Blackwell

TOPICS IN PHOSPHORUS CHEMISTRY

Volume 9

EDITORS

Edward J. Griffith, Monsanto Co., St. Louis, Missouri

Martin Grayson, John Wiley & Sons, New York, N.Y.

CONSULTING EDITORS

D. E. C. Corbridge, Research Department, Associated Companies Ltd., Harrogate, Yorks, England

R. F. Hudson, University of Kent, Canterbury, England

William E. McEwen, University of Massachusetts, Amherst, Massachusetts

John R. Van Wazer, Vanderbilt University, Nashville, Tennessee

New York • London • Sydney • Toronto

Topics in Phosphorus Chemistry

Volume 9

AUTHORS

H. M. Buck
E. H. J. M. Jansen
P. Schipper
O. Stelzer
A. E. R. Westman

An Interscience ® Publication

JOHN WILEY & SONS

Library of Congress Catalog Card Number: 64-17051
ISBN 0-471-32782-4

Printed in the United States of America

10 9 8 7 6 5 4 3 2 1

Preface

The chemistry of phosphorus in its fundamental and applied aspects has grown rapidly in the past decade. The progress being made in nearly every scientific discipline dealing with this key element has created a need for a continuous central forum in which progress may be reviewed rapidly on a larger, more critical scale and for a broader audience than the specialized journals permit.

It is the purpose of *Topics in Phosphorus Chemistry* to provide the general scientific reader as well as the specialist in phosphorus chemistry with a series of critical evaluations and reviews of progress in the diverse special areas of the science written by scientists actively engaged in work in the field. An attempt has been made to keep the articles timely and current to the extent that previously unpublished work can be included. Reviews are, at the same time, intended to give enough background to the specific subjects to allow the reader to cross disciplinary lines effectively. It is hoped that further developments in phosphorus chemistry will be stimulated by this approach.

No fixed pattern has been established for the *Topics in Phosphorus Chemistry*. A flexible attitude will be preserved and the course of the series will be dictated by the workers in the field. The editors of *Topics in Phosphorus Chemistry* not only welcome suggestions from our readers but eagerly solicit your advice and contributions. The series is dedicated to the phosphorus chemist in particular while providing all chemists and biochemists with a charted route into the various facets of phosphorus chemistry.

The Editors

Contents

TOPICS IN PHOSPHORUS CHEMISTRY

Volume 9

Transition Metal Complexes of Phosphorus Ligands

O. Stelzer

Lehrstuhl B. für Anorganische Chemie
der Technischen Universität Braunschweig

CONTENTS

The coordination chemistry of tricoordinate phosphorus compounds has in the last decade become a widespread area of research in chemistry. Of necessity it has not been possible to cover all relevant papers published on this topic in the period under review (1970 to the end of 1973). Only those publications that deal with the coordinated phosphorus ligands themselves or their influence on chemical and physical properties of their coordination compounds are cited and discussed.

Relevant to the general interest of this topic, several review articles that cover either the whole area [161, 1030] or only special aspects of it [162, 174, 469, 939, 940, 1228, 1229] have appeared in the literature.

However, with the exception of references 469 and 1228, the literature is covered only up to 1969 or 1970.

This review therefore is concerned mainly with the literature that has appeared since that time. In some cases references are made to work done before 1970 to show trends of interest and the development in this area of coordination chemistry.

The chemical nomenclature within this review corresponds to that used by the authors of the various papers.

List of Abbreviations

Me = methyl
Et = ethyl
Pr^i = isopropyl
Pr^n = normalpropyl
Bu^n = normalbutyl
Bu^t = tertiary butyl
Bu^s = secundary butyl
Bu^i = isobutyl
Cy = cyclohexyl
Ph = phenyl

I. Complexes of Phosphine, PH_3, Organophosphines, $R_{3-n}PH_n$ (n = 1, 2), and Related Ligands

The coordination chemistry of phosphine (PH_3) has been reviewed by Fluck [469] up to 1971. Booth [162] has prepared an excellent review of metal complexes of primary, secondary, and tertiary phosphines, $R_{3-n}PH_n$ (n = 0, 1, 2, R = organic group). The reader interested in the literature published before 1971 is also directed to earlier review articles written by Fluck and Novobilsky [470] and by Booth [161].

A. Complexes of Phosphine, PH_3, and Phosphides

Additional complexes of phosphine, PH_3, have been prepared and investigated.

Tetraphosphinenickel(O), $Ni(PH_3)_4$, has been prepared by reaction of gaseous PH_3 under pressure with di-1,5-cyclooctadienenickel at −40°C in a closed system

[1189]. $Ni(PH_3)_4$, like the known $Ni(CO)_3PH_3$ [1040], is unstable at room temperature. It was characterized by IR and Raman spectroscopy.

Mixed phosphine-triphenylphosphite-nickel(O) complexes, $Ni(PH_3)_x[P(OPh)_3]_{4-x}$ (x = 1-4) are formed by reaction of PH_3 with tetrakis(triphenylphosphite)nickel(O) or bis-1,5-cyclooctadienenickel(O) and subsequent treatment with a stoichiometric amount of the phosphite. In a similar fashion mixed complexes with PH_3 and triphenylphosphine as ligands [$Ni(PH_3)(Ph_3P)_3$ and $Ni(PH_3)_2(Ph_3P)_2$] have been synthesized [1092]. Replacement of one bidentate phosphorus ligand in the complex $[Ir(Ph_2PCH_2CH_2PPh_2)_2]X$ (X = Cl, BPh_4) by PH_3 produces $[Ir(PH_3)(Ph_2PCH_2CH_2PPh_2)]X$ at -70°C, but a reversible oxidative addition occurs at room temperature to give [*cis*-$IrH(PH_2)(Ph_2PCH_2CH_2PPh_2)_2$]X. The synthesis of the related complexes *trans*-$[IrH(PH_2)(Me_2PCH_2CH_2PMe_2)_2]X$ has also been reported [1092]. It is interesting to note that $MnBr(CO)_5$ reacts with phosphine at room temperature with replacement of one or two carbonyl molecules.

Fischer has reported the new complex eq-phosphinenonacarbonyldimanganese(O) [463].

The previously known compound π-$C_5H_5Mn(CO)_2PH_3$ [465] has been prepared by reaction of π-$C_5H_5Mn(CO)_2PCl_3$ with sodium borohydride in tetrahydrofuran-solution [602].

$$\pi\text{-}C_5H_5Mn(CO)_2PCl_3 + 3\ NaBH_4 \xrightarrow{THF} \pi\text{-}C_5H_5Mn(CO)_2PH_3 + 3\ NaCl + 3\ BH_3 \quad (1)$$

By using appropriate precursor complexes, for example, $M(CO)_3(MeCN)_3$ (M = Cr, Mo, W) PH_3 complexes may be formed in various reaction media, including water, alcohols, ethers, and benzene and its derivatives. Air must be excluded from the reaction mixture, but after formation most of the products are stable in air at 25°C [708].

Cleavage of the dinuclear rhenium complex $[Re(CO)_4I]_2$ with PH_3 gives $Re(CO)_4I(PH_3)$ in high (90%) yield [897].

Large-scale syntheses of $Cr(CO)_5PH_3$, *cis*-$M(CO)_4(PH_3)_2$, and fac-$M(CO)_3(PH_3)_3$ (M = Cr, Mo, W), based on reaction of phosphine with $NEt_4[CrCl(CO)_5]$, $M(CO)_4DTO$ (DTO = 2,2,7,7-tetramethyl-3,6-dithiaoctane), and $M(CO)_3(MeCN)_3$, have been reported by Guggenberger, Klabunde, and Schunn [550]. Hydrogen-deuterium exchange with D_2O adsorbed on the surface of alumina (used as column material for the chromatography) afforded some complexes of trideuterophosphine, PD_3; for example, $Cr(CO)_5PD_3$, *cis*-$Cr(CO)_4(PD_3)_2$, and fac-$Cr(CO)_3(PD_3)_3$.

From kinectic investigations on the catalytic oxidation of phosphine by copper(II) in the presence of potassium iodide a mechanism has been proposed that included the formation of a complex of copper(I) with phosphine $[CuCl_2PH_3]^-$ [1123].

Gaseous phosphine reacts smoothly with europium or ytterbium metal solutions in liquid ammonia at -78°C to form colored ammoniated dihydrogen phosphides,

$M(PH_2)_2 \cdot 7NH_3$, according to the equation

$$M + 2\,PH_3 \xrightarrow{\text{liq } NH_3} M(PH_2)_2 \cdot 7NH_3 + H_2 \qquad (2)$$

(M = Eu, Yb). The products were thermally decomposed and the resulting phosphides were studied by means of magnetic susceptibility, X-ray powder diffraction, and IR and Mössbauer spectroscopy [617]. Thermal decomposition of supported metal salt phosphine complexes has been used for the production of metal-coated substrates [999].

B. Complexes of Primary and Secondary Phosphines and Phosphides

Methyl- and dimethylphosphine react with titanium trichloride to give crystalline air-sensitive complexes $TiCl_3 \cdot 2$(alkylphosphine). The EPR spectra of toluene solutions which contain 3 moles of phosphine per mole of titanium trichloride show a 1:3:3:1 quartet at $g = 1.94$ at low temperatures. Three phosphine molecules therefore appear to be associated with one titanium atom [1075].

Dinitrogen complexes of rhenium, $ReCl(N_2)(Ph_2PH)_4$ and $ReCl(N_2)(PhPH_2)_2$-$(Ph_3P)_2$, are formed if [*N*-cinnamoylhydrazido(3-) *N'*,O] dichlorobis(triphenylphosphine)rhenium(V) is treated with diphenyl- or phenylphosphine [249].

Tetrakis(diphenylphosphine)nickel(O) has been prepared from bis(π-allyl)nickel, tetrakis(triphenylphosphine)nickel(O), or nickel powder or nickel bromide and diphenylphosphine [607]

$$Ni(\pi C_3H_5)_2 \xrightarrow{Ph_2PH} Ni(Ph_2PH)_4 \qquad (3)$$

$$Ni(Ph_3P)_4 \xrightarrow{Ph_2PH} Ni(Ph_2PH)_4 \qquad (4)$$

$$Ni \xrightarrow{Ph_2PH} Ni(Ph_2PH)_4 \qquad (5)$$

Four- and five-coordinate diamagnetic nickel complexes, $[NiX(Et_2PH)_3]^+$, $NiX_2(Et_2PH)_3$, and $[NiX(Et_2PH)_4]^+$ (X = Cl, Br, I, NCS), have been synthesized. The previously reported [633] six-coordinate complex $NiX_2(Et_2PH)_4$ should be formulated preferably as $[NiX(Et_2PH)_4]X$ in the solid state [1015].

A reinvestigation of the reaction between nickel bromide and diphenylphosphine revealed that the product formed was $Ni(Ph_2PH)_4$ [607,1248] and not a square planar complex containing the terminal diphenylphosphido groups formulated earlier [637]. Similar results have been obtained for the corresponding palladium complex [1248].

Dimeric phosphide-phosphine platinum complexes (*1*) have been prepared by the reaction of diphenylphosphine with $Pt(PhCN)_2X_2$ (X = Cl, Br, I) or Na_2PtX_4 (X = Cl, NCS).

$$
\begin{array}{ccccc}
X & & PPh_2 & & PPh_2H \\
\diagdown & \diagup & & \diagup & \\
& Pt & & Pt & \\
\diagup & & \diagdown & & \diagdown \\
HPh_2P & & PPh_2 & & X
\end{array}
$$

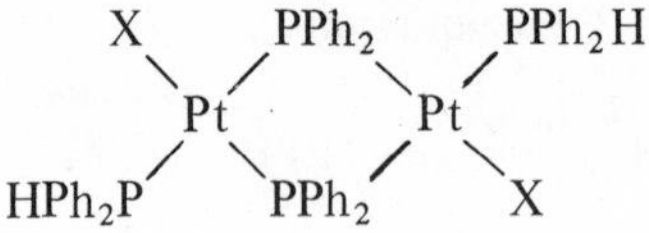

1

These complexes are expected to be potential anticancer agents [775].

Platinum metal complexes containing four or five diphenylphosphine ligands are reported [1045], e.g., *cis-* and *trans-*$MCl_2(Ph_2PH)_4$ (M = Ru, Os) and the salts of the cations $[Rh(Ph_2PH)_4]^+$, $[Ir(CO)(Ph_2PH)_4]^+$, and $[RuH(Ph_2PH)_5]^+$. The cationic complex $[Rh(Ph_2PH)_4]^+$, for example, may be synthesized by reaction of chlorotris(triphenylphosphine)rhodium(I) with diphenylphosphine and isolated as a tetrafluoroborate. Using *cis-* and *trans-*$RuCl_2(Ph_2PH)_4$ as starting materials, a variety of hydrido complexes of ruthenium could be prepared; for example, *cis-*$RuH_2(Ph_2PH)_4$, *trans-*$RuHX(Ph_2PH)_4$ (X = Cl, Br, I, SCN, $SnCl_3$). If *trans-*$RuCl_2(Ph_2PH)_4$ is treated with lithium methoxide in boiling methanol, *cis-*RuH_2-$(Ph_2PH)_4$ is obtained which, with lithium chloride in ethanol, gives *trans-*$RuHCl$-$(Ph_2PH)_4$ [1046]. In the series *trans-*$RuHX(Ph_2PH)_4$ the values of $\nu(Ru\text{-}H)$ give an order of trans influence of X, that increases from I to SCN as followes: I < Br < $SnCl_3$ < Cl < SCN.

Some of the corresponding iron complexes are reported as well [1046]. The H-NMR spectrum of $FeH_2(Ph_2PH)_4$ shows a close similarity to those reported by Meakin, Muetterties, Tebbe, and Jesson [876] for stereochemically nonrigid complexes, FeH_2P_4 (P = various phosphines, phosphonites, and phosphinites).

Chlorine hydrogen exchange in $\pi C_5H_5Mn(CO)_2(PhPCl_2)$, be use of sodium borohydride in tetrahydrofuran, produces the phenylphosphine complex πC_5H_5Mn-$(CO)_2(PhPH_2)$ [602]. Group VIB metal carbonyl derivatives of the diprimary bidentate ligand 1,2-diphosphinoethylene, $H_2PCH_2CH_2PH_2$ (P-en), have been prepared by replacement of weakly bonded ligands from $M(CO)_4$(bidentate) (M = Cr, Mo, W, bidentate = 2,2,7,7-tetramethyl-3,6-dithiaoctane or *N,N,-N',N'*-tetramethyl-1,3-propanediamine) [400]. In addition to the monomeric species $M(CO)_4$(P-en), insoluble polymeric powders of approximate stoichiometry $M(CO)_4$(P-en) were obtained.

Phosphido-phosphine complexes (*2*) are formed by treatment of Na_2PdCl_4 with trimethylenediphosphine in ethanol [131]

$$2\,Na_2PdCl_4 + 3\,H_2P(CH_2)_3PH_2 \longrightarrow [Pd_2(H_2P(CH_2)_3PH_2)_2(HP(CH_2)_3PH)]\,Cl_2$$
$$+ 4\,NaCl + 2\,HCl \tag{6}$$

$$
\left[
\begin{array}{ccccc}
& PH_2 & & PH & & PH_2 \\
& \diagup & \diagdown & | & \diagup & \diagdown \\
(CH_2)_3 & & Pd & (CH_2)_3 & Pd & (CH_2)_3 \\
& \diagdown & \diagup & | & \diagdown & \diagup \\
& PH_2 & & PH & & PH_2
\end{array}
\right]^{2+}
$$

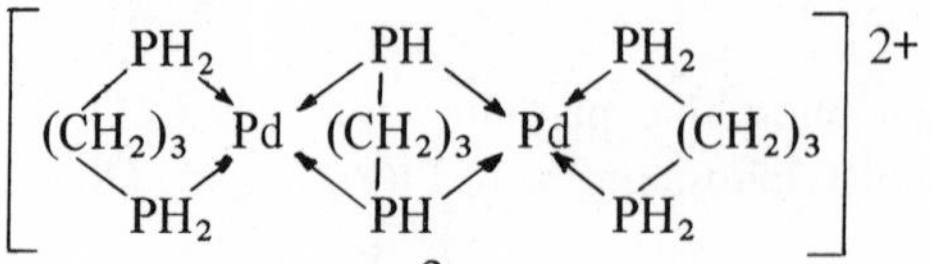

2

The displacement of carbon monoxide in various cobalt, nickel, and iron carbonyl complexes by bis(trifluoromethyl)phosphine led to a variety of substitution products with terminal $(CF_3)_2PH$ ligands or $(CF_3)_2P$ bridging units [399, 1039]. A complex reaction that occurs between $(CF_3)_2PH$ and $[\pi C_5H_5Co(CO)_2]$ gives $[Co(CO)_3(P(CF_3)_2)]_2$, $[(\pi C_5H_5)Co(P(CF_3)_2)]_2$, and $[(\pi\text{-}C_5H_5)Co(CO)(P(CF_3)_2)\text{-}Co(CO)_3]$. Concurrent with the formation of these complexes, reduction of the coordinated πC_5H_5 ligand is followed by the addition of phosphine or hydrogen to give $C_5H_7P(CF_3)_2$ and C_5H_8 [399].

Reactions of enneacarbonyldiiron, $Fe_2(CO)_9$, with various primary and secondary phosphines at 25°C produce the phosphine iron tetracarbonyl species $Fe(CO)_4L$ (L = $MePH_2$, $PhPH_2$, Ph_2PH, $(p\text{-}MeC_6H_4)_2PH$, Me_2PH, Et_2PH, PhMePH) as major product [1193]. Only small amounts of the dinuclear complexes $Fe_2(CO)_6(RR'P)_2$ are formed. On heating, or on irradiation, $Fe(CO)_4L$ species are converted to the dinuclear complexes. Phenylphosphine reacts with $Fe(CO)_5$ at 190°C to give $Fe_3\text{-}(CO)_9(PhP)_2$. Quite different results have been obtained in the reaction between bis(trifluoromethyl)phosphine and pentacarbonyliron or dodecacarbonyltriiron When $P(CF_3)_2H$ was heated with a slight excess of $Fe(CO)_5$ (100°C, 8 hr) in a sealed tube, a solid was precipitated from which by careful sublimation $Fe_2(CO)_6\text{-}[P(CF_3)_2]_2$ (3%) and $H_2Fe_2(CO)_6[P(CF_3)_2]_2$ (15% yield) could be isolated [397, 398]. The ^{1}H NMR spectrum of the latter compound showed the presence of two isomers (*3* and *4*).

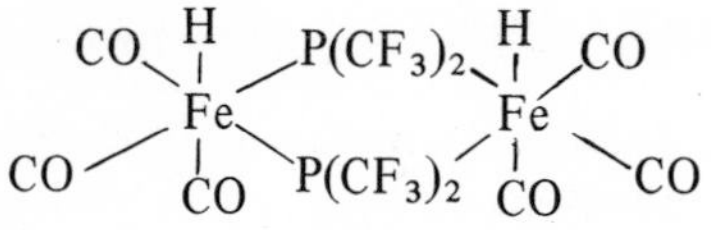

3
"*cis*"

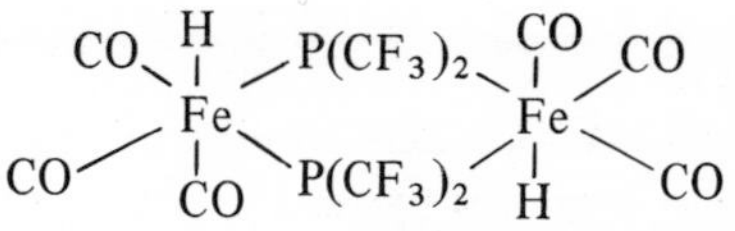

4
"*trans*".

Reaction of $P(CF_3)_2H$ with $Fe_2(CO)_9$, however, at 20°C gives $Fe(CO)_4[P(CF_3)_2H]$, which decomposes on heating to the dihydride [397]. In contrast to this rather complex reaction pattern, bis(trifluoromethyl)phosphine and trifluoromethylphosphine react in a normal manner with norbornadienetetracarbonylchromium or -molybdenum to give the cis-disubstituted products *cis*-$L_2M(CO)_4$ (L = $(CF_3)_2\text{-}PH$, CF_3PH_2, M = Cr, Mo) [945].

All experiments to protonate or deprotonate the three-membered heterocycle phosphirane have been unsuccessful so far. Bausch, Ebsworth, and Rankin [82] have shown, however, that it is possible to prepare transition metal carbonyl complexes of this ligand according to the following equations:

$$MeNC_5H_5^+ [Mo(CO)_5I]^- + (CH_2)_2PH \longrightarrow MeNC_5H_5I + Mo(CO)_5(CH_2)_2PH \quad (7)$$
$$C_7H_8Mo(CO)_4 + 2\,(CH_2)_2PH \longrightarrow C_7H_8 + Mo(CO)_4[(CH_2)_2PH]_2 \quad (8)$$
$$C_7H_8Mo(CO)_3 + 3\,(CH_2)_2PH \longrightarrow C_7H_8 + Mo(CO)_3[(CH_2)_2PH]_3 \quad (9)$$

Pentacarbonyl phosphirane molybdenum(O) may be deprotonated by potassium dihydrogenphosphide in dimethylether:

$$Mo(CO)_5(CH_2)_2PH + PH_2^- \longrightarrow [Mo(CO)_5(CH_2)_2P]^- + PH_3 \qquad (10)$$

Mercury bis(diorganophosphides), $Hg(PR_2)_3$, have been prepared for the first time [80]:

$$2\ Bu^t_2PH + HgBu^t_2 \longrightarrow Hg(Bu^t_2P)_2 + 2\ C_4H_{10} \qquad (11)$$

Irradiation with UV light in benzene causes decomposition of the remarkably stable mercury bis(*t*-butylphosphide) with precipitation of mercury:

$$Hg(Bu^t_2P)_2 \xrightarrow[C_6H_6]{h\nu} Bu^t_2P\text{-}PBu^t_2 + Hg \qquad (12)$$

By a similar route mercury bis(trifluoromethylphosphide) is formed from divinylmercury and bis(trifluoromethyl)phosphine [544]. Use of this compound has been proposed as a reagent for the introduction of the $P(CF_3)_2$ group.

The synthesis of other phosphido complexes, prepared by reactions of coordinated phosphine ligands, is reported in Chapter VIII.

II. Complexes of Tertiary Phosphines

The number of papers covered in this area, compared with that of other types of phosphine ligand, is by far the largest. Several reasons may account for this coverage: (a) There exist manifold well-established synthetic routes to a large variety of mono-, bi-, tri-, and polydentate tertiary phosphine ligands. (b) The hydrolytic and, in a number of cases, oxidative stability of these ligands is high. (c) For cocatalysts in homogeneous catalysis the phosphine ligands must be inert to the various substrates, a requirement that is met by a large number of tertiary phosphines. (d) The range of oxidation states of the metals involved in complex formation with these ligands is by far the more extensive. Although for phosphine ligand complexes with more electronegative groups (e.g., halogen, OR^-, NR_2) the oxidation number of the central metal is not greater than +2 in most cases, several complexes of tertiary phosphines R_3P (mono- or polydentate) have metals in higher oxidation states. This last statement obviously reflects the present theory of the transition metal phosphorus bond which, via π-backbonding, ascribes to the phosphorus ligands with electronegative substituents a special ability for stabilization of metals in low oxidation states.

For a discussion of the literature on the use of phosphine complexes in organic synthesis and homogeneous catalysis the reader is referred to Chapter VII.

Previous review articles to be mentioned are those of Booth [161, 162] and Robinson [1030].

A. *Tertiary Monodentate Phosphine Complexes*

1. LANTHANIDES AND ACTINIDES

There is little in the literature about phosphine complexes of these metals. Selbin, Ahmad, and Pribble [1097] for the first time prepared complexes of uranium(V) with phosphorus donors (e.g., Ph_3P and $Ph_2PCH_2CH_2PPh_2$) UCl_5PPh_3 and UCl_5-$Ph_2PCH_2CH_2PPh_2$:

$$UCl_5TCAC + Ph_3P \longrightarrow UCl_5Ph_3P + TCAC \qquad (13)$$

$$UCl_5TCAC + \left.\begin{matrix} Ph_2P \\ \\ Ph_2P \end{matrix}\right] \longrightarrow UCl_5(Ph_2P\diagup\!\!\diagdown\!\!\diagup PPh_2) + TCAC \qquad (14)$$

$$TCAC = \text{trichloroacryloylchloride}, Cl_2C{=}C\begin{matrix}\diagup Cl \\ \diagdown CO(Cl)\end{matrix}$$

The latter compound showed, up to 500°C (!), neither endothermic nor exothermic peaks under differential thermal analysis. Both compounds have been investigated by means of ESR, IR, and UV-visible spectroscopy. It was proved that the product was not a phosphine oxide or a phosphonium salt.

2. TITANIUM, ZIRCONIUM, AND HAFNIUM

Few complexes of monodentate tertiary phosphines with these metals have been reported in the literature. The presence of fast equilibria in solution of several $TiCl_4$-PR_3 systems involving 1:1 and 1:2 adducts and free tertiary phosphine has been suggested. At low temperature binuclear species of the type $Ti_2Cl_8(R_3P)$ may be present. Some 1:1 and 1:2 $TiCl_4$-R_3P adducts have been isolated [221]:

$$TiCl_4R_3P + TiCl_4 \rightleftharpoons Ti_2Cl_8R_3P \qquad (15)$$

$$Ti_2Cl_8R_3P + R_3P \rightleftharpoons 2\,TiCl_4R_3P \qquad (16)$$

$$TiCl_4R_3P + R_3P \rightleftharpoons TiCl_4(R_3P)_2 \qquad (17)$$

Reaction of methylphenylphosphines (Me_2PhP and $MePh_2P$) with the binuclear trivalent cyclopentadienyl complex $[(\pi C_5H_5)_2TiCl]_2$ yielded interesting complexes of titanium(III) [531]:

$$[(\pi C_5H_5)_2TiCl]_2 + 2L \longrightarrow 2(\pi C_5H_5)_2TiClL \qquad (18)$$

(L = Me_2PhP, $MePh_2P$). Analogous compounds in which L = Ph_3P and Cy_3P could not be prepared.

3. VANADIUM, NIOBIUM, AND TANTALUM

The preparation of the seven-coordinate main group mixed metal complexes $Ph_3SnM(CO)_5Ph_3P$, $EtHgM(CO)_5Ph_3P$, and $Ph_3PAuM(CO)_5Ph_3P$ by direct substitution of one carbonyl molecule according to

$$RM(CO)_6 + Ph_3P \longrightarrow RM(CO)_5Ph_3P + CO \tag{19}$$

(R = Ph_3Sn, EtHg, and Ph_3Au, M = V, Nb, Ta) has been described [368]. The triphenyltin derivatives of niobium and tantalum complexes are less susceptible to heterolytic cleavage than the corresponding vanadium ones.

1,2,5-Triphenylphosphole (TPP) reacts in dichloromethane solutions with niobium chloride, tantalum chloride, and bromide with formation of the complexes MCl_5TPP (M = Nb, Ta) and $TaBr_5TPP$. The reaction of the phosphole in these cases is rather surprising in view of its much weaker basicity compared with tertiary phosphines [it does not form complexes with silver(I), manganese(II), iron (II), cobalt(II), nickel(II), thorium(IV), and uranium(IV) chlorides, although phosphine complexes of these metals, except for the last two, are well established] [205].

Complexes of vanadium(IV) with ethylxanthate and phosphorus(III)-containing ligands, for example, Ph_3P, have been studied by ESR and IR spectroscopy. They contain one molecule of the phosphine ligand in the inner sphere [725].

Tri-valent bis(π-cyclopentadienyl)niobium hydrides, $[(\pi C_5H_5)_2NbHR_3P]$ and related derivatives $[(\pi C_5H_5)_2NbBrR_3P]$ have been obtained by treatment of the borohydride $(\pi C_5H_5)_2NbBH_4$ with tertiary phosphines R_3P (R_3 = Ph_3 or Me_2Ph) or $(\pi C_5H_5)_2NbHPR_3$ with *n*-butylbromide, respectively [794]. The hydrides $(\pi C_5H_5)_2NbHR_3P$ behave, as expected by analogy with $(\pi\text{-}C_5H_5)_2MoH_2$ and $(\pi C_5H_5)_2ReH$, as bases:

$$(\pi C_5H_5)_2NbHPR_3 \underset{-H^+}{\overset{+H^+}{\rightleftharpoons}} [(\pi C_5H_5)_2Nb(H)_2\text{-}PR_3]^+ \tag{20}$$

The ^1H NMR spectrum of the resulting cation shows a doublet assignable to the NbH_2 hydrogens at τ13.96 ppm, $^2J_{PH}$ 31.5 Hz.

Niobium and tantalum pentachloride form complexes with triphenyl and tri-*n*-butylphosphine of the general formula MCl_5R_3P (M = Nb, Ta) [388, 516]. Alcoholysis yields the substitution products, for example, $MCl_3(OEt)_2Ph_3P$ (M = Nb, Ta), and in ammonolysis the phosphine ligand is displaced by ammonia to give products with 6 or 7 molecules of NH_3 per atom of niobium or tantalum. The heating curve of $NbCl_5Bu^n{}_3P$ [214, 516] exhibits three endothermic stages at 410,

475, and 510°C. The complete analysis of the products formed suggests the following scheme of thermal decomposition:

$$NbCl_5Bu_3^nP \xrightarrow[410^\circ C]{-Bu^nCl} NbCl_4Bu_2^nP \xrightarrow[475^\circ C]{-2\ Bu^nCl} NbPCl_2 \xrightarrow[510^\circ C]{-Cl_2} \text{mixture of niobium phosphides} \quad (21)$$

Niobiumdichlorophosphate(III) is a dark brown viscous mass. It forms an extremely stable aquo complex, and no silver chloride is precipitated when silver nitrate is added to the aqueous solution. Cryoscopic determination of the molecular weight in benzene indicates polymerization, the degree of which is 7 at a concentration of 0.133 mass%.

4. CHROMIUM, MOLYBDENUM, AND TUNGSTEN

Complexes of Group VI metals (especially tungsten) with tertiary monodentate phosphine ligands continue to attract attention because of their use as homogeneous catalysts, the nonrigidity in some eight coordinate tungsten and molybdenum hydrides, and the interesting chemistry of complexes with π-bonded ligands (e.g., πC_5H_5 and πC_6H_6).

Chatt, Leigh, and Thankarajan [255] found that the substitution of carbon monoxide in Group VI metal carbonyls by tertiary phosphines in boiling ethanol is strongly catalyzed by sodium borohydride. It was found that the latter compound reduces the tendency of the hexacarbonyls to volatilize from the boiling solution. It is likely that the hydride attacks in an initial step at the carbon of the carbonyl group to give a relatively labile intermediate salt.

Substitution reactions between tris(*p*-fluorophenyl)phosphine and the appropriate transition metal carbonyl or their derivatives led to VIb metal carbonyl substitution products, $Cr(CO)_5L$, *trans*-$Cr(CO)_4L_2$, *trans*-$W(CO)_4L_2$, and *cis*-$Mo(CO)_4L_2$ [385]. Dinitrogen complexes $Mo(N_2)_2L_4$ (L = $MePh_2P$, Me_2PhP, Bu_3^nP) are formed if the complexes $MoCl_4L_2$ are reduced with sodium amalgam in tetrahydrofurane in the presence of two equivalents of L and under a stream of nitrogen [497].

Complexes of the type $M(CO)_5L$, $M(CO)_4L_2$ (cis and trans) and fac-$Mo(CO)_3L_3$ (M = Cr, Mo, W) have been prepared with phosphines containing bulky substituents such as *t*-butyl [1087, 1131] and *o*-, *m*-, and *p*-tolyl groups [166, 168]. With tri(*o*-tolyl)phosphine [166], in addition to the complexes $M(CO)_5L$ and $M(CO)_4L_2$, an interesting novel series of derivatives formulated as π-[(*o*-tolyl)$_3$P]-$M(CO)_3$ has been isolated. The proposed structure contains one π-bonded aromatic ring and an uncoordinate phosphorus (*1*). According to ^1H NMR measurements, the methyl groups on the two nonbonded *o*-tolyl groups are not equivalent (up to 60°C). Strong steric interaction between the methyl group and the adjacent carbonyl groups is thought to be the reason for this.

The reaction of hexacarbonylchromium with triphenylphosphine, tri(*o*-, *m*-, and *p*-tolyl)phosphine, at high temperature yields two series of complexes, $Cr(CO)_3L$ and $[Cr(CO)_2L]_2$ [168]. The first is analogous to the already mentioned molybdenum compound, whereas in the second the arylphosphines act as bridging ligands; one phosphorus and one aromatic π-system are the donor sites (*2*).

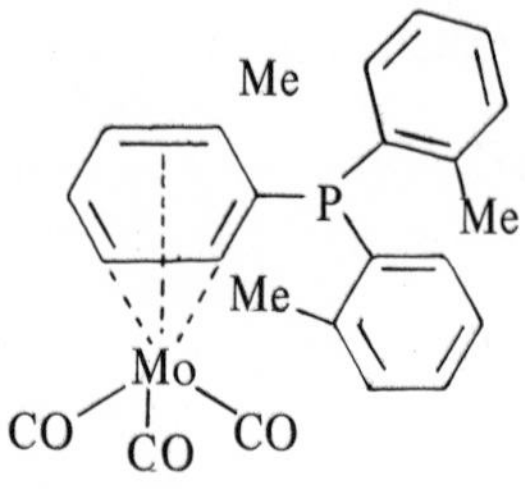

1

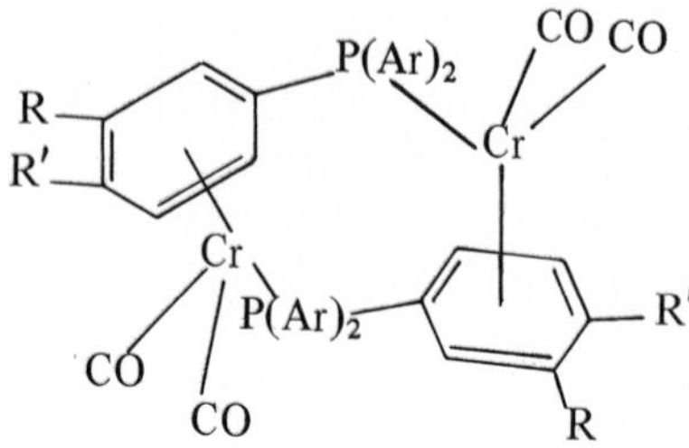

2

R = R′ = H, Ar = Ph
R = H, R′ = Me, Ar = *p*-tolyl
R = Me, R′ = H, Ar = *m*-tolyl

The reactions of triphenylphosphine with π-$C_6H_6M(CO)_3$ (M = Cr, Mo, W) which yield fac-$M(CO)_3(Ph_3P)_3$ have been reexamined [170]. Two solid state forms (α and β) with different molecular packing could be identified. Substitution of one carbonyl molecule in some azomethane complexes (*3*) by phosphine ligands causes a strong bathochromic effect in all solvents in the UV-visible spectra of these intensely colored compounds [1236]. The reaction of Ph_3P or Bu^n_3P with π-allyldicarbonylmolybdenum complexes, for example, (π-allyl)MoX-$(CO)_2(MeCN)_2$, has been reported. Phosphonium salts and tetra-substituted molybdenum carbonyls $Mo(CO)_2(MeCN)_2(R_3P)_2$ were obtained. Subsequent replacement of MeCN produced a series of interesting, highly substituted derivatives of hexacarbonylmolybdenum; for example, $Mo(CO)_2$(MeN=CHCH=NMe) $(R_3P)_2$ [477].

Methylthioethers Me-S-R′ (R′ = Me, allyl) can easily be replaced by organophosphines R_3P (R = Ph_3 and Bu^n_3) in the complexes $[(\pi C_5H_5)_2Mo(SR'Me)X]^+$-$PF_6^-$ (X = Cl, Br, R′ = Me; X = Cl, R′ = allyl). The complexes $[(\pi$-$C_5H_5)_2$-$MBrR_3P]^+PF_6^-$ (M = Mo, W, R = Ph) with sodium borohydride give the hydrides $[(\pi$-$C_5H_5)_2MHR_3P]^+PF_6^-$ [330].

Although the chemistry of various compounds containing the $\pi C_5H_5Mo(CO)_3$ and $\pi C_5H_5W(CO)_3$ moieties is well established, the analogous chromium compounds have received far less attention. The substituted derivatives $[\pi C_5H_5Cr$-$(CO)_2L]_2Hg$ were recently obtained by heating a solution of $[\pi$-$C_5H_5Cr(CO)_3]_2Hg$

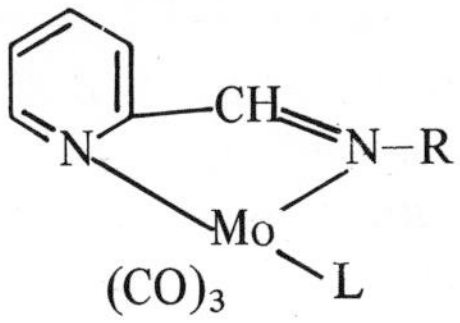

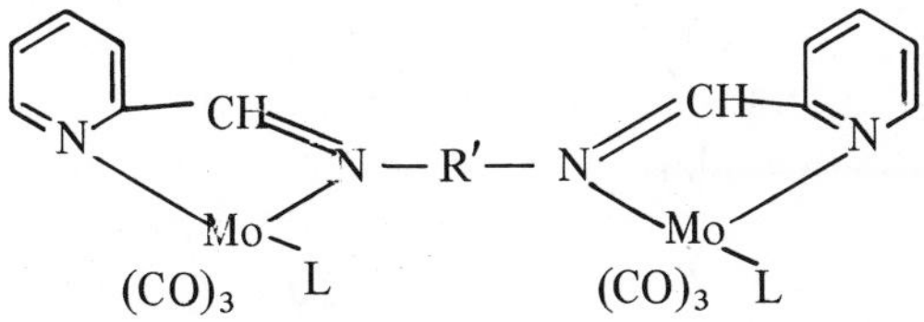

3

L = Ph_3P, Cy_3P, Bu_3P, Bu_2PhP
R = Et, Cy, $PhCH_2$, Ph, Me
R′ = p-C_6H_4, C_6H_4-C_6H_4

with an excess of the phosphine ligands L [L = Ph_3P, $P(OMe)_3$, and $P(OPh)_3$] in a suitable solvent [838]. The complexes $\pi C_5H_5Mo(CO)_2(L)SnMe_3$ [L = Ph_3P, $P(OPh)_3$, Me_2PhP] have been prepared by reaction of the anion $[\pi C_5H_5Mo(CO)_2$-$L]^-$ with Me_3SnCl. Substitution of one carbonyl in $\pi C_5H_5Mo(CO)_3SnMe_3$ by Ph_3P at 160°C provides a direct synthesis. All the $SnMe_3$ compounds liberate tetramethyltin at about 200°C with the formation of $[\pi C_5H_5Mo(CO)_2L]_2SnMe_2$ [495]. The corresponding tungsten complexes $\pi C_5H_5W(CO)_2LSnMe_3$ (L = Me_3P, Me_2PhP, $MePh_2P$, Ph_3P, etc.) have been prepared in a similar fashion to the molybdenum analogs [498]. For some new phosphine-substituted acyldicarbonyl π-cyclopentadienyl molybdenum complexes $[RCOMo(CO)_2(PhPMe_2)$-$\pi C_5H_5]$ it has been demonstrated by use of ^1H NMR spectroscopy that they seem to exist in solution only in the trans form ($>$99%) [332], whereas the corresponding alkyl complexes $RMo(CO)_2L$-πC_5H_5 [R = $PhCH_2$, Me, L = $P(OPh)_3$] have been shown to exist in cis and trans forms and undergo interconversion between the two [442].

King and Saran [699] have reported a series of very interesting reactions of transition metal cyanocarbon derivatives with triphenyl- and dimethylphenylphosphines and various other group V donors; for example, the reaction of $(CN)_2$-$C{=}CClMo(CO)_5\pi$-C_5H_5 with triphenylphosphine in boiling benzene results in the complete substitution of the three carbonyl ligands with two Ph_3P to yield the red-orange complex π-$C_5H_5MoCl[C{=}C(CN)_2](Ph_3P)_2$ containing the dicyanomethylenecarbene ligand. Neutral π-cyclopentadienyl nitrosyl molybdenum complexes $\pi C_5H_5Mo(NO)I_2L$ (L = Me_2PhP, $AsMe_2Ph$, $AsPh_3$ or p-MeO-C_6H_4NC) and the monocationic species $[\pi C_5H_5Mo(NO)IL_2]^+I^-$ (L = Me_2PhP) are formed by the reaction of the appropriate Lewis base with $[\pi C_5H_5Mo(NO)I_2]_2$. It is assumed that the structures of $\pi C_5H_5Mo(NO)I_2L$ and $[\pi\text{-}C_5H_5Mo(NO)IL_2]^+$ are based on that of 3:4 coordination (i.e. the πC_5H_5 ring occupies three coordination sites) [648].

If triphenylphosphine is refluxed in chloroform solution with $WX(NO)(CO)_4$, monosubstituted complexes of formula $WX(NO)(CO)_3Ph_3P$ (X = Cl, Br, and I) are initially formed [307]. Because the NO stretching frequency in these complexes (about 1680 cm^{-1}) is similar to that in $WX(NO)(CO)_4$ (at approximately 1710 cm^{-1}), it is suggested that the complexes $WX(NO)(CO)_3Ph_3P$ may be assigned to the meridional structure. Disubstituted compounds $WX(NO)(CO)_2L_2$ (L = Ph_3P, X = Cl, Br, I) are formed when the tricarbonyl derivatives are refluxed with excess Ph_3P. The *cis*(dicarbonyl)*cis*[bis(triphenylphosphine)] structure is believed to be the most likely (*4*):

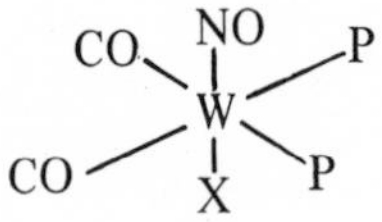

4

(P = Ph_3P, X = Cl, Br, I).

Replacement of carbonyl groups in the halocarbonyls of molybdenum and tungsten [$MoX_2(CO)_4$ and $WX_2(CO)_4$ (X = Cl, Br)] with triethylphosphine [1247] and meta- and paratolylphosphines [167] led to the complexes MX_2-$(CO)_3L_2$. The tritolylphosphine complexes could be readily converted into the corresponding dicarbonyls $MX_2(CO)_2L_2$. Treatment of halogenocarbonyls of the type $MX_2(CO)_4$ (M = Mo, W, X = Cl, Br) with 3 moles of Me_2PhP produced complexes of the type $MX_2(CO)_2(Me_2PhP)_3$ [911]. It is interesting to note that tris(*o*-tolyl)-phosphine does not form complexes of that kind. It reacts very slowly with $MoCl_2(CO)_4$, for example, with the formation of only small quantities (~5%) of $Mo(CO)_5(o\text{-tolyl})_3P$ and decomposition products. The halogen oxidation of *trans*-$Mo(CO)_4L_2$ (L = Ph_3P) is a rather complex reaction [167]. In the iodine oxidation $MoI_2(CO)_4$ is an intermediate; the ionic complex $[Ph_3PH]^+[MoI_3(CO)_3\text{-}Ph_3P]^-$ may be isolated if excess Ph_3P and chloroform as a solvent are employed. Similar observations were made when iodine was allowed to react with *trans*-$Mo(CO)_4[(p\text{-tolyl})_3P]_2$ and *trans*-$Mo(CO)_4[(m\text{-tolyl})_3P]_2$. These reactions which are summarized in eq. 22 are consistent with Scheme 1.

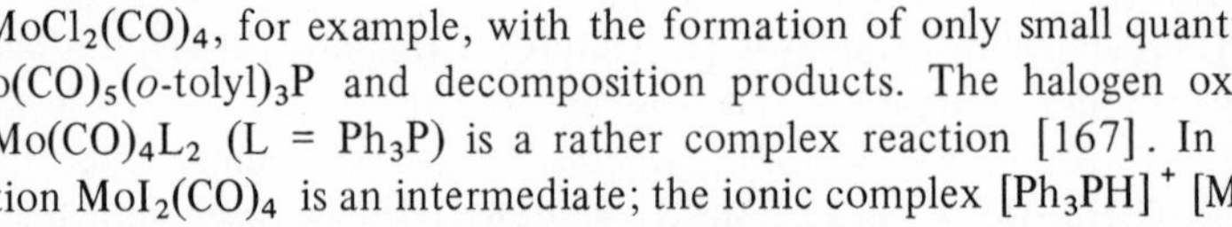

(22)

SCHEME 1. Some reactions of tungsten complexes (L = Me_2PhP or $AsMe_2Ph$); taken from [911].

$$WX_2(CO)_4 \xrightarrow{2L} WX_2(CO)_3L_2 \xrightarrow{L} WX_2(CO)_2L_3$$

$$X_2 \uparrow (a)$$

$$W(CO)_6 \xrightarrow{2L} W(CO)_4L_2 \xrightarrow[(a)]{X_2} WX_4L_2$$

$$3L \downarrow \qquad\qquad\qquad -L \updownarrow +L\,(b)$$

$$W(CO)_3L_3 \xrightarrow{I_2} [WI(CO)_3L_3]^+ \qquad WX_4L_3$$

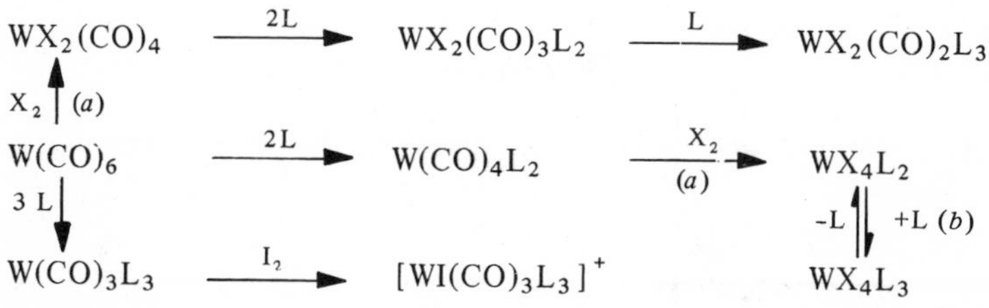

(a) X = Cl, Br
(b) equilibrium in solution

Halogen oxidation of complexes of the type $cis\text{-}M(CO)_4L_2$ [911] (M = Mo, W, L = Et_3P, Me_2PhP, etc.) produces seven coordinate complexes, $MX_2(CO)_3L_2$ (X = Cl, Br, I). $MoBr_2(CO)_3(Et_3P)_2$ loses carbon monoxide reversibly to give a blue complex, $MoBr_2(CO)_2(Et_3P)_2$. Chlorine oxidation of $cis\text{-}Mo(CO)_4(Me_2PhP)_2$ gives $Mo_2Cl_2(CO)_4(Me_2PhP)_4$. On heating the complexes $MoX_2(CO)_3(Me_2PhP)_2$ (X = Br, I), the dimers $Mo_2X_2(CO)_4(Me_2PhP)_4$ are formed. No identifiable products could be obtained by oxidation of $fac\text{-}M(CO)_3L_3$ with halogens.

If (an excess of bromine or cholorine is employed) for the oxidation of $cis\text{-}M(CO)_4L_2$ complexes, $trans\text{-}MX_4L_2$ (X = Br, Cl, L = Me_2PhP, M = Mo, W; X = Cl, L = Et_2PhP, M = W) are obtained. Schemes 1 and 2 summarize some reactions of molybdenum and tungsten complexes, including ligand replacement, addition, and oxidation.

Attempts to prepare the complexes WCl_4L_2 (L = tertiary phosphine or arsine) directly from tungsten(VI) chloride and the corresponding phosphine were unsuccessful. If amalgamated zinc in tetrahydrofuran is employed as the reducing agent in the presence of 2 moles of the ligand, a yield of roughly 40% may be obtained [217].

Aliphatic nitrile complexes $[WCl_4(RCN)_2]$ have been used for the synthesis of WCl_4L_2 (L = Ph_3P, Me_3P, and other group V donor ligands) under mild conditions [146]. Convenient new syntheses of tungsten(IV) halide adducts of the type WX_4L_2 (X = Cl, Br, L = Ph_3P and various other donors) have been reported [1065]; for example,

$$cis\text{-}WCl_4(SEt_2)_2 + 2\ Ph_3P \longrightarrow trans\text{-}WCl_4(Ph_3P)_2 + 2\ SEt_2 \qquad (23)$$

The trans structure of this complex has been ascertained by IR spectroscopy.

SCHEME 2. Some reactions of molybdenum complexes (L = Me_2PhP) except for (f) when L = Et_3P; reactions were also carried out with L = Et_2PhP and $AsMe_2Ph$; taken from [911]

MoX_4L_2 $MoX_2(CO)_2L_2$

$X_2 \uparrow$ (e)   +CO $\updownarrow$ –CO (f)

$Mo(CO)_4L_2 \xrightarrow[(a)]{X_2} MoX_2(CO)_3L_2 \xrightarrow[-2CO]{(c)} Mo_2X_4(CO)_4L_4$

(b)   L $\downarrow$ L (a)

$Mo(CO)_6 \xrightarrow[(d)]{X_2} MoX_2(CO)_4 \xrightarrow{3L} MoX_2(CO)_2L_3$

$Mo(CO)_3L_3 \xrightarrow[(b)]{X_2} [MoX(CO)_3L_3]^+$

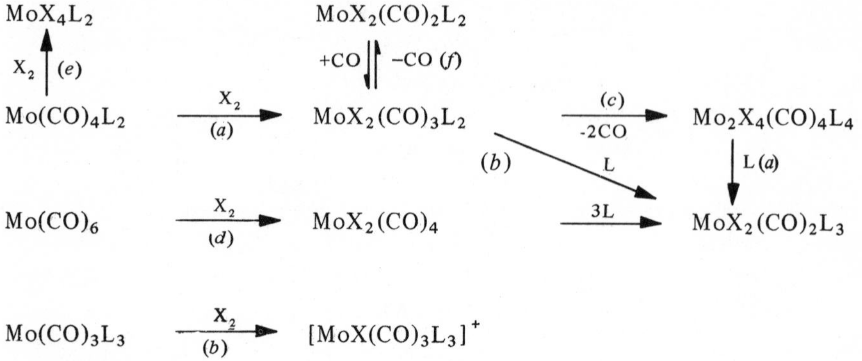

(a) X = Cl, Br, I
(b) X = I
(c) Spontaneous for X = Cl, on heating X = Br, I
(d) X = Cl, Br at –78°C
(e) X = Cl, Br at 20°C
(f) L = Et_3P, X = Cl

Reductive carbonylation of $MoCl_4(Ph_3P)_2$ in the presence of $EtAlCl_2$ yielded, after addition of Ph_3P, the molybdenum(II) complex $MoCl_2(CO)_3(Ph_3P)_2$; the corresponding tungsten compound could be obtained in a similar manner [98]. Halo metalcarbonyls as well as hexacarbonyls may be obtained in the absence of triphenylphosphine. Using sodium amalgam in tetrahydrofuran as a reducing agent and trans-WCl_4L_2 (L = tertiary phosphine), Bell, Chatt, and Leigh [89] obtained mixtures of substituted tungsten(0) carbonyls. Some of the products have been identified by their IR absorptions in the 2000 cm^{-1} region. If the same reaction is carried out in a nitrogen or hydrogen atmosphere, the dinitrogen complexes $W(N_2)_2(R_3P)_4$ (R_3 = Me_2Ph and Et_2Ph) or tungsten hydrido complexes are formed, respectively [88, 89] [e.g., $WH_6(Me_2PhP)_3$]. The major product from the reduction of tungsten(VI) chloride by sodium amalgam under argon in the presence of dimethylphenylphosphine was, however, trans-$WCl_2(Me_2PhP)_4$.

Reductive nitrosation [99] of molybdenum and tungsten halides yielded derivatives of dichlorodinitrosyl molybdenum or tungsten. These reactions may be represented by the following equations:

$$MCl_{5,6} \xrightarrow[\substack{EtAlCl_2 \\ hexane}]{NO} M(NO)_2Cl_2 \cdot Et_nAlCl_{3-n} \xrightarrow{Ph_3P} M(NO)_2Cl_2(Ph_3P)_2 \quad (24)$$

M = Mo, W, n = 0, 1;

$$MoCl_4(Ph_3P)_2 \xrightarrow[\text{hexane}]{NO + EtAlCl_2} Mo(NO)_2Cl_2(Ph_3P)_2 \qquad (25)$$

The reactions of anhydrous chromium(III) chloride with tertiary phosphines have been reinvestigated [103]. Only with Me_3P, Et_3P, and Bu^n_3P were reactions observed and only for Et_3P and Bu^n_3P could the green dimeric complexes $[CrCl_3(R_3P)_2]_2$ be isolated. On warming in solution, these complexes lose 1 mole of trialkylphosphine to form polymeric (probably halogen-bridged) complexes $[CrCl_3(R_3P)]_n$. The dimeric species, on reaction with Ph_4PCl or Ph_4AsCl, adds one chloride ion to give the ionic complexes $Ph_4P^+[CrCl_4(R_3P)_2]^-$ (or the corresponding Ph_4As salt), which are unstable in solution. No definite ligand arrangement may be derived from the analysis of the UV, IR, and dipole measurements for the neutral dimers, whereas a trans octahedral structure may be assigned to the ionic complexes.

Diamagnetic oxymolybdenum(IV) complexes with tertiary phosphines $MoOCl_2L_3$, $MoOCl_2L(LL)$, and $[MoOCl(LL)_2]X$ (L = alkylarylphosphine, LL = ditertiary phosphine, X = BF_4^-, BPh_4^-) have been prepared by reaction of $MoCl_4(EtCN)_2$ with the phosphines or from molybdenum pentachloride and the phosphines in ethanol[52, 215] or the more convenient synthesis using sodium molybdate in ethanol in the presence of HCl as a reagent [52]. The analogous tungsten complexes may be prepared by reaction of the appropriate phosphines (Me_2PhP, Et_2PhP, and $MePh_2P$) with $[NEt_4]^+[WCl_6]^-$, $WOCl_4$ or WCl_6 in wet alcohol [217]. For both tungsten and molybdenum complexes a meridional arrangement (5) of the phosphine ligands is proposed on the basis of an analysis of their ^1H NMR spectra.

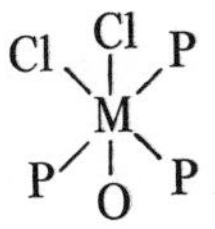

5

(P = Me_2PhP, M = Mo, W). The methyl protons in $WOCl_2(Me_2PhP)_3$ in the ^1H NMR spectra gave rise to two triplets at $\delta = -1.95, -1.77$ and a doublet at $\delta = 1.4$ ppm; the intensity ratios are 1:1:1. This may be rationalized in terms of the proposed structure with a strong trans P-P-coupling and without a plane of symmetry through the P-W-P axis.

Bisbenzenemolybdenum reacts readily with tertiary phosphines to form the monobenzene derivatives $\pi C_6H_6Mo(R_3P)_3$ [530] (R_3 = Me_2Ph, $MePh_2$), which may be protonated to form the monohydride cations $[\pi C_6H_6MoH(R_3P)_3]^+$ [530, 532]. NMR studies of this system have shown that solutions of the complexes $\pi C_6H_6Mo(MePh_2P)_3$ or $[\pi C_6H_6Mo(H)(MePh_2P)_3]^+$ in trifluoroacetic acid contain the molybdenum hydride derivative $[\pi C_6H_6MoH_2(Me_2PhP)_3]^{2+}$, which may be isolated as its hexafluorophosphate salt [530].

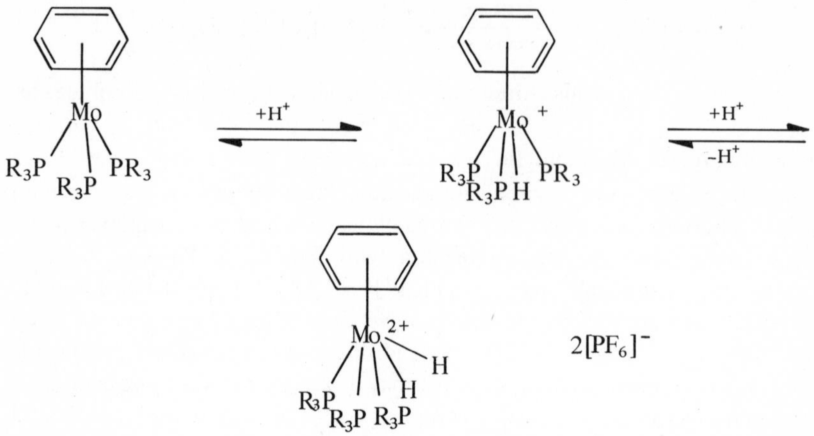

2[PF_6]^- (26)

(R_3 = Me_2Ph, $MePh_2$). Stereochemically nonrigid eight-coordinate molybdenum and tungsten tetrahydrides H_4ML_4 (M = Mo, W for L = $EtPh_2P$, $MePh_2P$, M = Mo for L = Et_3P, M = W for L = Me_2PhP, including bidentate phosphines and phosphonites as ligands) have been investigated in solution by ^1H NMR spectroscopy [873]. Possible alternatives to the tetrahedral jump mechanism [875, 876] are discussed to explain the temperature dependance of the ^1H NMR spectra. The temperature dependent ^{31}P NMR spectrum of $MoH_4(MePh_2P)_4$ in CH_2Cl_2 solution has been studied as well. It is assumed that the most likely mechanism involves, predominantly, a concerted motion of three of the hydrogen nuclei [654].

Stereochemical nonrigidity in the "crowded" molybdenum(II) complex πC_5H_5-$Mo(CO)(MePh_2P)_2Cl$ has been observed. At room temperature a rapid (with respect to the NMR time scale) intramolecular interconversion of cis and trans isomers occurs, which may be frozen at $-62°C$ [1270]. The kinetics of the related intramolecular exchange phenomenon in the complex $\pi C_5H_5MoH(CO)_2L$ (L = tertiary phosphines) has been studied by means of variable temperature NMR [670].

Triphenylchromium forms stable complexes with alkylarylphosphines (Bu^n_2PhP and Et_2PhP) of formulas $CrPh_3(Bu^n_2PhP)_2$ and $CrPh_3(Et_2PhP)_2$ [1074]. The tetrahydrofuran adduct $Ph_3Cr(THF)_3$ is used as starting material for the synthesis. The liberated tetrahydrofuran must be removed from the equilibrium by pumping on the reaction mixture. In contrast to the phosphine complexes of chromium(III) chloride [103] with analogous composition $[CrCl_3(R_3P)_2]_2$, these compounds are monomeric in benzene solution.

5. MANGANESE, TECHNETIUM, AND RHENIUM

Dicarbonylcyclopentadienyl (triphenylphosphine)manganese complexes with Et, $PhCH_2$, MeS, Cl, Br, I, and COOMe substituents in the cyclopentadienyl ring were synthesized from the corresponding derivatives of tricarbonylcyclopentadienylmanganese by replacement of a carbonyl ligand under the influence of UV irradiation. The chemical shift of the protons of the cyclopentadienyl ring depends on the nature of the substituents [679]. Tricarbonylmethylcyclopentadienylmanganese(I) yields two compounds when treated with triphenylphosphine in the presence of a sodium dispersion in tetrahydrofuran solution. One is the diamagnetic $\pi CH_3C_5H_4Mn(CO)_2Ph_3P$; the other is a novel paramagnetic manganese(O) compound, *cis*-$Mn(CO)_3(Ph_3P)_2$. The derivatives $BrMn(CO)_3(Ph_3P)_2$, $CH_3Mn(CO)_3(Ph_3P)_2$, and $Mn(CO)_3(Ph_3P)_2Mn(CO)_3(Ph_3P)_2 \cdot ½CHCl_3/½CHCl_3$ have been prepared from *cis*-$Mn(CO)_3(Ph_3P)_2$ [920].

Substitution of two carbonyl molecules in $[\pi C_5H_4RMn(NO)(CO)_2]PF_6$ by Me_2PhP, Ph_3P, and some other group V donors yielded the monocationic species $[\pi C_5H_4RMn(NO)L_2]^+$ (R = H, Me, L = Me_2PhP, Ph_3P, etc.) [647].

The racemization of the enantiomers (*8*) and (*9*), as well as the epimerization of the diastereomers (*6*) and (*7*), have been shown to proceed by first-order kinetics [201]:

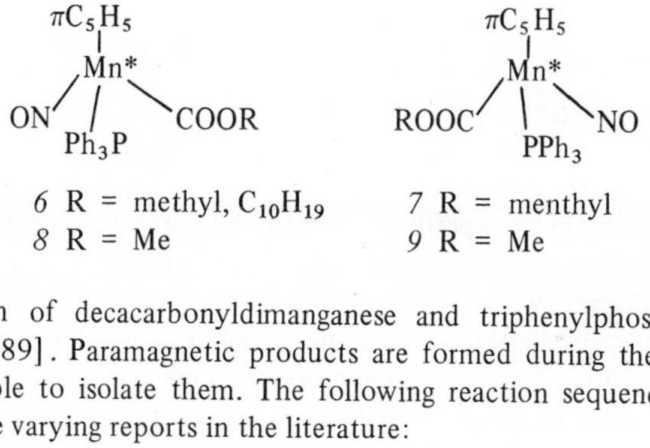

(27)

6 R = methyl, $C_{10}H_{19}$ *7* R = menthyl
8 R = Me *9* R = Me

The reaction of decacarbonyldimanganese and triphenylphosphine has been investigated [889]. Paramagnetic products are formed during the reaction but it was not possible to isolate them. The following reaction sequence is believed to account for the varying reports in the literature:

$$Mn_2(CO)_{10} + Ph_3P \longrightarrow Mn_2(CO)_9Ph_3P + CO \quad (28)$$

$$Mn_2(CO)_9Ph_3P + Ph_3P \longrightarrow (9,\ 10)\ Mn_2(CO)_8(Ph_3P)_2 + CO \quad (29)$$

$$Mn_2(CO)_8(Ph_3P)_2 + [H] \longrightarrow Mn(CO)_4Ph_3P + MnH(CO)_4Ph_3P \quad (30)$$

$$Mn(CO)_4Ph_3P + [H] \longrightarrow MnH(CO)_4Ph_3P \quad (31)$$

$$MnH(CO)_4Ph_3P + Ph_3P \longrightarrow MnH(CO)_3(Ph_3P)_2 + CO \quad (32)$$

[H] denotes hydrogen abstracted from solvent or triphenylphosphine. Replacement of carbon monoxide in pentacarbonylpentafluorophenylmanganese by various phosphorus(III) ligands led to a series of substitution products of the type $Mn(CO)_4C_6F_5L$ (L = Ph_3P, Bu_3P). In some cases it was possible to observe the coupling of phosphorus to para fluorine in the C_6F_5 ring [960].

Acetatobis(triphenylphosphine)dicarbonylmanganese(I), $Mn(CH_3CO_2)(CO)_2(Ph_3P)_2$ [374], has been accidentally synthesized by the reaction of $NaMn(CO)_5$ with Me_3SiCl, followed by the addition of triphenylphosphine and acetic acid. $Mn_2(CO)_9Ph_3P$ and $Mn(CO)_3(Ph_3P)_2Cl$ were also isolated from the reaction mixture, which suggests that the $[Mn_3(CO)_{14}]^-$ ion plays role as an intermediate.

Reactions of some triphenylphosphine substituted carbonyl anions, for example, $Na[Mn(CO)_4Ph_3P]$ and $Na[\pi C_5H_5Mo(CO)_2\,Ph_3P]$, with triphenylchlorosilane have been studied. No silylation of the transition metal species was observed in these cases, in contrast to the corresponding reaction with $Na[Co(CO)_3Ph_3P]$ [349]:

$$Na^+[Mn(CO)_4Ph_3P]^- \xrightarrow{Ph_3SiCl} Mn(CO)_3(Ph_3P)_2Cl + [Mn(CO)_4(Ph_3P)]_2 + Ph_3Si\text{-}O\text{-}SiPh_3 \qquad (33)$$

$$Na^+[Co(CO)_3Ph_3P]^- + Ph_3SiCl \longrightarrow \text{trans-}[Co(CO)_3Ph_3P(Ph_3Si)] + NaCl \qquad (34)$$

(31% yield, ν_{CO} = 1950 cm^{-1}). Perfluorocyclopentene reacts with the $[Mn(CO)_4Ph_3P]^-$ anion to give *trans*-$C_5F_7Mn(CO)_4Ph_3P$, in which a vinylic fluorine atom is replaced by the organometallic group [73]:

$$C_5F_8 + Na[Mn(CO)_4Ph_3P] \longrightarrow NaF + \qquad (35)$$

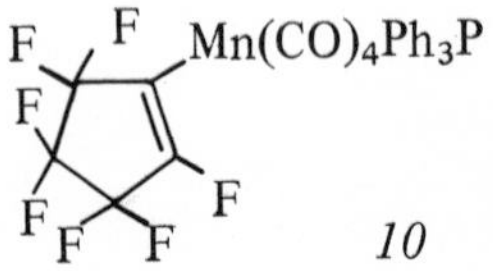

10

Treatment of pentacarbonyl-σ-(perfluoro-1-cyclopentenyl)manganese with triphenylphosphine gave a mixture of the cis and trans isomers of (*10*).

Anhydrous hydrazine with bromotricarbonylbis(dimethyphenylphosphine)manganese(I) and various halogenophosphinecarbonylrhenium(I) complexes gave isocyanato hydrazine tricarbonyl derivatives of manganese(I) and rhenium(I) [899, 900].

Stable σ-prop-2-ynyl complexes with the anions $[Mn(CO)_4Ph_3P]^-$ and $[Mn(CO)_3(Ph_3P)_2]^-$ have been reported [74]. Carboxylic acids form stable addition products, whereas acids HX (X = Cl, Br, I, CF_3CO_2) react to give $MnX(CO)_4Ph_3P$.

Manganese and rhenium pyrazolato borate complexes containing phosphines (Ph_3P, Me_3P, Me_2PhP, Pr^i_3P, Bu^n_3P, and Cy_3P) may be obtained by various synthetic routes [153, 1078] (see *11* and *12*).

Substitution reactions of decacarbonyldirhenium with dimethylphenylphosphine [1118] and bromopentacarbonylrhenium(I) with diethylphenylphosphine (including triphenylphosphite) [1023] have been studied. A marked temperature dependence was observed.

The reaction of the hydrido compound $ReH_3(Ph_3P)_4$ (in benzene suspension) with carbon monoxide leads to $ReH(CO)_2(Ph_3P)_3$ and $ReH(CO)_3(Ph_3P)_2$. The reaction of these compounds with iodine, acids, and CS_2 has been reported [474].

Zakharkin and L'vov have, for the first time, carried out the replacement of ligands in metalocarboranes [1283], for example.

$$Cs[\pi\text{-}(3)\text{-}1,2\text{-}B_9H_{11}C_2]Mn(CO)_3 + Ph_3P \xrightarrow{\text{UV irradiation}} CO + Cs[\pi\text{-}(3)\text{-}1,2\text{-}B_9H_{11}C_2]Mn(CO)_2Ph_3P \quad (36)$$

Dinitrogen complexes of rhenium(I) and rhenium(II) are obtained from the benzoylazo complexes $Re(N{=}NCOPh)Cl_2(Ph_3P)_xL_y$ ($L = Me_2PhP$, Ph_3P, $EtPh_2P$, Pr^nPh_2P and others, $x = 0, 2$; $y + x = 3$) with some phosphorus(III) ligands in boiling methanol [249]:

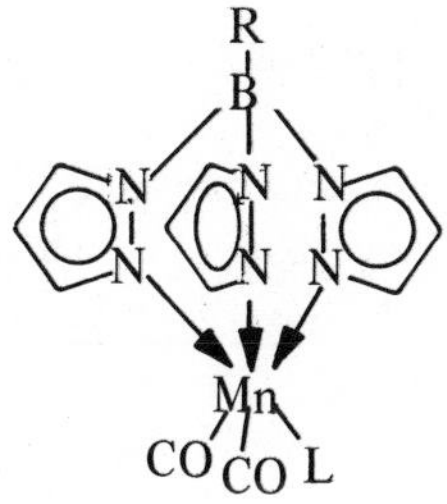

11

$L = Ph_3P, Bu^3_3P, Pr^i{}_3P, Me_3P, Cy_3P,$
$R = $ H, pyrazolyl

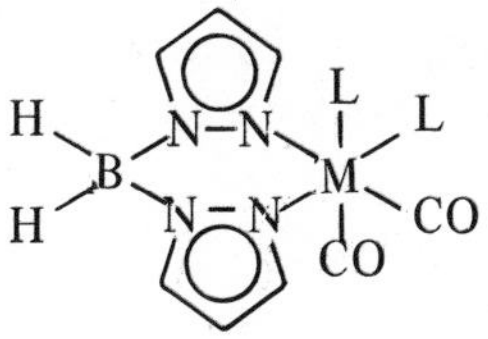

12

$L = Me_2PhP, M = Mn$
$L_2 = Ph_2PCH_2CH_2PPh_2, M = Mn$

On oxidation, the dinitrogen complexes form moderately stable paramagnetic cations; for example, $[ReCl(N_2)(Me_2PhP)_4]^+$. The complex *trans*-$[ReCl(N_2)(Me_2\text{-}PhP)_4]$ forms polynuclear dinitrogen complexes with a great variety of acceptor molecules [250]. Jabs and Herzog [642] have reported ligand exchange reactions in nitridodichloro-bis(triphenylphosphine)rhenium(V) $[ReNCl_2(Ph_3P)_2]$ with ni-

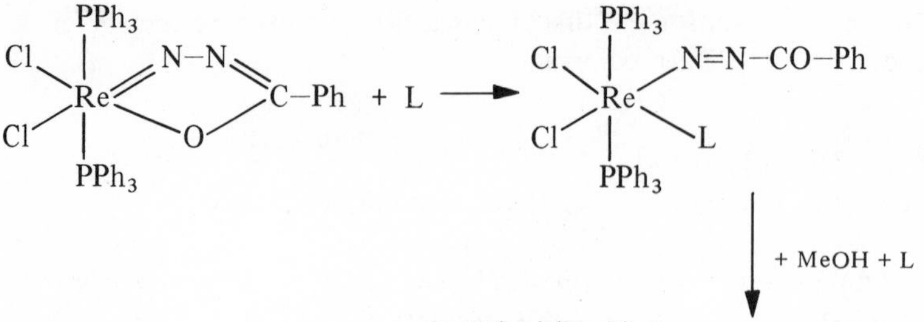

$$ReCl(N_2)(Ph_3P)_2L_2 \text{ or } ReCl(N_2)L_4 + PhCOOMe + HCl \ (+\ 2\ Ph_3P) \quad (37)$$

trogen donors (e.g., pyridine, α,α'-bipyridyl, 1,10-phenanthroline). Phenylhydrazine with mer-$ReCl_3(Me_2PhP)_3$ in refluxing ethanol yields two rhenium complexes, one possibly being a phenylhydrazinenitridorhenium(V) complex with trans phosphine ligands, which, when treated with dimethylphenylphosphine, gives a phenylazo derivative of rhenium(III) $[ReCl_2(N_2Ph)(Me_2PhP)_3]$, the second product of the reaction of mer-$ReCl_3(Me_2PhP)_3$ with phenylhydrazine [415].

The reaction of the β-rhenium(IV) chloride under strictly anhydrous conditions with various donor molecules [411] is more complex than reported earlier [322]. Triphenylphosphine in acetonitrile yields *trans*-$ReCl_4(Ph_3P)_2$, $[ReCl_3(Ph_3P)]_2$, and $ReCl_3(Ph_3P)_2 \cdot CH_3CN$. The products resulting from these reactions depend critically on the solvent [1237].

The coordination chemistry of rhenium(V) chloride has been investigated. With triphenylphosphine (L) as ligand, a series of products could be isolated; for example, $[ReCl_3L]_x$, $ReCl_4L_2$, and $[MeCOCH_2(CMe_2)Ph_3P]$ $[ReCl_5L]$, their composition being strongly dependent on the nature of the solvent employed in these reactions [493].

6. IRON, RUTHENIUM, AND OSMIUM

Braterman and Wallace [176] studied the reaction between enneacarbonyldiiron and *p*-tolylphosphine. The reactions (38) to (43) are believed to be consistent with the observations made on this system:

$$Fe_2(CO)_9(\text{solid}) \rightleftharpoons Fe(CO)_5 + Fe(CO)_4 \quad (\text{slow}) \qquad (38)$$

$$Fe(CO)_4 + L \longrightarrow Fe(CO)_4L \quad (\text{very fast}) \qquad (39)$$

$$Fe(CO)_4 + CO \longrightarrow Fe(CO)_5 \quad (\text{fast}) \qquad (40)$$

$$Fe(CO)_4 + O_2 \longrightarrow \text{oxidized products} + CO \quad (\text{slow under the conditions}) \qquad (41)$$

$$Fe_2(CO)_9(\text{solid}) \rightleftharpoons [\text{unknown}] + CO \qquad (42)$$

$$[\text{unknown}] + 2\,L \longrightarrow Fe(CO)_3L_2 + y\,Fe(CO)_5 + y\,Fe(CO)_4L \quad (43)$$
(y is probably zero)

Oxidative addition reactions of triphenylphosphine with dodecacarbonyltriosmium have been studied [172]. Some unusual new compounds, for example, $Os_3H(CO)_n(Ph_3P)(Ph_2PC_6H_4)$ (n = 8, 9), $Os_3H(CO)_7(Ph_2P)(Ph_3P)C_6H_4$, $Os_3(CO)_8(Ph_2P)Ph(PhPC_6H_4)$, and $Os_3(CO)_7(Ph_2P)_2C_6H_4$, could be isolated. Trinuclear ruthenium complexes of the type $Ru_3(CO)_{12-n}L_n$ (L = tertiary phosphine or arsine, n = 1 - 4) and their pyrolytic reactions in boiling decalin have been reported [199].

The expected Lewis base acid interaction of the trimethylsilyl or trifluorosilyl group, respectively, with the center of coordination (Fe) in carbonyliron complexes like $Fe(CO)_4(Me_3SiCH_2CH_2PMe_2)$ and $Fe(CO)_4(F_3SiCH_2CH_2PMe_2)$ could not be proved unambigously [545]. The related ligand tris(trimethylsilylmethyl)phosphine $(CH_2SiMe_3)_3P$ (L) reacts with carbonyliron to form the complexes $Fe(CO)_3L_2$ and $Fe(CO)_4L$ with a normal composition. The reaction of $[Co(CO)_3L_4]_2$ or $[Co(CO)_3L_2][Co(CO)_3L_2][Co(CO)_4]$ with halogenated solvents under evolution of CO form the emerald green complexe $[CoCl(CO)_2L]_2$, the first monophosphine dicarbonyl complex of cobalt (I) [619]; ruthenium complexes with this ligand have been reported as well [618].

It is found that tributylphosphine substitutes one carbonyl group in carbonyliron complexes $[FeBr_2(CO)_3(Me_2NAl)]_2$ with the unusual ligand $AlNMe_2$. The liberated CO molecule is inserted in the Al-N bond to form $[FeBr_2(CO)_2(Me_2N\text{-}(CO)Al)Bu^n_3P]_2$ [980].

Oxidative elimination reaction of stannic halides SnX_4 (X = Cl, Br) with triphenylphosphine-substituted tetracarbonyliron(O) complexes yields hexacoordinate iron derivatives of the $Fe(CO)_3(X)(Ph_3P)(SnX_3)$ type with an iron-tin bond [1159]. Enneacarbonyldiiron forms complexes with 3,6-diphenylpyridazine containing an iron-iron bond. Triphenylphosphine readily displaces one carbonyl ligand with formation of the monosubstituted derivative (*13*) [758].

The synthesis and the catalytic properties of carbonyltriphenylphosphineruthenium(II) complexes $RuH(CO)Cl(Ph_3P)_2$, $RuCl_2(CO)_2(Ph_3P)_2$ [646, 1266], and $RuCl_2(CO)_2(Ph_3P)_2$ have been published [646]; the latter compound is an efficient olefin isomerization catalyst.

Hydrogen-deuterium exchange in tricyclohexylphosphine was observed in the complexes $MHCl(CO)(Cy_3P)_2$ (M = Os, Ru) on reaction with deuterium gas and deuteriated ethanol [903].

New and improved syntheses of *cis*-$RuCl_2(CO)_2(Ph_3P)_2$, $OsHCl(CO)(Ph_3P)_3$, $OsH_2(CO)(Ph_3P)_3$, and $OsH_2(CO)_2(Ph_3P)_2$ use ethanolic potassium hydroxide or triethylamine as the basic reducing agent [7]. The bulky tertiary phosphines Bu^t_2PhP and Bu^t_2p-tolylP (L) with ruthenium trichloride form binuclear ruthenium(I) complexes $Ru_2Cl_2(CO)_4L_2$ [854].

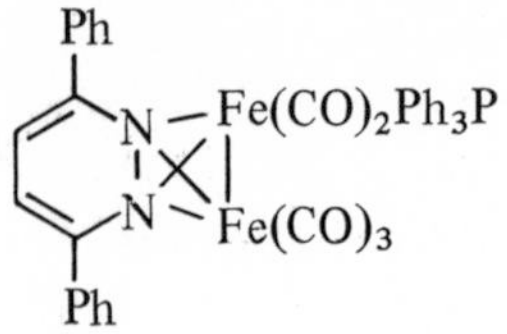

13

Pentacoordinate complexes $M(CO)_{5-n}(Ph_3P)_n$ (M = Ru, Os, n = 1, 2) are prepared by reacting the corresponding pentacarbonyl with Ph_3P. $M(CO)_3(Ph_3P)_2$ reacts with hydrogen in tetrahydrofuran at 130°C/120 atm to give $MH_2(CO)_2$-$(Ph_3P)_2$ [773]. Related hydridocarbonylruthenium(II), $RuHCl(CO)_2(Ph_3P)_2$, and carbonyl complexes *cis*- and *trans*-$RuCl_2(CO)_2(Ph_3P)_2$ have been synthesized [644]. Treatment of K_2OsCl_6 with tricyclohexylphosphine in alcohols leads to the hydridocarbonyl complex $OsHCl(CO)(Cy_3P)_2$ [901]. The reaction of sulfur dioxide and carbon disulfide with $OsHCl(CO)(Cy_3P)_2$ has been studied [906]. In contrast to the diastereomers and enantiomers $\pi C_5H_5Mn(NO)(Ph_3P)COOR$ [201] the related iron complexes $\pi C_5H_5Fe(CO)(Ph_3P)COOR$ (R = Me or menthyl) do not undergo racemization even at higher temperatures [203].

Reduction of iodo-π-cyclopentadienyliron [$\pi C_5H_5FeI(CO)_2$] with $LiAlH_4$ in the presence of phosphorus(III) ligands (L) like Me_3P, Me_2PhP, Ph_3P, $P(OMe)_3$, and $P(NMe_2)_3$ affords the corresponding derivatives $\pi C_5H_5FeH(CO)L$. With CCl_4 the chloro derivatives $\pi C_5H_5FeCl(CO)L$ are formed [671].

Hexafluorobut-2-yne may insert in the Ru-H bond of the cyclopentadienylruthenium complex $\pi C_5H_5RuH(Ph_3P)_2$ and the C–H bond of (*14*) formed primarily to give compounds of type *15*. [142].

The reaction of the bromosubstituted allene (*16*) with [$\pi C_5H_5Fe(CO)_2$]Na proceeds via an unstable acetyl intermediate (*17*) to the π-bonded allene complex (*18*). If the reaction is carried out in presence of tributylphosphine, the complex with the uncoordinated allene group (*19*) may be isolated [96]:

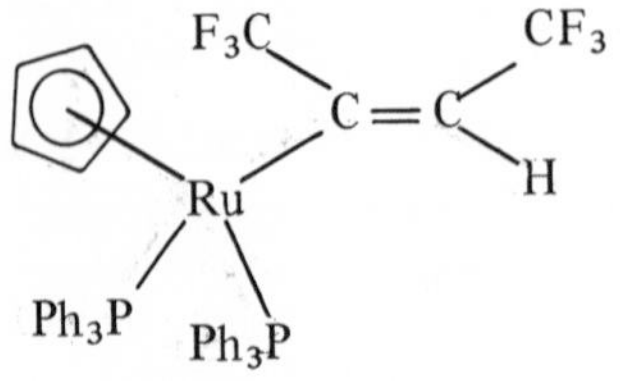

14

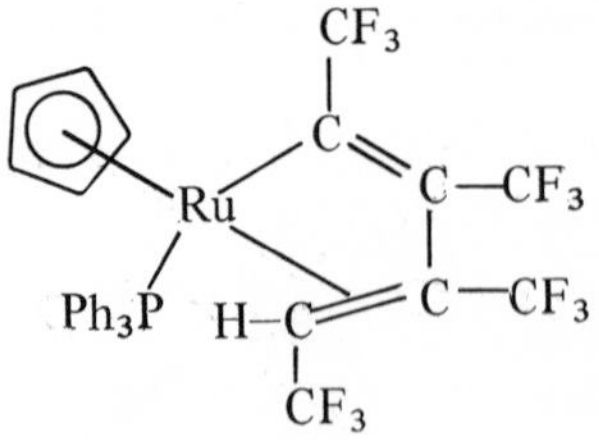

15

$$[\pi C_5H_5Fe(CO)_2]^- + \text{R-Br} \xrightarrow[6\text{-}12\,h]{25°C \text{ in THF}} \pi C_5H_5Fe(CO)_2\text{-R} \rightleftharpoons \pi C_5H_5Fe(CO)\text{-CO-R}\ (17) \qquad (44)$$

[for R, e.g., $(CH_2)_2CH{=}C{=}CH_2$ (*16*)]

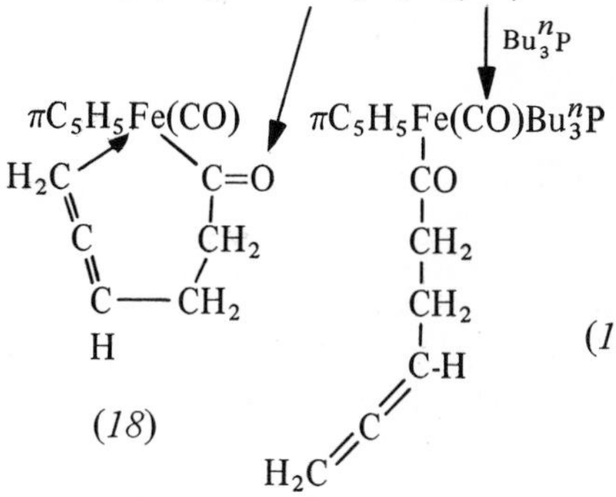

(*18*) (*19*)

Substitution reactions of $\pi C_5H_5Fe(CO)_2SnR_3$ (R = alkyl, aryl, halogens) with phosphorus(III) donors L (e.g., L = Ph_3P, Ph_2PCF_3, $MePh_2P$, Me_2PhP) have been investigated. The products isolated and studied were $\pi C_5H_5Fe(CO)LSnR_3$ [346, 723] and $\pi C_5H_5FeL_2SnR_3$ [346].

A series of new ruthenium(III) and ruthenium(II) complexes containing triphenylphosphine could be prepared by for example, $RuX_3(Ph_3P)_2So$ [So = Me_2SO, THF, RCN (R = Me, Ph, $PhCH_2$, $CH_2{=}CH$), Me_2CO, CS_2, etc., X = Cl, Br] using $RuX_3(MeOH)(Ph_3P)_2$ as starting material [1036, 1156]. Replacement of Ph_3P ligands in $RuCl_2(Ph_3P)_2$ by pyridine in excess yields the complexes $RuCl_2Ph_3P$-$(Py)_3$, $RuCl_2(Ph_3P)_2Py_2$, and $RuCl_2(Py)_4$ [996]. There are now some well established synthetic methods for ruthenium tertiary phosphine complexes of type $RuX_2(R_3P)_n$ (n = 3, 4, X = Cl, Br, R_3 = Ph_3, Ph_2PCl, Me_2Ph, $MePh_2$, Et_2Ph, $EtPh_2$) [47, 569].

Hydrated ruthenium trichloride $RuCl_3(H_2O)_3$ with dimethyl-(1-naphthyl)phosphine (L) gives mainly five coordinate complexes $RuCl_2L_3$ and a small quantity of the $[Ru_2Cl_3L_6]^+$ species which may be isolated as its BPh_4 salt. Reaction of L with osmium tetroxide in the presence of concentrated hydrochloric acid and alcohol yields maroon mer-$OsCl_3L_3$ and internally metallated $OsCl_2[Me_2P(C_{10}H_6)L_2]$ [409].

Anionic ruthenium(III) complexes $A[RuCl_4(R_3P)_2]$ (A = Me_4N^+, Ph_4As^+, R_3P = Ph_3P, Me_2PhP, Et_3P, etc.) have been synthesized by various methods; for example, by the "oxidation" of $RuCl_2(Ph_3P)_n$ (n = 3, 4) with concentrated hydrochloric acid [1135].

The blue solution of Ru(II) in methanol (which probably contains the anion $[Ru_5Cl_{12}]^{2-}$) may be used as a convenient starting material for the preparation of a variety of ruthenium(II) and ruthenium(III) complexes [501]. Zinc reduction of solutions of mer-$OsX_3(R_3P)_3$ (X = Cl, Br, R_3 = Me_2Ph, Et_2Ph, $EtPh_2$, $MePh_2$, Et_3) in anhydrous tetrahydrofuran under a nitrogen atmosphere produces dinitrogen complexes of osmium(II), $OsX_2(N_2)(R_3P)_3$ [253].

Cationic bisarylazo complexes $[Ru(N_2Ar)_2(Ph_3P)_2Cl]$ $[BF_4]$ are reported to be formed from $RuHCl(Ph_3P)_3$ and $[N_2Ar]$ $[BF_4]$ (Ar = *p*-MeC_6H_4 or *p*-$MeOC_6H_4$) with elimination of the hydrido group [805]. Complexes of osmium(II) containing besides various phosphine ligands, organic cyanides and isocyanides have been prepared [257, 259] and their reactions with sodium borohydride or appropriate primary amines investigated [259].

Treatment of iron(II) chloride dihydrate with sodium borohydride in ethanol in presence of tertiary phosphines R_3P (R_3 = $EtPh_2$, $BuPh_2$, $MePh_2$) under a nitrogen atmosphere results in the formation of dihydridodinitrogen complexes of iron(II), $FeH_2(N_2)(R_3P)_3$ [45]. These hydrido complexes react with carbon monoxide [45] and ethylene [132] with replacement of the weakly bonded dinitrogen. Elimination of hydrogen was observed when dihydridotetracarbonyliron was treated with triphenylphosphine. The products formed are $Fe(CO)_4Ph_3P$ and (if Ph_3P is used in excess) $Fe(CO)_3(Ph_3P)_2$ [447]. This is in sharp contrast to the analogous reactions of other hydridocarbonyls [160,583].

Dihydridotetrakis(triphenylphosphine)ruthenium, $RuH_2(Ph_3P)_4$ may be prepared from $RuCl_3$ or ruthenium(III) acetylacetonate, triphenylphosphine, and triethylaluminium in tetrahydrofuran. The corresponding rhodium complex, $RhH(Ph_3P)_4$ could be obtained in a similar manner. $RuH_2(Ph_3P)_4$ dissociates in solution into free triphenylphosphine and dihydridotris(triphenylphosphine)ruthenium [639]. Some reactions of these two and other related complexes are summarized in Scheme 3. In the exchange reaction with deuterium 24 to 25 hydrogen atoms per mole RuH_2-$(Ph_3P)_4$ were found to participate, i.e., all ortho hydrogens are involved [639]. It is worthwhile mentioning the complex $[RuH_2(N_2)L_3]_2B_{10}H_8$ as it contains dinitrogen as a bridging ligand between ruthenium and boron [718].

The hydrido complexes of osmium(II) and osmium(IV), OsH_2L_3L' and OsH_4-L_3 (L = L' = Me_2PhP, Et_2PhP, $MePh_2P$, $EtPh_2P$; L' = CO, L = Me_2PhP, Et_2-PhP, $MePh_2P$, $EtPh_2P$) have been prepared by various methods [90,778].

Robinson and Uttley published a convenient single-stage synthesis for a selection of ruthenium and osmium nitrosyl halide complexes, $MX_3(NO)(R_3P)_2$ (M = Ru, Os, X = Cl, Br, I, R_3 = Ph_3, p-$tolyl_3$, Et_3, $Pr^i{}_3$, $Bu^n{}_3$, $(p\text{-}C_6H_4Cl)_3$) with the phosphine ligands in trans positions [1032].

A rapid intramolecular phosphine exchange has been detected by means of NMR in hydridonitrosyltris(tertiary phosphine) complexes of ruthenium and osmium, $MH(NO)(R_3P)_3$ (M = Ru, Os, R_3 = Ph_3, $MePh_2$ for M = Ru in addition Pr^iPh_2, $CyPh_2$). Structure *20* was proposed for the high temperature isomer, being consistent with IR and NMR coupling constant data. The catalytic activity may be associated with the presence of isomer (*20*) in solution [1265]:

SCHEME 3

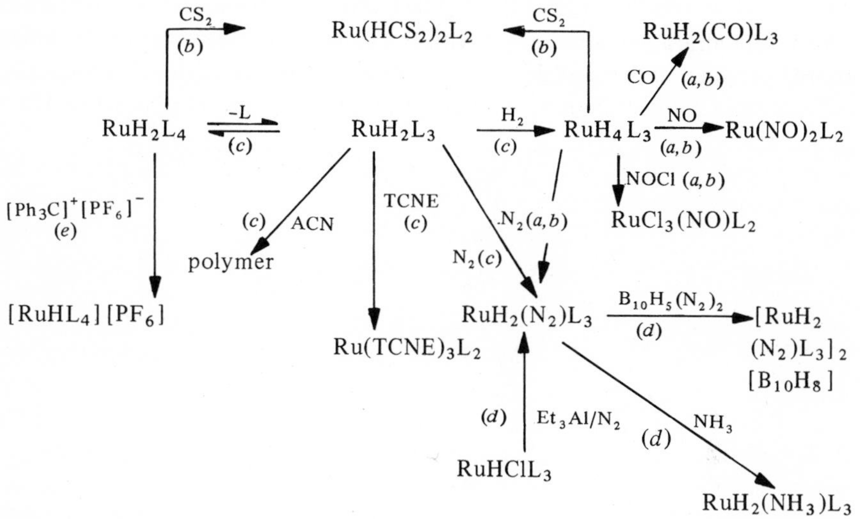

TCNE = tetracyanoethylene
ACN = Acrylonitrile
$B_{10}H_8(N_2)_2$ is the inner diazonium salt of $B_{10}H_{10}^{2-}$
L = Ph_3P
Letters in parentheses refer to corresponding references: (*a*) [429], (*b*) [577], (*c*) [639], (*d*) [718], (*e*) [1047]

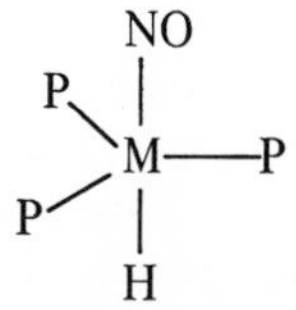

(*20*)

M = Ru, Os, P = phosphine ligands

Nitrosyltetrafluoroborate or -hexafluorophosphate react with *trans*-$Fe(CO)_3(Ph_3P)_2$ to give the cationic nitrosyl iron complex $[Fe(CO)_2(NO)(Ph_3P)_2]^+$. Treatment of $Ru_3(CO)_9(Ph_3P)_3$ with NO^+ in methanol or $OsCl(CO)(NO)(Ph_3P)_2$ with carbon monoxide in the presence of $NaBPh_4$ yields the corresponding ruthenium or osmium cations [660]. The reactions of tertiary phosphines (and arsines), for example, Me_2PhP, Me_2PhAs, with trichloronitrosyl-ruthenium leading to the complexes $RuCl_3(NO)L_2$ have been reinvestigated [1188].

The unusual complex (*21*) was obtained when tris(diphenylethylphosphine)dinitrogen-dihydridoiron was treated with 2,3-dimethylbutadiene under UV irradiation. Depending on the availability of an uncoordinated phosphorus atom in (*21*), the complex (*22*) may be formed photochemically with pentacarbonyliron [722].

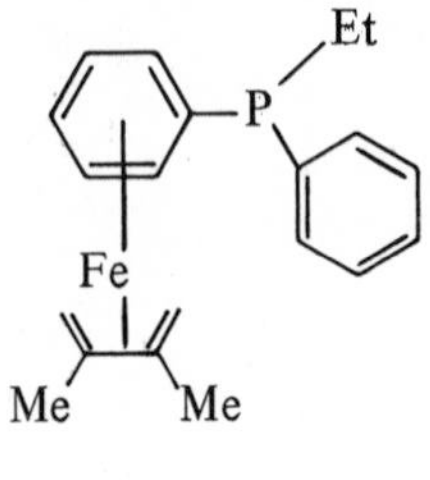

21

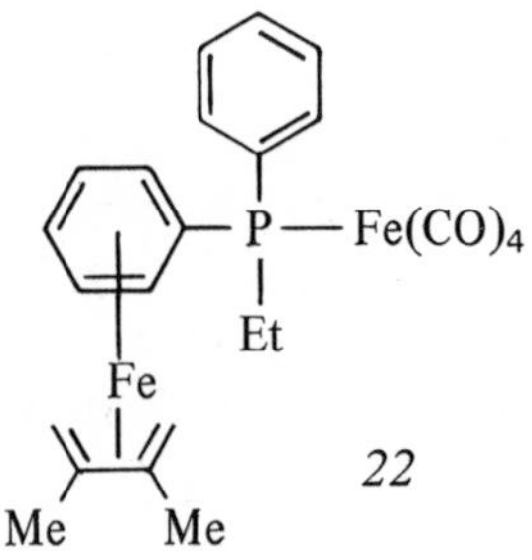

22

Similar bonding of phenylphosphines as in (*21*) and (*22*) was observed in some chromium, molybdenum and tungsten complexes [166, 168].

The complex acids $H_3[Fe(CN)_5L]H_2O$ have been synthesized by treatment of the corresponding sodium salts $Na_3[Fe(CN)_5L]$ (L = Ph_3P, Bu^n_3P, Cy_3P, etc.) with concentrated hydrochloric acid [970].

7. COBALT, RHODIUM, AND IRIDIUM

The coordination chemistry of cluster metal carbonyls continues to attract attention. Labroue and Poilblanc reported the synthesis of the first products of the $Co_4(CO)_{12-n}L_n$ series (L = Et_3P, n = 2, 3). The presence of isomeric species of $Co_4(CO)_{10}(Et_3P)_2$, $Co_4(CO)_9(Et_3P)_3$, $Co_4(CO)_{10}(Me_2PhP)_2$, and $Co_4(CO)_9(Me_2PhP)_3$ was detected by IR spectroscopy [761]. The temperature dependence of their ^1H NMR spectra is possibly due to steric nonrigidity in the cluster [318]. Pregaglia, Andreetta, Ferrari, Montrasi, and Ugo [1003] published the synthesis of cluster complexes of cobalt, $[Co(CO)_2L]_m$ (m is probably 3, L = Bu_3P, Bu_2PhP, Ph_3P). The reactions of the higher homolog cluster compound $Rh_4(CO)_{12}$ with phosphine (and arsine) ligands have been investigated and the complexes $Rh_4(CO)_{11}L$, $Rh_4(CO)_{10}L_2$ could be isolated [L = Ph_3P, p-tolyl$_3$P, $(p\text{-}FC_6H_4)_3P$, and 2L $Ph_2PCH_2CH_2PPh_2$]. Carbon monoxide under pressure, preferably in the presence of excess phosphine, splits these Rh_4-substituted clusters to give unstable dinuclear species, $Rh_2(CO)_6L_2$ [1257]. Degradation of the cluster occurs also if the hexanuclear derivatives of $Rh_6(CO)_{16}$, for example, $Rh_6(CO)_7L_9$ (L = Ph_3P), are treated with thiols, halogens, or carboxylic acids [657]. When the monosubstituted cluster $Co_4(CO)_{11}Ph_3P$ was treated with bis(pentafluorophenyl)-acetylene, besides the alkyne complex, a disubstituted derivative of octacarbonyldicobalt, $Co_2(CO)_6(Ph_3P)_2$, was obtained [392]. Reaction of carbon monoxide under high pressure (80 to 100 atm) with polynuclear phosphine-substituted iridium carbonyls, for example, $Ir_4(CO)_9L_3$ (L = Ph_3P or $(p\text{-tolyl})_3P$) and $Ir_4(CO)_8L_4$ (L = Et_3P or Bu^n_3P), involves replacement of coordinated phosphine and degradation of the cluster with formation of the dinuclear complexes $Ir_2(CO)_7L$ [1258]. The complexes $Ir_2(CO)_6L_2$ and $IrH(CO)_3L$ (L = Ph_3P) are obtained by refluxing $IrI(CO)L_2$ in benzene suspension with sodium ethoxide. Both complexes may react with hydrogen with the formation of the trihydride $IrH_3(CO)_2L$ [820].

Reactions of pentakis(t-butylisocyanide)cobalt cation with tertiary phosphines have been studied. Pentacoordinate compounds, for example, trans-$[Co(Bu^tNC)_3(Ph_3P)_2][PF_6]$, are formed [698] from $[Co(Bu^tNC)_5][PF_6]$ and Ph_3P. Five coordinate Co(I) isonitrile complexes containing tertiary phosphines as ligands are subject to rapid intra- and intermolecular ligand exchange [914]. Isonitrile complexes of rhodium and iridium $[M(CNR)_3(Ph_3P)_2]^+$ and $[M(CNR)_2(Ph_3P)_2]^+$ [360, 362] undergo oxidative addition reactions that produce $[M(CNR)_3(Ph_3P)_2X]^{2+}$, $[M(CNR)_2(Ph_3P)_2XY]^+$, and $[M(CNR)_2(Ph_3P)_2Y_2]^+$ (M = Rh, Ir, R = alkyl, aryl, X = halogen, methyl, acyl, allyl, HgCl, $SnCl_3$, Y = halogen) [362]. The hydrides $[MH(CNR)_3(Ph_3P)_2][PF_6]$ have been synthesized also [360]. The decomposition of methyltris(triphenylphosphine)cobalt, $[CoMe(Ph_3P)_3]$ and some of its reactions with olefins and acetylenes provide evidence for weakening the metal-carbon bond under the action of bases and olefins [419]. Phosphine-substituted derivatives of dicarbonyltricyanocobaltate(I), for example, $[Co(CN)_2(CO)_2Cy_3P]^+$, $[Co(CN)_2(CO)(Ph_3P)_2]^+$, and $Co(CN)(CO)_2(R_3P)_2$ (R_3= Ph_3, Et_3, Cy_3), have been reported [123]. By using the disulfito complex $Na_5[Co(SO_3)_2(CN)_4]$ as starting material tetracyanobis(triphenylphosphine)cobaltate(III), $Na[Co(CN)_4(Ph_3P)_2]$, may be synthesized [818].

The reactions of $[Co(CO)_4]_2Hg$ with various phosphorus(III) ligands in the absence of light gave $[Co(CO)_3L]_2Hg$ and, under more rigorous conditions, $[Co(CO)_2L_2]_2Hg$ (L = Et_3P, Bu^n_3P, Me_2PhP, Pr_2PhP, $MePh_2P$) [936].

A new series of tautomeric complexes containing nickel-cobalt bonds $\pi C_5H_5\text{-}Ni\text{-}Co(CO)_4L$ (L = Cy_2PhP, Ph_3P, $(p\text{-}FC_6H_4)_3P$) have been prepared by reaction of

πC_5H_5NiXL with $Na[Co(CO)_4]$ in tetrahydrofuran [836].

Substitution of the chloride ion in complexes $IrCl(CO)L_2$ (L = Ph_3P, Me_2PhP, $EtPh_2P$, Et_2PhP, Et_3P, Cy_3P, Pr^i_3P) by neutral ligands L and carbon monoxide in presence of sodium-tetraphenylborate led to cationic pentacoordinate complexes of iridium(I) [277,380]. The syntheses and interconversion reactions of complexes of type $[Ir(CO)_x(Me_2PhP)_{5-x}]^+$ (X = 1, 2, 3) are summarized in Scheme 4. Oxidative addition of hydrogen halides, hydrogen, chlorine, methyl halides, 2-methylallyl chloride, and mercury(II) halides to these five coordinate cationic species yielded hexacoordinate complexes of iridium(III) [381]. The dimeric compounds $[RhX(diene)]_2$ (diene = C_8H_{12} = cycloocta-1,5-diene) with Ph_3P or Bu^n_3P in methanolic solution in presence of $NaBPh_4$ give cationic species $[Rh(diene)L_2]^+$ (L = Bu^n_3P, Ph_3P) as their tetraphenylborate salts [555].

SCHEME 4

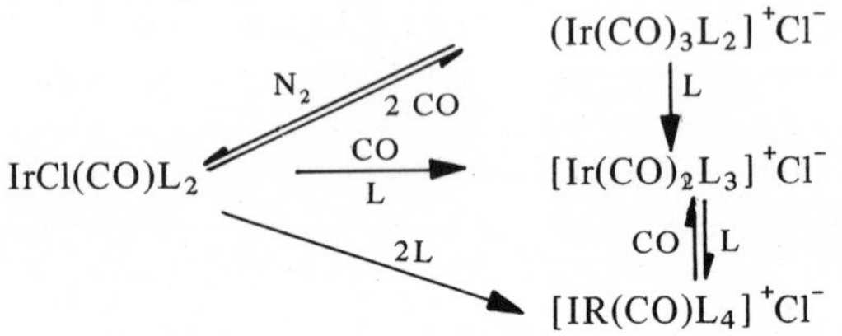

L = Me_2PhP.

Tetrakis(trimethylphosphine)hydridocobalt(I) [709] and the interesting compound tetrakis(trimethylphosphine)cobalt(O) $Co(Me_3P)_4$ [710] have been prepared from Co(II) halides either in aqueous solution or in tetrahydrofuran by using zinc dust or amalgamated sodium, respectively, as reducing agents:

$$CoX_2 + 4\,Me_3P + 2\,Na(Hg) \xrightarrow[20^\circ C]{THF} Co(Me_3P)_4 + 2\,NaX \qquad (45)$$

The reduction of $CoCl(R_3P)_3$ (R_3 = Ph_3, Et_2Ph) with sodium metal in an atmosphere of nitrogen gives bridged dinitrogen complexes $Co(R_3P)_3(N_2)Co(R_3P)_3$ of cobalt(O) or dinitrogen complexes $Na[Co(N_2)(R_3P)_3]$ of cobalt(-I) [46] depending on the reaction conditions employed.

Four coordinate iridium(I) cations $[Ir(MePh_2P)_4]^+$ have been prepared by reaction of silver perchlorate (or tetrafluoroborate) with $IrCl(CO)(Ph_3P)_2$ and subsequent ligand replacement (carbonyl and acetonritrile, which is used as solvent) in

$[Ir(CO)(MeCN)(Ph_3P)_2]^+$ by $MePh_2P$ [291]. Dicarbonylbis(phosphine)iridium(I) complexes, $[Ir(CO)_2L_2]$ [A] [444] (L = Ph_3P, $(o\text{-tolyl})_3P$, Pr^i_3P, Cy_3P, Bu^tPh_2P, A = BF_4^- or BPh_4^-), have been obtained from chlorotricarbonyliridium $[IrCl(CO)_3]_n$ and tertiary phosphines L in the presence of $AgBF_4$ or $NaBPh_4$. They absorb carbon monoxide to give cationic, five-coordinate compounds $[Ir(CO)_3L_2]^+$. Interesting reports have been made on the reactions of mer-$IrCl_3L_3$ (L = Me_2PhP, Et_2PhP) and *trans*-$IrCl(CO)(Ph_3P)_2$ with silver nitrate in acetone [237,1103]. Depending on the amount of silver nitrate used, two products, $Ir(CO)(NO_3)(Ph_3P)_2$ and $Ir(CO)(NO_3)_2(Ph_3P)_2Ag$ (with a silver-iridium bond), could be isolated [237]. The nitrato group in $IrX_2(NO_3)L_3$ is readily replaced by a variety of ligands Q (carbon monoxide, Me_2PhP, Me_2PhAs, NH_3, etc.) to form complexes of type $[IrX_2QL_3]^+$ (X = Cl, I, L as above) [1103].

Quite a large number of publications on the syntheses and reactions of chlorocarbonyliridium and -rhodium complexes have appeared in the literature. Only a selected number is given (Ir [261, 298, 306, 351, 765, 954, 1104, 1105], Rh [155, 213, 481, 482, 680, 815, 954, 1130, 1224, 1267]) and a few are discussed here.

The Lewis basicity of iridium(I) complexes, *trans*-$IrX(CO)L_2$ (X = Cl, F, L = Ph_3P, Ph_3As), $IrCl(CO)L_2$ (L = Et_2PhP, $MePh_2P$), and $IrCl(CS)(Ph_3P)_2$, toward a range of main group acceptors has been studied [467, 1095]. The relative affinities of some group III Lewis acids toward Ir(I) bases are as follows: $BF_3 > B(C_6F_5)_3 \gg B_2H_6$ and $Al_2Me_6 > GaMe_3 \gg BMe_3$ [1095]. For the iridium complex bases the following tentative order of relative basicity was proposed: $IrCl(CO)(Ph_3As)_2 > IrCl(CO)(Ph_3P)_2 > IrCl(CS)(Ph_3P)_2$ [467].

The cationic methyliridium complex $[IrClMe(CO)(Ph_3P)_2]$ $[SO_3F]$ was obtained by reaction of Vaska's compound, $IrCl(CO)(Ph_3P)_2$, with excess of $MeSO_3F$ [1151]. Oxidative addition of thiocyanogen with rhodium(I) and iridium(I) complexes, *trans*-$MX(CO)(Ph_3P)_2$ (M = Rh, Ir, X = Cl, NCO, NCS), gave $MX(CO)(SCN)_2(Ph_3P)_2$ [210]. In a "nucleophile-assisted" methyl migration the complex RhClI(CO)(COMe)-Ph_3P is formed from *trans*-$RhCl(CO)_2Ph_3P$ when treated with methyl iodide [1210].

Neutral methyl iridium complexes have been synthesized by reaction of methyllithium with chlorotris(triphenylphosphine)iridium(I) and some of its reactions, for example, with carbon monoxide, which at higher pressures affords acetyldicarbonyl complexes, $IrMeCO(CO)_2(Ph_3P)_2$, have been reported [1093].

Valentine and Valentine [1217] observed a photo-induced oxidative addition of 9,10-phenanthrenequinone to $IrCl(CO)(Ph_3P)_2$. This reaction results in a convenient synthesis of *o*-quinone complexes:

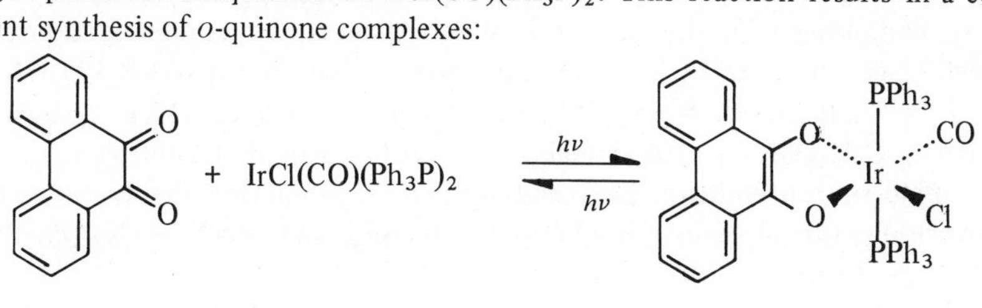

(46)

The metal complex $IrCl(CO)(Ph_3P)_2$ apparently serves to trap electronically excited states of *o*-quinones. Aryldiazo complexes of iridium may be obtained in a single-stage synthesis:

$$IrCl(CO)(Ph_3P)_2 \xrightarrow[ArN_2^+PF_6^-]{EtOH/ArCON_3} [IrCl(ArN_2)(Ph_3P)_2]^+PF_6^- \quad (47)$$

These aryldiazoniumiridium complexes react with carbon monoxide, isonitriles, phosphines, arsines, and stibines to form the corresponding five-coordinate species [581].

Carboxylato complexes of phosphine-substituted dimeric rhodium carbonyls, $Rh_2(CO)_3(OOCR)_2Ph_3P$ (R = Me, Et) [341], and phosphine-substituted tricarbonylacylcobalt complexes, $Co(CO)_3(XFHCCO)Ph_3P$ (X = F, H) [785], have been prepared according to the following equations:

$$[Rh(CO)_2(OOCMe)]_2 + Ph_3P \longrightarrow Rh_2(CO)_3(OOCMe)_2Ph_3P + CO \quad (48)$$

$$(XFHCCO)_2O + Na[Co(CO)_3Ph_3P] \longrightarrow XFHCCOONa + Co(CO)_3(XFHCCO)Ph_3P \quad (49)$$

(X = F, H).

The halogen oxidation of the compounds $Rh(LL)(CO)Ph_3P$ (LL = acac or 8-quinolinoyl) leads to octahedral complexes of rhodium (III) [1225]; for example,

$$Rh(acac)(CO)Ph_3P + I_2 \longrightarrow Rh(acac)(CO)I_2Ph_3P \quad (50)$$

An unusual disproportionation of nitric oxide has been observed on its reaction with $Co(NO)(Ph_3P)_3$ [1035]:

$$Co(NO)(Ph_3P)_3 + 7\,NO \longrightarrow Co(NO_2)(Ph_3P)(NO)_2 + 2\,Ph_3PO + 2\,N_2O + 0.5\,N_2 \quad (51)$$

In the reaction of $[Ir(NO)_2(Ph_3P)_2][PF_6]$ with carbon monoxide the coordinated nitrosyl ligand acts as an oxidant. The reaction products are $[Ir(CO)_3(Ph_3P)_2]$ $[PF_6]$, CO_2, and H_2O [656]. The trinuclear complex $[Co(NO)(CO)(CF_3)_3P]_3$ is formed by thermal decomposition of $Co(NO)(CO)_2(CF_3)_3P$ at 40°C [208].

Nitrosyl complexes of iridium and rhodium continue to attract interest. Convenient single-stage syntheses of halogenonitrosyl complexes, $MX_2(NO)(R_3P)_2$ (M = Rh, Ir, X = Cl, Br, I, R_3 = Et_3, Pr^i_3, Bu^n_3, $MePh_2$, Ph_3, $(p\text{-}ClC_6H_4)_3$, $(p\text{-}C_6H_4\text{-}Me)_3$, and $(p\text{-}C_6H_4OMe)_3$) [1031] make use of either *N*-methyl-*N*-nitrosotoluene-*p*-sulfonamide or pentylnitrite. Beck and Werner [85] reported the synthesis of dinitrosylrhodium complexes $[Rh(NO)_2(Ph_3P)_2]BF_4$ and $Rh(NO)_2(N_3)(Ph_3P)_2$

and a series of nitrosylrhodium complexes. Azido and isocyanatorhodium(I) complexes are used as starting materials for the reaction with $NOBF_4$ in alcoholic media. The iridium analog of $[Rh(NO)_2(Ph_3P)_2]BF_4$ has been prepared as well [1012]. $[IrCl(NO)(Ph_3P)_2]^+$, the cationic nitrosyl analog of Vaska's compound, is formed in a series of redox reactions with $IrCl(CO)(Ph_3P)_2$, $[Ir(NO)_2(Ph_3P)_2]^+$, or $IrHCl(NO)(Ph_3P)_2$ as starting materials [468, 1013]:

$$IrHCl(NO)L_2 \xrightarrow[-H_2\,[1013]]{+\,HY} [IrCl(NO)L_2]^+ \qquad (52)$$

$$[IR(NO)_2L_2]^+ \xrightarrow[[1013]]{+\,½\,Cl_2} [IrCl(NO)L_2]^+ \qquad (53)$$

$$IrCl(CO)L_2 \xrightarrow[[468]]{[1]\,RN_3\,[2]\,NOBF_4} [IrCl(NO)L_2]^+ \qquad (54)$$

$L = Ph_3P, Y = ClO_4, BF_4, PF_6, R =$ P-nitrobenzoyl.
References in brackets.
The reactivity pattern of nitrosyltris(triphenylphosphine)-iridium(-I) [1011] shows some similarities to that of the isoelectronic $Pt(Ph_3P)_4$.

Pseudotetrahedral Co(II) halide phosphine complexes, CoX_2L_2 and $[COX_3L]^-$ (X = Cl, Br, I, $L = Ph_3P$, Bu^n_3P), undergo complete solvolytic displacement of phosphine in coordinating solvents such as acetonitrile. Values of equilibrium constants are reported for the reaction

$$[CoX_3L]^- + X^- \rightleftharpoons [CoX_4]^{2-} + L \qquad (55)$$

(L and X as above). Phosphines replace halides with increasing facility in the order $Cl < Br < I$; the stability order for L is $Bu_3P > Ph_3P$ [1099]. The anions $[CoX_3L]^-$ (X as above, $L = Ph_3P$) are formed together with the cobalticenium cation by reaction of cobaltocene with triphenylphosphonium halides [1219]. Carbon monoxide reacts with complexes of the type CoX_2L_2 (L, e.g., Et_3P, Et_2PhP, Ph_3P, Cy_3P, X = Cl, Br, I) to produce five-coordinate adducts $CoX_2(CO)L_2$ and, on further reaction, cobalt(I) complexes, $CoX(CO)_2L_2$ [179]. Solution [1167] and solid-phase syntheses [1170] have been reported for $CoCl_2(Ph_3P)_2$. Coordinated trialkylphosphine is subject to slow but quantitative autoxidation in complexes such as CoX_2L_2, for example, $CoCl_2(Et_3P)_2$, to yield phosphine oxide complexes [1072]. The importance and the relative stability of five-coordinate geometries in Co(II) complexes (with and without phosphine ligands) and Ni(II) complexes have been discussed by Orioli [962] and Wood [1268].

A complete series of tertiary phosphine complexes of rhodium includes $RhCl_3L_3$, $RhClL_3$, $RhHCl_2L_3$, RhH_2ClL_3, $RhHCl_2L_2$, $[RhCl_3L_2]_2$, $RhCl(CO)L_2$, and $RhHCl_2(CO)L_2$ (L, e.g., Me_3P, Et_3P, Pr^i_3P, Pr^n_3P, Bu_3P, Cy_3P, and $Benzyl_3P$) [631]. A reinvestigation of the preparation of a chlorotris(*n*-butylphosphine)-

rhodium(III) complex, first described by Chatt and Shaw [260] as di-μ-chlorotetrachlorotetrakis(tri-*n*-butylphosphine)dirhodium, revealed that three different complexes, including di-μ- and tri-μ-chloro complexes of rhodium(III) had been formed [26]. Paramagnetic rhodium(II) complexes, *trans*-$RhCl_2L_2$, containing *t*-butylphosphines (L = Bu^t_2RP, R = Me, Et, Pr^n) and the complexes $RhHCl_2L_2$, RhH_2ClL_2 (L including Bu^t_3P), may be prepared from $RhCl_3$ trihydrate and L in alcoholic media [856, 857]. Paramagnetic tricyclohexylphosphinerhodium(II) complexes $RhX_2(Cy_3P)_2$(X = Cl, Br) (including some diamagnetic iridium complexes of this ligand) have been reported as well [902]. If, instead of rhodium(III) chloride, chloroiridous acid is used for the synthesis of *t*-butylphosphine complexes, a series of iridium(III) complexes, including $IrHCl_2(Bu^t_2RP)_2$ and $[Ir_2Cl_7(Bu^t_2Pr^n)_2]$ $[Bu^t_2Pr^nHP]$, may be synthesized. The hydride resonance of $IrHCl_2(Bu^t_2RP)_2$ at τ ca 60 ppm is the highest yet observed [858]. Photochemical isomerizations of some related iridium complexes, for example, mer-IrX_3-$(Et_3P)_3$ to fac-$IrX_3(Et_3P)_3$, have been studied [186].

Dimethyl(1-naphthyl)phosphine (L) is readily metalated in the 8-(peri) position by iridium(III) [417]. Hydride addition-elimination, outlined in Scheme 5., provides a mechanism for this reaction. 9-Phenyl-9-phospha-fluorene and substituted phospholes form complexes with rhodium(I) and rhodium(III) [604]. The phosphorus in the substituted phosphole system has poor donor properties.

The nature of $RhCl(Ph_3P)_3$ in solution and its dependence on various factors has been discussed in the literature [54, 770].

Dissociation according to

$$RhCl(Ph_3P)_3 \rightleftharpoons RhCl(Ph_3P)_2 + Ph_3P \tag{56}$$

is probably not extensive (i.e., less than 5% for $>10^{-2} M$ CH_2Cl_2 solutions) [770]. Methyltris(triphenylphosphine)rhodium with diphenylacetylene gives *trans*-(α-methylstilbene) in an insertion-type reaction [887].

SCHEME 5

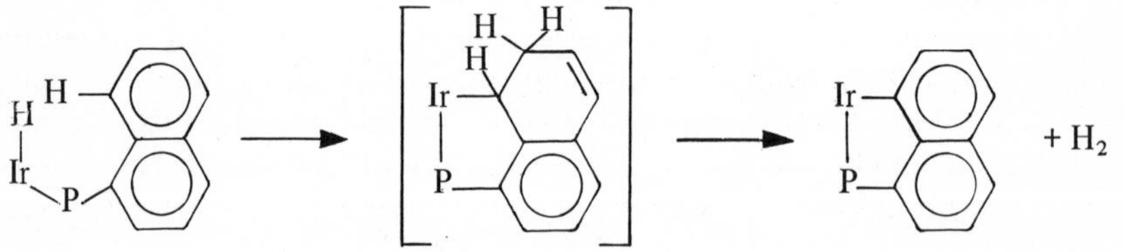

Syntheses and reactions of cyclopentadienyls of cobalt(I) and rhodium(I) or (III), for example $\pi C_5H_5Co(Ph_3P)_2$ and $\pi C_5H_5Rh(CO)(Me_2PhP)$ or $\pi C_5H_5RhBr_2$-(Me_2PhP), have been reported [959, 1277].

SCHEME 6

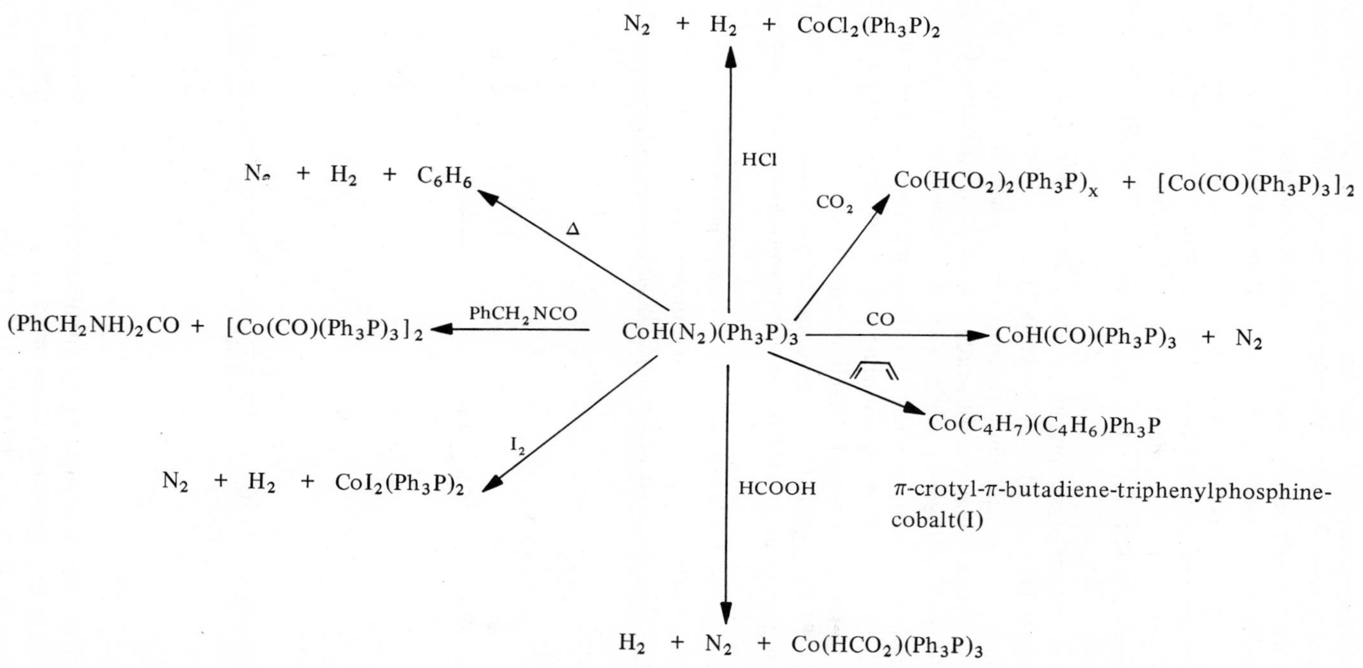
N2 + H2 + CoCl2(Ph3P)2
HCl
N2 + H2 + C6H6
Δ
Co(HCO2)2(Ph3P)x + [Co(CO)(Ph3P)3]2
CO2
(PhCH2NH)2CO + [Co(CO)(Ph3P)3]2
PhCH2NCO
CoH(N2)(Ph3P)3
CO
CoH(CO)(Ph3P)3 + N2
Co(C4H7)(C4H6)Ph3P
I2
N2 + H2 + CoI2(Ph3P)2
HCOOH
π-crotyl-π-butadiene-triphenylphosphine-
cobalt(I)
H2 + N2 + Co(HCO2)(Ph3P)3

Cationic tertiary phosphine complexes containing the olefinic ligands 1,5-cyclooctadiene and norbornadiene ($[Rh(diene)L_2]^+$ diene = 1,5-cyclooctadiene, norbornadiene, L = Ph_3P) are of interest as starting materials for a series of other cationic species of rhodium [1079] and iridium, some of which have catalytic properties for interconversions of unsaturated hydrocarbons into their isomers [529]. π-Diene derivatives of octacarbonyldicobalt, for example, (π-diene)Co_2-$(CO)_6$ (π-diene = norbornadiene, isoprene, 2,3-dimethyl-1,3-butadiene) with tertiary phosphines R_3P form the complexes $[Co(CO)_3R_3P]_2$ (R_3 = Ph_3, Bu_3, $MePh_2$) [937].

Preparation and catalytic properties of cobalt hydrides $CoH(CO)_{4-n}(Bu^n_3P)_n$, (n = 2, 3) have been reported by Pregaglia, Andreetta, Ferrari, and Ugo [1004].

The synthesis of hydridodinitrogentris(triphenylphosphine)cobalt(I), $CoH(N_2)$-$(Ph_3P)_3$ [894, 1041], and the trihydrido complex $CoH_3(Ph_3P)_3$ [1041] is now well established. The effect of substitution of the triphenylphosphine ligands by other phosphine ligands on the infrared N_2 stretching band has been studied [1272]. $CoH(N_2)(Ph_3P)_3$, besides having reversible reactions with dihydrogen and ammonia, shows a variety of irreversible reactions which are summarized in Scheme 6 [1026, 1272]. $Co(N_2)(Ph_3P)_3$ in benzene solution reacts with Bu^n_3P without liberation of gaseous nitrogen, but stepwise substitution of the triphenylphosphine ligands occurs [955].

trans-Chlorobis(triphenylphosphine)dinitrogeniridium reacts with carbon disulfide to give $IrCl(C_2S_5)(Ph_3P)_2$. Some reactions of this unusual complex are summarized in Scheme 7 [751] (together with tentatively proposed structures).

SCHEME 7

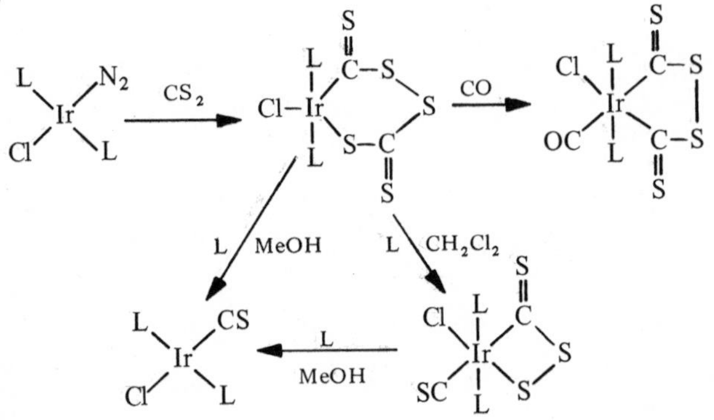

L = Ph_3P.

The iridium complex obtained by lithiumtetrahydridoaluminate reduction of mer-$IrCl_3(Et_2PhP)_3$ [previously formulated as a trihydride, $IrH_3(Et_2PhP)_2$] has been shown to be a pentahydride, $IrH_5(Et_2PhP)_2$ [827]. The pentahydrides

$IrH_5(R_3P)_2$ (R_3 = Et_2Ph, Et_3) react with a variety of neutral ligands L to give the trihydrides $IrH_3(R_3P)_2L$ (L = Ph_3P, $P(OMe)_3$, $SbPh_3$, $AsMe_2Ph$, CO, SMe_2, MeNC) [828, 831]. *O,O′*-Disubstituted dithiophosphate or *P,P*-disubstituted dithiophosphinates may function as bidentate ligands in hydridoiridium complexes, IrH_2L_2-(chel) (L = Ph_3P, chel = $(RO)_2PS_2^-$, $R_2PS_2^-$) [42]. It has been reported that five-coordinate iridium(III) hydrides of type $IrHCl_2(Bu^t_2RP)_2$ (R = Me, Et, Pr^n) form six coordinate adducts with a variety of π- or σ-bonding ligands [1106].

By reaction of di- and tetranuclear phosphine substituted iridium complexes, for example, $Ir_2(CO)_6(Ph_3P)_2$ and $Ir_4(CO)_8(Et_3P)_4$, with carbon monoxide/hydrogen mixed gases under pressure, hydridoiridium complexes may be formed [1259]. The results are summarized in equations (57) and (58).

$$IrH(CO)L_3 \xrightarrow[50\text{-}60^\circ C]{420\ atm,\ CO/H_2} IrH(CO)_2L_2 \xrightarrow[80\text{-}120^\circ C]{450\ atm,\ CO/H_2} IrH(CO)_3L$$

$$\underset{}{\overset{250\ atm,\ CO/H_2,\ 90\text{-}120^\circ C}{\rightleftharpoons}} Ir_2(CO)_6L_2 \qquad (57)$$

L = Ph_3P

$$Ir_4(CO)_8L_4 \xrightarrow[90\text{-}100^\circ C]{400\ atm,\ CO/H_2} Ir_4(CO)_9L_3 \xrightarrow[200^\circ C]{425\ atm,\ H_2} IrH(CO)_3L \qquad (58)$$

L = Et_3P, Bu^n_3P

Lithium diorganophosphido complexes were obtained when tris(tertiary phosphine)dinitrogencobalt complexes were treated with *n*-butyllithium [1124]:

$$Co(N_2)(Et_3P)_3 + Bu^nLi \rightarrow Li(Et_2P)Co(N_2)(Et_3P)_2 + C_6H_{14} \qquad (59)$$

$$Co(N_2)(Ph_3P)_3 + Bu^nLi \rightarrow Li(Ph_2P)Co(N_2)(Ph_3P)_2 + Bu^nPh \qquad (60)$$

$$Li(Ph_2P)Co(N_2)(Ph_3P)_2 + Bu^nLi \rightarrow Li_2(Ph_2P)_2Co(N_2)Ph_3P + Bu^nPh \qquad (61)$$

Cobalt(III) complexes with a tetradentate chelating system and a monodentate tertiary phosphine have been described. A five-coordinate square pyramidal structure is suggested for compounds of the type $[Co^{III}(chel)R_3P]^+ClO_4^-$ and $[Co^{III}(chel)(H_2O)R_3P]^+ClO_4^-$ (R_3 = Ph_3, $EtPh_2$, Bu_3, chel = *N,N′*-ethylenebis-(salicylideneiminato) (salen), *o*-phenylenebis(salicylideneiminato) (saloph), etc.) [1166].

There are several reports [1, 570, 1034] on cobalt and rhodium complexes of tertiary phosphines that contain the dimethylglyoximato (dmg) anion. The non-electrolytes $CoX(dmg)_2Ph_3P$ exist in two modifications (lamellar and acicular) with identical compositions [1]. Although the literature contains several examples of carbon monoxide transfer [22, 158] from one metal atom to another, only recently has the first report on nitrosyl transfer reactions appeared [238]; for example,

$$2\ Co(NO)(dmg)_2(MeOH) + CoCl_2L_2 + 2\ L \rightarrow Co(dmg)_2L + CoCl(dmg)_2L + CoCl(NO)_2L_2 \quad (62)$$

dmg = monoanion of dimethylglyoxime, L = Ph_3P.

Low-spin cobalt(II) complexes $Co(sacsac)_2L$ (L = Ph_3P, Ph_3As, Ph_3Sb, Ph_3Bi, sacsac = dithioacetylacetonate) have been obtained by reaction of equimolar amounts of L and $Co(sacsac)_2$ [1253].

The single-bridged hydroxo complex $[Co(salen)Bu_3P\text{-}OH\text{-}Co(salen)\text{-}Bu_3P]^+ClO_4^-$ (salen = *N,N'*-ethylenebis(salicylideneiminato)-tetradentate ligand) is formed if $[Co(salen)Bu_3P]^+ClO_4^-$ is treated with aqueous KOH. The complex formed in this reaction is the first example of a μ-hydroxo complex that contains the cobalt atoms with a tetradentate equatorial chelating system [1165].

Molecular hydrogen in the presence of tertiary phosphines reacts with Co(I) derivatives such as $[Co(chel)_3]^+ClO_4^-$ (chel = bipyridyl, *o*-phenanthroline) to give complexes of type (*23*) [886]:

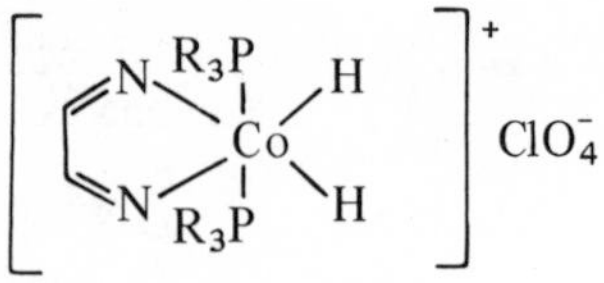

23

R_3 = Et_3, Pr_3^n, Bu_3^n, Et_2Ph

8. NICKEL, PALLADIUM, AND PLATINUM

Klein and Schmidbaur [716] have reported some interesting preparative methods for the synthesis of tetrakis(trimethylphosphine)nickel(O):

$$Ni^{2+} + 2\ OH^- + 5\ Me_3P \rightarrow Ni(Me_3P)_4 + Me_3PO + H_2O \quad (63)$$

$$5\ NiCl_2(Me_3P)_2 + 10\ OH^- \rightarrow 2\ Ni(Me_3P)_4 + 2\ Me_3PO + 3\ Ni(OH)_2 + 2\ H_2O + 10\ Cl^- \quad (64)$$

$$5\ NiCl_2(Me_3P)_2 + 10\ NaOSiMe_3 \rightarrow 2\ Ni(Me_3P)_4 + 2\ Me_3PO + 3\ Ni(OSiMe_3)_2 + 10\ NaCl + 2\ (Me_3Si)_2O \quad (65)$$

Tetrakis(triarylphosphine)palladium may be prepared from triarylphosphine and palladium(II) compounds, for example, Pd(II) acetylacetonate, under an inert gas in the presence of compounds with activated hydrogen atoms, such as amines, ketones, aldehydes, and nitriles [895]. Dissociation of tetrakis(arylmethylphosphine)palladium and -platinum compounds is strongly dependent on the nature of the phosphine ligand. Although the methyldiphenylphosphineplatinum complex loses one phosphine ligand below $-30°C$, the tetrakis complexes of dimethylphenylphosphine and dimethylpentafluorophenylphosphine show no evidence of dissociation [292].

The reactivity of Ni, Pd, and Pt(O) phosphine complexes toward a variety of substrates has been studied in a number of papers [209, 243, 271, 315, 499, 504, 779, 918, 1180, 1209]. Quite a few examples show the striking similarity of the reactivity pattern of carbenoids, and species like $Pt(Ph_3P)_2$ [1209]. $Pt(Ph_3P)_3$, which has a completely different chemical behavior [504], may be considered to be "Ph_3P poisoned" $Pt(Ph_3P)_2$. The order of stability of bis(triphenylphosphine)platinum(O)adducts, $Pt(Ph_3P)_2L$ for various Ls is roughly as follows [1209] : $O_2 > CO > SO_2 \gg CH{\equiv}CH \sim Ph_2P > H_2C{=}CH_2 \gg H_2 > N_2$. The smallest cyclic acetylene isolated so far is cyclooctyne, although some evidence exists for cycloheptyne and cyclohexyne as short-lived intermediates. Coordination to the $Pt(Ph_3P)_2$ moiety via reduction of the π-bond order relative to a linear C–C≡C–C arrangement stabilizes the last two cyclic acetylenes [111].

The ionic compounds $[(Ph_3P)_3Pt{-}N{=}N{-}C_6H_4{-}p{-}R]^+X^-$ (R = NO_2, F, H, OMe, Me, X = BF_4; R = NMe_2, NEt_2, X = BPh_4) have been obtained by interaction of $Pt(Ph_3P)_3$ with diazonium salts. When this complex is treated with hydrogen, nitrogen is evolved with formation of a cationic platinum hydride [243]

$$[(Ph_3P)_3Pt{-}N{=}N{-}C_6H_4{-}p{-}R]^+X^- + H_2 \xrightarrow[\text{suspension}]{\text{MeOH}} [PtH(Ph_3P)_3]^+X^- + C_6H_5R + N_2 \qquad (66)$$

R = OMe, Me, F, X = BF_4

Sulfur dioxide replaces one Ph_3P ligand in $M(Ph_3P)_4$ (M = Pd, Pt) to yield the dark purple air-sensitive derivatives $M(SO_2)(Ph_3P)_3$. Air oxidation of these complexes produces the sulfato derivatives $M(SO_4)(Ph_3P)_2{\cdot}H_2O$ [779]. The tertiary carbon-cyanocarbon bond in 1,1,1-tricyanoethane is cleaved on treatment with $Pt(Ph_3P)_4$ [209]:

$$Pt(Ph_3P)_4 + MeC(CN)_3 \rightarrow Pt(CN)[CMe(CN)_2](Ph_3P)_2 + 2\ Ph_3P \qquad (67)$$

A similar reaction was observed between $Pt(Ph_3P)_4$ and $C(CN)_4$ [84], in which only one type of C–C bond cleavage may occur:

$$C(CN)_4 + Pt(Ph_3P)_4 \rightarrow trans\text{-}Pt(CN)[C(CN)_3](Ph_3P)_2 + 2\ Ph_3P \qquad (68)$$

After carbonylation of tetrakis(triphenylphosphine)nickel(O) the substitution products $Ni(CO)_2(Ph_3P)_2$ and $Ni(CO)_3Ph_3P$ have been identified by IR spectroscopy [315]. Carbonyl complexes $Pt(CO)L_3$ ($L = Ph_3P$, $MePh_2P$) and $Pt(CO)_2L_2$ ($L = Ph_3P$, $EtPh_2P$, Pr^iPh_2P) may be synthesized in an analogous manner or by reductive carbonylation of alkalichloroplatinite in the presence of L [271]. The analogous palladium compound $Pd(CO)(Ph_3P)_3$ has been synthesized as well by reductive carbonylation in the presence of Ph_3P. Triethylaluminium and sodium-borohydride were employed as reducing agents. By use of equimolar quantities of palladium(II)acetylacetonate and triphenylphosphine the trinuclear cluster $Pd_3(CO)_3L_3$ was formed. Another trinuclear cluster, $Pd_3(CO)_3L_4$ ($L = Ph_3P$) [754] was created by heating $Pd(CO)L_3$.

Thermal decomposition of the disubstitution products $Pt(CO)_2L_2$ leads to trinuclear cluster compounds $Pt_3(CO)_3L_n$ (n = 4, $L = Ph_3P$, $Me\,Ph_2P$, $Et\,Ph_2P$ and n = 3, $L = Ph_2P{-}CH_2Ph$) which on further rearrangement and absorption of carbon monoxide give the tetranuclear species $Pt_4(CO)_5L_4$ and $Pt_4(CO)_5L_3$. The first type exists in two isomeric forms (*24*) and (*25*) [248]. New cluster carbonyls

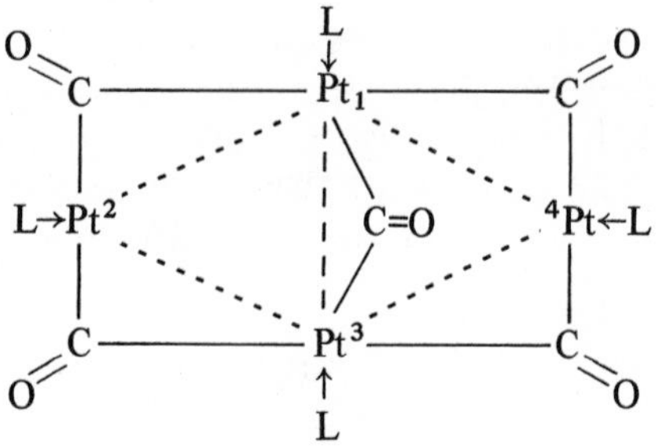

24

$L = Ph_3P, Me_2PhP.$

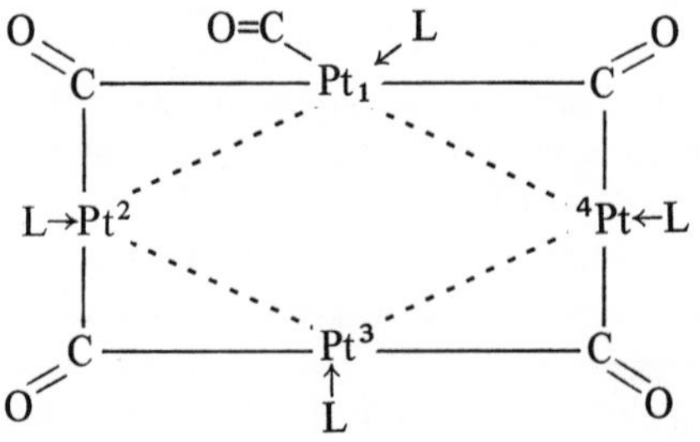

25

$L = Et_3P, Ph_2(CH_2Ph)P.$

(*26*) and (*27*) containing platinum and iron [198] have been obtained by reactions between Pt(II) or Pt(O) phosphine complexes, *cis*-$PtCl_2(Me_3P)_2$ or Pt(*trans*-stilbene)$(Ph_3P)_2$ and dodecacarbonyltriiron:

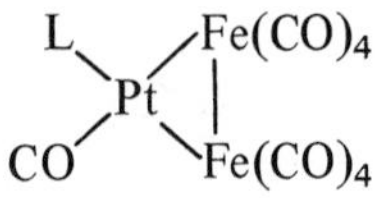

26

L = Me_3P

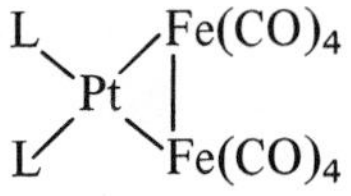

27

L = Ph_3P

Substitution products of tetrakis(alkylisocyanide)nickel(O), isoelectronic to tetracarbonylnickel, have been synthesized in liquid ammonia [926]

$$Ni(CNR)_4 + Ph_3P \xrightarrow[\text{ether}]{\text{liq } NH_3} Ni(CNR)_3Ph_3P + RNC \qquad (69)$$

R = Bu^n, Cy, Pr^i.

Treichel et al. [1196, 1197] have reported five-coordinate platinum(II) isocyanide complexes $[PtX(MeNC)_2(Ph_3P)_2]Y$ (X = Br, I, Y = BF_4^-, I^-). In solution a dissociation equilibrium involves free isocyanide, the pentacoordinate, and a tetracoordinate species. If the diamagnetic, dimeric palladium(I) complex $[PdI(Bu^tNC)_2]_2$ is treated with Ph_3P in toluene at room temperature $[PdI(Bu^tNC)Ph_3P]_2$, a dimeric Pd(I) phosphine complex is obtained [966]. In the complex $Pd(PhNC)_2$ (TCNE), which may be prepared from TCNE and $Pd(PhNC)_2$, triphenylphosphine replaces one PhNC ligand to produce the mixed derivative Pd(PhNC)(TCNE)-Ph_3P (TCNE = tetracyanoethylene) [164].

Duroquinone replacing the ethylene molecule in $M(C_2H_4)(Ph_3P)_2$(M = Ni, Pt) forms complexes $[M(Ph_3P)_2]_2$(duroquinone). Similar reactions have been observed with other quinones [1211]. Platinum complexes of the type Pt(quinone)$(Ph_3P)_2$ have also been synthesized [242]. Substitution of one phosphine ligand in diphenylacetylenebis(triphenylphosphine)platinum (O) by tetrachloro-*o*-benzoquinone yields $Pt(C_6Cl_4O_2)(PhC{\equiv}CPh)Ph_3P$. The acetylene in this complex is susceptible to nucleophilic attack [77]. The reactivity of coordinated acetylene in platinum complexes, for example, Pt(butyne-2)$(Ph_3P)_2$, with protonic acids HX has been studied [1198]. In addition to $PtX_2(Ph_3P)_2$, mixtures of *cis*- and *trans*-butene-2 have been obtained.

Alkynes with electronegative substituents, for example, CF_3 or C_2F_5, react with tetrakis(triphenylphosphine)platinum(O) with formation of the σ-bonded complex $Pt(C{\equiv}C{-}R_f)_2(Ph_3P)_2$ (R_f = CF_3, C_2F) [344] and not a π-bonded structure $Pt(HC{\equiv}C{-}R_f)(Ph_3P)_2$ as reported earlier [572].

Stable ortho-substituted aryl or perhalovinyl nickel complexes may be synthesized by reaction of $Ni(C_2H_4)(R_3P)_2$ (R_3 = Ph_3, Et_3) with aryl or vinyl halides [440].

Reduction of palladium acetylacetonate by ethoxydiethylaluminium in the presence of tertiary phosphines Ph_3P, Cy_3P, (o-tolyl O)$_3$P and ethylene provides a simple synthetic route to tertiary phosphine palladium(O) ethylene complexes [1221]. Active olefins (dimethyl fumarate) replace two Ph_3P molecules in $Pd(Ph_3P)_4$ to form Pd(dimethyl furmarate)$(Ph_3P)_2$ [554]. If nickel acetylacetonate is treated with trimethylaluminium in the presence of tricyclohexylphosphine and nitrogen, bis[bis-(tricyclohexylphosphine)nickel] dinitrogen is formed. According to molecular weight measurements and IR spectroscopic studies, it dissociates in solution and looses N_2 when argon is bubbled through the solutions [666]:

$$[Ni(Cy_3P)_2]_2N_2 \rightarrow Ni(N_2)(Cy_3P)_2 + Ni(Cy_3P)_2 \qquad (70)$$

$$Ni(N_2)(Cy_3P)_2 \rightarrow N_2 + Ni(Cy_3P)_2 \qquad (71)$$

In an interesting photochemical rearrangement the novel percyano-vinylplatinum complex $Pt[C(CN)=C(CN)_2](CN)(Ph_3P)_2$ is formed from tetracyanoethylenebis-(triphenylphosphine)platinum [1191]. The following mechanism has been proposed for this reaction:

$$Pt(TCNE)(Ph_3P)_2 \xrightarrow{h\nu} [Pt(Ph_3P)_2^{+} + TCNE\cdot^{-}]$$

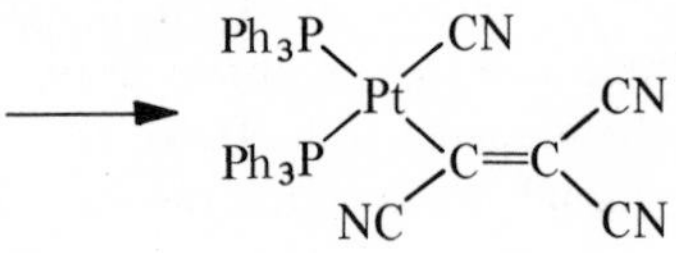

(72)

The versatile starting material $Pt(C_2H_4)(Ph_3P)_2$ may be employed for the synthesis of bis(triphenylphosphine)-1,2-dimethylcyclopropeneplatinum [1230]. Interaction between azobenzene, p-tolylphosphine, and bis(1,5-cyclooctadiene)nickel(O) affords bis(tri-p-tolylphosphine)azobenzenenickel(O), Ni(PhN=NPh)[(p-tolyl)$_3$P]$_2$, with π-bonded azobenzene [640]. Analogous complexes have been reported before by Klein and Nixon [715] and Otsuka, Yoshida, and Tatsuno [967]. π-Bonded trifluoroacetonitrile complexes of platinum may be obtained by reaction of trifluoroacetonitrile with Pt(*trans*-stilbene)$(Ph_3P)_2$ [145].

Hydride-bromide exchange in (π-allyl)triphenylphosphinenickel led to instable π-allylhydrido complexes, which, with excess of triphenylphosphite, form propene [150].

Head-to-tail addition of two butadiene molecules under the influence of *trans*-$NiBrMe(R_3P)_2$ resulted in the formation of unusual binuclear μ-(α,ω-octa-di-π-enyl)nickel compound (*28*) [223]:

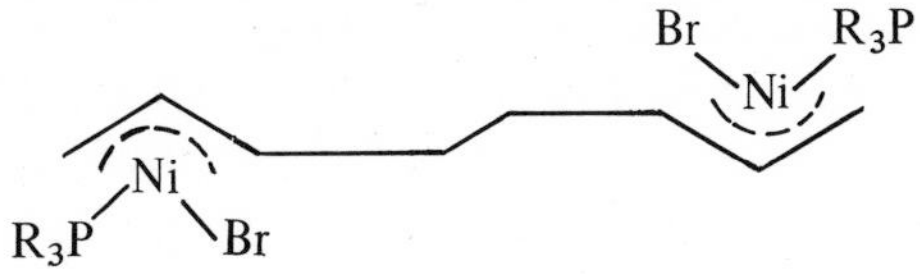

28

R = Pr^i.

A study has been made of the reaction between Ph_3P and $[(\pi\text{-crotyl})PdCl]_2$ in solution by using ^1H, ^{13}C, and ^{31}P NMR spectroscopy as well as electrodialysis. Rapid reactions involving various covalent and ionic species in which the allyl group is bonded to the palladium through a π- or a σ-bond [1122] take place.

Nucleophilic substitution of halide anions in πC_5H_5NiXL (L = Ph_3P, $MePh_2P$, Me_2PhP, Me_3P, $P(OPh)_3$, X = Cl, Br) with $Na[\pi C_5H_5Fe(CO)_2]$ (in tetrahydrofuran solution) produced the dinuclear species $\pi C_5H_5Ni(CO)_2Fe(\pi C_5H_5)L$ [1281]. Thioglycolic acid with $[\pi C_5H_5Ni(Bu^n{}_3P)_2]^+Cl^-$ [1056] yields tri-*n*-butylphosphine-π-cyclopentadienenickel mercaptide. Similar reactions for which, instead of thioglycolic acid, various mercaptides were used gave a variety of compounds $\pi C_5H_5Ni(Bu_3P)SR$ (R = H, Me, Et, etc.). The selenium and tellurium derivatives [1058, 1059] have been obtained in an analogous manner • $[\pi C_5H_5Ni(Bu_3P)_2]Cl$ in aqueous solution produced complexes of the type $[\pi C_5H_5Ni(Bu_3P)_2]X$ with various anions (X = NCO, NCS, N_3, ClO_4, ClO_3, NO_3, NO_2) [1057].

The replacement of πC_5H_5 groups in nickelocene with phosphines, resulting in tetrakis(phosphine)nickel(O), has been reported. In one case (for Bu^n_2PhP) it was possible to isolate the intermediate product $\pi C_5H_5Ni(Bu^n_2PhP)_2$ [1212] as deep blue paramagnetic crystals (μ_{eff} = 1.76).

A number of reports on the syntheses and properties of phosphine complexes of nickel, palladium, and platinum containing the Group IVb element transition metal bond E^{IV}–M (E^{IV} = C, Si, Ge, Sn, M = Ni, Pd, Pt) have been made. Methylnickel complexes with trimethylphosphine ligands $NiMe_2(Me_3P)_2$, $NiMe_2(Me_3P)_3$, $NiMeX(Me_3P)_2$ and tetrakis(trimethylphosphine)methylnickel halides $NiMeX(Me_3P)_4$ (X = Cl, Br, I) have been synthesized by Klein et al. [711, 712]. Some of these compounds show a remarkable thermal stability. Insertion of carbon monoxide into the nickel-carbon bond produces acetylnickel complexes, $NiX(MeCO)(Me_3P)_2$ [711].

Halogen abstraction from the nickel-chlorine bond and the *o*-bromophenyl substituent in *trans*-[chloro(*o*-bromophenyl)bis(triethylphosphine)nickel(II)] with lithium provided an interesting nickel heterocycle which (according to NMR investigations) has a structure with two C_6H_4 units bound to two nickel atoms [402]. In solution a dissociation equilibrium exists, as indicated in (73).

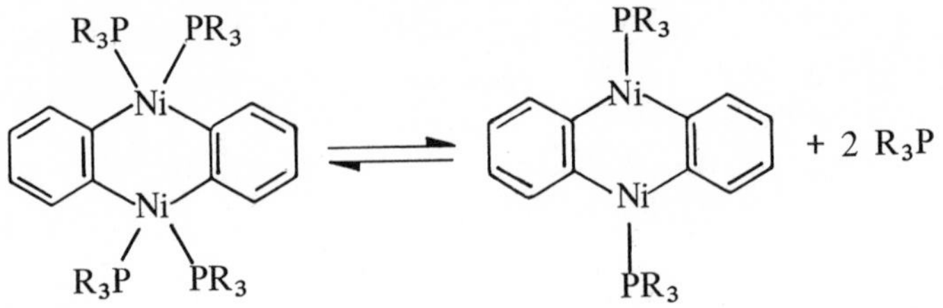

(73)

(R = Et).

Alkenylaryl compounds of nickel and palladium(II), *trans*-chloro(2-allylphenyl)-bis(triethylphosphine)metal(II) and *trans*-X(2-vinylphenyl)bis(triethylphosphine)-metal(II) (X = Cl, Br, I, NO_2, CN, NCS, NCO for M = Ni and X = Br for M = Pd) have been synthesized and the influence of the transition metal on ligand proton chemical shifts has been studied [890]. The different shifts are explained in terms of the paramagnetic anisotropy of the transition metal. Racemic tertiary phosphines L* have been resolved by a simple method that employs stereospecific reactions with an asymmetric palladium(II) complex [964]:

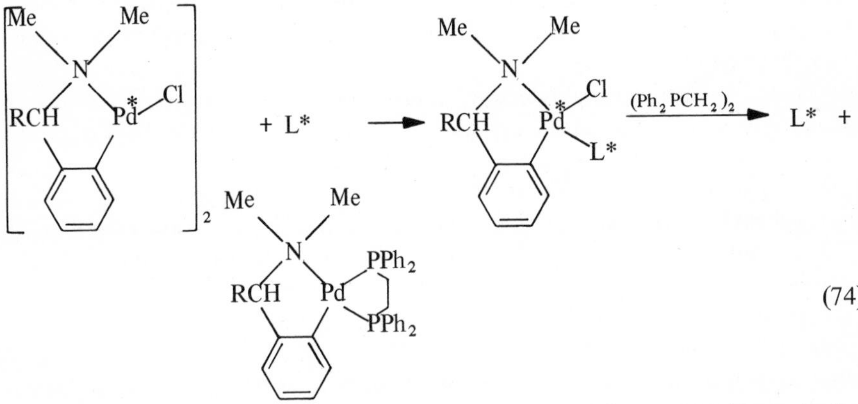

(74)

R = Me, L* = Ph(α-naphthyl)(*o*-tolyl)P, Ph(α-naphthyl)(*p*-$EtOC_6H_4$)P, including Ph_3P comparison.

Unsaturated σ-hydrocarbyl derivatives have been prepared by Lappert et al. [227]. The methods employed include R/Cl exchange between organometallics like CH_2=CHLi or $(CH_2=CH)_4Sn$ and *cis*-$PtCl_2(Et_2PhP)_2$ or oxidative addition of Me_3SnR (R = CF=CF_2) to Pt(O) phosphine complexes.

By using the amine elimination reaction between transition metal hydrides and trimethyl Group IV element dialkylamino compounds a number of complexes containing bonds between metals could be synthesised [228]; for example, $PtCl(SnMe_3)(Ph_3P)_2$:

$$R_3M^1\text{-}NMe_2 + M^2HL_n \longrightarrow HNMe_2 + R_3M^1\text{-}M^2L_n \quad (75)$$

M^1 = Si, Ge, Sn, M^2 = Pt, Pd. Dihydrogen and HCl elimination as a driving force have been utilized for the synthesis of transition metal phosphine complexes with Group IV element transition metal bonds [117, 118, 251, 514]

$$2\ Cl_3SiH + Pt(Ph_3P)_4 \rightarrow Pt(SiCl_3)_2(Ph_3P)_2 + 2\ Ph_3P + H_2 \quad (76)$$

$$PtHCl(R_3P)_2 + GeH_3Cl \rightarrow Pt(GeH_2Cl)H_2Cl(R_3P)_2 \rightarrow$$

$$\textit{trans}\text{-}PtCl(GeH_2Cl)(R_3P)_2 + H_2 \quad (77)$$

R = Et

$$\pi C_5H_5NiXL + GeHX_3 \rightarrow HX + \pi C_5H_5Ni(GeX_3)L \quad (78)$$

X = halide, L = Ph_3P, Et_3P, Pr^i_3P, Bu^n_3P, Et_2PhP.

The dihydrides *o*-$(HMe_2Si)_2C_6H_4$ and *o*-$(HMe_2Si)C_6H_4CH_2SiMe_2H$ react with $Pt(C_2H_4)(Ph_3P)_2$ to form five- and six-membered cyclic bis(silyl) complexes [420]; for example, (*29*):

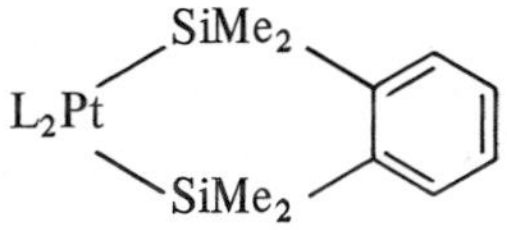

29

In a number of publications, some of which have already been mentioned, oxidative addition has proved to be a versatile synthetic principle. Further examples are the reaction between tetrakis(triphenylphosphine)platinum(O) and methyldichlorosilane [462] or triorganochlorostannanes R_3SnCl (R = Ph, Me) and tris(triphenylphosphine)nickel(O) [487]:

$$Pt(Ph_3P)_4 + 2\ SiCl_2HMe \rightarrow cis\text{-}Pt(SiCl_2Me)_2(Ph_3P)_2 + H_2$$

$$+ 2\ Ph_3P \quad (79)$$

$$Ni(Ph_3P)_3 + 2\ R_3SnCl \rightarrow Ni(R_3Sn)_2Cl_2(Ph_3P)_2 + Ph_3P \quad (80)$$

R = Ph, Me.

Nucleophilic replacement of the halogen of a transition metal-halogen bond by alkali metal derivatives of Group IV compounds, for example, Ph_3GeLi, X_3GeCs [514, 1174], or the insertion of $SnCl_2$ into the nickel-chlorine bond [515] represent alternative synthetic routes to transition metal Group IV element compounds.

Oxidative addition of HCl to palladium(O) complexes, for example, $Pd(CO)(Ph_3P)_3$, $Pd(Cy_3P)_2$, $Pd(Ph_3P)_4$, provides a simple preparative method for hydridochloropalladium complexes [753]. The reactivity of halobenzenes toward to $Pd(Ph_3P)_4$ decreases in the order PhI > PhBr > PhCl, but aryl chlorides with electron withdrawing groups in the para position to chlorine have been found to be quite reactive [466].

Sulfur-sulfur bond cleavage was observed when organic disulfides were treated with complexes $M(Ph_3P)_4$ (M = Pd, Pt). Complexes containing terminal and bridging mercaptido groups were isolated [1284].

Oxidative addition of bromine to *trans*-$PtCl_2LR_3P$ (R = Et, L = py, 3,5-Me_2-py, 4-Bu^tpy, Et_3P; R = Bu^n, L = py, py = pyridine) is followed by rapid halogen scrambling; the result is approximately statistical distribution of *trans*-$PtBr_xCl_{4-x}$-LEt_3P (x = 0 - 4) [582].

Tertiary phosphine hydride complexes of palladium may be obtained by reduction of *trans*-bis(tertiary phosphine)dichloropalladium by the hydridoborohydride *trans*-$MH(R_3P)_2(BH_4)$:

$$PdCl_2(R_3P)_2 + NiH(Cy_3P)_2(BH_4) \xrightarrow{-B_2H_6} MHCl(R_3P)_2 + MHCl(R_3P)Cy_3P + MHCl(Cy_3P)_2 \quad (81)$$

M = Ni, Pd, R = Et, Pr^n, Bu^n. Addition of Cy_3P to this mixture allows the isolation of *trans*-$PdHCl(Cy_3P)_2$ and $NiHCl(Cy_3P)_2$ as mixtures. These hydrides may be partly separated by recrystallization. If the nickel compound is used again this method may be considered as a cyclic route to $PdHCl(Cy_3P)_2$ [919].

The product formed by reaction of silver perchlorate and $PtHCl(Ph_3P)_2$, $PtH(ClO_4)(Ph_3P)_2$ has proved to be a useful intermediate for the synthesis of a range of hydridoplatinum complexes, $PtHY(Ph_3P)_2$ (Y = CO, C_2H_4, Ph_3P, NH_3, thiourea, pyridine) [492]. Dihydride complexes of tetracoordinate platinum(II), *trans*-PtH_2L_2 (L = Cy_3P, Cy_2Pr^iP, Cy_2EtP) [755] and six-coordinate platinum(IV), *cis,trans*-$PtH_2XY(Et_3P)_2$ (X, Y = Cl, Br, I) [34] may be prepared according to

$$Pt(acac)_2 + 2\ Cy_3P \xrightarrow[Et_2O]{AlR_3} trans\text{-}PtH_2(Cy_3P)_2 \quad (82)$$

R = Et

$$PtHY(Et_3P)_2 + HX \longrightarrow cis/trans\text{-}PtH_2XY(Et_3P)_2 \quad (83)$$

X, Y = Cl, Br, I.

$$2\ PtHCl(Et_3P)_2 + 2\ Cl_2 \longrightarrow PtH_2Cl_2(Et_3P)_2 + PtCl_4(Et_3P)_2 \quad (84)$$

Dumler and Roundhill [418] found by detailed IR and NMR study of the reaction between *trans*-$PtHCl(Ph_3P)_2$ and HCl, that the reaction product between $Pt(Ph_3P)_4$ and excess HCl originally claimed to be dihydridodichlorobis(triphenylphosphine) platinum(IV) [231] was really a different crystalline form of *trans*-$PtHCl(Ph_3P)_2$. Their results cast doubt on the existence of other dihydrides, $PtH_2X_2(R_3P)_2$.

Protonation of $Pt(Ph_3P)_4$ with carboxylic acids produces complexes of the type $[PtH(Ph_3P)_3]Y_2H$ [Y = CF_3COO, CF_2ClCOO, C_2F_5COO, CF_2COO, $(COO)_2H$] which contain the planar cation $[PtH(Ph_3P)_3]^+$ [1172]. In solutions of complex hydrides ligand exchange occurs which, as found by NMR, is dependent on the concentration of free ligand and temperature [3].

Dihalobis(tertiary phosphine)nickel(II) complexes have been investigated by a number of authors. A detailed physicochemical study has been carried out on the trans square planar complexes $NiX_2(Me_3P)_2$ (X = CN, NO_2, NCS, Cl, Br, I, Me_3P) [883]. Steric effects have been found to be relatively unimportant in affecting the thermodynamics of the square planar/tetrahedral equilibrium in dihalobis-(tertiary phosphine)-nickel(II) complexes [1006]. For cyclohexylphenylphosphine complexes $NiX_2(Cy_{3-n}Ph_nP)$ (X = Cl, Br, I, NCS, n = 0 - 3) the amount of paramagnetic tetrahedral isomer follows the order $Ph_3P > CyPh_2P > Cy_2PhP > Cy_3P$ for a given halide and I > Br > Cl for a given phosphine [1138]. Complexes of nickel(II) dihalides with alkyl- and aryl-substituted phosphines have been shown by thermogravimetric measurements to decompose in an oxygen-containing atmosphere at or above ~200°C, thus yielding solid phases based on nickel oxide and phosphorus pentoxide [898]. Substitution of the bromide ions in $NiBr_2(R_3P)_2$ (R = Et, Bu, Cy) by the pyrrole anion gives phosphine complexes of nickel(II) pyrrolate [1245]. Pentacoordinate nickel(II) complexes $NiX_2(Me_3P)_3$ (X = NCS, NO_2) [882, 884] show dynamic behavior in solution. A trigonal bipyramidal/square pyramidal equilibrium is believed to exist for $Ni(NO_2)_2(Me_3P)_3$, based on the temperature dependence of the ligand field spectra [881]. An interesting compound $NiI_2(CO)$-$(Me_3P)_2$, which probably has a bipyramidal structure, has been obtained by a number of synthetic methods [969]; for example,

$$Ni(CO)_2(Me_3P)_2 + I_2 \longrightarrow CO + NiI_2(CO)(Me_3P)_2 \qquad (85)$$

Hydrogen cyanide adds to $Ni(CN)_2(Pr_3P)_2$ in an equilibrium reaction involving the pentacoordinate species $Ni(CN)_2HCN(Pr_3P)_2$ [314].

Klein and Karsch reported the syntheses and spectroscopic investigations of the dinuclear methyl nickel complexes $[NiX(Me)L]_2$ (L = Me_3P, X = NH_2, p-NHC_6H_4Me, F) and L(Me)NiXYNi(Me)L (L = Me_3P, X = NMe_2, Y = OMe, F, Cl; X = F, Y = OMe, Cl; X = OMe, Y = Cl) [714].

For palladium(II) complexes *cis/trans*-PdX_2L_2 (X = Cl, Br, I), containing *t*-butylphosphine ligands, for example, Bu^t_2RP, Bu^tR_2P (R = Me, Et, Pr^i, Ph, *p*-tolyl) there exists a linear correlation between the chemical shift of the ^{31}P nuclei in the free phosphine and the coordination shift Δ on complexation [834]. Palladium(II) complexes of phosphine ligands with CH_2OH or CH_2OCOMe substituents exhibit a higher solubility in hydroxylic solvents (including water) than those of trialkyl phosphines [254].

Dimeric triphenylphosphine complexes of platinum(II), $Pt_2Cl_4(Ph_3P)_2$ are formed by the reaction of *cis*-$PtCl_2(Ph_3P)_2$ with platinum dichloride $PtCl_2$ [578]. The thermally induced conversion of trans dihalogenoplatinum(II) or palladium(II) tertiary phosphine carbene complexes to the cis compound (which is a rare example of trans to cis isomerization within platinum complexes) is slowed down as the *n*-alkyl chain length of the R-groups in the R_3P ligands trans to the carbene increases [245].

Mercuric halides form adducts with certain dihalogenobis(tertiary phosphine)-nickel, -palladium, or -platinum halides of the type $MX_2L_2HgX_2$ (M = Pd, Pt, L, e.g., Me_2PhP, Et_2PhP, Me_3P, Et_3P, Ph_3P, X = Cl, Br, I) [188]. (L = Et_3P, M = Ni, X = Cl) [173]. Magnetic properties, electronic spectra, and conductivity measurements revealed two bridging chlorines between the mercury and the transition metal present in these complexes.

Arene diazonium salts react with trans chlorohydridobis(triethylphosphine)-platinum(II) to give aryldiimide complexes which are precursors to new arylazo-platinum compounds and arylhydrazine complexes [973].

Specific deuteration at C-3 of the alkyl group in tripropyl- and tributylphosphine occurs if complexes $Pt_2Cl_4L_2$ (L = Pr_3^nP, Bu_3^nP) are treated with $MeCOOD/D_2O$ mixtures containing perchloric acid [855]. A mechanism proposed for this exchange requires the insertion of platinum into a C-H bond. Internal metalation the phosphine ligands, for example, Bu_2^t(*o*-tolyl)P, Bu^t(*o*-tolyl)$_2$P may be observed on complexation with palladium(II) [265] or platinum(II) halides (with various butylaryl phosphines) [264]. Internal metalation gives four- [451] or five-membered rings, the ease of reaction being increased for platinum in the order Cl < Br < I [264]. Cleavage of the palladium-carbon bond in such metalation products yields *o*-substituted arylphosphine palladium complexes [451]:

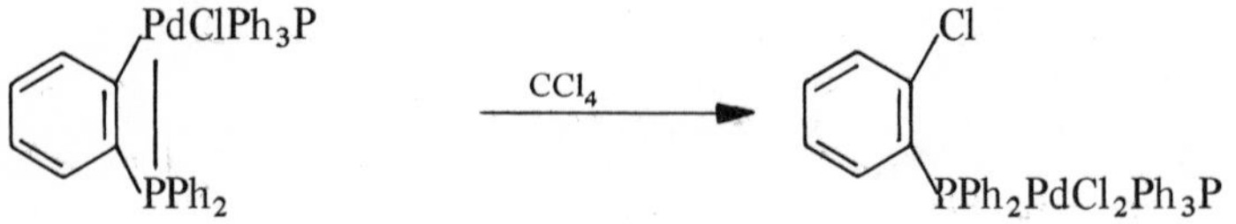

Abstraction of chloride with $AgPF_6$ from $PtCl(RCO)(Ph_3P)_2$ (R = Me, Ph, *p*-tolyl, *p*-$MeOC_6H_4$, *p*-$NO_2C_6H_4$) leads to cationic platinum carbonyl complexes $[Pt(CO)(R)(Ph_3P)_2]^+$. The alkyl or aryl group migration from carbon to platinum observed here is strongly solvent-dependent. In acetonitrile only the cationic acetyl complexes $[Pt(RCO)MeCN(Ph_3P)_2]^+$ are formed [752]. An interesting cleavage reaction occurs if 3,3,3-trifluoropropynyldiphenylphosphine is treated with MX_4^{2-} salts (M = Pd, Pt, X = Cl, SCN). Complexes of unsymmetric ditertiary phosphine $Ph_2PCH_2C(CF_3)$=$CHPPh_2$ are obtained [1116].

Phosphine complexes of nickel-, palladium-, and platinum-containing ligands with O, S, Se, or Te donor atoms were the subject of a series of publications. Dimeric methyl nickel complexes $(NiXMeL)_2$ (L = Me_3P, X = OH, OMe, OEt, $OSiMe_3$,

OPh, p-MeC$_6$H$_4$O-, O$_2$CH, O$_2$CMe) have been synthesized. An equilibrium exists between cis and trans isomers [713].

Some tertiary phosphine complexes of nickel and platinum(II) salts of dibasic oxoacids have been reported. Blindheim obtained bis(tributylphosphine)sulfato-nickel(II) by reaction of nickel(II) sulfate hydrate NiSO$_4$·6H$_2$O with Bu$_3^n$P in absolute ethanol. The sulfate anion acts as a bidentate ligand within a monomeric nonionic structure [147]. Oxalato and carbonato platinum complexes show interesting thermal and photochemical behavior [143, 144]. The oxalato platinum complex, when photolyzed in presence of H$_2$ gives, in a very unusual reaction, cluster compounds of platinum.

$$PtCO_3L_2 \xrightarrow[\text{reflux, 2h}]{\text{EtOH, dark}} (PtL_2)_2 + CO_2 + MeCHO + H_2O \qquad (87)$$

$$PtCO_3L_2 \xrightarrow[40°C]{\text{X/EtOH} \quad h\nu} PtXL_2 \qquad (88)$$

(a) L = Ph$_3$P, X = C$_2$H$_4$,
(b) L = MePh$_2$P, X = Ph$_3$P.

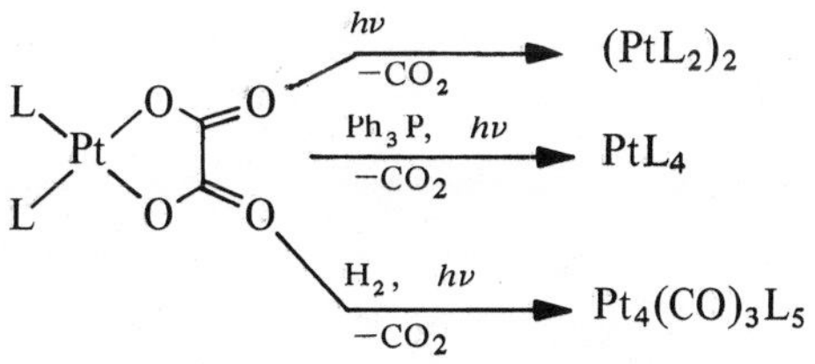

$$\xrightarrow[-CO_2]{h\nu} (PtL_2)_2 \qquad (89)$$

$$\xrightarrow[-CO_2]{Ph_3P, \ h\nu} PtL_4 \qquad (90)$$

$$\xrightarrow[-CO_2]{H_2, \ h\nu} Pt_4(CO)_3L_5 \qquad (91)$$

L = Ph$_3$P.

An interesting redox reaction occurs if nitric oxide is treated with tetrakis-(triphenylphosphine)platinum(O). A monomeric hyponitrite derivative of platinum-(II) is formed which probably has the structure (30) [244].

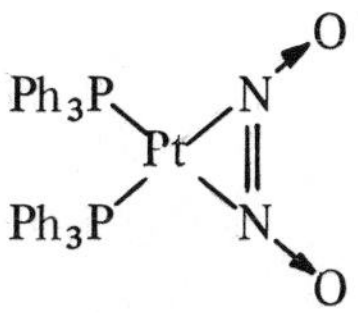

30

The reactivity of the azido groups in the azido metal complexes M(N$_3$)$_2$(Ph$_3$P)$_2$ (M = Pd, Pt) towards hydrogensulfide, disulfides, RSSR, or elemental sulfur has been studied [731]. The fate of the alkyl group in the dealkylation of organic

$$M(N_3)_2(Ph_3P)_2 + 2\ H_2S \longrightarrow M(SH)_2(Ph_3P)_2 + 2\ HN_3 \qquad (92)$$

M = Pd, Pt.

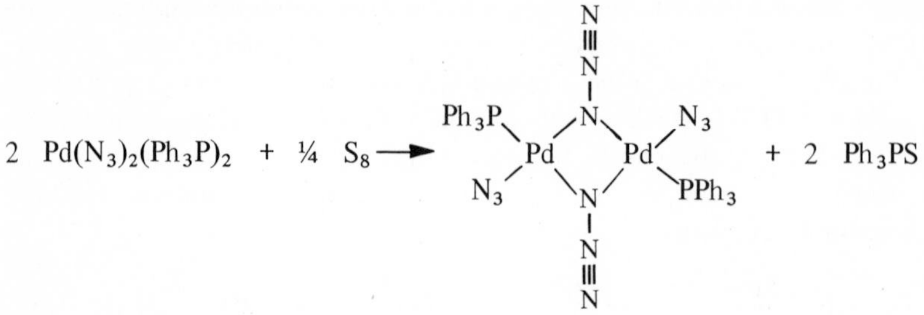

$$2\ Pd(N_3)_2(Ph_3P)_2 + \tfrac{1}{4}\ S_8 \longrightarrow \qquad\qquad\qquad + 2\ Ph_3PS \qquad (93)$$

azides by tetrakis(triphenylphosphine)palladium(O) or platinum(O) is unknown [587].

Reaction of dialkyldithiocarbamato ions with nickel(II) halides in the presence of the appropriate phosphine gives dialkyldithiocarbamatophosphine complexes $NiX(CS_2NEt_2)Ph_3P$ (X = Cl, Br) that were only characterized by elemental analysis and i.r. spectroscopy [869]. The reaction of $M(S-S)_2$ [M = Pd, Pt, (S-S) = S_2CNR_2 (R = Me, Et), S_2COR (R = Et, $PhCH_2$), $S_2P(OEt)_2$, S_2PR_2 (R = Me, Et, Ph)] with tertiary phosphines probably occurs stepwise. Unidentate/bidentate and ionic/bidentate coordination of the chelate ligands in the reaction products is observed [24].

Selenocarbamato complexes $M(OSeCNBu_2^n)_2(Ph_3P)_2$ (M = Pt, Pd), [1163] have been prepared as well. The selenocarbamato ligand is monodentate in the solid state, whereas in solution partial dissociation of triphenylphosphine offers the possibility of a bidentate coordination.

The anions $R_2PS_2^-$ (R = OEt, Ph) may be bidentate [23,1282], for example $Ni(R_2PS_2)Ph_3P$ and *31*, and probably monodentate in the same molecule (*32*) [23].

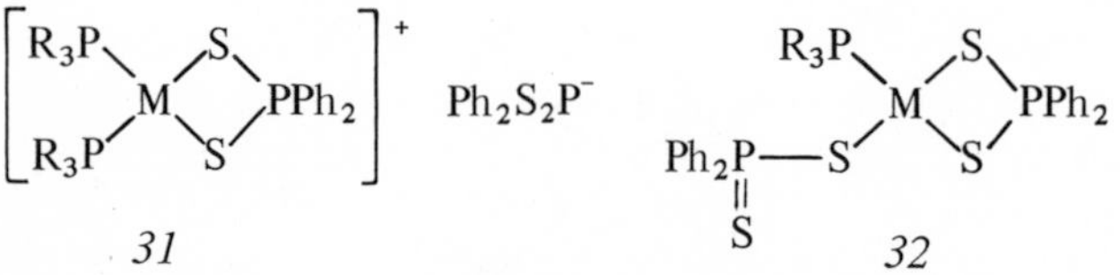

$$R_3 = Et_3,\ Me_2Ph,\ Ph_3,\ MePh_2 \qquad R_3 = Me_2Ph,\ MePh_2,\ Ph_3$$

M = Pd, Pt

Sayler, Beall, and Sieckhaus reported an unusual chelated *o*-carborane nickel complex $NiC_2B_{10}H_{10}(Ph_3P)_2$, which they obtained by reaction of 1,2-dilithio-*o*-carborane with $NiCl_2(Ph_3P)_2$ [1063]. Related platinum(II) and palladium(II) complexes have been synthesised by Bresadola, Frigo, Longato, and Rigatti [178].

9. COPPER, SILVER, AND GOLD

The preparation and properties of methylbis(triphenylphosphine)copper have recently been reported [896, 1273]. Insertion of carbon dioxide into the Cu-Me bond yields acetato copper(I) complexes [896]

$$CuMe(Ph_3P)_2(Et_2O)_{1/2} + CO_2 \longrightarrow Cu(MeCOO)(Ph_3P)_2 + \tfrac{1}{2}\ Et_2O \quad (94)$$

Cyclopentadienyltributylphosphinecopper(I) reacts smoothly with benzoylchloride to give 6-benzoyloxy-6-phenylfulvene [795], whereas with iodobenzenes arylation of the π-cyclopentadienyl ring, which yields arylcyclopentadienes, takes place [1234].

Monodentate tertiary phosphine complexes of the copper(I) halides $CuXL_m$ have been the subject of a number of investigations. Schmidbaur, Adlkofer, and Schwirten synthesized and studied trimethylphosphine complexes of copper(I) chloride (tetrameric $(CuClMe_3P)_4$, dimeric $[CuCl(Me_3P)_2]_2$, monomeric $CuCl(Me_3P)_3$, and probable ionic compounds $[Cu(Me_3P)_4]^+Cl^-$) by ^1H NMR, ^{35}Cl and ^{63}Cu NQR spectroscopy [1067]. Tricyclohexylphosphine (L) with copper(I) chloride, bromide, and iodide gives complexes of the type $CuXL_2$ (X = Cl, Br) and dimeric $(CuXL)_2$ (X = Cl, Br, I). Conductance measurements showed that $[CuL_2]ClO_4$ is a 1:1 electrolyte. Attempts to prepare $[CuL_4]^+ClO_4^-$ were unsuccessful [905]. Tri-*p*-tolylphosphine forms complexes of the type $CuX(p\text{-tolyl})_3P$ (X = Cl, Br, I) (which are tetramers according to molecular weight determination) if it is treated with CuX in a metal:ligand ratio of (1:1) in benzene solution [931].

Copper(I) complexes of type $CuX(biL)Ph_3P$ which contain bidentate (and monodentate) *N*-donor ligands (biL), for example, α,α'-bipyridyl, *o*-phenanthroline, can be prepared easily from the tetramers $[CuXPh_3P]_4$ and the appropriate ligands (biL) [649].

The behavior of tertiary phosphine complexes of copper(I) in chloroform solution [786] and the factors influencing the stereochemistry of four-coordinate phosphine-copper(I) complexes [787] have been studied. The relative stability of the complexes $CuClL_3$ increases in the order for L $Ph_3P \ll MePh_2P \simeq Me_2PhP$ [786]. Triphenylphosphine forms stable complexes $CuNO_3(Ph_3P)_2$ and $CuNO_3(Ph_3P)_3$ [650]. It has been shown in an X-ray crystallographic study that $CuNO_3(Ph_3P)_2$ contains tetrahedrally coordinated copper(I) with a bidentate nitrato group [885].

Triethyl- [56] and triphenylphosphine [961] are oxidized by copper(II) chloride. Copper(I) complexes $CuCl(Et_3P)_n$ (n = 1,3) in addition to "triethyldichlorophosphorane" (which has been shown to react immediately with CuCl to form complexes $Et_3PCl^+CuCl_2^-$ or $Et_3PCl^+Cu_2Cl_3^-$) or "triphenyldichlorophosphorane," may be isolated. The redox process occurring in the system copper(II) chloride-triphenylphosphine-acetone has been studied. Ph_3PO, $CuCl_2(Ph_3PO)_2$, $CuCl_2$-$(Ph_3PO)_4 \cdot 2H_2O$, and $Cu_4OCl_6(Ph_3PO)_4$ could be isolated by chromatography from the product obtained in a reaction with a 1:1 [copper(II) chloride:triphenylphosphine] molar ratio. $CuCl(Ph_3P)_3$ and $CuCl(Ph_3P)_2$ were obtained when a 1:4 [copper(II) chloride:triphenylphosphine] stoichiometry was employed [816].

In contrast to the well-known complexes of copper(I) with tertiary phosphines, there are only a few publications on copper(II) phosphine complexes [769, 1285, 1286]. They contain the hexafluoroacetylacetonate anion (hfac), for example, $Cu(hfac)_2L$ (L = Ph_3P, $MePh_2P$, Me_2PhP, Et_3P, Bu^n_3P) [1285, 1286], or the trifluoroacetylacetonate (tfac), for example $Cu(tfac)_2L$ (L as before including $P(OPh)_3$ [769]). These complexes are monomeric in chloroform solution, and, besides the copper hyperfine splitting, the ESR spectrum reveals phosphorus hyperfine splitting of 146-132 G.

Silver(I) chloride [1067], silverthiocyanate [329, 363], and silvercyanate [363] form complexes with tertiary phosphines of the general type $AgXL_n$. The values of *n*, depending on the nature of X and L, are summarized in Scheme 8.

SCHEME 8. Composition of silver(I) halide and pseudohalide complexes $AgXL_n$ (numbers of *n* are given)

X \ L	Me_3P	Et_2PhP	$EtPh_2P$	Bu^n_3P	Ph_3P
Cl	1-4	–	–	–	–
SCN	1	1, 2	3	1	2, 3
OCN	–	–	–	–	3

The solution structure of some coinage metal(I) phosphine complexes has been discussed by Muetterties and Alegranti [915]. Ligand exchange occurs in the systems $AuMe(Me_3P)/Me_3P$ [1071], $AuMe_3(Me_3P)/Me_3P$ [1071], and $AuMe/Me_3P/P(OMe)_3$ [1068]. These exchange reactions provide a valuable synthetic route to various organophosphinegold complexes [1070]. Tricyclohexylphosphinechlorogold(I), $AuClCy_3P$, for example, may be obtained by ligand exchange between Cy_3P and $AuClPh_3P$ [60]. Although methylgold with triorganophosphines (and phosphites) yields only 1:1 complexes, gold(I) halides, nitrate, and tetrafluoroborate also form 1:2 and 1:4 complexes [1068]. Addition of triorganophosphines (or the bidentate phosphine $Me_2PCH_2CH_2PMe_2$) to dimethylgold chloride produces complexes of the type $AuClMe_2(R_3P)$ or $[AuMe_2(R_3P)_2]^+Cl^-$ [1109].

Oxidative addition of MeI or HgX_2 to trimethylphosphinemethylgold did not result in the expected products; for example, $Au(Me_2I)Me_3P$ or $Au(MeX)(HgX)$-Me_3P. Instead, the gold(I) halide complexes $AuXMe_3P$ (X = I, X) and ethane or

HgXMe, respectively, were formed [1110]. A similar reductive elimination of ethane has been reported for cationic dimethylgold(III) complexes, $[AuMe_2L_2]Y$ [1107] (L = Me_3P, Me_2PhP, $MePh_2P$, Ph_3P, Y = Cl, I, ClO_4). Steric crowding facilitates reductive elimination [1107]. Bromo- and iodo-bis(pentafluorophenyl)triphenylphosphinegold(III) have a fairly high thermal stability which may be attributed to the stabilizing effect of the two electronegative substituents at the gold atom [1213]. Square planar organogold complexes, for example, $AuMe_2(SCN)Ph_3P$ [1137] and $[AuMe_2(Ph_3P)_2]ClO_4$ [1107, 1137] and the corresponding cyanato, selenocyanato and cyano analogs have been prepared [1136].

The cleavage of the gold-carbon bond in triphenylphosphinephenylgold(I) with various reagents (e.g., HCl, halogens, acetic acid, acetyl chloride, and trifluoroacetic anhydride) provides a useful synthetic route to a range of derivatives $AuXPh_3P$ [982]:

$$AuPhPh_3P \xrightarrow{hal_2} Au(hal)Ph_3P \qquad (95)$$

$$AuPhPh_3P \xrightarrow{MeCOOH} Au(MeCO_2)Ph_3P \qquad (96)$$

$$AuPhPh_3P \xrightarrow{MeCOCl} AuClPh_3P \qquad (97)$$

$$AuPhPh_3P \xrightarrow{(CF_3CO)_2O} Au(CF_3OCO)Ph_3P \qquad (98)$$

$$AuPhPh_3P \xrightarrow{HgCl_2} AuClPh_3P \qquad (99)$$

If phenyl or *p*-tolyltriphenylphosphinegold(I) are treated with HBF_4 in ether, the unusual compound $[AuAr(AuPh_3P)Ph_3P]^+BF_4^-$ (Ar = Ph, *p*-tolyl) with the two Au-atoms bond to the aromatic ring system is formed [526].

Phosphine-substituted gold clusters of the type $[Au_{11}(R_3P)_7]X_3$ (R = C_6H_4Cl, C_6H_4F, tolyl, X = I, SCN, CN) [229], $[Au_9L_8]X_3$ [L = tris(*p*-substituted phenylphosphine), X = NO_3, PF_6, picrate] [92], and $[Au(p\text{-tolyl})_3P]_6[BPh_4]_2$ [93] have been reported. The presence of two or three gold atoms in a positive oxidation state seems to be required for the stability of these clusters.

The complexes $AuLFe(CO)_3NO$ and $AuPh_3PFe(CO)_2(NO)L'$ [L = $P(OMe)_3$, Me_3P, Cy_2PhP, $CyPh_2P$, Ph_3P, $(p\text{-}ClC_6H_4)_3P$, $(p\text{-tolyl})_3P$, L′ = $P(OPh)_3$, Ph_3P, Et_2PhAs, possess an approximately linear L-Au-Fe-L′ system in a trigonal bipyramidal structure [236].

It is interesting to note that some triorganophosphinegold coordination complexes [e.g., $AuXR_3P$ (R = alkyl or aryl groups, X = Cl, *β-D*-glucopyranosylthiolate and related mercaptoethylglycosides)] have been shown to exert antiarthritic effects. The triethylphosphinegold complexes were most effective [808, 1154].

10. ZINC, CADMIUM, AND MERCURY

The coordination chemistry of these metals toward phosphorus(III) ligands has been little investigated during the period under review. Complex formation between trimethylphosphine and methylmercury chloride and mercury halides, cyanide, thiocyanide, acetate, and nitrate has been studied [1069]. Although HgClMe forms only a 1:1 complex, the mercury(II) halides, -acetate, -nitrate, -cyanide, and -thiocyanate are capable of bonding up to four phosphine molecules per mercury ion. With triphenylphosphine a 1:1 (dimeric) and a 1:2 complex of mercury(II) thiocyanate is obtained [364, 643]. As with the mercury(II) halide complexes of trimethylphosphine, $HgX_2(Me_3P)$ [1069] a dimeric structure is proposed for the 1:1 triphenylphosphine adducts of $Hg(SCN)_2$ [364]. The thermal decomposition of *p*-dimethylaminophenyldimethylphosphine (L) complexes, MX_2L_2 and MX_2L (M = Zn, Cd, Hg, X = Cl, Br, I), has been studied. The order of thermal stability is I > Br > Cl for MX_2L_2 but Br > Cl > I for complexes MX_2L [91].

Monomeric and dimeric tricyclohexylphosphine complexes MX_2L_2 and $(MX_2L)_2$ (X = Cl, Br, I, SCN, M = Zn, Cd, Hg) were reported as well. A distorted tetrahedral coordination around Hg or Cd was found for the complexes $(MX_2L)_2$ [904].

B. *Ditertiary Phosphines*

A large number of complexes containing ditertiary phosphines, $R_2P{-}X{-}PR_2$ (X being $(CH_2)_n$, CH=CH, C_6H_4, etc.) are known. Usually these ligands function as bidentate chelates, but complexes in which they serve as monodentate or bridging ligands have also been reported.

The following abbreviations for the bidentate ligands are used:

$Ph_2P(CH_2)_nPPh_2$	dppm (*n* = 1), dppe (*n* = 2), dppp (*n* = 3), dppb (*n* = 4)
$MePhP(CH_2)_2PMePh$	mppe
$Me_2P(CH_2)_2PMe_2$	dmpe
$Me_2P(CH_2)_3PMe_2$	dmpp
$(p\text{-tolyl})_2P(CH_2)_2P(p\text{-tolyl})_2$	dtpe
$Et_2P(CH_2)_2PEt_2$	depe
$Ph_2PCH{=}CHPPh_2$	dppv
$Ph_2PCH_2C{\equiv}CCH_2PPh_2$	dppby
$Ph_2PC{\equiv}CPPh_2$	dppac
$Ph_2PN(Et)PPh_2$	dppa
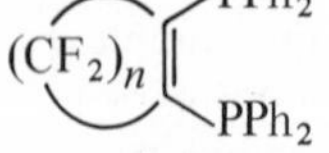	$f_4$fos (*n* = 2) $f_6$fos (*n* = 3) $f_8$fos (*n* = 4)

Coordination compounds of ditertiary phosphines for which the analogs with monodentate phosphines are known or in which the bidentate ligand does not exert a specific effect are mentioned only briefly.

The subject of this chapter has been reviewed by Levason and McAuliffe [774]. The literature covered includes 1970 and part of 1971 (references added in proof).

Rigo and Turco [1024] reported and summarized aspects of the coordination chemistry of (mainly) bidentate phosphine ligands with transition metal cyanides.

1. COMPLEXES OF THE DITERTIARY LIGANDS $R_2P(CH_2)_nPR_2$ (n = 1 - 4)

1,2-Bis(diphenylphosphino)ethane (dppe) forms stable complexes with $TiCl_3Me$ [297], and a convenient synthesis for $VH(CO)_4$dppe, $VI(CO)_4$dppe, and $V(CO)_4$-dppe has been demonstrated [369]. The predominant products from the reaction between bis(diphenylphosphino)methane (dppm) and the polymeric molybdenum nitrosyl halides $[MoX_2(NO)_2]_n$ (X = Cl, Br) were the complexes $MoX_2(NO)_2$dppm [169]. If hexacarbonylmolybdenum reacts with dppm, $Mo(CO)_4$dppm and subsequently *trans*-$Mo(CO)_2(dppm)_2$ [310] are formed. Dppm with tetracarbonylnitrosyltungsten halides forms the complexes *cis*-$WX(CO)_2(NO)$dppm and WX(CO)-(NO)$(dppm)_2$. The latter contains both mono- and bidentate dppm [308].

In complexes formed with π-allylchloropalladium $[\pi C_3H_5PdCl]_2$ dppm acts as a weak monodentate ligand (weaker than triphenylphosphine) or in some cases as a bridging unit that causes σ-π-rearrangement of the allyl group [632]

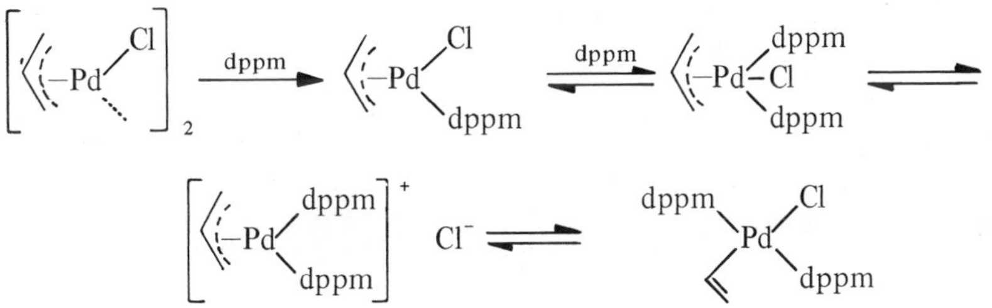

(100)

Lithiation of the CH_2 group in dppm yields $LiCH(Ph_2P)_2$ which reacts with $PdCl_2$ or π-allylpalladium chloride to give dimeric complexes $[PdClCH(Ph_2P)_2]_2$ and $[\pi C_3H_5PdCH(Ph_2P)_2]_2$ [632]. There is a marked difference between the reactions of dppm and dppe with hexacarbonylmolybdenum: dppm yields *trans*-$Mo(CO)_2$ $(dppm)_2$ and dppe gives *cis*-$Mo(CO)_2(dppe)_2$; dppm obviously exerts more steric crowding about the metal atom than dppe [309, 310].

Mononuclear dinitrogen complexes of molybdenum, *trans*-$Mo(N_2)_2(dppe)_2$ [51, 590] react with dihydrogen reversibly to yield the dihydrides $MoH_2(dppe)_2$, [*trans*-MoH_2(dppe)]$_2$-μ-dppe [590], whereas the complexes $Mo(N_2)_2$(dtme) (dtme = 1,2-bis(di-m-tolylphosphino)ethane) with hydrogen affords the tetrahydride MoH_4 $(dtme)_2$ [43]. Hydrido complexes of iron(I) $FeH(dppe)_2$ [484], and iron(II), ruthenium(II), and osmium(II), *trans*-$[MN(L)(depe)_2]$ BPh_4 [71] (L = CO, p-$MeOC_6H_4NC$, N_2, PhCN, M = Fe, Ru, Os) and the related dppe-complexes [500] have been reported. Dihydridobis-1,2-bis(diphenylphosphino)ethanemolybdenum(II) was prepared by reduction of $Mo(acac)_3$ (acac = 2,4-pentanedionate) with triisobutylaluminium in the presence of dppe under a hydrogen atmosphere [478]. Dppe methyl nickel complexes, $NiMe_2dppe$ [535], hydridonickel complexes, $[NiH(dppe)_2]Y$ (Y = BF_4^-, $AlCl_4^-$ HCl_2^-) [1091], and hydridopalladium complexes, $[PdH(R_3P)dppe]^+$-PF_6^- (R = Cy, Pr) [534], reveal a relatively high thermal stability.

Reactions of ditertiary phosphines (L = dppe, dppm, dppa) with a number of dinuclear iron complexes $[Fe(CO)_3SR]_2$ (R = Et, Me, Ph) have been reported. Products in which the ligand L is monodentate, for example, $Fe_2(CO)_5(SR)_2L$, and $[Fe(CO)_2LSR]_2$, bridges two iron atoms, $Fe(CO)_2(SR)_2Fe(CO)_2L$, or in which one ligand (L) is monodentate and the other bidentate, like in $Fe_2(CO)_3(SR)_2L_2$, could be isolated [376, 377].

Mague and Mitchener [814] prepared chlorocarbonyl complexes of ruthenium(II) and (III) with dppm ligands in addition to Rh(I) complexes of the asymmetrical chelate ligand 1,2-bis(methylphenylphosphino)ethane (mppe), $RhCl(mppe)_2$ [813]. In a series of papers Horner and Müller reported the synthesis and spectroscopic investigation of chiral and achiral mppe complexes of iron, cobalt, and nickel [610], ruthenium [608], and osmium [609].

Rhodium(I) and (III) complexes containing the ligand $Me_2PCH_2CH_2PMe_2$ (dmpe), $[RhXY(dmpe)_2]^+$, and $Rh(dmpe)_2Cl$ (X = Y = Cl, Br, H) have been synthesized and their chemical properties and structures investigated [218].

The NO stretching frequency in the nitrosylnickel complexes $NiCl(NO)(Ph_2P$-$(CH_2)_nPPh_2)$ are shifted to lower frequency as the length of the methylene chain $(CH_2)_n$ increases [588]; a similar trend was observed for ν_{NN} in the complexes $Mo(N_2)_2(Ph_2P(CH_2)_nPPh_2)$ [590].

Copper(I) halide complexes with dppe and dppm [847] with an CuX/L ratio of 2:1, 3:2, 4:3, 1:1, and 2:3 and the copper(I) trifluoroacetate complexes with dppe [393] have been reported. In $Cu(CF_3COO)(dppe)_2$ an essentially free CF_3COO^- ion is believed to be present. Zinc(II) halide complexes of dppe, ZnX_2dppe, and dppb, ZnX_2dppb (X = Cl, Br, I) were synthesized from zinc(II) halides and the ligand in alcoholic solutions [1049].

Monotertiary phosphines R_3P with molybdenum(V) oxide chloride yield complexes of the type $MoOCl_2(R_3P)_3$, whereas dppe produce the monomeric complexes $MoOCl_3dppe$. Complexes $[MoOCl(dppe)_2]^+$ and $MoOCl_2(THF)dppe$ (THF = tetrahydrofuran) have been prepared as well [216].

Various group VI metal carbonyl derivatives of unsymmetrical bis(tertiary phosphine) ligands, $Ph_2PCH_2CH_2PPhR$ (R = Me, Et, Pr^i, abbreviated by P∿P'), have been isolated. There are three possibilities of coordination of these ligands as indicated in (101) to (103), all of which have been realized [539]:

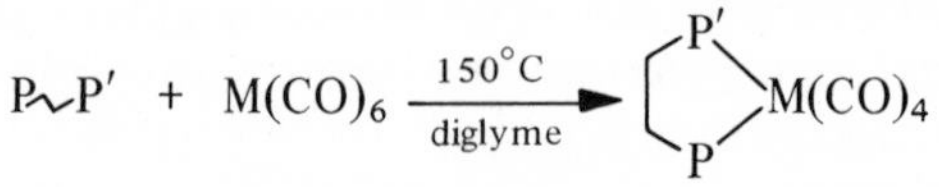

$$P\!\smallsmile\!P' \ + \ M(CO)_6 \ \xrightarrow[\text{diglyme}]{150°C} \ \left[\begin{array}{c} P' \\ P \end{array}\right]\!M(CO)_4 \qquad (101)$$

$$P\!\smallsmile\!P' \ + \ M(CO)_6 \ (\text{excess}) \ \xrightarrow[\text{diglyme}]{150°C} \ (CO)_5MP\!\smallsmile\!P'M(CO)_5 \qquad (102)$$

$$P\!\frown\!P' + W(CO)_5(PhNH_2) \xrightarrow[C_6H_6]{} P\!\frown\!P'W(CO)_5 \text{ or } P'\!\frown\!PW(CO)_5 \qquad (103)$$

The reaction between decacarbonyldimanganese or bromopentacarbonylmanganese and dppe or dppm provides a number of interesting substitution products; for example, $Mn_2Br_2(CO)_8$dppm or $Mn_2(CO)_8$dppm with a bridging dppm and Mn(CO)(dppe)$_2$ or MnBr(CO)(dppe)$_2$ [1021].

The ditertiary phosphines dppe and dmpe cleave the dimer $[\pi C_3H_5Mo(C_6H_6)Cl]_2$, thus giving the cationic species $[\pi C_3H_5Mo(C_6H_6)L]^+$ (L = dppe, dmpe) which reacts with nucleophiles R to yield neutral cyclohexadienylcomplexes $\pi C_3H_5Mo(C_6H_6R)L$ ($R = H^-$, Bu^- for dppe and dmpe, if L = dppe (CN^-) [533].

Reduction of the complexes $MoCl_4$dppe or $MoCl_4(THF)_3$ with sodium amalgam in the presence of dppe in tetrahydrofuran under a nitrogen atmosphere provides a simple preparative route to bis(dinitrogen) complexes of molybdenum $Mo(N_2)_2$-(dppe)$_2$. The corresponding diphenylmethylphosphine complexes may be obtained in a similar way [4]. Chatt et al. reported the reduction of N_2 within the complexes *trans*-$M(N_2)_2$(dppe)$_2$ (M = Mo, W) under the influence of hydrogen halides to yield the N_2H_2 complexes $MX_2(N_2H_2)$(dppe)$_2$ (M = Mo, W, X = Cl, Br). The mode of bonding of the N_2H_2 moiety is uncertain [252]. Oxidation of $Mo(N_2)_2$(dppe)$_2$ with iodine in methanol solution produced red $[Mo(N_2)_2(dppe)_2]^+I_3^-$, the first Mo(I) complex containing coordinated dinitrogen [496].

Aerial oxidation of *cis*-$Mo(CO)_2$(dppe)$_2$ in tetrahydrofuran in the presence of acids gave the cationic complex *trans*-$[Mo(CO)_2(dppe)_2]^+$ which was isolated as its perchlorate or bisulfate. The corresponding chromium compound was also synthesized [336].

An interesting complex of Re(IV) *trans*-ReO_2dppe, was obtained by Jabs et al. [641] in an attempt to recrystallize *trans*-ReO_2Cldppe from absolute nitromethane. The other reaction products were $ReOCl_4$ and oxidized dppe.

Seven coordinate complexes of the main group mixed metal complexes $Ph_3SnM(CO)_4$dppe (M = V, Nb, Ta) were synthesized by Davison and Ellis [368]. Magnetic moments and spectral measurements of the technetium(II) and (III) complexes (which are very rare with group V donors), for example TcX_2(dppe)$_2$ (X = Cl, Br, μ_{eff} = 2.05 or 2.28) and $[TcBr_2(dppe)_2]Br$ (μ_{eff} = 3.04), indicate octahedral coordination geometry. The reaction of the halogeno anions $TcCl_6^{2-}$ and $TcBr_6^{2-}$ with dppe in acidic solutions gave precipitates of the salts $[H_2dppe]^{2+}$ $[TcX_6]^{2-}$ (X = Cl, Br) rather than Tc(IV) complexes that may be obtained if Ph_3P is employed instead of dppe [454].

Treatment of $[\pi C_5H_5Fe(CO)]_2L$ (L = dppm, dppe, dppv, dppp, dppa) with iodine or Ag(I) salts leads to the formation of the cationic species $([\pi C_5H_5Fe(CO)]_2L)^+$ [556]. Stepwise coordination of dppe to the acetone solvated cation $[\pi C_5H_5Fe$-$(CO)_2(OCMe_2)]^+$ [193] has been observed. Depending on relative concentrations, this cationic complex reacts with dppe to give the 1:1 or 2:1 bridged complexes. Elimination of carbonyl occurs on irradiation with a sun lamp:

$$[\pi C_5H_5Fe(CO)_2Ac]^+ \xrightarrow[-Ac]{P-P} [\pi C_5H_5Fe(CO)_2(P-P)]^+ \xrightarrow[-CO]{h\nu}$$

$$[\pi C_5H_5Fe(CO)(P{-}P)]^+ \qquad (104)$$

$$[\pi C_5H_5Fe(CO)_2P-P)]^+ + [\pi C_5H_5Fe(CO)_2Ac]^+ \xrightarrow[-Ac]{}$$

$$[\pi C_5H_5Fe(CO)_2(P-P)Fe(CO)_2\pi C_5H_5]^{2+} \qquad (105)$$

P—P = dppe, Ac = Me_2CO.

Ionic and covalent derivatives of octacarbonyldicobalt, $[Co_2(CO)_4(L\text{-}L)_3][Co(CO)_4]_2$ (L-L = dppe, dtpe) and $[Co(CO)_2(L\text{-}L)]_2$ (L-L = dppe, dtpe) or $Co_2(CO)_6(L\text{-}L)$ (L-L = dppb), respectively, are obtained if $Co_2(CO)_8$ is treated with dppe, dppp or dtpe [1176].

Dppe forms stable cobalt(I) complexes of the type $CoX(dppe)_2$ (X = Br, ClO_4, BPh_4), whereas the bromide is tetracoordinate in polar solvents it is pentacoordinate in nonpolar solvents. The perchlorate and the tetraphenylborate are tetracoordinate both in solution and the solid state [948]. The analogous cyanide complex $Co(CN)(dppe)_2$ and $[Co(CN)(dppe)_2]^+$ of Co(I) and Co(II) have been synthesized and their reactivity toward dioxygen studied [1016]. The zwitterion complexes with the cation dicyanobis[1,2-bis(diphenylphosphino)ethane] cobalt-(III), $[Co(CN)_2(dppe)_2]^+$ (A^+) of general formula $A^+[MX_3]^-$ (M = Mn(II), Fe(II), Co(II), Ni(II), Zn(II), X = Cl, Br, SCN) may be considered pseudotetrahedral M(II) complexes in which the cobalt(III) cation is bound to the MX_3^- moieties by the nitrogen atom of one cyanide group [1017].

Spectrophotometric and conductometric data provide evidence of the existence of a lowspin five-coordinate species $[NiX(dppe)_2]^+$ (X = Cl, Br, I), which is in equilibrium with cis planar NiX_2dppe [907]. Analogous complexes with the ligand depe have been investigated. The five-coordinate complexes $[NiX(depe)_2]^+$-BPh_4^- are assigned a square pyramidal structure on the basis of their electronic spectra and an X-ray structure determination of $[NiI(depe)_2]I$ [33].

On the basis of IR and ^1H NMR spectroscopic studies Jonas and Wilke [663] postulated a hydrogen-bridged dinuclear species (*33*) containing the ditertiary phosphines $Cy_2P(CH_2)_nPCy_2$:

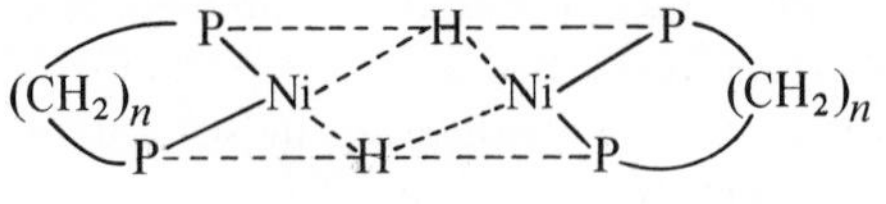

33

(n = 2, 3, 4).

Palladium and platinum complexes with terminal mercaptido groups RS (R = Ph, Me) of the type $M(RS)_2L$ (L = dppe) and their coordination compounds $M(SR)_2LM'(CO)_4$ (M' = Cr, Mo, W) have been prepared [177].

An unusual alkyl-P bond cleavage was observed during the reaction of dppe with (2-cupriobenzyl)dimethylamine (CuR) in the stoichiometric ratio 2:1 (dppe:CuR) [1222].

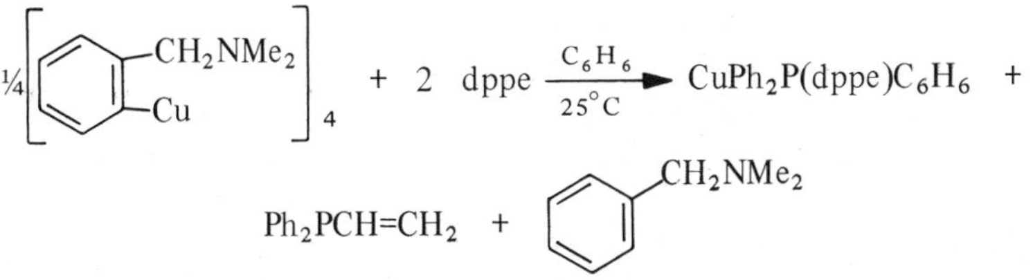

$$\frac{1}{4}\left[\begin{array}{c}\text{—CH}_2\text{NMe}_2\\ \text{—Cu}\end{array}\right]_4 + 2 \text{ dppe} \xrightarrow[25°C]{C_6H_6} \text{CuPh}_2\text{P(dppe)}C_6H_6 +$$

$$\text{Ph}_2\text{PCH=CH}_2 + \underset{}{\boxed{}}\text{—CH}_2\text{NMe}_2 \qquad (106)$$

Organocopper(I) complexes [224], for example, $(CuR)_2(dppe)_3$ (R = o-,p-tolyl, o-anisyl, phenylethynyl), $(CuR)_4(dppm)_3$, CuRdppe (R = phenylethynyl), which contain the ligands dppe and dppm, have been prepared by reaction of the appropriate phosphines with arylcopper compounds in ethereal solution.

Copper(I) [1050], silver(I) [1051], gold(I) [1051], and cadmium(II) [1048] halide complexes of the ditertiary phosphines dppe and dppb have been prepared. The complex CuCldppe in $PhNO_2$ has the ionic structure $[Cu(dppe)_2]^+ [CuCl_2]^-$ according to molar conductance measurements. Three types of silver(I) and gold(I) complexes were found: MXL, MXL_2, and $(MX)_2L_3$ (M = Ag, Au, X = Cl, Br, I, L = dppe, dppb).

$Ni(dppb)_2$ was more reactive than $Ni(dppp)_2$ toward cyanoalkenes and alkynes [316]. In the reaction between HCN and bis[1,4-bis(diphenylphosphino)butane]-nickel(O) a reactive unstable hydride species is formed initially. The final products are $[Ni(CN)_2dppb]_2$ and $Ni(CN)_2(dppb)_2$ [317].

$[\pi C_5H_5Fe(CO)_2]_2$ reacts with ditertiary phosphine ligands (L), $Ph_2P(CH_2)_n$-PPh_2 (n = 1, 2, 3) (including their arsenic analogs), cis-dppv, and dppa to give derivatives of the type $[\pi C_5H_5Fe(CO)]_2L$. Two terminal carbonyl groups in $[\pi C_5H_5Fe(CO)_2]_2$ are replaced [557]. Reversible oxidations with cyclic voltametry in ditertiary phosphine bridged derivatives of $[\pi C_5H_5Fe(CO)_2]_2$ have been studied; for example, with $[\pi C_5H_5Fe(CO)]_2dppp$ [453].

The complexes $\pi C_5H_5Ni(SPh)dppm$ and $\pi C_5H_5Ni(CN)dppm$ act as bidentate ligands, P and S or P and CN being the donors in their complexes; for example, πC_5H_5Ni-μ-SPh-μ-$(Ph_2PCH_2PPh_2)Mo(CO)_4$ [1055].

Ligand transfer reactions between nickelocene and bis(diphenylphosphino)-ethanenickel(II) complexes yield, for example, ionic $[\pi C_5H_5Nidppe]_2NiX_4$ (X = Cl, Br) and $[\pi C_5H_5NiX]_2dppe$ (X = I, CN) [667]. Dppe and dppb complexes of nickel(II) and cobalt(II) have been prepared [1052].

Several attempts to prepare stable five-coordinate nickel(II) complexes of dppe have generally been unsucessful [620, 802]. The more flexible aliphatic dmpp form the five-coordinate complexes $[MX(dmpp)_2]^+$ (M = Ni, X = Cl, Br, I, NCS;

M = Co, X = Cl, Br, I). Four-coordinate planar complexes NiX_2dmpp and six-coordinate complexes $[CoX_2(dmpp)_2]^+$ were also synthesized with this ligand [303].

2. COMPLEXES OF DITERTIARY PHOSPHINES WITH OLEFINIC, ACETYLENIC, OR AROMATIC BRIDGING UNITS

The rigid ligand *cis*-dppv has been used for the syntheses of complexes $[RhHX(dppv)_2][RhX_2(CO)_2]$ (X = Cl, Br) [813] and dinuclear iron complexes [377]. As one would expect, the trans isomer, *trans*-dppv, behaves as a nonchelating ligand; for instance, it forms the three-coordinate yellow Pt(*trans*-dppv)$_3$ and two-coordinate Pt(*trans*-dppv)$_2$ as well as the unusual complex $[Pt_2$(*trans*-dppv)$_4][PF_6]_4$ [690]. Palladium(II) complexes of the fluorine containing analog of dppv, *cis*-1,2-difluoro-1,2-bis(diphenylphosphino)ethylene were reported [273].

$[\pi\text{-}C_5H_5Fe(CO)_2]_2$ under UV irradiation with dppac produces monosubstituted diphosphine-bridged substitution products $[(\pi C_5H_5)_2Fe_2(CO)_3]_2$dppac, whereas with $\pi C_5H_5FeX(CO)_2$ (X = Cl, Br, I) complexes $[\pi C_5H_5FeX(CO)]_2$dppac are obtained [235]. 1,4-Bis(diphenylphosphino)butyne-2 affords only mono- and dibridged complexes; for example, $[Mo(CO)_2(COMe)(C_5H_5)]_2$dppby and $[Ni(CO)_2]_2$(dppby)$_2$ [684]. Tribridged complexes, which are well known for dppac, $(CuCl)_2$(dppac)$_3$ [234] being an example, could not be synthesized with dppby [684].

9,10-Bis(diphenylphosphino)phenanthrene (dpph), a ligand with a highly aromatic backbone and appropriate nickel(II) salts, produces complexes NiX_2dpph (X = Cl, Br, I, SCN) or $[Ni(dpph)_2](ClO_4)_2$ and the cationic species $[Ni(NO_3)(dpph)_2]^+$ [272].

A series of complexes containing the fluorocarbon-bridged unsaturated ditertiary phosphine ligands $f_4$fos, $f_6$fos, and $f_8$fos has been prepared. Derivatives of ruthenium carbonyls $[Ru_3(CO)_8(f_4fos)_2$, $Ru_2(CO)_6f_4fos$, $Ru(CO)_3f_6fos]$ [342], osmium carbonyls $[Os_3(CO)_{11}f_4fos$, $Os_3(CO)_{10}f_4fos$, $Os_2(CO)_6f_4fos$, $Os_3(CO)_8(f_4fos)_2]$ [337], enneacarbonyldiiron $[Fe_2(CO)_7f_6fos]$ [343], decacarbonyldimanganese and -rhenium $[Mn_2(CO)_8f_4fos$ and $Re_2(CO)_8f_4fos]$ [340], octacarbonyldicobalt $[Co_2(CO)_6f_4fos]$ [338], and halocarbonylrhodium, for example, $[Rh(L\text{-}L)_2]^+$[*cis*-$RhCl_2(CO)_2]^-$ (L-L = $f_4$fos, $f_6$fos) and $[Rh(L\text{-}L)_2]^+X^-$ (L-L = $f_4$fos, $f_6$fos, $f_8$fos, X = Cl, BPh_4) [345], have all been prepared.

C. Tri-, Tetra-, and Hexatertiary Phosphines

Mainly three types of tritertiary, four types of tetratertiary, and only one type of hexatertiary phosphine have been used as ligands in transition metal complexes. Their abbreviations are given in Scheme 9.

Two publications concerned with general aspects of the coordination chemistry of polytertiary phosphines, that is, systematics of the ligands and their metal complexes, synthetic routes, and coordination modes, must be mentioned [682, 683].

SCHEME 9

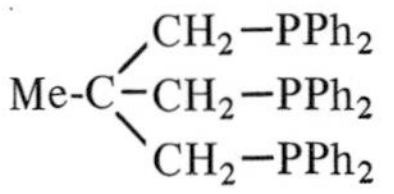

$$Me-C \begin{cases} CH_2-PPh_2 \\ CH_2-PPh_2 \\ CH_2-PPh_2 \end{cases}$$

$$T^1$$

$$Ph-P \begin{cases} (CH_2)_n \, PPh_2 \\ (CH_2)_n \, PPh_2 \end{cases}$$

$$T^2 \quad n \; = \; 2$$
$$T^3 \quad n \; = \; 3$$

$$T^4 \quad R \; = \; H$$

$$f_8 T^4 \quad R \; = \; F$$

$$C[CH_2PPh_2]_4 \qquad Q^1$$

$$\underset{Ph}{Ph_2P(CH_2)_2}\overset{Ph}{P}(CH_2)_2\overset{Ph}{P}(CH_2)_2PPh_2$$
$$Q^2$$

$$P[(CH_2)_2PPh_2]_3 \qquad Q^3$$

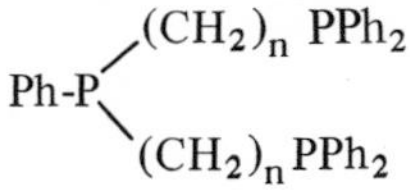

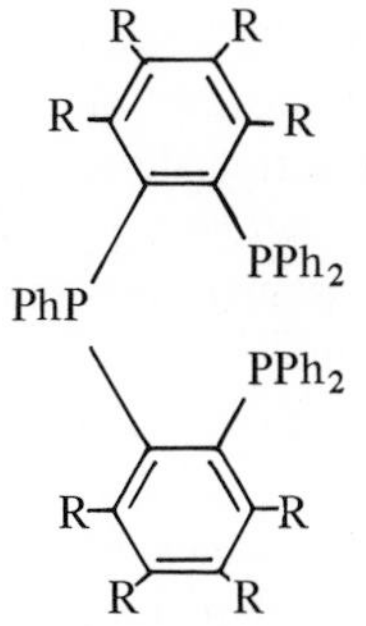

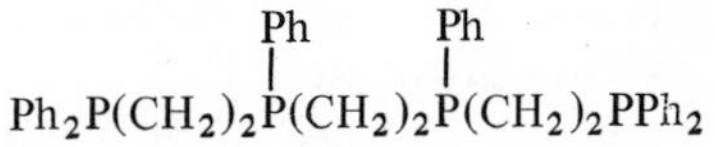

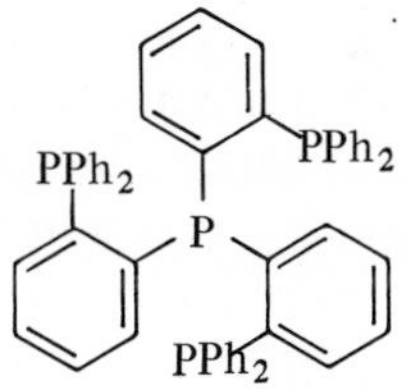

$$Q_4$$

$$[Ph_2P(CH_2)_2]_2P(CH_2)_2P[(CH_2)_2PPh_2]_2.$$

Hx

1. COMPLEXES OF THE TRITERTIARY PHOSPHINE T^1

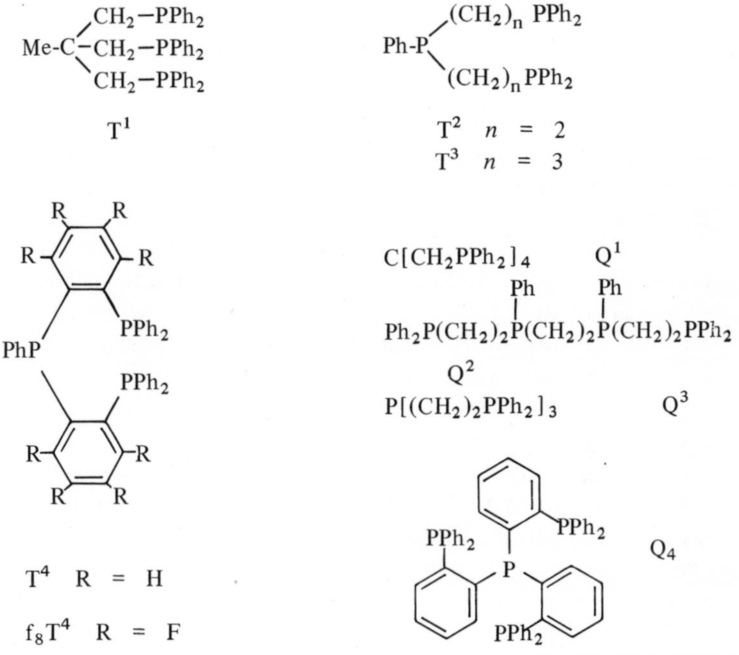

1,1,1-Tris(diphenylphosphinomethyl)ethane, T^1, reacts with cyclooctatetraene-tricarbonyliron, $C_8H_8Fe(CO)_3$, under mild conditions to yield $Fe(CO)_3T^1$ in which T^1 acts only as a bidentate ligand; under more rigorous conditions $Fe(CO)_2T^1$ is obtained [86]. Similarly, if T^1 interacts with diiodotetracarbonyliron $FeI_2(CO)_4$, $FeI_2(CO)_2T^1$ is obtained first in which T^1 is only bidentate; but, if this complex is treated with iodine, a tetraiodo complex, $FeI_4(CO)_2T^1$, which has two iodine atoms bound to the metal and two to the phosphorus atom of the noncoordinate branch of T^1 [87], may be isolated.

Rhodium(I) and iridium(I) complexes of T^1, $MCl(CO)T^1$ (M = Rh, Ir), have been synthesized [1113]. The reactivity of $Ru(CO)_2T^1$ obtained from the reaction between $Ru_3(CO)_{12}$ with T^1 in hot toluene is explained by kinetic lability of one of the Ru-P bonds; the result is a rapid equilibrium:

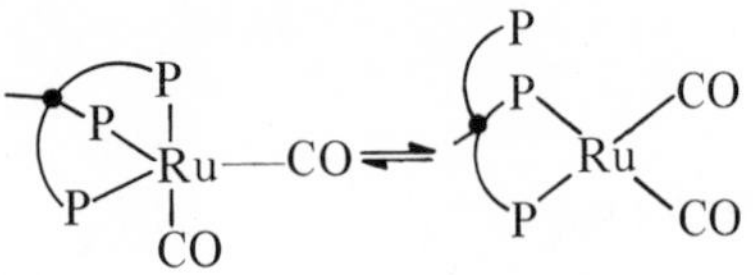

(107)

Whereas $Ru(CO)_2(Ph_3P)_3$ reacts rapidly with O_2 to form a dioxygen complex [240], with $Ru(CO)_2T^1$ and oxygen a carbonato complex, $Ru(CO)(CO_3)T^1$, is formed [1114].

The blue crystalline diamagnetic compounds of the formula $(Fe_2H_3T^1{}_2)Y$ (Y = Br, I, ClO_4, PF_6, BPh_4) have three bridge hydrogen bonds per molecule according to the ^{31}P NMR and an X-ray crystal structure determination [354]. Two isomers of $ReOCl_3T^1$ are formed in the reaction between perrhenate ion and T^1 in the presence of HCl and H_3PO_2. The blue isomer presumably is a six-coordinate, the green a seven-coordinate, rhenium complex. If sodium dithionite is used as a reducing agent for the reduction of $ReOCl_3T^1$ the Re(III) complex, $ReCl_3T^1$, is formed [366].

Four- and five-coordinate complexes of cobalt(II) and nickel(II), MX_2T^1 (X = Cl, Br, ClO_4, M = Co, Ni) have been reported [367]. In the perchlorate complexes the ClO_4^- ion functions as a monodentate ligand. A marked difference between T^1 and its arsenic analog is shown in the reaction between excess tetrafluoroethylene and the complexes $\overline{NiCF_2CF_2}L$ (L = T^1 or its As analog T^1_{as}). Only in the case of T^1_{as} does ring expansion occur to give $\overline{NiCF_2CF_2CF_2CF_2}L$ [843].

2. COMPLEXES OF THE TERTIARY PHOSPHINES T^2, T^3, AND T^4

Different synthetic routes for the preparation of these tritertiary phosphines have been developed by Cloyd and Meek [304] and King and Kapoor [692, 693].

The special arrangement of three P-donor atoms in T^2 naturally implies the formation of five [e.g., $Ni(CN)_2T^2$] and planar four-coordinate nickel(II) cations [304]. With halopentacarbonylmanganese(I) two isomers of fac-$MnX(CO)_3T^2$ (X = Cl, Br, I) in which T^2 is only bidentate may be isolated [327]. The potentially tridentate ligand T^2 forms polymetallic derivatives upon treatment with K_2PtCl_4 and $NaBH_4$ [690]. Napier and Meek [921] studied the effect of the chelate chain length on the coordination geometry in complexes of the tritertiary phosphines T^2 and T^3. The ligand field effect of the six-membered chelate ring is lower than that of the analogous five-membered ring.

Four-, five-, and six-coordinate rhodium complexes containing T^3, for example, $RhXT^3$ (X = Cl, Br, I), $RhClT^3 \cdot A$ (A = CO, BF_3, SO_2, O_2, S_2), and $RhClT^3(X)(Y)$ (XY = Cl_2, Br_2, I_2, HCl, H_2, MeI) have been synthesized [922, 923].

The complex $RhClT^3$ may be methylated by $MeFSO_3$ to give the unusual metal-alkyl cation $[RhClMeT^3]^+$ as its fluorosulfate [984]. As King, Kapoor, and Kapoor demonstrated in a large number of T^2 complexes of the metals Ni(O), Ni(II),

Pd(II), Co(II), Rh(I), Rh(III), Ir(III), Ru(II), Os(IV), Cr(O), Mo(O), W(O), Mn(I), and Fe(O), this ligand may act as monodentate [e.g., $Fe(CO)_4T^2$], bidentate [e.g., $Mo(CO)_4T^2$], or tridentate [e.g., $MnBr(CO)_2T^2$]. In all, it may bond in six possible ways (monoligate monometallic, biligate monometallic, triligate monometallic, biligate bimetallic, triligate bimetallic, and triligate trimetallic; for an explanation of these terms see [683]) [694]. Interchanging the ligands T^4 and f_8T^4 in the nickel complexes NiX_2L (X = Cl, Br, I, L = T^4, f_8T^4) has no pronounced effect on their electronic spectra [430].

3. TETRATERTIARY PHOSPHINE COMPLEXES OF Q^1, Q^2, Q^3, AND Q^4

The complex chemical behavior of the potentially four-dentate tetrakis(diphenyl-phosphinomethyl)methane Q^1 toward binuclear metal carbonyls $Co_2(CO)_8$ and $Mn_2(CO)_{10}$ has been investigated [433]. The results are summarized in (108):

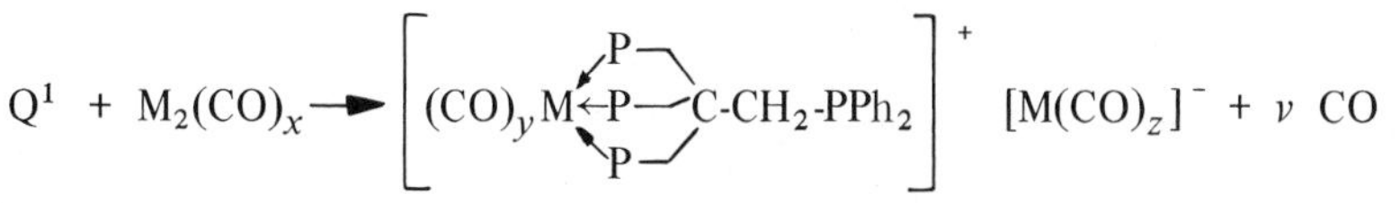

$$Q^1 + M_2(CO)_x \longrightarrow \left[(CO)_y M \overset{P}{\underset{P}{\overset{|}{\mathclap{}}}}P\!\!-\!\!C\text{-}CH_2\text{-}PPh_2 \right]^+ \quad [M(CO)_z]^- + v\ CO \qquad (108)$$

M = Co, x = 8, y = 2, z = 4, v = 2,
M = Mn, x = 10, y = 3, z = 5, v = 2.

The spiroheterocyclic complex (34) is formed if Q^1 is allowed to react with cyclo-pentadienetricarbonylchloromolybdenum, $\pi C_5H_5MoCl(CO)_3$ [432]:

$$\pi C_5H_5(CO)ClMo \overset{\displaystyle P(Ph_2)CH_2}{\underset{\displaystyle P(Ph_2)CH_2}{\diagdown\!\!\!\diagup}} C \overset{\displaystyle CH_2P(Ph_2)}{\underset{\displaystyle CH_2P(Ph_2)}{\diagup\!\!\!\diagdown}} MoCl(CO)\pi C_5H_5$$

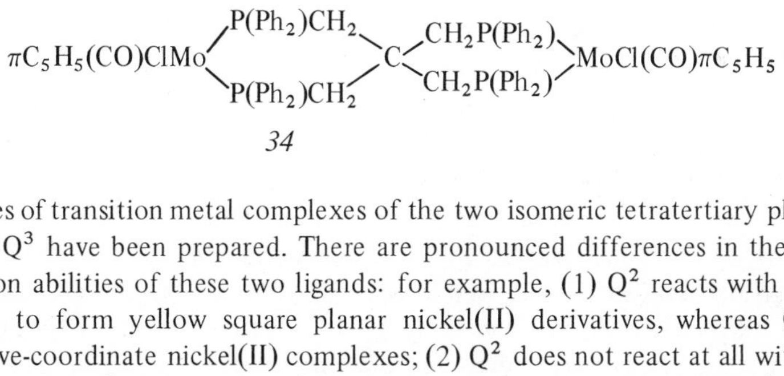

34

A series of transition metal complexes of the two isomeric tetratertiary phosphines Q^2 and Q^3 have been prepared. There are pronounced differences in the complex formation abilities of these two ligands: for example, (1) Q^2 reacts with nickel(II) chloride to form yellow square planar nickel(II) derivatives, whereas Q^3 yields violet five-coordinate nickel(II) complexes; (2) Q^2 does not react at all with ferrous chloride, and Q^3 forms the violet five-coordinate derivative $[FeClQ^3]^+$ [695].

$Co_2(CO)_8$ under the influence of Q^2 (and dppe or T^2) disproportionates to give cationic species of the type $[Co_2(CO)_4Q^2{}_3]^{2+}$ and $[Co(CO)Q^2{}_n]^+$ [983].

Bacci, Midollini, Stoppioni, and Sacconi have prepared complexes of the type $[MXQ^2]BPh_4$ (M = Fe, Co, Ni, X = Cl, Br, I). These complexes probably have a square pyramidal ligand arrangement. In the iron compounds a spin-equilibrium between a singlet ground state and a low-lying triplet state is observed [58].

Geometrical considerations for trigonal bipyramidal and octahedral species containing ligands of type Q^4 have shown that in many cases bond distances and bond angles must be severely distorted to form the complexes [371].

In complexes $[NiXQ^4]BPh_4$ (X = Cl, I) a change of the apical donor atom in the organic ligand Q^4 causes an anomalous ligand field effect. The unusual spectrochemical order P > As < Sb is explained by the compression of the apical bond caused on chelation [591]. Reductive carbonylation of the complexes $RuCl_2Q^4$ yields $Ru(CO)Q^4$, from which by halogen oxidation the ruthenium(II) complexes RuX_2Q^4 (X = Cl, I) may be obtained again [563].

An iridium(I) complex containing only phosphorus donor ligands $[Ir(Ph_3P)Q^4]^+ BPh_4^-$ has been reported [1227].

The relative locations of the phosphorus atoms in the hexatertiary phosphine Hx resemble the relative position of the O- and N-donor atoms in the well known hexadentate ethylenediaminetetraacetic acid (EDTA); for example, with nickel chloride, Hx forms the cations $[NiHx]^+$ which may be isolated as their hexafluorophosphate salts. A series of other metal Hx complexes containing the metal in various oxidation states may be synthesized [697].

D. *Tertiary Phosphines Containing Donor Groups Other Than Phosphorus*

Approximately six types of di-, tri-, or polydentate ligands contain other donor groups in addition to phosphorus. The donor sets include three-coordinate nitrogen, heavier analogs of phosphorus (Sb, As), sulfur, and selenium atoms, olefinic and acetylenic π systems, and nitrile groups.

1. P, N-DONOR SETS

Stable, mixed thiocyanate complexes of the type Pd(NCS)(SCN)L which contains one Pd-SCN and one Pd-NCS linkage were synthesized with the PN-Ligands $Ph_2P(CH_2)_nNMe_2$ (n = 2, 3) [879]. The terdentate ligand $Et_2N(CH_2)_2NR(CH_2)_2PPh_2$ (L) (R = H, Me) with the donor set NNP forms cobalt(II) complexes of general formula CoX_2L (X = halogen, NCS) [908]. "Hybrid" ligands like $Ph_2P(CH_2)_2NMe(CH_2)_2NMe(CH_2)_2PPh_2$ (L) (donor set PNNP) impose five-coordination on its complexes with nickel(II) and cobalt(II) [1043]. Chromium, molybdenum, and tungsten carbonyl complexes containing the "tripodlike" ligands of general formula (*35*) have been synthesized.

Ligands with the donor set N_2P_2 act as terdentate in the reaction with hexacarbonylchromium, -molybdenum, and -tungsten and yield monomeric compounds $M(CO)_3N_2P_2$, whereas with the donor set NP_3 only insoluble complexes $M(CO)_3NP_3$ could be obtained. The molybdenum compound $Mo(CO)_3NP_3$ may be converted to $Mo(CO)_2NP_3$ in which NP_3 is quadridentate [57].

$$\begin{array}{l} \quad\quad\quad\nearrow CH_2CH_2X \\ N-CH_2CH_2Y \\ \quad\quad\quad\searrow CH_2CH_2Z \end{array}$$

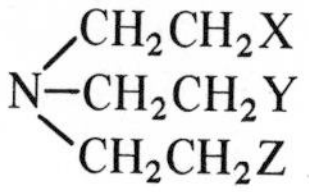

35

X = Y = Z = Ph_2P, donor set NP_3,
X = Y = Ph_2P, Z = NEt_2, donor set N_2P_2,
X = Y = NEt_2, Z = Ph_2P, donor set N_3P.

With hexacarbonyls of group VIb metals *o*-*N,N*-dimethylaminophenyl-diethyl-phosphine (donor set PN) forms penta- and tetracarbonyls $M(CO)_5PN$, $Mo(CO)_4$-$(PN)_2$, and $Mo(CO)_4PN$ [1033]. 2-Pyridyldiphenylphosphine (donor set PN) acts only as monodentate (either through phosphorus or nitrogen, depending on the hardness or softness of the transition metal) [39], whereas 8-chinolyldiphenyl-phosphine [Fe(II), Co(II), Ni(II), and Cu(II) complexes] [636] and phosphinamino-pyridines [9] may act as bidentate ligands.

2. P, As-, P, S-, AND P, Se-DONOR SETS

Five-coordinate cobalt(II) complexes, $[CoXL_2]ClO_4$ (X = Cl, Br, I, L = SP, SeP, AsP) [419a] and nickel(II) complexes of *cis*-(2-diphenylarsinovinyl)diphenylphos-phine (vasp) of the type NiX_2vasp, $[Ni(vasp)_2][ClO_4]_2$ (four-coordinate) and $[NiX(vasp)_2]Y$ (X = Cl, Br, I, NCS, Y = ClO_4, BPh_4) (five-coordinate) [272] have been reported. With the ligands SP, FSP, and SeP in complexes of the type $Pd(SCN)_2L$ two S-bonded thiocyanate ions are found [879].

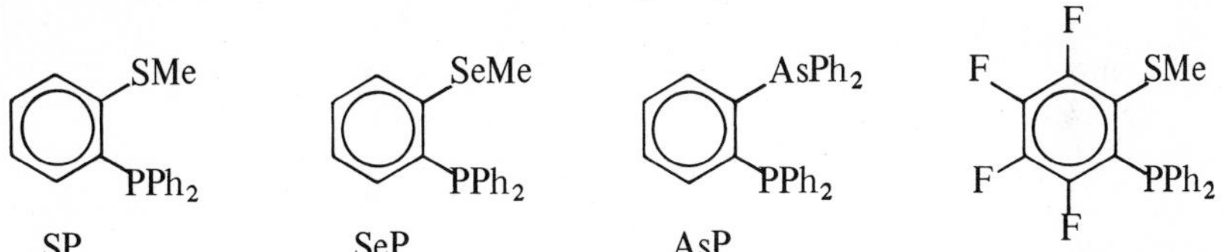

SP   SeP   AsP   FSP

The various trends in the ligand behavior of fluorocarbon-bridged AsP donor sets $\underline{LC}=CL'(CF_2)_{n/2}$ (L = Me_2As, Ph_2P, L' = Me_2As, Ph_2P, n = 4, 6, 8, but not all combinations) has been discussed by Chia, Cullen, and Harbourne [270].

Bis(2-diphenylarsinoethyl)phenylphosphine, $PhP(CH_2CH_2AsPh_2)_2$ (donor set As_2-P), does not form bimetallic or trimetallic derivatives under the conditions employed for the preparation of bimetallic and trimetallic derivatives of the tritertiary phosphine T^2. The lower donor ability of the arsenic donor atoms apparently are responsible [691]. The more flexible bis(3-dimethylarsinopropyl)phenylphosphine, $PhP(CH_2CH_2CH_2AsMe_2)_2$ (donor set As_2P), may act as tridentate ligand in a number of diamagnetic, distorted square pyramidal Ni(II) complexes NiX_2L (L = tridentate ligand as above, X = Cl, Br, I, NCS, CN) [803].

Dimeric complexes $[Ni_2(H_2O)L_3][ClO_4]_4$ [besides monomeric species NiX_2L (X = Br, I) and CoX_2L (X = Br, SCN)] have been reported for L = bis(*o*-diphenylarsinophenyl)phenylphosphine [596].

3. P, C=C -DONOR SETS

This topic has been reviewed in detail by Hall, Ling, and Nyholm. This article covers the literature through 1971 (and part of 1972) [564].

Most of the papers on the ligand properties of this kind of molecules are concerned with *o*-vinylphenylphosphines. The photochemical and thermal reactions between $Mo(CO)_6$, $Mo(CO)_4$(norbornadiene), $Mo(CO)_3$(cycloheptatriene), or $MoCl_2(CO)_4$ and the ligands $Ph_nP(CH_2CH_2CH{=}CH_2)_{3-n}$ (n = 0, 1, 2) are summarized in Scheme 10. Tri(3-butenyl)phosphine (tbp) functions as a tetradentate ligand

SCHEME 10 (taken from [488])

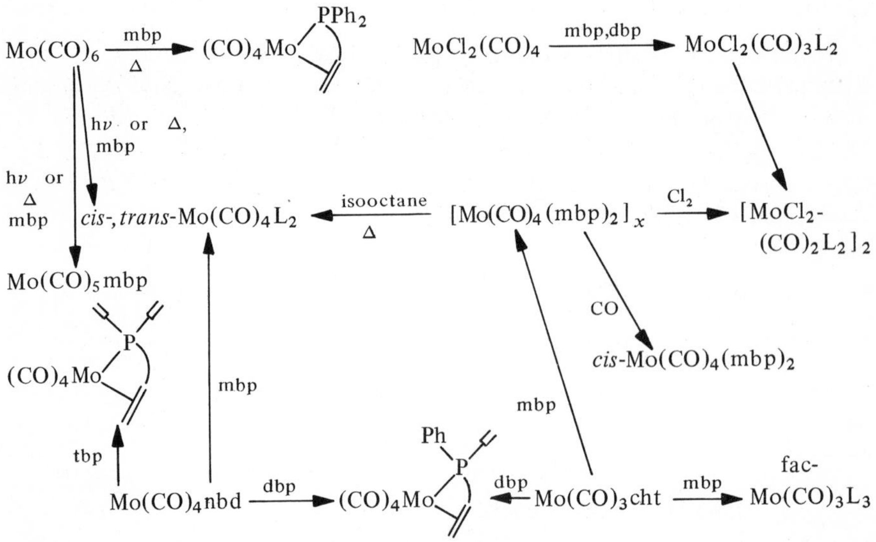

mbp = $Ph_2PCH_2CH_2CH{=}CH_2$,
dbp = $PhP(CH_2CH_2CH{=}CH_2)_2$,
tbp = $P(CH_2CH_2CH{=}CH_2)_3$,
nbd = norbornadiene,
cht = cycloheptatriene.

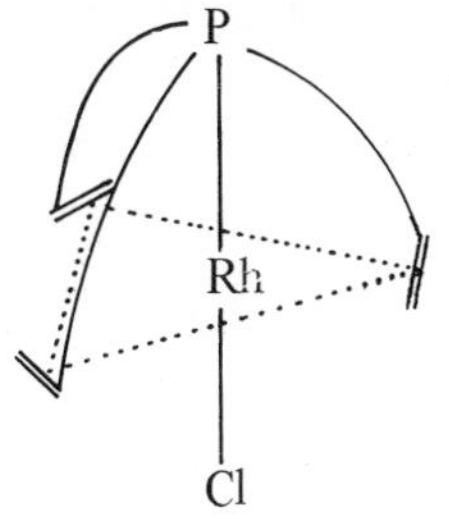

36

 = $(CH_2)_2$

in the complex RhCl(tbp). The structure proposed for this complex is shown in (*36*) [294]. Hydrogenation and isomerization of $RhCl(CO)(Ph_2P(CH_2)_n$-$CH{=}CH_2)_2$ (n = 0, 1, 2, 3) are controlled by steric requirements of the ligand [579]. (*o*-Vinylphenyl)diphenylphosphine [or *o*-styryldiphenylphosphine] (sp) forms Rh(I) complexes of composition $Rh(sp)_2Cl$ [104, 184], $[Rh(sp)_2]BPh_4$, $[Rh(sp)_2Q]BPh_4$ (Q = C_2H_4, SO_2, CS_2) [185], iron(O) and ruthenium(O) complexes, $M(CO)_3sp$, $M(CO)_2(sp)_2$ (M = Fe, Ru) [109], tetracarbonylchromium, -molybdenum and -tungsten complexes, $M(CO)_4sp$ (M = Cr, Mo, W) [112]. Reaction of $MH(CO)_5$ with sp gives, initially, $MH(CO)_4sp$ (M = Mn,Re); when heated, intermolecular hydride addition affords M-C σ-bonded chelate complexes [115]. Formation of a metal-carbon bond was also observed in the bromination of the linear P-coordinated AuBr(sp) [105]:

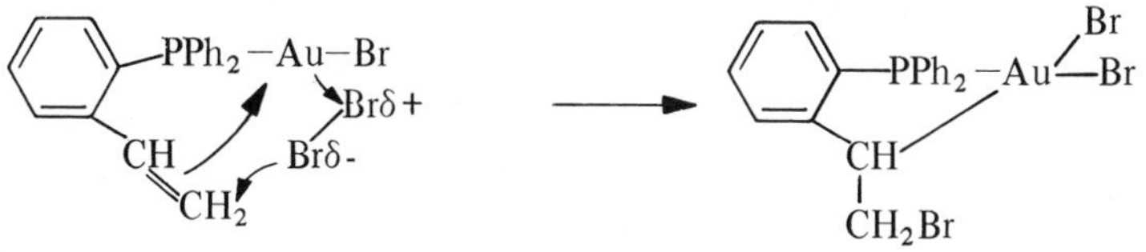

(109)

Insertion of the ortho-olefinic part of sp into the Pt-H bond in *trans*-$PtHCl(Ph_3P)_2$ gave the complex (*37*), which, according to IR spectral data, contains the Pt-CHMe

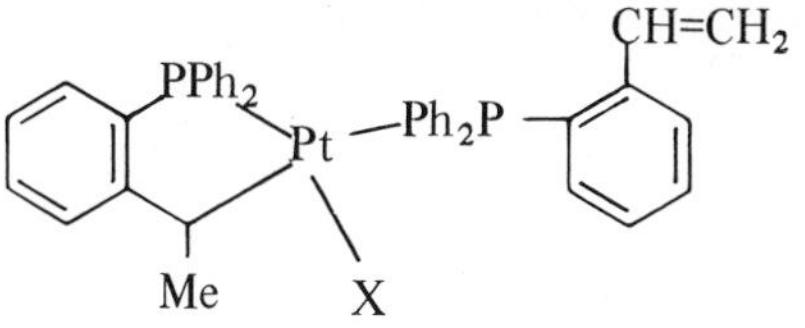

37

moiety [187]. Some (*o*-allylphenyl)diphenylphosphine (ap) complexes have been prepared, for example AuBr(ap) [105].

Reduction of vanadium(O) complexes $V(CO)_4L$ [L = (*o*-allyl-phenyl)diphenylphosphine (ap) or (*o-cis*-propenylphenyl)-diphenylphosphine (pp)] affords the anionic species $[V(CO)_5L]^-$ in which the potentially bidentate phosphines (ap and pp) are coordinated only through the phosphorus atom [630]. The oxidation of complexes $PtBr_2L$ (L = ap, sp) with elemental bromine produced compounds $PtBr_4L$. The bromination products probably have a metal-carbon σ-bond and may be formulated as dimers, for example, $(PtBr_3[CH(CH_2Br)C_6H_4Ph_2P\text{-}o])_2$ [107].

Hall and Nyholm studied the ligand properties of bis(*o*-vinylphenyl)phenylphosphine (bvpp) and tris-*o*-vinylphenyl-phosphine (tvpp) [565, 566]. In the five-coordinate rhodium(I) complexes of the As analog of tvpp a surprisingly facile displacement of the olefinic groups occurs with many ligands (e.g., CO, Ph_3P) [566].

o-Styryldiphenylphosphine is dimerized if reacted with rhodium(III) chloride in refluxing 2-methoxyethanol. A rhodium(I) complex of the tridentate ligand bis-1,3-[*o*-(diphenylphosphino)phenyl]-*trans*-1-butene is formed (*38*) which with hydrochloric acid/carbon monoxide gives a rhodium-carbon σ-bonded complex (*39*) [106]:

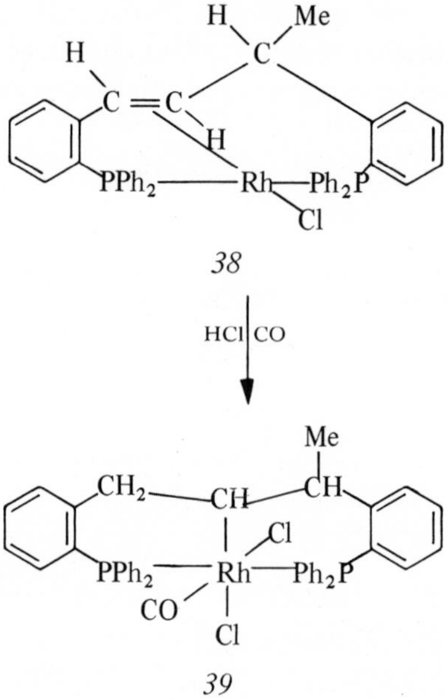

(110)

4. P, C≡C- DONOR SETS

A few derivatives of $Ph_2P{-}C{\equiv}CR$ (R = Me, Ph) [685, 1249], $Ph_2P{-}C{\equiv}C{-}PPh_2$ [1249], and $Ph_2PCH_2{-}C{\equiv}C{-}CH_2PPh_2$ [684], in which only phosphorus coordination occurs, have been described. Only recently Patel, Carty, and Hota demonstrated that under certain conditions both the alkyne and the phosphine moiety may act as donors if the phosphinoacetylenes $RR'PC{\equiv}CR''$ (R = R' = Ph, R'' = H, CF_3, Ph, Me, Bu^t; R = R' = C_6F_5, R'' = Ph, Me; R = Ph, R' = Me, R'' = Me) are allowed to react with octacarbonyldicobalt [975]. Because of the low steric demand of acetylenic substituents $-C{\equiv}C-R$ compared with alkyl and aryl groups in phosphines $R_2R'P$ (R_2 = Ph_2, MePh, R' = alkyl, aryl, acetylenic substituents), phosphinoacetylenes are capable of stabilizing the cis isomer of dihalobis(tertiary phosphine)palladium(II) complexes [1115]. Cleavage of the acetylenic bond in $Ph_2P-C{\equiv}C-CF_3$ was observed, when treated with MX_4^{2-}-salts (M = Pt, Pd, X = Cl, SCN). Complexes of the unsymmetrical ditertiary ligand $Ph_2P-CH_2-C(CF_3){=}CH-PPh_2$ were obtained [1116].

5. P, C≡N− DONOR SETS

The nearly unchanged CN stretching frequencies in nitrosylcobalt complexes, $Co(NO)(CO)_2L$ of bis(trifluoromethyl)cyanophosphine, $(CF_3)_2PCN$ or trifluoromethyldicyanophosphine, $CF_3P(CN)_2$, and tricyanophosphine, show that these ligands are coordinated by phosphorus rather than CN. No indication has been found that the CN group acts as a σ-donor system [1038].

Halopentacarbonylrhenium(I), $ReX(CO)_5$ (X = Cl, Br) reacts with one equivalent of (2-cyanoethyl)diphenylphosphine (PCN) to yield $Re_2X_2(CO)_6(PCN)_2$. PCN acts as a bridging function, as indicated in (111). The dimeric complexes react readily with donor solvents and σ-donor ligands to yield monomeric complexes [1139]:

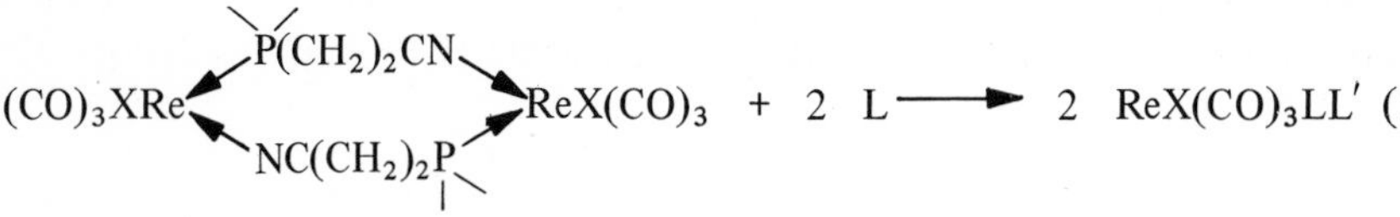

$$(CO)_3XRe \underset{NC(CH_2)_2P}{\overset{P(CH_2)_2CN}{\bigg\langle}} ReX(CO)_3 \ + \ 2 \ L \longrightarrow \ 2 \ ReX(CO)_3LL' \quad (111)$$

L' = C_5H_5N, Ph_3P, PCN, CO; L = PCN.

The proposed bridging system in the (2-cyanoethyl)diphenylphosphine complexes is similar to that in complexes of R_2PCN, $M_2(CO)_8(R_2PCN)_2$ (M = Cr, Mo, R_2 = Ph_2, CNPh, Me_2, $(EtO)_2$, $(Me_2N)_2$) [665], (R = CF_3) [947]).

The isocyanide $Ph_2PCH_2CH_2N{=}C$ (PNC), isomeric to $Ph_2PCH_2CH_2CN$, is believed to function also as a bridging ligand in complexes $Cr_3(CO)_{12}(PNC)_4$ [686].

o-Cyanophenyldiphenylphosphine (PCN) forms monomeric platinum(II) halide complexes $PtX_2(PCN)_2$ with PCN coordinated only by the phosphorus atom and trimeric species $[PtCl_2(PCN)]_3$ in which PCN acts as a bridging ligand with the nitrile group coordinated by the lone pairs of electrons on the nitrogen [977].

Recently the first examples of cyanophosphine complexes with a π-bonded nitrile group have been reported. They may be prepared by heating (o-cyanophenyl)-diphenylphosphine with halopentacarbonylmanganese(I) [978,979]:

$$MnX(CO)_5 + PCN \longrightarrow 2\ CO +$$ 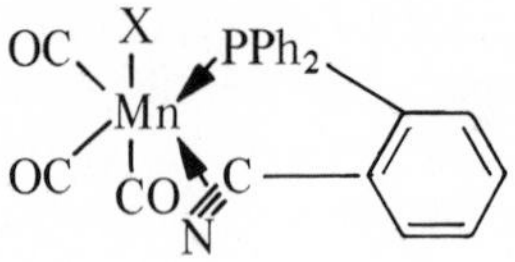 (112)

(X = ce, Br, I)

E. Various Phosphorus Ligands

The unusual ligand triferrocenylphosphine $(C_5H_5FeC_5H_4)_3P$ (TFP) reacts photochemically with $\pi C_5H_5Mn(CO)_3$ or $\pi C_6H_6Cr(CO)_3$ [933] or thermally with $M(CO)_6$ (M = Cr, Mo, W) [726, 994], $Fe(CO)_5$, $Mn_2(CO)_{10}$ [994], and $Co_2(CO)_8$ [726]:

$$TFP + \pi C_5H_5Mn(CO)_3 \xrightarrow{h\nu} CO + \pi C_5H_5Mn(CO)_2TFP \qquad (113)$$

$$TFP + M_o(CO)_m \xrightarrow[\Delta]{\text{diglyme}} CO + M_o(CO)_nTFP \qquad (114)$$

$o = 1,\ m = 6,\ n = 5,$ M = Cr, Mo, W,
$o = 1,\ m = 5,\ n = 4,$ M = Fe,
$o = 2,\ m = 10,\ n = 9,$ M = Mn.

The σ-donor ability seems to be greater for TFP than for Ph_3P. Complexes of phenyl-ferrocenylphosphines [1153] were also reported [726].

The tertiary phosphines Ph_2RP ($R = \pi C_5H_5(CO)_2M$, M = Fe, Ru) may function as ligands in the cationic species $[(\pi C_5H_5)_2M'M(CO)_4Ph_2P]^+$ (M′ = Ru, Fe) or the neutral mixed derivatives $\pi C_5H_5RuFe(CO)_5Ph_2P$ and $\pi C_5H_5RuFe(CO)_6Ph_2P$ [560]:

$$\pi C_5H_5M'Cl(CO)_2 + \pi C_5H_5M(CO)_2Ph_2P \longrightarrow (\pi\text{-}C_5H_5)_2M'M(CO)_4Ph_2P]^+Cl^- \qquad (115)$$

M = Fe, Ru,
M′ = Fe, Ru.

$$Fe_2(CO)_9 + \pi C_5H_5Ru(CO)_2Ph_2P \longrightarrow \pi C_5H_5RuFe(CO)_6Ph_2P + \pi C_5H_5RuFe(CO)_5Ph_2P \qquad (116)$$

The ligand properties of 2,4,6-triphenylphosphorine (TPP) [812] have been investigated by Deberitz and Nöth [378, 379], as well as by Fraser, Holah, Hughes, and Hui [473]. This ligand forms complexes in which either only the lone pair on phosphorus or the whole 6 π-electron system is involved. Some of its reactions are summarized in Scheme 11.

SCHEME 11

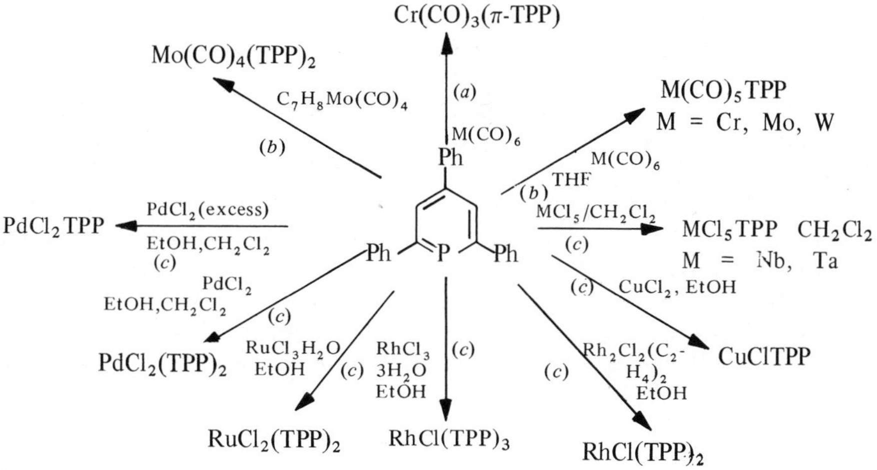

(*a*) [378], (*b*) [379], (*c*) [473].

1-Pentaboryl-bis(trifluoromethyl)phosphine with tetracarbonylnickel(O) forms a stable substitution product $Ni(CO)_3(CF_3)_2PB_5H_8$ (*40*), which has been identified by ^1H, ^{11}B, ^{19}F, and ^{31}P NMR spectroscopy [207]:

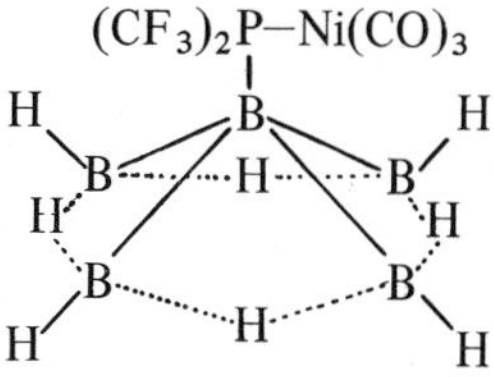

40

Complex formation of tris(carboxyethyl)phosphine, which is related to the complexons for example, $N(CH_2CH_2COOH)_3$, with various transition metal ions was found to be p_H-dependent. The stability constants of its complexes are lower than those of the nitrogen analogs [997].

III. Complexes of Organometalphosphines

Compared with triorganophosphines, organometalphosphines $(E^{IV}\ R_3)_{3-n}PR'_n$ [R = Me, R′ = Ph, Bu^t, Me, E^{IV} = C, Si, Ge, Sn, Pb, n = 0, 1, 2] should be expected to have a small σ-donor ability because of the participation of their phosphorus "lone pair" in a (p→d)π interaction between phosphorus and the group IV element [1080]. In contrast with these considerations a series of stable transition metal carbonyl complexes has been prepared. For the sake of comparison the lowest homolog of this series of organometalphosphines, tri-tert-butylphosphine, has been included in these investigations. The complexes $Ni(CO)_3L$ [1081, 1084, 1088], $Co(NO)(CO)_2L$ [1086], $Fe(CO)_4L$ [1084, 1089], $\pi C_5H_5Mn(CO)_2L$ [1085], and $M(CO)_5L$ (M = Cr, Mo, W) [1081, 1084] [L = $(E^{IV}Me_3)P$, E^{IV} = C, Si, Ge, Sn, Pb [1088], E^{IV} = C, Si, Ge, Sn [1085, 1086, 1089], L = $SnMe_3PBu^t_2$ [1084], M = Mo, L = $(SnMe_3)_2PMe$ [1081], M = Ni, Cr, Mo] have been prepared by standard synthetic methods. Although all these ligands have bulky substituents Bu^t, $SiMe_3$, $GeMe_3$, $SnMe_3$ (and in one case $PbMe_3$), it was possible to obtain di- and trisubstituted derivatives, *cis*-$M(CO)_4L_2$ [L = Bu^t_3P, $(GeMe_3)_3P$, $(SnMe_3)_3P$, M = Cr, Mo, W] [1083], fac-$M(CO)_3L_3$ [L = $(GeMe_3)_3P$, $(SnMe_3)_3P$, M = Cr, Mo] [1082]:

$$M(CO)_6 + 2L \xrightarrow[60^\circ C]{h\nu} 2\ CO + M(CO)_4L_2 \tag{117}$$

$$C_7H_8M(CO)_4 + 2\ L \longrightarrow C_7H_8 + M(CO)_4L_2 \tag{118}$$

$$M(CO)_3(MeCN)_3 + 3\ L \longrightarrow 3\ MeCN + M(CO)_3L_3 \tag{119}$$

C_7H_8 = norbornadiene,
L and M, as indicated above [1082, 1083].

These organometalphosphine complexes (for E^{IV} = Si, Ge, Sn, Pb) are of general interest because their facile synthesis and stability cast some doubt on the utility of the (p→d)π bonding concept for the explanation of stabilities and reactivities of these E^{IV} compounds. They are also of great interest as potential model substances for the investigation of reactions on coordinated ligands, especially as the $E^{IV}R_3$ groups are excellent leaving groups.

IV. Complexes of Phosphites, Phosphonites, and Phosphinites, $R'_{3-n}P(OR)_n$ (n = 1, 2, 3)

Quite a large number of publications on the coordination chemistry of these ligands is concerned with nickel, palladium, and platinum complexes. Only a few papers in the literature deal with the ligand properties of these types of phosphorus(III) donor in complexes with early transition metals.

A. *Scandium, Yttrium, Lanthanides, and Actinides*

Only one short report on uranium(IV) complexes treates of the triphenylphosphite ligand. The compound $UCl_4[P(OPh)_3]_2$ is formed by the reaction of uranyl(VI) acetate in hydrochloric acid-methanol mixtures with zinc and the subsequent addition of triphenylphosphite. The complex obtained in this way is quite stable but reduces Ag^+ ions to metallic silver and rapidly hydrolyzes in the presence of alkalies and ammonium hydroxide [1053].

B. Titanium, Zirconium, and Hafnium

Little has been published in this area during the period under review. Although not a phosphite, phosphinite, or phosphonite complex, the compound bis(cyclopentadienyl)bis-[bis(trifluoromethyl)phosphinoxy] titanium(IV) and its synthesis should be mentioned [1009]:

$$(\pi C_5H_5)_2TiCl_2 + 2\ (CF_3)_2POP(CF_3)_2 \longrightarrow 2\ (CF_3)_2PCl + (\pi C_5H_5)_2Ti[OP(CF_3)_2]_2 \qquad (120)$$

C. Vanadium, Niobium, and Tantalum

Phosphite, phosphinite, and phosphonite complexes of vanadium(IV) ethylxanthate have been investigated by means of ESR spectroscopy. The ligands studied were $P(OMe)_3$, $P(OEt)_3$, $P(OPr^n)_3$, $P(OPr^i)_3$, $P(OBu^n)_3$, $P(OBu^i)_3$, $PhP(OMe)_2$, $PhP(OBu^i)_2$, and Ph_2POEt (including Ph_3P). However, no complexes have been isolated or identified by other than ESR and IR spectroscopy [725].

D. Chromium, Molybdenum, and Tungsten

Mathieu and Poilblanc have prepared penta- and hexa-substituted zerovalent derivatives of group VI metal hexacarbonyls with the ligands F_2POPr^n, $FP(OMe)_2$, $P(OMe)_3$, $MeP(OMe)_2$, $Me_2P(OMe)$ (and Me_3P included) by extensive irradiation of mixtures of the ligands and metal carbonyls [864]. In addition mono-, di-, and trisubstituted derivatives, $M(CO)_{6-x}L_x$ [$M = Cr, Mo, W, x = 1, 2, 3, L = MeP(OMe)_2$] have been reported [664].

The ligand behavior of isomeric, monocyclic, ring-locked phosphites (*1*) and (*2*) toward carbonyls of group VIb metals has been studied. Evidence of stereoretention in the reaction of (*1*) and (*2*) with norbornadienetetracarbonylmolybdenum has been derived from a comparison of the carbonyl stretching frequencies of the complex *cis*-$Mo(CO)_4L_2$ (L corresponds to *1* or *2*) and their differences in basicity [1218].

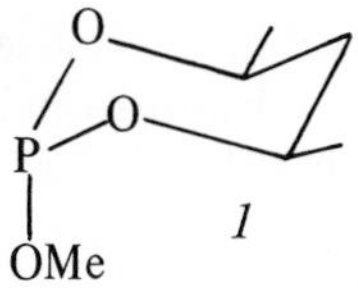

1

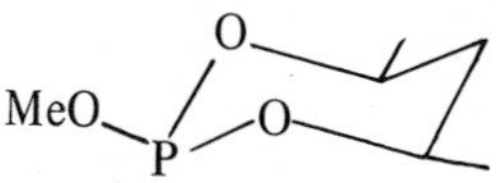

2

The constrained phosphite ester $P(OCH_2)_3CR$ (R = Me, Et, Pr) forms covalent complexes of the type $\pi C_5H_5MoX(CO)_2L$ and $\pi C_5H_5MoX(CO)L_2$ (X = Cl, Br, I, L = $P(OCH_2)_3CR$) [1129].

Depending on the conditions employed, triphenylphosphite reacts with phenyl derivatives of Mo and W cyclopentadienylcarbonyls, $\pi C_5H_5M(CO)_3Ph$ (M = Mo, W), either with carbonyl substitution and insertion or replacement of the cyclopentadienyl and phenyl rings [935]. Similar reactions for triphenylphosphine have been studied as well.

In complexes of chromium(I) with phosphites $P(OR)_3$ (R = Me, Et, Pr^n, Pr^i, Bu^n) and $PhP(OEt)_2$ of structure (*3*) the phosphorus ligand has a considerable influence on the stability [485].

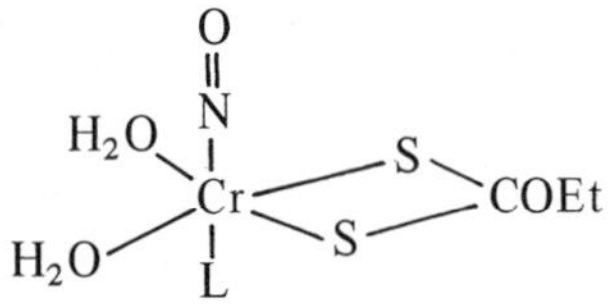

L = phosphite ligands

3

Derivatives of hexacarbonyltungsten with 2-substituted 1,3,2-benzodioxaphosphole have been prepared [494]. Tertiary alkylphosphites $P(OR)_3$ (R = Me, Et, Pr^i, Bu^n), $P(OCH_2)_3CMe$, and $P(OC_3H_5)_3$ with $[\pi C_5H_5Mo(CO)_3]_2$ yield unsymmetrical dinuclear substitution products $[\pi C_5H_5Mo(CO)_2[P(OR)_3]_2][\pi C_5H_5Mo(CO)_3]$ and symmetrical dinuclear substitution products $[\pi C_5H_5Mo(CO)_2P(OR)_3]_2$. An unusual Michaelis-Arbusov-type arrangement, via ionic intermediates provides $[\pi C_5H_5Mo(CO)_2[P(OR)_3]_2]^+[\pi C_5H_5Mo(CO)_3]^-$, complexes $\pi C_5H_5Mo(CO)_2P(OR)_3$-$P(O)(OR)_2$ of the P-bonded $P(O)(OR)_2$ ligand besides $\pi C_5H_5Mo(CO)_3R$ or $\pi C_5H_5MoR(CO)_2P(OR)_3$ [558, 562].

E. Manganese, Technetium, and Rhenium

The rate of hydrogen exchange [with CF_3COOD/D_2SO_4] in the cyclopentadienyl ring of the complexes $\pi C_5H_5Mn(CO)_2L$ (L = $P(OPh)_3$, Ph_3P, CO) is strongly dependent on the nature of L. If one CO is replaced by $P(OPh)_3$, the rate increases 70 times if Ph_3P is used 2000 times [1100]. The oxidation of various phosphite-substituted manganese carbonyl halides, for example, mer-*cis*-$Mn(CO)_2[PhP(OMe)_2]_3Br$, in which nitrogen dioxide as a single electron oxidizing agent was used, has been reported [1022].

Ligand exchange between triphenylphosphine hydrido complexes $ReH_3(Ph_3P)_4$ and triphenylphosphite affords a variety of partly substituted hydrido phosphite complexes; for example, $ReH_3(Ph_3P)_2[P(OPh)_3]_2$ [475].

F. Iron, Ruthenium, and Osmium

The temperature dependence of the ^{1}H NMR spectra of $FeH_2[P(OEt)_3]_4$ and $FeH_2[PhP(OEt)_2]_4$ is caused by their stereochemical nonrigidity. A polytopal rearrangement is thought to be operative [1168]. Replacement of triphenylphosphine by triphenylphosphite in the hydrido complexes $OsH_4(Ph_3P)_3$, $OsH_2(CO)(Ph_3P)_3$, and $OsHCl(CO)(Ph_3P)_3$ is followed by orthometalation, which yields either mono- or dicyclometalated osmium derivatives [8]:

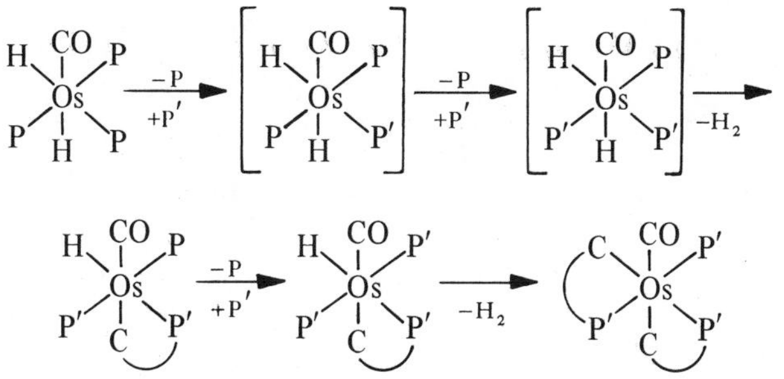

(121)

P = Ph_3P, P′ = $P(OPh)_3$.

The complex $FePt_2(CO)_5[P(OPh)_3]_3$ contains a triangular metal atom cluster and only terminally bound ligands [21].

Triarylphosphite complexes of osmium and ruthenium, $MX_2[P(OR)_3]_4$ (M = Ru, Os, X = Cl, Br, I), may be obtained by metathetical replacement reactions. On prolonged carbonylation, the dicarbonyl derivatives $MX_2(CO)_2[P(OR)_3]_2$ are formed. Nitrosyl and organoruthenium complexes, for example, $RuCl_3(NO)[P(OR)_3]_2$ and $RuCl(OPh)_2P(OC_6H_4)[P(OPh)_3]_3$ (with an *o*-metalated phenyl group), have been synthesized by using ruthenium triphenylphosphine complexes, for example, $RuHCl(Ph_3P)_3$, and excess triphenylphosphite [777]. Ruthenium(III) chloride reacts with excess of pure $P(OEt)_3$, Ph_2POMe, or $PhP(OEt)_2$ to give the ruthenium(II) complexes $RuCl_2L_3$ (L as indicated). Magnetic and EPR studies revealed the presence of ruthenium in the d^6 electronic state, implying a reduction or Ru(III) to Ru(II) on complex formation [655].

The effect of the reaction conditions on product formation (carbonyl substitution or insertion in the $Fe{-}C_{aryl}$ bond) between $\pi C_5H_5Fe(CO)_2Ar$ and phosphites (including phosphines) has been studied [934]. SO_2 insertion into the $Fe{-}C_{Me}$ bond in the phosphite complexes $\pi C_5H_5Fe(CO)Me[P(OR)_3]$ (R = Me, Bu^n, Ph) produces the corresponding S-sulfinates [1152]. Substitution of carbonyl groups, formation of ionic products, and Michaelis-Arbusov type rearrangements similar to those observed for cyclopentadienylcarbonylmolybolenum complexes [558, 562] occur on reaction between tertiary alkylphosphites and $\pi C_5H_5FeCl(CO)_2$ [559].

Nitrosonium hexafluorophosphate has been proposed as a "single electron oxidant" (e.g., for oxidation of $(\pi C_5H_5)_2Fe_2(CO)_3P(OPh)_3]$) to the salt $[\pi C_5H_5$-$Fe(CO)_2P(OPh)_2]PF_6$ [1020] without introducing a NO ligand in the oxidation products. Complexes of the type $[\pi C_5H_5Fe[P(OPh)_3]_2L]^+A^-$ [L = MeCN, EtCN, $ClCH_2CN$, Me_2NCN, CO, SO_2, C_2H_4, Et_3P, $P(OPh)_3$, A = PF_6, BF_4] have been prepared and some of their reactions [536] are summarized in Scheme 1.

SCHEME 1

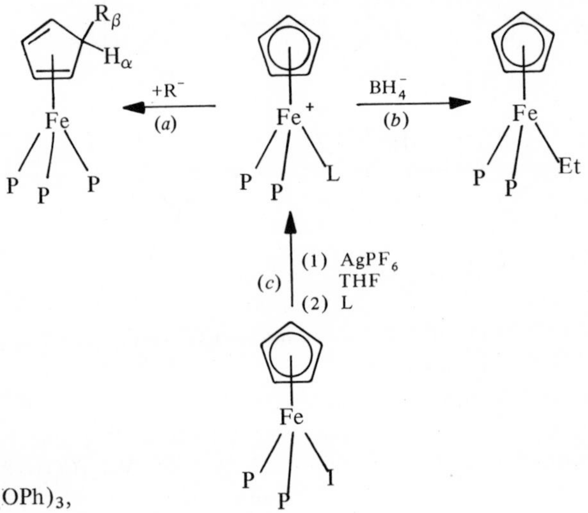

(a) L = $P(OPh)_3$,
(b) L = C_2H_4,
(c) L = RCN, SO_2, C_2H_4, $P(OPh)_3$,
R = Ph, H,
P = $P(OPh)_3$.

A new type of isomerism for transition metal carbonyl complexes (π-ring)MX(CO)L, in which L is a phosphine or a phosphite ligand, has been observed. Restricted rotation about the M—P or P—O—C bonds is thought to be the reason for the existence of different isomers [190]. Controlled pyrolysis of the cluster compound $Ru_3(CO)_9[P(OPh)_3]_3$ [200] yields $Ru_2H(CO)_3[P(OC_6H_4)(OPh)_2]_2[OP(OPh)_2]$ in which the two ruthenium atoms are bridged by one of the two orthometalated $P(OPh)_3$ ligands and one $OP(OPh)_2$ group [197, 200].

The synthesis and halogen oxidation of the tetrastannane derivative πC_5H_5Fe-$[P(OPh)_3]_2Sn_4Me_9$ have been reported [706, 707]. The disubstitution products $Fe(CO)_3L_2$ [L = $P(OMe)_3$ or, for the sake of comparison, Me_3P] react with mercury(II) halides HgX_2 (X = Cl, Br, I) to form 1:1 [for both L], 1:2 (X = Br, I), and 1:4 (X = Cl) adducts [for L = $P(OMe)_3$]. These adducts may be regarded as donor-acceptor complexes which undergo dissociation in polar solvents to give ionic species [386].

Oxidative elimination reactions involving *trans*-$Os(CO)_3[P(OMe)_3]_2$ and tetrafluoroethylene, trifluoroethylene, and chlorotrifluoroethylene show a marked difference in their results, compared with those of the analogous phenyldimethylphosphine complexes, *trans*-$Os(CO)_3(Me_2PhP)_3$ [312].

G. Cobalt, Rhodium, and Iridium

Orthometalation has been observed in triphenylphosphite complexes of rhodium(I) [76] and iridium(I) [11, 12, 102] as in those of osmium [8]. Poilblanc et al. [53, 760] reported a general examination of the reactions of trimethylphosphite with octacarbonyldicobalt or dodecacarbonyltetracobalt. Partial substitution of carbonyl groups in $Rh_4(CO)_{12}$ by phosphorus ligands [Ph_3P, $P(OCH_2)_3CEt$] leads to $Rh_4(CO)_9L_3$ and $Rh_4(CO)_8L_4$, whereas $Rh_6(CO)_{16}$ under similar conditions yields $Rh_6(CO)_{10}L_6$. If the reaction is carried out in the presence of carbon monoxide, a breakdown of the cluster occurs [159].

Stereochemical rigidity could be reached only at low temperatures for a series of cations $[M[P(OR)_3]_5]^+$ [R = Me, Et and in one case Bu^n, M = Co, Rh, Ir(I), Ni(II)Pd(II)]. The observed (and calculated) A_2B_3 type ^{31}P NMR spectra are in agreement with a trigonal bipyramidal structure for these cations [653].

Half-sandwich type compounds, $\pi C_5H_5Co[P(OR)_3]_2$, are obtained by reaction of cobaltocene, $(\pi C_5H_5)_2Co$, with tertiary phosphites $P(OR)_3$ (R = Me, Et, Ph) [573].

Five-coordinate complexes of cobalt(II) of the type $[CoXL_4]BPh_4$ and $Co(NCS)_2$-L_3 [X = Cl, Br, I, N_3, NCS, L = $PhP(OEt)_2$] have been prepared [125]. The electronic spectra in solution provide evidence for a dissociation equilibrium of the $[CoXL_4]^+$ cationic complexes involving $[CoXL_3]^+$ and L.

Ligand exchange studies in the hydrogenation catalysts $MX(CO)L_2$ [M = Ir, Rh, L = Ph_3P, $P(OPh)_3$] reveal that the rhodium complexes are much more dissociated than their iridium homologs [1150]. The NMR spectra of phosphite and phosphonate complexes of rhodium (including palladium and platinum) have been discussed and an order for the trans influence on platinum-phosphorus bonds for uncharged and anionic ligands, derived [27].

Triphenylphosphite with $IrH_3(Ph_3P)_3$ gives the hydrido complex $IrH[P(OPh)_3]_4$ which, by protonation or ligand replacement, gives *trans*-$[IrH_2[P(OPh)_3]_4]^+$, $IrH(CO)[P(OPh)_3]_3$, and $Ir(NO)[P(OPh)_3]_3$ [513]. The preparation of a series of triarylphosphite cobalt and nickel complexes of the type $CoH[P(OAr)_3]_4$ and $Ni[P(OAr)_3]_4$ where Ar = Ph, *m*-tolyl, *p*-tolyl, *p*-ClC_6H_4 have been reported and some features of their reactivity discussed [776].

H. Nickel, Palladium, and Platinum

The chemistry of zerovalent tetrakis(triarylphosphite)nickel complexes has been studied in view of their catalytic properties. For the synthesis of $Ni[P(OR)_3]_4$ (R = Me, p-$MeOC_6H_4$,Ph) nickel(II) halides NiX_2 (X = Cl, Br, I) [580, 672] or nickel(II) acetylacetonate [809] were used as starting materials and hydrogen [580], metals (Zn, Cu, Th) [672], or sodium borohydride [809], as reducing agents. Ni(O) and Pt(O) complexes of the cyclic 5,5-dimethyl-2R-1,3,2-dioxaphosphorinanes with the ligand in chair ring form and an axial phenoxy group have been prepared [972].

Tolman published a series of papers concerned with the formation and decomposition of hydridotetrakis(triethylphosphite)nickel [1183], its reaction with 1,3-butadiene [1184], NMR study of these reactions [1185], and the mechanism of the $[NiH(P(OEt)_3)_4]^+$ catalyzed olefin isomerization [1186]. Jesson et al. did similar protonation reactions of $Ni[P(OEt)_3]_4$ [414].

Reaction between $Pt(Ph_2POBu^n)_4$ or $Pt(Ph_2POMe)_4$ and protonic acids gave only labile hydrides. Bis(diphenylphosphinato)bis(hydroxydiphenylphosphine)platinum(II) has been prepared by different routes; the simplest used $Pt(Ph_2POBu^n)_4$ as starting material [724]. Some reactions of this complex and related compounds are summarized in Scheme 2.

SCHEME 2

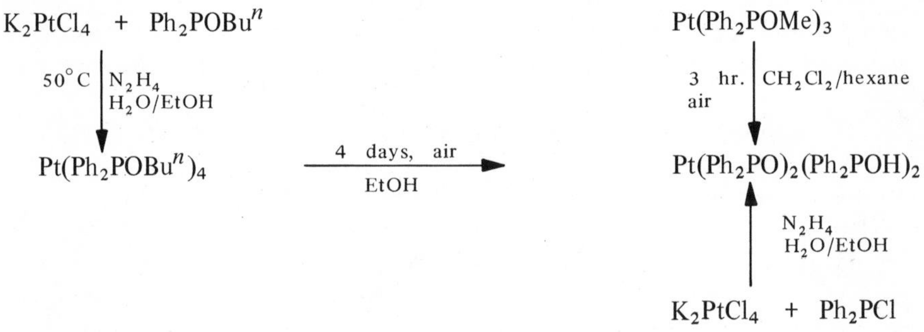

The synthesis and various interconversion reactions of diphenylphosphinato, and hydroxydiphenyl- as well as dialkyl- and diphenylphosphonato complexes of platinum(II) and palladium(II) have been reported by Dixon and Rattray [395] and Pidcock and Waterhouse [987]. With palladium, no bridging of the two metal atoms by the diphenylphosphinato group has been observed; in platinum compounds Ph_2PO bridges (P and O being donor atoms) are stable [6, 987].

Phosphites with bulky alkyl groups form three-coordinate nickel(O) complexes; for example, tris(tri-*o*-tolylphosphite)nickel(O) [522] and tris(2-biphenylylphosphite)nickel(O) [584]. The coordinatively unsaturated tris(tri-*o*-tolylphosphite)-nickel(O) may add one more molecule of tri-*o*-tolylphosphite to form the coordinatively saturated tetrakis(tri-*o*-tolylphosphite)nickel(O) [522].

On addition of alkyl nitriles to solutions of $Ni[P(O\text{-}o\text{-tolyl})_3]_3$ the equilibrium (122), which includes nitrile complexes of zerovalent nickel, is established [1179]:

$$MeCN + NiL_3 \rightleftharpoons NiMeCNL_3 \quad (fast) \qquad (122)$$

Reduction of nickel(II) acetylacetonate by triethylaluminium in the presence of tri-*o*-tolylphosphite and ethylene produced ethylene[bis(tri-*o*-tolylphosphite)] nickel(O) in high yield [1096]. Addition of tri-*o*-tolylphosphite to a solution of C_2H_4Ni-$[P(O\text{-}o\text{-tolyl})_3]_2$ liberates ethylene and forms tris(tri-*o*-tolylphosphite)nickel(O) and subsequently tetrakis(tri-*o*-tolylphosphite)nickel(O) [1096]. Reaction of nickel(II) acetylacetonate with triphenylphosphite, triethylaluminium, and carbon monoxide or the carbonylation of $Ni[P(OPh)_3]_4$ yields tris(triphenylphosphite)monocarbonyl nickel [958].

Addition of RCN (R = Me, Ph, Bu, *trans*-EtCH=CH, etc.) to $Ni[P(OC_6H_3MeR^1\text{-}2,x)_3]_3$ provided useful hydrocyanation catalysts $Ni(RCN)[P(OC_6H_3MeR^1\text{-}2,x)_3]_3$ (R^1, e.g., H, Me-4, Me-5, Cl-4) [521].

Cationic complexes of platinum and palladium of the types $[ML^1_5]^{2+}$, $[ML^2_4]^{2+}$, $[MClL^1_3]^+$ (M = Pd, Pt) in which L^1 stands for $P(OR)_3$, L^2 for $PhP(OR)_2$, and Ph_2POR (R = alkyl, e.g., Me) have been investigated [323, 324]. The tendency to accommodate five phosphorus donor ligands in cations of the form $[ML_5]^{2+}$ decreases in the order $P(OR)_3 > PhP(OR)_2 > Ph_2POR$, as expected from steric considerations. Similar trends have been observed for other metals (e.g., Rh, Ir) [323].

The reaction of various triarylphosphites $P(OAr)_3$ (Ar = Ph, *p*-ClC_6H_4, *o*-, *m*-, *p*-tolyl) with the salts Na_2MCl_4 (M = Pd, Pt) in ethanol or MX_2(cycloocta-1,5-diene) in benzene produces cis and trans complexes of type $MX_2[P(OAr)_3]_2$ [5]. Ginzburg, Sentemov, and Troitskaya studied the complex formation of K_2Pd-$(SCN)_4$ with trialkylphosphites in absolute ethanol. According to their work, the trialkylphosphite complexes of palladium(II) can be arranged in the following order of decreasing ligand field strength [510]: $P(OEt)_3 \gtrsim P(OPr^n)_3 \approx P(OPr^i)_3 > P(OBu^n)_3 \approx P(OBu^i)_3 \approx P(OBu^s)_3$.

The ^1H NMR spectra of some trimethylphosphite complexes of platinum(II), $PtX_2[P(OMe)_3]_2$ (X = Cl, Br, I, NCO, N_3, Ph) have been reported and analyzed [276].

Complex formation between nickel(II) iodide and trialkylphosphites takes place stepwise. The resulting complexes contain four molecules of trialkylphosphite per nickel atom. The thermal instability constants have been evaluated [511]. The chain length and structure of the hydrocarbon group R in trialkylphosphites $P(OR)_3$ affects their coordinating abilities. Increasing the chain length beyond four carbon atoms, however, has no further influence on the ligand field strength [512].

The introduction of thiourea into the inner coordination sphere of the platinum-(IV) derivatives $PtCl_4[P(OR)_3]_2$ (R = Et, Pr^i) was accompanied by hydrolysis of the trialkylphosphite ligands [1199]. *o*-Metalation reactions involving triarylphosphite derivatives of palladium(II) and platinum(II) take place at elevated temperatures with the elimination of hydrogen halide and the formation of the complexes (*4*) [6].

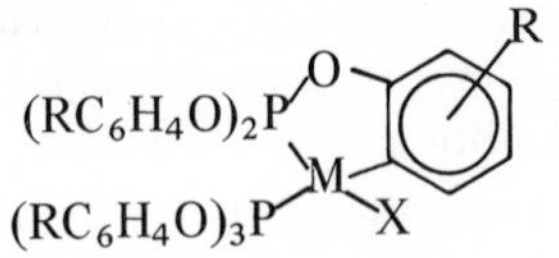

4

M = Pd, Pt,
X = Cl, Br, I,
R = H, Cl, Me.

I. Copper, Silver

Reduction of the triisopropylphosphitecopper(I) complex, $CuClP(OPr^i)_3$ with triethylstannane yields a stable hydridocopper complex:

$$CuClP(OPr^i)_3 + Et_3SnH \longrightarrow CuHP(OPr^i)_3 + Et_3SnCl \qquad (123)$$

With excess triisopropylphosphite the complex $3CuH \cdot 2P(OPr^i)_3$ may be obtained [821].

The solution structure of silver(I) triethylphosphite complexes has been investigated by means of NMR spectroscopy, conductivity, and molecular weight studies. At room temperature an averaged shift which is temperature-dependent is observed in the NMR spectra. The ligand lability in $[AgL_3]^+$ complexes is greater than in $[AgL_2]^+$ complexes (L = $P(OEt)_3$) [916].

V. Complexes of Halophosphines, $R_{3-n}PX_n$ (n = 1, 2, 3)

By far the largest number of publications that have appeared in the literature on this general topic during the period under review are concerned with fluorophosphines. The coordination chemistry of these ligands has been reviewed by Nixon [939, 940] and Clark and Busch have discussed stereochemical rearrangements in trifluorophosphine complexes [296]. Timms' elegant synthesis of trifluorophosphine complexes by reaction of transition metal vapors and trifluorophosphine [1177] deserves mention.

A. Vanadium, Niobium, and Tantalum

The first trifluorophosphine complex of a group Vb metal, π-allylpentakis(trifluorophosphine)tantalum(O), $\pi C_3H_5Ta(PF_3)_5$, has been synthesized by reaction of tetrakis(π-allyl)tantalum and PF_3 under pressure [735]. Replacement of two carbonyl groups in $C_5H_5V(CO)_4$ by piperidinodifluorophosphine produces the yellow disubstituted derivative $C_5H_5V(CO)_2(PF_2NC_5H_{10})_2$ [700].

Only phosphonium salts, for example, $[RBu^tPCl_2][MCl_6]$ (M = Nb, Ta, R = Cl, Me) [206] were formed in a Kinnear-Perren type reaction between the phosphorus chlorides PCl_3 and $MePCl_2$ and niobium or tantalum chloride (tungsten hexachloride included) in presence of *t*-butylchloride.

B. Chromium, Molybdenum, and Tungsten

Ligand replacement between tris(π-allyl)chromium and PF_3 gives dark green tris(π-allyl)trifluorophosphinechromium, which is slowly converted into hexakis(trifluorophosphine)chromium. The corresponding molybdenum and tungsten analogs may be obtained by reaction between molybdenum(V) or tungsten(VI) chloride and PF_3 under pressure and in the presence of copper [734].

$$(\pi C_3H_5)_3Cr \xrightarrow[Et_2O,\,PF_3]{20\text{-}30at} (\pi C_3H_5)_3CrPF_3 \xrightarrow[Et_2O,PF_3]{20\text{-}30at} Cr(PF_3)_6 + 1.5\ C_6H_{10} \qquad (124)$$

$$MCl_n + n\ Cu + 6\ PF_3 \xrightarrow{200\ at, 250^\circ C} M(PF_3)_6 + n\ CuCl \qquad (125)$$

$n = 5$ (Mo), $n = 6$ (W).

Various cyclopentadienylmolybdenum complexes with the dialkylaminodifluorophosphines Me_2NPF_2, Et_2NPF_2, and $C_5H_{10}NPF_2$, for example, $C_5H_5Mo(CO)_2(COMe)Me_2NPF_2$, $C_5H_5Mo(CO)(C_3H_5)Et_2NPF_2$, $C_5H_5MoCl(CO)_2Et_2NPF_2$, $C_5H_5MoCl(CO)(R_2NPF_2)_2$ (R = Me, Et, C_5H_{10}) and the complex with the asymmetric molybdenum atom, $C_5H_5MoCl(CO)(Ph_3P)Me_2NPF_2$ have been reported. The reaction between $C_5H_5Mo(CO)_3CH_3$ and $C_5H_{10}NPF_2$ results in a loss of the methyl and the cyclopentadienyl groups to give *cis*-$Mo(CO)_2(C_5H_{10}NPF_2)_4$ [700]. Convenient synthetic routes to dialkylaminodifluorophosphine metal pentacarbonyls, $M(CO)_5Et_2NPF_2$ (M = Cr, Mo, W), and their reactions with anhydrous HBr have been worked out by Douglas and Ruff [407].

In an unusual reaction the bis(trifluoromethyl)fluorophosphine complex $W(CO)_5(CF_3)_2PF$ was formed when hexacarbonyltungsten was heated in a sealed tube with an approximately equimolar quantity of $(CF_3)_4P_2$ [396].

Carbonylmolybdenum complexes, $Mo(CO)_5L$, of halophosphines, for example, PCl_3, Me_2NPCl_2, $(Me_2N)_2PCl$, PI_3, Me_2NPI_2, have been obtained by treatment of $Mo(CO)_5(Me_2N)_3P$ with hydrogen halides HX (X = Cl, Br, I) under appropriate conditions [601]:

$$Mo(CO)_5(Me_2N)_3P + 6\ HCl \xrightarrow[\text{in an autoclave}]{20^\circ C} Mo(CO)_5PCl_3 + 3\ [Me_2NH_2]Cl \quad (126)$$

$$Mo(CO)_5(Me_2N)_3P + 4\ HI \longrightarrow Mo(CO)_5Me_2NPI_2 + 2\ [Me_2NH_2]I \quad (127)$$

C. *Manganese, Technetium, and Rhenium*

An extremely fast intramolecular positional exchange in the pentacoordinate complex molecules of the substitution series $Mn(CO)_{4-n}(NO)(PF_3)_n$ has been observed in the ^{19}F NMR spectra. Solvolysis studies showed that isomerization occurs among the compositional species [1205].

Dinitrogenfluorophosphinerhenium(I) complexes, $ReCl(N_2)(Me_2NPF_2)_4$ and $ReCl(N_2)(PF_3)_2(Ph_3P)_2$, were synthesized from the benzoylazo complexes, $ReCl_2(N{=}NCOPh)(Ph_3P)_xL_y$ (L = monodentate ligand or ½ bidentate ligand, x = 0, 2, $x + y = 3$) [249].

Reaction of trifluorophosphine with photochemically generated trinitrosylcarbonylmanganese yields trinitrosyltrifluorophosphinemanganese $Mn(NO)_3PF_3$ [333].

Cleavage of the P—N bond in cyclopentadienyldicarbonylbis-(diethylamino)-phenylphosphinemanganese(I) by hydrogen halides HX (X = Cl, Br, I) gave the corresponding dihalogenophenylphosphine complexes. Fluorine/halogen exchange produced fluorophosphine complexes; for example, $\pi C_5H_5Mn(CO)_2PhPF_2$ and $\pi C_5H_5Mn(CO)_2(Et_2N)PhPF$ [603].

D. *Iron, Ruthenium and Osmium*

Tetracarbonylfluorophosphineiron derivatives [406] are formed in the reaction between enneacarbonyldiiron and the appropriate fluorophosphines according to

$$R_nPF_{3-n} + Fe_2(CO)_9 \longrightarrow Fe(CO)_4R_nPF_{3-n} + Fe(CO)_5 \quad (128)$$

(n = 1, R = Ph, Et_2N; n = 2, R = Et_2N). On treatment with anhydrous HCl or HBr, mixed halogenophosphine adducts could be obtained in which Br or Cl could be exchanged by standard methods for SCN or N_3. Carbonylnickel reacts with $Fe(CO)_4PF_2Br$ in refluxing hexane to give the bis-μ-difluorophosphido complex $[Fe(CO)_3PF_2]_2$. On the basis of the ^{19}F NMR spectrum (two doublets were observed), the structure (*1*) was proposed [405].

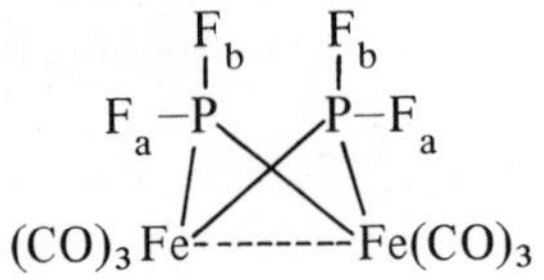

1

π-Allyltris(trifluorophosphine)iron halides, $\pi C_3 H_5 FeX(PF_3)_3$ (X = Br, I), have been synthesized photochemically:

$$Fe(PF_3)_5 + C_3 H_5 X \xrightarrow[\text{ether}]{h\nu} \pi C_3 H_5 FeX(PF_3)_3 + 2 PF_3 \qquad (129)$$

(X = Br, I). If allylchloride is used instead of allylbromide or -iodide, the diene complex π-hexa-1,3-dienetris(trifluorophosphine)iron(0), $\pi C_6 H_{10} Fe(PF_3)_3$, is obtained [737].

Depending on the relative partial pressures of PF_3 or CO, dodecacarbonyltriruthenium forms two series of compounds. High PF_3 pressure produces $Ru(CO)_{5-x}$-$(PF_3)_x$ (x = 3, 4, 5) as main products. At high carbon monoxide pressure, however, the complexes $Ru_3(CO)_{12-y}(PF_3)_y$ (y = 0 – 6) are main products [1206].

Halogen oxidation of pentakis(trifluorophosphine)iron results in the formation of cis octahedral dihalogenotetrakis(trifluorophosphine)iron complexes [741]. The formally analogous reaction between pentakis(trifluorophosphine)iron and dihydrogen, which must be photochemically conducted, yields dihydridotetrakis(trifluorophosphine)iron [739, 742] in addition to the dinuclear species $(PF_3)_3 Fe(PF_2)_2$-$Fe(PF_3)_3$:

$$Fe(PF_3)_5 \xrightarrow[\text{H}_2,\ \text{Pt} \quad \text{asbestos}]{h\nu,\ -20°C,\ \text{ether}} FeH_2(PF_3)_4 \ + \ (PF_3)_3 Fe \genfrac{}{}{0pt}{}{PF_2}{PF_2} Fe(PF_3)_3 \qquad (130)$$

Temperature-dependent ^1H, ^{19}F, and ^{31}P NMR spectra of the dihydride species indicate a rearrangement equilibrium. The potassium salt $K_2 Fe(PF_3)_4$ shows tetrahedral (T_d) symmetry, whereas the monopotassium salt $K[HFe(PF_3)_4]$ has a trigonal bipyramidal structure distorted toward tetrahedral, as shown by Mössbauer spectroscopy [740].

Photochemical reaction between conjugated dienes and pentakis(trifluorophosphine)iron(0) in ether yields π-dienetris(trifluorophosphine)iron(0) complexes. The mode of bonding the dienes is discussed on the basis of the ^1H NMR spectra of these complexes [738]. On ultraviolet irradiation of pentakis(trifluorophosphine)-iron and cyclopentadiene ($C_5 H_6$) in ether solutions, two PF_3 ligands in $Fe(PF_3)_5$ are replaced by $C_5 H_6$ and π-cyclopentadienetris(trifluorophosphine)iron(O), πC_5-$H_6 Fe(PF_3)_3$ is formed. It is converted into hydrido-π-cyclopentadienylbis(trifluorophosphine)iron, $\pi C_5 H_5 FeH(PF_3)_2$ by weak bases [736].

King, Zipperer, and Ishaq [700] have prepared various aminofluorophosphine derivatives of π-cyclopentadienylcarbonyliron. PF_3 substitution products of dienetricarbonyliron complexes (diene)$Fe(CO)_x(PF_3)_{3-x}$ (diene = butadiene, 1,3-cyclohexadiene) show fluxional behavior [1239, 1240].

Photochemically induced replacement of carbonyl groups in trimethylenemethanetricarbonyliron by trifluorophosphine yields all possible compounds of the type $(TMM)Fe(CO)_{3-x}(PF_3)_x$ (TMM = trimethylenemethane). A hindered motion of the PF_3 group in $(TMM)Fe(CO)_2 PF_3$ relative to the trimethylenemethane ligand at lower temperature is considered to be an explanation of the temperature dependence of the ^1H NMR spectra [295].

E. *Cobalt, Rhodium, and Iridium*

The hydridotetrakis(trifluorophosphine)metal complexes, $MH(PF_3)_4$ (M = Co, Rh, Ir), may be employed as versatile starting materials for the synthesis of a large number of complexes. Some of these reactions are summarized in Scheme 1.

SCHEME 1

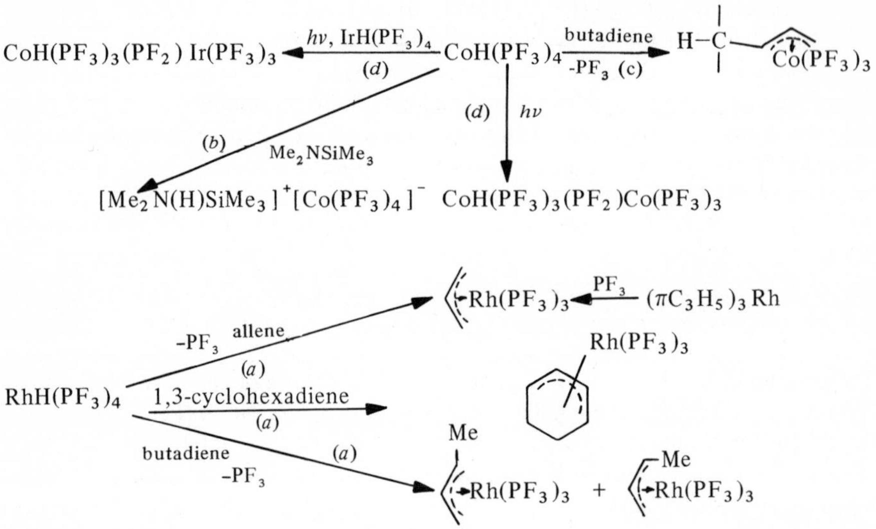

$$2\ IrH(PF_3)_4 \xrightarrow[(e)]{h\nu} (PF_3)_4Ir{-}Ir(PF_3)_4 + H_2$$

(*a*) [302], (*b*) [592], (*c*) [745], (*d*) [746], (*e*) [743].

"Reductive fluorophosphination" of iridium trichloride at 80 to 200 at PF_3 pressure yields the unusual compound $PIr_3(PF_3)_9$, μ_3-phosphido-nonakis(trifluorophosphine)triiridium in which a three-coordinate phosphorus functions as a threefold bridge (*2*) [746].

2

Syn- and anti-π-allylic complexes of cobalt(I) containing trifluorophosphine have been prepared by making use of syn- [1026,1231] or anti-$\pi C_4H_7(C_4H_6)CoPh_3P$ [1231] in the reaction with trifluorophosphine [220]. Because of the low nucleophilicity of the complex anion $[Co(PF_3)_4]^-$, alkylation according to

$$[Co(PF_3)_4]^- + RX \longrightarrow RCo(PF_3)_4 + X^- \qquad (131)$$

(R = Me, Et, C_7H_7(tropylium), X = BF_4) is possible only if strong alkylating agents like oxonium salts are employed [744].

Oxidative addition reactions of pentamethylcyclopentadienyl-bis(trifluorophosphine)rhodium have been studied [689]

$$Me_5C_5Rh(PF_3)_2 + YI \longrightarrow Me_5C_5RhYI(PF_3) + PF_3 \qquad (132)$$

(Y = I, CF_3, C_2F_5, n–C_3F_7, n–C_7F_{15}).

SCHEME 2

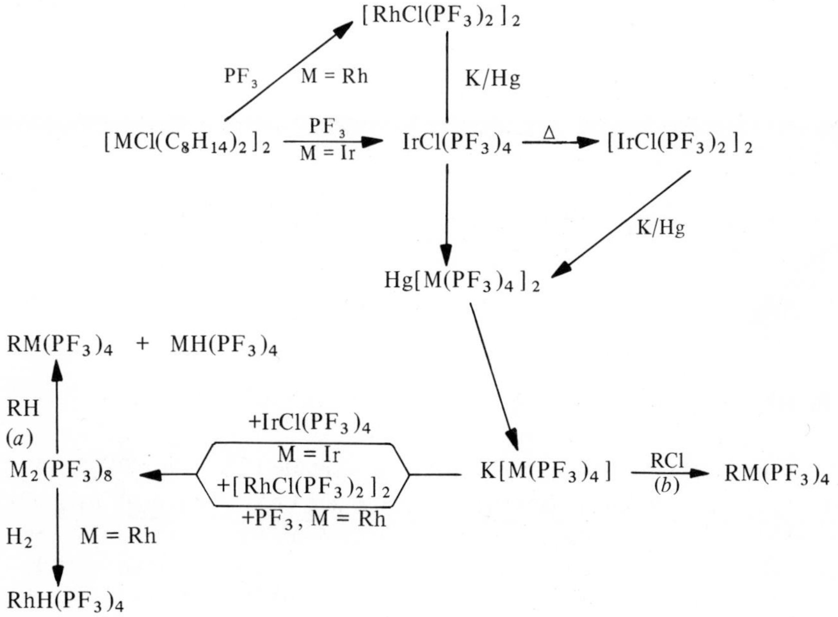

M = Rh, Ir,
(*a*) R = Ph_3Si, $(EtO)_3Si$, Cl_3Si, Ph_3Ge,
(*b*) R = $AuPh_3P$, Ph_3Sn.

Replacement of carbonyl groups in the complexes $[RhX(CO)_2]_2$ by trifluorophosphine affords (with excess PF_3) the binuclear trifluorophosphine complexes of rhodium(I), $[RhX(PF_3)_2]_2$ (X = Cl, Br, I) [946], (X = Cl) [113], or mixed trifluorophosphinecarbonyl complexes $Rh_2Cl_2(CO)_{4-x}(PF_3)_x$ (x = 1, 2, 3) (if $[RhCl(CO)_2]_2$ is treated with smaller amounts of PF_3) [946]. Scrambling and ligand exchange reactions with these complexes have been studied by means of ^{19}F NMR spectroscopy [300, 946].

Bennett and Patmore reported another route to $[RhCl(PF_3)_2]_2$. They observed pronounced ligand mobility in this complex which is contrasted by the stability of π-cyclopentadienylbis(trifluorophosphine)rhodium with respect to ligand exchange [108]. Scheme 2 contains some of the reactions of rhodium(I) trifluorophosphine complexes, including those of the iridium analogs.

Dimethylaminodifluorophosphinerhodium(I) [114, 301] and diethylaminodifluorophosphinerhodium(I) [110, 114] complexes have been obtained by ligand replacement reactions (Scheme 3). Nowlin and Cohn prepared pentacoordinate

SCHEME 3

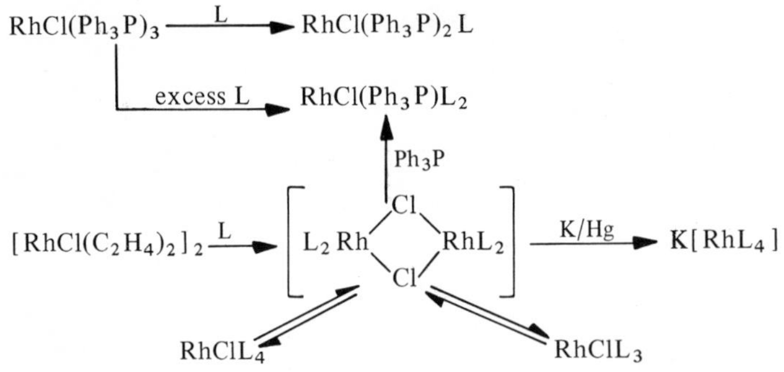

$L = Me_2NPF_2$ [301],
$L = Et_2NPF_2$ [110]

Co(II) complexes with dimethylaminodifluorophosphine $CoX_2(Me_2NPF_2)_3$ (X = Br, I) [951] and bis(dimethylamino)fluorophosphine $CoX_2[(Me_2N)_2PF]_3$ [952] (X = I isolated, X = Cl, Br spectroscopic evidence). ΔH and ΔS values have been obtained for the equilibrium between the low-spin five-coordinate complex and a high-spin, pseudotetrahedral complex of the type $CoX_2[(Me_2N)_2PF]_2$ (X = Cl, Br, I).

The first complexes of dialkylfluorophosphines with metals in a positive oxidation state (i.e., > 1) have been synthesized by the reaction between di(t-butyl)-fluorophosphine and cobalt halides [1132]:

$$CoX_2 + 2\ Bu^t_2PF \longrightarrow \underset{\text{pseudotetrahedral}}{CoX_2(Bu^t_2PF)_2} \qquad (133)$$

($X = Cl, Br$).

F. Nickel, Palladium, and Platinum

Reactions of transition metal complexes with $Ni(PF_3)_4$, tetrakis(trifluorophosphine)nickel, provide a new pathway for the preparation of trifluorophosphine metal complexes [687, 688]. Some examples given in the schematic equations (134) to (137), include other transition metal trifluorophosphine complexes:

$$[Me_5C_5RhCl_2]_2 \xrightarrow{Ni(PF_3)_4} Me_5C_5Rh(PF_3)_2 \qquad (134)$$

$$MnBr(CO)_5 \xrightarrow{Ni(PF_3)_4} Mn_2(CO)_{10-x}(PF_3)_x \qquad (135)$$

($x = 0.5 - 1.5$).

$$ReBr(CO)_5 \xrightarrow{Ni(PF_3)_4} ReBr(CO)_3(PF_3)_2 \qquad (136)$$

$$\pi C_5H_5MoCl(CO)_3 \xrightarrow{Ni(PF_3)_4} \pi C_5H_5)_2Mo_2(CO)_5PF_3\,\pi[C_5H_5Mo(CO)_2PF_3]_2 \qquad (137)$$

Ligand replacement of PF_3 in $Ni(PF_3)_4$ with PH_3 under pressure gave $Ni(PF_3)_3PH_3$ [1190].

Fluorophosphine-induced ligand coupling of π-allylnickel and -palladium complexes has been reported [403].

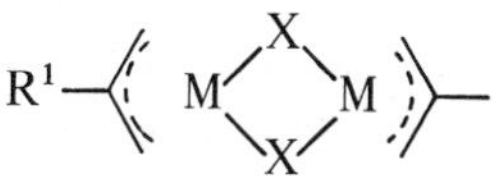
$$+ 8\ RPF_2 \longrightarrow 2\ CH_2{=}CR^1CH_2X + 2\ M(RPF_2)_4 \qquad (138)$$

$M = Ni$, $R^1 = H$, $X = Br$, $R = Me_2N$,
$M = Pd$, $R^1 = H, Me$, $X = Cl$, $R = F, Me_2N, Et_2N$.

Direct reaction between platinum(II) chloride and PF_3, CF_3PF_2, or $(CF_3)_2PF$ or potassium-tetrachloroplatinite and $(CF_3)_2PF$ yields the corresponding fluorophosphine complexes of zerovalent platinum [944]. This reaction represents an extension of the work of Kruck et al. [733].

Nickel(II) halide complexes of bis(t-butyl)fluorophosphine, $NiX_2(Bu^t_2PF)_2$ (X = Cl, Br, I), the first examples of nonnoble metal complexes of alkylfluorophosphines with metals in oxidation state higher than one, have been prepared. They have a square planar trans ligand arrangement. The temperature dependent ^{19}F NMR spectra are explained by a cis-trans equilibrium and hindered rotation of the phosphine ligands with respect to the Ni–P axis [1132].

The reaction between copper butylacetylide and PCl_3 produced a complex of composition $BuC{\equiv}CCuPCl_3$ [1119]. It is believed to be structurally related to the corresponding complexes of tertiary phosphines.

Ligand redistribution reactions have been observed and monitored by means of ^{19}F NMR spectroscopy for the complexes $Ni(CO)_2(C_5H_{10}NPF_2)_2$ ($C_5H_{10}N$ = piperidyl), $Ni(CO)_2(Me_2NPF_2)_2$, and $Ni(CO)_2(Et_2NPF_2)_2$. The species identified are $Ni(CO)_3(R_2NPF_2)$, $Ni(CO)_2(R_2NPF_2)_2$, and $Ni(CO)(R_2NPF_2)_3$ [943].

VI. Complexes of Aminophosphines, $R'_{3-n}P(NR_2)_n$ (n = 1, 2, 3), Diphosphines, $R_2P{-}PR_2$, and Related Ligands

The number of papers that have appeared in the literature on this topic is small by comparison with papers on the coordination chemistry of triorganophosphines. Verkade and Coskran [1229] in their comprehensive review have covered the literature on this topic up to 1970.

Dialkylaminodiphenylphosphines (L), Ph_2PNR_2 (R = Me, Et, Pr, Bu) form complexes of the type $Mo(CO)_5L$ and *cis*- or *trans*-$Mo(CO)_4L_2$. The attempted synthesis of dinuclear complexes with a μ-phosphido and a μ-amino bridge in the same molecule was unsuccessful [49].

Monosubstituted hexacarbonyl metal derivatives $M(CO)_5L$ (M = Cr, Mo, W) of tris(dimethylamino)phosphine and $Fe(CO)_4L$ complexes have been reported by King and Korenowski [696]. In the reactions between $C_7H_8M(CO)_4$ (C_7H_8 = norbornadiene), $C_7H_8M(CO)_3$ (C_7H_8 = cycloheptatriene) (M = Cr, Mo, W) or *cis*-$(C_6H_8)_2M(CO)_2$ (C_6H_8 = cyclohexadiene, M = Mo, W), and $P(NMe_2)_3$ the first step is probably the complete displacement of the coordinated olefin(s). If, however, the $P(NMe_2)_3$ ligands are in cis positions, further reactions occur which include isomerization of the cis to the trans isomer and substitution of one $P(NMe_2)_3$ ligand for a carbon monoxide molecule that has been generated by decomposition of other carbonyl species [696].

Cleavage of the P–N bond in the complexes $\pi C_5H_5Mn(CO)_2PhP(NEt_2)_2$ by HX (X = Cl, Br, I) gave the corresponding halophosphine complexes $\pi C_5H_5Mn(CO)_2PhPX_2$ [603].

N,N-Dimethyl-*N'*-(diphenylphosphino)hydrazine (DPH) functions in most cases as a unidentate ligand with the phosphorus atom as donor. Complexes of this ligand with palladium and platinum halides and mercury(II) chloride, for example, $PdCl_2(DPH)$, $PdCl_2(DPH)_2$, $[PdCl(DPH)_2]ClO_4$, $PtCl_2(DPH)_2$, $PtCl_4(DPH)$, $HgCl_2(DPH)$, have been synthesized as well. $PdCl_2(DPH)$ is dimeric in solution with chloro bridges, whereas in $[PdCl(DPH)_2]ClO_4$ one ligand DPH forms a chelate ring that uses both N and P as donor atoms; the other DPH is monodentate [10].

Some reports have been made on complexes of tetraorganodiphosphines. Issleib and Giesder prepared a number of transition metal halide complexes of tetraethyldiphosphine of the type $MX_nEt_2PPEt_2$ (n = 2, MX_2 = $FeCl_2$, $FeBr_2$, FeI_2, $CoBr_2$, $NiCl_2$, $NiBr_2$, NiI_2, CdI_2, $HgBr_2$; n = 1, MX = CuCl, CuBr). No cleavage of the P–P bond was observed on complex formation with tetraethyldiphosphine [634]. Similar behavior was found for tetramethyldiphosphine. However, the composition of its complexes with transition metal halides is more variable. Complexes of the following type could be isolated: $(M^nX_n)_2(Me_2PPMe_2)_3$ (M = Ti^{3+}, V^{3+}, Cr^{3+}, Fe^{2+}, Co^{2+}, Ni^{2+}, Cu^+), $MX_2(Me_2PPMe_2)_2$ (M = Co^{2+}, Ni^{2+}, Hg^{2+}), and $MX_2(Me_2PPMe_2)$ (M = Fe^{2+}, Ni^{2+}, Zn^{2+}, Cd^{2+}, Hg^{2+}) (X = Cl, Br, I). The low solubility of most of these complexes makes the elucidation of their structures quite difficult [635].

Structure (*1*) was assigned to a complex formed by reaction between $Fe_2(CO)_9$

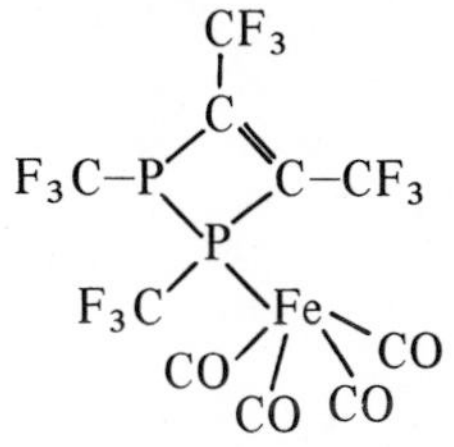

1

and the diphosphacyclobutene $(CF_3)_2P_2C_2(CF_3)_2$. If (*1*) is treated with $Fe_2(CO)_9$ the triiron complex $(CF_3)_2P_2C_2(CF_3)_2Fe_3(CO)_{10}$ is obtained for which two structures, (*2*) and (*3*), have been proposed. Small quantities of this complex were

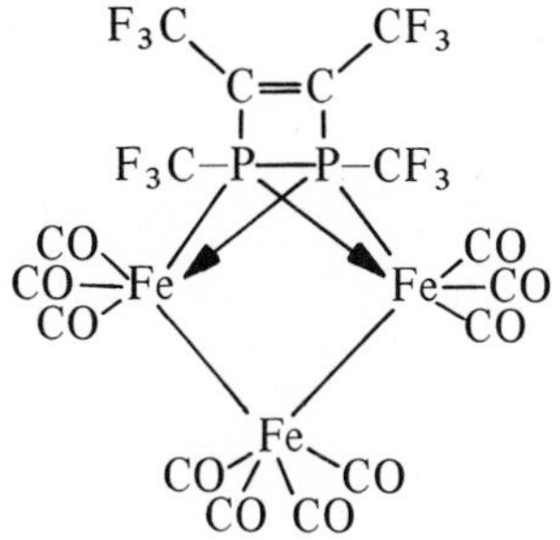

2

also formed in the direct reaction between $(CF_3)_2P_2C_2(CF_3)_2$ and $Fe_2(CO)_9$ [328]. Tetrakis(trifluoromethyl)cyclotetraphosphine $(CF_3P)_4$ with $Fe_2(CO)_9$ gives a compound of empirical composition $Fe(CO)_3(CF_3P)_2$ to which structure (*4*) was

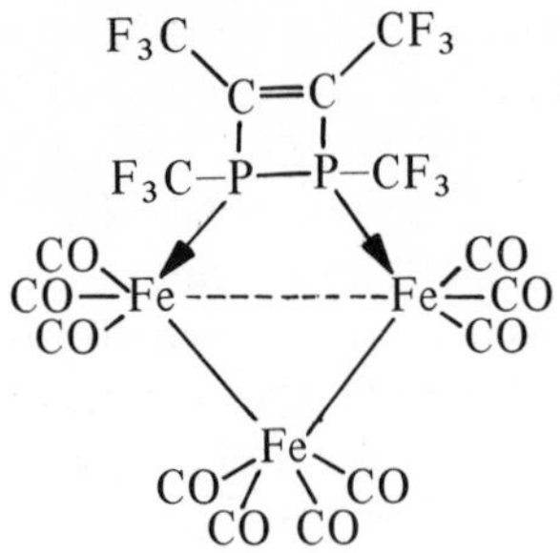

3

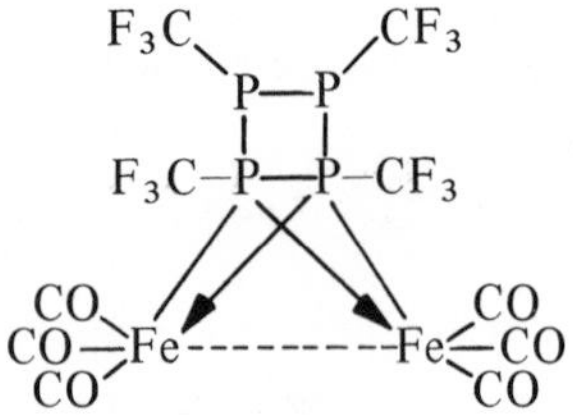

4

assigned being consistent with the ^{19}F NMR, mass spectrum, and the IR spectrum in the carbonyl stretching vibrations [328].

An unusual phosphido carbonylcobalt cluster $Co_2(CO)_6P_2$ was isolated in small yields (1 to 10%) from the reaction mixture obtained after treating sodium tetracarbonylcobaltate(−I) with phosphorus trichloride or bromide in tetrahydrofuran at 25° C. On the basis of analogies with the IR spectra of $Co_2(CO)_6$(acetylene) and $Co_2(CO)_5$(acetylene)R_3P in the carbonyl stretching vibration region, structure (*5*) was proposed for this phosphido complex, It may be considered as a

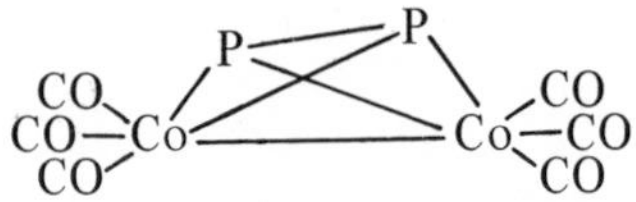

5

complex of the (unstable) P_2 molecule [1232].

VII. Homogeneous and Heterogeneous Catalysis by Transition Metal Phosphine Complexes

It is not within the scope of this review to give a complete or nearly complete representation of the literature that has appeared on this topic in the period under review. Only a limited number of catalytic processes is cited and listed in Table I. A few that seem to be of general interest are discussed in detail. For a review article on homogeneous catalyzed hydrogenation, its kinetics and mechanism, the reader is referred to [1141]. For a more general treatment of the kinetics of hydrogenation by $IrX(CO)L_2$ (X = halide, L = tertiary phosphines) see [1142].

Of great interest for general application in synthetic chemistry are those catalysts that achieve or promote stereospecific reactions. For that purpose transition metal phosphine complexes are used in which centers of chirality are on the phosphorus atoms [719,957,1164,1275] or in their substituents [353,704,910]. Such stereospecific reactions, which yield products up to 90% of optical purity, are mainly hydrogenations or hydrosilylations of prochiral ketones [957,1164] or olefinic carboxylic acids and their derivatives [353, 668, 719, 910]. The work of Horner and Siegel [612, 613], which, in a systematic way, demonstrates the use of asymmetric induction in homogeneous hydrogenation catalyzed by complexes of chiral tertiary phosphines, deserves to be mentioned. These authors proposed to use the homogeneous hydrogenation by chiral rhodium phosphine complexes (of known configuration) for the determination of the absolute configurations of the hydrogenation products [612]. Some asymmetric hydrosilylation reactions of prochiral olefins in which transition metal phosphine complexes (with centers of chirality on phosphorus [1275] or on a carbon [704] of the substituents on phosphorus) were used have been reported as well; for example,

$$\mathrm{MePhC{=}CH_2 + MeSi\underset{\bullet}{H}Cl_2 \xrightarrow[39\%]{NiR_3P^*} MePhCHCH_2SiMeHCl + MePhCHCH_2SiMeCl_2} \qquad (139)$$

[R_3P^* = (+)(R)-($PhCH_2$)MePhP (81% optical purity).]. To give the reader an idea of the synthetic use of such catalyzed asymmetric syntheses some examples of hydrogenation products of α-acylaminoacrylic acids are listed in Table II.

Wilke and co-workers [151] reported the catalyzed asymmetric codimerization of ethylene and 1,3-cyclooctadiene:

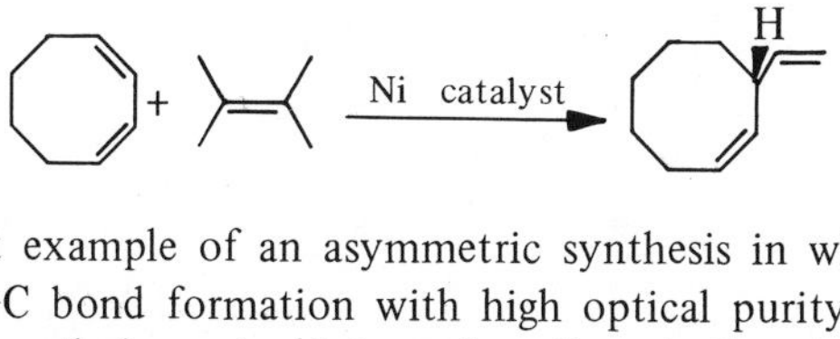

(140)

This was the first example of an asymmetric synthesis in which the chirality is introduced by C–C bond formation with high optical purity. The catalysts that were active were menthyl, myrtanyl, bornyl, or 2-methylbutylphosphine complexes of the type $\pi C_3H_5NiX(R_3P)\cdot AlX_3$ [151].

Another interesting development in catalysis is the synthesis and use of metal-containing polymeric catalysts which may be obtained, for example, by reaction of $PhPCl_2$ with polyvinylalcohol and subsequent treatment with $CoCl_2$, $RhCl(Ph_3P)_3$, or Na_2PtCl_4 [31].

TABLE I. Chemical reactions catalyzed by transition metal phosphine complexes

Substrate	Catalyst	Products, Comments	Refs.
Polymerization			
1,3-Butadiene	$RuCl_2(C_{12}H_{18})Ph_3P$ $RuCl_2(C_{12}H_{18})Bu_3P$ $RuCl_2(C_{12}H_{18}) + 2\ Ph_3P$	*cis*-1,4-, *trans*-1,4 and 1,2-units in the polybutadiene, their contents in the polymers is dependent on the ratio Ru complex: phosphine	599
1,3-Butadiene	$CoX_2(Et_3P)_2 + Et_2AlCl$ (X = Cl, Br, I)	1,2-Polybutadiene	626
Ethylene, butadiene	$RuHCl(Ph_3P)_3$	Polyethylene, polybutadiene, kinetic study	645
Propylene	Catalysts supported on SiO_2, Al_2O_3 or Al_2O_3/SiO_2, treatment of these with $NiCl_2$ followed by reaction with Ph_3P and Et_3Al or $EtAlCl_2$	C_6 olefins, polypropylene	595
Methylmethacrylate	$Ni[P(OPh)_3]_4$, initiation by CCl_4	Polymethylmethacrylate, kinetic study of polymerization	66
Methylmethacrylate	Dichloro(dodeca-2,6,10-triene-1,12-diyl)ruthenium(IV) + Ph_3P	Polymethylmethacrylate, radical polymerization	600
1,3,5,7-Tetramethyl-1,3,5,7-tetravinyl-cyclotetrasiloxane and 1,3,5,7-tetramethyl-cyclotetrasiloxane	$Pd(Ph_3P)_4$	Polymers containing cyclotetrasiloxane rings bound by ethylene groups	1201

Dimerization, codimerization, oligomerization, and cooligomerization			
Olefins (generally)	$NiCl(o\text{-}ClC_6H_4)(Et_3P)_2$ $NiCl(F_2C{=}CF)(Ph_3P)_2$ $NiCl(2,5\text{-}Cl_2C_6H_3)(Et_3P)_2$	Dimers	441
1,3-Butadiene	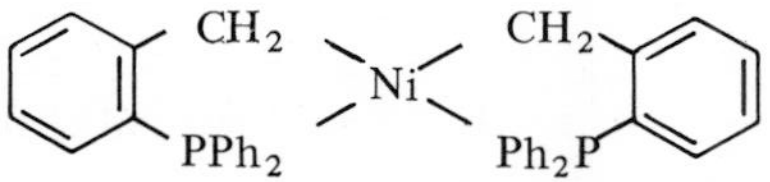	10% cyclooctadiene, 50% cyclododecatriene	790
Isoprene	$NiX_2(Ph_3P)_2$ + EtMgBr (X = Cl,Br)	2,6-Dimethylocta-1,3,6-triene	1241
Isoprene	$PdBr_2(Ph_2PCH_2CH_2PPh_2)$ + PhONa + PhOH	Five linear dimers identified, distribution of the isomeric dimers determined by molar ratio isoprene: PhOH	1157
Ethylene, propylene	$(1\text{-}C_{10}H_7)NiBr(Ph_3P)_2 + BF_3{\cdot}Et_2O$ $1\text{-}C_{10}H_7$ = 1-naphthyl	2-Methyl-1-butene, 2-methyl-2-butene, 3-methyl-1-butene, *cis*-2-pentene, *trans*-2-pentene, 1-pentene	849
Ethylene, styrene	$RNiBr(Ph_3P)_2 + BF_3{\cdot}Et_2O$ R = *o*-tolyl, 1-naphthyl, mesityl	3-Phenyl-1-butene	674
Ethylene, 1,3-butadiene	$CoCl_2[Ph_2P(CH_2)_nPPh_2]_m + Pr^nMgCl$ + EtOH (n = 2,3,4, m = 1,2)	1,4-*cis*-Hexadiene	669
Propylene, 1,3-butadiene	$(\pi C_3H_5PdCl)_2$, $AlCl_3$, R_3P R = Cy, Bu^n, Ph, OPh, OEt	C_7-Dienes, e.g., *trans*-2-methylhexadiene-1,4, *trans,trans*-heptadiene-2,5, *trans,cis*-heptadiene-2,5, *trans*-heptadiene-1,5	638

TABLE I. Chemical reactions catalyzed by transition metal phosphine complexes (continued)

Substrate	Catalyst	Products, Comments	Refs.
Ethylene and vinyl cyclic hydrocarbons, e.g., styrene ethylene	$Ni(PA_nX_{3-n})_4$ + AlX_3 or $AlHX_2$ + alkyllithium or aryllithium A = alkyl, cycloalkyl, aryl, aralkyl, alkoxy, aryloxy, H, X = halogen, n = 0, 1, 3	Phenylbutenes 2.2% 1-butene, 69.5% *trans*-butene, 28.3% *cis*-butene	629
1,3-Butadiene, active methylene compounds, $R^1R^2CH_2$	$PdBr_2(Ph_2PCH_2CH_2PPh_2)_2$	$R^1R^2CHCH(Me)CH{=}CH_2$, $R^1R^2CHCH_2CH{=}CHMe$	1158
Phenols and conjugated dienes, e.g., PhOH, 1,3-butadiene	$Ni[P(OPh)_3]_4$	$PhOCH_2CH{=}CH(CH_2)_3CH{=}CH_2$ 54% selectivity, dimerization addition reaction	275
PhOH, 1,3-butadiene	$Rh(OPh)(Ph_3P)_3$	$PhOCH_2CH{=}CH(CH_2)_3CH{=}CH_2$ 85% selectivity	391
Hydrogenation			
Olefins, e.g., cyclohexene	$CoH_3(Ph_3P)_3$ or $CoH(CO)(Ph_3P)_3$	Alkanes or cyclohexane	589
Olefins, acetylenes	$MCl(Ph_3P)_3$ (M = Rh, Ir)	Alkanes, reactivity studies Influence of phosphine ligands	1144,1145 611
	$MCl(CO)(Ph_3P)_2$ (M = Rh, Ir)	Influence of central metal atom on the reaction rate of hydrogenation	1148,1149
Olefins, dienes, cycloalkenes, e.g., 1,3-butadiene	$CoCl_2(Ph_3P)_2$	Alkanes, cycloalkanes, alkenes Butane (49%), 2-butene (46%). 1-butene (5%)	1252

1-Hexene	$IrCl(C_8H_{14})_2 + Ph_3P$	*trans*-2-Hexene (26%), *cis*-2-hexene (4%), *cis*-3-hexene (<1%), hexane (68%)	1220
1-Hexene	$Ru_2(OAc)_4 \cdot 2\ Ph_3P$	Selective hydrogenation, 1-hexene is hydrogenated but not *cis*-2-pentene and inner alkenes	1263
1-Hexene	$RhCl(Ph_3P)_3$, $RhCl_3(Ph_3P)_3$	Hexane (95%)	389
Cycloalkenes	$RhBr(Ph_3P)_3$, $RhCl(C_8H_{14})_2 + R_3P$ $R_3 = EtPh_2$, $Ph(C_5H_{10}N)_2$, $Ph_2C_5H_{10}N$	Cycloalkanes, kinetics and deuteration experiments	621
Styrene	$CoHN_2(Ph_3P)_3$ + NaNp (Np = naphthalene	Ethylbenzene	1203
Styrene	$RhCl(Ph_3P)_3$	Ethylbenzene	1121
1,3-Cyclohexadiene hexene	$CuClR_3P + SnCl_2$ (R = Ph, OPh, $PhCH_2$)	Cyclohexane hexane	1171
Acrylonitrile	$RhCl(Ph_3P)_3$ or $RhX(R_3P)_3$ (X = Cl, Br, I, R = alkyl, aryl)	EtCN (100%)	390
α,β-Unsaturated carbonyl compounds	$RuCl_2(Ph_3P)_3$	Saturated carbonyl compounds, hydrogen transfer from alcohols, e.g., $PhCH_2OH$	1054
MeCOPh, butanone, pinacolone	bis(R-ethylmethylphenylphosphine)-(2,5-norbornadiene)rhodium(I) hexafluorophosphate	Optically active alcohols	1164

TABLE I. Chemical reactions catalyzed by transition metal phosphine complexes (continued)

Substrate	Catalyst	Products, Comments	Refs.
Atropic acid	$Rh_2Cl_2(C_8H_{14})_4$ + Et_3N + L L = $Ph_2PCH_2C(H)(MeO)-C(H)(OMe)CH_2PPh_2$ L(+)- and D(−) form or 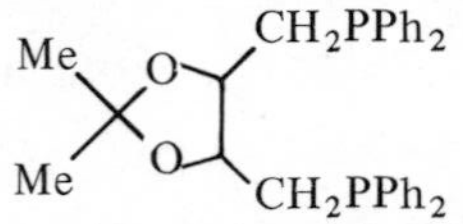	Hydratropic acid (64% optical yield)	668
(E)-β-Methylcinnamic acid	Chlorotris(neomenthyldiphenylphosphine)rhodium(I) in C_6H_6/EtOH/NEt_3	3-Phenylbutanoic acid (80%) 61% enantiomeric excess of the S isomer	910
(E)-α-Methylcinnamic acid		(R)-2-Methyl-3-phenylpropanoic acid (52%)	
α,β-Unsaturated nitro compounds			575
3,4-(Methylenedioxy)-β-nitrostyrene	$RhCl(Ph_3P)_3$	2-[3,4-(Methylenedioxy)phenyl] nitroethane	
2,5-Dimethoxy-β-methyl-β-nitrostyrene	$RhCl_3[(4\text{-}C_6H_5C_6H_4)(1\text{-}C_{10}H_7)-(C_6H_5)P]_3$	2-(2,5-Dimethoxyphenyl)-2-methyl-nitroethane	

Unsaturated prochiral acids, e.g., α-acetamidocinnamic acid	Optically active diphosphine Rh catalyst, e.g., $[RhCl(1,5\text{-hexadiene})]_2$ + 2 L(L = *o*-anisylmethylphenylphosphine	Optically active aminoacids	353,719
Hydrosilylation			
Terminal olefins C_nH_{2n}/$HMeSiCl_2$	$NiCl_2(dmpf)$ (dmpf = $Fe(C_5H_4PMe_2)_2$	$C_nH_{2n+1}SiMeClH$ + $C_nH_{2n+1}SiMeCl_2$ n = 5 80% 18% n = 6 87% 18% n = 7 55% 28% n = 8 79% 16%	756
Terminal olefins/$MeSiHCl_2$	$Pt(Ph_3P)_4$	Selective addition to the terminal double bond	461
$PhMeC{=}CH_2/MeCl_2SiH$	$NiCl_2(R_3P^*)_2$ (R_3P^* = (+)–(R)–$(PhCH_2)MePhP$, 81% optical purity)	Asymmetric hydrosilylation $PhMeCHCH_2$-$SiMeHCl$ + $PhMeCHCH_2SiMeCl_2$ 95% + pure 8% yield 31% yield	1275
1-Hexene/$MeSiHCl_2$	$NiBr_2(Me_3SiCH_2PBu_2)_2$	$Me(CH_2)_3CH(SiMeHCl)Me$ (5%) $Me(CH_2)_5SiMeHCl$ (20%) $Me(CH_2)_3CH(SiMeCl_2)Me$ (10%) $Me(CH_2)_5SiMeCl_2$ (65%)	247
Olefins, 1,3-dienes/$SiHCl_3$, e.g., styrene	$PdCl_2(PhCN)_2$ + chiral phosphine, e.g., menthyldiphenylphosphine	$CH_3CHPh(SiCl_3)$ (87%), after conversion to the methylated product 5.1% enantiomeric excess of S-isomer	704

TABLE I. Chemical reactions catalyzed by transition metal phosphine complexes (continued)

Substrate	Catalyst	Products, Comments	Refs.
Acetylenes, olefins/R_2SiHCl (R_2 = Cl_2, MeCl, Me_2) e.g., 1-octene/$MeSiHCl_2$, 1,3-cyclo-octadiene/$MeSiHCl_2$	$NiCl_2(Me_3P)_2$ $NiCl_2$(dmpf) dmpf = $Fe(C_5H_4PMe_2)_2$	Reactivity decreases in the order $SiHCl_3$ > $MeSiHCl_2$ > Me_2SiHCl ≫ Me_3SiH $C_8H_{17}SiMeHCl$ (50%) + $C_8H_{17}SiMeCl_2$ (50%) 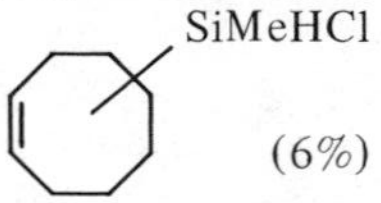 (6%) 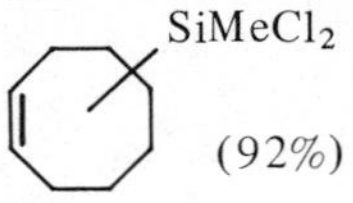 (92%)	703
2-Pentene/phenyldimethyl-silane	$RhCl(Ph_3P)_3$, $RhCl(CO)(Ph_3P)_2$, $RhH(CO)(Ph_3P)_3$	$C_5H_{11}SiPhMe_2$	246
Styrene/trichlorosilane	$NiCl_2(Bu^n_3P)_2$	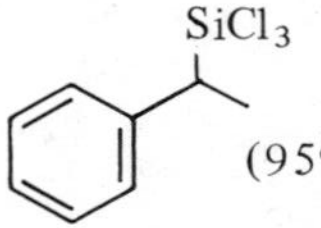 (95%) + 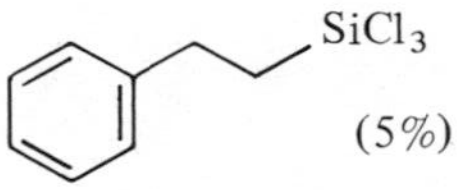 (5%)	100
PhRCO/$MeCl_2SiH$ R = alkyl groups	$[PtCl_2R_3P^*]_2$ (R_3P^* = (+)(R)–(PhCH$_2$)MePhP)	PhRCH(OSiMeCl$_2$), R = Bu^t: yield 33%, optical (S) yield: 18.6%	1274
Ketones/$PhMe_2SiH$, e.g., PhCOMe	$(RhClQ)_2$ + (S)(−)($PhCH_2$)MePhP Q = 1,5-hexadiene, cyclooctadiene	Reaction via asymmetric Rh(I) complex, after hydrolysis gives (R)(+)-1-phenylethanol in 43.1% optical yield	957
Acrylonitrile/dimethylchlorosilane	$RhH(CO)(Ph_3P)_3$ + Bu_3N or $Co_2(CO)_8$	MeCH(CN)$SiMe_2Cl$	246

Isomerization			
1-Heptene	$RhH(CO)(Ph_3P)_3$	Double-bond migration under UV irradiation	1140,1143
1-Hexene, n-pentenes, e.g., 1-pentene	$RhH(CO)(Ph_3P)_3$	*cis*- and *trans*-2-Pentene	1271
1-Heptene, 2-*trans*-heptene, 2-*cis*-heptene	$RhH(CO)(Ph_3P)_3$	Kinetic study	1146,1147
cis-1,4-Hexadiene	NiL_4/HCl mixtures, L, e.g., Ph_3P, $P(OPh)_3$, $P(OEt)_3$, $P(p\text{-}MeOC_6H_4O)_3$, $Pd(Ph_3P)_4$, $Pt(Ph_3P)_4$, $NiBr(R_3P)_2$/ $NaBH_4$ (R = Bu^n, Ph)	Skeletal isomerization to give mixtures of 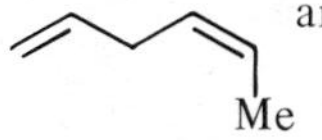 and 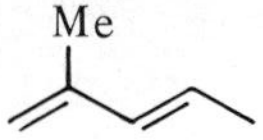	520
1,4-Diarylbutenes and 2-methyl-1,4-diphenylbutenes, e.g. *trans*-1,4-Diphenylbutene-2	$RuCl_2(Ph_3P)_3$	Double-bond migration, the electronic nature of the *p*-substituents of the phenyl groups affects the ratio terminal: inner olefin *trans*-$PhCH_2CH{=}CHCH_2Ph$, *cis*-$PhCH_2CH{=}CHCH_2Ph$, *trans*-$PhCH{=}CHCH_2CH_2Ph$	148
Carbonylation, decarbonylation			
Oxoprocess	$RhCl(CO)L_2$ (L = Ph_3P, Pr^i_3P, Cy_3P, $(PhCH_2)_3P$, $[o\text{-}(m\text{-},p\text{-})tolyl]_3P$		598
Oxoprocess	$MCl_xH_y(CO)_zL_t$ M = Rh, Ni, Pt; $x + y = 1$, for M = Rh; $x + y = 0$, for M = Ni, Pt, $x + y + z + t = 4$, L = aliphatic phosphite, aliphatic or aromatic phosphine (for $y = 1$, $z = 0$)		126

TABLE I. Chemical reactions catalyzed by transition metal phosphine complexes (continued)

Substrate	Catalyst	Products, Comments	Refs.
e.g., 1-Hexene, CO/H_2 100°C/12 at	$RhCl(CO)[P(OEt)_3]_2$	74% heptanal, 64% 1-hexene conversion	
Vinylchloride, CO in methanol	$Pd(Ph_3P)_4$	Methylacrylate	1169
Propene, $CO + H_2$	$Co_2(CO)_6(Bu^n_3P)_2$	Hydroformylation, reaction mechanism is discussed	985
Unsaturated compounds, e.g., 1-dodecene, $CO + H_2$, 184°C/85 at	Co-octanoate + eicosyl(octahydropentalenyl)phosphine, KOH	98.5% completion with 14% conversion to saturated hydrocarbons, the rest to primary alcohols (60.4% *n*-tridecanol), no aldehyde obtained	1223
Heptene, H_2	$RhCl(CO)(Ph_3P)_2$, Et_3N	30% of hexene hydroformylated in 12 h	1262
Allylic alcohols $RCH{=}CHCH_2OH$	$RhCl(Ph_3P)_3$	RCH_2CH_3 (major) and $RCH{=}CH_2$ (minor)	435
Aldehydes, e.g., 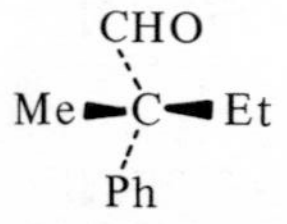 (−)(R)-isomer	$RhCl(Ph_3P)_3$	Stereoselective decarbonylation 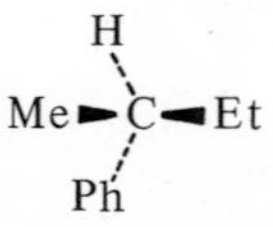 93% retention of configuration	1235

Oxidation			
Bu_3P	$Pd(Ph_2PCH_2CH_2PPh_2)_2$	Bu_3PO is obtained at 25 to 45°C	348
Ph_3P	$Ru(CO)(NO)(NCS)(Ph_3P)_2$ $IrX(CO)(Ph_3P)_2$	Ph_3PO, room temperature, O_2 pressure 1 at Ph_3PO	524 1160
Cyclohexene	Complexes of Ph_3P with Rh, Ir, Pt or Au,	Cyclohexene oxide, 2-cyclohexene-1-one, 2-cyclohexene-1-ol	480
Various catalytic processes			
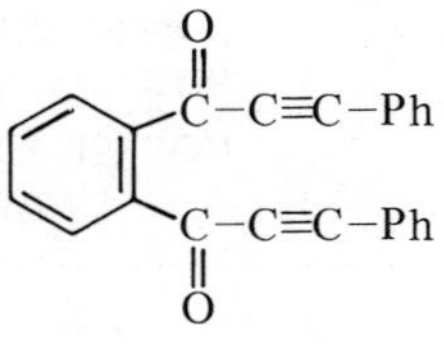	$RhCl(Ph_3P)_3$	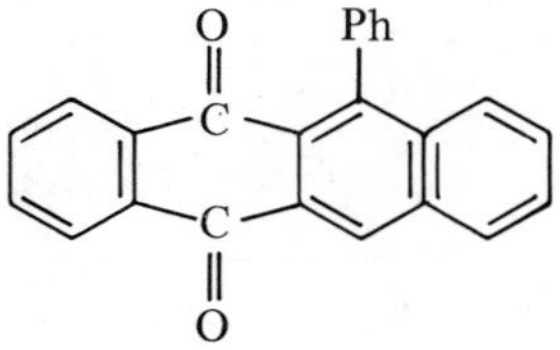	912
$BuPH_2/CH{\equiv}CPr$ $P(OEt)_2(O)H$/4-octyne	$Ni(CO)_2(Ph_3P)_2$	$BuPHCH{=}CHPr$ *cis*-$P(OEt)_2(O)CPr{=}CHPr$	782
$EtMgBr/PhCl$ $PhMgBr/ClCH{=}CHCl$	$NiCl_2(Ph_2PCH_2CH_2PPh_2)$	Cross coupling, PhEt $PhCH{=}CHPh$	1161
C_2D_4/C_2H_4 D_2/H_2	$NiX_2(Ph_3P)_2$ (X = halide)	Isotopic exchange catalysed via nickel hydride complex	37
Ph_2CH_2	$RhCl(CO)(Ph_3P)_2$	Benzophenone, benzhydrol	459

TABLE II. Hydrogenation of α-acylaminoacrylic acid [719]

Chiral Phosphine			Substrate		Product
o-Anisyl–P(R^1)–R^2			$R^3CH{=}C(NHCOR^4)CO_2H$		$R^3CH_2CH(NHCOR^4)CO_2H$
R^1	R^2	Approximate Optical Purity (%)	R^3	R^4	Optical Purity (%)
Me	Ph	95	3-MeO-4-(OH)C_6H_3	Ph	58
Cyclohexyl	Me	95	3-MeO-4-(OH)C_6H_3	Ph	87[a]
Cyclohexyl	Me	95	3-MeO-4-(OH)C_6H_3	Ph	90[b]
Cyclohexyl	Me	90	3-MeO-4-(AcO)C_6H_3	Me	77[c]
Cyclohexyl	Me	95	3-MeO-4-(AcO)C_6H_3	Me	85[a]
Cyclohexyl	Me	95	3-MeO-4-(AcO)C_6H_3	Me	88[b]
Cyclohexyl	Me	95	Ph	Me	85[a]
Cyclohexyl	Me	95	Ph	Ph	85[a]

a) Solvent 95% ethanol,
b) Solvent propan-2-ol,
c) 0.05 triethylamine

VIII. Reactions of Coordinated Ligands

An increasing number of papers deal with reactions of coordinated ligands for the synthesis of new types of ligand.

Deprotonation and subsequent alkylations of phosphine metal complexes have been studied [1194, 1195]:

$$M(CO)_5Ph_2PH \xrightarrow[THF]{LiBu} M(CO)_5Ph_2PLi \xrightarrow{MeI} M(CO)_5MePh_2P \quad (141)$$

(M = Cr, Mo, W),

$$Mo(CO)_4(Ph_2PH)_2 \xrightarrow[THF]{2\ LiBu} Mo(CO)_4(Ph_2PLi)_2 \xrightarrow{MeI} Mo(CO)_4(MePh_2P)_2 \quad (142)$$

$$Fe(CO)_4RPhPH \xrightarrow[THF]{LiBu} Fe(CO)_4RPhPLi \xrightarrow{MeI} Fe(CO)_4MePhRP \quad (143)$$

(R = Me, Ph),

$$MeC_5H_4Mn(CO)_2Ph_2PH \xrightarrow[THF]{LiBu} MeC_5H_4Mn(CO)_2Ph_2PLi \xrightarrow{MeI} MeC_5H_4Mn(CO)_2MePh_2P \quad (144)$$

$$Fe_2(CO)_6(PhPH)_2 \xrightarrow[THF]{LiMe} Fe_2(CO)_6(PhPLi)_2 \xrightarrow{MeI} Fe_2(CO)_6(MePhP)_2 \quad (145)$$

$$[C_5H_5Fe(CO)_2PhPH_2]PF_6 \xrightarrow[THF,\ -78°C]{LiBu} \underset{\mathit{1}}{[C_5H_5Fe(CO)_2(PhPLi_2)]PF_6} \xrightarrow[-78°C]{MeI} [C_5H_5Fe(CO)_2Me_2PhP]PF_6 \quad (146)$$

$$\mathit{1} \xrightarrow{25°C} \underset{\mathit{2}}{[C_5H_5Fe(CO)_2PPh]Li} + LiPF_6 \quad (147)$$

$$\mathit{2} \longrightarrow (PhP)_x + [C_5H_5Fe(CO)_2]Li \xrightarrow{MeI} C_5H_5Fe(CO)_2Me \quad (148)$$

Deprotonation of tetrakis(diphenylphosphine)nickel(0), $Ni(Ph_2PH)_4$, with butyllithium, phenyllithium, or sodium amide and subsequent addition of *n*-butylbromide, did not yield tetrakis(butyldiphenylphosphine)nickel(0) [607]. Deprotonation of phosphine complexes may lead to dimerization via μ-phosphido groups with carbonyl loss and intramolecular rearrangements [1194]:

$$[Mn(CO)_4Ph_2PH]\,Br \xrightarrow[THF]{LiBu^n} [Mn(CO)_4Ph_2PLi]\,Br \xrightarrow[6\ days]{-LiBr} [Mn(CO)_4(Ph_2P)]_2 \quad (149)$$

$$[C_5H_5Fe(CO)Ph_2PH]\,Br \xrightarrow[THF]{LiBu^n}$$

$$[C_5H_5Fe(CO)Ph_2PLi]\,Br \xrightarrow[2\ h]{-LiBr}$$

$$[C_5H_5Fe(CO)Ph_2P]_2 \quad (150)$$

$$2M(CO)_5R_2PLi \xrightarrow{-2\ CO} [M_2(CO)_8(R_2P)_2]\,Li \xrightarrow{oxidation}$$

$$M_2(CO)_8(R_2P)_2 \quad (151)$$

(M = Cr, Mo, W).

Reaction between complexes of diphenylphosphine and various complex transition metal halides under the influence of amines or Grignard reagents produced mixed transition metal complexes with two μ-diphenylphosphido or one μ-carbonyl and one μ-diphenylphosphido group [1279,1280]; for example,

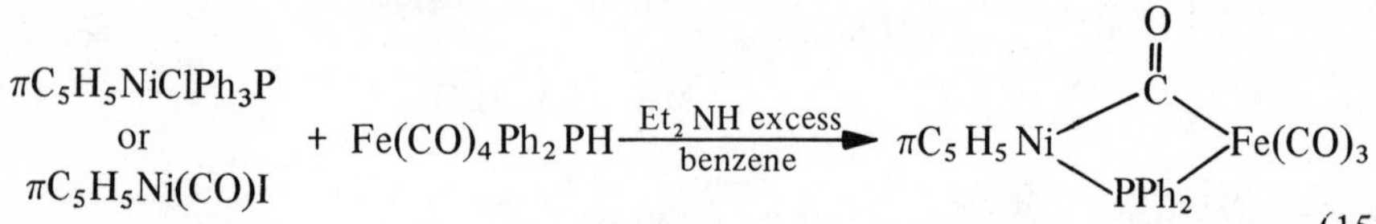

$$(152)$$

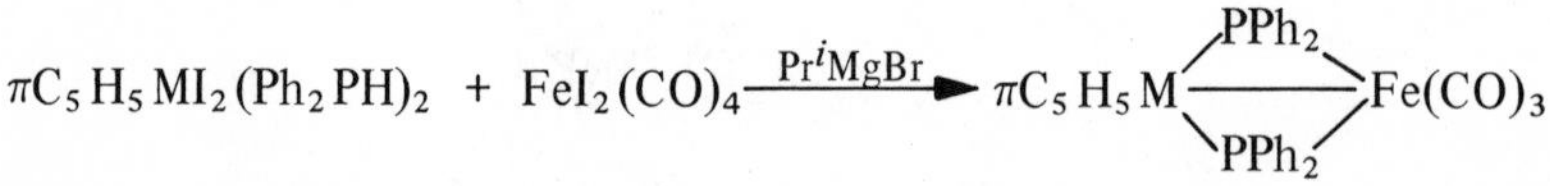

$$\pi C_5H_5MI_2(Ph_2PH)_2 + FeI_2(CO)_4 \xrightarrow{Pr^iMgBr} \pi C_5H_5M \begin{array}{c} PPh_2 \\ \diagdown \\ PPh_2 \end{array} Fe(CO)_3 \quad (153)$$

(M = Co, Rh)

Photochemical reaction of metal hexacarbonyl $M(CO)_6$ (M = Cr, Mo, W) with tetramethyldiphosphine yields the organometallic bases $M(CO)_5Me_2PPMe_2$. Pyrolysis and reaction with more hexacarbonyl involves the second noncoordinated phosphorus lone pair [183].

$$M(CO)_5Me_2PPMe_2 + M'(CO)_6 \xrightarrow[THF]{h\nu}$$

$$+ CO \quad M(CO)_5Me_2PPMe_2M'(CO)_5 \xrightarrow[benzol]{250°C} 2\ CO +$$

$$(CO)_4M \begin{array}{c} PMe_2 \\ \diagup \\ \diagdown \\ PMe_2 \end{array} M'(CO)_4 \quad (154)$$

(M, M' = Cr, Mo, W).

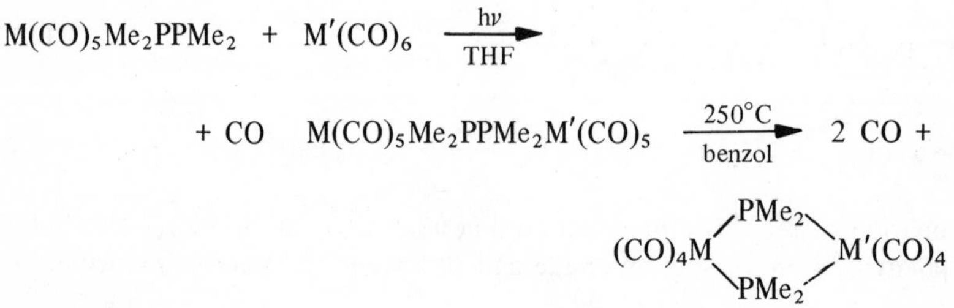

Nucleophilic replacement of fluoride ions in PF_3 transition metal complexes by OH^-, OR^-, or R_2NH leads to complexes with difluorophosphine ligands $R'PF_2$ ($R' = O^-$, OR, R_2N), of which the complexes containing the PF_2O^- ligand are of special interest [732].

Hydrolysis of diphenyl- and dimethylchlorophosphinepentacarbonylmolybdenum produces, for example, complexes of diphenyl- and dimethylphosphinous acid in addition to diphosphoxane complexes. The reactions studied by Kraihanzel et al. [728], including those with other nucleophiles, for example, hydrogensulfide, ethanethiol, ammonia, and amines [729], are summarized in Scheme 1.

SCHEME 1

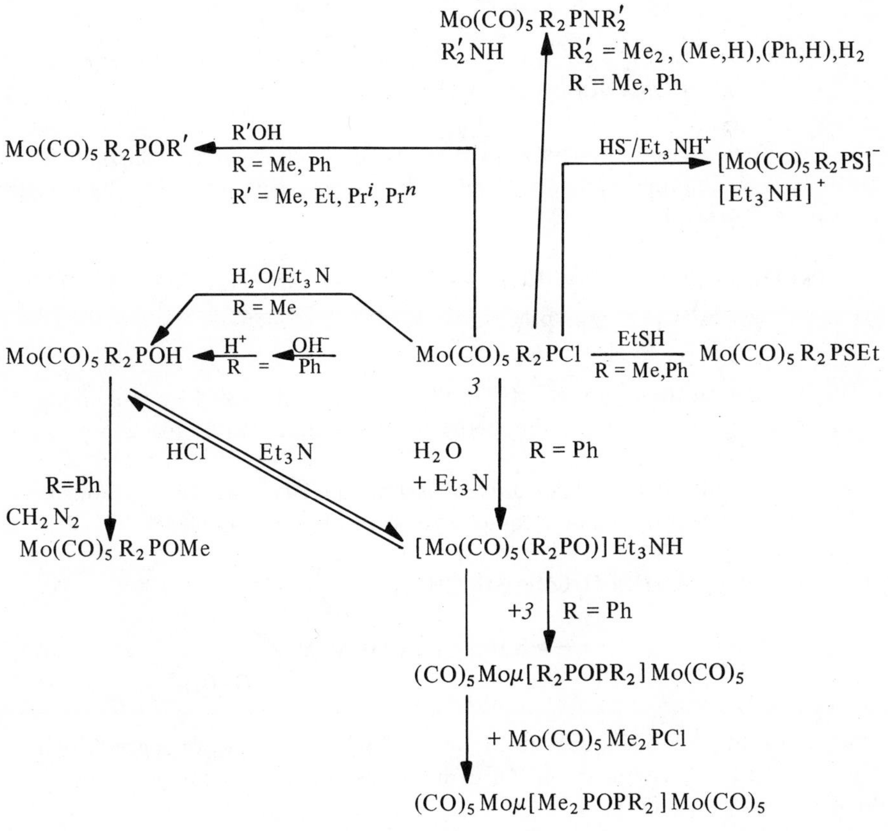

Methanolysis of tetrakis(phosphorus trichloride)nickel(0) in hydrocarbon solvent is, in contrast to the reactions reported above, a slow, heterogenous oxidation-reduction process [59] which may be summarized by (155) to (157):

$$Ni(PCl_3)_4 + 17\ MeOH \longrightarrow [Ni(MeOH)_6]Cl_2 + 3\ HP(O)(OMe)_2 + H_2P(O)OMe + 6\ HCl + 4\ MeCl \qquad (155)$$

Accompanying steps:

$$HP(O)(OMe)_2 + HCl \longrightarrow HP(O)(OH)OMe + MeCl \qquad (156)$$

$$H_2P(O)OMe \longrightarrow x\ PH_3 + \text{other products} \qquad (157)$$

Halogen-hydride exchange between sodium borohydride or lithium aluminumhydride in tetrahydrofuran or ether and $\pi C_5H_5Mn(CO)_2PCl_3$ or $\pi C_5H_5Mn(CO)_2$-Ph_2PCl has been proposed as an alternative synthetic route to phosphine complexes [602]:

$$\pi C_5H_5Mn(CO)_2Ph_nPCl_{3-n} + (3-n)\ NaBH_4 \xrightarrow{THF} (3-n)\ NaCl + (3-n)\ BH_3 + \pi C_5H_5Mn(CO)_2Ph_nPH_{3-n} \qquad (158)$$

If alkali metal-transition metal carbonylates are employed as nucleophiles in reactions with chlorophosphine complexes, dinuclear complexes bridged by a single phosphine may be obtained [424]:

$$M(CO)_5Me_2PCl + NaM'(CO)_5 \xrightarrow{-NaCl} (CO)_5M{-}PMe_2{-}M'(CO)_5 \qquad (159)$$

(M = Cr, Mo, W, M′ = Mn, Re)

The bromine in the complex $Fe(CO)_4PF_2Br$ may be displaced by the ambident anion of 2,2,5,5-tetrakis(trifluoromethyl)-4-oxazolidone with formation of a P–N–C–O chelate ring [81].

Cyclic phosphite ligand (1,3,2-dioxaphosphorinane) complexes are formed by alcoholysis of dichlorophosphine complexes with diols; for example,

$$Mo(CO)_5MePCl_2 + HOCH_2CMe_2CH_2OH \longrightarrow 2\ HCl + (CO)_5MoMeP\begin{matrix} O{-}CH_2 \\ O{-}CH_2 \end{matrix}CMe_2 \qquad (160)$$

Analysis of the ^1H NMR spectra of this and related complexes suggests that a common chair form, with the metal carbonyl positioned equatorially, is the predominant conformer [79].

By treating mono- and diketones with $Cr(CO)_5PH_3$, $Mo(CO)_4(PH_3)_2$, and $Cr(CO)_{4-n}(PH_3)_{n+2}$ in the presence of H_2O or a base, bidentate phosphine ligands (*3*) could be synthesized [705].

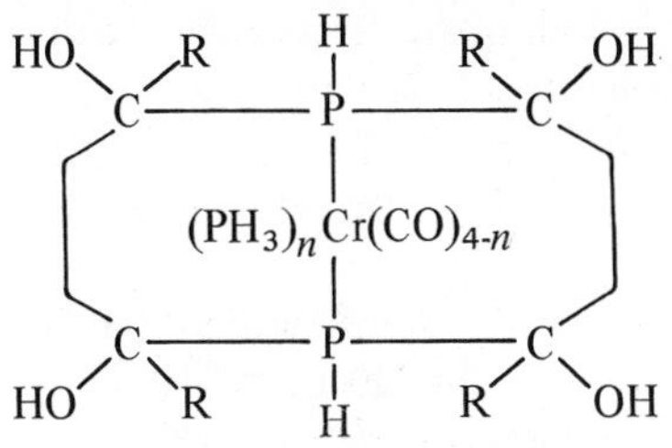

4

R = Me, n = 0, 1

For the synthesis of novel transition metal phosphine complexes some authors have made use of the triorgano-group IV element moiety as an excellent leaving group [1090, 1215]:

$$M(CO)_5P(Me_3Sn)_3 + 3\ Ph_2PCl \longrightarrow M(CO)_5P(Ph_2P)_3 + 3\ Me_3SnCl \quad (161)$$

(M = Cr, Mo, W),

$$W(CO)_5PhP(SnMe_3)_2 + 2\ Me_2ECl \longrightarrow 2\ SnMe_3Cl + (CO)_5WPPh(EMe_2)_2 \quad (162)$$

(E = P, As). In contrast to these reactions, boron trichloride or -bromide does not split the P–Sn bond of the coordinated, triorganometal-substituted phosphine ligand. Instead, the tin-carbon bond is cleaved and a diorganomonohalogeno metal-substituted transition metal complex is formed [949]:

$$M(CO)_5Ph_2PSnMe_3 + BX_3 \longrightarrow M(CO)_5Ph_2PSnMe_2X + MeBX_2 \quad (163)$$

(M = Cr, Mo, W, X = Cl, Br).

The reactivity of the metal-coordinated allyl ligand in $[\pi\text{-allylPd}(R_3P)_2]^+$ has been used for the synthesis of the otherwise inaccessible compound $Pd(Ph_3P)_3$; $Pd[P(OPh)_3]_3$ may be prepared in a similar manner [757]:

$$3\ L + (\pi\text{-methallyl})PdCl \xrightarrow[-PhCH_2NH_3Cl]{PhCH_2NH_2} PdL_3 + PhCH_2NHC_4H_6 + PhCH_2N(C_4H_6)_2 \quad (164)$$

($L = Ph_3P, P(OPh)_3$).

Nucleophilic additions to coordinated isocyanides with OH^-, SH^-, NHR^- [717], or alcohols [258] yield carbene complexes; for example,

$$cis\text{-}PtCl_2(CNR)L + R'OH \longrightarrow cis\text{-}PtCl_2[C(RNH)(OR')]L \quad (165)$$

($L = Ph_3P$, $EtPh_2P$, Et_2PhP, Et_3P, $R = p\text{-}MeOC_6H_4$, $p\text{-}NO_2C_6H_4$, $R' = Et$).

Nonionic, carbamoyl-carbonyl complexes, $\pi C_5H_4RFe(CO)(CONH_2)L$, have been prepared by reaction of the appropriate carbonyl compound with liquid ammonia [431]:

$$[\pi C_5H_4RFe(CO)_2L]X + 2NH_3 \xrightarrow{\text{liq } NH_3} NH_4X + \pi C_5H_4RFe(CO)(CONH_2)L \quad (166)$$

($R = H$, $CHPh_2$, $L = CO$, Ph_3P, Et_3P, $X = PF_6^-$, BPh_4^-, Cl^-, I^-). In an analogous reaction (alkoxycarbonyl)iridium compounds $IrI_2(CO)(CO_2R)L_2$ or $[IrI(CO)_2(CO_2R)L]_2$ were obtained from $IrI(CO)_2L_2$ or $[Ir(CO)_3L]_2$ with iodine and alcohols [819]:

$$[Ir(CO)_3L]_2 + 2\,I_2 \longrightarrow [IrI(CO)_3L]_2^{2+} + 2\,I^- \quad (167)$$

$$[IrI(CO)_3L]_2^{2+} + 2\,ROH \longrightarrow [IrI(CO)_2(CO_2R)L]_2 + 2\,H^+ \quad (168)$$

($L = Ph_3P$, $R = Me, Et$).

Alkoxide ion attack on the nitrosyl group of $[IrCl_3(NO)(Ph_3P)_2]^+$ gave alkyl nitrite complexes $IrCl_3(RONO)L_2$ ($L = Ph_3P$, $R = Me, Et, Pr$) [1012].

Borodko, Broitman, Kachapina, Shilov, and Ukhin reported the partial reduction of coordinated N_2 to N_2H_4 in the molecular nitrogen complex isolated from the system $FeCl_3(Ph_3P)_2 + Pr^iMgCl + N_2$ [163].

Reduction of coordinated dinitrogen was also achieved by hydrogen halides in the complexes $Mo(N_2)_2(dppe)_2$. The diimine derivatives $MoX_2(N_2H_2)(dppe)_2$ were obtained [252]. It was reported that the N_2 ligand in $CoH(N_2)(Ph_3P)_3$ may be reduced to ammonia by a rather complex reaction medium containing naphthalene, lithium, and titanium tetrachloride in tetrahydrofurane [1232a].

Triethylaluminum forms a 2:1 adduct with *cis*-$Mo(N_2)_2L_4$, *trans*-$Mo(N_2)_2(Ph_2PCH_2CH_2PPh_2)_2$ *cis*- and *trans*-$Mo(CO)_2(Ph_2PCH_2CH_2PPh_2)_2$, *cis*-$Mo(CO)_2L_4$ ($L = Me_2PhP$), and *trans*-$Mo(N_2)L'_4$ ($L' = Et_2PhP$). The Et_3Al is bound to the oxygen of the carbonyl group or terminal N of the N_2 group [44].

Coordinated ethylene may alkylate ammonia. If *cis*-dichloroethylenetriphenylphosphine platinum(II) were allowed to react with ammonia, the ethyl amines $EtNH_2$, Et_2NH, and Et_3N, after hydrolysis, could be isolated [387].

Nucleophilic replacement reactions on coordinated trifluorophosphine in the complexes $Fe(NO)_2(PF_3)_2$ and $Co(NO)(PF_3)_3$ have been studied. If alcoholates and phenolate are employed as nucleophiles, the corresponding phosphite complexes, for example, $Fe(NO)_2(P(OR)_3)_2$ and $Co(NO)(P(OR)_3)_3$ (R = Me, Et, Pr^i, Bu^n, Ph), are obtained. Controlled hydrolysis of $Fe(NO)_2(PF_3)_2$ and $Co(NO)(PF_3)_3$ produced the fluorophosphonato complexes $[Fe(NO)_2(PF_2O)_2]^{2-}$, $[Co(NO)(PF_3)_2(PF_2O)]^-$, and $[Co(NO)(PF_3)(PF_2O)_2]^{2-}$ [747]. Tetrakis(difluorophosphonato) platinate(II) and palladate(II) have been recently prepared in a interesting Michaelis-Arbusov type reaction between allyldifluorophosphite and *cis*- or *trans*-$(C_5H_5N)_2PtCl_2$ [546].

IX. Thermodynamic, Kinetic, and Mechanistic Studies

The number of papers concerned with this topic is quite large. It was not possible, therefore, to give detailed and complete information on this subject within the limits of the present review article. Peloso [980a] has published an excellent review article on the kinetics of reactions of nickel, palladium, and platinum complexes which is concerned with transition metal phosphine complexes.

A. *Reactions and Exchange Equilibria Involving Carbon Monoxide or Phosphine in Transition Metal Carbonyls and Their Substitution Reactions*

Kinetics and mechanisms of substitution reactions on transition metal carbonyls and their derivatives have been the subject of much recent work.

The volume of activation $\Delta V^{\neq}$ has been shown to be a valuable mechanistic criterion for substitution (or decomposition) reactions of the type [189]

$$M(CO)_n \xrightarrow{\text{slow}} CO + M(CO)_{n-1} \xrightarrow{\text{L, fast}} \text{products} \qquad (169)$$

The decomposition of $Mo(CO)_5$(amine) complexes in the presence of Lewis bases (L) to form substituted $Mo(CO)_5L$ species (L = phosphines, arsines) has been discussed [358]. Of particular interest is the nature of the intermediate in these reactions because it is also believed to be present in both the photochemical [1140] and thermal [1246] substitution reactions of group VI hexacarbonyls. ^{13}CO labeling experiments are expected to lead to a differentiation between the two possible geometries, trigonal bipyramidal and square pyramidal, of the intermediate $Mo(CO)_5$ [359].

The photochemical formation of mixed phosphine-amine molybdenum and tungsten tetracarbonyls, $M(CO)_4LL'$ (M = Mo, W, L, L′ = phosphine, amine), has been studied. Isotopic substitution studies revealed that the principal product obtained from irradiation of $Mo(CO)_5Ph_3P$ in the presence of amines, *cis*-$Mo(CO)_4$(amine)Ph_3P, results from an isomerization process [1094]. Kinetic investigations of the amine substitution in pentacarbonylaminemolybdenum(0) complexes, $Mo(CO)_5A$, by phosphorus(III) donors L (e.g., Ph_3P) gave for k_1, the first-order dissociative process rate constant, the following ranking of A: cyclohexylamine > piperidine > quinuclidine. $\Delta H\neq$ and $\Delta S\neq$ are about 25 kcal/mole and 3 eu respectively. The overall rate law may be expressed as follows [326]:

$$\frac{d[Mo(CO)_5A]}{dt} = (\frac{k_1}{\gamma} + k_2[L])[Mo(CO)_5A]$$

Thermal decomposition and substitution reactions of *cis*-$Mo(CO)_4(amine)Ph_3P$ (amine = $C_5H_{10}NH$, C_5H_5N) have been studied. The first-order rate law may be written as $\nu = k[Mo(CO)_4(amine)Ph_3P]$ at low concentrations of the entering substituent L, whereas at higher concentrations of L a second-order term has to be taken into account:

$$\nu = [k + k'[L]]\,[Mo(CO)_4(amine)Ph_3P]$$

The quantitative order of preference of the ligands L for the five-coordinate intermediate $Mo(CO)_4Ph_3P$ was the following: CO (0.31) < $AsPh_3$ (0.47) < Bu^n_3P (0.68) < $C_5H_{10}NH$ (1) < $SbPh_3$ (1.11) < Ph_3P (1.47) ~ C_5H_5N < $P(OCH_2)_3CEt$ (2.38) [625].

Careful kinetic investigation of tertiary phosphine substitution in di-μ-nitrosyldecacarbonyltriruthenium [950] revealed that replacing the terminal carbonyl ligands in $Ru_3(CO)_{12}$ with bridging nitrosyl groups has no effect on the mechanism of the reaction.

$$Ru_3(CO)_{10}(NO)_2 + Ph_3P \xrightarrow{k_1} CO + Ru_3(CO)_9(NO)_2Ph_3P \quad (170)$$

$$Ru_3(CO)_9(NO)_2Ph_3P + Ph_3P \xrightarrow{k_2} CO + Ru_3(CO)_8(NO)_2(Ph_3P)_2 \quad (171)$$

The nitrosyl groups in $Ru_3(NO)_2(CO)_{10}$ do not serve as "electron wells" to stabilize coordinatively supersaturated transition states [373, 891].

Dodecacatricarbonylruthenium reacts in a bimolecular manner with various phosphorus and arsenic donor ligands. The activation parameters have been obtained, and the rate constants are strongly dependent on the nature of the ligands. This indicates a high degree of bond making in the transition state [998].

For the disubstitution reaction

$$Co(CO)_2(NO)L + L \longrightarrow Co(CO)(NO)L_2 + CO \quad (172)$$

dissociative and associative paths (involving pentacoordinate species) are competitive. The order of reactivity for the associative path for various Ls is $P(OMe)_3$ > $P(OBu^n)_3$ > $P(OEt)_3$ > $P(OPr^i)_3$ > Bu^n_3P > Ph_3P > $P(OPh)_3$, whereas for the dissociative path the L may be ranked as $Ph_3P > Bu_3^nP > P(OPr^i)_3 > P(OPh)_3$ [226]. Ligand replacement in the complexes $Co(CO)_2(NO)L$ by carbon monoxide

$$Co(CO)_2(NO)L + CO \rightleftharpoons Co(CO)_3(NO) + L \qquad (173)$$

follows a two-term rate law, $v = k_{-1}[Co(CO)_2(NO)L] + k_{-2}[Co(CO)_2(NO)L][CO]$; the equilibrium constant and the rate constant vary considerably with the σ-donor abilities of the ligands L (L = Ph_3P, Ph_3As, Ph_3Sb [628].

Kinetic studies of the chelation reaction

$$M(CO)_5 Ph_2 P(CH_2)_n PPh_2 \longrightarrow CO + (CO)_4 M \overset{PPh_2}{\underset{PPh_2}{<>}} (CH_2)_n \qquad (174)$$

(M = Cr, W, n = 1, 2, 3) revealed that the smaller the potential chelate ring, the faster the reaction. The positive entropy of activation and its large variation (+6 to 71 $K^{-1}mole^{-1}$) are compatible with a concerted process in the transition state [311].

Solvent effects in substitution reactions of cyclopentadienylhalocarbonyliron and -ruthenium complexes by phosphite ligands have been studied. The rate constants vary with solvent polarity:

$$k_{octane} > k_{DNB} > k_{xylene} > k_{CHCl_3} > k_{PhNO_2}$$

| dielectric constant | 1.96 | 3.08 | 2.3 | 4.8 | 34.8 |

(DNB = di-n-butylether)

Replacing iron in the complexes with its heavier congener ruthenium causes a sharp decrease in reactivity [191].

A marked difference in mechanism, rate, and activation parameters has been found for the carbonyl substitution reactions by phosphine ligands (Ph_3P, $P(OPh)_3$, $P(OMe)_3$) of the π-tetrahydroindenyl complex $\pi C_9H_{11}MoCl(CO)_3$ and the π-indenyl complex $\pi C_9H_7MoCl(CO)_3$ [1251].

Reaction of various phosphorus(III) ligands with π-cyclopentadienylhalotricarbonylmolybdenum(II) yields $\pi C_5H_5MoX(CO)_2L$ by a dissociative mechanism. The rates of the dissociative processes decrease in the order Cl > Br > I. Basic phosphines, for example, Bu_3^nP and Bu_2^nPhP, also give fac-$Mo(CO)_3L_3$ and cis-$Mo(CO)_4$-L_2 in temperature- and solvent-dependent amounts in a first-order reaction involving $\pi C_5H_5MoCl(CO)_3$ and L [1250].

Contrary to the results for the manganese series, the halogenopentacarbonylrhenium compounds undergo successive mono- and disubstitution by Ph_3P; the final products are always the disubstituted compounds $ReX(CO)_3(Ph_3P)_2$. The rhenium compounds react in the monosubstitution step about 60 times more slowly than the corresponding manganese compounds. The rate constants of various $ReX(CO)_5$ compounds vary as follows: $Cl > Br > I$ [192]. Steric and stereochemical limitations of higher substitution of bromocarbonylmanganese by various phosphines and phosphites have been discussed by Reimann and Singleton [1019].

Two competing mechanisms have been proposed for the reaction between triphenylgermyl- or triphenylstannylpentacarbonylmanganese(0) and phosphines L ($L = Ph_3P$) [401] (Scheme 1).

SCHEME 1

Five-coordinate intermediate   Seven-coordinate intermediate

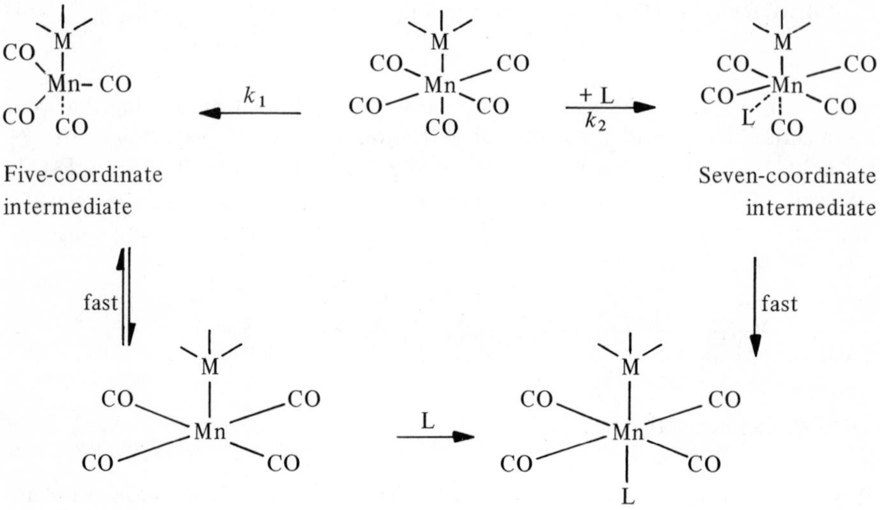

$L = Ph_3P$, $M = Ge, Sn$

For the reactions of tetracarbonyl(2,2′-dipyridyl)tungsten(0) with Lewis bases, a modification of the mechanism proposed by Graham and Angelici [525] has to be taken into account. Initial formation of a five-coordinate intermediate is involved with 2,2′-dipyridyl functioning as a monodentate ligand. Subsequent attack by L affords the precursor of the products formed in this reaction [880]:

$$\underset{N}{\overset{N}{\bigcirc}}W(CO)_4 \underset{k_4}{\overset{k_3}{\rightleftharpoons}} \overset{\frown}{N}\ N W(CO)_4 \xrightarrow[L]{k_5} \overset{\frown}{N}\ N W(CO)_4 L \xrightarrow{fast} \text{products}$$

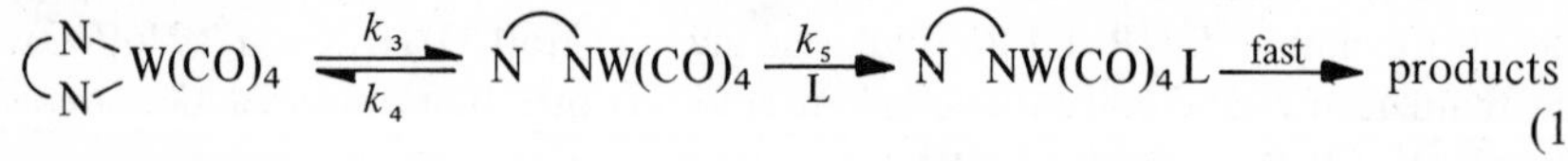

(175)

(NN = 2,2′-dipyridyl).

Under forcing conditions, the reaction between o-phenanthrolinetetracarbonylmolybdenum and Lewis bases, for example, $P(OEt)_3$, yields $Mo(CO)_2(phen)L_2$ and CO. The overall reaction consists of two steps [615]. Shrader, Ross, Jernigan, and Dobson [1112] investigated the kinetics of the second step:

$$Mo(CO)_4(phen) + L \xrightarrow{k_1} CO + Mo(CO)_3(phen)L \qquad (176)$$

$$Mo(CO)_3(phen)L + L \xrightarrow{k_2} CO + Mo(CO)_2(phen)L_2 \qquad (177)$$

The relative first-order rates of reaction (177) are dependent on the nature of L. They decrease as follows [1112]: $CO > P(OPh)_3 > P(OEt)_3 > Bu^n_3P$.

The kinetics of the reactions of h^4-1,2,3,6-$C_8H_{12}Fe(CO)_3$ with tertiary phosphines and phosphites have been examined:

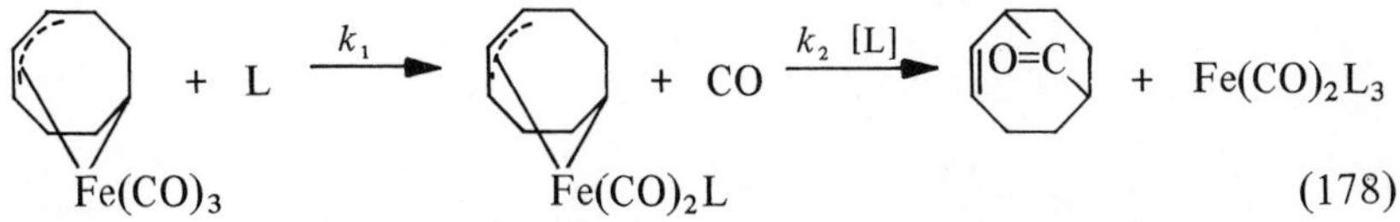

(178)

1

With relatively poor nucleophiles [e.g., Ph_3P, $P(OPh)_3$] only product (*1*) is formed by a CO dissociative mechanism, whereas with more nucleophilic phosphines the reaction may proceed by two paths, the rate law for which may be expressed as $k_{obs} = k_1 + k_2[L]$ [658].

Faraone, Cusmano, and Pietropaolo [443] studied the kinetics of reactions of cyclooctatetraenetricarbonylruthenium with various phosphorus(III) ligands. The C_8H_8 ligand facilitates the bimolecular attack by the phosphines.

The rate of decarbonylation of the acetyl complex $\pi C_5H_5Mo(CO)_2(COMe)L$ (L = Ph_3P, $(p\text{-tolyl})_3P$, $(p\text{-OMeC}_6H_4)_3P$, $MePh_2P$, Me_2PhP, Bu^n_3P, Cy_3P, Pr^iPh_2-P)

$$\pi C_5H_5Mo(CO)_2(COMe)L \longrightarrow CO + \pi C_5H_5Mo(CO)_2(Me)L \qquad (179)$$

depends markedly on the nature of the organic groups attached to phosphorus. Steric and electronic factors contribute to the observed rate differences (see Table III) [78].

TABLE III. Rate constants for the decarbonylation of $\pi C_5H_5Mo(CO)_2(COMe)L$ at 60°C, using MeCN as solvent and 5 to 10 molar excess of ligand to inhibit dissociation

L	$k(sec^{-1})(10^6)$	L	$k(sec^{-1})(10^6)$
Ph_3P	13.2	Me_2PhP	3.6
$(p\text{-tolyl})_3P$	10.9	Bu^n_3P	3.9
$(p\text{-MeOC}_6H_4)_3P$	9.6	Cy_3P	46.4
$MePh_2P$	6.8	Pr^iPh_2P	23.9

On treatment of $[IrCl_2(CO)_2Et]_2$ with various phosphorus ligands L (L = Me_2PhP, $MePh_2P$, $P(OMe)_3$, $PhP(OMe)_2$, Ph_2POMe), the propionyl complexes $IrCl_2(CO)(COEt)L_2$ (*2*), which subsequently rearrange to a more stable isomeric form (*3*), are obtained. Kinetics and mechanism of this rearrangement have been studied [1269].

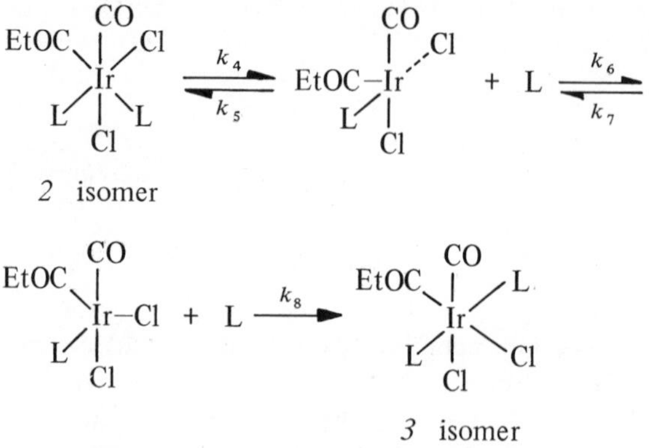

(180)

A strictly second-order rate law (first-order in each reagent) has been found for the substitution of carbonyl in bis(mercaptotricarbonyliron) complexes, for example, $Fe_2(CO)_6S_2C_6H_3Me$, by various phosphorus(III) donor ligands R_3P (R_3 = Ph_3, $(OPh)_3$, Bu^n_3, $(OMe)_3$, Ph_2H, Ph_2Et, Cy_3, etc.) [434].

B. *Reactions Involving Elimination of Molecular Nitrogen, Hydrogen or Halogen*

Substitution of molecular nitrogen in complexes $OsX_2(N_2)L_3$ (X = Cl, Br, L = Me_2PhP, Et_2PhP) and $OsX_2(N_2)L_2L'$ (L = Me_2PhP, L′ = $PhP(OMe)_2$, X = Cl, Br) by various phosphorus ligands proceeds by a reaction first-order with respect to complex concentration and independent of phosphorus ligand concentration. A dissociative (S_N1) mechanism is compatible with the kinetic results [844]. The rate of reductive elimination of molecular hydrogen from $[IrH_2(CO)_2L_2]BPh_4$ (L = tertiary phosphine or arsine)

$$[IrH_2(CO)_2L_2]^+ + L' \underset{}{\overset{k_1}{\rightleftharpoons}} [Ir(CO)_2L_2L']^+ + H_2 \qquad (181)$$

is independent of the nature and concentration of the substituting ligand L′ but decreases for various L in the series $Ph_3P > Ph_3As > MePh_2P > EtPh_2P > Et_2PhP > Et_3P > Cy_3P \cong Pr^i_3P$ [871].

Paramagnetic iron nitrosyl complexes, $FeX(NO)_2L$ (X = Cl, Br, I, L = Ph_3P, Bu^n_3P, $P(OPr^i)_3$, Ph_3As), undergo halogenic substitution with Ph_3P, $P(OPr^i)_3$, $P(OPh)_3$, $P(OMe)_3$, the rate of which depends on the halogen and on the concentration and nature of reagent [992]:

$$FeX(NO)_2Ph_3P + Ph_3P \underset{k_{-1}}{\overset{k_1}{\rightleftharpoons}} FeX(NO)_2(Ph_3P)_2 \xrightarrow{k_2}$$

$$Fe(NO)_2(Ph_3P)_2 + X \text{ (reacts with } Ph_3P) \tag{182}$$

C. Exchange, Dissociation, and Isomerization Reactions Involving Phosphine Ligands and Their Complexes

The kinetics of the substitution reactions of $Ni(PF_3)_4$ and $Pt(PF_3)_4$ with cyclohexylisonitrile have been investigated and an S_N1 mechanism has been established:

$$M(PF_3)_4 \longrightarrow M(PF_3)_3 + PF_3 \tag{183}$$

$$M(PF_3)_3 + L \longrightarrow M(PF_3)_3L \tag{184}$$

The activation energies for dissociation of the metal-phosphorus bond in the complexes $M(PF_3)_3$ show the following trend: Ni > Pd < Pt [662].

Tolman demonstrated that steric effects are much more important than electronic effects in determining the exchange equilibria among phosphorus ligands L′ on Ni(O) complexes NiL_4, where L and L′ represent different phosphorus donor molecules [1182]. The ligands used for these investigations may be ranked in a series of stability of the corresponding complexes: $PhP(OEt)_2 \gtrsim P(OCH_2)_3MeC = P(OCH_2)_3EtC = Me_2PCH_2CH_2PMe_2 = P(OMe)_3 = P(OEt)_3 = P(OCH_2CH_2Cl)_3 \gtrsim P(OCH_2CCl_3)_3 \gg P(o\text{-}p\text{-}C_6H_4CN)_3 \gtrsim P(OPh)_3 = P(O\text{-}p\text{-}C_6H_4Me)_3 = P(O\text{-}p\text{-}C_6H_4\text{-}OMe)_3 = Me_3P = Me_2PCF_3 \gtrsim Ph_2POEt \gg Et_3P = Bu^n_3P \sim EtPh_2P \sim Ph_3P \sim P(O\text{-}o\text{-}C_6H_4Me)_3 \gg Pr^i_3P > Bu^t_3P > P(o\text{-}C_6H_4Me)_3, P(C_6F_5)_3$.

In order to evaluate the relative importance of σ-donor and π-acceptor properties for the bonding in transition metal complexes, Gibbs free energy values and the equilibrium constants for the equilibria (185) to (188) have been determined [673]:

$$2\ Ni(CO)_3L \overset{K_1}{\rightleftharpoons} Ni(CO)_2L_2 + Ni(CO)_4 \tag{185}$$

$$Ni(CO)_2[P(CF_3)_3]_2 + L \overset{K_2}{\rightleftharpoons} Ni(CO)_2P(CF_3)_3L + P(CF_3)_3 \tag{186}$$

$$Ni(CO)_2P(CF_3)_3L + L \overset{K_3}{\rightleftharpoons} Ni(CO)_2L_2 + P(CF_3)_3 \tag{187}$$

$$Ni(CO)_2L_2 + Ni(CO)_2[P(CF_3)_3]_2 \overset{K_4}{\rightleftharpoons} 2\ Ni(CO)_2P(CF_3)_3L \tag{188}$$

($L = MeP(CF_3)_2, EtP(CF_3)_2, Bu^iP(CF_3)_2, Et_2PCF_3$).

A mechanism has been proposed for the reaction of olefin(tetramethyl-1,4-

benzoquinone)nickel(O) complexes, Ni(diene)($C_{10}H_{12}O_2$) (diene = cycloocta-1,5-diene, dicyclopentadiene, norbornadiene, cyclooctatetraene, bezoquinone) with trimethylphosphite, involving initial bimolecular attack of the nickel complex to form an intermediate with monodentate olefin [986].

The relative affinities of tertiary phosphines (and arsines) L for ruthenium in exchange equilibria involving $RuHCl(CO)(R_2PhP)_2L$ (R = Et, Pr^n, Bu^n) have been evaluated by ^1H NMR spectroscopy. Affinity increases as shown in the series $Et_2PhAs < Me_2PhAs < Bu_2^nPhP \approx Pr_2^nPhP \approx Et_2PhP < Et_3P < P(OEt)_3 \approx Me_2PhP < PhP(OMe)_2$ [408].

A rate law $v = [k_1 + k_2(\text{halide})] [PtX_2L_2]$ has been found for the isotopic exchange in methanol between chloride $^{36}Cl^-$ or bromide $^{82}Br^-$ and the corresponding halogenides in the complexes *trans*-PtX_2L_2 [X = Cl, Br, $L_2 = (Et_3P)_2$, $Et_3P(NH_3)$, $Et_3P(Et_2NCH_2Ph)$]. k_1 can be considered as characterizing a solvation step [233].

The dependence of the reactivity and reaction mechanism on an increasing steric hindrance induced by groups R (R = Ph, *o*-tolyl, mesityl) in substitution reactions of complexes *trans*-$PtClR(Et_3P)_2$ has been investigated [445].

Kinetic and equilibrium studies on the systems bis(1,2-dicyano-1,2-dithioethylene)-iron(III)/organic base (e.g., Ph_3P, Ph_3PO) in acetonitrile revealed that the equilibrium constants of the reaction

$$[Fe(mnt)_2]^- + L \rightleftharpoons [FeL(mnt)_2]^- \qquad (189)$$

(mnt = 1,2-dicyano-1,2-dithioethylene, L = e.g., Ph_3P, Ph_3PO) for Ph_3P and Ph_3PO are of the same magnitude [1278]. The redistribution equilibrium

$$2\ CoI_2(Ph_3P)(Ph_3PO) \rightleftharpoons CoI_2(Ph_3PO)_2 + CoI_2(Ph_3P)_2 \qquad (190)$$

with an equilibrium constant $K = (3.9 \pm 0.3) \cdot 10^{-2}$ at 25°C is shifted to the right with increasing temperature. The mixed complex $CoI_2(Ph_3P)(Ph_3PO)$ may be isolated as a solid crystalline compound [1025].

For the dissociation of tris(triphenylphosphine)chlororhodium(I) in benzene solution the equilibrium constant has been determined spectrophotometrically to be $(1.4 \pm 0.4) \cdot 10^{-4}$ [*m*] at 25°C [41].

$$RhCl(Ph_3P)_3 \overset{K}{\rightleftharpoons} RhCl(Ph_3P)_2 + Ph_3P \qquad (191)$$

The heats of the reactions (192) to (195) have been obtained by calorimetric measurements:

$$[RhCl(\text{olefin})_n]_2 + 2\ P \longrightarrow Rh_2Cl_2P_2(\text{olefin})_n + n\ \text{olefin} \qquad (192)$$

$$Rh_2Cl_2P_2(\text{olefin})_n + 2\ P \longrightarrow [RhClP_2]_2 + n\ \text{olefin} \qquad (193)$$

$$[RhClP_2]_2 + 2\ P \longrightarrow 2\ RhClP_3 \qquad (194)$$

$$Rh(acac)(olefin) + 2\ P \longrightarrow Rh(acac)P_2 + olefin \qquad (195)$$

(olefin = norbornadiene, 1,5-cyclooctadiene, 1,3,5,7-cyclooctatetraene, dicyclopentadiene, benzoquinone, for all these $n = 1$; cis-cyclooctene, $n = 2$; olefin = 1,5-cyclooctadiene, 1,3,5,7-cyclooctatetraene for reaction (195); P in all cases = $P(OPh)_3$). Values for $-\Delta H$ increase in the order [values in parentheses are the ΔH values for reaction (193) in kcal/mole]

norbornadiene >	1,5-cyclooctadiene >	1,3,5,7-cyclooctatetraene >	dicyclopentadiene
(−22.6 ± 0.6)	(−23.0 ± 0.6)	(−25.2 ± 0.6)	(−26.3 ± 0.7)

These displacement energies are similar to the heats of adsorption on metallic surfaces [974].

The kinetics of the reaction of $RuCl_2(Ph_3P)_3$ and oxygen showed that the rate of oxygen uptake decreased with increasing amounts of free triphenylphosphine present in benzene solutions. The oxidation of cyclohexene by molecular oxygen at 60°C and atmospheric pressure catalyzed by $RuCl_2(Ph_3P)_3$ is a radical reaction, the products being 2-cyclohexen-1-one, 2-cyclohexen-1-ol, cyclohexene oxide, and polymers [241].

The thermodynamics and kinetics of the planar tetrahedral interconversion of 26 dihalobis(diarylmethylphosphine)nickel(II) complexes have been studied. The rate of structural change increases in the order Br < Cl < I [993].

A consecutive displacement mechanism for the tertiary phosphine catalyzed cis-trans isomerization of the platinum(II) complexes PtX_2L_2 [X = Cl, I, L = Me_2PhP, $Me_2(o\text{-tolyl})P$] via an ionic intermediate $[PtXL_3]^+X^-$ could be established, as indicated in Scheme 2 [313].

SCHEME 2. Consecutive displacement mechanism for L′ catalyzed cis-trans isomerization of PtX_2L_2

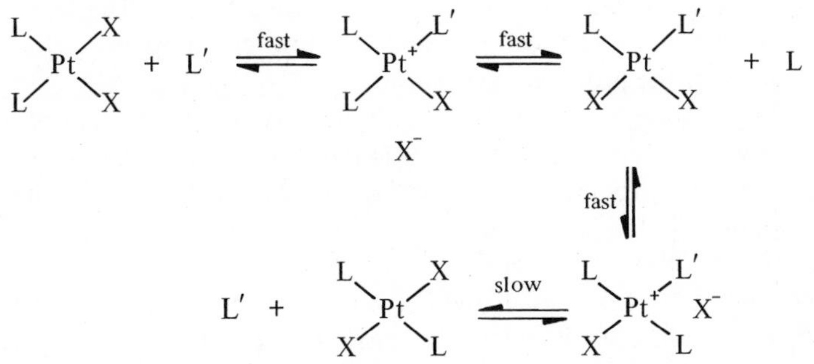

The solvation of platinum complexes has been studied by determination of the solvent dependency of the phosphine catalyzed cis-trans isomerization rate constant in *cis*-$PtCl_2(Bu_3^nP)_2$ [553]. Some solvation numbers and equilibrium constants for solvent association with *cis*-$PtCl_2(Bu_3^nP)_2$ according to

$$PtCl_2(Bu_3P)_2 + nS \rightleftharpoons PtCl_2S_n(Bu_3P)_2 \tag{196}$$

$$K = \frac{[PtCl_2S_n(Bu_3P)_2]}{[S]^n[PtCl_2(Bu_3P)_2]}$$

(S = solvent), are listed in Table IV.

TABLE IV. Solvation numbers and equilibrium constants for solvent association with *cis*-$PtCl_2(Bu_3^nP)_2$ (taken from [553])

Solvent $S^{(a)}$	$n^{(b)}$	$K^{(b)}$
$MeNO_2$	1.00 ± 0.01	21. ± 1.0
MeCN	1.05 ± 0.05	16.8 ± 0.4
Et_2O	1.07 ± 0.03	3.5 ± 0.2
$CHCl_3$	0.99 ± 0.04	0.284 ± 0.005
MeOH	2.04 ± 0.5	110 ± 2

(a) Added to cyclohexane in low concentration.
(b) Standard deviations from least-square treatment of equation $1/k_2' = 1/k_2 + [S]^n K/k_2$, where k_2 refers to solutions without polar solvents S and k_2' with polar solvent added to cyclohexane solutions.

The uncatalyzed cis-trans isomerization of chloro(*o*-tolyl)bis(triethylphosphine)-platinum(II), *cis*-$PtCl(o\text{-tolyl})(Et_3P)_2$, in protic solvents follows a nonsynchronous mechanism. The rate-determining step involves Pt-Cl bond breaking and the formation of a three-coordinate intermediate [446]. This was the first example of a dissociative path for the cis-trans isomerization of a Pt(II) complex in the presence of a much more favorable bimolecular path for substitution. The large negative entropy of activation (−32.5 cal/deg) and the low enthalpy of activation (9.7 kcal/mole) supports the proposed mechanism.

Bis(diethyldithiophosphato)nickel(II) with tributylphosphine forms a stable diamagnetic five-coordinate adduct $Ni[PS_2(EtO)_2]_2Bu_3P$. For the replacement of Bu_3P with bidentate ligands (e.g., $Ph_2PCH_2CH_2PPh_2$) a dissociative mechanism has been proposed [1155].

By use of ^1H NMR techniques the thermodynamics and kinetics of cis-trans isomerization of cyclopentadienylmolybdenum complexes $\pi C_5H_5Mo(CO)_2RL$ (L = Ph_3P, Bu_3^nP, Me_2PhP, $P(OMe)_3$, $P(OPh)_3$, R = H, D, Me, $PhCH_2$, Cl, Br, I) have been evaluated

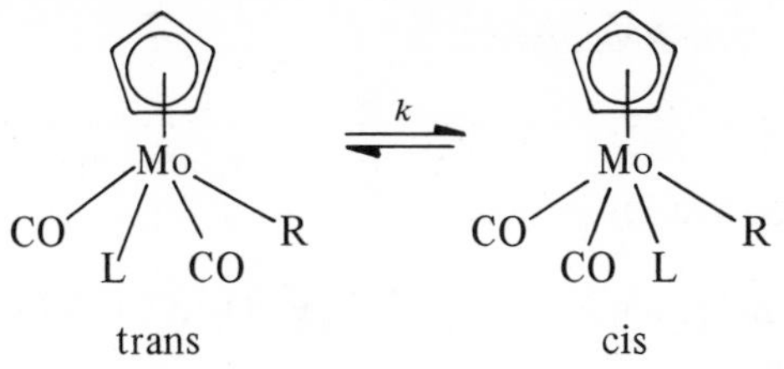

(197)

The equilibrium constant K [cis]/[trans] increases in the following order for R and L. R: $PhCH_2 < Me < H, D < I < Br < Cl$; L: $Me_2PhP \sim Ph_3P \sim Bu^n_3P < P(OMe)_3 < P(OPh)_3$. An intermediate of approximately trigonal bipyramidal structure with the cyclopentadienyl ring in an apical position has been proposed [442]. The rate of phosphine exchange in $\pi C_5H_5Mn(COOMe)(NO)Ph_3P$ is equal to the rate of racemization of the (+) and (−) isomers in the same solvent. A dissociative mechanism, which involves a relatively stable achiral planar intermediate, $\pi C_5H_5Mn(NO)$-(COOMe), has been proposed [202].

D. *Intramolecular Rearrangements*

Shapley and Osborn [1101] discussed rapid intramolecular rearrangements in pentacoordinate transition metal compounds with d^8 electronic configuration. The physicochemical method employed for investigation of these dynamic processes was variable temperature NMR spectroscopy. Free energies of activation, $\Delta G_c{\neq}$ obtained for some of these compounds, are listed in Table V.

TABLE V. Free energies of activation $\Delta G_c{\neq}$ for intramolecular rearrangements within pentacoordinate transition metal complexes of d^8 electronic configuration in kcal/mole (taken from [1101]).

$Ir(COD)RL_2$ (a)	R = Me	R = H
L_2 = 2 Me_2PhP	16.3	13.3
2 $MePh_2P$	>16.9 (b)	14.1
2 Ph_3P	⩾14.0 (b)	17.7
2 $EtPh_2P$		17.6
2 Pr^iPh_2P		>20.6 (c)
2 $CyPh_2P$		>20.1 (c)
dppe	<9.0	
dppp	13.4	
dppb	>16.5	
1,2-bis(dimethylarsino)-benzene	11.9	
2 Me_2PhAs	>15.5	

(*a*) COD = 1,5-cyclooctadiene
(*b*) $\Delta G_c{\neq}$ value at coalescence temperature of vinyl resonances
(*c*) Lower limit because internal exchange intervenes

E. *Kinetics of Catalytic Reactions*

Increasing basicity of the substituents R on phosphorus in the complexes *trans*-$IrCl(CO)(R_3P)_2$ enhances the rate of formation and the stability of dioxygen adducts, $IrCl(CO)(O_2)(R_3P)_2$

$$IrCl(CO)L_2 + O_2 \underset{k_{-1}}{\overset{k_2}{\rightleftharpoons}} O_2IrCl(CO)L_2 \quad (198)$$

(L = tertiary phosphine (or arsine); $K = k_2/k_{-1} = [O_2IrCl(CO)L_2]/[IrCl(CO)L_2][O_2]$). Steric and electronic factors have to be considered for the interpretation of the kinetic and thermodynamic parameters of these reactions [1226]. The mechanism of the tris(triphenylphosphine)platinum(O) catalyzed oxidation of triphenylphosphine [139] may be illustrated by (199) to (201):

$$Pt(Ph_3P)_3 + O_2 \xrightarrow{k_1} PtO_2(Ph_3P)_2 + Ph_3P \quad (199)$$

$$PtO_2(Ph_3P)_2 + Ph_3P \xrightarrow{k_2} [PtO_2(Ph_3P)_3] \xrightarrow[\text{fast}]{/2\ Ph_3P} Pt(Ph_3P)_3 + 2\ Ph_3PO \quad (200)$$

$$2\ Ph_3P + O_2 \xrightarrow{Pt(Ph_3P)_3} 2\ Ph_3PO \quad (201)$$

A more detailed study has confirmed this mechanism [571].

Qualitative measurements of the homogeneous oxidation of triphenylphosphine and diphenylmethylphosphine by the dioxygen complex, $[Ir(CO)(O_2)(MePh_2P)_3]$-ClO_4 have shown that it is an ineffective catalyst [274]. The uptake of O_2 by $RuCl_2(Ph_3P)_3$ in benzene [241] is at least five times faster than by $[Ir(CO)$-$(MePh_2P)_3]^+$ [274].

Kinetic studies in which IR, UV, and ^{31}P NMR spectroscopy were used revealed that the reactions between $CoH(N_2)(Ph_3P)_3$ or $CoH(CO)(Ph_3P)_3$ and substituted olefins that give adducts or polymers are first-order with respect to the cobalt complexes [750]. Expressions for the pseudo first-order rate constants were given. The homogeneous hydrogenation of cyclohexene by $CoH(CO)_2(Bu_3P)_2$ under 30 atm of hydrogen at 60°C follows the rate law:

$$\text{rate} = \frac{k'K_2[H_2][cat][S]}{1 + K_2[S] + [R_3P]/K_1}$$

where [S], $[H_2]$, [cat], and $[R_3P]$ are concentrations of olefin, hydrogen, catalyst $CoH(CO)_2(Bu_3P)_2$, and free phosphine, respectively, k' is the rate constant for the rate-determining step, and K_1 and K_2 are equilibrium constants for the dissociation of phosphine, Bu_3P, or the association of olefin with the coordinative unsaturated species $CoH(CO)_2R_3P$ [455].

A new type of hydrogen migration (1,3-shift) in a dinuclear intermediate (*4*) is presumably responsible for the isomerization of 1-hexene to *cis*-2-hexene catalyzed by $Co(N_2)(Ph_3P)_3$ [727].

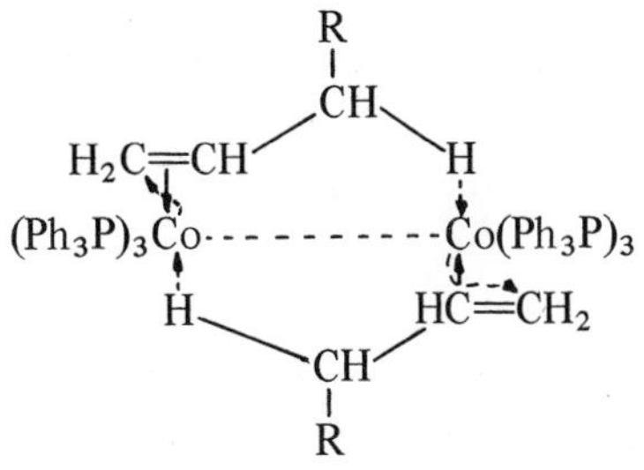

(*4*)

Proposed structure of reaction, intermediate of "1,3-hydrogen shift"

Hydrosilylation of olefins catalyzed by nickel(O) and nickel(II) phosphine complexes has been compared with respect to rate of product formation. The nickel(II) complexes exhibit a substantially longer induction period [702].

The kinetics and mechanism of catalytic homogeneous hydrogenation of olefins which use chlorocarbonylbis(triphenylphosphine)iridium(I) or hydridocarbonyl-tris(triphenylphosphine)-iridium(I) have been studied in detail [211, 212, 1142].

The rate of isomerization of the product formed in the reaction between dienes and hydridotetrakis(triethylphosphite)nickel was found to be dependent on both structural and electronic factors. The reactivity of $CoH[P(OEt)_3]_4$ compared with the isoelectronic $[NiH[P(OEt)_3]_4]^+$ is much lower [1185].

Catalytic cycloaddition of allene, $CH_2{=}C{=}CH_2$, has been effected with a variety of nickel(O) phosphorus systems; for example, $Ni(COD)(Bu^n_3P)_2$ (COD = cycloocta-1, 5-diene), $Ni(C_2H_4)(Ph_3P)_2$. These metal-assisted cycloadditions, which lead to the cyclotrimer (*5*), cyclotetramer (*6*), or cyclopentamer (*7*), proceed by multistep reaction paths over Ni(O) templates [965].

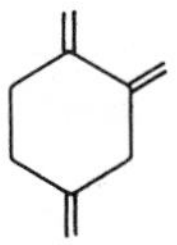

5

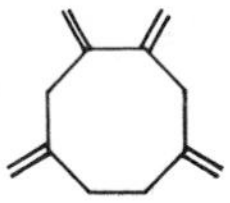

6

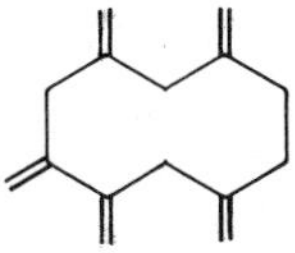

7

F. Solid-State and Thermal Decomposition Reactions

Solid-state reactions in the carbonylchlorobis(triphenylphosphine)iridium(I)-hydrogen chloride system have been investigated. The rate-controlling step of the reaction takes place at the phase surface with an activation energy of 37 kJ/mole

$$[\mathrm{IrCl(CO)(Ph_3P)_2}]_s + \mathrm{HCl}_g \rightleftharpoons \underset{\text{trans chloro isomer}}{\mathrm{Ir(H)Cl_2(CO)(Ph_3P)_2}} \quad (202)$$

takes place at the phase surface with an activation energy of 37 kJ/mole and an entropy change of $-184\ \mathrm{K^{-1}\ mole^{-1}}$. The reverse reaction has an activation energy of 51 kJ/mole and entropy change of $-176\ \mathrm{K^{-1}\ mole^{-1}}$. The reduced time plots $t/t_{0.5}$ versus α $t_{0.5}$ = time when $\alpha = 0.5$, α = proportion of the complex reacted) suggest a two-dimensional process [65].

The thermal decomposition of neophyl(tributylphosphine)copper(I) or -silver(I) is a free radical process for which the silver and copper complexes probably share a common decomposition mechanism [1256]. Kinetic data for the thermal decomposition of *trans*-tricarbonylhalobis(triphenylphosphine)manganese, $\mathrm{MnX(CO)_3(Ph_3P)_2}$ (X = Cl, Br, I), have been obtained by means of IR spectroscopy. One mechanism proposed for these reactions [1125] is

$$trans\text{-}\mathrm{MnX(CO)_3(Ph_3P)_2} \xrightarrow{k_1} \mathrm{MnX(CO)_3Ph_3P} + \mathrm{Ph_3P} \quad (203)$$

$$trans\text{-}\mathrm{MnX(CO)_3(Ph_3P)_2} + \mathrm{MnX(CO)_3Ph_3P} \xrightarrow{\text{fast}} cis\text{-}\mathrm{MnX(CO)_4Ph_3P} + \mathrm{MnX(CO)_2(Ph_3P)_2}\ (?) \quad (204)$$

Brown, Puddephat, and Upton [194] studied the kinetics of the thermal elimination of ethane from iodotrimethylbis(dimethylphenylphosphine)platinum(IV)

$$\mathrm{PtIMe_3(Me_2PhP)_2} \longrightarrow \mathrm{C_2H_6} + trans\text{-}\mathrm{PtIMe(Me_2PhP)_2} \quad (205)$$

Using selectively deuterated derivatives (*8*), they were able to show that the reductive elimination is an intramolecular process [194].

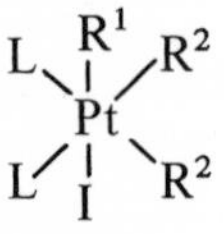

(*8*)

L = $\mathrm{Me_2PhP}$	% $\mathrm{C_2H_6}$	% $\mathrm{MeCD_3}$	% $\mathrm{C_2D_6}$
$R^1 = R^2 = \mathrm{Me}$	100	0	0
$R^1 = \mathrm{CD_3}$, $R^2 = \mathrm{Me}$	40	60	0
$R^1 = \mathrm{Me}$, $R^2 = \mathrm{CD_3}$	0	68	32
$R^1 = R^2 = \mathrm{CD_3}$	0	0	100

The thermal decomposition of di-*n*-butyl-bis(triphenylphosphine)platinum(II) proceeds via an intramolecular process that involves (1) dissociation of one molecule of triphenylphosphine, (2) elimination of 1-butene and its concomitant transfer to the vacant coordination site of the platinum hydride species, formed simultaneously and (3) final reductive elimination of *n*-butane from an intermediate with both hydride and butyl bonded to platinum [1255].

G. *Kinetics of Some Reactions on Coordinated Ligands*

The reversible formation of alkoxy-carbonyl platinum complexes according to (206) has been investigated kinetically for a series of alcohols ROH (R = Me, Et, Pr^n, Pr^i, $PhCH_2$, $CH_2(OMe)CH_2$. A mechanism proposed for these reactions

$$[PtCl(CO)(Ph_3P)_2]^+ + ROH \underset{k_{-1}}{\overset{k_1}{\rightleftharpoons}} PtCl(Ph_3P)_2COOR + H^+ \qquad (206)$$

involves a nucleophilic attack of the alcohol on the carbon atom of the coordinated CO ligand [219].

Cationic metal carbonyl complexes of manganese and rhenium react with primary aliphatic amines with formation of carbamoyl complexes

$$[trans\text{-}M(CO)_4L_2]^+ + 2\ RNH_2 \rightleftharpoons trans\text{-}M(CO)_3L_2(CONHR) + RNH_3^+ \qquad (207)$$

The rates of these base catalyzed reactions follow the rate law $v = k_3[Mo(CO)_4\text{-}L_2^+][RNH_2]^2$, with k_3 decreasing in the order $MePh_2P > Me_2PhP > Ph_3P$ for various ligands L [40].

The rate of hydrogen deuterium exchange of the cyclopentadienyl protons in $\pi C_5H_5Mn(CO_2)R_3P$ is significantly dependent on the nature of the groups R attached to phosphorus. The carbonyl stretching vibration and the logarithm of the rate constant correlate with the Taft inductive constants σ_R^* [509].

X. Spectroscopic Investigations

A. *NMR and NQR Spectroscopy*

In order to represent the literature on this topic systematically, it has become necessary to subdivide this heading along functional lines.

1. NMR SHIFTS AND COUPLING CONSTANTS

Under this subheading reference is made to those publications that, for a related series of complexes, collect and interpret shift values and coupling constants.

NMR Shifts. For complexes of the type *trans*-$PdCl_2L_2$ (L = tertiary phosphine) [832] and *trans*-$RhCl(CO)L_2$ [L = tertiary phosphine, $P(OMe)_3$] [826] approximately linear correlations have been found between the changes($\Delta\delta$) in the ^{31}P NMR shifts on coordination and the shifts of the corresponding free ligand:

$$trans\text{-}PdCl_2L_2{:}\Delta\delta = -(0.085 \pm 0.047)\delta - (40.04 \pm 1.40)$$

$$trans\text{-}RhCl(CO)L_2{:}\Delta\delta = -(0.355 \pm 0.026)\delta - (35.88 \pm 0.69)$$

Similar relationships have been found for the mutually trans phosphines (P^1 and P^3) and the phosphine in cis position within complexes of the type mer-MCl_3L_3 (M = Rh, Ir, L = Et_3P, Bu^n_3P, Me_2PhP, Et_2PhP, $MePh_2P$) [830] and for complexes of the type *trans*- or *cis*-$RuCl_2(CO)_2L_2$ (L = various teriary phosphines, some of which contain *t*-butyl groups) [503]. In the complexes *cis*- and *trans*-$PdCl_2L_2$, for a particular ligand L (L = teriary phosphines $R_nPh_{3\text{-}n}P$, n = 0, 1, 2, 3, R = Me, Et, Pr, Bu), the chemical shift δ ^{31}P of the cis isomer is downfield from that of the trans isomer [542].

Heteronuclear double resonance experiments have been used to determine ^{31}P (and ^{195}Pt) chemical shifts as well as signs and magnitudes of coupling constants in an extended series (44 complexes) of six- and four-coordinate *trans*-bis(triethylphosphine)platinum complexes. Nearly all six-coordinate complexes studied had negative ^{31}P chemical shifts, whereas four-coordinate had positive values of δ ^{31}P (i.e., to high field) [35].

Mathieu, Lenzi, and Poilblanc studied the variation of ^{31}P NMR shifts with *n* for the complexes $M(CO)_{m\text{-}n}(PA_3)_n$ (M = Ni, m = 4; M = Cr, Mo, W, m = 6; $1<n<4$; A = F, Cl, OMe, SMe, NMe_2, Me, Et). They found a great similarity in the behavior of ligands differing with respect to steric crowding [e.g., PF_3, Me_3P, $P(SMe)_3$, $P(OCH_2)_3CEt$] and rationalized the characteristic shifts in terms of $\sigma + \pi$ charge transfer interactions [863]. In the complexes mer-$IrCl_3(Ph_nR_{3\text{-}n}P)_3$(R = Me, Et, Pr, Bu, n = 0, 1, 2) all phosphorus shifts, with the exception of the dimethylphenylphosphine compound, are upfield from the uncomplexed ligand; the shift of the phosphorus atom trans to a chlorine (P_2) is in all cases upfield from that of the mutually trans P atoms ($P_{1,3}$) (see Table VI) [541]. Analogous observations have been made for the meridional isomers of $M(CO)_3L_3$ (M = Cr, Mo, W, L = phosphine ligands) [863].

Inductive effects or electronegativities of the anions X in complexes $\pi C_5H_5NiXPh_3P$ (X = Me, Et, Pr^n, Pr^i, Bu^n, Bu^s, CH_2SiMe_3, CH_2CMe_3, $CHPhSiMe_3$, Ph, CH_2Ph, $SnPh_3$, $PbPh_3$, $SnCl_3$, NCO, NCS, CN, NO_2, Cl, Br, I) gave only an unsatisfactory correlation with the ^{31}P chemical shifts [1173, 1175]. The correlation between the wavelength of the stronger electronic (*d-d*) absorption and δ ^{31}P, which could be expected according to the theory of Buckingham and Stephens [204], did not hold for about half the compounds studied [1173].

The theory of contact and pseudocontact shift in the NMR spectra of tetrahedral complexes of Ni(II) has been developed. The pseudocontact shift may in some cases be quite large and has to be taken into account in any analysis of NMR shifts [807].

By analysis of the isotropic ^1H NMR shifts for the *o*-, *m*-, *p*-phenyl, and methyl protons in complexes $MX_n(MePh_2P)_{4\text{-}n}$ (X = Br, I, M = Co, n = 0, 1, 2; M = Ni, n =

TABLE VI. ^{31}P NMR data for mer-$IrCl_3L_3$ compounds (chemical shift in ppm relative to H_3PO_4) [541]

L	$\delta^{(a)}$ Free Ligand	$\delta P_{1,3}{}^{(b)}$	$\delta P_2{}^{(c)}$	$\Delta P_{1,3}{}^{(d)}$	$\Delta P_2{}^{(d)}$
Bu_3P	+32.3	+37.6	+39.9	+ 5.3	+ 7.6
Pr_3P	+33	+37.4	+40.0	+ 4.4	+ 7.0
Et_3P	+20.4	+33.0	+35.5	+12.6	+15.1
Bu_2PhP	+26.2	+37.9	+41.4	+11.7	+15.2
Pr_2PhP	+27.7	+37.8	+41.1	+10.1	+13.4
Et_2PhP	+16	+33.3	+36.7	+17.3	+20.7
Me_2PhP	+46	+40.4	+49.5	− 5.6	+ 3.5
$BuPh_2P$	+17.1	+29.5	+35.0	+12.4	+17.9
$PrPh_2P$	+17.6	+29.9	+35.1	+12.3	+17.5
$EtPh_2P$	+12	+27.5	+31.6	+15.5	+19.6
$MePh_2P$	+26	+41.8	+55.2	+15.8	+29.2

(*a*) S. O. Grim, W. McFarlane, and E. F. Davidoff, *J. Org. Chem.*, *32*, 781 (1967).
(*b*) $P_{1,3}$ are mutually trans phosphines.
(*c*) P_2 is trans to chlorine.
(*d*) Δ (coordination chemical shift) = δ complex − δ free ligand.

1, 2) La Mar, Sherman, and Fuchs [768] were able to assign values to these complexes for the relative covalencies by using the *p*-H spin density calculated by McConnels relationship [421, 766] ($\rho = A2S/Q$, where $Q = -63$ MHz, $2S$ = total number of unpaired spins, and A = hyperfine coupling constant). The covalencies increase slightly on reducing the divalent species (see Table VII). On the basis

TABLE VII. Hyperfine coupling constants, spin densities, and relative covalencies for $MBr_n(MePh_2P)_{4-n}$ (taken from [768]).

Complex	d^m	$\mu^{(a)}$	(A) $10^{-5\,(b)}$	$\rho \cdot 10^3$	2 S	Relative Covalency$^{(c)}$
$CoBr_2(MePh_2P)_2$	d^7	4.39	0.96	4.6	3	1.00
$CoBr(MePh_2P)_3$	d^8	3.03	1.06	3.4	2	1.10
$Co(MePh_2P)_4$	d^9	1.71	1.39	2.2	1	1.45
$NiBr_2(MePh_2P)_2$	d^8	3.35$^{(d)}$	1.91	6.1	2	1.98
$NiBr(MePh_2P)_3$	d^9	1.7	2.21	3.5	1	2.25

(*a*) Magnetic moment, in BM at 25°C.
(*b*) Hyperfine coupling constant in Hertz.
(*c*) Relative covalency with the Co(II) complex covalency, arbitrarily set at 1.00.
(*d*) Taken from G. N. La Mar and E. O. Sherman, *J. Amer. Chem. Soc.*, *92*, 2691 (1970).

of their results, La Mar, Sherman, and Fuchs conclude that ($d_\pi \to d_\pi$) bonding is present in the zerovalent complex, whereas for the divalent species ($d_\pi \to d_\pi$) bonding is unimportant; the spin transfer occurs mainly by σ-π nonorthogonality.

^{19}F NMR shifts of meta- and parafluorophenylgold(I) complexes $AuPh^fL$ (Ph^f = *m*-, *p*-fluorophenyl group) have been used to rank the phosphorus ligands L according to their π-acceptor properties [938]: $P(OPh)_3 > (C_6F_5)Ph_2P > Ph_3P \approx EtPh_2P \approx (p\text{-tolyl})_3P > Me_2PhP \approx Et_3P \approx Bu_3P$.

^{59}Co NMR chemical shifts in complexes $CoX(NO)_2PZ_3$ [X = Cl, Br, I, PZ_3 = various phosphorus(III) ligands] are dependent on the covalency of the Co-X bond (increasing in the order Cl < Br < I). The influence of the ligands PZ_3 on ^{59}Co chemical shifts is mainly determined by π-interactions between π_P-acceptor orbitals and metal orbitals of appropriate symmetry [1018].

^{183}W chemical shifts are slightly different for *cis*- and *trans*-bis(tributylphosphine)-tetracarbonyltungsten (cis isomer δ ^{183}W = 1965 ppm, trans isomer δ ^{183}W = 2021 ppm relative to WF_6) [537].

^{195}Pt chemical shifts have been determined in a series of 44 *trans*-bis(triethylphosphine)platinum complexes. No simple additivity rules exist. The ^{195}Pt chemical shifts for different isomers of the same compound are different; for example, (values in ppm) [35]:

$PtBr_2HI(Et_3P)_2$,	trans-trans isomer	+ 743
	cis-trans isomer	+ 313
$PtCl_2(Et_3P)_2$,	trans isomer	+ 946
	cis isomer	+ 414
$PtHCl(Et_3P)_2$,	trans isomer	0 (reference)

Coupling constants $J(^{31}P-{}^1H)$. Hildenbrand and Dreeskamp made a comparative determination of various coupling constants $J(^{31}P-{}^1H)$ and $J(^{31}P-{}^{13}C)$ in free and complexed phosphorus ligands. The changes of the values on complexation are explained in terms of differing hybridization [593].

$^3J(^{31}P-{}^1H)$ is found to be proportional to δ $^1H(OCH_2)$ in $M(CO)_xL_y$ complexes of polycyclic phosporus ligands L, $P(OCH_2)_3P$, $P(OCH)_3(CH_2)_3$ (x = 3, 4, y = 1, 2, M = Fe; x = 1, 2, 3, y = 3, 2, 1, M = Ni; x = 4, 5, y = 2, 1, M = Cr, Mo, W) [30]. The absolute signs for $^3J(^{31}P-{}^1H)$ nuclear spin coupling constants in the methyl platinum complexes $PtMe(NO_3)Ph_2PCH_2CH_2PPh_2$, $PtMeXC_6H_4PPh_2(CH{=}CH_2)$ (X = Cl, Me) are positive for phosphorus atoms cis to the methyl group and negative for a phosphorus atom in the trans position to the methyl group [101].

The two methyl groups of *cis*-$\pi C_5H_5MoX(CO)_2(Me_2PhP)$ are nonequivalent in the ^1H NMR spectrum but equivalent for the trans isomer [867].

Orthophenyl hydrogen interactions in bis[1,2-bis(diphenylphosphino)ethane] metal complexes $[IrX_2(dppe)_2]Cl$ (X = O, S, Se), *cis*-$M(CO)_2(dppe)_2$ (M = Cr, Mo,

W), $[IrC_3S_2(dppe)_2]Cl$, and *cis*-$RuMe(dppe)_2Cl$ have been observed. All these complexes show a 1:2:1 triplet of intensity 4 protons or, for the last two compounds, two triplets each of intensity 2 protons. The presence of such triplets in the ^1H NMR spectra of six-coordinate bis(dppe) complexes may be taken as evidence for a cis geometry, whereas their absence does not indicate trans geometry [506].

The interpretation of small coupling constants in the complexes *trans*-$\pi C_5H_5WX(CO)_2R_3P$ as primarily "through space" in origin has been criticized [677].

Metal-Phosphorus Coupling Constants. Metal-phosphorus coupling constants have been determined for various metals and phosphorus(III) ligands in a variety of complexes.

$^{199}Hg - ^{31}P$ spin-spin coupling constants, obtained for a series of 20 phosphine and phosphite complexes of mercury(II), HgX_2L_2 [X = Cl, Br, I, SCN, CN, L = Bu_3P, Oc_3P, $P(OEt)_3$], $(HgX_2L)_2$ [X = Cl, I, L = $P(OEt)_3$], $HgClP(O)(OEt)_2$, and $Hg[P(O)(OEt)_2]_2$, are quite sensitive to ligand electronegativity. They ranged from 3960 Hz [for $HgI_2(Oc_3P)_2$] to 12830 Hz [for $HgClP(O)(OEt)_2$] (Oc = *n*-octyl) [1276].

Because of a rapid intermolecular ligand exchange, no cadmium-phosphorus coupling is observed. However, on cooling, the satellites due to $^{111}Cd-^{31}P$ and $^{113}Cd - ^{31}P$ coupling are clearly detectable. AT −90°C, $^1J(^{111}Cd - ^{31}P)$ and $^1J(^{113}Cd - ^{31}P)$ coupling was observed for all complexes studied (see Table VIII) [822].

TABLE VIII. ^{31}P chemical shifts at 36.43 MHz with respect to 85% H_3PO_4 and cadmium-phosphorus coupling constants at −90°C in CH_2Cl_2 for the complexes CdI_2L_2 (taken from [822])

L	^{31}P [ppm] ± 0.1	$^1J(^{111}Cd-^{31}P)$ [Hz] ± 0.2	$^1J(^{113}Cd-^{31}P)$ [Hz] ± 0.2
Me_2PhP	+34.1	1236.3	1293.3
$MePh_2P$	+23.9	1123.8	1175.6
Et_3P	+10.2	1308.3	1368.4
Et_2PhP	+11.3	1238.8	1295.7
$EtPh_2P$	+10.6	1143.3	1195.7

The ^{31}P NMR spectra of the complexes $Fe(CO)_4Et_nPh_{3-n}P$ (n = 1, 2, 3) have been studied with complete decoupling of the protons. Several pairs of weak satellites due to $^1J(^{57}Fe-^{31}P)$ and $J(^{13}C-^{31}P)$ were observed. $^1J(^{57}Fe-^{31}P)$ increases as n decreases. Some values of $^1J(^{57}Fe-^{31}P)$ are reported: 25.9 ± 0.4 Hz for $Fe(CO)_4Et_3P$, 26.5 ± 0.4 Hz for $Fe(CO)_4Et_2PhP$, and 27.4 ± 0.4 Hz for $Fe(CO)_4EtPh_2P$ [824].

The first measurement of $^{1}J(^{187}Os-^{31}P)$ has been reported by Gill, Mann, Masters, and Shaw for the complexes $OsH_4(Et_2PhP)_3$ (166 Hz) and *cis*-$OsCl_2(CO)_2(Bu^tPr_2^nP)_2$ (149.8 Hz) [502].

The determination of $^{103}Rh-^{31}P$ spin coupling constants in the complexes mer-$RhCl_3(R_nPh_{3-n}P)_3$ (R = alkyl, n = 1, 2, 3) shows that $^{1}J(^{103}Rh-^{31}P)$ is always larger for a phosphine trans to another phosphorus as compared to $^{1}J(^{103}Rh-^{31}P)$ for one trans to halogen [540].

$^{183}W-^{31}P$ nuclear spin coupling constants have been reported for quite an extended series of compounds [464, 675, 859]. For the complexes $W(CO)_5X_3P$ (X = F, Cl, Br, I) a linear correlation of $^{1}J(^{183}W-^{31}P)$ with Sanderson electronegativity values exists [464]. $^{183}W-^{31}P$ coupling may be observed through three bonds in the complex $W(CO)_5Ph_2PCH_2PPh_2$ [675].

For the coupling constants $^{1}J(^{195}Pt-^{31}P)$ (for P trans to X) in the four-coordinate complexes, $[PtX(Et_3P)_3]ClO_4$, and *cis*-$PtX_2(Bu_3P)_2$, the magnitudes are varying in the order $ONO_2 > N_3$, NCO, NCS $> NO_2 >$ CN, and usually Cl $>$ Br $>$ I. The effects induced by anionic ligands trans to P are larger than those by anionic ligands cis to the phosphine ligands [861]. In six-coordinate complexes containing two trans positioned triethylphosphine ligands $^{1}J(^{195}Pt-^{31}P)$ is larger and the more hydride ligands and the fewer halide ligands are coordinated to platinum. The signs of $^{1}J(^{195}Pt-^{31}P)$ are positive [35]. Cis and trans influence series of L on $|^{1}J(^{195}Pt-^{31}P)|$ in the complexes *cis*- and *trans*-$PtCl_2LR_3P$ and $[PtClL(R_3P)_2]Cl$ (R_3P = Bu_3P, L = $P(OMe)_3$, $P(OPh)_3$, Et_3P, Ph_3P, Me_2PhP, Bu_3^nP) have been obtained [28]. The correlation of $^{1}J(^{195}Pt-^{31}P)$ with bond lengths supports the interpretation of the coupling constants in terms of the *s*-orbital bond order (see Table IX [860].

Phosphorus-Phosphorus Coupling Constants. Factors influencing sign and magnitude of $^{31}P-^{31}P$ spin-spin coupling constants in coordination compounds have been discussed on the basis of the Pople-Santry molecular orbital model [128, 945, 956]. The compounds studied were *cis*- and *trans*-$M(CO)_4L_2$ (M = Cr, Mo, W), *cis*- and *trans*-PdX_2L_2 (X = Cl, I), and *trans*-$Fe(CO)_3L_2$ [L = Me_3P, $P(NMe_2)_3$, $P(OMe)_3$] [128] [L = PH_3, R_3P, $P(NR_2)_3$, $P(OR)_3$, PF_3] [956], *cis*-$M(CO)_4L_2$ [L = $RP(CF_3)_2$, Q_2PCF_3 (M = Cr, Q = H, M = Mo, Q = Cl, Br, H, R = Cl, Br, I, H)] [945]. In addition, $^{2}J(^{31}P-^{31}P)$ has been obtained for a series of Rh(III) complexes, mer-$RhCl_3L_3$ (L = Bu_3P, Pr_3^iP, Me_3P, Et_3P, Bu_2PhP, Pr_2PhP, Et_2PhP, Me_2PhP, $BuPh_2P$, $PrPh_2P$, $EtPh_2P$, $MePh_2P$) [540] and the iridium complexes $IrCl_2X(Me_2PhP)(Ph_2PCH_2CH_2PPh_2)$ (X = NO_3, NO_2, NCO, N_3, NCS, Cl, Br, I, pyridine, CO) and related (partial tertiary arsine substituted) compounds [829].

The coupling constants $^{2}J(^{31}P-^{31}P)$ are usually much smaller for cis complexes of second- and third-row transition metals than for the trans isomers, and this has been used widely as a stereochemical criterion [28, 796, 798, 1000]. Care has to be taken when stereochemical assignments are made on the basis of NMRP-P coupling constants for first-row transition metal complexes. Johnson, Lynden-Bell, and

TABLE IX. Coupling constants $^1J(^{195}\text{Pt}-^{31}\text{P})$ and bond length l (Pt-P) in platinum(II) complexes (taken from [860])

Complex	$^1J(^{195}\text{Pt}-^{31}\text{P})$ [Hz]	Complex	l (Pt–P) [Å]
cis-PtCl$_2$(PhNC)Et$_3$P	3049	cis-PtCl$_2$(EtNC)Et$_2$PhP	2.244(8)
cis-PtCl$_2$(Et$_3$P)$_2$	3520	cis-PtCl$_2$(Me$_3$P)$_2$	2.247(7)
trans-PtHCl(Et$_3$P)$_2$	2723	trans-PtHCl(EtPh$_2$P)$_2$	2.267(8)
trans-PtCl$_2$Et$_3$P[$\overline{\text{CN(Ph)CH}_2\text{CH}_2\text{N}}$Ph]	2440	trans-PtCl$_2$Et$_3$P[$\overline{\text{CN(Ph)CH}_2\text{CH}_2\text{N}}$Ph]	2.29(1)
trans-PtCl$_2$(Et$_3$P)$_2$	2400	trans-PtCl$_2$(Et$_3$P)$_2$	2.289(18)
trans-PtBr$_2$(Bu$_3$P)$_2$	2327	trans-PtBr$_2$(Et$_3$P)$_2$	2.315(4)
sym-trans-Pt$_2$Cl$_4$(Bu$_3$P)$_2$	3800	sym-trans-Pt$_2$Cl$_4$(Pr$_3$P)$_2$	2.230(9)

Nixon found that the P-P coupling constant for *cis*-$Cr(CO)_4(PF_3)_2$ (77.0 Hz) is larger than for the trans isomer (34.0 Hz) [661]. Additional problems are caused by the fact that $^2J(^{31}P-^{31}P)$(cis) and $^2J(^{31}P-^{31}P)$(trans) may have opposite signs [128, 661, 956]. It is interesting to note that for complexes $M(CO)_4L_2$ (M = Cr, Mo, W) $^2J(^{31}P-^{31}P)$(trans) increases positively with the electronegativity of the groups attached to phosphorus, whereas $^2J(^{31}P-^{31}P)$(cis) increases in magnitude but the sign is negative [956].

The dependence of $^1J(^{31}P-^{183}W)$ on σ- or π-bonding properties has been discussed. On the basis of the calculation of bond angles and hybridization at phosphorus in a series of complexes $Cr(CO)_5L$ [L = $P(OPh)_3$, Ph_3P] and *trans*-$Cr(CO)_4P(OPh)_3$, *cis*-$PtCl_2(Me_3P)_2$, and *trans*-$PtBr_2(Et_3P)_2$, it is believed that variation in *s*-character is not the determining factor for the variation of $J(^{31}P-^{183}W)$ in some of the corresponding tungsten complexes [543].

2. ANALYSIS OF NMR SPECTRA

To keep within the limits of this review article only a selected number of papers believed to be representative and relevant to this topic are discussed.

Mann [823] reported the analysis of NMR spectra of the type $[AMX_n]_2$ in order to explain a novel five-line pattern in the ^1H NMR spectra of *trans*-bis(di-*t*-butylphosphine) metal complexes [182]. The ^1H NMR spectrum of the platinum-bonded methyl groups in *cis*-$PtMe_2(Me_3P)_2$ may not be treated as an $[AX_3]_2$ spin type but rather as $[AR_3X_9]_2$. An analysis of this spin system has been given [519]. A partial analysis of the ^1H NMR spectra of the planar $[AX_n]_4$ spin systems represented by the complexes *trans*-$[IrCl_2(Me_3P)_4][NO_3]$ and $[Pt(Me_3P)_4][BF_4]_2$ [518] follows the procedure of Finer and Harris [460] and assumes that coupling between protons on different phosphines is zero. The ^1H NMR spectra of the unique phosphine ligands (i.e., in cis position to the other two trans phosphines) in the complexes mer-$MoOX_2(Me_2PhP)_3$ and mer-$OsX_2L(Me_2PhP)_3$ (X = anionic, L = neutral ligand) show anomalies that may be explained by near-zero shift differences for the ^{31}P nuclei. Calculations of $[AM_3S_3]_2BX$ spectra, ^{31}P spectra at 24.29 MHz and ^1H spectra at 220 MHz confirmed this explanation [797].

The deuterium decoupled ^1H NMR spectrum of di-*n*-butyl-2,2-d_2-bis(triphenylphosphine)platinum(II) has been analyzed [1254] by using a slightly modified version of the program LAOCN3 [165].

A theoretical treatment of NMR spectra of the $[AX_n]_3$ spin system has been presented for C_{3v} symmetry and $|J_{AX}| \gg |J_{AA}| \gg |J_{AX'}| > J_{XX} = 0$ in order to explain the ^{19}F NMR spectra of fac trisubstituted fluorophosphine molybdenum complexes [576]. Only a partial analysis of the ^{19}F NMR spectra of NiL_4 complexes [L = PF_3, CF_3PF_2, $(CF_3)_2PF$, CCl_3PF_2, CH_2ClPF_2, Me_2NPF_2, $C_6H_4O_2PF$] could be effected because of insufficient resolution of all lines and the large number of wave functions involved. The values $^1J(^{31}P-^{19}F)$, $^3J(^{31}P-^{19}F)$, $^2J(^{31}P-^{31}P)$, and $^4J(^{31}P-^{19}F)$ were obtained [799, 800].

The analysis of the ^{19}F NMR spectra of tris(trifluoro)phosphine metal complexes with approximate C_{3v} symmetry, $\pi C_3H_5Rh(PF_3)_3$, $\pi C_6H_6Cr(PF_3)_3$, and Rh(NO)-$(PF_3)_3$, representing spin systems of the type $[AX_3]_3$ [576], revealed the values for $^1J(^{31}P-^{19}F)$, $^2J(^{31}P-^{31}P)$, $^3J(^{31}P-^{19}F)$, and $^4J(^{19}F-^{19}F)$ [942].

3. DYNAMIC PROCESSES STUDIED BY NMR SPECTROSCOPY

Stereochemistry and stereochemical nonrigidity in transition metal hydrido complexes, detected by NMR spectroscopy (and other spectroscopic methods), have been discussed in detail by Jesson [652].

The temperature-dependence of the ^1H NMR hydride spectrum of the complexes MH_2L_4 (M = Fe, Ru, L = $P(OMe)_3$, $P(OEt)_3$, $PhP(OEt)_2$, $P(OPr^i)_3$, $PhP(OPr^i)_2$, $P(OCH_2)_3CEt$, $Ph_2PCH_2CH_2PPh_2$, Me_2PhP, Et_2PhP, $MePh_2P$, Ph_2POMe indicates a polytopic rearrangement including vibrational populations of a pseudotetrahedral ML_4 state with hydride tunneling of face and edge positions [872]. Hydrogen atom traverse of faces in the ML_4 tetrahedral substructure has been proposed as a mechanism to explain the observed stereochemical nonrigidity of the five-coordinate complexes MHL_4^{n-} (n = 0 for M = Co, Rh, Ir; n = 1 for M = Ru, Os) [874]. Seven-coordinate trihydridorhenium complexes $[ReH_3(Ph_2PCH_2CH_2PPh_2)_2$, $ReH_3(Ph_2PCH_2CH_2PPh_2)(Ph_3P)_2$, $ReH_3(Ph_2AsCH_2CH_2AsPh_2)(Ph_3P)_2$, including $ReH_3(Ph_2AsCH_2CH_2AsPh_2)_2]$ show fluxional behavior indicated by the 300-MHz ^1H NMR spectra. The stereochemical nonrigidity may be accounted for by vibrational deformation of a given isomer to a polytopal isomer in which hydrogen atom exchange may then take place [507].

Rapid intramolecular rearrangements have been observed in a series of pentacoordinate transition metal compounds; for example, $IrR(COD)L_2$ [1102], (R = H, Me, COD = 1,5-cyclooctadiene, L = Me_2PhP, $MePh_2P$, Ph_3P), $Fe(CO)_3(Me_2PCH_2CH_2PMe_2)$ [13], and $C_4H_6Fe(CO)_x(PF_3)_{3-x}$ (C_4H_6 = 1,3-butadiene, x = 0, 1) [1240]. The mechanisms discussed for the rearrangements include the Berry "pseudo rotation" [124] and Ugi's "turnstile rotation" [1208].

The spin-free ⇌ spin-paired equilibrium has been studied

$$\text{tetrahedral } NiX_2L_2 \underset{k_s}{\overset{k_t}{\rightleftharpoons}} \text{planar } NiX_2L_2 \qquad (208)$$

X = Cl, Br, I, L = $MePh_2P$, $EtPh_2P$, Pr^nPh_2P, Bu^nPh_2P [767]; X = Cl, Br, L = R_3P, R_2PhP, RPh_2P, R = cyclopropyl, cyclohexyl, and, in one case, Bu^t, for example, $NiBr_2(Bu^tPh_2P)_2$ [1006]. A complex dependence of the rate of isomerization on L and X exists [767], whereas steric factors seem to be relatively unimportant in affecting the thermodynamics of the structural equilibrium [1006]. Thermodynamic data for the cis-trans equilibria of the compounds $PdCl_2(Me_2PhP)_2$ and $PdCl_2(MePh_2P)_2$ in 11 different solvents have been obtained by variable temperature ^1H NMR spectroscopy. In most solvents the cis isomer is thermodynamically more stable. Only the trans isomers could be detected in benzene, toluene, and 1,1,2-trichloroethane [1010].

An intramolecular exchange process which involves B_3H_8-Cu bond breakage has been proposed to explain the observed differing temperature-dependence of triphenylphosphine and triphenylphosphite complexes of type (*1*) [83].

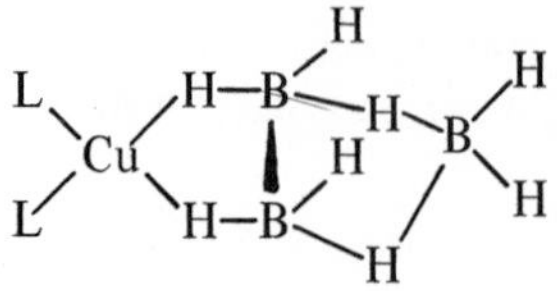

(*1*)

[L = Ph_3P, $P(OPh)_3$]. "Thermal" decoupling and apparent rapid intramolecular exchange in $Cu(Ph_3P)_2BH_4$ is thought to be responsible for the temperature-dependence of the ^1H NMR spectrum [523].

By studying the temperature-dependence of the ^{13}C and ^1H NMR spectra of $[Os(CO)(NO)(C_2H_4)(Ph_3P)_2]PF_6$ direct experimental evidence for a mode of rotation of the olefin molecule, indicated in (*2*), has been obtained for the first time [659].

Temperature-dependent ^{31}P or ^{19}F NMR revealed the existence of different rotamers with respect to the M-P axis for complexes *trans*-$MCl(CO)(RBu^t_2P)_2$ (M = Rh, Ir, R = Me, Et, Pr^n) [833] or *trans*-$NiCl_2(Bu^t_2PF)_2$ [1132].

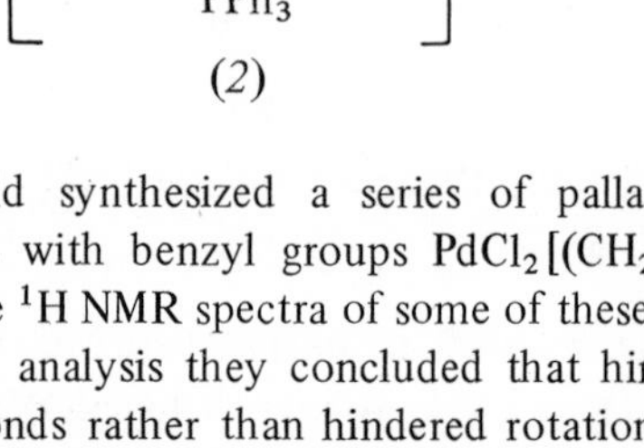

(*2*)

Nelson and Redfield synthesized a series of palladium chloride complexes containing phosphines with benzyl groups $PdCl_2[(CH_2Ph)_nPh_{3-n}P]_2$ (n = 1, 2, 3) and investigated the ^1H NMR spectra of some of these complexes at 60, 100 and 220 MHz. From their analysis they concluded that hindered rotation about the phosphorus carbon bonds rather than hindered rotation about the Pd-P bond has to be taken into account to explain the observed spectral pattern [932].

Rapid intra- and intermolecular $SnCl_2$ transfer has been found for the complexes (*3*) and (*4*) with added $SnCl_2$ by analysis of the temperature-dependence of the ^1H NMR spectra [1044].

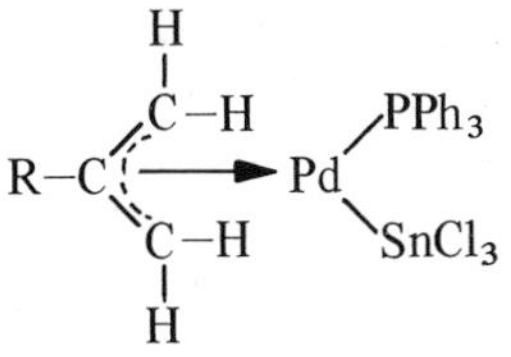

3

(R = Me, H),

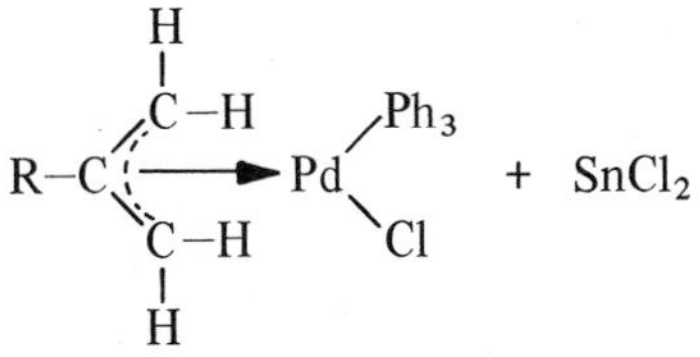

4

In an elegant experiment Mawby, Wright, and Ewing showed that no rapid intramolecular rearrangement through a planar transition state is occurring for $Co(CO)(NO)(Me_2PhP)_2$ (which should lead to change in line shape and multiplicity of the Me signals) [868].

The applications of "dynamic nuclear magnetic resonance," that is, temperature-dependent NMR spectra (calculation of rate constants and Arrhenius parameters) have been reviewed by Sergeev [1098].

Intermolecular dynamic processes have also been studied by NMR spectroscopy. From observed isotropic shifts of the CH_2N and MeN protons of $[CoBr_3Ph_3P]$-$[Bu^n_3NMe]$ and $[CoBr_3Ph_3P]$ $[Bu^n_4N]$ information about ion pair geometry and association constants in various solvents of different dielectric constants has been obtained [781, 1162].

Decoupling of phosphorus-hydrogen spin-spin interactions on addition of paramagnetic complexing reagents of the first-row transition metals may be observed in the ^1H and ^{31}P NMR spectra of trialkyl phosphites (and phosphates). It is interesting to note that decoupling of phosphorus-hydrogen is always observed at a lower concentration ratio of paramagnetic complexing reagent:phosphite in the phosphorus spectrum than in the proton spectrum. Paramagnetic line broadening in the ^{31}P spectra before true chemical decoupling may be responsible for this effect [437, 438].

4. CARBON-13 NMR SPECTROSCOPY

Improved techniques, that is, improved signal-to-noise ratio by pulsed Fourier spectroscopy, noise decoupling in compounds containing C-H bonds, etc., have made ^{13}C NMR spectroscopy a useful technique for the investigation of inorganic and/or metal organic compounds.

Farnell, Randall, and Rosenberg have made a functional group survey for the series $\pi C_5H_5Fe(CO)_2$ X (X = Cl, Br, I, CN, COMe, Me) and some other related species [e.g. $Mo(CO)_5P(OPr^i)_3$]. They found that all the $\delta^{31}C$ values varied significantly with X [448].

^{13}C shifts and coupling constants have been reported for a series of substitution products of transition metal carbonyls by various authors [175, 483, 503, 825]. Correlations between ^{13}C shift values and Cotton-Kraihanzel force constants are satisfactory; their interpretation, however, is not in all cases straightforward [483, 503, 825].

The changes in $^1J(^{183}W-^{13}C)$ are thought to be a measure of the trans effect of the ligands L [$Et_3P < P(OMe)_3 < CO$] in substitution products of $W(CO)_6$ [825].

The coupling constant $^1J(^{95}Mo-^{13}C)$ in $Mo(CO)_6$ [68 H_2] has been measured for the first time [825]. The phosphorus-carbon monoxide coupling constants vary in a similar way to phosphorus-phosphorus coupling constants [175]. ^{13}C NMR parameters, including the $^{195}Pt-^{13}C$ coupling constant [594 ± 5 H_2] have been reported for *cis*-$PtMe_2(Me_2PhP)_2$ and for the corresponding arsenic analog [263].

The suggestion that ^{13}C NMR spectroscopy could be a more applicable technique than ^1H NMR spectroscopy in determining stereochemistry of bis(phosphine) metal derivatives [835] was questioned by Axelson and Holloway [55] because $^2J(^{31}P-^1H)$ is often of comparable magnitude to $^1J(^{31}P-^{13}C)$ and the ^1H and ^{13}C spectra are of different types. ^{13}C contact shift studies of triphenylphosphine complexed with nickel(II) acetylacetonate reveal that the skew conformation of the phenyl ring (the lone pair orbital on the P atom is nearly parallel to the plane of the phenyl ring) may account for the σ-spin delocalization [909].

^{13}C isotropic shifts have been used as a probe of mechanisms of spin transmission in metal complexes. Spin density is delocalized into the σ- and π-frameworks of triphenylphosphine complexed to nickel(II) or cobalt(II) acetylacetonates. Chemically induced "phosphorus decoupling" could be observed in the ^{13}C NMR spectrum of the two triphenylphosphine complexes [404].

5. NQR SPECTROSCOPY

NQR spectroscopy has rarely been employed in the study of structural problems in transition metal complexes of trivalent phosphorus compounds [140, 479, 1067].

Bishop, Cullen, and Gerry [140] reported ^{35}Cl NQR data for the complexes $Ni(PCl_3)_4$ (27.98 MHz at 77°K and 27.47 MHz at 295°K) and fac-$Mo(CO)_3(PCl_3)_3$ (28.22 MHz at 77°K). The ^{35}Cl NQR frequencies in a series of compounds MCl_2L_2 (L = Bu^n_3P, Et_2PhP, Bu^n_2PhP, Et_3P, M = Ni, Pd, Pt) have been measured (Table X). The ^{35}Cl signal at 15.99 MHz in $NiCl_2(Bu_3P)_2$ was the first to be reported in a nickel(II) complex of type $NiCl_2L_2$ [479].

TABLE X. ^{35}Cl NQR frequencies and the estimated charge on Cl atoms of some d^8 complexes of the type MCl_2L_2 (taken from [479])

	ν_Q [MHz]	(±0.01 MHz)	Charge on Cl Atoms $-(1-\sigma)$
$NiCl_2(Bu^n_3P)_2$	15.99 m		- 0.71
$PdCl_2(Bu^n_3P)_2$	18.37 w	18.50 w	- 0.66
	18.58 w	18.63 w	
cis-$PtCl_2(Bu^n_3P)_2$	17.73 w	17.79 w	- 0.68
	17.89 w	17.96 vw	
cis-$PtCl_2(Et_2PhP)_2$	17.82 m	17.99 w	- 0.68
cis-$PtCl_2(Bu^n_2PhP)_2$	18.33 w		- 0.67
trans-$PtCl_2(Et_3P)_2$	20.99 w		- 0.62
trans-$PtCl_2(Bu^n_3P)_2$	20.90 w	21.04 w	- 0.62
	(doublet)	21.08 w	
trans-$PtCl_2(Bu^n_2PhP)_2$	21.32 w	21.64 w	- 0.61
$[PdCl_2(Bu^n_3P)]_2$	12.77 m	19.50 m	
$[PtCl_2(Pr^n_3P)]_2$	15.46 m	22.36 m	

B. ESR Spectroscopy

The complexes formed by reaction of tertiary phosphines R_3P (R_3 = Et_3, Me_2Ph, Pr_3, Bu_3, $EtPh_2$, Ph_3) with vanadyl chloride, $VOCl_2$, have been studied [585, 586]. The increasing line width with increasing molecular volume of the penta-coordinate complexes $VOCl_2(R_3P)_2$ can be described by an equation derived by Kivelson et al. [50, 1264].

Low-spin nitrosyl chromium(I) complexes ($3d^5$, S = ½) with ligands containing tricoordinate phosphorus of types (*5*) and (*6*) have been investigated by ESR. [L = $P(OMe)_3$, $P(OEt)_3$, $P(OPr^i)_3$, $P(OPr^n)_3$, $P(OBu^n)_3$, $PhP(OEt)_2$ C_6H_4ClP-

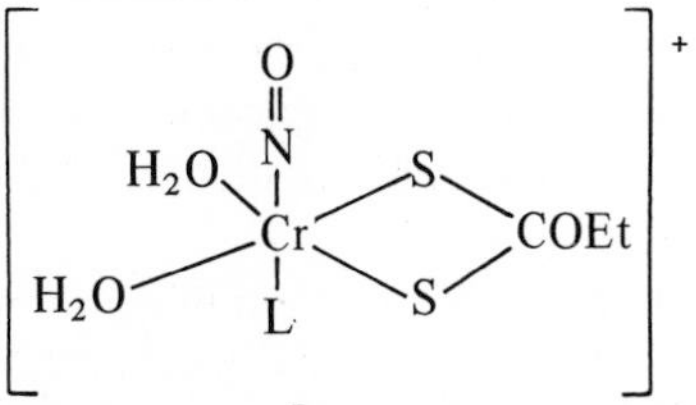

5

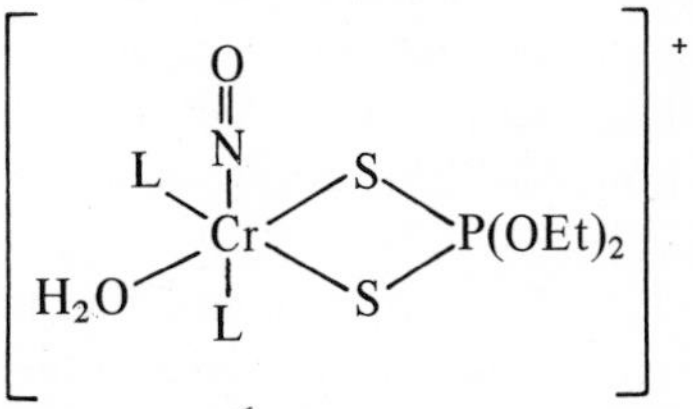

6

$(OEt)_2$, $(EtS)_3P$, $EtP(SEt)_2$, $Bu^nP(SEt)_2$, Et_2PSPh]. The isotropic nature of $A^{31}P$ (additional hyperfine coupling constant) of the axial ligands in (*5*) and (*6*) provides evidence that the unpaired electron enters the *s*-orbital of the phosphorus atom, whereas delocalization to vacant *d*-orbitals on phosphorus is not at all significant [486].

Both donor and acceptor properties of the ligand L influence the spin densities at phosphorus in the dinitrosyliron complexes, $Fe(NO)_2XL$ (L = primary, secondary, and tertiary phosphines, aminophosphines, phosphites, phosphonites, halophosphines, etc.) [730, 1073]. The isotropic g-values may be used to rank some of the ligands according to their π-acceptor abilities: $P(OEt)_3 > PhP(OEt)_2 > Et_3P > PhP(NEt_2)_2 > P(NEt_2)_3$. The ESR spectra of these complexes clearly indicate that the aminophosphines are bound through phosphorus to the transition metal [1073].

From detailed study of the ESR and electronic spectra and the bulk magnetism of diarylbis(diethylphenylphosphine)cobalt(II) complexes the relative ligand field strengths of the aryl groups could be arranged as follows: mesityl ~ 2-methylnaphthyl > C_6F_5 ~ C_6Cl_5 with aryl > phosphine. From mesityl to C_6Cl_5 the symmetry changes from D_{2h} to D_{4h}. Approximate configuration energies have been calculated from the powder ESR spectral data [120].

The ^{31}P hyperfine interaction constant (AHFS) in the complexes formed by the reaction between phosphorus(III) donor molecules [$P(OPh)_3$, $ClP(OPh)_2$, Cl_2POPh, PCl_3] and low-spin cobalt(II) bis(dithio-α-diketonate) (proposed structure *7*) shows a significant dependence on the nature of the groups attached to phosphorus [1200].

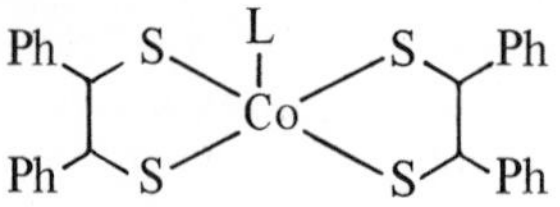

L = PCl_3, Cl_2POPh, $ClP(OPh)_2$, $P(OPh)_3$

7

ESR studies of a series of five-coordinate low-spin cobalt(II) complexes, CoX_2L_3 (X = Cl, Br, I, L = Me_2NPF_2, $(Me_2N)_2PF$), revealed that their structure is intermediate between square pyramidal and trigonal bipyramidal. As indicated by the total cobalt hyperfine interaction for these complexes, the complex $CoI_2[(Me_2N)_2PF]_3$ possesses a higher degree of covalency than $CoI_2[(Me_2N)PF_2]_3$ [953].

No hyperfine structure could be resolved in the EPR spectrum of osmium-doped trichlorotris(triphenylphosphine)rhodium(III). The g-tensor has the principal values 3.32, 1.44, and 0.32 at 77°K, which have been used to calculate the electronic structure. The effective value of the spin orbit coupling is 2650 ± 100 cm^{-1} the orbital reduction factor, 0.80 ± 0.01 [594].

In a series of bis(trifluoroacetylacetonato)copper(II) phosphine or phosphite complexes, $Cu(tfac)_2R_3P$ (R_3 = Bu^n_3, Et_3, Ph_3, Me_2Ph, $MePh_2$, $(PhO)_3$) the phosphorus hyperfine coupling constant decreases with increasing basicity of the ligand in the order $Et_3P > Bu^n_3P > Me_2PhP \gg MePh_2P \gg Ph_3P > P(OPh)_3$ [769]. Copper(II) complexes of the type $Cu(hfac)_2L$ (hfac = hexafluoroacetylacetonate, L =

Ph_3P, $MePh_2P$, Me_2PhP, Et_3P, Bu^n_3P, Ph_3As) have been studied by ESR spectroscopy [1285, 1286]. In $Cu(hfac)_2Me_2PhP$ the four-line spectrum of copper(II) ($I = \frac{3}{2}$) is split into eight lines by coupling with the phosphorus. No correlation exists between the copper hyperfine splitting and the isotropic shifts/linewidths of ^{19}F NMR resonance for the complexes studied [1285].

C. Infrared and Raman Spectroscopy

Intensity measurements of CO or CN stretching vibrations in the Raman and IR spectra of transition metal complexes have been employed to evaluate the relative importance of σ- and π-interactions in the metal-ligand bonds [36, 624, 720, 721, 1066]. The intensity of IR bands due to Fe-CO bonding modes in complexes $Fe(CO)_4L$ and $Fe(CO)_3L_2$ ($L = Me_3P$, Me_3As, Me_3Sb) has been discussed as well [134]. Darensbourg measured the integrated IR intensities of the dinitrogen and carbonyl stretching vibrations in complexes of the type $OsX_2(Y_2)(R_2PhP)_3$ ($Y_2 = N_2$, CO, R = Me, Et, OMe, X = Cl, Br). With these data, dipole moment derivatives involving OsN or OsC and NN or CO stretching motions could be calculated. From the effective group dipole moment derivatives it was concluded that carbon monoxide is a better σ donor and π acceptor than N_2 [357].

Tolman proposed a method for determining the electron donor-acceptor properties of phosphorus(III) ligands based on the A_1 carbonyl stretching frequency of $Ni(CO)_3L$ in CH_2Cl_2. An expression, $\nu(CO)A_1 = 2056.1 + \sum_{i=1}^{3} \chi_i$ [cm^{-1}], has been given to calculate the $\nu(CO)A_1$ of 70 compounds of the type $Ni(CO)_3L$ studied with substituent parameters χ_i [1181].

By use of the Cotton-Kraihanzel force field a more exact direct method has been applied for calculating the CO stretching force constants of the substituted metal carbonyls *cis*-$M(CO)_4L_2$ (M = Cr, Mo, W, L = phosphine or amine) [383, 384].

Isotopic substitution, ^{58}Ni and ^{62}Ni, ^{104}Pd and ^{110}Pd, in nickel(II) and palladium(II) halide phosphine complexes MX_2L_2 [M = Ni, Pd, X = Cl, Br, I, L_2 = two monodentate or one bidentate phosphorus(III) donor ligand], produced reliable assignments of the metal-phosphorus and metal-halogen stretching bands [1111, 1207, 1238]. According to Boormann and Carty [157], the Ni-X stretching frequencies are useful in distinguishing the cis and trans configurations of square planar nickel(II) phosphine halide complexes. For nickel bromide complexes it has been shown that the Ni-Br and Ni-P stretching vibrations of the tetrahedral complexes occur generally at lower wave numbers than those of the planar complexes [238].

Cis and trans influences of the ligands L in complexes of the type *cis*-$[PtX(dppe)L]^+ClO_4^-$ (X = Cl, Br, L = Me_3CNC, $P(OPh)_3$, $P(OMe)_3$, Et_3P) and *cis*-$[PtX(dmpe)L]^+ClO_4^-$ (X = Cl, Br, L = Me_3CNC, $P(OPh)_3$, Et_3P) could be derived from metal-chlorine stretching frequencies [61]. The trans effect of triphenylphosphine and triphenylstibine has been evaluated by a comparative study of the IR spectra of *trans*-[$PtCl(NH_3)_2L$]NO_3 (L = Ph_3P, Ph_3Sb) in the 500 to 200 cm^{-1} region. The ν(Pt-Cl) values are believed to demonstrate that the σ-electron donating and the π-electron accepting capacities of Ph_3Sb are much lower than those of Ph_3P [491].

Infrared spectral studies in the region of 700 to 300 cm^{-1} of a series of octahedral metal carbonyls, $M(CO)_5L$ (L = various amines and phosphorus(III) donor ligands, M = Cr, Mo, W), revealed that the metal-carbon stretching frequency is lowest for "hard" bases and correlates well with the complex reactivity involving rate-determining dissociation of carbon monoxide [195].

For the complexes $\pi C_5H_5Mo(CO)_2RCOL$ (R = Me, Et, σC_3H_5, CH_2Ph, L = Ph_3P, $P(OCH_2)_3CMe$), $\pi C_5H_5MoCl(CO)_2Ph_3P$, and $\pi C_5H_5WH(CO)_2P(OPh)_3$ assignments for the metal-carbon-oxygen deformation, metal-carbon stretching metal-ring and metal-phosphorus stretching vibrations have been proposed [331]. The protonation of arenedicarbonyltriphenylphosphinechromium by trifluoroacetic acid may be followed by monitoring the IR absorption bands and their intensities in the CO stretching frequency region (2100 to 1800 cm^{-1}) [789]:

$$Cr(CO)_2ArL + H^+ \rightleftharpoons [Cr(CO)_2HArL]^+ \quad (209)$$

(L = Ph_3P, Ar = C_6H_6, MeC_6H_5, $MeOC_6H_5$, $Me_3C_6H_3$, $Me_2NC_6H_5$).

Restricted rotation within the phosphorus-containing ligand in complexes $\pi C_3H_4XCo(CO)_2Y_3P$ (X = H, 1-Me, 2-Me, 2-Cl, Y = OPh, OMe, Ph) causes conformational isomerism indicated by the multiplicity of the ν(CO) absorption in the IR spectra [299].

Phosphine derivatives of cyclopentadienyltricarbonylmanganese interact reversibly with Lewis acids (including $SnCl_4$) in CH_2Cl_2 solution:

$$\pi C_5H_5Mn(CO)_2L^1 + n\ SnCl_4 \rightleftharpoons \pi C_5H_5Mn(CO)_2L^1\ n\ SnCl_4 \quad (210)$$

$$\pi C_5H_5Mn(CO)L^2{}_2 + n\ Y \rightleftharpoons \pi C_5H_5Mn(CO)L^2{}_2\ n\ Y \quad (211)$$

(Y = $SnCl_4$, $SbCl_3$, $HgCl_2$, $GeCl_4$, L^1 = CO, $P(OPh)_3$, Ph_3P, Ph_3As, Ph_3Sb, $(p\text{-tolyl})_3P$, $P(C_6H_4OMe)_3$, Pr^i_3P, Cy_3P, L_2 = $(Ph_3P)_2$, dppm, dppe, dppp). These equilibria have been studied by observing the changes of the IR spectra in the CO stretching frequency region [508].

The metal-phosphorus stretching frequencies of the complexes *cis*-$MnI(CO)_4Me_3P$, *cis*-$MnI(CO)_4PH_3$, *cis*-$MnI(CO)_3(PH_3)_2$, $Ni(CO)_3PH_3$, $Fe(CO)_4PH_3$, *cis*-$FeI_2(CO)_3$-PH_3, and $FeI_2(CO)_2(PH_3)_2$ could be assigned by careful Raman and IR spectroscopic studies [135]. An axial position for the substituent Ph_3P has been proposed for $MnCo(CO)_8Ph_3P$ on the basis of a detailed analysis of the IR spectra in the CO stretching frequency region, including the unsubstituted parent molecules $MCo(CO)_9$ (M = Mn, Re) [1064].

By using partly deuterated and substituted cyclopentadienyl derivatives assignments could be made of the low-frequency vibrations of the π-cyclopentadienyl-metal and metal-carbonyl moieties in the cyclopentadienylmetal carbonyls $\pi C_5H_5Mn(CO)_3$, $\pi C_5H_5MoX(CO)_3$, $\pi C_5H_5FeX(CO)_2$ (X = halide), and their phosphine substitution derivatives [971].

According to Raman and IR spectroscopic studies of $Fe(CO)(PF_3)_4$ in the solid state at low temperature, only one isomer exists, whereas at room temperature two isomers of C_{3v} and C_{2v} symmetry are present. By a comparative study of liquid $Fe(CO)(PF_3)_4$ and $CoH(PF_3)_4$ (C_{3v} symmetry) it was possible to obtain the difference in enthalpy between the two iron complexes (+ 0.52 ± 0.05 kcal/mole) [136].

Although not a transition metal complex, the single-crystal Raman studies on trigonal bipyramidal $InCl_3(Me_3P)_2$ should be mentioned; they afforded the assignment of ν_s(InP) (135 cm^{-1}) [968].

The high-pressure reaction of molecular hydrogen with $Ru(CO)_4Ph_3P$ [and $Ru(CO)_5$] has been followed by IR spectroscopy in the CO stretching frequency region. A reversible equilibrium seems to exist between $RuH_2(CO)_3Ph_3P$ and $Ru(CO)_4Ph_3P$ and hydrogen [1260].

T_d symmetry for $KCo(PF_3)_4$ and C_{3v} symmetry for $CoH(PF_3)_4$ have been established by Raman and IR spectroscopy [97].

Force constants have been calculated on the basis of a nonrigorous model for the complexes $[Co(CO)_3L]_2$, $[Co(CO)_3L]_2Hg$ (L = Et_3P, Bu^n_3P, Et_2PhP, Ph_2PH, $MePh_2P$) and their implications discussed [837]. Loutellier and Bigorgne [791] reported symmetry coordinates, the F- and G-matrices, and the equations for symmetry force constants for the complexes of the type $Ni(CO)_{4-n}(PA_3)_n$ (A = F, OMe, Me, n = 1 - 4). On the basis of these results they assigned the Raman and IR bands [792] and calculated the symmetry force constants [793].

Vibrational assignments and force constant calculations have also been given for tetrakis(trifluorophosphine)nickel, -palladium, and -platinum [422], and tetrakis-(trichlorophosphine)nickel [423]. Cyvin and Müller calculated the mean amplitudes of vibration for tetrakis(trifluorophosphine)nickel(O) [350].

Bands in the far infrared region of some tetrakis(bicyclic phosphite) complexes of nickel(O), palladium(O), platinum(O), copper(I), and silver(I) could be assigned to metal-phosphorus stretching vibrations [676].

It has been demonstrated by a laser Raman study that coordination of a phosphine donor ligand with the Pd atom in palladium allyls does not cause conversion of the π-allyl group to the σ-bonded form but that cleavage of the Pd-Cl bridge occurs [772].

Although $PtHCl(Ph_3P)_2$ in different solvents is present only in the trans form, there are three different crystalline modifications with quite different multiplicities, intensities, and positions of the IR bands in the region 2400 to 2000 cm^{-1} [ν(Pt-H)] and 900 to 800 cm^{-1} [δ(Pt-H) [305].

Far infrared studies in the range 600 to 120 cm^{-1} of complexes $PdX(N\text{-}C)L_2$

obtained by bridge-splitting reactions between Ph_3P or Et_3P and palladium(II) halogeno bridged complexes $[PdX(N\text{-}C)]_2$ [X = Cl, Br, (N-C) = azobenzene-2-C,N, *N,N*-dimethylbenzylamine-2-C,N, 2-methoxy-3-*N,N*-dimethylaminopropyl] established that the ν(PdCl) (range 300 to 160 cm^{-1}) for compounds with a trans alkyl group are generally lower than the corresponding ones for a trans aryl group [334]. Vibrational spectra of trimethylphosphine (and trimethylarsine) complexes AuXL, $[Au(Me_3P)_2]^+$, $[MX_3L]^-$, $M_2X_4L_2$, MX_2L_2 (cis and trans), $[M_2X_2L_4]^{2+}$, $[MXL_3]^+$, $[ML_4]^{2+}$ (M = Pd, Pt, X = Cl, Br, I) have been reported and assignments suggested in some cases [416]. The IR and Raman spectra of the mercury(II) cations, $[HgXMe_3P]^+$ (X = Cl, Br, I, CN, Me), have been measured and assignments made for all stretching and CPC deformation vibrations of the $HgXPC_3$ skeleton [517].

D. *ESCA and UV Photoelectron Spectroscopy*

Trifluorophosphine complexes of iron, chromium, nickel, platinum, and rhodium have been studied by UV photoelectron spectroscopy [528, 597, 913, 941] ([913] includes the complex $\pi C_5H_5Mn(CO)_2PF_3$). The first ionization bands are generally attributed to metal molecular orbitals of largely *d* character. From comparison of the PE spectra of the complexes and free ligands it was concluded that the bonding in the PF_3 ligand itself is unchanged on coordination to the metal [528, 597]. This was interpreted in terms of π-backbonding, compensating for charge migration by the σ-donation through the "lone pair" of the ligand PF_3 [528, 597]. The phosphorus lone pair energies vary from PF_3 to $Pt(PF_3)_3$ in the series $PF_3 < Cr(PF_3)_6 < Fe(PF_3)_5 < Ni(PF_3)_4 < RhH(PF_3)_4 < Pt(PF_3)_3$. This ranking reflects the order of σ-donation of PF_3 in these complexes [941].

As for trifluorophosphine complexes, the binding energy of the electrons on the triphenylphosphine ligands in their nickel, palladium, and cadmium complexes is similar to that of the free ligand. The opposing effects of the σ-bonding (P → metal donation) and π-bonding (metal → P donation) result in a zero or almost zero charge drift from the ligands to the coordination center [141].

The X-ray photoelectron spectra are believed to be consistent with a conformational isomerism that involves change in the formal oxidation state of the nitrosyl ligand in the complexes $CoCl_2(NO)L_2$ (L = Me_2PhP, $MePh_2P$) [1001].

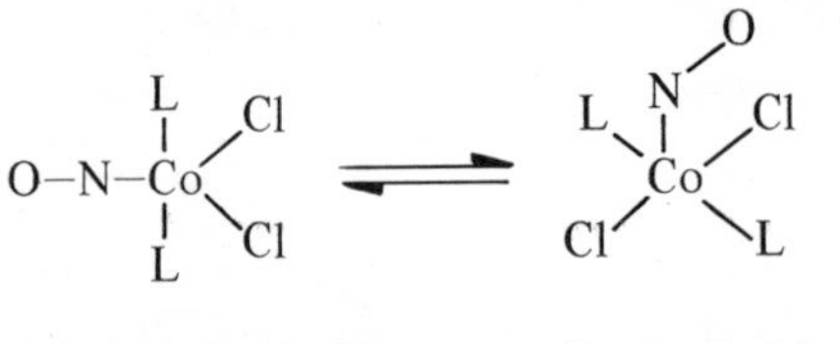

(212)

Linear O–N–Co Co(I)–NO^+ isomer   Bent O–N–Co Co(III)–NO^- isomer

Holsboer, Beck, and Bartunik [606] found for triphenylphosphineiridium complexes $[IrX(NO)(Ph_3P)_2]^{n+}$ [X = CO, Ph_3P, Cl_2, HCl for n = 0; X = $(Ph_3P)H$, (CO)Cl, (CO)I for n = 1] that the electron density (measurable by the N 1s binding energies) on the NO^+ is not necessarily smaller than on the NO^- group. These authors also reported Ir $4f_{5/2}$ and Ir $4f_{3/2}$ core binding energies for a series of IrX(CO)-$(Ph_3P)_2$ type complexes (X e.g. Cl, Br, N_3, SCN, NO, $C(CN)_3$, N_4CCF_3). The almost unchanged charge distribution around the iridium atom on variation of X is compensated by the carbonyl ligand. The tetracyanoethylene, fumaronitrile and acrylonitrile adducts of $IrCl(CO)(Ph_3P)_2$ were investigated as well by X-ray photoelectron spectroscopy [605].

For an extended series of rhodium(I) and rhodium(III) complexes the binding energies of the Rh $3d_{5/2}$, N 1s, Cl $2p_{3/2}$ and in some cases S $2p_{3/2}$ electrons have been determined. As indicated by the binding energies of Rh $3d_{5/2}$ levels for the series $RhX(Ph_3P)_3$, the ability of the substituents X to withdraw electron density from rhodium decreases in the order NO > I > Br > Cl [930].

N 1s electrons of the dinitrogen ligand in a dinitrogen complex of iron, $FeH_2(N_2)$-$(Ph_3P)_3$, gave only a single broad peak of half width 2.4 eV, which is indicative of two unresolved peaks [471].

E. Mössbauer Spectroscopy

The approximate invariance of $|\Delta E_Q|$, the ^{57}Fe quadrupole splitting, in pentacoordinate phosphine (and arsine) derivatives of $Fe(CO)_5$ indicates that replacement of CO may be regarded only as a perturbation of the electronic structure of $Fe(CO)_5$. The sign of the electric field gradient (EFG) at the iron nucleus and the (estimated) asymmetry parameters η for $Fe(CO)_5$, $Fe(CO)_5Ph_3P$, $Fe(CO)_3(Ph_3P)_2$, $Fe_2(CO)_8$dppe, and $Fe(CO)_3$dppe have been obtained by examination of the Mössbauer quadrupole interaction in an applied magnetic field. The sign of ΔE_Q and η may be used as diagnostic of a lack of axial symmetry in derivatives of $Fe(CO)_5$ [293]. The ^{57}Fe quadrupole splitting at room temperature in *trans*-$FeCl_2(depe)_2$ is + 1.29mm/s [67]. The effect of phosphorus(III) ligands on the Mössbauer parameters (isomer shifts and quadrupole splitting) of an extended range of complexes, $Fe(CO)_4L$, $Fe(CO)_3L_2$ [171, 232], $FeX_2(CO)_2L_2$, $FeX_2(CO)_3L$ [X = Cl, Br, I for $FeX_2(CO)_2L_2$, X = Br, I for $FeX_2(CO)_3L$, L = various phosphorus(III) donor ligands] [69], has been studied and the trends in isomer shift values interpreted in terms of variable σ- and π-interactions in the Fe-P bond. In the disubstituted series $Fe(CO)_3L_2$, δ(Fe), the isomer shift of ^{57}Fe, has been found to vary with L as follows: $Ph_3Sb > Ph_3As > Ph_3P$ [171].

Mössbauer spectra of low-spin iron(II) phosphine complexes, for example, *trans*-$FeX_2(depe)_2$ [X_2 = Cl_2, Br_2, I_2, $(NCO)_2$, $(NCS)_2$, $(N_3)_2$, $SnCl_3(Cl)$, H(Cl), H(I)], $Na_3[Fe(CN)_5Ph_3P]$, $Fe(NCS_2)L$ [L = tris(*o*-diphenylphosphinophenyl)phosphine], have been reported. An inverse correlation exists between the partial centre shift (PCS) values (calculated relative to stainless steel) and the spectrochemical ranking of the ligands [70].

Further iron complexes studied by ^{57}Fe Mössbauer spectroscopy are πC_5H_5FeX-

(dppe) (X = H, Cl, $SnCl_3$, $SnBr_3$, SnI_3, $SnMe_3$, Me), $[\pi C_5H_5Fe(CO)dppe]^+$ [870], $Fe(NO)_2(L\text{-}L)$ [L-L = $Ph_2PC{=}CPPh_2(CF_2)_n$, n = 2, 3, 4], $FeX(NO)_2L$ (X = Br, I, $Fe(CO)(NO)_2L$ [L = Ph_3P, $MePh_2P$, Ph_3As, $P(OPh)_3$] [339].

Mössbauer centre shifts indicated that the relative bonding properties ($\sigma + \pi$) of the ligands L in the complexes *trans*-$[MHL(depe)_2]^+$ BPh_4^- (M = Fe, Ru, Os) increase in the order [71] MeCN < PhCN ~ N_2 < $P(OPh)_3$ ~ $P(OMe)_3$ < Me_3CNC < *p*-$MeOC_6H_4NC$ < CO. The dinitrogen complex has the most positive value for the ^{57}Fe quadrupole splitting but is one of the poorest ($\sigma + \pi$) ligands. The π-acceptor properties seem to be more important for N_2 than for carbon monoxide [68].

Satisfactory correlation between ^{193}Ir isomer shift values and carbonyl stretching frequencies for the molecular adducts of trans-bis(triphenylphosphine)chlorocarbonyliridium, $IrCl(CO)Ph_3P\cdot A$ (A = O_2, H_2, HCl, MeI, I_2, Cl_2) was found with the exception of the O_2 adduct [1261]. The ^{193}Ir Mössbauer data for the complexes $IrX(CO)(Ph_3P)_2$ (XX = Cl, Br, N_3, SCN, NO) are almost insensitive to variation of X [606, 1261].

The effect of the substitution of two carbonyl groups for the bidentate phosphine dppe on the ^{119}Sn quadrupole splittings is very small in complexes of the type $\pi C_5H_5FeL_2SnX_3$ [L_2 = $(CO)_2$, dppe, X = Cl, Br, I, Me] [870]. The presence of Sn(II) in the complexes $[CoCl(dppe)_2]SnX_3$ (X = Cl, Br) could be proved by ^{119}Sn Mössbauer spectroscopy [1128]. The isomer shifts obtained for these complexes are similar to those found for Sn(II) in $KSnCl_3\cdot H_2O$ and $SnCl_2\cdot 2H_2O$ [452].

F. Electronic Spectroscopy

In this chapter only those papers that are mainly concerned with electronic spectra and their interpretation are discussed. A rigorous and perhaps subjective selection was necessary in order to stay within the limits of this review.

The *d-d*-transitions in the electronic spectra of the rhenium complexes $[ReX_5Ph_3P][Ph_3PH]$ (X = Cl, Br) of C_{4v} symmetry, $ReX_4(Ph_3P)_2$ (X = Cl, Br), and $[ReX_2(dppe)_2]X$ (X = Cl, Br) have been correlated. By using point charge model calculations an ordering of the monoelectronic energy levels of the chloro derivatives could be obtained [230].

The charge transfer bands of a series of octahedral complexes mer-MX_3L_3 [M = Re, Ru, Os, X = Cl, Br, L = Et_2PhP, Me_2PhP, Pr^n_3P, Bu^n_2PhP), fac-$OsCl_3(Bu^n_2PhP)_3$, and *trans*-MX_4L_2 (M = Re, Os, Ir, Pt, X = Cl, Br, L = Me_2PhP, Pr^n_3P, Et_3P] were reported by Leigh and Mingos [771]. The lowest energy charge transfer band was assigned to the π orbital(ligand) → t_{2g} ligand-metal transition and its position varies but little with different phosphine ligands. From crystal polarized spectra of some pseudotetrahedral Co(II) complexes, for example, $CoX_2(Ph_3P)_2$ (X = Br, I) the sequence of components derived from T_d symmetry was found to be $^4T_1$ (F): $B_1 < A_2 < B_2$ and $^4T_1$(P): $B_2 < A_2 < B_1$ [1187].

On the basis of a relative energy diagram of the metal-carbon bond in the planar complexes *trans*-$M(C{\equiv}CR)_2L_2$ (M = Ni, Pd, Pt, L = Et_3P, Me_3P, R = H, Me, CH_2F, $CH_2{=}CH$, $HC{\equiv}C$, Ph, $MeC{\equiv}C$), the lowest frequency charge transfer bands were assigned to transitions from bonding π orbitals of alkynyl groups [$b_{1g}(\pi)$ or $b_{3g}(\pi)$] to $b_{3u}(\pi^*)$ or $b_{1u}(\pi^*)$ orbitals which contain the metal *p* orbitals [850].

The observed polarization spectra of tetraphenylarsonium triiodotriphenylphosphinenickelate(II) between 380 and 77°K are not consistent with an axial symmetry (C_{3v}) of the $NiPI_3$ chromophore. This is ascribed to a low-symmetry component of the ligand field [127].

The electronic spectra (and structure) of the tetrahedrally distorted low-spin nickel(II) complex $NiI_2Ph_2P(CH_2)_2O(CH_2)_2O\text{-}(CH_2)_2PPh_2$ has been reported by Sacconi and Dapporto [1042].

Assignments of ligand field and charge transfer bands for planar tetracoordinate Ni(II) complexes $NiX_2(Me_3P)_2$ (X = CN, NO_2, NCS, Cl, Br, I, Me_3P) were given by Merle, Dartiguenave and Dartiguenave [883].

Ligand effects, band assignments, and temperature-dependence in the electronic spectra of low-spin pentacoordinate nickel(II) complexes of trimethylphosphine [881, 882, 884] and nickel, palladium, and platinum(II) complexes of monodentate and quadridentate phosphine ligands [370] have been discussed. From the temperature-dependent electronic spectrum of $Ni(NO_2)_2(Me_3P)_3$ evidence for a square pyramidal (C_{2v}) $\rightleftharpoons$ trigonal bipyramidal (C_{2v}) equilibrium was deduced [881]. The temperature-dependence of the ν_1 band ($^1A_1' \rightarrow {}^1E'$ transition) in a series of five-coordinate nickel, palladium, and platinum(II) complexes, $Ni(CN)_2(R_2PhP)_3$ (R = Me, OEt) and $[MX(\text{tripod ligand})]^+$ [X = unidentate anion, tripod ligand, e.g., $(o\text{-}Ph_2PC_6H_4)_3P$] could be caused by solute-solvent interaction or a dynamic Jahn-Teller effect in a regular trigonal bipyramidal structure [372].

Pressure effects (blue shifts of ν_1) on the ligand field spectra of five-coordinate nickel(II) and cobalt(II) complexes MX_2L_3 (X = Cl, Br, I, L = Me_3P, Ph_2PH, M = Co, Ni) have been observed [456].

High external pressures cause spin pairing in the solid state of the complex $NiBr_2(Ph_2PCH_2Ph)_2$; the green (high-spin) species is transformed to the red (low-spin) species [457].

Rhodium(III) and iridium(III) complexes of bis(diphenylphosphino)ethane $[MCl_2\text{-}(dppe)_2]Cl$ (M = Rh, Ir) show emissions that are assigned as metal localized phosphorescence ($^3E \rightarrow {}^1A_1$) [48]. Photoluminescence of some phosphine d^{10} metal complexes, $[M_rX_n(AR^1R^2L)_m]_p$ [A = phenyl or substituted phenyl, R^1, R^2 = alkyl, aryl, alkoxy, L = P, As, M = Ni(O), Pd(O), Pt(O), Cu(I), Ag(I), Zn(II), Cd(II), Hg(II), X = various anions, e.g., halides and pseudohalides, *m, r, n, p* = integers, 1 - 4, 1 or 2, 0 - 3, and 1 - 4, respectively], has been observed. In some cases the luminescent properties can be used for identification of complexes; for example,

$Pt(Ph_3P)_3$ shows a bright red and $Pt(Ph_3P)_4$ a bright orange luminescence [1288].

An approach to direct evaluation of π-bonding in metal carbonyls $Cr(CO)_6$ and $Cr(CO)_5L$ and $Cr(CO)_5$ (L = amines, phosphines) has been presented. By use of a simple MO diagram the electronic spectral data provide direct information about the relative importance of π-bonding in these complexes [319].

G. *Various Methods*

The Faraday effect (magneto-optical rotation) has been studied for various nickel(O) complexes; for example, $Ni[P(OR)_{3-x}F_x]_4$ (R = Et, Pr^n, Bu^n, x = 0, 1, 2) [1061], $Ni(R_{3-x}PF_x)_4$ (for R = OPh, x = 0, 1, 2, 3; for R = Ph, x = 0, 2) [888], $Ni(R_{3-x}PCl_x)_4$ (for R = OPh, x = 0, 1, 2, 3; for R = Ph, x = 1, 2) [1062], and $Ni(PX_3)_4$ (X = F, Cl, Br) [1060]. The actual magnetic rotation of the Ni-P bonds was used within a theoretical formalism to discriminate the σ- and π-contributions.

McCleverty et al. [361, 788, 804] studied the voltametric properties of phosphorus(III) donor complexes of chromium, molybdenum, and tungsten carbonyls [788, 804]. Bis(tertiary phosphine)tris(isonitrile) Co(I) complexes [361] have been investigated as well. The half-wave potentials, $E_{1/2}$, are dependent on the π-acceptor σ-donor properties of the phosphorus(III) ligands.

The mass spectra of monosubstituted group VIB carbonyls, $M(CO)_5L$ [M = Cr, Mo, W, L = $P(OPh)_3$, $Ph_2PC_6F_5$, Ph_3P, $P(pC_6H_4F)_3$, (m-tolyl)$_3$P, including nitrogen, arsenic, and antimony donor ligands], have been obtained and the fragmentation patterns, studied [154]. The composition of the cobalt carbonyl cluster compounds, $Co_4S(CO)_9(Me_2P)_2$ and $Co_3S_2(CO)_7Me_2P$, could be elucidated from the analysis of their mass spectra [927].

Appearence potential measurements on transition metal carbonyl complexes of tris(dimethylamino)phosphine (L) permitted the estimation of the average bond dissociation energies of the metal-carbonyl bond (Table XI) [1037].

TABLE XI. Bond energies of metal carbonyl complexes of $P(NMe_2)_3$ (L)

Compound	Bond Energy (eV) of the First M–CO Bond
$Cr(CO)_5L$	1.0
$Cr(CO)_4L_2$	3.0
$Mo(CO)_5L$	2.1
$Mo(CO)_4L_2$	4.3
$W(CO)_4L_2$	4.8
$Fe(CO)_4L$	0.4
$Fe(CO)_3L_2$	2.0

XI. Crystal and Molecular Structures and Electron Diffraction Studies

There is a considerably increasing number of papers on crystal and molecular structures of transition metal phosphine complexes. This may be because of the commercial availability of improved X-ray diffractometers and high-capacity computers. X-ray analysis no longer seems to be considered as a rather time-consuming physicochemical method but as a quick technique that provides direct structural information. In Table XII compounds are listed for which X-ray analysis has been performed. Some structural parameters, for example, space-group, metal-phosphorus bond distances, and bond angles are given. In order to stay within the limits of this review, only a few structures are discussed here in detail.

The crystal and molecular structures of the PH_3 complexes *cis*-$Cr(CO)_4L_2$, *cis*-$Cr(CO)_2L_4$, and fac-$Cr(CO)_3L_3$ (L = PH_3) have been studied [550, 622, 623]. The Cr-P bond distances in *cis*-$Cr(CO)_4(PH_3)_2$ [550] are shorter than the "normal single" bond distance of 2.58 [assuming radii of 1.48 Å for Cr(O) and 1.10 Å for P]. This may be attributed to dπ dπ-bonding or to the good σ-donor properties of the sterically small PH_3 molecule.

The complex $NiBr_3L_2{\cdot}0.5NiBr_2L_2{\cdot}C_6H_6$ (L = Me_2PhP) needs to be mentioned because it contains complex molecules of different coordination geometry at Ni in the same unit cell [1127].

The structure of eight-coordinate tetrahydridotetrakis(methyldiphenylphosphine)-molybdenum, $MoH_4(MePh_2P)_4$, has been determined. The coordination polyhedron is a distorted variant of a D_{2d}-$\bar{4}$ $2m$ dodecahedron (see Fig. 1). All four hydride

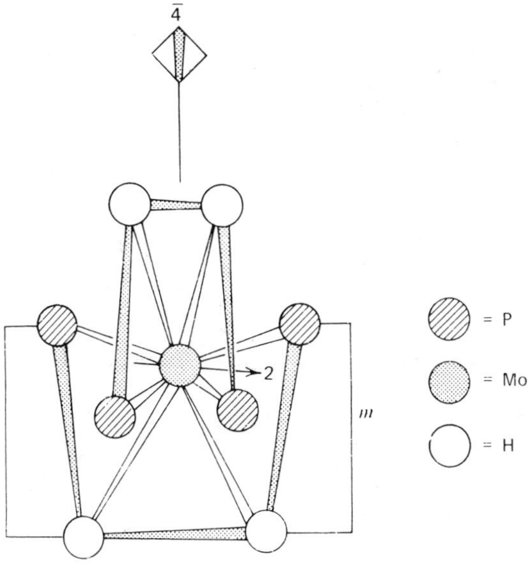

Fig. 1.

TABLE XII. Crystal and molecular structures of transition metal complexes containing phosphorus(III) ligands. The numbers in parentheses are the standard deviations in bond lengths or angles. When more than one metal-phosphorus distance are listed, the different d(M-P) values are indicated by P(1), P(2), etc., instead of d[M-P(1)], d[M-P(2)], etc. The symbol A-B denotes the A-B distance; ∡ABC the bond angle at B between A and C.

Ligand L	Complex	Space Group	Metal-P Distances d(M-P)[Å]	Comments	Refs.
PH_3	*cis*-$Cr(CO)_4L_2$	$Pna2_1$	2.349(2)	$Cr\text{-}C_{eq}$ 1.847(4), $Cr\text{-}C_{ax}$ 1.914(7)	550
	fac-$Cr(CO)_3L_3$	$P2_1/m$	2.346(4)	Cr-C 1.84(1)	623
	cis-$Cr(CO)_2L_4$	C2/c	P(1,2) 2.282(4) P(3,4) 2.338(4)	Cr-C 1.817(7), P(1) trans to P(2), P(3,4) trans to CO	622
	$FeI_2(CO)_2L_2$	$P2_1/a$	P(1) 2.24(2) P(2) 2.27(2)	Distorted octahedron, P(1) trans to P(2), I(1) cis to I(2), CO(1) cis to CO(2), ∡P(1)FeP(2) 173.2(7), ∡C(1)FeC(2) 99(4), ∡I(1)FeI(2) 93.0(3)	137
Me_3P	$[IrCl(CO)(C_3Ph_3)L_2]$ $BF_4 \cdot CH_2Cl_2$	$P2_12_12_1$	P(1) 2.360(5) P(2) 2.369(5)	Distorted octahedral, Ls in trans position, C_3Ph_3 monopositive bidentate propenylium-1,3-diyl group	1202
Et_3P	*trans*-$PdCl(C_{12}H_9N_2)L_2$	$P2_1/c$ ($C_{2h}5$)	av 2.306(5)	$C_{12}H_9N_2$ = σ-C bonded 2-(phenylazo)-phenyl group, Pd-C 1.994(15)	1243
	trans-$PdHClL_2$	$P2_1/c$	P(1) 2.310(4) P(2) 2.306(3)	Pd-Cl 2.427(5)	1077
	trans-$PtI(CMeNC_6H_4Cl)L_2$	$P2_1/c$	P(1) 2.312(3) P(2) 2.321(3)	Square planar	1233
	trans-$Pt(NO_2)_2L_2$	$Pca2_1$	2.32(1) 2.30(1)	Pt-N 2.11(2), 2.09(2)	527
	πC_5H_5CuL	$P2_1/m$	2.136(9)	Av Cu-C 2.238(26)	382
	$Hg[Fe(CO)_2(NO)L]_2$	Molecule $\bar{3}(S_6)$	2.223(3)	Hg-Fe 2.534(2)	1134

Pr^n_3P	*cis*-$LClPt(SEt)_2PtClL$	Pbca	2.267(7) 2.257(7)	PtS_2Pt bridge bent, dihedral angle 130	567
	α-$Pt_2Cl_2(SCN)_2L_2$	$P2_1/c$	2.244(4)	Pt-S 2.327(5), μ-thiocyanato groups, S trans to Cl, N trans to P	538
	β-$Pt_2Cl_2(SCN)_2L_2$	$P2_1/n$	2.262(4)	Pt-S 2.408(4), μ-thiocyanato groups, S trans to P, N trans to Cl	538
Pr^i_3P	$Ir(C_2H_4)_2Pr^i_2PC_3H_6L$	$P4_12_12$	mean 2.400	Pentacoordinate, internal metallated L, $Pr^i_2PC_3H_6$; contains four-membered ring IrPCC	981
Bu^n_3P	$\pi C_5H_5MoI(CO)_2L$	$P4_2/n$ C^4_{4h}	2.50(1)	Square-pyramid-type geometry, P trans to CO	449
	$RhCl_3P(OMe)_3L_2$	$P2_1/n$	P(1) 2.400 (L) P(2) 2.379 (L) 2.199 $P(OMe)_3$	Octahedral, Ls trans to one another, 3 Cl meridional	25
	trans-$Pt(C_6F_5S)_2L_2$	C2/c	2.329(8)	Approximately square planar, the Pt atom 0.28 Å out of PtS_2P_2 square	450
Me_2PhP	$Mn_2(CO)_9L$	C2/c	2.239	L axially attached, staggered $Mn(CO)_4$ units, Mn-Mn 2.904, Mn-C 1.76	764
	$ReCl_2(N_2Ph)L_3$		P(1) 2.446(4) P(2) 2.453(4) P(3) 2.397(4)	Octahedral, P(1) and P(2) mutually trans, P(3) cis to P(1,2), Re-Cl(1) 2.470(4), Cl(1) trans to P(3), Re-Cl(2) 2.448(4), Cl(2) trans to N_2Ph	415
	$MoCl_2(CO)_2L_3$	Pbca	P(1) 2.500(14) P(2) 2.528(15) P(3) 2.623(16)	C_s symmetry, seven-coordinate, P(2), P(3) in trans position, Mo-P(3) unusually long	865
	$MoCl_4L_3 \cdot EtOH$	$P2_1/c$	P(1) 2.572(3) P(2) 2.577(4) P(3) 2.583(4)	Seven-coordinate, capped octahedron, distorted, C_{3v} symmetry, capped face formed by the three P atoms, three equivalent Mo-Cl 2.448, one extra Mo-Cl 2.399(3)	839

TABLE XII. cont. Crystal and molecular structures of transition metal complexes containing phosphorus(III) ligands. The numbers in parentheses are the standard deviations in bond lengths or angles. When more than one metal-phosphorus distance are listed, the different d(M-P) values are indicated by P(1), P(2), etc., instead of d[M-P(1)], d[M-P(2)], etc. The symbol A-B denotes the A-B distance; ∡ABC the bond angle at B between A and C.

Ligand L	Complex	Space Group	Metal-P Distances d(M-P)[Å]	Comments	Refs.
	$MoBr_4L_3$	P_3^-	Av 2.580(7)	Seven-coordinate, capped octahedron, approximately C_{3v} symmetry, capped face formed by three P atoms, another by three Br atoms, the fourth Br occupies the unique capping position, Mo-Br 2.560(5), Mo-Br(unique) 2.425(7)	413
	$MoOCl_2L_3$ blue isomer	Pbca	P(1) 2.500(3) P(2) 2.541(3) P(3) 2.558(3)	Meridional octahedral, Cls cis to one another, P(2) trans to P(3), P(1) trans to Cl, Mo-O, 1.676(7), Mo-Cl(trans P) 2.464(3), Mo-Cl (trans 0) 2.551(3)	840, 841
	$MoOCl_2L_3$ green isomer (L = Et_2PhP)	$P2_1/c$	P(1) 2.521(5) P(2) 2.582(6) P(3) 2.556(6)	As above, Mo-O 1.803(11), Mo-Cl(trans P) 2.479(5), Mo-Cl(trans O) 2.426(6), distorted isomer	256
	$Fe_3(CO)_9L_3$	Pbca	P(1) 2.242(9) P(2) 2.232(9) P(3) 2.236(9)	Skeleton of $Fe_3(CO)_2$ retained, P(2), P(3) attached to Fe(2) and Fe(3) respectively, which are CO bridged; all Ps coplanar with the Fe_3 triangle	1008
	$NiBr_3L_2 \cdot 0.5NiBr_2L_2 \cdot C_6H_6$	C_i^1-$P\bar{1}$	Trans planar $NiBr_2L_2$ 2.251(3)	Two molecules $NiBr_3L_2$, one $NiBr_2L_2$ and two molecules C_6H_6 in the unit cell, trans planar $NiBr_2L_2$: Ni-Br 2.297-	1127

			trigonal bipyramidal P(1) 2.263(3) P(2) 2.273(3)	(2), trigonal bipyramidal: Ni-Br 2.349(2), 2.375(2), 2.339(2)	
	trans-[PtMe]CMe(OMe)]-$L_2]^+[PF_6]$	Pnma	2.283(3)	The two L's are related by reflection in a mirrorplane containing the Me carbon, Pt, carbene carbon, Pt-Me 2.13(2), Pt-C(carbene) 2.13(2)	1133
	$[Ir(CO)_3L_2]ClO_4$	$P2_1/c$	P(1) 2.34(2) P(2) 2.29(2)	Trigonal bipyramidal, P(1), P(2) in axial position; cation has approximate C_2 symmetry	1007
	$IrCl(CO)L_3$	$P2_1/c$	P(1) 2.342(5) P(2) 2.320(6) P(3) 2.330(6)	Distorted trigonal bipyramidal, P(2), P(3) axial, ∡P(2)IrP(3) 165.1(2), ∡ClIrP(1) 96.3(2), ∡C(1)IrCl 129.5(6)	262
	$IrMe(C_8H_{12})L_2$	Pbca	P(1) 2.316(5) P(2) 2.329(5)	Trigonal bipyramidal, P(1), P(2) in equatorial, Me in axial position, $Ir\text{-}Me_{ax}$ 2.202(22), ∡P(1)IrP(2) 101.5(2)	278, 281
	$PtCl_2(HgCl_2)L_2$	$P2_1/c$	P(1) 2.256(7) P(2) 2.253(7)	Two bridging Cl atoms between Hg and Pt, square planar about Pt atom, distorted tetrahedral about Hg	62
$MePh_2P$	$MnH(CO)_3L_2$	$C2/c$	P(1) 2.253 P(2) 2.257	Distorted square pyramid, $Mn(CO)_3$ portion coplanar within 0.01 Å, P(1), P(2) in trans position to each other, ∡P(1)MnP(2) 175	763
	$Mn_2(CO)_8L_2$	$P2_1/c$	2.23	The two L are attached axially, $Mn(CO)_4$ groups staggered, Mn-Mn 2.90, ∡MnMnC 86	762
	MoH_4L_4	$P2_1/c$	P(1) 2.433(2) P(2) 2.503(8)	Eight-coordinate, distorted variant of D_{2d}-$\bar{4}2m$ dodecahedron, the four Hs	549

TABLE XII. cont. Crystal and molecular structures of transition metal complexes containing phosphorus(III) ligands. The numbers in parentheses are the standard deviations in bond lengths or angles. When more than one metal-phosphorus distance are listed, the different d(M-P) values are indicated by P(1), P(2), etc., instead of d[M-P(1)], d[M-P(2)], etc. The symbol A-B denotes the A-B distance; ∡ABC the bond angle at B between A and C.

Ligand L	Complex	Space Group	Metal-P Distances d(M-P)[Å]	Comments	Refs.
				forming a flattened tetrahedron, Mo-H 1.70(3); the molecule has nearly idealized $C_2(2)$ symmetry leading to two Mo-P distances	
	$[Ru(NO)(\mu Ph_2P)L]_2$	$P2_1/c$	2.315(4) (L) 2.304(4) (μ-phosphido)	Distorted tetrahedron, NO terminally, Ru-Ru 2.629(2)	1014
	trans-$Ni(\sigma C_6F_5)(\sigma C_6Cl_5)L_2$	$C2/c$	2.230(2)	Square planar, ∡PNiP 176.73(7)	287, 288
	trans-$NiBr(\sigma C_6F_5)L_2$	$C2/c$	P(1) 2.2164(13) P(2) 2.2148(13)	Distorted square planar, ∡P(1)NiP(2) 175.10(5)	284
	trans-$Ni(\sigma C_6F_5)_2L_2$	Pbca	2.2061	Square planar, precise C_i symmetry, each of the σC_6F_5 rings lies at an angle of 86.13 to the NiP_2C_2 plane	289
	$Cu(NO_3)L_3$	$P2_{1/n}$	P(1) 2.259(2) P(2) 2.315(2) P(3) 2.270(2)	Tetrahedral, NO_3 acts as monodentate ligand, Cu-O 2.206(6), ∡P(1)CuO 107.4(2), ∡P(2)CuO 87.9(2), ∡P(3)CuO 96.3(2)	862
	$[Cu(NCS)L_2]_2$	$P\bar{1}$	P(1) 2.27(1) P(2) 2.24(1)	μ-Thiocyanato groups highly distorted, tetrahedral geometry at Cu, ∡P(1)Cu(1)P(2) 123.0(1), $Cu_2(SCN)_2$ unit: planar eight-membered ring	490

$EtPh_2P$	$IrCl(CO)(O_2)L_2$	$P2_1/a$	2.366(7)	Distorted trigonal bipyramidal, P atoms in trans axial position, ∡PIrP 174,5(1)	1244
Ph_3P	$Cr(CO)_5L$	$P\bar{1}$	2.422(1)	Cr-C(trans P) 1.845(4), Cr-C(cis) 1.878(4), 1.894(4), 1.882(5), 1.867(4)	995
	$[Cr(CO)_2Ph_2P(\pi C_6H_5)]_2$	$P\bar{1}$		Bridging by P and πC_6H_5 group, Cr-Cr 4.657	1029
	$MoBr_2(CO)_2L_2$	$P\bar{1}$	P(1) 2.55(2) P(2) 2.47(2)	Nonoctahedral! ∡P(1)MoP(2) 127.8(4), ∡BrMoBr 83.1(2), ∡CMoC 119.4(24)	412
	$Mn(CO)_2(MeCO_2)L_2$	$P2_1/c$	P(1) 2.260(3) P(2) 2.275(3)	Octahedral, the two Ls mutually trans, $MeCO_2$ group bidentate, ∡P(1)MnP(2) 177.4(1)	374
	$\pi C_5H_5Mn(CO)_2L$	$P\bar{1}$	2.236(3)		75
	$MnCl(CO)_4L$	$P\bar{1}$	2.398(4)	Octahedral, Cl and P in cis position, ∡ClMnP 86.55(14), Mn-Cl 2.358(5)	1214
	$Mn(CO)_4MeL$ Molecule 1 Molecule 2	$P2_12_12_1$	 2.311(18) 2.315(18)	Asymmetric unit contains two molecules; molecule 1: P atom and Me group in cis position; molecule 2: superposition of cis and trans isomers	866
	$Re(CO)_2(HCS_2)L_2$	$P\bar{1}$	P(1) 2.426(4) P(2) 2.412(4)	Distorted octhedral, approximately C_{2v}, P(1) trans to P(2), CO groups cis to one another, $(HCS_2)^-$ bidentate, ∡P(1)ReP(2) 177.6(1)	16
	$ReCl_3(MeCN)L_2$	$P2_1/n$	P(1) 2.47(1) P(2) 2.48(1)	Distorted octahedral; the two Ls are mutually trans, the three Cls in meridional position to one another	411
	$\pi C_5H_5Fe(CO)(\sigma\text{-}\alpha C_4H_3S)L$	Pc	2.22(1)	$\sigma\text{-}\alpha C_4H_3S$ = σ-(α-thienyl), octahedral coordination	38
	RuH_2L_4	$P\bar{1}$			627

TABLE XII. cont. Crystal and molecular structures of transition metal complexes containing phosphorus(III) ligands. The numbers in parentheses are the standard deviations in bond lengths or angles. When more than one metal-phosphorus distance are listed, the different d(M-P) values are indicated by P(1), P(2), etc., instead of d[M-P(1)], d[M-P(2)], etc. The symbol A-B denotes the A-B distance; ∡ABC the bond angle at B between A and C.

Ligand L	Complex	Space Group	Metal-P Distances d(M-P)[Å]	Comments	Refs.
	$RuH(CHO_2)L_3$	$P2_1/c$	P(1) 2.350 P(2) 2.364	Highly distorted octahedral, formiate ligand bidentate, P(1) trans to P(2), P(3) trans to O	551
	$Ru(\pi C_3H_5)_2L_2$	$P\bar{1}$	P(1) 2.342(4) P(2) 2.344(3)	Coordination relative to two P and two central C atoms of the π-allyl groups tetrahedral, ∡P(1)-RuP(2) 109.9	1120
	$[RuCl(NO)_2L_2][PF_6]\cdot C_6H_6$	$P2_1/c$	P(1) 2.431(6) P(2) 2.419(6)	Distorted square pyramid, NO in apical position bond in bent, NO in basal plane in linear manner, P(1) trans to P(2), ∡P(1)RuP(2) 159.6(2)	988
	$RuH(NO)L_3$	$P2_1/n$	P(1) 2.342(3) P(2) 2.347(3) P(3) 2.328(3)	Distorted C_{3v} symmetry, ∡P(1)RuP(2) 119.0(1), ∡P(1)RuP(3) 111.5(1), ∡P(2)RuP(3) 113.3(1)	989, 991
	$OsCl_2(NO)(HgCl)L_2$	$R\bar{3}$	P(1) 2.39(2) P(2) 2.40(2)	Octahedral coordinated P(1) trans to P(2)	119
	$[Os(OH)(NO)_2L_2][PF_6]$	$P2_1/a$	P(1) 2.42(1) P(2) 2.45(1)	Tetragonal pyramid, P(1) trans to P(2), apical NO bond in bent manner, basal NO linearly bond	1242
	$Co(CO)_2(NO)L$	$P\bar{1}$	2.224(3)	Distorted tetrahedral	17
	$Co(CO)(NO)L_2$	$P2/c$	2.230(3)	Distorted tetrahedral, ∡PCoP 114.1(2)	17

$RhH(AsPh_3)L_3 \cdot \frac{1}{2}C_6H_6$	T_h^6-Pa3	2.36(6)	Random occupation of the four heavy atom positions by three P and one As atom	63
Rh(CO)(oxq)L	$P\bar{1}$ or P1	2.27	Square planar, oxq = 8-hydroxyquinoline, CO trans to O of oxq, P trans to N of oxq	759
$RhHCl(SiCl_3)L_2 \cdot xSiHCl_3$	C_i^1-$P\bar{1}$	P(1) 2.344(4) P(2) 2.332(4)	Highly distorted trigonal bipyramid, P(1) trans to P(2) axial, ∡P(1)RhP(2) 161.7(1), H trans to Cl	917
$Rh_2(C_4H_7N_2O_2)_4L_2 \cdot H_2O \cdot Pr^nOH$	$P2_1/c$	Rh(1)-P(1) 2.430(5) Rh(2)-P(2) 2.447(5)	$C_4H_7N_2O_2^-$ = dimethylglyoximato ion, distorted octahedron, close to D_{2d} symmetry, Rh(1)-Rh(2) 2.936(2)	239
$RhCl(C_4H_7N_2O_2)_2L$	$P2_1/c$	2.327(1)	Octahedral, Cl trans to P, ∡ClRhP 178.25(5), the $Rh(C_4H_7N_2O_2)_2$ is not completely planar	321
$[Rh(CO)L_2]_2 \cdot 2CH_2Cl_2$	C2/c	P(1) 2.321(2) P(2) 2.327(2)	Two bridging CO groups, Rh-Rh 2.630(1), indicative of metal-metal bond, the two Rh atoms are five-coordinate	352, 1117
$[Ir(NO)_2L_2][ClO_4]$	C_{2h}^6-C2/c	2.3393(3)	approximately tetrahedral	891
$IrCl_2(NO)L_2$	C_{2h}^6-I2/a	2.367(2)	Distorted square pyramidal, NO nonlinear bonded	892
$IrI(Me)(NO)L_2$	C_{2h}^6-I2/a	2.348(3)	Distorted square pyramidal, P atoms in trans position	893
$Ir(NO)L_3$	P3	Ir(1)-P(1) 2.321(6) Ir(2)-P(2) 2.316(8) Ir(3)-P(3) 2.290(10)	Three independent molecules in the unit cell, each C_3 symmetry, absolute configuration of the crystal is R,S,S	20
$Ir_2(CO)_2(tdt)_3L_2$	$B\bar{1}$	Ir(1)-P(1)	tdt = toluene-3,4-dithiolato anion,	678

TABLE XII. cont. Crystal and molecular structures of transition metal complexes containing phosphorus(III) ligands. The numbers in parentheses are the standard deviations in bond lengths or angles. When more than one metal-phosphorus distance are listed, the different d(M-P) values are indicated by P(1), P(2), etc., instead of d[M-P(1)], d[M-P(2)], etc. The symbol A-B denotes the A-B distance; ∡ABC the bond angle at B between A and C.

Ligand L	Complex	Space Group	Metal-P Distances d(M-P)[Å]	Comments	Refs.
			2.380(10) Ir(2)-P(2) 2.397(11)	octahedral coordination about each Ir atom, Ir(1)(CO)(tdt)L, Ir(2)(CO)(tdt)$_2$L, one S from each tdt ligand functions as a bridge to Ir(1)	
	[Ir(CO)(Ph_2P)L]$_2$	P2$_1$/n	2.279 (L) 2.301 (Ph_2P) 2.306	Distorted tetrahedral, Ir-Ir 2.554	853
	Ir(CO)(NO)L_2	P2$_1$/c C^5_{2h}	P(1) 2.324(2) P(2) 2.323(2)	Distorted tetrahedron, ∡P(1)IrP(2) 103.9(1), ∡CIrN 128.7(4)	1002
	[Ir(CO)$_2$(CS)L_2][PF_6] ·Me_2CO	P2$_1$/m	P(a) 2.367(5) P(b) 2.372(4)	Two crystallographically independent cations a and b, distorted trigonal bipyramidal, ∡P(a)IrP(a) 175.8(2) ∡P(b)IrP(b) 173.2(2)	458
	Ni(C_2H_4)L_2	triclinic	P(1) 2.157(4) P(2) 2.147(4)	Strongly distorted tetrahedron, ∡P(1)NiP(2) 110.5(2), ∡CNiC 42.1(5)	266
	Ni(NO)(N_3)L_2	P2$_1$/c C^5_{2h}	2.257(2)	Pseudotetrahedral, ∡PNiP 120.52(8) ∡N_3NiNO 128.8(3)	436
	πC_5H_5Ni(CF_3)L	P2$_1$/c C^5_{2h}	P(a) 2.148(8) P(b) 2.172(9)	Two crystallographic independent molecules in the unit cell with different conformations of the Ph groups in L	285
	Ni(CO)(CF_3)$_2$$L_2$	P2$_1$/n		CO group sideways coordinated	325

trans-PdI_2L_2	$P2_1/c$	2.331(2)	Trans square planar, ∢IPdP 87.1(1)	375
Pt(TNCE)L_2	$P2_1/c$	P(1) 2.291(9) P(2) 2.288(8)	TNCE = tetracyanoethylene, distorted square planar, ∢P(1)PtP(2) 101.4(3), ∢CPtC 41.5(13)	152
$Pt(CS_2)L_2$	$P2_1/n$	P(1) 2.346(10) P(2) 2.240(15)	Strongly distorted square planar, ∢P(1)PtP(2) 107.1(5), dihedral angle (between CSPt and P(1)PtP(2) plane): 6°, P(1) "trans" C, P(2) "trans" S	851
$[Pt(S_2CF)L_2][HF_2]$	$P\bar{1}$	P(1) 2.269(4) P(2) 2.296(4)	Irregular square planar, ∢P(1)PtP(2) 99.1, ∢S(1)PtS(2) 74.7, Pt-S(1) 2.340(5), S(1) trans to P(1), Pt-S(2) 2.322(4), S(2) trans to P(2)	439
$[Pt(C_3Ph_3)L_2][PF_6]$ $\cdot C_6H_6$	$P2_1/n$	P(1) 2.285(3) P(2) 2.322(3)	C_3Ph_3 = triphenylcyclopropenyl, C_3 ring perpendicular to P(1)PtP(2) plane, ∢P(1)PtP(2) 104.5(1)	806
$Pt(C_4H_8)L_2$	$P\bar{1}$	P(1) 2.279(5) P(2) 2.285(6)	C_4H_8 = four CH_2 groups of the puckered tetrahydroplatinole ring, square planar coordination, ∢P(1)PtP(2) 98.8(2), ∢CPtC 80.9(8)	133
$Pt(C_2H_4)L_2$	$P2_1/a$	P(1) 2.265(2) P(2) 2.270(2)	Planar dihedral angle [P(1)PtP(2)] [CPtC] 1.6°, ∢P(1)PtP(2) 111.6(1), ∢CPtC 39.7(4)	267
$Pt[Cl_2C{=}C(CN)_2]L_2$	$P2_1/c$	P(1) 2.260(6) P(2) 2.339(6)	$Cl_2C(1){=}C(2)(CN)_2$, Pt-C(1) 2.00(2), Pt-C(2) 2.10(2), C-C 1.42(3), ∢P(1)PtP(2) 102.0(2), P(2) "cis" to C(2)	801
$Pt(Cl_2C{=}CCl_2)L_2$	$P2_1/c$	P(1) 2.292(7) P(2) 2.278(8)	Trigonal coordination (if the olefin is counted as a monodentate ligand), ∢P(1)PtP(2) 100.6(2), ∢CPtC 47.1(1), dihedral angle	472

TABLE XII. cont. Crystal and molecular structures of transition metal complexes containing phosphorus(III) ligands. The numbers in parentheses are the standard deviations in bond lengths or angles. When more than one metal-phosphorus distance are listed, the different d(M-P) values are indicated by P(1), P(2), etc., instead of d[M-P(1)], d[M-P(2)], etc. The symbol A-B denotes the A-B distance; ∡ABC the bond angle at B between A and C.

Ligand L	Complex	Space Group	Metal-P Distances d(M-P)[Å]	Comments	Refs.
				[P(1)PtP(2)] [CPtC] 12.3(1.5)	
	$Cu_2Cl_2L_3 \cdot C_6H_6$	$P\bar{1}$	P(1) 2.19(1) P(2) 2.27(1) P(3) 2.25(1)	Two chloride ions are bridging, Cu(1) planar, Cu(2) distorted tetrahedral, dihedral angle [ClCu(2)Cl] [P(1)Cu(2)P(2)] 94.3(1)	780
	$Cu_2Cl_2L_3$	$P2_1/c$	P(1) 2.183(4) P(2) 2.236(5) P(3) 2.245(5)	Dihedral angle 86.2(3)	18
	$[Cu(N_3)L_2]_2$	$P2_1/c$	Cu(1)-P(1) 2.271(4) Cu(1)-P(2) 2.250(3) Cu(2)-P(1′) 2.269(4) Cu(2)-P(2′) 2.266(3)	Dimer, two μ-azido groups, tetrahedral coordination at Cu, eight-membered ring $(N_3)_2Cu_2$, N_3 groups tilted by 38° with respect to one another	1287
	$[CuClL]_4$	Pbcn		Molecule has precise C_2 symmetry, Cu_4Cl_4 distorted cube, at each Cu distorted tetrahedral coordination	283
	$[CuBrL]_4 \cdot 2CHCl_3$	$C2/c$		Molecule has precise C_i symmetry, Cu_4Br_4 defining a "step" configuration, two Cu(I) have tetrahedral, two Cu(I) trigonal planar coordination geometry	283

	$CuBrL_2 \cdot \frac{1}{2}C_6H_6$	$P\bar{1}$	P(1) 2.282(3) P(2) 2.263(3)	Trigonal planar, ∡P(1)CuP(2) 126.0(1)	365
	$Cu_6H_6L_6 \cdot HCONMe_2$	Pbca D_{2h}^{15}	*av* 2.240(17)	Octahedral cluster of Cu atoms, each Cu apically bonded to one L	282
	$[Ag(SCN)L_2]_2$	$P\bar{1}$	Av 2.48		616
	$[AuC(CF_3)L]_2$	$P2_1/c$	2.28(1)	LAu-C=C-AuL moiety planar C_{2v} symmetry, Au-C 2.05(6)	505
	$Au(\sigma C_6F_5)L$	$P2_1/c$	2.27(1)	Linear coordination, ∡$C(C_6F_5)AuP$ 178(1)	64
	$AuCl_3L$	$P\bar{1}$	2.335(4)	Nearly square planar coordination, ∡Cl(1)AuP 93.8(1), ∡Cl(1)AuCl(3) 88.5(1), ∡Cl(2)AuP 89.1(1) ∡Cl(1)AuCl(2) 172.7(2)	72
	$CdCl_2L_2$	$Pna2_1$	P(1) 2.635(5) P(2) 2.632(6)	Pseudotetrahedral coordination, ∡P(1)CdP(2) 107.6(2) ∡ClCdCl 113.9(2)	222
	$Hg(SCN)_2L_2$	$P2_1/c$	P(1) 2.489(3) P(2) 2.487(3)	Distorted tetrahedral, ∡P(1)HgP(2) 118.1(1), ∡SHgS 96.7(1)	817
(*p*-tolyl)$_3$P	$[Au_6L_6][BPh_4]_2$	$P\bar{1}$	P(1) 2.274(8) P(2) 2.311(11) P(3) 2.293(8)	Centrosymmetric Au_6 octahedron distorted, each Au coordinates one L	95
$(pFC_6H_4)_3P$	$Au_{11}I_3L_7$	R3	3 classes 2.21(1) - 2.29(1)	Au_{11} cluster icosahedron in which one triangular face is replaced by one Au atom	94
$(pClC_6H_4)_3P$	$Au_{11}I_3L_7$	$R\bar{3}c$			19
Cy_3P	$Ni(\pi C_2H_4)_2L$	$P2_1/n$	2.196(2)		749
Ph_2PCH_2Ph	*trans*-$NiBr_2L_2$ tetrahedral $NiBr_2L_2$	$P\bar{1}$	2.263(7) P(1) 2.316(7) P(2) 2.314(8)	Unit cell contains three single molecules, one with trans square planar configuration; the other two are tetrahedral, trans: Ni-Br 2.305(3), tetrahedral: Ni-Br av 2.355	681

TABLE XII. cont. Crystal and molecular structures of transition metal complexes containing phosphorus(III) ligands. The numbers in parentheses are the standard deviations in bond lengths or angles. When more than one metal-phosphorus distance are listed, the different d(M-P) values are indicated by P(1), P(2), etc., instead of d[M-P(1)], d[M-P(2)], etc. The symbol A-B denotes the A-B distance; ∡ABC the bond angle at B between A and C.

Ligand L	Complex	Space Group	Metal-P Distances d(M-P)[Å]	Comments	Refs.
$(Ph_2P)_2CH_2$	$Fe(CO)_3L$	$P\bar{1}$	P(1) 2.209(3) P(2) 2.225(3)	Coordination geometry intermediate between trigonal bipyramidal and square pyramidal, ∡P(1)FeP(2) 73.5(1)	320
	$Mo(CO)_4L$	$P2_1/c$	P(1) 2.535(3) P(2) 2.501(2)	Coordination approximately octahedral, ∡P(1)MoP(2) 67.3(1)	269
	$CuL \cdot \frac{2}{3} C_6H_5Me$	$P2_1/c$	av 2.317(5)	Trimeric, two Cu atoms trigonal planar coordinated to three P atoms; one Cu atom bridges two methine groups of two L	225
	Cu_2I_2L	Pbca		Crystals contain centrosymmetric dimetric units $[Cu_2I_2L]_2$, tri- and four-coordinate Cu atoms	848
$(Cy_2P)_2CH_2$	NiL_2	$P2_1/c$	P(1) 2.214(2) P(2) 2.215(2) P(3) 2.193(2) P(4) 2.218(2)	Distorted tetrahedron, ∡P(1)NiP(2) 77.7 (four-membered ring), ∡P(3)NiP(4) 77.3 (four-membered ring)	748
$(Ph_2P)_2C_2H_4$	$Cr(CO)_4L$	Pbca	2.360(2)	Distorted octahedron, ∡PCrP 83.41(8), Cr-C(trans P) 1.831(7), Cr-C(trans C) 1.884(7)	116
	$[MoOClL_2]$ $[ZnCl_3(OCMe_2)]$	Pbca	Av 2.57	Octahedral geometry, O trans to Cl atom	2

trans-$Mo(N_2)_2L_2$	$P\bar{1}$	P(1) 2.500(4) P(2) 2.450(5)	Distorted octahedron, ∡P(1)MoP(2) 80.9(1), P(1) trans P(1′), P(1), P(2) are the two P atoms of the same L	1204
$\pi C_5H_5MoCl(CO)L$	Pbca	P(1) 2.498(2) P(2) 2.435(3)	Sandwich type, CO cis to Cl atom, ∡P(1)MoP(2) 75.2(1)	335
$MoBr_2(CO)_3L$	$P2_1/c$	P(1) 2.500(4) P(2) 2.618(5)	Mo atom seven-coordinate, distorted capped octahedron, capped face contains two CO and one P atom	410
ReH_3L_2	$P2/c$	P(1) 2.34(2) P(2) 2.35(2)	ReL_2 has C_2 symmetry, three H atoms occupy equatorial position of a slightly distorted pentagonal bipyramid	14
$[Ru(NO)L_2][BPh_4]$ $\cdot COMe_2$	$P2_1/n$	Av 2.39(1)	Approximately trigonal bipyramidal, the two L bridging axial and equatorial positions, NO linearly coordinated	990, 991
$[CoClL_2][SnCl_3]$ red	$P2_1/c$ C_{2h}^5	P(1) 2.291(2) P(2) 2.254(2) P(3) 2.283(2) P(4) 2.274(2)	Tetragonal pyramid with apical Cl atom	1126
green	C_i^1-$P\bar{1}$	P(1) 2.258(5) P(2) 2.252(5) P(3) 2.268(5) P(4) 2.253(6)	Trigonal bipyramidal, two P in axial [P(1), P(3)], two P [P(2), P(4)], and one Cl atom in equatorial position	
$[RhL_2]ClO_4$	$P2_1/c$	Av 2.306	Square planar, ∡PRhP(intraligand) av 82.7	568
$[Ir(S_2)L_2]Cl\cdot MeCN$	$P\bar{4}2_1c$	P(1) 2.371(5) P(2) 2.331(4) P(3) 2.363(5) P(4) 2.337(4)	Approximately trigonal bipyramidal, S_2 π-bonded equatorial, P(1), P(3) axial, P(2), P(4) equatorial	156
$IrMe(C_8H_{12})L$	Pnma	2.308(3)	C_8H_{12} = cycloocta-1,5-diene, distorted trigonal bipyramidal; L occupies two	279

TABLE XII. cont. Crystal and molecular structures of transition metal complexes containing phosphorus(III) ligands. The numbers in parentheses are the standard deviations in bond lengths or angles. When more than one metal-phosphorus distance are listed, the different d(M-P) values are indicated by P(1), P(2), etc., instead of d[M-P(1)], d[M-P(2)], etc. The symbol A-B denotes the A-B distance; ∡ABC the bond angle at B between A and C.

Ligand L	Complex	Space Group	Metal-P Distances d(M-P)[Å]	Comments	Refs.
				equatorial positions, ∡PIrP 84.9(2)	
	[Ni(CH_2CMeCH_2)L] Br	$P\bar{1}$	P(1) 2.178(4) P(2) 2.180(4)	Five-coordinate π-allyl Ni species, square pyramidal; L occupies two basal sites, ∡P(1)NiP(2) 88.8(1)	286
	Pd$(SCN)_2$L	$P2_12_12_1$	P(1) 2.260(4) P(2) 2.245(4)	Square planar, P(1) trans to S, P(2) trans to N	122
	$Cu_2Cl_2L_3 \cdot 2C_3H_6O$	$P2_1/c$	P(1) 2.291(5) P(2) 2.311(4) P(3) 2.284(6) (bridge)	Tetrahedral coordination of Cu, (one Cl and three P), two L being chelating, one L bridging	15
	$Cu_2(N_3)_2L_3$	Pbca	P(1) 2.317(6) P(2) 2.295(6) P(3) 2.319(5)	Distorted tetrahedral, P(2), P(3) part of a five-membered chelate ring at Cu(1) or Cu(2), P(1) part of a nonchelate L, connecting Cu(1) and Cu(2)	489
$(MePhP)_2C_2H_4$	$PdCl_2L_2 \cdot EtOH$	Pccn	2.333(3)	Meso anti-trans form; i.e., equal ligands at opposite sides of the plane through the P atoms, ∡PPdP(intraligand) 85.0(1)	547
$(Ph_2P)_2C_3H_6$	IrMe(C_8H_{12})L	$C2/c$	P(1) 2.309(4) P(2) 2.337(4)	C_8H_{12} = cycloocta-1,5-diene, distorted trigonal bipyramid; L occupies two equatorial sites, ∡P(1)IrP(2) 93.4(1)	280
$PhP(C_2H_4\text{-}PPh_2)_2$	$MnBr(CO)_3LCr(CO)_5$		P(1) 2.328(12)	P(3) connected to Cr; the Ph ring	138,

	α-isomer	Pbcn	P(2) 2.301(13) P(3) 2.41(1)	attached to the central P can take a position cis or trans to the Br, relative to the five-membered chelate ring and this gives rise to the two isomers	1076
	β-isomer	Cc	P(1) 2.369(7) P(2) 2.346(7) P(3) 2.376(7)		
$MeC(CH_2Ph_2P)_3$	NiIL	$Pn2_1a$	P(1) 2.220(4) P(2) 2.211(4) P(3) 2.210(4)	Distorted tetrahedron, ∡P(1)NiP(2) 92.2(1), ∡P(1)NiP(3) 94.4(1), ∡P(2)NiP(3) 97.7(1)	355
	$Ni(C_2F_4)L$	$P2_1/c$	P(1) 2.28(1) P(2) 2.26(1) P(3) 2.21(1)	Distorted tetrahedron, ∡PNiP (av) 92.5	196
$(o\text{-}Ph_2PC_6H_4)_3P$	$[CoClL][BPh_4]$	$P2_1nb$	P(1) 2.280(4) P(2) 2.318(4) P(3) 2.261(4) P(4) 2.057(4)	Distorted trigonal bipyramidal, P(4) axial, threefold axial symmetry distorted by Jahn-Teller effect	149
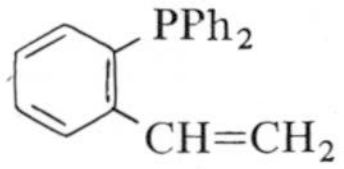	$Mn(CO)_4L$	$P2_1/n$	2.279(3)	octahedron, Mn-C(HCMe) 2.21(1), ∡PMnC 80.4	1028
	$Fe(CO)_2L_2$	$B2_1/c$	P(1) 2.205(2) P(2) 2.239(2)	Approximately trigonal bipyramidal, equatorial position: two CO, one π-vinyl, P(2) unidentate ligand L	1027
$(\sigma\text{-vinyl}C_6H_4)_3P$	RhBrL	R3c	2.167(11)	Trigonal bipyramid, Br, Rh, P atoms on a C_3 axis, the three o-vinyl groups occupy equatorial positions	928, 929
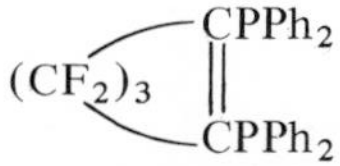	$[RhL_2][cis\text{-}RhCl_2(CO)_2]$	$I4_1/a$	P(1) 2.282(6) P(2) 2.300(6)	Each ion nearly square planar with C_2 axis, P(1) cis P(2)	425
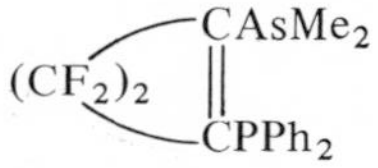	$Fe_2(CO)_4L_2$	C2/c	P(1) 2.270(2) P(2) 2.239(2)	Fe atoms in different environments, Fe(1): $P_2As_2Fe(CO)$, Fe(2): $[\pi(C{=}C)](CO)_3$	426
	$Fe_3(CO)_9L$	$P2_12_12_1$	2.48	Me_2As group cleaved from L	428
	$Fe(CO)_4L$	$P2_1/c$	2.224(3)	Ligand coordinated through P	427

TABLE XII. cont. Crystal and molecular structures of transition metal complexes containing phosphorus(III) ligands. The numbers in parentheses are the standard deviations in bond lengths or angles. When more than one metal-phosphorus distance are listed, the different d(M-P) values are indicated by P(1), P(2), etc., instead of d[M-P(1)], d[M-P(2)], etc. The symbol A-B denotes the A-B distance; ∢ABC the bond angle at B between A and C.

Ligand L	Complex	Space Group	Metal-P Distances d(M-P)[Å]	Comments	Refs.
$(oSeC_6H_4)Ph_2P$	NiL_2	$P2_1/c$	2.177(3)	Square planar, ∢SeNiP(intraligand) 88.4(I)	347
$Me_2N(CH_2)_3Ph_2P$	Pd(SCN)(NCS)L	$P2_1/c$	2.243(2)	Square planar, isothiocyanate trans to P,thiocyanatetrans to N	290
$Ph_2P(C_2H_4)NH$-$(C_2H_4)NEt_2$	$Co(NCS)_2L$	$P\bar{1}$	2.32(1)	Trigonal bipyramid distorted toward elongated square pyramid	963
$Ph_2P(C_2H_4)NMe$-$(C_2H_4)NEt_2$	$Co(NCS)_2L$	Cc	2.40(1)	Trigonal bipyramidal, distorted toward tetrahedron	963
$Ph_2P(C_2H_4)NMe(C_2H_4)$ $Ph_2P(C_2H_4)NMe$	$[CoBrL][PF_6]$	$P2_1/c$	2.22(1) av	Elongated square pyramid, Br in apical position, Co-Br 2.534(6)	130
	[NiBrL]Br ·0.5BuOH	$P2_1/c$	2.21(1) av	Elongated square pyramid, Br in apical position, Ni-Br 2.807(4)	130
$N(C_2H_4NEt_2)$-$(C_2H_4PPh_2)_2$	[CoIL]I	$P2_1/c$	P(1) 2.446(6) P(2) 2.416(6)	Trigonal bipyramidal, distorted toward tetrahedron, Co-1 2.647(3)	129
	[NiIL]I	$P2_1/c$	P(1) 2.209(3) P(2) 2.256(3)	Trigonal bipyramidal, distorted toward elongated square pyramid, Ni-I 2.550(2)	129
$N(C_2H_4OMe)$-$(C_2H_4PPh_2)_2$	$[CoClL][PF_6]$	P1	P(1) 2.414(7) P(2) 2.438(8)	Trigonal bipyramidal, distorted toward tetrahedron, ∢P(1)CoP(2) 117.3(2)	356
$N(C_2H_4PPh_2)_3$	[CoIL]I	$P2/c$	P(1) 2.229(8) P(2) 2.266(8)	Distorted square pyramidal, basal plane formed by 2P, N and I, one P	877

			P(3) 2.279(8)	at the apex [P(3)]	
	$[CoClL][PF_6]$	$P2_1/c$	2.37(1) mean	Distorted trigonal bipyramid, Co-N 2.675(10)	394
	NiL	$C2/c$	P(1) 2.117(3) P(2) 2.121(3) P(3) 2.118(3)	Trigonal bipyramid, ∡NNiP 90	878
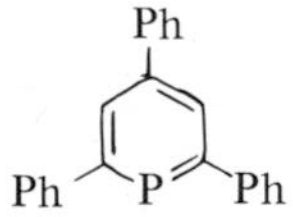	$Cr(CO)_5L$	$P2_1/c$	2.372(4)	P-bonded phosphorine, Cr-C(1) 1.83(2), Cr-C(2) 1.85(2), Cr-C(3) 1.86(2), Cr-C(4) 1.83(2), Cr-C(5) 1.82(2), C(5) trans to P	1216
$Ph_2P\text{-}C{\equiv}CCF_3$	$Co_4(CO)_{10}L_2$	$P2_1/n$	Co(3)-P(1) 2.236 Co(4)-P(2) 2.229	Acetylene unit and phosphorus(III) function as ligands	614
$(Ph_2PNEt)_2PhP$	$Mo(CO)_3L$	C_c	P(1) 2.425(4) P(2) 2.527(4) P(3) 2.477(4)	Distorted octahedron, ∡P(1)MoP(2) 65.0(1), ∡P(1)MoP(3) 65.3(1)	268
$PhCH_2P$	*cis*-$NiCl_2L_2$	$P2_1/c$	2.154(2) av	Phospholene rings puckered	810
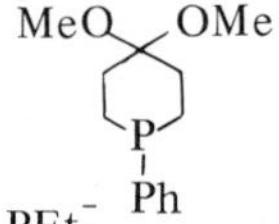	*trans*-$NiCl_2L_2$	$P2_1/c$	2.250(2)	Trans square planar, phosphorinane ligands in chair conformation	811
PEt_2^-	$[Mo(CO)_4L]_2$	$C2/c$	Mo(1) 2.51(1) Mo(2) 2.49(1)	μ-PEt_2, octahedral coordination, planar Mo_2P_2 skeleton, ∡MoPMo 75.4(3) ∡Mo(1)Mo(2)P 52.7(3), ∡Mo(1)Mo(2)P′ 52.0(3)	784, 925
	$[W(CO)_4L]_2$	$C2/c$	W(1) 2.54(2) W(2) 2.58(1)	μ-PEt_2	783, 784
P_2Et_4	$[Mo(CO)_5]_2L$	Pbca		Octahedral at Mo, P-P 2.21	924
PPh_2^-	$[Ni(CO)_2L]_2$ Molecule 1	$P\bar{1}$	Ni(1) 2.185(3)	Unit cell contains two crystallographically different centrosymmetric	651

TABLE XII. cont. Crystal and molecular structures of transition metal complexes containing phosphorus(III) ligands. The numbers in parentheses are the standard deviations in bond lengths or angles. When more than one metal-phosphorus distance are listed, the different d(M-P) values are indicated by P(1), P(2), etc., instead of d[M-P(1)], d[M-P(2)], etc. The symbol A-B denotes the A-B distance; ∡ABC the bond angle at B between A and C.

Ligand L	Complex	Space Group	Metal-P Distances d(M-P)[Å]	Comments	Refs.
			Ni(2) 2.196(3)	molecules, planar di-μ-phosphido bridge	
	Molecule 2		Ni(1) 2.197(3) Ni(2) 2.191(3)		
	[Rh[FeL$(CO)_2$-$(\pi C_5H_4Me]_2$][PF_6]	Pbca	Fe(1)-P(1) 2.241(4) Fe(2)-P(2) 2.224(4) Rh-P(1) 2.239(4) Rh-P(2) 2.241(4)	μ-Ph_2P and μ(CO) groups, distorted octahedron at Rh	561
	$Fe_2(CO)_6$(C≡CPh)L	$P2_1/c$	Fe(1) 2.213(2) Fe(2) 2.224(2)	Coordinated acetylide functions as one-electron donor to Fe(1) via σ-bond and as two-electron donor to Fe(2) via π-bond	976
$P(p\text{-tolyl})_2^-$	$Fe_2(CO)_6$(OH)L	C2/c			1192
$P(OCH_2)_3P$	$Fe(CO)_3L_2$	Pnma	P(1) 2.116(4) P(2) 2.190(4)	Trigonal bipyramid, PC_3 eclipsed with respect to the $Fe(CO)_3$ unit; PO_3 is staggered relative to $Fe(CO)_3$, P(1) "phosphite" side of L, P(2) "phosphine" side of L	29
$P(OMe)_3$	$Ru(\pi C_4H_7)_2L_2$	$P2_1/a$	P(1) 2.210(5) P(2) 2.233(7)	πC_4H_7 = π-2-methylallyl, octahedrally coordinated Ru, Ls mutually cis, ∡P(1)RuP(2) 97.9(2)	846

	$\pi C_5H_5MoClC_2(CN)_2L_2$	$P2_1/n$	P(1) 2.461(2) P(2) 2.470(2)	distorted square pyramidal	701
	$\pi C_5H_5MoI(CO)_2L$	$P2_12_12_1$	2.406(9)	CO groups mutually trans	574
	$\pi C_5H_4MeMoI(CO)_2L$	$P2_12_12_1$	2.388(8)	CO groups mutually trans	574
$P(OPh)_3$	$Cr(CO)_5L$	$P\bar{1}$	2.309(1)	Slightly distorted octahedral, Cr-C(1) 1.892(5), Cr-C(2) 1.895(4), Cr-C(3) 1.904(6), Cr-C(4) 1.892(4), Cr-C(5) 1.861(4) C(t) trans to P	995
	trans-$Cr(CO)_4L_2$	$P2_1/c$	2.252(1)	Octahedral, Cr-C(1) 1.875(7) Cr-C(2) 1.881(6)	1005
	$IrCl[P(OPh)_2$-$(OC_6H_4)]_2L$	$P\bar{1}$	P(1) 2.291(3) P(2) 2.268(3) P(3) 2.192(3)	Two Ph rings orthometalated, Ir atom in a distorted octahedral environment (3 P, 2 *o*-C, 1 Cl), P(ligand) = P(1)	552
P(*o*-tolyl-O)$_3$	$Ni(C_2H_3CN)L_2$	$P2_1/c$	P(1) 2.121(4) P(2) 2.096(4)	Trigonally coordinated Ni atom	548
	$Ni(C_2H_4)L_2$	Pc	P(1) 2.093(5) P(2) 2.098(4)	Trigonally coordinated Ni atom	548
$PhP(OEt)_2$	$CoHL_4$	$P\bar{1}$	P(1) 2.115(2) P(2) 2.103(2) P(3) 2.126(1) P(4) 2.128(1)	Aproximately trigonal bipyramidal, Co-H(axial) 1.38(54), P(4) axial	1178
PF_3	$CoHL_4$	$C2/c$	2.052(5)	P-F 1.549(12), distorted tetrahedron of PF_3 groups about Co, ∡PCoP 101.8(3), 118.0(2), 108.2(2), 109.7(2), H-atom not located	476
Bu^t_2PF	*trans*-$NiBr_2L_2$	$P2_1/n$	2.232(3)	Trans square planar, the Ni,P,F plane forms a small dihedral angle (3.8°) with that of the Br, Ni, P atoms	1108
Et_2NPF_2	*cis*-$RhCl(Ph_3P)L_2$	$P2_1/c$	P(1) 2.215(3) P(2) 2.136(3) P(3) 2.352(3)	Square planar, the two Ls are mutually cis, P(1) cis P(2), P(1) trans to Ph_3P (P(3)), P(2) trans to Cl	110

TABLE XII. cont. Crystal and molecular structures of transition metal complexes containing phosphorus(III) ligands. The numbers in parentheses are the standard deviations in bond lengths or angles. When more than one metal-phosphorus distance are listed, the different d(M-P) values are indicated by P(1), P(2), etc., instead of d[M-P(1)], d[M-P(2)], etc. The symbol A-B denotes the A-B distance; ∡ABC the bond angle at B between A and C.

For Further X-ray Studies See	Ref.
cis-mer-$MoOCl_2(Et_2PhP)_3$	842
$[Ph_2C_4H_2N_2][Fe_2(CO)_5Ph_3P]$	758
$[Fe_2H_3(T^1)_2PF_6 \cdot 1.5CH_2Cl_2$	354
$Ru_2H(CO)_3[P(OC_6H_4)(OPh)_2]_2[OP(OPh)_2]$	197
$[\pi C_5H_5)_2Co][CoI_3Ph_3P]$	1219
$CoX(dmg)_2Ph_3P$ (X = Cl, Br, I)	1
$CoCl_2(NO)(MePh_2P)_2$	1001
$[CoCl(dppe)_2]SnCl_3$, 2 forms	1128
$Rh_2(MeCO_2)_2(dmg)_2(Ph_3P)_2$	570
$[Ir(Ph_3P)Q^4][BPh_4]$	1227

$[IrS_2(dppe)_2]Cl$	506
$Ni(C_2B_{10}H_{10})(Ph_3P)_2$	1063
$Ni(PhN{=}NPh)[(p\text{-tolyl})_3P]_2$	640
μ-(α,ω-octa-di-π-enyl)[Nidppe]$_2$	223
μ-(α,ω-octa-di-π-enyl)[$NiBr(Pr^i{}_3P)]_2$	223
$NiI_2[Ph_2P(CH_2)_2O(CH_2)_2O(CH_2)_2PPh_2]$	1042
trans-$Pd(CNS)_2(Ph_2PC{\equiv}CBu^t)_2$	121
$Pt(C_3H_2Me_2)(Ph_3P)_2$	1230
trans-$PtCl(R)(Et_2PhP)_2$ (R = $HC{=}CH_2$, $C{\equiv}CPh$)	227
Pt(cycloheptyne)$(Ph_3P)_2$	111

dmg = dimethylglyoximato
$C_3H_2Me_2$ = 1,2-dimethylcyclopropene
dppe = 1,2-bis(diphenylphosphino)ethane
Q^4 = *tris*-(o-diphenylphosphinophenyl)phosphine
T^1 = 1,1,1-*tris*(diphenylphosphinomethyl)ethane

hydrogen atoms were located and refined to an average Mo-H distance of 1.70(3) Å [549].

Another quite interesting transition metal hydride structure is represented by hexameric hydridotriphenylphosphinecopper(I), $Cu_6H_6(Ph_3P)_6$·DMFA (DMFA = dimethylformamide). It contains a Cu_6 octahedron, each copper is apically bonded to a triphenylphosphine ligand. The presence of hydride ligands could not, however, in this case be ascertained from crystallographic studies [282].

Tetramethylene metallocyclic species are thought to be intermediates in the olefin metathesis reaction:

$$2\ CH_2{=}CHR \rightleftharpoons CH_2{=}CH_2 + CHR{=}CHR \qquad (213)$$

In this context the crystal and molecular structure of bis(triphenylphosphine)tetramethyleneplatinum(II) is quite interesting, with respect to conformational features of the tetrahydridoplatinole ring. These are indicated in Fig. 2 [133].

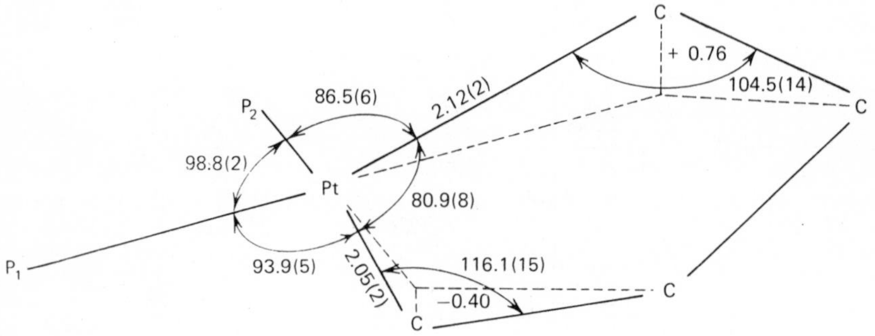

Fig. 2. Bond lengths in angstroms, bond angles in degrees.

The dependence of metal-ligand bond lengths on the nature of the ligand groups has been discussed by Mason and Randaccio [852] for several complexes of third-row transition metal halides with tertiary phosphines MCl_4L_2 (M = W, Re, Os, Ir, Pt). These authors defined theoretical metal bond radii as the distances from the metal to the centroid of overlap density of the metal ligand bonds. The calculated covalent radii of the metals with respect to phosphorus and chlorine reflect the observed bond lengths metal-phosphorus and metal-chlorine.

In contrast to the quite extended use of X-ray analysis, only a few electron diffraction studies have been made on transition metal complexes of phosphorus(III) ligands. The compounds investigated were exclusively PF_3 complexes and are listed in Table XIII.

TABLE XIII. Electron diffraction data on complexes of trifluorophosphine. The numbers in parantheses are the standard deviations. A-B stands for the distance A-B, ∢ABC for the bond angle at B between A and C

Ligand	Complex	Metal-P Distance [Å]	FPF Bond Angles	Comments	Refs.
PF_3	$Mo(CO)_5L$	2.369(10)	99.5(5)	P-F 1.557(4), Mo-C 2.063(6)	180
	NiL_4	2.099(3)	99.3	P-F 1.561(3), ∢PNiP 109.47 (not refined)	32
		2.116(10)	98.4(8)	P-F 1.561(5)	845
	PtL_4	2.230(10)	98.9(7)	P-F 1.546(6)	845
	$Rh(NO)L_3$	2.245(5)	100.1(7)	P-F 1.558(3), Rh-N 1.858(18), N-O 1.149(19), ∢PRhP 110.4(5)	181

Acknowledgments

I wish to thank Dr. J. A. Gibson (Braunschweig) for carefully reading the manuscript and Professor R. Schmutzler (Braunschweig) for his interest and helpful comments. I am indebted also to my wife Elisabeth for her assistance in the literature search and careful typing of the manuscript.

References

1. Ablov, A. V., A. M. Gol'dman, O. A. Bologa, Yu. A. Simonov, and M. M. Botoshanskii, *Zh. Neorg. Khim.*, **16**, 2187 (1971).
2. Adam, V. C., U. A. Gregory, and B. T. Kilbourn, *Chem. Commun.*, **1970**, 1400.
3. Adlard, M. W., and G. Socrates, J. C. S. *Chem. Comm.*, **1972**, 17.
4. Adrian, G. T., and C. D. Seibold, *Inorg. Chem.*, **12**, 2544 (1973).
5. Ahmad, N., E. W. Ainscough, T. A. James, and S. D. Robinson, *J.C.S. Dalton*, **1973**, 1148.
6. Ahmad, N., E. W. Ainscough, T. A. James, and S. D. Robinson, *J.C.S. Dalton*, **1973**, 1151.
7. Ahmad, N., S. D. Robinsom, and M. F. Uttley, *J.C.S. Dalton*, **1972**, 843.
8. Ainscough, E. W., T. A. James, S. D. Robinson, and J. N. Wingfield, *J. Organomet. Chem.*, **60**, C63 (1973).
9. Ainscough, E. W., and L. K. Peterson, *Inorg. Chem.*, **9**, 2699 (1970).
10. Ainscough, E. W., L. K. Peterson, and D. E. Sabourin, *Can. J. Chem.*, **48**, 401 (1970).
11. Ainscough, E. W., and S. D. Robinson, *Chem. Commun.*, **1970**, 863.
12. Ainscough, E. W., S. D. Robinson, and J. J. Levison, *J. Chem. Soc. (A)*, **1971**, 3413.
13. Akhtar, M., P. D. Ellis, A. G. MacDiarmid, and J. D. Odom, *Inorg. Chem.*, **11**, 2917 (1972).
14. Albano, V. G., and P. L. Bellon, *J. Organomet. Chem.*, **37**, 151 (1972).
15. Albano, V. G., P. L. Bellon, and G. Ciani, *J.C.S. Dalton*, **1972**, 1938.
16. Albano, V. G., P. L. Bellon, and G. Ciani, *J. Organomet. Chem.*, **31**, 75 (1971).
17. Albano, V. G., P. L. Bellon, and G. Ciani, *J. Organomet. Chem.*, **38**, 155 (1972).
18. Albano, V. G., P. L. Bellon, G. Ciani, and M. Manassero, *J.C.S. Dalton*, **1972**, 171.
19. Albano, V. G., P. L. Bellon, M. Manassero, and M. Sansoni, *Chem. Commun.*, **1970**, 1210.
20. Albano, V. G., P. Bellon, and M. Sansoni, *J. Chem. Soc. (A)*, **1971**, 2420.
21. Albano, V. G., G. Ciani, M. I. Bruce, G. Shaw, and F. G. A. Stone, *J. Organomet. Chem.*, **42**, C99 (1972).
22. Alexander, J., and A. Wojcicki, *Inorg. Chem.*, **12**, 74 (1973).
23. Alison, J. M. C., and T. A. Stephenson, *Chem. Commun.*, **1970**, 1092.
24. Alison, J. M. C., and T. A. Stephenson, *J.C.S. Dalton*, **1973**, 254.
25. Allen, F. H., G. Chang, K. K. Cheung, T. F. Lai, L. M. Lee, and A. Pidcock, *Chem. Commun.*, **1970**, 1297.
26. Allen, F. H., and K. M. Gabuji, *Inorg. Nucl. Chem. Lett.*, **7**, 833 (1971).
27. Allen, F. H., A. Pidcock, and C. R. Waterhouse, *J. Chem. Soc. (A)*, **1970**, 2087.
28. Allen, F. H., and S. N. Sze, *J. Chem. Soc. (A)*, **1971**, 2054.
29. Allison, D. A., J. Clardy, and J. G. Verkade, *Inorg. Chem.*, **11**, 2804 (1972).
30. Allison, D.A., and J. G. Verkade, *Phosphorus*, **2**, 257 (1973).
31. Allum, K. G., and R. D. Hancock, Ger. Offen. 2,003,294, 03 Dec 1970, Brit. Appl. 30 Jan 1969, through *Chem. Abstr.*, **74**, 54804m (1971).

32. Almenningen, A., B. Andersen, and E. E. Astrup, *Acta Chem. Scand.*, **24**, 1579 (1970).
33. Alyea, E. C., and D. W. Meek, *Inorg. Chem.*, **11**, 1029 (1972).
34. Anderson, D. W. W., E. A. V. Ebsworth, and D. W. H. Rankin, *J.C.S. Dalton*, **1973**, 854.
35. Anderson, D. W. W., E. A. V. Ebsworth, and D. W. H. Rankin, *J.C.S. Dalton*, **1973**, 2370.
36. Anderson, P. W., and T. L. Brown, *J. Organomet. Chem.*, **32**, 343 (1971).
37. Ando, N., K. Maruya, T. Mizoroki, and A. Ozaki, *J. Catal.*, **20**, 299 (1971).
38. Andrianov, V. G., G. N. Sergeeva, Yu. T. Struchkov, K. N. Anisimov, N. E. Kolobova, and A. S. Beschastnov, *Zh. Strukt. Khim.*, **11**, 168 (1970).
39. Ang, H. G., W. E. Kow, and K. F. Mok, *Inorg. Nucl. Chem. Lett.*, **8**, 829 (1972).
40. Angelici, R. J., and R. W. Brink, *Inorg. Chem.*, **12**, 1067 (1973).
41. Arai, H., and J. Halpern, *Chem. Commun.*, **1971**, 1571.
42. Araneo, A., F. Bonati, and G. Minghetti, *Inorg. Chim. Acta*, **4**, 61 (1970).
43. Archer, L. J., and T. A. George, *J. Organomet. Chem.*, **54**, C25 (1973).
44. Aresta, M., *Gazz. Chim. Ital.*, **102**, 781 (1972).
45. Aresta, M., P. Giannoccaro, M. Rossi, and A. Sacco, *Inorg. Chim. Acta*, **5**, 203 (1971).
46. Aresta, M., C. F. Nobile, M. Rossi, and A. Sacco, *Chem. Commun.*, **1971**, 781.
47. Armit, P. W., and T. A. Stephenson, *J. Organomet. Chem.*, **57**, C80 (1973).
48. Arnold, G. S., W. L. Klotz, W. Halper, and M. K. DeArmond, *Chem. Phys. Lett.*, **19**, 546 (1973).
49. Atkinson, L. K., and D. C. Smith, *J. Organomet. Chem.*, **33**, 189 (1971).
50. Atkins, R. W., and D. Kivelson, *J. Chem. Phys.*, **44**, 169 (1966).
51. Atkinson, L. K., A. H. Mawby, and D. C. Smith, *Chem. Commun.*, **1971**, 157.
52. Atkinson, L. K., A. H. Mawby, and D. C. Smith, *Chem. Commun.*, **1970**, 1399.
53. Attali, S., and R. Poilblanc, *Inorg. Chim. Acta*, **6**, 475 (1972).
54. Augustine, R. L., and J. F. VanPeppen, *Chem. Commun.*, **1970**, 497.
55. Axelson, D. E., and C. E. Holloway, *J.C.S. Chem. Comm.*, **1973**, 455.
56. Axtell, D. D., and J. T. Yoke, *Inorg. Chem.*, **12**, 1265 (1973).
57. Bacci, M., and S. Midollini, *Inorg. Chim. Acta*, **5**, 220 (1971).
58. Bacci, M., S. Midollini, P. Stoppioni, and L. Sacconi, *Inorg. Chem.*, **12**, 1801 (1973).
59. Bachman, D. F., E. D. Stevens, T. A. Lane, and J. T. Joke, *Inorg. Chem.*, **11**, 109 (1972).
60. Bailey, J., *J. Inorg. Nucl. Chem.*, **35**, 1921 (1973).
61. Bailey, J., M. J. Church, and M. J. Mays, *J. Coord. Chem.*, **3**, 63 (1973).
62. Baker, R. W., M. J. Braithwaite, and R. S. Nyholm, *J.C.S. Dalton*, **1972**, 1924.
63. Baker, R. W., B. Ilmayer, P. J. Pauling, and R. S. Nyholm, *Chem. Commun.*, **1970**, 1077.
64. Baker, R. W., and P. J. Pauling, *J.C.S. Dalton*, **1972**, 2264.
65. Ball, M. C., and J. M. Pope, *J.C.S. Dalton*, **1973**, 1802.
66. Bamford, C. H., and E. O. Hughes, *Proc. R. Soc. (London), A*, **326**, 469 (1972).
67. Bancroft, G. M., R. E. B. Garrod, and A. G. Maddock, *J. Chem. Soc (A)*, **1971**, 3165.
68. Bancroft, G. M., R. E. B. Garrod, A. G. Maddock, M. J. Mays, and B. E. Prater, *J. Am. Chem. Soc.*, **94**, 647 (1972).
69. Bancroft, G. M., and E. T. Libbey, *J.C.S. Dalton*, **1973**, 2103.
70. Bancroft, G. M., M. J. Mays, and B. E. Prater, *J. Chem. Soc (A)*, **1970**, 956.
71. Bancroft, G. M., M. J. Mays, B. E. Prater, and F. P. Stefanini, *J. Chem. Soc. (A)*, **1970**, 2146.
72. Bandoli, G., D. A. Clemente, G. Marangoni, and L. Cattalini, *J.C.S. Dalton*, **1973**, 886.
73. Banks, R. E., R. N. Haszeldine, M. Lappin, and A. B. P. Lever, *J. Organomet. Chem.*, **29**, 427 (1971).
74. Bannister, W. D., B. L. Booth, R. N. Haszeldine, and P. L. Loader, *J. Chem. Soc. (A)*, **1971**, 930.
75. Barbeau, C., K. S. Dichmann, and L. Ricard, *Can. J. Chem.*, **51**, 3027 (1973).
76. Barefield, E. K., and G. W. Parshall, *Inorg. Chem.*, **11**, 964 (1972).

77. Barlex, D. M., R. D. W. Kemmitt, and G. W. Littlecott, *J. Organomet. Chem.*, **43**, 225 (1972).
78. Barnett, K. W., T. G. Pollman, and T. W. Solomon, *J. Organomet. Chem.*, **36**, C23 (1973).
79. Bartish, C. M., and C. S. Kraihanzel, *Inorg. Chem.*, **12**, 391 (1973).
80. Baudler, M., and A. Zarkadas, *Chem. Ber.*, **105**, 3844 (1972).
81. Bauer, D. P., W. M. Douglas, and J. K. Ruff, *J. Organomet. Chem.*, **57**, C19 (1973).
82. Bausch, R., E. A. V. Ebsworth, and D. W. H. Rankin, *Angew. Chem.*, **83**, 111 (1971).
83. Beall, H., C. H. Bushweller, and M. Grace, *Inorg. Nucl. Chem. Lett.*, **7**, 641 (1971).
84. Beck, W., K. Schorpp, C. Oetker, R. Schlodder, and H. S. Smedal, *Chem. Ber.*, **106**, 2144 (1973).
85. Beck, W., and K. von Werner, *Chem. Ber.*, **106**, 868 (1973).
86. Behrens, H., H. D. Feilner, and E. Lindner, *Z. Anorg. Allg. Chem.*, **385**, 321 (1971).
87. Behrens, H., E. Lindner, and H. D. Feilner, *Z. Anorg. Allg. Chem.*, **385**, 325 (1971).
88. Bell, B., J. Chatt, and G. J. Leigh, *Chem. Commun.*, **1970**, 842.
89. Bell, B., J. Chatt, and G. J. Leigh, *J.C.S. Dalton*, **1972**, 2492.
90. Bell, B., J. Chatt, and G. J. Leigh, *J.C.S. Dalton*, **1973**, 997.
91. Bell, N. A., and L. A. Nixon, *Thermochim. Acta*, **3**, 61 (1971).
92. Bellon, P. L., F. Cariati, M. Manassero, L. Naldini, and M. Sansoni, *Chem. Commun.*, **1971**, 1423.
93. Bellon, P. L., M. Manassero, L. Naldini, and M. Sansoni, *J.C.S. Chem. Comm.*, **1972**, 1035.
94. Bellon, P., M. Manassero, and M. Sansoni, *J.C.S. Dalton*, **1972**, 1481.
95. Bellon, P., M. Manassero, and M. Sansoni, *J.C.S. Dalton*, **1973**, 2423.
96. Benaim, J., J. Y. Merour, and J. L. Roustan, *Tetrahedron Lett.*, **1971**, 983.
97. Benazeth, S., A. Loutellier, and M. Bigorgne, *J. Organomet. Chem.*, **24**, 479 (1970).
98. Bencze, L., *J. Organomet. Chem.*, **37**, C37 (1972).
99. Bencze, L., *J. Organomet. Chem.*, **56**, 303 (1973).
100. Bennett, E. W., and P. J. Orenski, *J. Organomet. Chem.*, **28**, 137 (1971).
101. Bennett, M. A., R. Bramley, and I. B. Tomkins, *J.C.S. Dalton*, **1973**, 166.
102. Bennett, M. A., and R. Charles, *Aust. J. Chem.*, **24**, 427 (1971).
103. Bennett, M. A., R. J. H. Clark, and A. D. J. Goodwin, *J. Chem. Soc. (A)*, **1970**, 541.
104. Bennett, M. A., and E. J. Hann, *J. Organomet. Chem.*, **29**, C15 (1971).
105. Bennett, M. A., K. Hoskins, W. R. Kneen, R. S. Nyholm, P. B. Hitchcock, R. Mason, G. B. Robertson, and A. D. C. Towl, *J. Am. Chem. Soc.*, **93**, 4591 (1971).
106. Bennett, M. A., R. N. Johnson, and I. B. Tomkins, *J. Organomet. Chem.*, **54**, C48 (1973).
107. Bennett, M. A., W. R. Kneen, and R. S. Nyholm, *J. Organomet. Chem.*, **26**, 293 (1971).
108. Bennett, M. A., and D. J. Patmore, *Inorg. Chem.*, **10**, 2387 (1971).
109. Bennett, M. A., G. B. Robertson, I. B. Tomkins, and P. O. Whimp, *Chem. Commun.*, **1971**, 341.
110. Bennett, M. A., G. B. Robertson, T. W. Turney, and P. O. Whimp, *Chem. Commun.*, **1971**, 762.
111. Bennett, M. A., G. B. Robertson, P. O. Whimp, and T. Yoshida, *J. Am. Chem. Soc.*, **93**, 3797 (1971).
112. Bennett, M. A., and I. B. Tomkins, *J. Organomet. Chem.*, **51**, 289 (1973).
113. Bennett, M. A., and T. W. Turney, *Aust. J. Chem.*, **26**, 2321 (1973).
114. Bennett, M. A., and T. W. Turney, *Aust. J. Chem.*, **26**, 2335 (1973).
115. Bennett, M. A., and R. Watt, *Chem. Commun.*, **1971**, 94.
116. Bennett, M. J., F. A. Cotton, and M. D. LaPrade, *Acta Cryst.*, **B 27**, 1899 (1971).
117. Bentham, J. E., S. Cradock, and E. A. V. Ebsworth, *Chem. Commun.*, **1969**, 528.
118. Bentham, J. E., and E. A. V. Ebsworth, *Inorg. Nucl. Chem. Lett.*, **6**, 671 (1970).
119. Bentley, G. A., K. R. Laing, W. R. Roper, and J. M. Waters, *Chem. Commun.*, **1970**, 998.

120. Bentley, R. B., F. E. Mabbs, W. R. Smail, M. Gerloch, and J. Lewis, *J. Chem. Soc. (A)*, **1970**, 3003.
121. Beran, G., A. J. Carty, P. C. Chieh, and H. A. Patel, *J.C.S. Dalton*, **1973**, 488.
122. Beran, G., and G. J. Palenik, *Chem. Commun.*, **1970**, 1354.
123. Bercaw, J., G. Guastalla, and J. Halpern, *Chem. Commun.*, **1971**, 1594.
124. Berry, S., *J. Chem. Phys.*, **32**, 933 (1960).
125. Bertacco, A., U. Mazzi, and A. A. Orio, *Inorg. Chem.*, **11**, 2547 (1972).
126. Bertoux, J., J. P. Martinaud, and R. Poilblanc, Ger. Offen. 2,039,938, 25 Feb 1971, Fr. Appl. 12 Aug 1969, through *Chem. Abstr.*, **74**, 99461p (1971).
127. Bertini, I., D. Gatteschi, and F. Mani, *Inorg. Chem.*, **11**, 2464 (1972).
128. Bertrand, R. D., F. B. Ogilvie, and J. G. Verkade, *J. Am. Chem. Soc.*, **92**, 1908 (1970).
129. Bianchi, A., P. Dapporto, G. Fallani, C. A. Ghilardi, and L. Sacconi, *J.C.S. Dalton*, **1973**, 641.
130. Bianchi, A., C. A. Ghilardi, C. Mealli, and L. Sacconi, *J.C.S. Chem. Comm.*, **1972**, 651.
131. Bianco, V. D., and S. Doronzo, *J. Organomet. Chem.*, **30**, 431 (1971).
132. Bianco, V. D., S. Doronzo, and M. Aresta, *J. Organomet. Chem.*, **42**, C63 (1972).
133. Biefeld, C. G., H. A. Eick, and R. H. Grubbs, *Inorg. Chem.*, **12**, 2166 (1973).
134. Bigorgne, M., *J. Organomet. Chem.*, **24**, 211 (1970).
135. Bigorgne, M., A. Loutellier, and M. Pankowski, *J. Organomet. Chem.*, **23**, 201 (1970).
136. Bigorgne, M., and J. B. Tripathi, *J. Mol. Struct.*, **10**, 449 (1971).
137. Birck, J. L., Y. LeCars, N. Baffier, J. J. Legendre, and M. Huber., *C.R. Acad. Sci.*, Ser. C, **273**, 880 (1971).
138. Bird, P. H., N. J. Coville, I. S. Butler, and M. L. Schneider, *Inorg. Chem.*, **12**, 2902 (1973).
139. Birk, J. P., J. Halpern, and A. L. Pickard, *J. Am. Chem. Soc.*, **90**, 4491 (1968).
140. Bishop, J. K. B., W. R. Cullen, and M. C. L. Gerry, *Can. J. Chem.*, **49**, 3910 (1971).
141. Blackburn, J. R., R. Nordberg, F. Stevie, R. G. Albridge, and M. M. Jones, *Inorg. Chem.*, **9**, 2374 (1970).
142. Blackmore, T., M. I. Bruce, F. G. A. Stone, R. E. Davis, and A. Garza, *Chem. Commun.*, **1971**, 852.
143. Blake, D. M., and R. Mersecchi, *Chem. Commun.*, **1971**, 1045.
144. Blake, D. M., and C. J. Nyman, *J. Am. Chem. Soc.*, **92**, 5359 (1970).
145. Bland, W. J., R. D. W. Kemmitt, and R. D. Moore, *J.C.S. Dalton*, **1973**, 1292.
146. Blight, D. G., D. L. Kepert, R. Mandyczewsky, and K. R. Trigwell, *J.C.S. Dalton*, **1972**, 313.
147. Blindheim, U., *Inorg. Chim. Acta*, **4**, 507 (1970).
148. Blum, J., and Y. Becker, *J.C.S. Perkin II*, **1972**, 982.
149. Blundell, T. L., and H. M. Powell, *Acta Cryst.*, **B 27**, 2304 (1971).
150. Bönnemann, H., *Angew. Chem.*, **82**, 699 (1970).
151. Bogdanovic, B., B, Henc, B. Meister, H. Pauling, and G. Wilke, *Angew. Chem.*, **84**, 1070 (1972).
152. Bombieri, G., E. Forsellini, C. Panattoni, R. Graziani, and G. Bandoli, *J. Chem. Soc. (A)*, **1970**, 1313.
153. Bond, A., and M. Green, *J. Chem. Soc (A)*, **1971**, 682.
154. Bond, S. T., and N. V. Duffy, *J. Inorg. Nucl. Chem.*, **35**, 3241, (1973).
155. Bondarenko, I. B., N. A. Buzina, Yu. S. Varshavskii, M. I. Gel'fman, V. V. Razumovskii, and T. G. Cherkasova, *Zh. Neorg. Khim.*, **16**, 3071 (1971).
156. Bonds, Jr. W. D., and J. A. Ibers, *J. Am. Chem. Soc.*, **94**, 3413 (1972).
157. Boorman, P. M., and A. J. Carty, *Inorg. Nucl. Chem. Lett.*, **4**, 101 (1968).
158. Booth, B., M. Else, R. Fields, H. Goldwhite, and R. Haszeldine, *J. Organomet. Chem.*, **14**, 417 (1968).

159. Booth, B. L., M. J. Else, R. Fields, and R. N. Haszeldine, *J. Organomet. Chem.*, **27**, 119 (1971).
160. Booth, B. L., and R. N. Haszeldine, *J. Chem. Soc. (A)*, **1966**, 157.
161. Booth, G., *Advan. Inorg. Radiochem.*, **6**, 1 (1964).
162. Booth, G., *Organophosphorus Compounds*, Vol 1, G. M. Kosolapoff and L. Maier, Eds., Wiley, New York, 1972, p. 433
163. Borodko, Yu. G., M. O. Broitman, L. M. Kachapina, A. E. Shilov, and L. Yu. Ukhin, *Chem. Commun.*, **1971**, 1185.
164. Boschi, T., P. Uguagliati, and B. Crociani, *J. Organomet. Chem.*. **30**, 283 (1971).
165. Bothner-By and A. A., S. M. Castellano, in D. F. DeTar, Ed., *Computer Programs for Chemistry*, Vol. 1, Benjamin, New York, 1968, p10.
166. Bowden, J. A., and R. Colton, *Aust. J. Chem.*, **24**, 2471 (1971).
167. Bowden, J. A., and R. Colton, *Aust. J. Chem.*, **25**, 17 (1972).
168. Bowden, J. A., and R. Colton, *Aust. J. Chem.*, **26**, 43 (1973).
169. Bowden, J. A., R. Colton, and C. J. Commons, *Aust. J. Chem.*, **25**, 1393 (1972).
170. Bowden, J. A., R. Colton, and C. J. Commons, *Aust. J. Chem.*, **26**, 655 (1973).
171. Bowen, L. H., P. E. Garrou, and G. G. Long, *Inorg. Chem.*, **11**, 182 (1972).
172. Bradford, C. W., and R. S. Nyholm, *J.C.S. Dalton*, **1973**, 529.
173. Braithwaite, M. J., and R. S. Nyholm, *J. Inorg. Nucl. Chem.*, **35**, 2237 (1973).
174. Braterman, P. S., *Organomet. Chem. Rev. B*, **9**, 1 (1972).
175. Braterman, P. S., D. W. Milne, E. W. Randall, and E. Rosenberg, *J.C.S. Dalton*, **1973**, 1027.
176. Braterman, P. S., and W. J. Wallace, *J. Organomet. Chem.*, **30**, C17 (1971).
177. Braterman, P. S., V. A. Wilson, and K. K. Joshi, *J. Organomet. Chem.*, **31**, 123 (1971).
178. Bresadola, S., A. Frigo, B. Longato, and G. Rigatti, *Inorg. Chem.*, **12**, 2788 (1973).
179. Bressan, M., B. Corain, P. Rigo, and A. Turco, *Inorg. Chem.*, **9**, 1733 (1970).
180. Bridges, D. M., G. C. Holywell, D. W. H. Rankin, and J. M. Freeman, *J. Organomet. Chem.*, **32**, 87 (1971).
181. Bridges, D. M., D. W. H. Rankin, D. A. Clement, and J. F. Nixon, *Acta Cryst. B*, **28**, 1130 (1972).
182. Bright, A, B. E. Mann, C. Masters, B. L. Shaw, R. M. Slade, and R. E. Stainbank, *J. Chem. Soc. (A)*, **1971**, 1826.
183. Brockhaus, M., F. Staudacher, and H. Vahrenkamp, *Chem. Ber.*, **105**, 3716 (1972).
184. Brookes, P. R., *J. Organomet. Chem.*, **42**, 459 (1972).
185. Brookes, P. R., *J. Organomet. Chem.*, **43**, 415 (1972).
186. Brookes, P. R., C. Masters, and B. L. Shaw, *J. Chem. Soc. (A)*, **1971**, 3756.
187. Brookes, P. R., and R. S. Nyholm, *Chem. Commun.*, **1970**, 169.
188. Brookes, P. R., and B. L. Shaw, *J.C.S. Dalton*, **1973**, 783.
189. Brower, K. R., and T. Chen, *Inorg. Chem.*, **12**, 2198 (1973).
190. Brown, D. A., H. J. Lyons, and A. R. Manning, *Inorg. Chim. Acta*, **4**, 428 (1970).
191. Brown, D. A., H. J. Lyons, and R. T. Shane, *Inorg. Chim. Acta*, **4**, 621 (1970).
192. Brown, D. A., and R. T. Shane, *J. Chem. Soc. (A)*, **1971**, 2088.
193. Brown, M. L., T. J. Meyer, and N. Winterton, *Chem. Commun.*, **1971**, 309.
194. Brown, M. P., R. J. Puddephatt, and C. E. E. Upton, *J. Organomet. Chem.*, **49**, C61 (1973).
195. Brown, R. A., and G. R. Dobson, *Inorg. Chim. Acta*, **6**, 65 (1972).
196. Browning, J., and B. R. Penfold, *J.C.S. Chem. Comm.*, **1973**, 198.
197. Bruce, M. I., J. Howard, I. W. Nowell, G. Shaw, and P. Woodward, *J.C.S. Chem. Comm.*, **1972**, 1041.
198. Bruce, M. I., G. Shaw, and F. G. A. Stone, *J.C.S. Dalton*, **1972**, 1082.
199. Bruce, M. I., G. Shaw, and F. G. A. Stone, *J.C.S. Dalton*, **1972**, 2094.

200. Bruce, M. I., G. Shaw, and F. G. A. Stone, *J.C.S. Dalton,* **1973**, 1667.
201. Brunner, H., and H. D. Schindler, *Chem. Ber.,* **104**, 2467 (1971).
202. Brunner, H., and H. D. Schindler, *Z. Naturforsch.* **26b**, 1220 (1971).
203. Brunner, H., and E. Schmidt, *J. Organomet. Chem.,* **50**, 219 (1973).
204. Buckingham, A. D., and P. J. Stephens, *J. Chem. Soc.,* **1964**, 2747.
205. Budd, D., R. Chuchman, D. G. Holah, A. N. Hughes, and B. C. Hui, *Can. J. Chem.,* **50**, 1008 (1972).
206. Bullock, J. I., F. W. Parrett, and N. J. Taylor, *J.C.S. Dalton,* **1973**, 522.
207. Burg, A. B., and I. B. Mishra, *J. Organomet. Chem.,* **24**, C33 (1970).
208. Burg, A. B., and I. H. Sabherwal, *Inorg. Chem.,* **9**, 974 (1970).
209. Burmeister, J. L., and L. M. Edwards, *J. Chem. Soc. (A),* **1971**, 1663.
210. Burmeister, J. L., and E. T. Weleski, Jr., *Syn. Inorg. Met-Org. Chem.,* **2**, 295 (1972).
211. Burnett, M. G., and R. J. Morrison, *J. Chem. Soc. (A),* **1971**, 2325.
212. Burnett, M. G., R. J. Morrison, and C. J. Strugnell, *J.C.S. Dalton,* **1973**, 701.
213. Busetto, L., G. Carturan, A. Palazzi, and U. Belluco, *J. Chem. Soc (A),* **1970**, 474.
214. Buslaev, Yu. A., M. A. Glushkova, and M. M. Ershova, U.S.S.R. 277,756, 05 Aug 1970, Appl. 04 Aug 1969, through *Chem. Abstr.,* **74**, 33169t (1971).
215. Butcher, A. V., and J. Chatt, *J. Chem. Soc. (A),* **1970**, 2652.
216. Butcher, A. V., and J. Chatt, *J. Chem. Soc. (A),* **1971**, 2356.
217. Butcher, A. V., J. Chatt, G. J. Leigh, and P. L. Richards, *J.C.S. Dalton,* **1972**, 1064.
218. Butter, S. A., and J. Chatt, *J. Chem. Soc. (A),* **1970**, 1411.
219. Byrd, J. E., and J. Halpern, *J. Am. Chem. Soc.,* **93**, 1634 (1971).
220. Cairns, M. A., and J. F. Nixon, *J. Organomet. Chem.,* **51**, C27 (1973).
221. Calderazzo, F., S. A. Losi, and B. P. Susz, *Helv. Chim. Acta,* **54**, 1156 (1971).
222. Cameron, A. F., K. P. Forrest, and G. Ferguson, *J. Chem. Soc. (A),* **1971**, 1286.
223. Cameron, T. S., M. L. H. Green, H. Munakata, C. K. Prout, and M. J. Smith, *J. Coord. Chem.,* **2**, 43 (1972).
224. Camus, A., and N. Marsich, *J. Organomet. Chem.,* **21**, 249 (1970).
225. Camus, A., N. Marsich, G. Nardin, and L. Randaccio, *J. Organomet. Chem.,* **60**, C39 (1973).
226. Cardaci, G., S. M. Murgia, and G. Reichenbach, *Inorg. Chim. Acta,* **4**, 118 (1970).
227. Cardin, C. J., D. J. Cardin, M. F. Lappert, and K. W. Muir, *J. Organomet. Chem.,* **60**, C70 (1973).
228. Cardin, D. J., S. A. Keppie, and M. F. Lappert, *J. Chem. Soc. (A),* **1970**, 2594.
229. Cariati, F., and L. Naldini, *Inorg. Chim. Acta,* **5**, 172 (1971).
230. Cariati, F., A. Sgamellotti, F. Morazzoni, and V. Valenti, *Inorg. Chim. Acta,* **5**, 531 (1971).
231. Cariati, F., R. Ugo, and F. Bonati, *Inorg. Chem.,* **5**, 1128 (1966).
232. Carroll, W. E., F. A. Deeney, J. A. Delaney, and F. J. Lalor, *J.C.S. Dalton,* **1973**, 718.
233. Carturan, G., and D. S. Martin, Jr., *Inorg. Chem.,* **9**, 258 (1970).
234. Carty, A. J., and A. Efraty, *Inorg. Chem.,* **8**, 543 (1969).
235. Carty, A. J., A. Efraty, T. W. Ng, and T. Birchall, *Inorg. Chem.,* **9**, 1263 (1970).
236. Casey, M., and A. R. Manning, *J. Chem. Soc. (A),* **1971**, 2989.
237. Cash, D. N., and R. O. Harris, *Can. J. Chem.,* **49**, 3821 (1971).
238. Caulton, K. G., *J. Am. Chem. Soc.,* **95**, 4076 (1973).
239. Caulton, K. G., and F. A. Cotton, *J. Am. Chem. Soc.,* **93**, 1914 (1971).
240. Cavit, B. E., K. R. Grundy, and W. R. Roper, *J.C.S. Chem. Comm.,* **1972**, 60.
241. Cenini, S., A. Fusi, and G. Capparella, *J. Inorg. Nucl. Chem.,* **33**, 3576 (1971).
242. Cenini, S., R. Ugo, and G. LaMonica, *J. Chem. Soc. (A),* **1971**, 416.
243. Cenini, S., R. Ugo, and G. LaMonica, *J. Chem. Soc. (A),* **1971**, 3441,
244. Cenini, S., R. Ugo, G. LaMonica, and S. D. Robinson, *Inorg. Chim. Acta,* **6**, 182 (1972).

245. Cetinkaya, B., E. Cetinkaya, and M. F. Lappert, *J.C.S. Dalton,* **1973**, 906.
246. Chalk, A. J., *J. Organomet. Chem.,* **21**, 207 (1970).
247. Chandra, G., Ger. Offen, 2,302,231, 02 Aug 1973, Brit. Appl. 2646, 19 Jan 1972, through *Chem. Abstr.,* **79**, 105400q (1973).
248. Chatt, J., and P. Chini, *J. Chem. Soc. (A),* **1970**, 1538.
249. Chatt, J., J. R. Dilworth, and G. J. Leigh, *J.C.S. Dalton,* **1973**, 612.
250. Chatt, J., J. R. Dilworth, J. G. Leigh, and R. L. Richards, *Chem. Commun.,* **1970**, 955.
251. Chatt, J., C. Eaborn, and P. N. Kapoor, *J. Chem. Soc. (A),* **1970**, 881.
252. Chatt, J., G. A. Heath, and R. L. Richards, *J.C.S. Chem. Comm.,* **1972**, 1010.
253. Chatt, J., G. J. Leigh, and R. L. Richards, *J. Chem. Soc. (A),* **1970**, 2243.
254. Chatt, J., G. J. Leigh, and R. M. Slade, *J.C.S. Dalton,* **1973**, 2021.
255. Chatt, J., G. J. Leigh, and N. Thankarajan, *J. Organomet. Chem.,* **29**, 105 (1971).
256. Chatt, J., L. Manojlovic-Muir, and K. W. Muir, *Chem. Commun.,* **1971**, 655.
257. Chatt, J., D. P. Melville, and R. L. Richards, *J. Chem. Soc. (A),* **1971**, 1169.
258. Chatt, J., R. L. Richards, and G. H. D. Royston, *Inorg. Chim. Acta,* **6**, 669 (1972).
259. Chatt, J., R. L. Richards, and G. H. D. Royston, *J.C.S. Dalton,* **1973**, 1433.
260. Chatt, J., and B. L. Shaw, *J. Chem. Soc.,* **1964**, 2508.
261. Chen, J. Y., and J. Halpern, *J. Am. Chem. Soc.,* **93**, 4939 (1971).
262. Chen, J. Y., J. Halpern, and J. Molin-Case, *J. Coord. Chem.,* **2**, 239 (1973).
263. Cheney, A. J., B. E. Mann, and B. L. Shaw, *Chem. Commun.,* **1971**, 431.
264. Cheney, A. J., B. E. Mann, B. L. Shaw, and R. M. Slade, *J. Chem. Soc. (A),* **1971**, 3833.
265. Cheney, A. J., and B. L. Shaw, *J.C.S. Dalton,* **1972**, 860.
266. Cheng, P. T., C. D. Cook, C. H. Koo, S. C. Nyburg, and M. T. Shiomi, *Acta Cryst. B,* **27**, 1904 (1971).
267. Cheng, P. T., and S. C. Nyburg, *Can. J. Chem.,* **50**, 912 (1972).
268. Cheung, K. K., T. F. Lai, and S. Y. Lam, *J. Chem. Soc. (A),* **1970**, 3345.
269. Cheung, K. K., T. F. Lai, and K. S. Mok, *J. Chem. Soc. (A),* **1971**, 1644.
270. Chia, L. S., W. R. Cullen, and D. A. Harbourne, *Can. J. Chem.,* **50**, 2182 (1972).
271. Chini, P., and G. Longoni, *J. Chem. Soc. (A),* **1970**, 1542.
272. Chow, K. K., M. T. Halfpenny, and C. A. McAuliffe, *J.C.S. Dalton,* **1973**, 147.
273. Chow, K. K., and C. A. McAuliffe, *Inorg. Nucl. Chem. Lett.,* **9**, 1189 (1973).
274. Choy, V. J., and C. J. O'Connor, *J.C.S. Dalton,* **1972**, 2017.
275. Chung, H., and W. Keim, U.S. 3,636,162, 18 Jan 1972, Appl. 653,032, 13 Jul 1967, through *Chem. Abstr.,* **76**, 85545y (1972).
276. Church, M. J., and M. J. Mays, *J. Inorg. Nucl. Chem.,* **33**, 253 (1971).
277. Church, M. J., M. J. Mays, R. N. F. Simpson, and F. P. Stefanini, *J. Chem. Soc. (A),* **1970**, 2909.
278. Churchill, M. R., and S. A. Bezman, *Inorg. Chem.,* **11**, 2243 (1972).
279. Churchill, M. R., and S. A. Bezman, *Inorg. Chem.,* **12**, 260 (1973).
280. Churchill, M. R., and S. A. Bezman, *Inorg. Chem.,* **12**, 531 (1973).
281. Churchill, M. R., and S. A. Bezman, *J. Organomet. Chem.,* **31**, C43 (1971).
282. Churchill, M. R., S. A. Bezman, J. A. Osborn, and J. Wormald, *Inorg. Chem.,* **11**, 1818 (1972).
283. Churchill, M. R., and K. L. Kalra, *J. Am. Chem. Soc.,* **95**, 5772 (1973).
284. Churchill, M. R., K. L. Kalra, and M. V. Veidis, *Inorg. Chem.,* **12**, 1656 (1973).
285. Churchill, M. R., and T. A. O'Brien, *J. Chem. Soc. (A),* **1970**, 161.
286. Churchill, M. R., and T. A. O'Brien, *J. Chem. Soc. (A),* **1970**, 206.
287. Churchill, M. R., and M. V. Veidis, *J. Chem. Soc. (A),* **1971**, 3463.
288. Churchill, M. R., and M. V. Veidis, *Chem. Commun.,* **1970**, 1099.
289. Churchill, M. R., and M. V. Veidis, *J.C.S. Dalton,* **1972**, 670.
290. Clark, G. R., and G. J. Palenik, *Inorg. Chem.,* **9**, 2754 (1970).

291. Clark, G. R., C. A. Reed, W. R. Roper, B. W. Skelton, and T. N. Waters, *Chem. Commun.*, **1971**, 758.
292. Clark, H. C., and K. Itoh, *Inorg. Chem.*, **10**, 1707 (1971).
293. Clark, M. G., W. R. Cullen, R. E. B. Garrod, A. G. Maddock, and J. R. Sams, *Inorg. Chem.*, **12**, 1045 (1973).
294. Clark, P. W., and G. E. Hartwell, *Inorg. Chem.*, **9**, 1948 (1970).
295. Clark, R. J., M. R. Abraham, and M. A. Busch, *J. Organomet. Chem.*, **35**, C33 (1972).
296. Clark, R. J., and M. A. Busch, *Acc. Chem. Res.*, **6**, 246 (1973).
297. Clark, R. J. H., and A. J. McAlees, *J. Chem. Soc. (A)*, **1970**, 2026.
298. Clarke, B., M. Green, and F. G. A. Stone, *J. Chem. Soc. (A)*, **1970**, 951.
299. Clarke, H. L., and N. J. Fitzpatrick, *Inorg. Nucl. Chem. Lett.*, **9**, 75 (1973).
300. Clement, D. A., and J. F. Nixon, *J.C.S. Dalton*, **1972**, 2553.
301. Clement, D. A., and J. F. Nixon, *J.C.S. Dalton*, **1973**, 195.
302. Clement, D. A., J. F. Nixon, and B. Wilkins, *J. Organomet. Chem.*, **37**, C43 (1972).
303. Cloyd, J. C., Jr., and D. W. Meek, *Inorg. Chim. Acta*, **6**, 480 (1972).
304. Cloyd, J. C., Jr., and D. W. Meek, *Inorg. Chim. Acta*, **6**, 607 (1972).
305. Collamati, I., A. Furlani, and G. Attioli, *J. Chem. Soc. (A)*, **1970**, 1694.
306. Collman, J. P., N. W. Hofman, and J. W. Hosking, *Inorg. Synth.* **12**, 8 (1970).
307. Colton, R., and C. J. Commons, *Aust. J. Chem.*, **26**, 1487 (1973).
308. Colton, R., and C. J. Commons, *Aust. J. Chem.*, **26**, 1493 (1973).
309. Colton, R., and J. J. Howard, *Aust. J. Chem.*, **22**, 2535, 2543 (1969).
310. Colton, R., and J. J. Howard, *Aust. J. Chem.*, **23**, 223 (1970).
311. Connor, J. A., J. P. Day, E. M. Jones, and G. K. McEwen, *J.C.S. Dalton*, **1973**, 347.
312. Cooke, M., M. Green, and T. A. Kuc, *J. Chem. Soc. (A)*, **1971**, 1200.
313. Cooper, D. G., and J. Powell, *J. Am. Chem. Soc.*, **95**, 1102 (1973).
314. Corain, B., *Coord. Chem. Rev.*, **8**, 159 (1972).
315. Corain, B., M. Bressan, and G. Favero, *Inorg. Nucl. Chem. Lett.*, **7**, 197 (1971).
316. Corain, B., M. Bressan, and P. Rigo, *J. Organomet. Chem.*, **28**, 133 (1971).
317. Corain, B., P. Rigo, and G. Favero, *Inorg. Chem.*, **10**, 2329 (1971).
318. Cotton, F. A., *Inorg. Chem.*, **5**, 1083 (1966).
319. Cotton, F. A., W. T. Edwards, F. C. Rauch, M. A. Graham, R. N. Perutz, and J. J. Turner, *J. Coord. Chem.*, **2**, 247 (1973).
320. Cotton, F. A., K. I. Hardcastle, and G. A. Rusholme, *J. Coord. Chem.*, **2**, 217 (1973).
321. Cotton, F. A., and J. G. Norman, Jr., *J. Am. Chem. Soc.*, **93**, 80 (1971).
322. Cotton, F. A., W. R. Robinson, and R. A. Walton, *Inorg. Chem.*, **6**, 223 (1967).
323. Couch, D. A., and S. D. Robinson, *Inorg. Nucl. Chem. Lett.*, **9**, 1079 (1973).
324. Couch, D. A., and S. D. Robinson, *Chem. Commun.*, **1971**, 1508.
325. Countryman, R., and B. R. Penfold, *J. Cryst. Mol. Struct.*, **2**, 281 (1972).
326. Covey, W. D., and T. L. Brown, *Inorg. Chem.*, **12**, 2820 (1973).
327. Coville, N. J., and I. S. Butler, *J. Organomet. Chem.*, **57**, 355 (1973).
328. Cowley, A. H., and K. E. Hill, *Inorg. Chem.*, **12**, 1446 (1973).
329. Cox, J. L., and J. Howatson, *Inorg. Chem.*, **12**, 1205 (1973).
330. Crabtree, R. H., A. R. Dias, M. L. H. Green, and P. J. Knowles, *J. Chem. Soc. (A)*, **1971**, 1350.
331. Craig, P. J., *Can. J. Chem.*, **48**, 3089 (1970).
332. Craig, P. J., and J. Edwards, *J. Organomet. Chem.*, **46**, 335 (1972).
333. Crichton, O., and A. J. Rest, *Inorg. Nucl. Chem. Lett.*, **9**, 391 (1973).
334. Crociani, B., T. Boschi, R. Pietropaolo, and U. Belluco, *J. Chem. Soc. (A)*, **1970**, 531.
335. Cross, J. H., and R. H. Fenn, *J. Chem. Soc. (A)*, **1970**, 3019.
336. Crossing, P. F., and M. R. Snow, *J. Chem. Soc. (A)*, **1971**, 610.
337. Crow, J. P., and W. R. Cullen, *Inorg. Chem.*, **10**, 1529 (1971).

338. Crow, J. P., W. R. Cullen, W. Harrison, and J. Trotter, *J. Am. Chem. Soc.,* **92**, 6339 (1970).
339. Crow, J. P., W. R. Cullen, F. G. Herring, J. R. Sams, and R. L. Tapping, *Inorg. Chem.,* **10**, 1616 (1971).
340. Crow, J. P., W. R. Cullen, and F. L. Hou, *Inorg. Chem.,* **11**, 2125 (1972).
341. Csontos, G., B. Heil, and L. Marko, *J. Organomet. Chem.,* **37**, 183 (1972).
342. Cullen, W. R., and D. A. Harbourne, *Inorg. Chem.,* **9**, 1839 (1970).
343. Cullen, W. R., D. A. Harbourne, B. V. Liengme, and J. R. Sams, *Inorg. Chem.,* **9**, 702 (1970).
344. Cullen, W. R., and F. L. Hou, *Can. J. Chem.,* **49**, 3404 (1971).
345. Cullen, W. R., and J. A. J. Thompson, *Can. J. Chem.,* **48**, 1730 (1970).
346. Cullen, W. R., J. R. Sams, and J. A. J. Thompson, *Inorg. Chem.,* **10**, 843 (1971).
347. Curran, R., J. A. Cunningham, and R. Eisenberg, *Inorg. Chem.,* **9**, 2749 (1970).
348. Curry, J. D., U.S. 3,760,000, 18 Sep 1973, Appl. 124,452, 15 Mar 1971, through *Chem. Abstr.,* **79**, 126609e (1973).
349. Curtis, M. D., *Inorg. Nucl. Chem. Lett.,* **6**, 859 (1970).
350. Cyvin, S. J., and A. Müller, *Acta Chem. Scand.,* **25**, 1149 (1971).
351. Dammann, C. B., J. L. Hughey, D. C. Jicha, T. J. Meyer, P. E. Rakita, and T. R. Weaver, *Inorg. Chem.,* **12**, 2206 (1973).
352. Dammann, C. B., P. Singh, and D. J. Hodgson, *J.C.S. Chem. Comm.,* **1972**, 586.
353. Dang, T. P., and H. B. Kagan, *Chem. Commun.,* **1971**, 481.
354. Dapporto, P., G. Fallani, S. Midollini, and L. Sacconi, *J. Am. Chem. Soc.,* **95**, 2021 (1973).
355. Dapporto, P., G. Fallani, S. Midollini, and L. Sacconi, *J.C.S. Chem. Comm.,* **1972**, 1161.
356. Dapporto, P., G. Fallani, and L. Sacconi, *J. Coord. Chem.,* **1**, 269 (1971).
357. Darensbourg, D. J., *Inorg. Chem.,* **10**, 2399 (1971).
358. Darensbourg, D. J., and T. L. Brown, *Inorg. Chem.,* **7**, 1679 (1968).
359. Darensbourg, D. J., M. Y. Darensbourg, and R. J. Dennenberg, *J. Am. Chem. Soc.,* **93**, 2807 (1971).
360. Dart, J. W., M. K. Lloyd, R. Mason, and J. A. McCleverty, *J.C.S. Dalton,* **1973**, 2046.
361. Dart, J. W., M. K. Lloyd, R. Mason, J. A. McCleverty, and J. Williams, *J.C.S. Dalton,* **1973**, 1747.
362. Dart, J. W., M. K. Lloyd, J. A. McCleverty, and R. Mason, *Chem. Comm.,* **1971**, 1197.
363. Dash, R. N., and D. V. Ramana Rao, *Z. Anorg. Allg. Chem.,* **393**, 309 (1972).
364. Davis, A. R., C. J. Murphy, and R. A. Plane, *Inorg. Chem.,* **9**, 423 (1970).
365. Davis, P. H., R. L. Belford, and I. C. Paul, *Inorg. Chem.,* **12**, 213 (1973).
366. Davis, R., and J. E. Fergusson, *Inorg. Chim. Acta,* **4**, 16 (1970).
367. Davis, R., and J. E. Fergusson, *Inorg. Chim. Acta,* **4**, 23 (1970).
368. Davison, A., and J. E. Ellis, *J. Organomet. Chem.,* **36**, 113 (1972).
369. Davison, A., and J. E. Ellis, *J. Organomet. Chem.,* **36**, 131 (1972).
370. Dawson, J. W., H. B. Gray, J. E. Hix, Jr., J. R. Preer, and L. M. Venanzi, *J. Am. Chem. Soc.,* **94**, 2979 (1972).
371. Dawson, J. W., B. C. Lane, R. J. Mynott, and L. M. Venanzi, *Inorg. Chim. Acta,* **5**, 25 (1971).
372. Dawson, J. W., L. M. Venanzi, J. R. Preer, J. E. Hix, Jr., and H. B. Gray, *J. Am. Chem. Soc.,* **93**, 778 (1971).
373. Day, J. P., D. Diemente, and F. Basolo, *Inorg. Chim. Acta,* **3**, 363 (1969).
374. Dean, W. K., G. L. Simon, P. M. Treichel, and L. F. Dahl, *J. Organomet. Chem.,* **50**, 193 (1973).
375. Debaerdemaeker, T., A. Kutoglu, G. Schmid, and L. Weber, *Acta Cryst. B,* **29**, 1283 (1973).

376. De Beer, J. A., and R. J. Haines, *J. Organomet. Chem.*, **36**, 297 (1972).
377. De Beer, J. A., R. J. Haines, R. Greatrex, and N. N. Greenwood, *J. Chem. Soc. (A)*, **1971**, 3271.
378. Deberitz, J., and H. Nöth, *Chem. Ber.*, **103**, 2541 (1970); *Chem. Ber.*, **106**, 2222 (1973).
379. Deberitz, J., and H. Nöth, *J. Organomet. Chem.*, **49**, 453 (1973).
380. Deeming, A. J., and B. L. Shaw, *J. Chem. Soc. (A)*, **1970**, 2705.
381. Deeming, A. J., and B. L. Shaw, *J. Chem. Soc. (A)*, **1970**, 3356.
382. Delbaere, L. T. J., D. W. McBride, and R. B. Ferguson, *Acta Cryst. B*, **26**, 515 (1970).
383. Delbeke, F. T., E. G. Claeys, and G. P. Van der Kelen, *J. Organomet. Chem.*, **25**, 219 (1970).
384. Delbeke, F. T., E. G. Claeys, G. P. Van der Kelen, and Z. Eeckhaut, *J. Organomet. Chem.*, **25**, 213 (1970).
385. Delbeke, F. T., and G. P. Van der Kelen, *J. Organomet. Chem.*, **21**, 155 (1970).
386. Demerseman, B., G. Bouquet, and M. Bigorgne, *J. Organomet. Chem.*, **35**, 341 (1972).
387. De Renzi, A., G. Paiaro, A. Panunzi, and L. Paolillo, *Gazz. Chim. Ital.*, **102**, 281 (1972).
388. Desnoyers, J., and R. Rivest, *Can. J. Chem.*, **43**, 1879 (1965).
389. Dewhirst, K. C., U.S. 3,489,786, 13 Jan 1970, Appl. 10 Dec 1964, through *Chem. Abstr.*, **72**, 89833f (1970).
390. Dewhirst, K. C., U.S. 3,639,439, 01 Feb 1972, Appl. 417,482, 10 Dec 1964, through *Chem. Abstr.*, **77**, 34708s (1972).
391. Dewhirst, K. C., W. Keim, and H. E. Thyret, U.S. 3,502,725, 24 Mar 1970, Appl. 20 Oct 1966, through *Chem. Abstr.*, **72**, 111023r (1970).
392. Dickson, R. S., and G. R. Tailby, *Aust. J. Chem.*, **23**, 229 (1970).
393. Dines, M. B., *Inorg. Chem.*, **11**, 2949 (1972).
394. Di Vaira, M., and A. B. Orlandini, *Inorg. Chem.*, **12**, 1292 (1973).
395. Dixon, K. R., and A. D. Rattray, *Can. J. Chem.*, **49**, 3997 (1971).
396. Dobbie, R. C., *Inorg. Nucl. Chem. Lett.*, **9**, 191 (1973).
397. Dobbie, R. C., M. J. Hopkinson, and D. Whittaker, *J.C.S. Dalton*, **1972**, 1030.
398. Dobbie, R. C., and D. Whittaker, *Chem. Commun.*, **1970**, 796.
399. Dobbie, R. C., and D. Whittaker, *J.C.S. Dalton*, **1973**, 2427.
400. Dobson, G. R., and A. J. Rettenmaier, *Inorg. Nucl. Chem. Lett.*, **6**, 327 (1970).
401. Dobson, G. R., and E. P. Ross, *Inorg. Chim. Acta.*, **5**, 199 (1971).
402. Dobson, J. E., R. G. Miller, and J. P. Wiggen, *J. Am. Chem. Soc.*, **93**, 554 (1971).
403. Dodd, H. T., and J. F. Nixon, *J. Organomet. Chem.*, **32**, C67 (1971).
404. Doddrell, D., and J. D. Roberts, *J. Am. Chem. Soc.*, **92**, 6839 (1970).
405. Douglas, W. M., and J. K. Ruff, *Inorg. Chem.*, **11**, 901 (1972).
406. Douglas, W. M., and J. K. Ruff, *J. Chem. Soc. (A)*, **1971**, 3558.
407. Douglas, W. M., and J. K. Ruff, *Syn. Inorg. Met-Org. Chem.*, **2**, 151 (1972).
408. Douglas, P. G., and B. L. Shaw, *J. Chem. Soc. (A)*, **1970**, 1556.
409. Douglas, P. G., and B. L. Shaw, *J.C.S. Dalton*, **1973**, 2078.
410. Drew, M. G. B., *J.C.S. Dalton*, **1972**, 1329.
411. Drew, M. G. B., D. G. Tisley, and R. A. Walton, *Chem. Commun.*, **1970**, 600.
412. Drew, M. G. B., I. B. Tomkins, and R. Colton, *Aust. J. Chem.*, **23**, 2517 (1970).
413. Drew, M. G. B., J. D. Wilkins, and A. P. Wolters, *J.C.S. Chem. Comm.*, **1972**, 1278.
414. Drinkard, W. C., D. R. Eaton, J. P. Jesson, and R. V. Lidsey, Jr., *Inorg. Chem.*, 9, 392 (1970).
415. Duckworth, V. F., P. G. Douglas, R. Mason, and B. L. Shaw, *Chem. Commun.*, **1970**, 1083.
416. Duddell, D. A., P. L. Goggin, R. J. Goodfellow, M. G. Norton, and J. G. Smith, *J. Chem. Soc. (A)*, **1970**, 545.
417. Duff, J. M., and B. L. Shaw, *J.C.S. Dalton*, **1972**, 2219.

418. Dumler, J. T., and D. M. Roundhill, *J. Organomet. Chem.*, **30**, C35 (1971).
419. D'yachkovskii, F. S., N. E. Khrushch, and A. E. Shilov, *Zh. Obshch. Khim.*, **40**, 1726 (1970).
419a. Dyer, G., and D. W. Meek, *J. Am. Chem. Soc.*, **89**, 3983 (1967).
420. Eaborn, C., T. N. Metham, and A. Pidcock, *J. Organomet. Chem.*, **54**, C3 (1973).
421. Eaton, D. R., and W. D. Philips, *Adv. Magn. Resonance*, **1**, 103 (1965).
422. Edwards, H. G. M., and L. A. Woodward, *Spectrochim. Acta*, **26A**, 897 (1970).
423. Edwards, H. G. M., and L. A. Woodward, *Spectrochim. Acta*, **26A**, 1077 (1970).
424. Ehrl, W., and H. Vahrenkamp, *Chem. Ber.*, **104**, 3261 (1971).
425. Einstein, F. W. B., and C. R. S. M. Hampton, *Can. J. Chem.*, **49**, 1901 (1971).
426. Einstein, F. W. B., and R. D. G. Jones, *Inorg. Chem.*, **12**, 255 (1973).
427. Einstein, F. W. B., and R. D. G. Jones, *J.C.S. Dalton*, **1972**, 442.
428. Einstein, F. W. B., and R. D. G. Jones, *J.C.S. Dalton*, **1972**, 2563.
429. Eliades, T. I., R. O. Harris, and M. C. Zia, *Chem. Commun.*, **1970**, 1709.
430. Eller, P. G., and D. W. Meek, *Inorg. Chem.*, **11**, 2518 (1972).
431. Ellermann, J., H. Behrens, and H. Krohberger, *J. Organomet. Chem.*, **46**, 119 (1972).
432. Ellermann, J., and W. Uller, *Z. Naturforsch.*, **25b**, 756 (1970).
433. Ellermann, J., and W. Uller, *Z. Naturforsch.*, **25b**, 1353 (1970).
434. Ellgen, P. C., and J. N. Gerlach, *Inorg. Chem.*, **12**, 2526 (1973).
435. Emery, A., A. C. Oehlschlager, and A. M. Unrau, *Tetrahedron. Lett.*, **1970**, 4401.
436. Enemark, J. H., *Inorg. Chem.*, **10**, 1952 (1971).
437. Engel, R., *Chem. Commun.*, **1970**, 133.
438. Engel, R., and L. Gelbaum, *J.C.S. Perkin I*, **1972**, 1233.
439. Evans, J. A., M. J. Hacker, R. D. W. Kemmitt, D. R. Russell, and J. Stocks, *J.C.S. Chem. Comm.*, **1972**, 72.
440. Fahey, D. R., *J. Am. Chem. Soc.*, **92**, 402 (1970).
441. Fahey, D. R., U.S. 3,686,245, 22 Aug 1972, Appl. 50,958, 29 Jun 1970, through *Chem. Abstr.*, 77, 140298t (1972).
442. Faller, J. W., and A. S. Anderson, *J. Am. Chem. Soc.*, **92**, 5852 (1970).
443. Faraone, F., F. Cusmano, and R. Pietropaolo, *J. Organomet. Chem.*, **26**, 147 (1971).
444. Faraone, F., P. Piraino, and R. Pietropaolo, *J.C.S. Dalton*, **1973**, 1625.
445. Faraone, G., V. Ricevuto, R. Romeo, and M. Trozzi, *Inorg. Chem.*, **9**, 1525 (1970).
446. Faraone, G., V. Ricevuto, R. Romeo, and M. Trozzi, *J. Chem. Soc. (A)*, **1971**, 1877.
447. Farmery, K., and M. Kilner, *J. Chem. Soc. (A)*, **1970**, 634.
448. Farnell, L. F., E. W. Randall, and E. Rosenberg, *Chem. Commun.*, **1971**, 1078.
449. Fenn, R. H., and J. H. Cross, *J. Chem. Soc. (A)*, **1971**, 3312.
450. Fenn, R. H., and G. R. Segrott, *J. Chem. Soc. (A)*, **1970**, 2781.
451. Fenton, D. M., U.S. 3,720,697, 13 Mar 1973, Appl. 119,836, 01 Mar 1971, through *Chem. Abstr.*, 78, 136410b (1973).
452. Fenton, D. E., and J. J. Zuckerman, *Inorg. Chem.*, **8**, 17777 (1969).
453. Ferguson, J. A., and T. J. Meyer, *Chem. Commun.*, **1971**, 1544.
454. Fergusson, J. E., and J. H. Hickford, *Aust. J. Chem.*, **23**, 453 (1970).
455. Ferrari, G. F., A. Andreetta, G. F. Pregaglia, and R. Ugo, *J. Organomet. Chem.*, **43**, 213 (1972).
456. Ferraro, J. R., and K. Nakamoto, *Inorg. Chem.*, **11**, 2290 (1972).
457. Ferraro, J. R., K. Nakamoto, J. T. Wang, and L. Lauer, *J.C.S. Chem. Comm.*, **1973**, 266.
458. Field, J. S., and P. J. Wheatley, *J.C.S. Dalton*, **1972**, 2269,
459. Fine, L. W., M. Grayson, and V. H. Suggs, *J. Organomet. Chem.*, **22**, 219 (1970).
460. Finer, E. G., and R. K. Harris, *J. Chem. Soc. (A)*, **1969**, 1972.
461. Fink, W., *Helv. Chim. Acta*, **54**, 1304 (1971).
462. Fink, W., and A. Wenger, *Helv. Chim. Acta*, **54**, 2186 (1971).

463. Fischer, E. O., and W. A. Herrmann, *Chem. Ber.*, **105**, 286 (1972).
464. Fischer, E. O., L. Knauss, R. L. Keiter, and J. G. Verkade, *J. Organomet. Chem.*, **37**, C7 (1972).
465. Fischer, E. O., E. Louis, W, Bathelt, and J. Müller, *Chem. Ber.*, **102**, 2547 (1969).
466. Fitton, P., and E. A. Rick, *J. Organomet. Chem.*, **28**, 287 (1971).
467. Fitzgerald, R. J., N. Y. Sakkab, R. S. Strange, and V. P. Narutis, *Inorg. Chem.*, **12**, 1081 (1973).
468. Fitzgerald, R. J., and H. M. Wu Lin, *Inorg. Chem.*, **11**, 2270 (1972).
469. Fluck, E., *Fortschr. Chem. Forsch.*, **35**, 1 (1973).
470. Fluck, E., and V. Novobilsky, *Fortschr. Chem. Forsch.*, **13**, 125 (1969).
471. Folkeson, B., *Acta Chem. Scand.*, **27**, 1441 (1973).
472. Francis, J. N., A. McAdam, and J. A. Ibers, *J. Organomet. Chem.*, **29**, 131 (1971).
473. Fraser, M., D. G. Holah, A. N. Hughes, and B. C. Hui, *J. Heterocycl. Chem.*, **9**, 1457 (1972).
474. Freni, M., D. Giusto, and P. Romiti, *J. Inorg. Nucl. Chem.*, **33**, 4093 (1971).
475. Freni, M., and P. Romiti, *Inorg. Nucl. Chem. Lett.*, **6**, 167 (1970).
476. Frenz, B. A., and J. A. Ibers, *Inorg. Chem.*, **9**, 2403 (1970).
477. Friedel, H., I. W. Renk, and H. Tom Dieck, *J. Organomet. Chem.*, **26**, 247 (1971).
478. Frigo, A., G. Puosi, and A. Turco, *Gazz. Chim. Ital.*, **101**, 637 (1971).
479. Fryer, C. W., and J. A. S. Smith, *J. Chem. Soc. (A)*, **1970**, 1029.
480. Fusi, A., R. Ugo, F. Fox, A. Pasini, and S. Cenini, *J. Organomet. Chem.*, **26**, 417 (1971).
481. Gallay, J., D. De Montauzon, and R. Poiblanc, *C.R. Acad. Sci., Ser. C*, **273**, 988 (1971).
482. Gallay, J., D. De Montauzon, and R. Poiblanc, *J. Organomet. Chem.*, **38**, 179 (1972).
483. Gansow, O. A., B. Y. Kimura, G. R. Dobson, and R. A. Brown, *J. Am. Chem. Soc.*, **93**, 5922 (1971).
484. Gargano, M., P. Giannoccaro, M. Rossi, and A. Sacco, *J.C.S. Chem. Comm.*, **1973**, 233.
485. Garif'yanov, N. S., A. D. Troitskaya, A. I. Razumov, P. A. Gurevich, and O. I. Kondrat'eva, *Zh. Obshch. Khim.*, **41**, 710 (1971).
486. Garif'yanov, N. S., A. D. Troitskaya, A. I. Razumov, I. V. Ovchinnikov, P. A. Gurevich, and O. I. Kondrat'eva, *Zh. Neorg. Khim.*, **17**, 1346 (1972).
487. Garrou, P. E., and G. E. Hartwell, *J.C.S. Chem. Comm.*, **1972**, 881.
488. Garrou, P. E., and G. E. Hartwell, *J. Organomet. Chem.*, **55**, 331 (1973).
489. Gaughan, A. P., R. F. Ziolo, and Z. Dori, *Inorg. Chem.*, **10**, 2776 (1971).
490. Gaughan, A. P., R. F. Ziolo, and Z. Dori, *Inorg. Chim. Acta*, **4**, 640 (1970).
491. Gavrilova, I. V., M. I. Gel'fman, N. V. Ivannikova, N. V. Kiseleva, and V. V. Razumovskii, *Zh. Neorg. Khim.*, **18**, 194 (1973).
492. Gavrilova, I. V., M. I. Gel'fman, N. V. Ivannikova, and V. V. Razumovskii, *Zh. Neorg. Khim.*, **16**, 1124 (1971).
493. Gehrke, H., Jr., and G. Eastland, *Inorg. Chem.*, **9**, 2722 (1970).
494. George, A. D., and T. A. George, *Inorg. Chem.*, **11**, 892 (1972).
495. George, T. A., *Inorg. Chem.*, **11**, 77 (1972).
496. George, T. A., and C. D. Seibold, *J. Am. Chem. Soc.*, **94**, 6859 (1972).
497. George, T. A., and C. D. Seibold, *J. Organomet. Chem.*, **30**, C13 (1971).
498. George, T. A., and C. D. Turnipseed, *Inorg. Chem.*, **12**, 394 (1973).
499. Gerlach, D. H., A. R. Kane, G. W. Parshall, J. P. Jesson, and E. L. Muetterties, *J. Am. Chem. Soc.*, **93**, 3543 (1971).
500. Giannoccaro, P., M. Rossi, and A. Sacco, *Coord. Chem. Rev.*, **8**, 77 (1972).
501. Gilbert, D. J., D. Rose, and G. Wilkinson, *J. Chem. Soc. (A)*, **1970**, 2765.
502. Gill, D. F., B. E. Mann, C. Masters, and B. L. Shaw, *Chem. Commun.*, **1970**, 1269.
503. Gill, D. F., B. E. Mann, and B. L. Shaw, *J.C.S. Dalton*, **1973**, 311.
504. Gillard, R. D., R. Ugo, F. Cariati, S. Cenini, and F. Bonati, *Chem. Commun.*, **1966**, 869.

505. Gilmore, C. J., and P. Woodward, *Chem. Commun.*, **1971**, 1233.
506. Ginsberg, A. P., and W. E. Lindsell, *Inorg. Chem.*, **12**, 1983 (1973).
507. Ginsberg, A. P., and M. E. Tully, *J. Am. Chem. Soc.*, **95**, 4749 (1973).
508. Ginzburg, A. G., B. V. Lokshin, V. N. Setkina, and D. N. Kursanov, *J. Organomet. Chem.*, **55**, 357 (1973).
509. Ginzburg, A. G., V. N. Setkina, and D. N. Kursanov, *Izv. Akad. Nauk SSSR, Ser. Khim.*, **1971**, 177.
510. Ginzburg, G. D., V. V. Sentemov, and A. D. Troitskaya, *Zh. Neorg. Khim.*, **17**, 1397 (1972).
511. Ginzburg, G. D., A. D. Troitskaya, and I. M. Babina, *Zh. Neorg. Khim.*, **18**, 1576 (1973).
512. Ginzburg, G. D., E. A. Zgadzai, N. S. Kolyubakina, P. A. Kirpichnikov, N. A. Mukmeneva, V. A. Pishchulina, and A. D. Troitskaya, *Zh. Neorg. Khim.*, **16**, 1923 (1971).
513. Giusto, D., and G. Cova, *Gazz. Chim. Ital.*, **101**, 519 (1971).
514. Glockling, F., and A. McGregor, *J. Inorg. Nucl. Chem.*, **35**, 1481 (1973).
515. Glockling, F., and S. R. Stobart, *MTP International Review of Science*, Vol. 6, Part 2, H. J. Emeleus, Ed., Butterworth, London, 1972, p. 63.
516. Glushkova, M. A., M. M. Ershova, N. A. Ovchinnikova, and Yu. A. Buslaev, *Zh. Neorg. Khim.*, **17**, 144 (1972).
517. Goggin, P. L., R. J. Goodfellow, S. R. Haddock, and J. G. Eary, *J.C.S. Dalton*, **1972**, 647.
518. Goggin, P. L., R. J. Goodfellow, J. R. Knight, M. G. Norton, and B. F. Taylor, *J.C.S. Dalton*, **1973**, 2220.
519. Goodfellow, R. J., M. J. Hardy, and B. F. Taylor, *J.C.S. Dalton*, **1973**, 2450.
520. Gosser, L. W., and G. W. Parshall, *Tetrahedron Lett.*, **1971**, 2555.
521. Gosser, L. W., and C. A. Tolman, Ger. Offen. 2,237,704, 15 Feb 1973, US Appl. 168,353, 02 Aug 1971, through *Chem. Abstr.*, **78**, 113465j (1973).
522. Gosser, L. W., and C. A. Tolman, *Inorg. Chem.*, **9**, 2350 (1970).
523. Grace, M., H. Beall, and C. H. Bushweller, *Chem. Commun.*, **1970**, 701.
524. Graham, B. W., K. R. Laing, C. J. O'Connor, and W. R. Roper, *J.C.S. Dalton*, **1972**, 1237.
525. Graham, J. R., and R. J. Angelici, *J. Am. Chem. Soc.*, **87**, 5590 (1965).
526. Grandberg, K. I., T. V. Baukova, E. G. Perevalova, and A. V. Nesmeyanov, *Dokl. Akad. Nauk SSSR*, **206**, 1355 (1972).
527. Graziani, R., G. Bombieri, and E. Forsellini, *Inorg. Nucl. Chem. Lett.*, **8**, 701 (1972).
528. Green, J. C., D. I. King, and J. H. D. Eland, *Chem. Commun.*, **1970**, 1121.
529. Green, M., and T. A. Kuc, Ger. Offen. 2,153,332, 27 Apr 1972, Brit Appl. 50,729/70, 26 Oct 1970, through *Chem. Abstr.*, **77**, 37215q (1972).
530. Green, M. L. H., J. Knight, L. C. Mitchard, G. G. Roberts, and W. E. Silverthorn, *Chem. Commun.*, **1971**, 1619.
531. Green, M. L. H., and C. R. Lucas, *J.C.S. Dalton*, **1972**, 1000.
532. Green, M. L. H., L. C. Mitchard, and W. E. Silverthorn, *J. Chem. Soc. (A)*, **1971**, 2929.
533. Green, M. L. H., L. C. Mitchard, and W. E. Silverthorn, *J.C.S. Dalton*, **1973**, 2177.
534. Green, M. L. H., and H. Munakata, *Chem. Commun.*, **1971**, 549.
535. Green, M. L. H., and M. J. Smith, *J. Chem. Soc. (A)*, **1971**, 639.
536. Green, M. L. H., and R. N. Whiteley, *J. Chem. Soc. (A)*, **1971**, 1943.
537. Green, P. J., and T. H. Brown, *Inorg. Chem.*, **10**, 206 (1971).
538. Gregory, U. A., J. A. J. Jarvis, B. T. Kilbourn, and P. G. Owston, *J. Chem. Soc. (A)*, **1970**, 2770.
539. Grim, S. O., J. Del Gaudio, C. A. Tolman, and J. P. Jesson, *Inorg. Nucl. Chem. Lett.*, **9**, 1083 (1973).
540. Grim, S. O., and R. A. Ference, *Inorg. Chim. Acta*, **4**, 277 (1970).
541. Grim, S. O., and R. A. Ference, *J. Coord. Chem.*, **2**, 225 (1973).
542. Grim, S. O., and R. L. Keiter, *Inorg. Chim. Acta*, **4**, 56 (1970).

543. Grim, S. O., H. J. Plastas, C. L. Huheey, and J. E. Huheey, *Phosphorus,* **1**, 61 (1971).
544. Grobe, J., and R. Demuth, *Angew. Chem.*, **84**, 1153 (1972).
545. Grobe, J., and U. Möller, *J. Organomet. Chem.*, **36**, 335 (1972).
546. Grosse, J., and R. Schmutzler, *Z. Naturforsch.*, **28b**, 515 (1973).
547. Groth, P., *Acta Chem. Scand.*, **24**, 2785 (1970).
548. Guggenberger, L. J., *Inorg. Chem.*, **12**, 499 (1973).
549. Guggenberger, L. J., *Inorg. Chem.*, **12**, 2295 (1973).
550. Guggenberger, L. J., U. Klabunde, and R. A. Schunn, *Inorg. Chem.*, **12**, 1143 (1973).
551. Gusev, A. I., G. G. Aleksandrov, and Yu. T. Struchkov, *Zh. Strukt. Khim.*, **14**, 685 (1973).
552. Guss, J. M., and R. Mason, *J.C.S. Dalton,* **1972**, 2193.
553. Haake, P., and R. M. Pfeiffer, *J. Am. Chem. Soc.*, **92**, 5243 (1970).
554. Hagihara, N., and S. Takahashi, Japan. 71 11,166, 22 Mar 1971, Appl. 15 Mar 1967, through *Chem. Abstr.*, **75**, 36349h (1971).
555. Haines, L. M., *Inorg. Chem.*, **9**, 1517 (1970).
556. Haines, R. J., and A. L. Du Preez, *Inorg. Chem.*, **11**, 330 (1972).
557. Haines, R. J., and A. L. Du Preez, *J. Organomet. Chem.*, **21**, 181 (1970).
558. Haines, R. J., A. L. Du Preez, and I. L. Marais, *J. Organomet. Chem.*, **28**, 97 (1971).
559. Haines, R. J., A. L. Du Preez, and I. L. Marais, *J. Organomet. Chem.*, **28**, 405 (1971).
560. Haines, R. J., A. L. Du Preez, and C. R. Nolte, *J. Organomet. Chem.*, **55**, 199 (1973).
561. Haines, R. J., R. Mason, J. A. Zubieta, and C. R. Nolte, *J.C.S. Chem. Comm.*, **1972**, 990.
562. Haines, R. J., and C. R. Nolte, *J. Organomet. Chem.*, **24**, 725 (1970).
563. Halfpenny, M. T., and L. M. Venanzi, *Inorg. Chim. Acta,* **5**, 91 (1971).
564. Hall, D. I., J. H. Ling, and R. S. Nyholm, *Structure and Bonding,* **15**, 3 (1973).
565. Hall, D. I., and R. S. Nyholm, *J. Chem. Soc. (A),* **1971**, 1491.
566. Hall, D. I., and R. S. Nyholm, *J.C.S. Dalton,* **1972**, 804.
567. Hall, M. C., J. A. J. Jarvis, B. T. Kilbourn, and P. G. Owston, *J.C.S. Dalton,* **1972**, 1544.
568. Hall, M. C., B. T. Kilbourn, and K. A. Taylor, *J. Chem. Soc. (A),* **1970**, 2539.
569. Hallman, P. S., T. A. Stephenson, and G. Wilkinson, *Inorg. Syntheses,* **12**, 237 (1970).
570. Halpern, J., E. Kimura, J. Molin-Case, and C. S. Wong, *Chem. Commun.*, **1971**, 1207.
571. Halpern, J., and A. L. Pickard, *Inorg. Chem.*, **9**, 2798 (1970).
572. Harbourne, D. A., and F. G. A. Stone, *J. Chem. Soc. (A),* **1968**, 1765.
573. Harder, V., J. Müller, and H. Werner, *Helv. Chim. Acta,* **54**, 1 (1971).
574. Hardy, A. D. U., and G. A. Sim, *J.C.S. Dalton,* **1972**, 1900.
575. Harman, R. E., J. L. Parsons, and S. K. Gupta, *Org. Prep. Proc.*, **2**, 25 (1970).
576. Harris, R. K., J. R. Woplin, and R. Schmutzler, *Ber. Bunsenges. Phys. Chem.*, **75**, 134 (1971).
577. Harris, R. O., N. K. Hota, L. Sadavoy, and J. M. C. Yuen, *J. Organomet. Chem.*, **54**, 259 (1973).
578. Hartley, F. R., and G. W. Searle, *Inorg. Chem.*, **12**, 1949 (1973).
579. Hartwell, G. E., and P. W. Clark, *Chem. Commun.*, **1970**, 1115.
580. Hattori, S., and M. Katsuta, Japan. Kokai 72 20,124, 27 Sep 1972, Appl. 71 10,881, 02 Mar 1971, through *Chem. Abstr.*, 78, 6051z (1973).
581. Haymore, B. L., and J. A. Ibers, *J. Am. Chem. Soc.*, **95**, 3052 (1973).
582. Heaton, B. T., and K. J. Timmins, *J.C.S. Chem. Comm.*, **1973**, 931.
583. Heck, R. F., *J. Am. Chem. Soc.*, **85**, 657 (1963).
584. Heimbach, P., H. Selbeck, and E. Troxler, *Angew. Chem.*, **83**, 731 (1971).
585. Henrici-Olive, G., and S. Olive, *Angew. Chem.*, **82**, 955 (1970).
586. Henrici-Olive, G., and S. Olive, *J. Am. Chem. Soc.*, **93**, 4154 (1971).
587. Hessett, B., J. H. Morris, and P. G. Perkins, *Inorg. Nucl. Chem. Lett.*, **7**, 1149 (1971).
588. Hidai, M., M. Kokura, and Y. Uchida, *Bull. Chem. Soc. Jap.*, **46**, 686 (1973).

589. Hidai, M., T. Kuse, T. Hikita, Y. Uchida, and A. Misono, *Tetrahedon Lett.*, **1970**, 1715.
590. Hidai, M., K. Tominari, and Y. Uchida, *J. Am. Chem. Soc.*, **94**, 110 (1972).
591. Higginson, B. R., C. A. McAuliffe, and L. M. Venanzi, *Inorg. Chim. Acta,* **5**, 37 (1971).
592. Highsmith, R. E., J. R. Bergerud, and A. G. MacDiarmid, *Chem. Commun.*, **1970**, 48.
593. Hildenbrand, K., and H. Dreeskamp, *Z. Naturforsch.*, **28b**, 226 (1973).
594. Hill, N. J., *J.C.S. Faraday Trans.*, 2, **1972**, 427.
595. Hill, T., Ger. Offen. 2,004,361, 10 Dec 1970, Brit. Appl. 30 Jan 1969, through *Chem. Abstr.*, **74**, 54350k (1970).
596. Hill, W. E., J. Dalton, and C. A. McAuliffe, *J.C.S. Dalton,* **1973**, 143.
597. Hillier, I. H., V. R. Saunders, M. J. Ware, P. J. Bassett, D. R. Lloyd, and N. Lynaugh, *Chem. Commun.*, **1970**, 1316.
598. Himmele, W., A. Werner, and F. J. Müller, Ger. Offen. 1,957,300, 19 May 1971, Appl. 14 Nov 1969, through *Chem. Abstr.*, **75**, 36336b (1971).
599. Hiraki, K., and H. Hirai, *Macromolecules,* **3**, 382 (1970).
600. Hiraki, K., S. Kaneko, and H. Hirai, *Polym. J.*, **2**, 225 (1971).
601. Höfler, M., and W. Marre, *Angew. Chem.*, **83**, 174 (1971).
602. Höfler, M., and M. Schnitzler, *Chem. Ber.*, **104**, 3117 (1971).
603. Höfler, M., and M. Schnitzler, *Chem. Ber.*, **105**, 1133 (1972).
604. Holah, D. G., A. N. Hughes, and B. C. Hui, *Can. J. Chem.*, **50**, 3714 (1972).
605. Holsboer, F., W. Beck, and H. D. Bartunik, *Chem. Phys. Lett.*, **18**, 217 (1973).
606. Holsboer, F., W. Beck, and H. D. Bartunik, *J.C.S. Dalton,* **1973**, 1828.
607. Horner, L., and K. Kunz, *Chem. Ber.*, **104**, 717 (1971).
608. Horner, L., and E. Müller, *Phosphorus,* **2**, 73 (1972).
609. Horner, L., and E. Müller, *Phosphorus,* **2**, 77 (1972).
610. Horner, L., and E. Müller, *Phosphorus,* **2**, 117 (1972).
611. Horner, L., and H. Siegel, *Liebigs Ann. Chem.*, **751**, 135 (1971).
612. Horner, L., and H. Siegel, *Phosphorus,* **1**, 199 (1971).
613. Horner, L., and H. Siegel, *Phosphorus,* **1**, 209 (1972).
614. Hota, N. K., H. A. Patel, A. J. Carty, M. Methew, and G. J. Palenik, *J. Organomet. Chem.*, **32**, C55 (1971).
615. Houk, L. W., and G. R. Dobson, *Inorg. Chem.*, **5**, 2119 (1966); *J. Chem. Soc. (A),* **1966**, 317.
616. Howatson, J., and B. Morosin, *Cryst. Struct. Commun.*, **2**, 51 (1973).
617. Howell, J. K., and L. L. Pytlewski, *Inorg. Nucl. Chem. Lett.*, **6**, 681 (1970).
618. Hsieh, A. T. T., J. D. Ruddick, and G. Wilkinson, *J.C.S. Dalton,* **1972**, 1966.
619. Hsieh, A. T. T., and G. Wilkinson, *J.C.S. Dalton,* **1973**, 867.
620. Hudson, M. J., R. S. Nyholm, and M. H. B. Stiddard, *J. Chem. Soc. (A),* **1968**, 40.
621. Hussey, A. S., and Y. Takeuchi, *J. Org. chem.*, **35**, 643 (1970).
622. Huttner, G., and S. Schelle, *J. Cryst. Mol. Struct.*, **1**, 69 (1971).
623. Huttner, G., and S. Schelle, *J. Organomet. Chem.*, **19**, P9 (1969).
624. Hyde, C. L., and D. J. Darensbourg, *Inorg. Chem.*, **12**, 1075 (1973).
625. Hyde, C. L., and D. J. Darensbourg, *Inorg. Chem.*, **12**, 1286 (1973).
626. Ichikawa, M., Y. Takeuchi, and A. Kogure, Ger. 1,770,545, 21 Jan 1971, Japan. Appl. 02 Jun 1967, through *Chem. Abstr.*, **74**, 112614n (1971).
627. Immirzi, A., and A. Luccarelli, *Cryst. Struct. Commun.*, **1**, 317 (1972).
628. Innorta, G., G. Reichenbach, and A. Foffani, *J. Organomet. Chem.*, **22**, 731 (1970).
629. International Synthetic Rubber Co. Ltd., Fr. 1,599,774, 28 Aug 1970, Brit. Appl. 20 Sep 1967, through *Chem. Abstr.*, **75**, 35368b (1971).
630. Interrante, L. V., and G. V. Nelson, *J. Organomet. Chem.*, **25**, 153 (1970).
631. Intille, G. M., *Inorg. Chem.*, **11**, 695 (1972).
632. Issleib, K., H. P. Abicht, and H. Winkelmann, *Z. Anorg. Allg. Chem.*, **388**, 89 (1972).

633. Issleib, K., and G. Döll, *Z. Anorg. Allg. Chem.*, **305**, 1 (1960).
634. Issleib, K., and U. Giesder, *Z. Anorg. Allg. Chem.*, **379**, 9 (1970).
635. Issleib, K., U. Giesder, and H. Hartung, *Z. Anorg. Allg. Chem.*, **390**, 239 (1972).
636. Issleib, K., and K. Hörnig, *Z. Anorg. Allg. Chem.*, **389**, 263 (1972).
637. Issleib, K., and E. Wenschuh, *Z. Anorg. Allg. Chem.*, **305**, 15 (1960).
638. Ito, T., T. Kawai, and Y. Takami, *Tetrahedron Lett.*, **1972**, 4775.
639. Ito, T., S. Kitazume, A. Yamamoto, and S. Ikeda, *J. Am. Chem. Soc.*, **92**, 3011 (1970).
640. Ittel, S. D., and J. A. Ibers, *J. Organomet. Chem.*, **57**, 389 (1973).
641. Jabs, W., and S. Herzog, *Z. Chem.*, **12**, 268 (1972).
642. Jabs, W., and S. Herzog, *Z. Chem.*, **12**, 297 (1972).
643. Jain, S. C., and R. Rivest, *Inorg. Chim. Acta*, **4**, 291 (1970).
644. James, B. R., and L. D. Markham, *Inorg. Nucl. Chem. Lett.*, **7**, 373 (1971).
645. James, B. R., and L. D. Markham, *J. Catal.*, **27**, 442 (1972).
646. James, B. R., L. D. Markham, B. C. Hui, and G. L. Rempel, *J.C.S. Dalton*, **1973**, 2247.
647. James, T. A., and J. A. McCleverty, *J. Chem. Soc. (A)*, **1970**, 850.
648. James, T. A., and J. A. McCleverty, *J. Chem. Soc. (A)*, **1971**, 1596.
649. Jardine, F. H., L. Rule, and A. G. Vohra, *J. Chem. Soc. (A)*, **1970**, 238.
650. Jardine, F. H., A. G. Vohra, and F. J. Young, *J. Inorg. Nucl. Chem.*, **33**, 2941 (1971).
651. Jarvis, J. A. J., R. H. B. Mais, P. G. Owston, and D. T. Thompson, *J. Chem. Soc. (A)*, **1970**, 1867.
652. Jesson, J. P., in *Transition Metal Hydrides*, E. L. Muetterties, Ed., Marcel Dekker, New York, 1971, p. 75.
653. Jesson, J. P., and P. Meakin, *Inorg. Nucl. Chem. Lett.*, **9**, 1221 (1973).
654. Jesson, J. P., E. L. Muetterties, and P. Meakin, *J. Am. Chem. Soc.*, **93**, 5261 (1971).
655. Jezowska-Trzebiatowska, B., H. Rataiczak, P. Soboda, and R. Tyka, *Bull. Acad. Pol. Sci., Ser. Sci. Chim.*, **20**, 869 (1972).
656. Johnson, B. F. G., and S. Bhaduri, *J.C.S. Chem. Comm.*, **1973**, 650.
657. Johnson, B. F. G., J. Lewis, and P. W. Robinson, *J. Chem. Soc. (A)*, **1970**, 1100.
658. Johnson, B. F. G., J. Lewis, and M. V. Twigg, *J. Organomet. Chem.*, **52**, C31 (1973).
659. Johnson, B. F. G., and J. A. Segal, *J.C.S. Chem. Comm.*, **1972**, 1312.
660. Johnson, B. F. G., and J. A. Segal, *J. Organomet. Chem.*, **31**, C79 (1971).
661. Johnson, T. R., R. M. Lynden-Bell, and J. F. Nixon, *J. Organomet. Chem.*, **21**, P15 (1970).
662. Johnston, R. D., F. Basolo, and R. G. Pearson, *Inorg. Chem.*, **10**, 247 (1971).
663. Jonas, K., and G. Wilke, *Angew. Chem.*, **82**, 295 (1970).
664. Jones, C. E., and K. J. Coskran, *Inorg. Chem.*, **10**, 55 (1971).
665. Jones, C. E., and K. J. Coskran, *Inorg. Chem.*, **10**, 1664 (1971).
666. Jolly, P. W., K. Jonas, C. Krüger, and Y. H. Tsay, *J. Organomet. Chem.*, **33**, 109 (1971).
667. Kaempfe, L. A., and K. W. Barnett, *Inorg. Chem.*, **12**, 2578 (1973).
668. Kagan, H., and T. P. Dang, Ger. Offen. 2,161,200, 22 Jun 1972, Fr. Appl. 70 44,632, 10 Dec 1970, through *Chem. Abstr.*, **77**, 114567k (1972).
669. Kagawa, T., and H. Hashimoto, *Bull. Chem. Soc. Jap.*, **45**, 2586 (1972).
670. Kalck, P., R. Pince, R. Poilblanc, and J. Roussel, *J. Organomet. Chem.*, **24**, 445 (1970).
671. Kalck, P., and R. Poiblanc, *C.R. Acad. Sci., Ser. C*, **274**, 66 (1972).
672. Kane, N. J., and J. B. Thompson, U.S. 3,631,191, 28 Dec 1971, Appl. 26, 770, 08 Apr 1970, through *Chem. Abstr.*, **76**, 87990p (1972).
673. Kang, D. K., and A. B. Burg, *Inorg. Chem.*, **11**, 902 (1972).
674. Kawata, N., K. Maruya, T. Mizoroki, and A. Ozaki, *Bull. Chem. Soc. Jap.*, **44**, 3217 (1971).
675. Keiter, R. L., and L. W. Cary, *J. Am. Chem. Soc.*, **94**, 9232 (1972).
676. Keiter, R. L., and J. G. Verkade, *Inorg. Chem.*, **9**, 404 (1970).

677. Kennedy, J. D., W. McFarlane, and D. S. Rycroft, *Inorg. Chem.*, **12**, 2742 (1973).
678. Khare, G. P., and R. Eisenberg, *Inorg. Chem.*, **11**, 1385 (1972).
679. Khatami, A. I., A. G. Ginzburg, M. N. Nefedova, V. N. Setkina, and D. N. Kursanov, *Zh. Obshch. Khim.*, **42**, 2665 (1972).
680. Kiji, J., and J. Furukawa, *Chem. Commun.*, **1970**, 977.
681. Kilbourn, B. T., and H. M. Powell, *J. Chem. Soc. (A)*, **1970**, 1688.
682. King, R. B., *Acc. Chem. Res.*, **5**, 177 (1972).
683. King, R. B., *J. Coord. Chem.*, **1**, 67 (1971).
684. King, R. B., and A. Efraty, *Inorg. Chim. Acta*, **4**, 123 (1970).
685. King, R. B., and A. Efraty, *Inorg. Chim. Acta*, **4**, 319 (1970).
686. King, R. B., and A. Efraty, *J. Am. Chem. Soc.*, **93**, 564 (1971).
687. King, R. B., and A. Efraty, *J. Am. Chem. Soc.*, **93**, 5260 (1971).
688. King, R. B., and A. Efraty, *J. Am. Chem. Soc.*, **94**, 3768 (1972).
689. King, R. B., and A. Efraty, *J. Organomet. Chem.*, **36**, 371 (1972).
690. King, R. B., and P. N. Kapoor, *Inorg. Chem.*, **11**, 1524 (1972).
691. King, R. B., and P. N. Kapoor, *Inorg. Chim. Acta*, **6**, 391 (1972).
692. King, R. B., and P. N. Kapoor, *J. Am. Chem. Soc.*, **91**, 5191 (1969).
693. King, R. B., and P. N. Kapoor, *J. Am. Chem. Soc.*, **93**, 4158 (1971).
694. King, R. B., P. N. Kapoor, and R. N. Kapoor, *Inorg. Chem.*, **10**, 1841 (1971).
695. King, R. B., R. N. Kapoor, M. S. Saran, and P. N. Kapoor, *Inorg. Chem.*, **10**, 1851 (1971).
696. King, R. B., and T. F. Korenowski, *Inorg. Chem.*, **10**, 1188 (1971).
697. King, R. B., and M. S. Saran, *Inorg. Chem.*, **10**, 1861 (1971).
698. King, R. B., and M. S. Saran, *Inorg. Chem.*, **11**, 2112 (1971).
699. King, R. B., and M. S. Saran, *J. Am. Chem. Soc.*, **95**, 1817 (1973).
700. King, R. B., W. C. Zipperer, and M. Ishaq, *Inorg. Chem.*, **11**, 1361 (1972).
701. Kirchner, R. M., J. A. Ibers, M. S. Saran, and R. B. King, *J. Am. Chem. Soc.*, **95**, 5775 (1973).
702. Kiso, Y., M. Kumada, K. Maeda, K. Sumitani, and K. Tamao, *J. Organomet. Chem.*, **50**, 311 (1973).
703. Kiso, Y., M. Kumada, K. Tamao, and M. Umeno, *J. Organomet. Chem.*, **50**, 297 (1973).
704. Kiso, Y., K. Yamamoto, K. Tamao, and M. Kumada, *J. Am. Chem. Soc.*, **94**, 4373 (1972).
705. Klabunde, U., U.S. 3,702,336, 07 Nov 1972, Appl. 97,009, 10 Dec 1970, through *Chem. Abstr.*, 78, 72373g (1973).
706. Kläui, W., and H. Werner, *J. Organomet. Chem.*, **54**, 331 (1973).
707. Kläui, W., and H. Werner, *J. Organomet. Chem.*, **60**, C19 (1973).
708. Klanberg, F. K., U.S. 3,695,853, 03 Oct 1972, Appl. 704,267, 09 Feb 1968, through *Chem. Abstr.*, **79**, 116689s (1973).
709. Klein, H. F., *Angew. Chem.*, **82**, 885 (1970).
710. Klein, H. F., *Angew. Chem.*, **83**, 363 (1971).
711. Klein, H. F., *Angew. Chem.*, **85**, 403 (1973).
712. Klein, H. F., and H. H. Karsch, *Chem. Ber.*, **105**, 2628 (1972).
713. Klein, H. F., and H. H. Karsch, *Chem. Ber.*, **106**, 1433 (1973).
714. Klein, H. F., and H. H. Karsch, *Chem. Ber.*, **106**, 2438 (1973).
715. Klein, H. F., and J. F. Nixon, *Chem. Commun.*, **1971**, 42.
716. Klein, H. F., and H. Schmidbaur, *Angew. Chem.*, **82**, 885 (1970).
717. Knebel, W. J., and P. M. Treichel, *Chem. Commun.*, **1971**, 516.
718. Knoth, W. H., *J. Am. Chem. Soc.*, **94**, 104 (1972).
719. Knowles, W. S., M. J. Sabacky, and B. D. Vineyard, *J.C.S. Chem. Comm.*, **1972**, 10.
720. Koenig, M. F., and M. Bigorgne, *Adv. Raman. Spectrosc.* **1**, 563 (1972).

721. Koenig, M. F., and M. Bigorgne, *Spectrochim. Acta,* **28A**, 1693 (1972).
722. Koerner von Gustorf, E., I. Fischler, J. Leitich, and H. Dreeskamp, *Angew. Chem.,* **84**, 1143 (1972).
723. Kolobova, N. E., V. V. Skripkin, and K. N., Anisimov, *Izv. Akad. Nauk SSSR, Ser. Khim.,* **1970**, 2225.
724. Kong, P. C., and D. M. Roundhill, *Inorg. Chem.,* **11**, 749 (1972).
725. Koshkina, G. N., I. V. Ovchinnikov, and A. D. Troitskaya, *Zh. Obshch. Khim.,* **43**, 956 (1973).
726. Kotz, J. C., and C. L. Nivert, *J. Organomet. Chem.,* **52**, 387 (1973).
727. Kovacs, J., G. Speier, and L. Marko, *Inorg. Chim. Acta,* **4**, 412 (1970).
728. Kraihanzel, C. S., and C. M. Bartish, *J. Am. Chem. Soc.,* **94**, 3572 (1972).
729. Kraihanzel, C. S., and C. M. Bartish, *J. Organomet. Chem.,* **43**, 343 (1972).
730. Kramolowsky, R., and J. Schmidt, *Z. Naturforsch.,* **25b**, 1487 (1970).
731. Kreutzer, B., P. Kreutzer, and W. Beck, *Z. Naturforsch.,* **27b**, 461 (1972).
732. Kruck, T., *Chimia,* **24**, 375 (1970).
733. Kruck, T., and K. Baur, *Z. Anorg. Allg. Chem.,* **364**, 192 (1969).
734. Kruck, T., H. L. Diedershagen, and A. Engelmann, *Z. Anorg. Allg. Chem.,* **397**, 31 (1973).
735. Kruck, T., and H. U. Hempel, *Angew. Chem.,* **83**, 437 (1971).
736. Kruck, T., and L. Knoll, *Chem. Ber.,* **105**, 3783 (1972).
737. Kruck, T., and L. Knoll, *Z. Naturforsch.,* **28b**, 34 (1973).
738. Kruck, T., L. Knoll, and J. Laufenberg, *Chem. Ber.,* **106**, 697 (1973).
739. Kruck, T., and R. Kobelt, *Chem. Ber.,* **105**, 3765 (1972).
740. Kruck, T., and R. Kobelt, *Chem. Ber.,* **105**, 3772 (1972).
741. Kruck, T., R. Kobelt, and A. Prasch, *Z. Naturforsch.,* **27b**, 344 (1972).
742. Kruck, T., and A. Prasch, Ger. Offen. 1,921,465, 25 Mar 1971, Appl. 26 Apr 1969, through *Chem. Abstr.,* **74**, 113910t (1971).
743. Kruck, T., G. Sylvester, and I. P. Kunau, *Angew. Chem.,* **83**, 725 (1971).
744. Kruck, T., G. Sylvester, and I. P. Kunau, *Z. Anorg. Allg. Chem.,* **396**, 165 (1973).
745. Kruck, T., G. Sylvester, and I. P. Kunau, *Z. Naturforsch.,* **28b**, 28 (1973).
746. Kruck, T., G. Sylvester, and I. P. Kunau, *Z. Naturforsch.,* **28b**, 38 (1973).
747. Kruck, T., J. Waldmann, M. Höfler, G. Birkenhäger, and C. Odenbrett, *Z. Anorg. Allg. Chem.,* **402**, 16 (1973).
748. Krüger, C., and Y. H. Tsay, *Acta Cryst. B,* **28**, 1941 (1972).
749. Krüger, C., and Y. H. Tsay, *J. Organomet. Chem.,* **34**, 387 (1972).
750. Kubo, Y., A. Yamamoto, and S. Ikeda, *J. Organomet. Chem.,* **60**, 165 (1973).
751. Kubota, M., and C. R. Carey, *J. Organomet. Chem.,* **24**, 491 (1970).
752. Kubota, M., R. K. Rothrock, and J. Geibel, *J.C.S. Dalton,* **1973**, 1267.
753. Kudo, K., M. Hidai, T. Murayama, and Y. Uchida, *Chem. Commun.,* **1970**, 1701.
754. Kudo, K., M. Hidai, and Y. Uchida, *J. Organomet. Chem.,* **33**, 393 (1971).
755. Kudo, K., M. Hidai, and Y. Uchida, *J. Organomet. Chem.,* **56**, 413 (1973).
756. Kumada, M., Y. Kiso, and M. Umeno, *Chem. Commun.,* **1970**, 611.
757. Kuran, W., and A. Musco, *J. Organomet. Chem.,* **40**, C47 (1972).
758. Kuzmina, L. G., N. G. Bokii, Yu. T. Struchkov, A. V. Arutyunyan, L. V. Rybin, and M. I. Rybinskaya, *Zh. Strukt. Khim.,* **12**, 875 (1971).
759. Kuzmina, L. G., Yu. S. Varshavskii, N. G. Bokii, Yu. T. Struchkov, and T. G. Cherkasova, *Zh. Strukt. Khim.,* **12**, 653 (1971).
760. Labroue, D., and R. Poilblanc, *C.R. Acad. Sci., Ser. C,* **271**, 1585 (1970).
761. Labroue, D., and R. Poilblanc, *Inorg. Chim. Acta,* **6**, 387 (1972).
762. Laing, M., T. Ashworth, P. Sommerville, E. Singleton, and R. Reimann, *J.C.S. Chem. Comm.,* **1972**, 1251.

763. Laing, M., E. Singleton, and G. Kruger, *J. Organomet. Chem.*, **54**, C30 (1973).
764. Laing, M., E. Singleton, and R. Reimann, *J. Organomet. Chem.*, **56**, C21 (1973).
765. Lam, C. T., and C. V. Senoff, *J. Organomet. Chem.*, **57**, 207 (1973).
766. La Mar, G. N., *Inorg. Chem.*, **10**, 2633 (1971).
767. La Mar, G. N., and E. O. Sherman, *J. Am. Chem. Soc.*, **92**, 2691 (1970).
768. La Mar, G. N., E. O. Sherman, and G. A. Fuchs, *J. Coord. Chem.*, **1**, 289 (1971).
769. Leh, F., and K. M. Chan, *Bull. Chem. Soc. Jap.*, **45**, 2709 (1972).
770. Lehman, D. D., D. F. Shriver, and I. Wharf, *Chem. Commun.*, **1970**, 1486.
771. Leigh, G. J., and D. M. P. Mingos, *J. Chem. Soc. (A)*, **1970**, 587.
772. Leites, L. A., V. T. Aleksanyan, S. S. Bukalov, and A. Z. Rubezhov, *Chem. Commun.*, **1971**, 265.
773. L'Eplattenier, F., and F. Calderazzo, U.S. 3,597,461, 03 Aug 1971, Appl. 13 Aug 1968, through *Chem. Abstr.*, **75**, 110424x (1971).
774. Levason, W., and C. A. McAuliffe, *Advances in Inorganic Chemistry and Radiochemistry*, H. J. Emeleus and A. G. Sharpe, Eds., Vol. 14, Academic, New York, 1972, p. 173.
775. Levason, W., C. A. McAuliffe, and B. Riley, *Inorg. Nucl. Chem. Lett.*, **9**, 1201 (1973).
776. Levison, J. J., and S. D. Robinson, *J. Chem. Soc. (A)*, **1970**, 96.
777. Levison, J. J., and S. D. Robinson, *J. Chem. Soc. (A)*, **1970**, 639.
778. Levison, J. J., and S. D. Robinson, *J. Chem. Soc. (A)*, **1970**, 2947.
779. Levison, J. J., and S. D. Robinson, *J.C.S. Dalton*, **1972**, 2013.
780. Lewis, D. F., S. J. Lippard, and P. S. Welcker, *J. Am. Chem. Soc.*, **92**, 3805 (1970).
781. Lim, Y. Y., and R. S. Drago, *J. Am. Chem. Soc.*, **94**, 84 (1972).
782. Lin, K. C., U.S. 3,673,285, 27 Jun 1972, Appl. 876,139, 12 Nov 1969, through *Chem. Abstr.*, **77**, 101890k (1972).
783. Linck, M. H., *Cryst. Struct. Commun.*, **2**, 379 (1973).
784. Linck, M. H., and L. R. Nassimbeni, *Inorg. Nucl. Chem. Lett.*, **9**, 1105 (1973).
785. Lindner, E., H. Stich, K. Geibel, and H. Kranz, *Chem. Ber.*, **104**, 1524 (1971).
786. Lippard, S. J., and J. J. Mayerle, *Inorg. Chem.*, **11**, 753 (1972).
787. Lippard, S. J., and G. J. Palenik, *Inorg. Chem.*, **10**, 1322 (1971).
788. Lloyd, M. K., J. A. McCleverty, D. G. Orchard, J. A. Connor, M. B. Hall, I. H. Hillier, E. M. Jones, and G. K. McEwen, *J.C.S. Dalton*, **1973**, 1743.
789. Lokshin, B. V., V. I. Zdanovich, N. K. Baranetskaya, V. N. Setkina, and D. N. Kursanov, *J. Organomet. Chem.*, **37**, 331 (1972).
790. Longoni, G., P. Chini, and F. Canziani, Ger. Offen. 2,148,925, 06 Apr 1972, Ital. Appl. 30,358A/70, 30 Sep. 1970, through *Chem. Abstr.*, **78**, 437181 (1973).
791. Loutellier, A., and M. Bigorgne, *J. Chim. Phys. Physicochim. Biol.*, **67**, 78 (1970).
792. Loutellier, A., and M. Bigorgne, *J. Chim. Phys. Physicochim. Biol.*, **67**, 99 (1970).
793. Loutellier, A., and M. Bigorgne, *J. Chim. Phys. Physicochim. Biol.*, **67**, 107 (1970).
794. Lucas, C. R., and M. L. H. Green, *J.C.S. Chem. Comm.*, **1972**, 1005.
795. Lundin, R., C. Moberg, R. Wahren, and O. Wennerström, *Acta Chem. Scand.*, **26**, 2045 (1972).
796. Lupin, M. S., and B. L. Shaw, *J. Chem. Soc. (A)*, **1968**, 741.
797. Lynden-Bell, R. M., G. G. Mather, and A. Pidcock, *J.C.S. Dalton*, **1973**, 715.
798. Lynden-Bell, R. M., J. F. Nixon, J. Roberts, J. R. Swain and W. McFarLane, *Inorg. Nucl. Chem. Lett.*, **7**, 1187 (1971).
799. Lynden-Bell, R. M., J. F. Nixon, and R. Schmutzler, *Colloques Internationaux du Centre National de la Recherche Scientifique*, **191**, 283 (1970).
800. Lynden-Bell, R. M., J. F. Nixon, and R. Schmutzler, *J. Chem. Soc. (A)*, **1970**, 565.
801. McAdam, A., J. N. Francis, and J. A. Ibers, *J. Organomet. Chem.*, **29**, 149 (1971).
802. McAuliffe, C. A., and D. W. Meek, *Inorg. Chem.*, **8**, 904 (1969).
803. McAuliffe, C. A., M. O. Workman, and D. W. Meek, *J. Coord. Chem.*, **2**, 137 (1972).

804. McCleverty, J. A., D. G. Orchard, J. A. Connor, E. M. Jones, J. P. Lloyd, and P. D. Rose, *J. Organomet. Chem.*, **30**, C75 (1971).
805. McCleverty, J. A., and R. N. Whiteley, *Chem. Commun.*, **1971**, 1159.
806. McClure, M. D., and D. L. Weaver, *J. Organomet. Chem.*, **54**, C59 (1973).
807. McGarvey, B. R., *J. Am. Chem. Soc.*, **94**, 1103 (1972).
808. McGusty, E. R., B. M. Sutton, and D. T. Walz, Ger. Offen. 2,061,181, 16 Jun 1971, US Appl. 12 Dec 1969, through *Chem. Abstr.*, **75**, 52818r (1971).
809. McLaughlin, J. R., *Inorg. Nucl. Chem. Lett.*, **9**, 565 (1973).
810. McPhail, A. T., R. C. Komson, J. F. Engel, and L. D. Quin, *J.C.S. Dalton*, **1972**, 874.
811. McPhail, A. T., and J. C. H. Steele, Jr., *J.C.S. Dalton*, **1972**, 2680.
812. Märkl, G., *Angew. Chem.*, **78**, 907 (1966); G. Märkl, F. Lieb, and A. Merz, *Angew. Chem.*, **79**, 947 (1967).
813. Mague, J. T., *Inorg. Chem.*, **11**, 2558 (1972).
814. Mague, J. T., and J. P. Mitchener, *Inorg. Chem.*, **11**, 2714 (1972).
815. Maisonnat, A., P. Kalck, and R. Poilblanc, *C.R. Acad. Sci., Ser. C*, **276**, 1263 (1973).
816. Makanova, D., G. Ondrejovic, and J. Gazo, *Proc. Conf. Coord. Chem., 3rd*, **1971**, 215.
817. Makhija, R. C., A. L. Beauchamp, and R. Rivest, *J.C.S. Dalton*, **1973**, 2447.
818. Maki, N., and K. Ohshima, *Bull. Chem. Soc. Jap.*, **43**, 3970 (1970).
819. Malatesta, L., M. Angoletta, and G. Caglio, *J. Chem. Soc. (A)*, **1970**, 1836.
820. Malatesta, L., M. Angoletta, and F. Conti, *J. Organomet. Chem.*, **33**, C43 (1971).
821. Malykhina, I. G., M. A. Kazankova, and I. F. Lutsenko, *Zh. Obshch. Khim.*, **41**, 2103 (1971).
822. Mann, B. E., *Inorg. Nucl. Chem. Lett.*, **7**, 595 (1971).
823. Mann, B. E., *J. Chem. Soc. (A)*, **1970**, 3050.
824. Mann, B. E., *Chem. Commun.*, **1971**, 1173.
825. Mann, B. E., *J.C.S. Dalton*, **1973**, 2012.
826. Mann, B. E., C. Masters, and B. L. Shaw, *J. Chem. Soc. (A)*, **1971**, 1104.
827. Mann, B. E., C. Masters, and B. L. Shaw, *Chem. Commun.*, **1970**, 703.
828. Mann, B. E., C. Masters, and B. L. Shaw, *Chem. Commun.*, **1970**, 846.
829. Mann, B. E., C. Masters, and B. L. Shaw, *J.C.S. Dalton*, **1972**, 48.
830. Mann, B. E., C. Masters, and B. L. Shaw, *J.C.S. Dalton*, **1972**, 704.
831. Mann, B. E., C. Masters, and B. L. Shaw, *J. Inorg. Nucl. Chem.*, **33**, 2195 (1971).
832. Mann, B. E., C. Masters, B. L. Shaw, R. M. Slade, and R. E. Stainbank, *Inorg. Nucl. Chem. Lett.*, **7**, 881 (1971).
833. Mann, B. E., C. Masters, B. L. Shaw, and R. E. Stainbank, *Chem. Commun.*, **1971**, 1103.
834. Mann, B. E., B. L. Shaw, and R. M. Slade, *J. Chem. Soc. (A)*, **1971**, 2976.
835. Mann, B. E., B. L. Shaw, and R. E. Stainbank, *J.C.S. Chem. Comm.*, **1972**, 151.
836. Manning, A. R., *J. Organomet. Chem.*, **40**, C73 (1972).
837. Manning, A. R., and J. R. Miller, *J. Chem. Soc. (A)*, **1970**, 3352.
838. Manning, A. R., and D. J. Thornhill, *J. Chem. Soc. (A)*, **1971**, 637.
839. Manojlovic-Muir, L., *Inorg. Nucl. Chem. Lett.*, **9**, 59 (1973).
840. Manojlovic-Muir, L., *J. Chem. Soc. (A)*, **1971**, 2796.
841. Manojlovic-Muir, L., *Chem. Commun.*, **1971**, 147.
842. Manojlovic-Muir, L., and K. W. Muir, *J.C.S. Dalton*, **1972**, 686.
843. Maples, P. K., M. Green, and F. G. A. Stone, *J.C.S. Dalton*, **1973**, 388.
844. Maples, P. L., F. Basolo, and R. G. Pearson, *Inorg. Chem.*, **10**, 765 (1971).
845. Marriott, J. C., J. A. Salthouse, M. J. Ware, and J. M. Freeman, *Chem. Commun.*, **1970**, 595.
846. Marsh, R. A., J. Howard, and P. Woodward, *J.C.S. Dalton*, **1973**, 778.
847. Marsich, N., A. Camus, and E. Cebulec, *J. Inorg. Nucl. Chem.*, **34**, 933 (1972).
848. Marsich, N., G. Nardin, and L. Randaccio, *J. Am. Chem. Soc.*, **95**, 4053 (1973).

849. Maruya, K., T. Mizoroki, and A. Ozaki, *Bull. Chem. Soc. Jap.*, **46**, 993 (1973).
850. Masai, H., K. Sonogashira, and N. Hagihara, *Bull. Chem. Soc. Jap.*, **44**, 2226 (1971).
851. Mason, R., and A. I. M. Rae, *J. Chem. Soc. (A)*, **1970**, 1767.
852. Mason, R., and L. Randaccio, *J. Chem. Soc. (A)*, **1971**, 1150.
853. Mason, R., I. Sotofte, S. D. Robinson, and M. F. Uttley, *J. Organomet. Chem.*, **46**, C61 (1972).
854. Mason, R., K. M. Thomas, D. F. Gill, and B. L. Shaw, *J. Organomet. Chem.*, **40**, C67 (1972).
855. Masters, C., *J.C.S. Chem. Comm.*, **1973**, 191.
856. Masters, C., W. S. McDonald, G. Raper, and B. L. Shaw, *Chem. Commun.*, **1971**, 210.
857. Masters, C., and B. L. Shaw, *J. Chem. Soc. (A)*, **1971**, 3679.
858. Masters, C., B. L. Shaw, and R. E. Stainbank, *J.C.S. Dalton*, **1972**, 664.
859. Mather, G. G., and A. Pidcock, *J. Chem. Soc. (A)*, **1970**, 1226.
860. Mather, G. G., A. Pidcock, and G. J. N. Rapsey, *J.C.S. Dalton*, **1973**, 2095.
861. Mather, G. G., G. J. N. Rapsey, and A. Pidcock, *Inorg. Nucl. Chem. Lett.*, **9**, 567 (1973).
862. Mathew, M., G. J. Palenik, and A. J. Carty, *Can. J. Chem.*, **49**, 4119 (1971).
863. Mathieu, R., M. Lenzi, and R. Poilblanc, *Inorg. Chem.*, **9**, 2030 (1970).
864. Mathieu, R., and R. Poilblanc, *Inorg. Chem.*, **11**, 1858 (1972).
865. Mawby, A., and G. E. Pringle, *J. Inorg. Nucl. Chem.*, **34**, 517 (1972).
866. Mawby, A., and G. E. Pringle, *J. Inorg. Nucl. Chem.*, **34**, 877 (1972).
867. Mawby, R. J., and G. Wright, *J. Organomet. Chem.*, **21**, 169 (1970).
868. Mawby, R. J., G. Wright, and D. Ewing, *J. Organomet. Chem.*, **23**, 545 (1970).
869. Maxfield, P. L., *Inorg. Nucl. Chem. Lett.*, **6**, 693 (1970).
870. Mays, M. J., and P. L. Sears, *J.C.S. Dalton*, **1973**, 1873.
871. Mays, M. J., R. N. F. Simpson, and F. P. Stefanini, *J. Chem. Soc. (A)*, **1970**, 3000.
872. Meakin, P., L. J. Guggenberger, J. P. Jesson, D. H. Gerlach, F. N. Tebbe, W. G. Peet, and E. L. Muetterties, *J. Am. Chem. Soc.*, **92**, 3482 (1970).
873. Meakin, P., L. J. Guggenberger, W. G. Peet, E. L. Muetterties, and J. P. Jesson, *J. Am. Chem. Soc.*, **95**, 1467 (1973).
874. Meakin, P., J. P. Jesson, F. N. Tebbe, and E. L. Muetterties, *J. Am. Chem. Soc.*, **93**, 1797 (1971).
875. Meakin, P., E. L. Muetterties, and J. P. Jesson, *J. Am. Chem. Soc.*, **94**, 5271 (1972).
876. Meakin, P., E. L. Muetterties, F. N. Tebbe, and J. P. Jesson, *J. Am. Chem. Soc.*, **93**, 4701 (1971).
877. Mealli, C., P. L. Orioli, and L. Sacconi, *J. Chem. Soc. (A)*, 2691 (1971).
878. Mealli, C., and L. Sacconi, *J.C.S. Chem. Comm.*, **1973**, 886.
879. Meek, D. W., P. E. Nicpon, and V. I. Meek, *J. Am. Chem. Soc.*, **92**, 5351 (1970).
880. Memering, M. N., and G. R. Dobson, *Inorg. Chem.*, **12**, 2490 (1973).
881. Merle, A., M. Dartiguenave, and Y. Dartiguenave, *Bull. Soc. Chim. Fr.*, **1972**, 87.
882. Merle, A., M. Dartiguenave, and Y. Dartiguenave, *C.R. Acad. Sci., Ser. C*, **272**, 2046 (1971).
883. Merle, A., M. Dartiguenave, and Y. Dartiguenave, *J. Mol. Struct.*, **13**, 413 (1972).
884. Merle, A., M. F. Obier, M. Dartiguenave, and Y. Dartiguenave, *C.R. Acad. Sci., Ser. C*, **272**, 1956 (1971).
885. Messmer, G. G., and G. J. Palenik, *Can. J. Chem.*, **47**, 1440 (1969).
886. Mestroni, G., A. Camus, and C. Cocevar, *J. Organomet. Chem.*, **29**, C17 (1971).
887. Michman, M., and M. Balog, *J. Organomet. Chem.*, **31**, 395 (1971).
888. Micoud, M. H., J. M. Savariault, and P. Cassoux, *Bull. Soc. Chim. Fr.*, **1972**, 3374.
889. Miller, J. R., and D. H. Myers, *Inorg. Chim. Acta*, **5**, 215 (1971).

890. Miller, R. G., R. D. Stauffer, D. R. Fahey, and D. R. Parnell, *J. Am. Chem. Soc.,* **92**, 1511 (1970).
891. Mingos, D. M. P., and J. A. Ibers, *Inorg. Chem.,* **9**, 1105 (1970).
892. Mingos, D. M. P., and J. A. Ibers, *Inorg. Chem.,* **10**, 1035 (1971).
893. Mingos, D. M. P., W. T. Robinson, and J. A. Ibers, *Inorg. Chem.,* **10**, 1043 (1971).
894. Misono, A., *Inorg. Syntheses,* **12**, 12 (1970).
895. Miyamoto, S., Japan. Kokai. 72 18,836, 18 Sep. 1972, Appl. 71 07,815, 20 Feb 1971, through *Chem. Abstr.,* **77**, 152359b (1972).
896. Miyashita, A., and A. Yamamoto, *J. Organomet. Chem.,* **49**, C57 (1973).
897. Moedritzer, K., *Syn. Inorg. Met.-Org. Chem.,* **2**, 121 (1972).
898. Moedritzer, K., and R. E. Miller, *Thermochim. Acta,* **1**, 87 (1970).
899. Moelwyn-Hughes, J. T., A. W. B. Garner, and A. S. Howard, *J. Chem. Soc. (A),* **1971**, 2361.
900. Moelwyn-Hughes, J. T., A. W. B. Garner, and A. S. Howard, *J. Chem. Soc. (A),* **1971**, 2370.
901. Moers, F. G., *Chem. Commun.,* **1971**, 79.
902. Moers, F. G., J. A. M. De Jong, and P. M. H. Beaumont, *J. Inorg. Nucl. Chem.,* **35**, 1915 (1973).
903. Moers, F. G., and J. P. Langhout, *Recl. Trav. Chim. Pays-Bas,* **91**, 591 (1972).
904. Moers, F. G., and J. P. Langhout, *Recl. Trav. Chim. Pays-Bas,* **92**, 996 (1973).
905. Moers, F. G., and P. H. Op Het Veld, *J. Inorg. Nucl. Chem.,* **32**, 3225 (1970).
906. Moers, F. G., R. W. M. Ten Hoedt, and J. P. Langhout, *Inorg. Chem.,* **12**, 2196 (1973).
907. Morassi, R., and A. Dei, *Inorg. Chim. Acta,* **6**, 314 (1972).
908. Morassi, R., F. Mani, and L. Sacconi, *Inorg. Chem.,* **12**, 1246 (1973).
909. Morishima, I., T. Yonezawa, and K. Goto, *J. Am. Chem. Soc.,* **92**, 6651 (1970).
910. Morrison, J. D., R. E. Burnett, A. M. Aguiar, C. J. Morrow, and C. Phillips, *J. Am. Chem. Soc.,* **93**, 1301 (1971).
911. Moss, J. R., and B. L. Shaw, *J. Chem. Soc. (A),* **1970**, 595.
912. Müller, E., and E. Langer, *Tetrahedron. Lett.,* **1970**, 993.
913. Müller, J., K. Fenderl, and B. Mertschenk, *Chem. Ber.,* **104**, 700 (1971).
914. Muetterties, E. L., *J.C.S. Chem. Comm.,* **1973**, 221.
915. Muetterties, E. L., and C. W. Alegranti, *J. Am. Chem. Soc.,* **92**, 4114 (1970).
916. Muetterties, E. L., and C. W. Alegranti, *J. Am. Chem. Soc.,* **94**, 6386 (1972).
917. Muir, K. W., and J. A. Ibers, *Inorg. Chem.,* **9**, 440 (1970).
918. Mukhedar, A. J., M. Green, and F. G. A. Stone, *J. Chem. Soc. (A),* **1970**, 947.
919. Munakata, H., and M. L. H. Green, *Chem. Commun.,* **1970**, 881.
920. Nakayama, H., *Bull. Chem. Soc. Jap.,* **43**, 2057 (1970).
921. Nappier, T. E., Jr., and D. W. Meek, *Inorg. Chim. Acta,* **7**, 235 (1973).
922. Nappier, T. E., Jr., and D. W. Meek, *J. Am. Chem. Soc.,* **94**, 306 (1972).
923. Nappier, T. E., Jr., D. W. Meek, R. M. Kirchner, and J. A. Ibers, *J. Am. Chem. Soc.,* **95**, 4194 (1973).
924. Nassimbeni, L. R., *Inorg. Nucl. Chem. Lett.,* **7**, 187 (1971).
925. Nassimbeni, L. R., *Inorg. Nucl. Chem. Lett.,* **7**, 909 (1971).
926. Nast, R., and H. Schulz, *Chem. Ber.,* **103**, 785 (1970).
927. Natile, G., S. Pignataro, G. Innorta, and G. Bor, *J. Organomet. Chem.,* **40**, 215 (1972).
928. Nave, C., and M. R. Truter, *Chem. Commun.,* **1971**, 1253.
929. Nave, C., and M. R. Truter, *J.C.S. Dalton,* **1973**, 2202.
930. Nefedov, V. I., E. F. Shubochkina, I. S. Kolomnikov, I. B. Baranovskii, V. P. Kukolev, M. A. Golubnichaya, L. K. Shubochkin, M. A. Porai-Koshits, and M. E. Vol'pin, *Zh. Neorg. Khim.,* **18**, 845 (1973).
931. Negoiu, M., and P. Spacu, *An. Univ. Bucuresti, Chim.,* **19**, 73 (1970), through *Chem.*

Abstr., **75**, 104627z (1971).
932. Nelson, J. H., and D. A. Redfield, *Inorg. Nucl. Chem. Lett.*, **9**, 807 (1973).
933. Nesmeyanov, A. N., D. N. Kirsanov, V. N. Setkina, B. D. Vil'cheskaya, N. R. Baranetskaya, A. I. Krilova, and L. A. Glushenko, *Dokl. Akad. Nauk SSSR*, **199**, 1336 (1971).
934. Nesmeyanov, A. N., L. G. Makarova, and I. V. Polovyanyuk, *J. Organomet. Chem.*, **22**, 707 (1970).
935. Nesmeyanov, A. N., N. A. Ustynyuk, L. V. Bogatyreva, and L. G. Makarova, *Izv. Akad. Nauk SSSR, Ser. Khim.*, **1973**, 62.
936. Newman, J., and A. R. Manning, *J.C.S. Dalton*, **1972**, 241.
937. Newman, J., and A. R. Manning, *J.C.S. Dalton*, **1973**, 1593.
938. Nichols, D. I., *J. Chem. Soc. (A)*, **1970**, 1216.
939. Nixon, J. F., *Adv. Inorg. Chem. Radiochem.*, **13**, 363 (1970).
940. Nixon, J. F., *Endeavour*, **32**, 19 (1973).
941. Nixon, J. F., *J.C.S. Dalton*, **1973**, 2226.
942. Nixon, J. F., *J. Fluorine Chem.*, **3**, 179 (1973/1974).
943. Nixon, J. F., M. Murray, and R. Schmutzler, *Z. Naturforsch.*, **25b**, 110 (1970).
944. Nixon, J. F., and M. D. Sexton, *J. Chem. Soc. (A)*, **1970**, 321.
945. Nixon, J. F., and J. R. Swain, *J.C.S. Dalton*, **1972**, 1038.
946. Nixon, J. F., and J. R. Swain, *J.C.S. Dalton*, **1972**, 1044.
947. Nixon, J. F., and J. R. Swain, *J. Organomet. Chem.*, **21**, P13 (1970).
948. Nobile, C. F., M. Rossi, and A. Sacco, *Inorg. Chim. Acta*, **5**, 698 (1971).
949. Nöth, H., and S. N. Sze, *J. Organomet. Chem.*, **43**, 249 (1972).
950. Norton, J. R., and J. P. Collman, *Inorg. Chem.*, **12**, 476 (1973).
951. Nowlin, T., and K. Cohn, *Inorg. Chem.*, **10**, 2801 (1971).
952. Nowlin, T., and K. Cohn, *Inorg. Chem.*, **11**, 560 (1972).
953. Nowlin, T., S. Subramanian, and K. Cohn, *Inorg. Chem.*, **11**, 2907 (1972).
954. O'Connor, C., *J. Inorg. Nucl. Chem.*, **32**, 2299 (1970).
955. Ötvös, I., G. Speier, and L. Marko, *Acta Chim. Acad. Sci. Hung.*, **66**, 27 (1970).
956. Ogilvie, F. B., J. M. Jenkins, and J. G. Verkade, *J. Am. Chem. Soc.*, **92**, 1916 (1970).
957. Ojima, I., T. Kogure, and Y. Nagai, *Chem. Lett.*, 541 (1973).
958. Olechowski, J. R., *J. Organomet. Chem.*, **32**, 269 (1971).
959. Oliver, A. J., and W. A. G. Graham, *Inorg. Chem.*, **9**, 243 (1970).
960. Oliver, A. J., and W. A. G. Garham, *Inorg. Chem.*, **9**, 2578 (1970).
961. Ondrejovic, G., D. Makanova, D. Valigura, and J. Gazo, *Z. Chem.*, **13**, 193 (1973).
962. Orioli, P. L., *Coord. Chem. Rev.*, **6**, 285 (1971).
963. Orlandini, A. B., C. Calabresi, C. A. Ghilardi, P. L. Orioli, and L. Sacconi, *J.C.S. Dalton*, **1973**, 1383.
964. Otsuka, S., A. Nakamura, T. Kano, and K. Tani, *J. Am. Chem. Soc.*, **93**, 4301 (1971).
965. Otsuka, S., K. Tani, and T. Yamagata, *J.C.S. Dalton*, **1973**, 2491.
966. Otsuka, S., Y. Tatsuno, and K. Ataka, *J. Am. Chem. Soc.*, **93**, 6705 (1971).
967. Otsuka, S., T. Yoshida, and Y. Tatsuno, *Chem. Commun.*, **1971**, 67.
968. Ozin, G. A., *J. Chem. Soc. (A)*, **1970**, 1307.
969. Pankowski, M., and M. Bigorgne, *J. Organomet. Chem.*, **35**, 397 (1972).
970. Papp, S., and T. Schönweitz, *J. Inorg. Nucl. Chem.*, **32**, 697 (1970).
971. Parker, D. J., *J. Chem. Soc. (A)*, **1970**, 1382.
972. Parrott, D. W., and D. G. Hendricker, *J. Coord. Chem.*, **2**, 235 (1973).
973. Parshall, G. W., *Inorg. Syntheses*, **12**, 26 (1970).
974. Partenheimer, W., and E. F. Hoy, *J. Am. Chem. Soc.*, **95**, 2840 (1973).
975. Patel, H. A., A. J. Carty, and N. K. Hota, *J. Organomet. Chem.*, **50**, 247 (1973).
976. Patel, H. A., R. G. Fischer, A. J. Carty, D. V. Naik, and G. J. Palenik, *J. Organomet. Chem.*, **60**, C49 (1973).

977. Payne, D. H., and H. Frye, *Inorg. Nucl. Chem. Lett.*, **8**, 73 (1972).
978. Payne, D. H., and H. Frye, *Inorg. Nucl. Chem. Lett.*, **9**, 505 (1973).
979. Payne, D. H., Z. A. Payne, R. Rohmer, and H. Frye, *Inorg. Chem.*, **12**, 2540 (1973).
980. Petz, W., and G. Schmid, *J. Organomet. Chem.*, **35**, 321 (1972).
980a. Peloso, A., *Coord. Chem. Rev.*, **10**, 123 (1973).
981. Perego, G., G. Del Piero, M. Cesari, M. G. Clerici, and E. Perrotti, *J. Organomet. Chem.*, **54**, C51 (1973).
982. Perevalova, E. G., T. V. Baukova, E. I. Goryunov, and K. I. Grandberg, *Izv. Akad. Nauk SSSR, Ser. Khim.*, **1970**, 2148.
983. Petersen, R. L., and K. L. Watters, *Inorg. Chem.*, **12**, 3009 (1973).
984. Peterson, J. L., T. E. Nappier, Jr., and D. W. Meek, *J. Am. Chem. Soc.*, **95**, 8195 (1973).
985. Piacenti, F., M. Bianchi, E. Benedetti, and A. Frediani, *J. Organomet. Chem.*, **23**, 257 (1970).
986. Pidcock, A., and G. G. Roberts, *J. Chem. Soc. (A)*, **1970**, 2922.
987. Pidcock, A., and C. R. Waterhouse, *J. Chem. Soc. (A)*, **1970**, 2080.
988. Pierpont, C. G., and R. Eisenberg, *Inorg. Chem.*, **11**, 1088 (1972).
989. Pierpont, C. G., and R. Eisenberg, *Inorg. Chem.*, **11**, 1094 (1972).
990. Pierpont, C. G., and R. Eisenberg, *Inorg. Chem.*, **12**, 199 (1973).
991. Pierpont, C. G., A. Pucci, and R. Eisenberg, *J. Amer. Chem. Soc.*, **93**, 3050 (1971).
992. Pignataro, S., G. Distefano, and A. Foffani, *J. Am. Chem. Soc.*, **92**, 6425 (1970).
993. Pignolet, L. H., W. DeW. Horrocks, Jr., and R. H. Holm, *J. Am. Chem. Soc.*, **92**, 1855 (1970).
994. Pittman, C. U., Jr., and G. O. Evans, *J. Organomet Chem.*, **43**, 361 (1972).
995. Plastas, H. J., J. M. Stewart, and S. O. Grim., *Inorg. Chem.*, **12**, 265 (1973).
996. Poddar, R. K., and U. Agarwala, *J. Inorg. Nucl. Chem.*, **35**, 567 (1973).
997. Podlaha, J., and J. Podlahova, *Collect. Czech. Chem. Commun.*, **38**, 1730 (1973).
998. Poë, A. J., and M. V. Twigg, *J. Organomet. Chem.*, **50**, C39 (1973).
999. Potrafke, E. M., U.S. 3,625,755, 07 Dec 1971, Appl. 14 Apr 1969, through *Chem. Abstr.*, **76**, 62691d (1972).
1000. Powell, J., and B. L. Shaw, *J. Chem. Soc. (A)*, **1968**, 211.
1001. Pratt Brock, C., J. P. Collman, G. Dolcetti, P. H. Farnham, J. A. Ibers, J. E. Lester, and C. A. Reed, *Inorg. Chem.*, **12**, 1304 (1973).
1002. Pratt Brock, C., and J. A. Ibers, *Inorg. Chem.*, **11**, 2812 (1972).
1003. Pregaglia, G. F., A. Andreetta, G. F. Ferrari, G. Montrasi, and R. Ugo, *J. Organomet. Chem.*, **33**, 73 (1971).
1004. Pregaglia, G. F., A. Andreetta, G. F. Ferrari, and R. Ugo, *J. Organomet. Chem.*, **30**, 387 (1971).
1005. Preston, H. S., J. M. Stewart, H. J. Plastas, and S. O. Grim, *Inorg. Chem.*, **11**, 161 (1972).
1006. Que, L., Jr., and L. H. Pignolet, *Inorg. Chem.*, **12**, 156 (1973).
1007. Raper, G., and W. S. McDonald, *Acta Cryst. B*, **29**, 2013 (1973).
1008. Raper, G., and W. S. McDonald, *J. Chem. Soc. (A)*, **1971**, 3430.
1009. Reagan, W. J., and A. B. Burg, *Inorg. Nucl. Chem. Lett.*, **7**, 741 (1971).
1010. Redfield, D. A., and J. H. Nelson, *Inorg. Chem.*, **12**, 15 (1973).
1011. Reed, C. A., and W. R. Roper, *J. Chem. Soc. (A)*, **1970**, 3054.
1012. Reed, C. A., and W. R. Roper, *J.C.S. Dalton*, **1972**, 1243.
1013. Reed, C. A., and W. R. Roper, *J.C.S. Dalton*, **1973**, 1014.
1014. Reed, J., A. J. Schultz, C. G. Pierpont, and R. Eisenberg, *Inorg. Chem.*, **12**, 2949 (1973).
1015. Rigo, P., and M. Bressan, *Inorg. Chem.*, **11**, 1314 (1972).
1016. Rigo, P., and M. Bressan, *Inorg. Nucl. Chem. Lett.*, **9**, 527 (1973).
1017. Rigo, P., B. Longato, and G. Favero, *Inorg. Chem.*, **11**, 300 (1972).
1018. Rehder, D., and J. Schmidt, *Z. Naturforsch.*, **27b**, 625 (1972).

1019. Reimann, R. H., and E. Singleton, *J.C.S. Dalton,* **1973**, 841.
1020. Reimann, R. H., and E. Singleton, *J. Organomet. Chem.,* **32**, C44 (1971).
1021. Reimann, R. H., and E. Singleton, *J. Organomet. Chem.,* **38**, 113 (1972).
1022. Reimann, R. H., and E. Singleton, *J. Organomet. Chem.,* **57**, C75 (1973).
1023. Reimann, R. H., and E. Singleton, *J. Organomet. Chem.,* **59**, 309 (1973).
1024. Rigo, P., and A. Turco, *Coord. Chem. Rev.,* 8, 175 (1972).
1025. Rimbault, J., and R. Hugel, *Inorg. Nucl. Chem. Lett.,* **9**, 1 (1973).
1026. Rinze, P. V., and H. Nöth, *J. Organomet. Chem.,* **30**, 115 (1971).
1027. Robertson, G. B., and P. O. Whimp, *J.C.S. Dalton,* **1973**, 2454.
1028. Robertson, G. B., and P. O. Whimp, *J. Organomet. Chem.,* **49**, C27 (1973).
1029. Robertson, G. B., and P. O. Whimp, *J. Organomet. Chem.,* **60**, C11 (1973).
1030. Robinson, S. D., *MTP Int. Rev. Sci.,* **6**, 121 (1972), H. J. Emeleus, Ed., Butterworths, London, University Park Press, Baltimore, 1972.
1031. Robinson, S. D., and M. F. Uttley, *J. Chem. Soc. (A),* **1971**, 1254
1032. Robinson, S. D., and M. F. Uttley, *J.C.S. Dalton,* **1972**, 1.
1033. Ros, R., M. Vidali, and G. Rizzardi, *Inorg. Chim. Acta,* **4**, 562 (1970).
1034. Roshchupkina, O. S., I. P. Rudakova, T. A. Pospelova, A. M. Yurkevich, and Yu. G. Borod'ko, *Zh. Obshch. Khim.,* **40**, 466 (1970).
1035. Rossi, M., and A. Sacco, *Chem. Commun.,* **1971**, 694.
1036. Ruiz-Ramirez, L., T. A. Stephenson, and E. S. Switkes, *J.C.S. Dalton,* **1973**, 1770.
1037. Saalfeld, F. E., J. J. De Corpo, and M. V. McDowell, *J. Organomet. Chem.,* **44**, 333 (1972).
1038. Sabherwal, I. H., and A. B. Burg, *Inorg. Chem.,* **11**, 3138 (1972).
1039. Sabherwal, I. H., and A. B. Burg, *Inorg. Chem.,* **12**, 697 (1973).
1040. Sabherwal, I. H., and A. B. Burg, *Inorg. Nucl. Chem. Lett.,* **5**, 259 (1969).
1041. Sacco, A., and M. Rossi, *Inorg. Syntheses,* **12**, 18 (1970).
1042. Sacconi, L., and P. Dapporto, *J. Am. Chem. Soc.,* **92**, 4133 (1970).
1043. Sacconi, L., and A. Dei, *J. Coord. Chem.,* **1**, 229 (1971).
1044. Sakakibara, M., Y. Takahashi, S. Sakai, and Y. Ishii, *J. Organomet. Chem.,* **27**, 139 (1971).
1045. Sanders, J. R., *J. Chem. Soc. (A),* **1971**, 2991.
1046. Sanders, J. R., *J.C.S. Dalton,* **1972**, 1333.
1047. Sanders, J. R., *J.C.S. Dalton,* **1973**, 743.
1048. Sandhu, S. S., R. Dass, and M. P. Gupta, *Indian J. Chem.,* 8, 458 (1970).
1049. Sandhu, S. S., R. Dass, and M. P. Gupta, *J. Indian. Chem. Soc.,* **47**, 1137 (1970).
1050. Sandhu, S. S., and R. S. Sandhu, *Indian J. Chem.,* 8, 189 (1970).
1051. Sandhu, S. S., and R. S. Sandhu, *Indian J. Chem.,* 9, 482 (1971).
1052. Sandhu, S. S., R. S. Sandhu, M. P. Gupta, and C. R. Kanekar, *Indian J. Chem.,* **9**, 1142 (1971).
1053. Sarkar, S., *J. Indian Chem. Soc.,* **49**, 185 (1972).
1054. Sasson, Y., and J. Blum, *Tetrahedron Lett.,* **1971**, 2167.
1055. Sato, F., T. Uemura, and M. Sato, *J. Organomet. Chem.,* **56**, C27 (1973).
1056. Sato, M., F. Sato, and T. Yoshida, *J. Organomet. Chem.,* **27**, 273 (1971).
1057. Sato, M., F. Sato, and T. Yoshida, *J. Organomet. Chem.,* **31**, 415 (1971).
1058. Sato, M., and T. Yoshida, *J. Organomet. Chem.,* **39**, 389 (1972).
1059. Sato, M., and T. Yoshida, *J. Organomet. Chem.,* **51**, 231 (1973).
1060. Savariault, J. M., P. Cassoux, and F. Gallais, *C.R. Acad. Sci., Ser. C,* **271**, 477 (1970).
1061. Savariault, J. M., P. Cassoux, and J. F. Labarre, *J. Chim. Phys. Physicochim. Biol.,* **67**, 235 (1970).
1062. Savariault, J. M., M. H. Micoud, P. Cassoux, and J. F. Labarre, *Bull. Soc. Chim. Fr.,* **1971**, 2413.

1063. Sayler, A. A., H. Beall, and J. F. Sieckhaus, *J. Am. Chem. Soc.*, **95**, 5790 (1973).
1064. Sbrignadello, G., G. Bor, and L. Maresca, *J. Organomet. Chem.*, **46**, 345 (1972).
1065. Schaefer King, M. A., and R. E. McCarley, *Inorg. Chem.*, **12**, 1972 (1973).
1066. Schlodder, R., S. Vogler, and W. Beck, *Z. Naturforsch.*, **27b**, 463 (1972).
1067. Schmidbaur, H., J. Adlkofer, and K. Schwirten, *Chem. Ber.*, **105**, 3382 (1972).
1068. Schmidbaur, H., and R. Franke, *Chem. Ber.*, **105**, 2985 (1972).
1069. Schmidbaur, H., and K. G. Räthlein, *Chem. Ber.*, **106**, 2491 (1973).
1070. Schmidbaur, H., and A. Shiotani, *Chem. Ber.*, **104**, 2821 (1971).
1071. Schmidbaur, H., A. Shiotani, and H. F. Klein, *Chem. Ber.*, **104**, 2831 (1971).
1072. Schmidt, D. D., and J. T. Yoke, *J. Am. Chem. Soc.*, **93**, 637 (1971).
1073. Schmidt, J., *Z. Naturforsch.*, **27b**, 600 (1972).
1074. Schmiedeknecht, K., W. Reichardt, and W. Seidel, *Z. Chem.*, **11**, 432 (1971).
1075. Schmulbach, C. D., C. H. Kolich, and C. C. Hinckley, *Inorg. Chem.*, **11**, 2841 (1972).
1076. Schneider, M. L., N. J. Coville, and I. S. Butler, *J.C.S. Chem. Comm.*, **1972**, 799.
1077. Schneider, M. L., and H. M. M. Shearer, *J.C.S. Dalton*, **1973**, 354.
1078. Schoenberg, A. R., and W. P. Anderson, *Inorg. Chem.*, **11**, 85 (1972).
1079. Schrock, R. R., and J. A. Osborn, *J. Am. Chem. Soc.*, **93**, 2397 (1971).
1080. Schumann, H., *Angew. Chem.*, **81**, 970 (1969).
1081. Schumann, H., and U. Arbenz, *J. Organomet. Chem.*, **22**, 411 (1970).
1082. Schumann, H., and W. W. Du Mont, *J. Organomet. Chem.*, **49**, C25 (1973).
1083. Schumann, H., and J. Kuhlmey, *J. Organomet. Chem.*, **42**, C57 (1972).
1084. Schumann, H., L. Rösch, and O. Stelzer, *J. Organomet. Chem.*, **21**, 351 (1970).
1085. Schumann, H., O. Stelzer, J. Kulmey, and U. Niederreuther, *J. Organomet. Chem.*, **28**, 105 (1971).
1086. Schumann, H., O. Stelzer, and U. Niederreuther, *Chem. Ber.*, **103**, 1391 (1970).
1087. Schumann, H., O. Stelzer, and U. Niederreuther, *J. Organomet. Chem.*, **16**, P64 (1969).
1088. Schumann, H., O. Stelzer, U. Niederreuther, and L. Rösch, *Chem. Ber.*, **103**, 1383 (1970).
1089. Schumann, H., O. Stelzer, U. Niederreuther, and L. Rösch, *Chem. Ber.*, **103**, 2350 (1970).
1090. Schumann, H., and E. von Deuster, *J. Organomet. Chem.*, **40**, C27 (1972).
1091. Schunn, R. A., *Inorg. Chem.*, **9**, 394 (1970).
1092. Schunn, R. A., *Inorg. Chem.*, **12**, 1573 (1973).
1093. Schwartz, J., and J. B. Cannon, *J. Am. Chem. Soc.*, **94**, 6226 (1972).
1094. Schwenzer, G., M. Y. Darensbourg, and D. J. Darensbourg, *Inorg. Chem.*, **11**, 1967 (1972).
1095. Scott, R. N., D. F. Shriver, and D. D. Lehman, *Inorg. Chim. Acta*, **4**, 73 (1970).
1096. Seidel, W. C., and C. A. Tolman, *Inorg. Chem.*, **9**, 2354 (1970).
1097. Selbin, J., N. Ahmad, and M. J. Pribble, *J. Inorg. Nucl. Chem.*, **32**, 3249 (1970).
1098. Sergeev, N. M., *Russ. Chem. Rev.*, **42**, 339 (1973).
1099. Sestili, L., C. Furlani, and G. Festuccia, *Inorg. Chim. Acta*, **4**, 542 (1970).
1100. Setkina, V. N., A. G. Ginzburg, N. V. Kislyakova, and D. N. Kursanov, *Izv. Akad. Nauk SSSR, Ser. Khim.*, **1971**, 434.
1101. Shapeley, J. R., and J. A. Osborn, *Acc. Chem. Res.*, **6**, 305 (1973).
1102. Shapeley, J. R., and J. A. Osborn, *J. Am. Chem. Soc.*, **92**, 6976 (1970).
1103. Shaw, B. L., and R. M. Slade, *J. Chem. Soc. (A)*, **1971**, 1184.
1104. Shaw, B. L., and R. E. Stainbank, *J. Chem. Soc. (A)*, **1971**, 3716.
1105. Shaw, B. L., and R. E. Stainbank, *J.C.S. Dalton*, **1972**, 223.
1106. Shaw, B. L., and R. E. Stainbank, *J.C.S. Dalton*, **1972**, 2108.
1107. Shaw, C. F., J. W. Lundeen, and R. S. Tobias, *J. Organomet. Chem.*, **51**, 365 (1973).
1108. Sheldrick, W. S., and O. Stelzer, *J.C.S. Dalton*, **1973**, 926.

1109. Shiotani, A., and H. Schmidbaur, *Chem. Ber.*, **104**, 2838 (1971).
1110. Shiotani, A., and H. Schmidbaur, *J. Organomet. Chem.*, **37**, C24 (1972).
1111. Shobotake, K., and K. Nakamoto, *J. Am. Chem. Soc.*, **92**, 3332 (1970).
1112. Shrader, D., E. P. Ross, R. T. Jernigan, and G. R. Dobson, *Inorg. Chem.*, **9**, 1286 (1970).
1113. Siegl, W. O., S. J. Lapporte, and J. P. Collman, *Inorg. Chem.*, **10**, 2158 (1971).
1114. Siegl, W. O., S. J. Lapporte, and J. P. Collman, *Inorg. Chem.*, **12**, 674 (1973).
1115. Simpson, R. T., and A. J. Carty, *J. Coord. Chem.*, **2**, 207 (1973).
1116. Simpson, R. T., S. Jacobson, A. J. Carty, M. Mathew, and G. J. Palenik, *J.C.S. Chem. Comm.*, **1973**, 388.
1117. Singh, P., C. B. Damman, and D. J. Hodgson, *Inorg. Chem.*, **12**, 1335 (1973).
1118. Singleton, E., J. T. Moelwyn-Hughes, and A. W. B. Garner, *J. Organomet. Chem.*, **21**, 449 (1970).
1119. Sladkov, A. M., and I. R. Goldina, *Izv. Akad. Nauk SSSR, Ser. Khim.*, **1970**, 2644.
1120. Smith, A. E., *Inorg. Chem.*, **11**, 2306 (1972).
1121. Smith, G. V., and R. J. Shuford, *Tetrahedron Lett.*, **1970**, 525.
1122. Sokolov, V. N., G. M. Khvostic, I. Ya. Poddubnyi, and G. P. Kondratenkov, *J. Organomet. Chem.*, **54**, 361 (1973).
1123. Sokol'skii, D. V., Ya. A. Dorfman, and L. S. Ernestova, *Zh. Prikl. Khim.*, **45**, 1344 (1972).
1124. Speier, G., and L. Marko, *J. Organomet. Chem.*, **21**, P46 (1970).
1125. Spendjian, H. K., and I. S. Butler, *Inorg. Chem.*, **9**, 1268 (1970).
1126. Stalick, J. K., P. W. R. Corfield, and D. W. Meek, *Inorg. Chem.*, **12**, 1668 (1973).
1127. Stalick, J. K., and J. A. Ibers, *Inorg. Chem.*, **9**, 453 (1970).
1128. Stalick, J. K., D. W. Meek, B. Y. K. Ho, and J. J. Zuckerman, *J.C.S. Chem. Comm.*, **1972**, 630.
1129. Stanclift, W. E., and D. G. Hendricker, *J. Organomet. Chem.*, **50**, 175 (1973).
1130. Steele, D. F., and T. A. Stephenson, *J.C.S. Dalton*, **1972**, 2161.
1131. Stelzer, O., and R. Schmutzler, *J. Chem. Soc. (A)*, **1971**, 2867.
1132. Stelzer, O., and E. Unger, *J.C.S. Dalton*, **1973**, 1783.
1133. Stepaniak, R. F., and N. C. Payne, *J. Organomet. Chem.*, **57**, 213 (1973).
1134. Stephens, F. S., *J.C.S. Dalton*, **1972**, 2257.
1135. Stephenson, T. A., *J. Chem. Soc. (A)*, **1970**, 889.
1136. Stocco, F., G. C. Stocco, W. M. Scovell, and R. S. Tobias, *Inorg. Chem.*, **10**, 2639 (1971).
1137. Stocco, G. C., and R. S. Tobias, *J. Am. Chem. Soc.*, **93**, 5057 (1971).
1138. Stone, P. J., and Z. Dori, *Inorg. Chim. Acta*, **5**, 434 (1971).
1139. Storhoff, B. N., *J. Organomet. Chem.*, **43**, 197 (1972).
1140. Strohmeier, W., *Angew. Chem.*, **76**, 873 (1964).
1141. Strohmeier, W., *Fortschr. Chem. Forsch.*, **25**, 71 (1972).
1142. Strohmeier, W., *J. Organomet. Chem.*, **32**, 137 (1971).
1143. Strohmeier, W., *J. Organomet. Chem.*, **60**, C60 (1973).
1144. Strohmeier, W., and R. Endres, *Z. Naturforsch.*, **25b**, 1068 (1970).
1145. Strohmeier, W., and R. Endres, *Z. Naturforsch.*, **26b**, 730 (1971).
1146. Strohmeier, W., and W. Rehder-Stirnweiss, *J. Organomet. Chem.*, **22**, C27 (1970).
1147. Strohmeier, W., and W. Rehder-Stirnweiss, *J. Organomet. Chem.*, **26**, C22 (1971).
1148. Strohmeier, W., W. Rehder-Stirnweiss, and R. Fleischmann, *Z. Naturforsch.*, **25b**, 1480 (1970).
1149. Strohmeier, W., W. Rehder-Stirnweiss, and R. Fleischmann, *Z. Naturforsch.*, **25b**, 1481 (1970).
1150. Strohmeier, W., W. Rehder-Stirnweiss, and G. Reischig, *J. Organomet. Chem.*, **17**, 393 (1971).
1151. Strope, D., and D. F. Shriver, *J. Am. Chem. Soc.*, **95**, 8197 (1973).

1152. Su, S. R., and A. Wojcicki, *J. Organomet. Chem.*, **27**, 231 (1971).
1153. Sullivan, C. E., and W. E. McEwen, *Org. Prep. Proced.*, **2**, 157 (1970).
1154. Sutton, B. M., E. McGusty, D. T. Walz, and M. J. Di Martino, *J. Med. Chem.*, **15**, 1095 (1972).
1155. Sweigart, D. A., and P. Heidtmann, *J.C.S. Chem. Comm.*, **1973**, 556.
1156. Switkes, E. S., L. Ruiz-Ramirez, T. A. Stephenson, and J. Sinclair, *Inorg. Nucl. Chem. Lett.*, 8, 593 (1972).
1157. Takahashi, K., G. Hata, and A. Miyake, *Bull. Chem. Soc. Jap.*, **46**, 600 (1973).
1158. Takahashi, K., A. Miyake, and G. Hata, *Chem. Ind. (London)*, **1971**, 488.
1159. Takano, T., *Bull. Chem. Soc. Jap.*, **46**, 522 (1973).
1160. Takao, K., Y. Fujiwara, T. Imanaka, and S. Teranishi, *Bull. Chem. Soc. Jap.*, **43**, 1153 (1970).
1161. Tamao, K., K. Sumitani, and M. Kumada, *J. Am. Chem. Soc.*, **94**, 4374 (1972).
1162. Tan, T. C., and Y. Y. Lim., *Inorg. Chem.*, **12**, 2203 (1973).
1163. Tanaka, K., and T. Tanaka, *Inorg. Nucl. Chem. Lett.*, **9**, 429 (1973).
1164. Tanaka, M., Y. Watanabe, T. Mitsudo, H. Iwane, and Y. Takegami, *Chem. Lett.*, **1973**, 239.
1165. Tauzher, G., and G. Costa, *J. Inorg. Nucl. Chem.*, **34**, 2676 (1972).
1166. Tauzher, G., G. Mestroni, A. Puxeddu, R. Costanzo, and G. Costa, *J. Chem. Soc. (A)*, **1971**, 2504.
1167. Tayim, H. A., S. K. Thabet, and M. U. Karkanawi, *Inorg. Nucl. Chem. Lett.*, 8, 235 (1972).
1168. Tebbe, F. N., P. Meakin, J. P. Jesson, and E. L. Muetterties, *J. Am. Chem. Soc.*, **92**, 1068 (1970).
1169. Temkin, O. N., O. L. Kaliya, G. K. Shestakov, S. M. Brailovskii, R. M. Flid, and A. P. Aseeva, *Kinet. Katal.*, **11**, 1592 (1970).
1170. Thabet, S. K., H. A. Tayim, and M. U. Karkanawi, *Inorg. Nucl. Chem. Lett.*, 8, 211 (1972).
1171. Thatcher, J. G., and W. R. Deever, U.S. 3,732,329, 08 May 1973, Appl. 102,166, 28 Dec 1970, through *Chem. Abstr.*, **79**, 4982a (1973).
1172. Thomas, K., J. T. Dumler, B. W. Renoe, C. J. Nyman, and D. M. Roundhill, *Inorg. Chem.*, **11**, 1795 (1972).
1173. Thomson, J., and M. C. Baird, *Can. J. Chem.*, **51**, 1179 (1973).
1174. Thomson, J., and M. C. Baird, *Inorg. Chim. Acta*, 7, 105 (1973).
1175. Thomson, J., D. Groves, and M. C. Baird, *J. Magn. Resonance*, **5**, 281 (1971).
1176. Thornhill, D. J., and A. R. Manning, *J.C.S. Dalton*, **1973**, 2086.
1177. Timms, P. L., *J. Chem. Soc. (A)*, **1970**, 2526.
1178. Titus, D. D., A. A. Orio, R. E. Marsh, and H. B. Gray, *Chem. Commun.*, **1971**, 322.
1179. Tolman, C. A., *Inorg. Chem.*, **10**, 1540 (1971).
1180. Tolman, C. A., *Inorg. Chem.*, **11**, 3128 (1972).
1181. Tolman, C. A., *J. Am. Chem. Soc.*, **92**, 2953 (1970).
1182. Tolman, C. A., *J. Am. Chem. Soc.*, **92**, 2956 (1970).
1183. Tolman, C. A., *J. Am. Chem. Soc.*, **92**, 4217 (1970).
1184. Tolman, C. A., *J. Am. Chem. Soc.*, **92**, 6777 (1970).
1185. Tolman, C. A., *J. Am. Chem. Soc.*, **92**, 6785 (1970).
1186. Tolman, C. A., *J. Am. Chem. Soc.*, **94**, 2994 (1972).
1187. Tomlinson, A. A. G., C. Bellitto, O. Piovesana, and C. Furlani, *J.C.S. Dalton*, **1972**, 350.
1188. Townsend, R. E., and K. J. Coskran, *Inorg. Chem.*, **10**, 1661 (1971).
1189. Trabelsi, M., A. Loutellier, and M. Bigorgne, *J. Organomet. Chem.*, **40**, C45 (1972).
1190. Trabelsi, M., A. Loutellier, and M. Bigorgne, *J. Organomet. Chem.*, **56**, 369 (1973).
1191. Traverso, O., V. Carassiti, M. Graziani, and U. Belluco, *J. Organomet. Chem.*, **57**, C22

(1973).
1192. Treichel, P. M., W. K. Dean, and J. C. Calabrese, *Inorg. Chem.*, **12**, 2908 (1973).
1193. Treichel, P. M., W. K. Dean, and W. M. Douglas, *Inorg. Chem.*, **11**, 1609 (1972).
1194. Treichel, P. M., W. K. Dean, and W. M. Douglas, *J. Organomet. Chem.*, **42**, 145 (1972).
1195. Treichel, P. M., W. M. Douglas, and W. K. Dean, *Inorg. Chem.*, **11**, 1615 (1972).
1196. Treichel, P. M., and W. J. Knebel, *J. Coord. Chem.*, **2**, 67 (1972).
1197. Treichel, P. M., W. J. Knebel, and R. W. Hess, *J. Am. Chem. Soc.*, **93**, 5724 (1971).
1198. Tripathy, P. B., B. W. Renoe, K. Adzamli, and D. M. Roundhill, *J. Am. Chem. Soc.*, **93**, 4406 (1971).
1199. Troitskaya, A. D., and Z. L. Shmakova, *Zh. Neorg. Khim.*, **16**, 1113 (1971).
1200. Troitskaya, A. D., Yu. V. Yablokov, A. V. Ryzhmanova, B. I. Kudryavtsev, and E. I. Semenova, *Zh. Neorg. Khim.*, **17**, 3122 (1972).
1201. Tsuji, J., M. Hara, and K. Ohno, Ger. Offen. 1,942,798, 23 Apr 1970, Japan. Appl. 26 Aug - 26 Sep 1968, through *Chem. Abstr.*, **73**, 15497y (1970).
1202. Tuggle, R. M., and D. L. Weaver, *Inorg. Chem.*, **11**, 2237 (1972).
1203. Tyrlik, S., *J. Organomet. Chem.*, **50**, C46 (1973).
1204. Uchida, T., Y. Uchida, M. Hidai, and T. Kodama, *Bull. Chem. Soc. Jap.*, **44**, 2883 (1971).
1205. Udovich, C. A., and R. J. Clark, *J. Organomet. Chem.*, **25**, 199 (1970).
1206. Udovich, C. A., and R. J. Clark, *J. Organomet. Chem.*, **36**, 355 (1972).
1207. Udovich, C. A., J. Takemoto, and K. Nakamoto, *J. Coord. Chem.*, **1**, 89 (1971).
1208. Ugi, I., D. Marquarding, H. Klusacek, P. Gillespie, and F. Ramirez, *Acc. Chem. Res.*, **4**, 288 (1971).
1209. Ugo, R., G. La Monica, F. Cariati, S. Cenini, and F. Conti, *Inorg. Chim. Acta*, **4**, 390 (1970).
1210. Uguagliati, P., A. Palazzi, G. Deganello, and U. Belluco, *Inorg. Chem.*, **9**, 724 (1970).
1211. Uhlig, E., and R. Münzberg, *Z. Chem.*, **13**, 142 (1973).
1212. Uhlig, E., and H. Walther, *Z. Chem.*, **11**, 23 (1971).
1213. Uson, R., P. Royo, and A. Laguna, *Inorg. Nucl. Chem. Lett.*, **7**, 1037 (1971).
1214. Vahrenkamp, H., *Chem. Ber.*, **104**, 449 (1971).
1215. Vahrenkamp, H., *Chem. Ber.*, **105**, 3574 (1972).
1216. Vahrenkamp, H., and H. Nöth, *Chem. Ber.*, **106**, 2227 (1973).
1217. Valentine, J. S., and D. Valentine, Jr., *J. Am. Chem. Soc.*, **92**, 5795 (1970).
1218. Van de Griend, L. J., and J. G. Verkade, *Inorg. Nucl. Chem. Lett.*, **9**, 1137 (1973).
1219. Van den Akker, M., R. Olthof, F. Van Bolhuis, and F. Jellinck, *Recl. Trav. Chim. Pays-Bas*, **91**, 75 (1972).
1220. Van der Ent, A., H. G. A. M. Cuppers, Neth. Appl. 70 01,018, 27 Jul 1971, Appl. 23 Jan 1970, through *Chem. Abstr.*, **75**, 151336 (1971).
1221. Van der Linde, R., and R. O. De Jongh, *Chem. Commun.*, **1971**, 563.
1222. Van Koten, G., and J. G. Noltes, *J.C.S. Chem. Comm.*, **1972**, 452.
1223. Van Winkle, J. L., Brit. 1,191,815, 13 May 1970, US Appl. 26 Feb 1968, through *Chem. Abstr.*, **73**, 66020v (1970).
1224. Varshavskii, Yu. S., T. G. Cherkasova, and N. A. Buzina, *J. Organomet. Chem.*, **56**, 375 (1973).
1225. Varshavskii, Yu. S., T. G. Cherkasova, and N. A. Buzina, *Zh. Neorg. Khim.*, **17**, 2208 (1972).
1226. Vaska, L., and L. S. Chen, *Chem. Commun.*, **1971**, 1080.
1227. Venanzi, L. M., R. Spagna, and L. Zambonelli, *Chem. Commun.*, **1971**, 1570.
1228. Verkade, J. G., *Coord. Chem. Rev.*, **9**, 1 (1972).
1229. Verkade, J. G., and K. J. Coskran, *Organic Phosphorus Compounds*, **Vol. 2**, G. M. Kosolapoff and L. Maier, Eds., Wiley, New York, 1972, p. 1.

1230. Visser, J. P., A. J. Schipperijin, J. Lucas, D. Bright, and J. J. De Boer, *Chem. Commun.*, **1971**, 1266.
1231. Vitulli, G., L. Porri, and A. L. Segre, *J. Chem. Soc. (A)*, **1971**, 3246.
1232. Vizi-Orosz, A., G. Palyi, and L. Marko, *J. Organomet. Chem.*, **60**, C25 (1973).
1232a. Volpin, M. E., V. S. Lenenko, and V. B. Shur, *Izv. Akad. Nauk SSSR, Ser. Khim.*, **1971**, 463.
1233. Wagner, K. P., P. M. Treichel, and J. C. Calabresse, *J. Organomet. Chem.*, **56**, C33 (1973).
1234. Wahren, R., *J. Organomet. Chem.*, **57**, 415 (1973).
1235. Walborsky, H. M., and L. E. Allen, *Tetrahedron Lett.*, **1970**, 823.
1236. Walther, D., *Z. Anorg. Allg. Chem.*, **396**, 46 (1973).
1237. Walton, R. A., *Inorg. Chem.*, **10**, 2534 (1971).
1238. Wang, J. T., C. Udovich, K. Nakamoto, A. Quattrochi, and J. R. Ferraro, *Inorg. Chem.*, **9**, 2675 (1970).
1239. Warren, J. D., M. A. Busch, and R. J. Clark, *Inorg. Chem.*, **11**, 452 (1972).
1240. Warren, J. D., and R. J. Clark, *Inorg. Chem.*, **9**, 373 (1970).
1241. Watanabe, S., K. Suga, and H. Kikuchi, *Aust. J. Chem.*, **23**, 385 (1970).
1242. Waters, J. M., and K. R. Whittle, *Chem. Commun.*, **1971**, 518.
1243. Weaver, D. L., *Inorg. Chem.*, **9**, 2250 (1970).
1244. Weininger, M. S., I. F. Taylor, Jr., and E. L. Amma, *Chem. Commun.*, **1971**, 1172.
1245. Wenschuh, E., and G. Mehner, *Z. Chem.*, **10**, 73 (1970).
1246. Werner, H., *Angew. Chem.*, **80**, 1017 (1968).
1247. Westland, A. D., and N. Muriithi, *Inorg. Chem.*, **12**, 2356 (1973).
1248. Weston, C. W., G. W. Bailey, J. H. Nelson, and H. B. Jonassen, *J. Inorg. Nucl. Chem.*, **34**, 1752 (1972).
1249. Wheelock, K. S., J. H. Nelson, and H. B. Jonassen, *Inorg. Chim. Acta*, **4**, 399 (1970).
1250. White, C., and R. J. Mawby, *Inorg. Chim. Acta*, **4**, 261 (1970).
1251. White, C., and R. J. Mawby, *J. Chem. Soc. (A)*, **1971**, 940.
1252. White, R. C. and J. G. Thatcher, U.S. 3,692,864, 19 Sep 1972, Appt. 76,991, 30 Sep 1970, through *Chem. Abstr.*, **78**, 71371t (1973).
1253. White, J. F., and M. F. Farona, *Inorg. Chem.*, **10**, 1080 (1971).
1254. Whitesides, G. M., and J. F. Gaasch, *J. Organomet. Chem.*, **33**, 241 (1971).
1255. Whitesides, G. M., J. F. Gaasch, and E. R. Stedronsky, *J. Am. Chem. Soc.*, **94**, 5258 (1972).
1256. Whitesides, G. M., E. J. Panek, and E. R. Stedronsky, *J. Am. Chem. Soc.*, **94**, 232 (1972).
1257. Whyman, R., *J.C.S. Dalton*, **1972**, 1375.
1258. Whyman, R., *J. Organomet. Chem.*, **24**, C35 (1970).
1259. Whyman, R., *J. Organomet. Chem.*, **29**, C36 (1971).
1260. Whyman, R., *J. Organomet. Chem.*, **56**, 339 (1973).
1261. Wickman, H. H., and W. E. Silverthorn, *Inorg. Chem.*, **10**, 2333 (1971).
1262. Wilkinson, G., Brit. 1,219,763, 20 Jan 1971, Appl. 02 Feb 1967, through *Chem. Abstr.*, **74**, 87398k (1971).
1263. Wilkinson, G., Ger. Offen. 2,034,908, 28 Jan 1971, Brit. Appl. 14 Jul 1969, through *Chem. Abstr.*, **74**, 88127h (1971).
1264. Wilson, R., and D. Kivelson, *J. Chem. Phys.*, **44**, 154, 4440, 4445 (1966).
1265. Wilson, S. T., and J. A. Osborn, *J. Am. Chem. Soc.*, **93**, 3068 (1971).
1266. Winzer, A., and R. Griebel, Ger. (East) 76,674, 12 Oct 1970, Appl. 30 Aug 1969, through *Chem. Abstr.*, **75**, 110425y (1971).
1267. Winzer, A., and R. Griebel, *Z. Chem.*, **12**, 181 (1972).
1268. Wood, J. S., *Progress in Inorganic Chemistry*, S. J. Lippard, Ed., Vol. 16, Wiley, New York, 1972, p. 227 ff.

1269. Wright, G., R. W. Glyde, and R. J. Mawby, *J.C.S. Dalton,* **1973**, 220.
1270. Wright, G., and R. J. Mawby, *J. Organomet. Chem.,* **29**, C29 (1971).
1271. Yagupsky, M., and G. Wilkinson, *J. Chem. Soc. (A),* **1970**, 941.
1272. Yamamoto, A., S. Kitazume, L. S. Pu, and S. Ikeda, *J. Am. Chem. Soc.,* **93**, 371 (1971).
1273. Yamamoto, A., A. Miyashita, T. Yamamoto, and S. Ikeda, *Bull. Chem. Soc. Jap.,* **45**, 1583 (1972).
1274. Yamamoto, K., T. Hayashi, and M. Kumada, *J. Organomet. Chem.,* **46**, C65 (1972).
1275. Yamamoto, K., Y. Uramoto, and M. Kumada, *J. Organomet. Chem.,* **31**, C9 (1971).
1276. Yamasaki, A., and E. Fluck, *Z. Anorg. Allg. Chem.,* **396**, 297 (1973).
1277. Yamazaki, H., and N. Hagihara, *Bull. Chem. Soc. Jap.,* **44**, 2260 (1971).
1278. Yandell, J. K., and N. Sutin, *Inorg. Chem.,* **11**, 448 (1972).
1279. Yasufuku, K., and H. Yamazaki, *Bull. Chem. Soc. Jap.,* **43**, 1588 (1970).
1280. Yasufuku, K., and H. Yamazaki, *Bull. Chem. Soc. Jap.,* **46**, 1502 (1973).
1281. Yasufuku, K., and H. Yamazaki, *J. Organomet. Chem.,* **38**, 367 (1972).
1282. Yoon, N., M. J. Incorvia, and J. I. Zink, *J.C.S. Chem. Comm.,* **1972**, 499.
1283. Zakharkin, L. I., and A. I. L'vov, *Zh. Obshch. Khim.,* **41**, 1880 (1971).
1284. Zanella, R., R. Ros, and M. Graziani, *Inorg. Chem.,* **12**, 2736 (1973).
1285. Zelonka, R. A., and M. C. Baird, *Can. J. Chem.,* **50**, 1269 (1972).
1286. Zelonka, R. A., and M. C. Baird, *Chem. Comm.,* **1971**, 780.
1287. Ziolo, R. F., A. P. Gaughan, Z. Dori, C. G. Pierpont, and R. Eisenberg, *J. Am. Chem. Soc.,* **92**, 738 (1970).
1288. Ziolo, R. F., S. Lipton, and Z. Dori, *Chem. Commun.,* **1970**, 1124.

Addenum

This addendum to the foregoing review covers the literature in *Chemical Abstracts,* Vols. 80 and 81 under the key words phosphine to phosphorus. In order to keep within reasonable limits, only the titles or some short comments for the cited publications are given.

I. Reviews

1. C. A. McAuliffe (Edit.), *Transition Metal Complexes of Phosphorus, Arsenic, and Antimony Ligands,* MacMillan, London, 1973.
   A comprehensive multiauthor review which covers the literature up to 1972.
2. P. Rigo and A. Turco, *Coord. Chem. Rev., 13,* 133 (1974).
   Cyanide phosphine complexes.
3. P. Uguagliati, G. Deganello, and U. Belluco, *Inorg. Chim. Acta, 9,* 203 (1974).
   A short review concerned with the system μ-dichlorotetracarbonyldirhodium tertiary phosphines.

II. Complexes of Tertiary Phosphines

A. Tertiary Monodentate Phosphine Complexes

3. *Vanadium, Niobium, and Tantalum*

4. C. Santini-Scampucci and J. G. Riess, *J. C. S. Dalton,* **1973**, 2436.

Preparation of monomethylniobium(V) and -tantalum(V) halide complexes, for example, $MeNbCl_4 \cdot L$ L = Ph_3P including $[Ph_2P{-}CH_2]_2$).

4. *Chromium, Molybdenum, and Tungsten*

5. U. Schubert and E. O. Fischer, *Chem. Ber., 106,* 3882 (1973).
6. E. O. Fischer, G. Kreis, F. R. Kreissl, C. G. Kreiter, and J. Müller, *Chem. Ber., 106,* 3910 (1973).
7. E. O. Fischer and H. Fischer, *Chem. Ber., 107,* 657 (1974).
8. H. Fischer and E. O. Fischer, *Chem. Ber., 107,* 673 (1974).
9. H. Fischer, E. O. Fischer, and F. R. Kreissl, *J. Organomet. Chem., 64,* C41 (1974).
10. H. Fischer, E. O. Fischer, C. G. Kreiter, and H. Werner, *Chem. Ber., 107,* 2459 (1974).
Chemistry of phosphine substituted VIb transition metal carbene complexes.
11. P. Hackett, P. S. O'Neill, and A. R. Manning, *J. C. S. Dalton, 1974,* 1625.
Reaction of Ph_3P with $[(\pi\text{-}C_5H_5)Cr(CO)_3]_2$; the only product is $[(\pi\text{-}C_5H_5)Cr(CO)_2Ph_3P]_2$.
12. N. G. Connelly, *J. C. S. Dalton, 1973,* 2183.
Cationic nitrosyl complexes of molybdenum and tungsten, $MoCl(CO)(NO)(Ph_2PCH_2)_2L$ (L = Ph_3P, $P(OPh)_3$, $P(OMe)_3$) and $[Mo(CO)(NO)L_2]PF_6$ [L = $(Ph_2PCH_2)_2$].
13. B. D. Dombek and R. J. Angelici, *J. Am. Chem. Soc., 95,* 7516 (1973).
The complexes *trans*-$Cr(CO)_4(CS)Ph_3P$, *trans*-$W(CO)_4(CS)Ph_3P$, and *trans*-$W(CO)_4$-(CS)py were prepared from the $M(CO)_5(CS)$ complexes.
14. C. Miniscloux, G. Martino, and L. Sajus, *Bull. Soc. Chim. Fr., 1973,* 2179.
Tris(acetonitrile)hexachlorodimolybdenum; synthesis and substitution reactions. Complexes of the type $MoCl_4L_2$ (L = Ph_3P, Cy_3P) were synthesized.
15. N. J. Copper and M. L. H. Green, *J. C. S. Chem. Comm., 1974,* 208.
Me_2PhP reacted with $[(\pi\text{-}C_5H_5)_2M(CH_2{=}CH_2)Me]^+PF_6$ (M = W) gave $[(\pi\text{-}C_5H_5)_2W(CH_2PMe_2Ph)H]^+PF_6^-$ whereas for M = Mo $[(\pi\text{-}C_5H_5)_2MoMe(Me_2PhP)]^+PF_6^-$ was obtained.
16. F. Pennella, *Inorg. Synth., 15,* 42 (1974).
Tetrahydridotetrakis(methyldiphenylphosphine)molybdenum(IV).
17. D. Walther, *Z. Anorg. Allg. Chem., 405,* 8 (1974).
Substitution of CO by Ph_3P or $Bu^n{}_3P$ lowers the stability of the benzil bis(phenylimine) complexes of molybdenum, $Mo(CO)_4L$ (L = $[p{-}R{-}C_6H_4(p\text{-}R^1C_6H_4N{:})C]_2$, R, R^1 = H, Me, OMe).
18. J. M. Burlitch and T. W. Theyson, *J. C. S. Dalton, 1974,* 828.
$Tl[Co(CO)_3Ph_3P]$ has been synthesized and spectroscopically identified.

5. *Manganese, Technetium, and Rhenium*

19. R. H. Reimann and E. Singleton, *J. C. S. Dalton, 1974,* 808.
fac-$[Mn(CO)_3(MeCN)_3]PF_6$, prepared by refluxing $MnBr(CO)_5$ in MeCN, reacts with phosphines and phosphites (L) to give *fac*-$[Mn(CO)_3(MeCN)_2L]^+$, *fac*-$[Mn(CO)_3(MeCN)L_2]^+$, *cis*-$[Mn(CO)_2(MeCN)_2L_2]^+$, and *mer-cis*-$[Mn(CO)_2(MeCN)L_3]^+$
20. A. N. Nesmeyanov, E. G. Perevalova, L. I. Leonteva, and L. I. Khomik, *Izv. Akad. Nauk SSSR, Ser. Khim., 1974,* 938.
The carbonyl group in $Mn(CO)_3(\pi\text{-}C_5H_5)Ni(Ph_3P)(\pi\text{-}C_5H_5)$ may be replaced by Ph_3P.
21. H. Brunner and W. Rambold, *Angew. Chem., 85,* 1118 (1973).

22. H. Brunner and M. Lappus, *Z. Naturforsch., 29b,* 363 (1974).
Optically active manganese carbonyl or manganese carbonyl nitrosyl complexes.
23. J. Chatt, J. R. Dilworth, H. P. Gunz, and G. J. Leigh, *J. Organomet. Chem.,* 64, 245 (1974).
New synthetic routes to mono-, di-, and tricarbonyl halo complexes of rhenium(I) and rhenium(II) with tertiary phosphine, *trans*-$ReX(CO)_3(Ph_3P)_2$ (X = Cl, Br, I).
24. G. LaMonica, S. Cenini, and M. Freni, *J. Organomet. Chem., 76,* 355 (1974).
$Re(CO)_2(NO)(Ph_3P)_2$ with aroyl azides gives isocyanate complexes Re(CO)-$(RCONCO)(NO)(Ph_3P)_2$.
25. D. J. Darensbourg, M. Y. Darensbourg, D. Drew, and H. L. Conder, *J. Organomet. Chem., 74,* C33 (1974).
Synthesis and spectral properties of manganese pentacarbonyl phosphine and phosphite cation derivatives.
26. R. W. Adams, J. Chatt, N. E. Hooper, G. J. Leigh, *J. C. S. Dalton, 1974,* 1075.
Synthesis and reactions of five coordinate paramagnetic $ReCl_2(NO)(Ph_3P)_2$.
27. G. La Monica, M. Freni, and S. Cenini, *J. Organomet. Chem., 71,* 57 (1974).
New Nitrosyl derivatives of rhenium, $Re(CO)_2(NO)(Ph_3P)_2$ and $ReH(NO)_2(Ph_3P)_2$.
28. M. Herberhold and A. Razavi, *J. Organomet. Chem., 67,* 81 (1974).
Replacement of CO by Ph_3P and $P(OMe)_3$ in $Mn(CO)(NO)_3$ affords complexes $Mn(NO)_3Ph_3P$ and $Mn(NO)_3P(OMe)_3$.
29. A. E. Fenster and I. S. Butler, *Inorg. Chem., 13,* 915 (1974).
Thiocarbonyl and carbondisulfide complexes of manganese are formed if $\pi C_5H_5Mn(CO)_2(C_8H_{14})$ (C_8H_{14} = *cis*-cyclooctene) is treated with CS_2 and Ph_3P.
30. F. A. Cotton, B. A. Frenz, and J. R. Ebner, *J. C. S. Chem. Comm., 1974,* 4.
New class of binuclear rhenium(II) halide species containing a strong metal-metal bond. Chemistry and structure of complexes of the type $Re_2X_4(R_3P)_4$.
31. U. Mazzi, G. De Paoli, G. Rizzardi, and L. Magon, *Inorg. Chim. Acta, 10,* L2 (1974).
Complexes of technetium(III) and (IV) with dimethylphenylphosphine, *trans*-$TcX_4(Me_2PhP)_2$ and *mer*-$TcX_3(Me_2PhP)_3$.

6. *Iron, Ruthenium, and Osmium*

32. A. J. Deeming and M. Underhill, *J. C. S. Dalton, 1973,* 2727.
Carbon-hydrogen cleavage of trimethyl and triethylphosphine complexes of osmium, $Os_3(CO)_{11}R_3P$ and $Os_3(CO)_{10}(R_3P)_2$ (R = Me, Et).
33. C. G. Cooke and M. J. Mays, *J. Organomet. Chem., 74,* 449 (1974).
Reactions of $[FeCo_3(CO)_{12}]^-$ and $[MnFe_2(CO)_{12}]^-$ with a number of monodentate P-donor ligands (L). Complexes of the type $[FeCo_3(CO)_{11}L]^-$ and $[MnFe_2(CO)_{11}$-$L]^-$ are isolated.
34. A. J. Deeming, R. E. Kimber, and M. Underhill, *J. C. S. Dalton, 1973,* 2589.
Benzyne complexes of osmium derived from dimethylphenylphosphine or dimethylphenylarsine.
35. J. A. Carroll, D. R. Fisher, G. W. Rayner-Canham, and D. Sutton, *Can. J. Chem., 52,* 1914 (1974).
Reaction of aryldiazoniumtetrafluoroboarates with $Fe(CO)_3(R_3P)_2$.
36. G. Cardaci and A. Foffani, *J. C. S. Dalton, 1974,* 1808.
1-3η allyl and but-3-enoyl iron carbonyl nitrosyl complexes with phosphine ligands.
37. J. Evans, D. V. Howe, and B. F. G. Johnson, *J. Organomet. Chem., 61,* C 48 (1973).
Addition of triphenylphosphine and pyridine to cyclohexadienyl- and cycloheptadienyltricarbonyl iron cations.

38. A. Vessieres and P. Dixneuf, *Tetrahedron Lett., 1974,* 1499.
$Fe(CO)_3$ complexes with PhCH=CHCOR (R = H, Me, Ph) react with Me_2PhP or $P(OMe)_3$ (L) to give the corresponding $Fe(CO)_3L$ complexes.
39. W. Schäfer, A. Zschunke, H. J. Kerrinnes, and U. Langbein, *Z. Anorg. Allg. Chem., 406,* 105 (1974).
40. W. Schäfer, H. J. Kerrinnes, and U. Langbein, *Z. Anorg. Allg. Chem., 406,* 101 (1974).
The synthesis and characterization of phosphine-stabilized iron diolefin complexes.
41. T. Blackmore, M. I. Bruce, and F. G. A. Stone, *J. C. S. Dalton, 1974,* 106.
Cyclopentadienylruthenium phosphine complexes, $(\pi\text{-}C_5H_5)Ru[\textit{cis}\text{-}C(CF_3)\text{=}CH\text{-}CF_3](Ph_3P)_2$ and $(\pi\text{-}C_5H_5)Ru\text{-}[\textit{trans}\text{-}C(CO_2Me)\text{=}CH(CO_2Me)](Ph_3P)_2$.
42. H. Kono and Y. Nagai, *Chem. Lett., 1974,* 931.
Hydrosilylation of $RuH_2(Ph_3P)_4$ with RR^1R^2SiH gave $RuH_3(SiRR^1R^2)(Ph_3P)_3$ (R, R^1, R^2 = alkyl, aryl, cyclohexyl, hydrogen, and ethoxy groups).
43. W. H. Knoth, *Inorg. Synth., 15,* 31 (1974).
Dinitrogen and hydrogen complexes of ruthenium, $RuH_2(N_2)(Ph_3P)_3$, and $RuH_4(Ph_3P)_3$.
44. B. F. G. Johnson and J. A. Segal, *J. C. S. Dalton, 1974,* 981.
Preparation and study of the osmium cation $[OsH_2(CO)(NO)L_2]^+$ (L = Ph_3P, Cy_3P).
45. F. G. Moers, R. W. M. Ten Hoedt, and J. P. Langhout, *J. Organomet. Chem., 65,* 93 (1974).
Carbonyl tricyclohexylphosphine complexes of ruthenium(II) and osmium(II), $RuHCl(CO)(Cy_3P)_2$ and $RuCl_2(CO)(Cy_3P)_2$.
46. P. W. Armit and T. A. Stephenson, *J. Organomet. Chem., 73,* C33 (1974).
Syntheses of di- and tri-μ-halide complexes of ruthenium(II) containing carbonyl and tertiary phosphine ligands.
47. M. M. T. Khan and R. K. Andal, *Proc. Chem. Symp., 2,* 133 (1972).
Synthesis of the dioxygen complexes $RuCl_2(O_2)(AsPh_3)L_2$ (L = $MePh_2P$, Pr_2PhP, Ph_3P), $RuHCl(O_2)L_3$ (L = $AsMePh_2$, $AsPh_3$), and $RuCl_2(O_2)(HgCl)(MePh_2P)_2$.
48. L. Ruiz-Ramirez and T. A. Stephenson, *J. C. S. Dalton, 1974,* 1640.
Reaction of benzyltriphenylphosphonium bicyclo-[2,2,1]-hepta-2,5-dienecarbonyltrichlororuthenate with Lewis bases.
49. M. M. T. Khan and S. S. Ahmed, *Proc. Chem. Symp., 2,* 155 (1972).
Complexes of osmium(III) and osmium(IV), $OsCl_3(Ph_3P)_2$ and *trans*-$OsCl_4L_2$ (L = Pr_2PhP, $MePh_2P$, $SbPh_3$).
50. D. Pawson and W. P. Griffith, *Inorg. Nucl. Chem. Lett., 10,* 253 (1974).
Suspensions of $(Bu_4N)(MCl_4N)$ (M = Os, Ru) in acetone with R_3P give $MCl_3(NPR_3)(R_3P)_2$ (R_3 = Et_2Ph, $EtPh_2$, $MePh_2$, Et_3).
51. R. K. Poddar and U. Agarwala, *J. Inorg. Nucl. Chem., 35,* 3769 (1973).
Reactions of dichlorobis(triphenylphosphine)ruthenium and tetrachlorotetrakis-(triphenylarsine)diruthenium with various donor molecules.
52. M. M. T. Khan and S. Vancheesan, *Proc. Chem. Symp., 2,* 137 (1972).
Preparation of $RuCl_2(HgCl)(Ph_3E)_3$, $RuCl(SnCl_3)(Ph_3E)_3$, and $RuCl_2(GeEt_2Cl_2)(Ph_3E)_3$ (E = P, As, Sb).
53. S. D. Robinson and M. F. Uttley, *J. C. S. Dalton, 1973,* 1912.
Carboxylato(triphenylphosphine) derivatives of ruthenium, osmium, rhodium, and iridium, $MH(O_2CR)(Ph_3P)_3$ (M = Ru, Os) and $RuX(O_2CR)(CO)(Ph_3P)_2$ (X = H, Cl).
54. A. Spencer and G. Wilkinson, *J. C. S. Dalton, 1974,* 786.
$Ru(CO)(CO_2Me)_2(Ph_3P)_2$ and $Ru(MeNC)_2(CO_2Me)_2(Ph_3P)_2$ were prepared.
55. D. J. Cole-Hamilton and T. A. Stephenson, *J. C. S. Dalton, 1974,* 739.
The preparation of the complexes RuX_2L_2 (X = R_2PS_2, Me_2NCS_2; R = Me, Et, Ph;

$L = Ph_3P, MePh_2P, Me_2PhP, P(OPh)_3$) is reported.
56. H. Brunner and J. Strutz, *Z. Naturforsch., 29b,* 446 (1974).
Reactions of optically active iron acyl and ester compound, (+) $-C_5H_5Fe(CO)(Ac)Ph_3P$ are reported.

7. *Cobalt, Rhodium, and Iridium*

57. L. Malatesta, M. Angoletta, and G. Caglio, *J. Organomet. Chem., 73,* 265 (1974).
Preparation and reactions of bis[triphenylphosphinetricarbonyliridium].
58. M. R. Churchill, J. J. Hackbarth, A. Davison, D. D. Traficante, and S. S. Wreford, *J. Am. Chem. Soc., 96,* 4041 (1974).
Oxidative addition of 1- or 2-bromopentaborane to trans-$IrCl(CO)(Me_3P)_2$ leads to the formation of 2-$[IrBr_2(CO)(Me_3P)_2]B_5H_8$.
59. L. Dahlenburg and R. Nast, *J. Organometal. Chem., 71,* C49 (1974).
Alkyl (or aryl) bis(triphenylphosphine)carbonyliridium and their oxygen adducts.
60. P. Piraino, F. Faraone, R. Pietropaolo, *Atti Accad. Peloritana Pericolanti, Cl. Sci. Fis., Mat. Natur., 52,* 269 (1972), through *Chem. Abstr., 80,* 133605z (1974).
$[IrCl(CO)_3]_n$ reacts with tertiary phosphines in the presence of $AgBF_4$ or $NaBPh_4$ to give cationic complexes $[Ir(CO)_2L_2]^+$ (L = Ph_3P, $P(o\text{-}MeC_6H_4)$, $Pr^i{}_3P$, Cy_3P, Bu^tPh_2P).
61. P. J. Fraser, W. R. Roper, and F. G. A. Stone, *J. Organomet. Chem., 66,* 155 (1974).
Transesterification reactions of dicarbonylbis(triphenylphosphine)methoxycarbonyl iridium(I).
62. M. Bressan and P. Rigo, *J. C. S. Chem. Comm., 1974,* 553.
This new synthetic route to phosphine nitrosyl complexes of cobalt, $Co(NO)(R_3P)_3$ uses $Co(NO_3)_2 \cdot 6\ H_2O$ as a starting material in the reaction with R_3P (R_3 = Ph_2H, PhHCy, Cy_2H, Me_2Ph, $MePh_2$).
63. Y. Kubo, A. Yamamoto, and S. Ikeda, *J. Organometal. Chem., 59,* 353 (1973).
Reactions of $CoH(N_2)(Ph_3P)_2$ and $CoMe(Ph_3P)_3$ with substituted olefins.
64. E. Perrotti and M. Clerici, Ger. Offen., 2,359,552, 12 Jun 1974, Ital. Appl. 32,211 A/72, 29 Nov 1972; through *Chem. Abstr. 81,* 91736f (1974).
Phosphine σ- and π-complexes of iridium and rhenium, e.g. $IrH_5(Pr^i{}_3P)_2$.
65. Yu. S. Varshavskii, T. G. Cherkasova, N. A. Buzina, and V. A. Kormer, *J. Organomet. Chem., 77,* 107 (1974).
The synthesis of $Rh(CO)(acac)Ph_3P$ and $Rh(CO)(oxq)Ph_3P$ (acac= acetylacetonate, oxq = 8-hydroxyquinoline).
66. I. S. Kolomnikov, H. Stepowska, S. Tyrlik, and M. E. Volpin, *Zh. Obshch, Khim., 44,* 1743 (1974).
Acrolein with $CoH(N_2)(Ph_3P)_2$ gives either $Co(OCOEt)_2(Ph_3P)_2$ or $Co(CO_3)(CO_2Et)(Ph_3P)_2$ depending on reaction conditions.
67. P. V. Rinze, *Angew. Chem., 86,* 351 (1974).
2,3-Dimethyl-1,3-butadienehydrobis(triphenylphosphine)cobalt(I), 1-4-η-CH_2=CMeCMe=$CH_2CoH(Ph_3P)_2$.
68. M. Nonoyama, *J. Organomet. Chem., 74,* 115 (1974).
The 8-methylquinoline (mq) complex $RhCl_2(mq) \cdot 1/3$ (Hmq$\cdot$HCl) reacts with Bu_3P to yield $RhCl_2(mq)Bu_3P$.
69. R. L. Bennett, M. I. Bruce, and R. C. F. Gardener, *J. C. S. Dalton, 1973,* 2653.
Oxidative addition reactions of the complex, pentafluorophenylbis(triphenylphosphinecarbonyl iridium(I).

70. Y. Wakatsuki and H. Yamazaki, *J. Organomet. Chem.*, *64*, 393 (1974).
Reactions of π-cyclopentadienylbis(triphenylphosphine)rhodium(I), (π-C_5H_5)Rh-$(Ph_3P)_2$ with alkyl halides, olefins, acetylenes, CS_2, and S.
71. R. J. Foot and B. T. Heaton, *J.C.S. Chem. Comm.*, *1973*, 838
2-Vinylpyridine is metalated by $Rh_2X_6(Bu_3P)_4$ (X = Cl, Br) to give

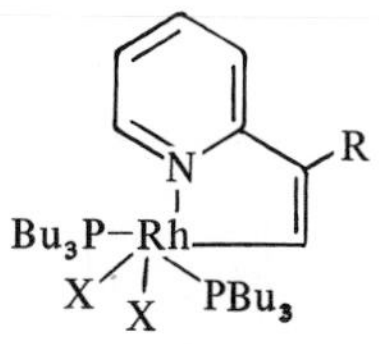

72. C. S. Cundy, M. F. Lappert, and R. Pearce, *J. Organomet. Chem.*, *59*, 161 (1973).
Treating $RhCl(Ph_3P)_3$ with Me_3CCH_2Li, Me_3SiCH_2MgCl, $(Me_3CCH_2)_4Zr$, $(Me_3Si$-$CH_2)_4Ti$, or $(Me_3SiCH_2)_4Zr$ gave Me_4C or Me_4Si and Rh(o-$C_6H_4PPh_2$) $(Ph_3P)_2$ in equimolar amounts.
73. D. Strope and D. F. Shriver, *Inorg. Chem.*, *13*, 2652 (1974).
$MeOSO_2F$ or $MeOSO_2CF_3$ undergo oxidative addition with $IrCl(CO)(Ph_3P)_2$ to give the highly labile products $IrCl(CO)(Me)X(Ph_3P)_2$ (X = OSO_2F or OSO_2CF_3).
74. F. Faraone, R. Pietropaolo, G. G. Troilo, and P. Piraino, *Inorg. Chim. Acta*, *7*, 792 (1973).
Cationic π-cyclopentadienylnitrosyl rhodium(I) complexes have been obtained in the reaction between (π-C_5H_5)Rh(CO)L (L = Ph_3P, Me_2PhP, Cy_3P) and $NOPF_6$.
75. G. K. N. Reddy and B. R. Ramesh, *Proc. Chem. Symp.*, *2*, 143 (1972).
76. G. K. N. Reddy and B. R. Ramesh, *J. Organomet. Chem.*, *67*, 443 (1974).
Cationic carbonyl complexes of rhodium(I), $[Rh(CO)_2L_3]ClO_4$ (L = Ph_3P, $SbPh_3$, $AsMePh_2$, $AsPrPh_2$).
77. C. A. L. Becker, *Synth. React. Inorg. Metal-Org. Chem.*, *4*, 213 (1974).
Preaparation of tris(*p*-chlorophenylisocyanide)bis(trimethyl phosphite)cobalt(I) tetrafluoroborate and tris-(*p*-chlorophenylisocyanide)bis(triphenylphosphine)cobalt-(I).
78. E. Bordignon, U. Croatto, U. Mazzi, and A. A. Orio, *Inorg. Chem.*, *13*, 935 (1974).
Mixed ligand complexes of five coordinate cobalt(I) with carbonyls, phosphines, and isocyanides, $CoI(CO)_2L_2$, $[Co(CO)_2L_3]^+$, $[Co(CNR)_2L_3]^+$, and $[Co(CNR)_3L_2]^+$ (L = $P(OMe)_3$, $PhP(OEt)_2$).
79. D. J. Cole-Hamilton and T. A. Stephenson, *J. C. S. Dalton*, *1974*, 1818.
mer-$RhCl_3(Me_2PhP)_3$ with L^-(L = S_2CNMe_2, S_2PMe_2, S_2PPh_2, S_2COEt) in methanol for 10 min gave *mer*-$RhCl_2L(Me_2PhP)_3$ containing a unidentate dithio-acid group.
80. S. Hasegawa, K. Itoh, and Y. Ishii, *Inorg. Chem.*, *13*, 2675 (1974).
Reaction between $RhCl(Ph_3P)_3$ and PhCONCO afforded [RhCl(PhCONCO) $(Ph_3P)_2]_2$. Some further reactions of this complex are reported (including analogous reactions with $Pd_2[(CO)(PhCH{=}CH)_2]_3$).
81. M. F. Lappert and A. J. Oliver, *J. C. S. Dalton*, *1974*, 65.
82. P. J. Fraser, W. R. Roper, and F. G. A. Stone, *J. C. S. Dalton*, *1974*, 760.
Carbene complexes containing stabilizing phosphine ligands.
83. M. Kubota and C. J. Curtis, *Inorg. Chem.*, *13*, 2277 (1974).
$CSCl_2$ has been used as a starting material for the synthesis of thiocarbonyl complexes, for example, $IrCl_3(CS)(Ph_3P)_2$, $[IrCl_2(CO)(CS)(Ph_3P)_2]PF_6$.

84. M. Doyle and R. Poilblanc, *C. R. Acad. Sci., Ser. C, 278,* 159 (1974).
Oxidative addition reaction of $[RhCl(CO)PZ_3]_2$ (Z = Me, OMe).

85. K. Kawakami, Y. Ozaki, and T. Tanaka, *J. Organomet. Chem., 69,* 151 (1974).
Oxidative addition reactions of carbon diselenide and areneselenols to some iridium(I) and platinum(O) complexes, trans-$IrCl(CO)(Ph_3P)_2$ and PtL_4 (L = Ph_3P, $MePh_2P$). The CSe_2 adducts have configurations similar to the corresponding CS_2 adducts.

86. P. R. Brookes and B. L. Shaw, *J. C. S. Dalton, 1974,* 1702.
Mercury(II) halide adducts of tertiary phosphine and arsine complexes of rhodium(III) and iridium(III).

87. J. S. Valentine, *J. C. S. Chem. Comm., 1973,* 857.
New hydridoiridium(III) isomer of chlorocarbonylbis(triphenylphosphine)iridium,

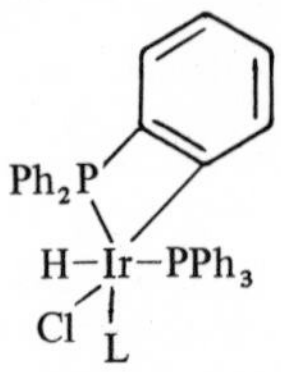

88. S. A. Gardner and M. D. Rausch, *Inorg. Chem., 13,* 997 (1974).
Grignard reaction of $(\pi\text{-}C_5H_5)MI(CF_3)Ph_3P$ (M = Co, Rh, Ir) with MeMgI gave both $(\pi\text{-}C_5H_5)MMeI(Ph_3P)$ and $(\pi\text{-}C_5H_5)MI_2Ph_3P$.

89. C. A. Tolman, P. Meakin, D. I. Lindner, and J. P. Jesson, *J. Am. Chem. Soc., 96,* 2762 (1974).
Triarylphosphine, hydride, and ethylene complexes of rhodium(I) chloride.

90. G. Speier and F. Ungvary, *Acta Chim. Acad. Sci. Hung., 82,* 25 (1974).
Reaction of $Co(N_2)(Ph_3P)_3$ or $CoH(N_2)(Ph_3P)_3$ and $CoX_2(Ph_3P)_2$ (X = Cl, Br, I) with Ph_3P gave $CoX(Ph_3P)_3$.

91. A. Maisonnat, P. Kalck, and R. Poilblanc, *J. Organomet. Chem., 73,* C36 (1974).
Treating the dinuclear complexes $[RhCl(CO)(C_2H_4)]_2$ and $[RhSR(CO)_2]_2$ with nucleophiles gives dinuclear penta-coordinated Rh(I) complexes, for example, $[Rh(SR)(CO)R_3P]_2$ (R = Ph, Bu^t; R_3 = Me_3, Me_2Ph, $(NMe_2)_3$, $(OMe)_3$).

92. G. Aklan, C. Foerster, E. Hergovich, G. Speier, and L. Marko, *Veszpremi Vegyip. Egyet. Kozlem, 12,* 131 (1973); through *Chem. Abstr., 80,* 77750d (1974).
Reaction of CoX_2L_2 complexes with CO (L = tertiary phosphine, X = Cl, Br, I).

93. H. L. M. Van Gaal, F. G. Moers, and J. J. Steggerda, *J. Organomet. Chem., 65,* C43 (1974).
Adducts of chlorobis(tricyclohexylphosphine)rhodium.

94. L. Vaska and J. Peone, jr., *Inorg. Synth., 15,* 64 (1974).
Fluoro complexes of rhodium(I) and iridium(I), $RhF(CO)(Ph_3P)_2$ and *trans*-$IrF(CO)(Ph_3P)_2$ were prepared from the corresponding chloride complexes.

95. L. Vaska and W. V. Miller, *Inorg. Synth., 15,* 72 (1974).
trans-Carbonyl(cyanotrihydridoborato)bis(triphenylphosphine)rhodium(I).

96. N. Ahmad, J. J. Levison, S. D. Robinson, and M. F. Uttley, *Inorg. Synth., 15,* 45 (1974).
Preparation of $MHCl(CO)L_3$, $MH_2(CO)L_3$, $MCl_3(NO)L_2$ (M = Os, Ru), $IrCl_2(NO)L_2$, $RhCl_2(NO)L_2$, $Ru(CO)_3L_2$, $Ru(NO)_2L_2$, $OsH_2(CO)_2L_2$, OsH_4L_3, $RhHL_4$, $RhH(CO)L_3$, $Rh(NO)L_3$ (L = Ph_3P).

97. J. Peone, jr., B. R. Flynn, and L. Vaska, *Inorg. Synth., 15,* 68 (1974).

Covalent perchlorato complexes of iridium and rhodium, *trans*-$M(CO)(OClO_3)(Ph_3P)_2$ (M = Ir, Rh) were prepared by reaction of the corresponding chloride complexes with $AgClO_4$.

8. *Nickel, Palladium, and Platinum*

98. R. G. Pearson, W. Louw, and J. Rajaram, *Inorg. Chim. Acta, 9,* 251 (1974).
Reaction of secondary alkyl halides with complexes of the type PtL_n (n = 3, 4, L = alkylaryl phosphines).
99. F. Glockling, T. McBride, and R. J. Pollock, *J. C. S. Chem. Comm., 1973,* 650.
Platinum phosphine cluster compounds, $[Pt(Ph_2P)(C_6H_4Ph_2P)]_n$ (n = 2, 3), were obtained by pyrolysis of $Pt(RC_6H_4)_2(Ph_3P)_2$.
100. P. W. Jolly and K. Jonas, *Inorg. Synth., 15,* 29 (1974).
(μ-Dinitrogen-*N,N'*)bis[bis(tricyclohexylphosphine)-nickel(O)].
101. M. E. Jason, J. A. McGinnety, and K. B. Wiberg, *J. Am. Chem. Soc., 96,* 6531 (1974).
Preparation, structure, and reactions of an organometallic [2,2,1]propellane, the bis(triphenylphosphine)platinum complex of $\Delta^{1,4}$-bicyclo[2,2,0]hexene.
102. C. A. Tolman, D. H. Gerlach, J. P. Jesson, and R. A. Schunn, *J. Organomet. Chem., 65,* C23 (1974).
Reaction of nitrogen with triethylphosphine complexes of zero-valent nickel, palladium, and platinum. Evidence for a nickel dinitrogen complex, $Ni(N_2)(Et_3P)_3$.
103. G. R. Newkome and G. L. McClure, *J. Am. Chem. Soc., 96,* 617 (1974).
Synthesis of a stable metallabicycle, bis(triphenylphosphine)] bis(2-pyridyl)methyllene] platinumdichlorocobalt(II).
104. B. F. G. Johnson, J. Lewis, J. D. Jones, and K. A. Taylor, *J. C. S. Dalton, 1974,* 34.
Reactions of $M(MePh_2P)_4$ (M = Pt, Pd) with vinyl chlorides $C_2H_nCl_{4-n}$ ($n = 0 - 2$) gave a series of chlorovinyl complexes.
105. T. Inglis and M. Kilner, *Nature* (London), *239,* 13 (1972).
Studies of palladium and platinum triphenylphosphine complexes with CO at high pressures.
106. Yu. S. Varshavskii and N. V. Kiseleva, Brit. 1,338,741, 28. Nov 1973, Appl. 30,048/72, 27 Jun 1972; through *Chem. Abstr. 80,* 122897v (1974).
Platinum(O) or palladium(O) triphenylphosphine complexes have been obtained by reduction of M_2PtCl_6 (M = H, K) or $PdCl_2$ by hydrazine in pressence of Ph_3P.
107. S. Miyamoto, Japan. 73 38,688, 19 Nov 1973, Appl. 70 92,143, 21 Oct 1970; through *Chem. Abstr., 81,* 25806c (1974).
Tetrakis(triarylphosphine)palladium complexes were prepared by reaction of β-diketone palladium complexes with a triarylphosphine.
108. W. Beck, W. Rieber, S. Cenini, F. Porta, and G. La Monica, *J. C. S. Dalton, 1974,* 298.
Reactions of toluene-*p*-sulfonyl azide and isocyanate with low valent transition metal complexes, $Pd(CO)(Ph_3P)_3$, $M(CO)(NO)(Ph_3P)_2$ (M = Rh, Ir), $Pt(Ph_3P)_4$, and $Pt(CO)_{4-n}(Ph_3P)_n$ (n = 2, 3).
109. C. S. Cundy, *J. Organomet. Chem., 69,* 305 (1974).
The preparation, properties, and some oxidative addition reactions of the complexes $Ni(Et_3P)_4$ and Ni(1,5-cyclooctadiene)$(Bu_3P)_2$ are described.
110. S. D. Ittel and J. A. Ibers, *J. Organomet. Chem., 74,* 121 (1974).
A series of Ni(O) stilbene complexes was prepared by the reaction of triarylphosphines or Bu^tNC with bis(1,5-cyclooctadiene)nickel(O). The structure of

bis(tri-*p*-tolylphosphine)(*trans*-stilbene)nickel(O) hemitetrahydrofuranate, $Ni[P(p\text{-}MeC_6H_4)_3]_2[PhCH{=}CHPh]\cdot 0.5THF$ was determined.

111. J. Browning, M. Green, J. L. Spencer, and F. G. A. Stone, *J. C. S. Dalton, 1974,* 97.
Reaction of Pt(*trans*-stilbene)$(Et_3P)_2$ with $C_6(CF_3)_6$ gives $Pt[C_6(CF_3)_6](Et_3P)_2$.

112. L. Cassar and A. Giarrusso, *Gazz. Chim. Ital., 103,* 793 (1973).
Stereochemistry of the oxidative addition of vinyl halides to nickel(O) complexes and related reactions.

113. K. Tanaka, Y. Kawata, and T. Tanaka, *Chem. Lett., 1974,* 831.
$NiCl_2L_2$ (L = Ph_3P, $MePh_2P$, 1/2 $(Ph_2P{-}CH_2)_2$, 1/2 $(Ph_2PCH{=}CHPPh_2)$ reacts with CO in presence of MeSNa as reducing agent to give $Ni(CO)_2L_2$ and Me_2S_2.

114. K. Schorpp and W. Beck, *Z. Naturforsch., 28b,* 738 (1973).
Reactions of ethylenebis(triphenylphosphine)platinum(O) with ketene derivatives.

115. H. A. Tayim and N. S. Akl, *J. Inorg. Nucl. Chem., 36,* 1071 (1974).
Oxidative substitution reactions of $Pt(Ph_3P)_4$ with HX (phenol, substituted phenols, and thiophenols) gave $PtX_2(Ph_3P)_2$.

116. J. H. Nelson, A. W. Verstuyft, J. D. Kelly, and H. B. Jonassen, *Inorg. Chem., 13,* 27 (1974).
Reactions of palladium(O) phosphine complexes with terminal α-hydroxyacetylenes.

117. I. V. Gavrilova, M. I. Gel'fman, and V. V. Razumovskii, *Zh. Neorg. Khim., 19,* 2490 (1974).
Interaction of hydroperchloratobis(triphenylphosphine) platinum with amines gives dimeric $[PtL(Ph_3P)_2]ClO_4$ (L = a bridging ligand; HL = NH_3, $MeNH_2$, Me_2NH).

C. Eaborn, B. Ratcliff, and A. Pidcock, *J. Organomet. Chem., 65,* 181 (1974).
$Pt(C_2H_4)(Ph_3P)_2$ reacts with silicon hydrides in an oxidative addition to yield hydrido(silyl)bis(triphenylphosphine)platinum(II) complexes.

119. P. M. Treichel and K. P. Wagner, *J. Organomet. Chem., 61,* 415 (1973).
Treating PtR_2L_2 (R = Me, Ph, L = $MePh_2P$, Ph_3P, Et_3P, Me_2PhP) with isocyanides gave either insertion or substitution reactions.

120. E. Uhlig and H. Walther, *Z. Anorg. Allg. Chem., 409,* 89 (1974).
$Ni(\pi\text{-}C_5H_5)_2$ (C_5H_5 = cyclopentadienyl) reacts with Bu_3P or $BuPh_2P$ to give $Ni(Bu_3P)_4$ or $Ni(BuPh_2P)_4$, whereas with Bu_2RP (R = Ph, *p*-tolyl) the complexes $\pi\text{-}C_5H_5Ni(Bu_2RP)_2$ were obtained.

121. H. F. Klein, H. H. Karsch, and W. Buchner, *Chem. Ber., 107,* 537 (1974).
The methyltetrakis(trimethylphosphine)nickel cation.

122. C. J. Wilson, M. Green, and R. J. Mawby, *J. C. S. Dalton, 1974,* 421.
One-step reduction of $PtCl_2$ to the cluster complexes $Pt_4(CO)_5L_4$ (L = Ph_3P, $Ph_2(o\text{-}MeC_6H_4)P$, $Ph(o\text{-}MeC_6H_4)_2P$) followed by oxidative addition with MeI gave PtIMe(CO)L.

123. C. Eaborn, A. Pidcock, and B. Ratcliff, *J. Organomet. Chem., 66,* 23 (1974).
The reactions between silicon hydrides and dimethylbis(phosphine)platinum(II) complexes afford Pt-silyl compounds.

124. J. A. Duff, B. L. Shaw, and B. L. Turtle, *J. Organomet. Chem., 66,* C18 (1974).
Dihapto-[(8-methylene)-1-naphthyl] bis(triphenylphosphine) platinum(II).

125. V. I. Sokolov, V. V. Bashilov, L. M. Anishchenko, and O. A. Reutov, *J. Organomet. Chem., 71,* C41 (1974).
Reaction between organomercurials and phosphineplatinum(O) complexes. New versatile synthesis of σ-bonded organoplatinum(II) compounds, $PtRX(Ph_3P)_2$ R = *p*-tolyl, $(p\text{-}MeOC_6H_5)_2C{=}CH$, MeO_2CCH_2, etc., X = Cl, Br).

126. G. Oehme and H. Pracejus, *Z. Chem., 14,* 24 (1974).
Oxidative addition of ClCH=CHR and BrCH=CHCN to $Pd(Ph_3P)_4$; CH_2=CClCN reacts with $Pd(Ph_3P)_4$ to give $Pd(CH_2=CClCN)(Ph_3P)_2$.
127. R. Uson, P. Royo, and J. Fornies, *Synth. React. Inorg. Metal-Org. Chem., 4,* 157 (1974).
Cationic pentafluorophenyl complexes of palladium, $[Pd(C_6F_5)L(Ph_3P)_2]ClO_4$ (L = Ph_3P, Et_3P, $Bu^t{}_3P$, pyridine, $AsPh_3$, etc.
128. H. C. Clark and K. Von Werner, *Synth. React. Inorg. Metal-Org. Chem., 4,* 355 (1974).
Preparation and reactivity of some alkoxycarbonyl complexes of palladium, platinum, and iridium, for example, trans-$MX(CO_2Me)L_2$ (M = Pd, X = Cl, L = Ph_3P, Me_2PhP; M = Pd, X = Br, L = Ph_3P; M = Pt, X = Cl, L = Ph_3P, $MePh_2P$, Me_2PhP).
129. E. A. Allen and W. Wilkinson, *J. Inorg. Nucl. Chem., 36,* 1663 (1974).
Cis- and *trans*-complexes of palladium with phenyldialkylphosphines.
130. W. J. Cherwinski, B. F. G. Johnson, and J. Lewis, *J. C. S. Dalton, 1974,* 1405.
$Pt[(PhCH=CH)_2CO]_2$ with L(Ph_3P, Et_3P, $MePh_2P$, $AsPh_3$, $AsEt_3$) gives $Pt[(PhCH=CH)_2CO]L_2$ in which Pt is coordinated to one double bond of $(PhCH=CH)_2CO$.
131. D. G. Holah, A. N. Hughes, B. C. Hui, and K. Wright, *Can. J. Chem., 52,* 2990 (1974).
Reactions of sodium tetrahydroborate and cyanotrihydroborate with divalent Co, Ni, Cu, Pd, and Pt chlorides in the presence of tertiary phosphines.
132. K. A. Jensen and E. Huge-Jensen, *Acta. Chem. Scand., 27,* 3605 (1973).
Addition of CSe_2 to $Pt(Ph_3P)_3$ and $Ni(CO)_2(Ph_3P)_2$.
133. B. Cetinkaya, M. F. Lappert, and J. McMeeking, *J. C. S. Dalton, 1973,* 1975.
Pt(II), Rh(I), and Ir(I) halide phosphine complexes undergo halogen substitution reactions with $LiN=C(CF_3)_2$ or $Me_3SnN=C(CF_3)_2$.
134. J. R. Dilworth and A. S. Kasenally, *J. Organomet. Chem., 60,* 203 (1973).
Dibenzoyl- and diacetylhydrazido(2-) platinum complexes.
135. K. J. Klabunde, J. Y. F. Low, and H. F. Efner, *J. Am. Chem. Soc., 96,* 1984 (1974).
Metal-atom synthesis of organonickel and -palladium complexes of triethylphosphine.
136. Y. Yamamoto and H. Yamazaki, *Inorg. Chem., 13,* 438 (1974).
Single and multiple insertion of isocyanide into palladium-carbon σ-bonds.
137. D. Negoiu and L. Paruta, *Rev. Roum. Chim., 18,* 2059 (1973).
Synthesis and properties of mixed stibine-phosphine complexes.
138. D. Negoiu, C. Parlog, D. Sandulescu, *Rev. Roum. Chim., 19,* 387 (1974).
Complexes of the type $M(N_2)L_n$ (M = Co, Ni, Fe, L = triarylphosphine, -arsine, -stibine, n = 2, 3).
139. K. R. Dixon, K. C. Moss, and M. A. R. Smith, *Can. J. Chem., 52,* 692 (1974).
Stereospecific synthesis of neutral and cationic tertiary phosphine complexes of platinum.
140. S. Baba, T. Ogura, and S. Kawaguchi, *Bull. Chem. Soc. Jap., 47,* 665 (1974).
Reactions of bis(acetylacetonato)palladium(II) with triphenylphosphine and nitrogen bases.
141. K. Jakob, *Z. Chem., 13,* 475 (1973).
Preparation of bis(*o*-chlorobenzyl)bis(trialkylphosphine)nickel(II) complexes.
142. T. G. Appleton, H. C. Clark, and L. E. Manzer, *J. Organomet. Chem., 65,* 275 (1974).
Oxidative addition reactions of I_2, MeI, CF_3I with trans-$PtRI(Me_2PhP)_2$ (R = Me, Ph) and the stereochemistry of the resulting Pt(IV) compounds.
143. P. J. Fraser, W. R. Roper, and F. G. A. Stone, *J. C. S. Dalton, 1974,* 102.

Cationic cyclic carbene complexes of Ir(III), Ni(II), Pd(II), and Pt(II), for example, $[IrCl_2(CO)(RH)(Me_2PhP_2]^+$ and $[PtCl(RH)(Et_3P)_2]^+BF_4^-$.

144. R. L. Hassel and J. L. Burmeister, *Inorg. Chim. Acta, 8,* 155 (1974).
Effects of geometric isomerism on the linkage isomers formed by dithiocyanotobis(triphenylphosphosphine or -arsine)platinum(II) complexes.

145. J. M. Duff, B. E. Mann, B. L. Shaw, and B. Turtle, *J. C. S. Dalton, 1974,* 139.
Internal metallations of dimethyl(1-naphthyl)phosphine in Pd and Pt complexes.

146. K. Tanaka and T. Tanaka, *Inorg. Nucl. Chem. Lett., 20,* 605 (1974).
Synthesis and reactions of $Ni[SeC(O)NEt_2]_2L_2$ ($L = Ph_3P, (Ph_2PCH_2)_2$) with CO.

147. R. Uson, P. Royo, and J. Gimeno, *Rev. Acad. Cienc. Exactas, Fis.-Quim. Natur. Zaragoza, 28,* 355 (1973); through *Chem. Abstr., 81,* 49793x (1974).
Reactions of $TiBr(C_6F_5)_2$ with $PtCl_2(Et_3P)_2$ to give Pt(IV) complexes.

148. A. R. Siedle, D. McDowell, and L. J. Todd, *Inorg. Chem., 13,* 2735 (1974).
The reactive anions $B_9H_{13}^{2-}$, $B_9H_{11}S^{2-}$, and $B_9H_9S^{2-}$ are used to produce new derivatives $Pd(B_9H_9S)(Ph_3P)_2$.

149. M. Gorio, B. C. Castellani, M. Cola, and A. Perotti, *J. Inorg. Nucl. Chem., 36,* 1168 (1974).
Oxidation of substituted phosphines and phosphine complexes of nickel(II) with sodium chlorite and chlorine dioxide.

150. F. Sato, M. Etoh, and M. Sato, *J. Organometal. Chem., 70,* 101 (1974).
Preaparation and properties of organonickel derivatives of 1-substituted tetrazoline-5-thiones,

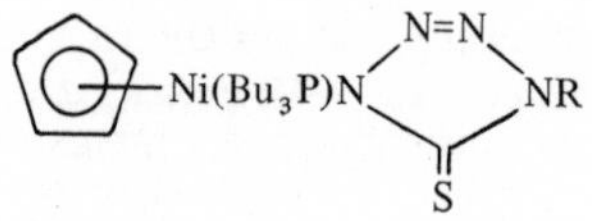

151. K. Schorpp and W. Beck, *Chem. Ber., 107,* 1371 (1974).
Oxidative addition reactions of nitroalkanes to tetrakis(triphenylphosphine)-platinum.

152. W. Beck, K. Schorpp, and C. Oetker, *Chem. Ber., 107,* 1380 (1974).
Reactions of trans-difulminatobis(triphenylphosphine)-platinum.

153. B. T. Heaton and D. J. A. McCaffrey, *J. Organometal. Chem., 70,* 455 (1974).
Reactions of 2-(alkenyl)pyridines with $Pt_2X_4(Et_3P)_2$ (X = Cl, Br).

154. D. R. Fahey, U.S. 3,808,246, 30 Apr 1974, Appl. 310,809, 30 Nov 1972; through *Chem. Abstr. 81,* 25810z (1974).
2,2′-Bis[chlorobis(triethylphosphine)nickel] biphenyl.

155. S. Bresadola, B. Longato, and F. Morandini, *J. C. S. Chem. Comm., 1974,* 510.
Reacting 1-lithiumcarboranes with *trans*-$PtHCl(Et_3P)_2$ gives *cis*- or *trans*-Pt(II) hydrido dicarba-closo-dodecaborane complexes.

9. *Copper, Silver, and Gold*

156. T. Ikariya and A. Yamamoto, *J. Organomet. Chem., 72,* (1974).
Preparation and properties of ligand-free copper alkyls coordinated with 2,2′-bipyridyl and tircyclohexylphosphine.

157. M. I. Bruce and A. P. P. Ostazewski, *J. C. S. Dalton, 1973,* 2433.
CuCl in presence of CO reacts with potassium hydridotris-(pyrazol-1–yl)borate (L) to give the complexes Cu(CO)L, in which the CO may be replaced by a variety of phosphine ligands.

158. D. J. Saturnino and P. R. Girardot, *Inorg. Chem., 13,* 2914 (1974).

The compounds $Cu(Ph_3P)_3FeCl_3X$ (X = Cl, Br) are prepared by the reaction of $CuX(Ph_3P)_3$ with $FeCl_3$.
159. A. N. Nesmeyanov, E. G. Perevalova, T. V. Baukova, and K. I. Grandberg, *Izv. Akad. Nauk SSSR, Ser. Khim., 1973,* 2641.
Triphenylphosphine complex of cyclopentadienyl(manganesedicarbonyltriphenylphosphine)gold.
160. A. Laguna, P. Royo, and R. Uson, *Rev. Acad. Cienc. Exactas, Fis.-Quim. Natur. Zaragoza, 28,* 71 (1973); through *Chem. Abstr., 80,* 60012s (1974).
Preparation and properties of pentafluorophenyl derivatives of gold(III), for example, $AuX(C_6F_5)_2Ph_3P$ (X = SCN, CN, NO_2).
161. A. Tamaki and J. K. Kochi, *J. Organomet. Chem., 64,* 411 (1974).
162. A. Johnson and R. J. Puddephatt, *Inorg. Nucl. Chem. Lett., 9,* 1175 (1973).
Oxidative addition reactions of methylgold compounds, AuMeL (L = Me_3P, Me_2PhP, Ph_3P).
163. R. Roulet, Nguyen Quang Lan, W. R. Mason, and G. P. Fenske, Jr., *Helv. Chim. Acta, 56,* 2405 (1973).
Oxidation-reduction reactions of tetrachloroaurate(III) anion with triphenyl derivatives of group V elements.
164. K. I. Grandberg, E. I. Smyslova, and A. N. Kosina, *Izv. Akad. Nauk SSSR, Ser. Khim., 1973,* 2787.
Reaction of vinyl(triphenylphosphine)gold.
165. J. C. Bommer and K. W. Morse, *J. Am. Chem. Soc., 96,* 6222 (1974).
Copper(I) and silver(I) complexes with O-ethylboranocarbonate, for example, $Cu(H_3BCO_2Et)(Ph_3P)_2$.
166. B. M. Sutton and J. Weinstock, U.S. 3,787,568, 22 Jan 1974, Appl. 249,280, 01 May 1972; through *Chem. Abstr., 80,* 112654z (1974).
167. E. R. McGusty and B. M. Sutton, U.S. 3,792,165, 12 Feb 1974, Appl. 137,961, 27 Apr 1971; through *Chem. Abstr., 80,* 112666e (1974).
168. E. R. McGusty and B. M. Sutton, U.S. 3,784,687, 08 Jan 1974, Appl. 150,734, 07 Jun 1971; through *Chem. Abstr., 80,* 100212r (1974).
Antiarthritic properties of phosphinegold complexes.

B. Ditertiary phosphines

169. J. A. Connor and G. A. Hudson, *J. Organomet. Chem., 73,* 351 (1974).
170. J. A. Connor, G. K. McEwen, and C. J. Rix, *J. C. S. Dalton, 1974,* 589.
171. J. A. Connor, G. K. McEwen, and C. J. Rix, *J. Less-Common Metals, 36,* 207 (1974).
172. J. A. Connor, G. K. McEwen, and C. J. Rix, *Chem. Uses Molybdenum, Proc. Conf.,* 1st, 111 (1973, Pub. 1974); through *Chem. Abstr. 81,* 180450a (1974).
Synthesis, spectroscopic studies and oxidation of molybdenum (and in some cases manganese) complexes of $Me_2PCH_2CH_2PMe_2$.
173. C. Miniscloux, G. Martino, and L. Sajus, *Bull. Soc. Chim. Fr., 1973,* 2183.
Syntheses of phosphinemolybdenum complexes containing molecular nitrogen by reduction of low-valent molybdenum halides in presence of $(Ph_2PCH_2)_2$ and N_2.
174. T. Ito, T. Kokubo, T. Yamamoto, A. Yamamoto, and S. Ikeda, *J. C. S. Chem. Comm., 1974,* 136.
π-Ethylene complex of molybdenum(O) and hydrido(acetylacetonato) complex of molybdenum(II) with diphosphine ligands.
175. J. Dehand and J. F. Nennig, *Inorg. Nucl. Chem. Lett., 10,* 875 (1974).
Heteronuclear triangular cluster compounds with platinum-cobalt bonds.

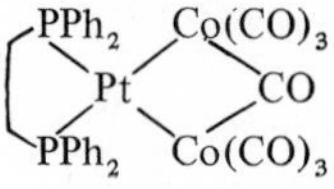

176. S. O. Grim, J. Del Gaudio, R. P. Molenda, C. A. Tolman, and J. P. Jesson, *J. Am. Chem. Soc., 96,* 3416 (1974).
177. S. O. Grim, W. L. Briggs, R. C. Barth, C. A. Tolman, and J. P. Jesson, *Inorg. Chem., 13,* 1095 (1974).
Group VI metal carbonyl complexes of unsymmetrical bidentate ligands.
178. M. Hidai, K. Tominari, Y. Uchida, and A. Misono, *Inorg. Synth., 15,* 25 (1974).
trans-Bis(dinitrogen)bis[ethylenebis(diphenylphosphine)] molybdenum(O).
179. H. Behrens, W. Topf, and J. Ellermann, *J. Organomet. Chem., 63,* 349 (1973).
Treating the CO-bridged complex $[Mo(CO)_3Q]_2$ (Q = bipyridyl, phenanthroline) with multidentate ligands L (L = Ph_2PPPh_2, $Ph_2P(CH_2)_nPPh_2$, n = 1,2,3, Ph_2-$PCH_2CH_2NEt_2$) gives $Mo(CO)_3QL$.
180. S. S. Sandhu and A. K. Mehta, *J. Organomet. Chem., 77,* 45 (1974).
Photochemically induced substitution in Group VI metal carbonyls with ditertiary phosphines (1,2-bis(diphenylphosphino)ethane, 1,4-bis(diphenylphosphino)butane).
181. D. Fenske and H. J. Becher, *Chem. Ber., 107,* 117 (1974).
2,3-Bis(diphenylphosphino)maleic anhydride and diphenylphosphino derivatives of cyclobutenedione as ligands in metal carbonyls.
182. D. J. Darensbourg and D. Madrid, *Inorg. Chem., 13,* 1532 (1974).
$ReCl(N_2)(Ph_2PCH_2CH_2PPh_2)_2$ reacts with $InCl_3$ in the absence of O to give $[ReCl(N_2)(Ph_2PCH_2CH_2PPh_2)_2]Cl$.
183. M. J. Mays and B. E. Prater, *Inorg. Synth., 15,* 21 (1974).
trans-(Dinitrogen)bis[ethylenebis(diethylphosphine)hydridoiron(II) tetraphenylborate.
184. W. G. Peet and D. H. Gerlach, *Inorg. Synth., 15,* 38 (1974).
$FeH_2[C_2H_4(Ph_2P)_2]_2 \cdot 2$ PhMe was prepared from $FeCl_2$ and the phosphine ligand using $NaBH_4$ as reducing agent.
185. P. Piraino, F. Faraone, and R. Pietropaolo, *Inorg. Nucl. Chem. Lett., 9,* 1237 (1973).
The complex $IrCl(CO)_2(Ph_2PCH_2CH_2PPh_2)$ may be obtained from $[IrCl_2(CO)_2]$-$AsPh_4$ and the bidentate ligand.
186. F. Pruchnik, *Inorg. Nucl. Chem. Lett., 10,* 661 (1974).
Reactions of dichloro-2-methylallyrhodium(III) with 1,2-bis(diphenylphosphino)-ethane.
187. T. Fukumoto, Y. Matsumura, and R. Okawara, *J. Organomet. Chem.,* **69,** 437 (1974).
Cobalt carbonyl complexes of $R_2PCH_2PR_2$ ligands.
188. N. J. De Stefano, D. K. Johnson, and L. M. Venanzi, *Angew. Chem., 86,* 133 (1974).
Square planar Ni, Pd, and Pt complexes of 2,11-bis-(diphenylphosphinomethyl)-benzo[c]-phenanthrene.
189. W. E. Hill, W. Levason, and C. A. McAuliffe, *Inorg. Chem., 13,* 244 (1974).
Nickel(II) and cobalt(II) complexes containing bis-(diphenylphosphino)-*o*-carborane.
190. K. K. Chow, W. Levason, and C. A. McAuliffe, *Inorg. Chim. Acta, 7,* 589 (1973).

191. K. K. Chow and C. A. McAuliffe, *Inorg. Chim. Acta, 10,* 197 (1974).
Cobalt, nickel, and palladium complexes of *trans*-1,2-bis(diphenylphosphino)-ethylene and nickel(II) complexes of bis(diphenylphosphino)methane.

192. A. J. Pryde, B. L. Shaw, and B. Weeks, *J. C. S. Chem. Comm., 1973,* 947.
Large ring compounds involving trans-bonding bidentate ligands.

193. K. Suzuki, H. Yamamoto, and S. Kanie, *J. Organomet. Chem., 73,* 131 (1974).
The reaction of *trans*-$PtCl(CH_2CN)(Ph_3P)_2$ with Br^-, I^-, $Ph_2PCH_2CH_2PPh_2$, $Ph_2PCH_2CH_2AsPh_2$, and *cis*-$Ph_2PCH=CHPPh_2$ was studied.

194. W. Levason and C. A. McAuliffe, *Inorg. Chim. Acta, 8,* 25 (1974).
Silver(I) complexes of $Ph_2P(CH_2)_nPPh_2$ (n = 1, 2, 3) and *cis*-1,2-bis(diphenyl-phosphino)ethylene.

195. A. Camus, N. Marsich, G. Nardin, and L. Randaccio, *J. Organomet. Chem., 60,* C-39 (1973).
Properties and structure of the product of the reaction of aryl copper compounds with bis(diphenylphosphino)methane.

196. R. T. Sane and P. R. Kulkarni, *Curr. Sci., 43,* 42 (1974).
Preparation of new derivatives of triphenylphosphinegold(1+) tetracarbonylco-baltate(1−) with some bidentate ligands.

197. M. M. T. Khan and S. S. Ahmed, *Proc. Chem. Symp., 2,* 149 (1972).
Synthesis of the complexes $OsCl(SnCl_3)L_2$ (L = $Ph_2(CH_2)_nPPh_2$ (n = 1, 2)), OsCl-$(HgCl)(Ph_2PCH_2PPh_2)$, and $[OsCl(HgCl)Ph_2PCH_2CH_2PPh_2]HgCl_3$.

198. L. F. Warren and M. A. Bennett, *J. Am. Chem. Soc., 96,* 3340 (1974).
Stabilization of high formal oxidation states of the first-row transition metal series by *o*-phenylenebis(dimethylphosphine).

C. Tri-, Tetra, and Hexatertiary Phosphines

199. I. S. Butler and N. J. Coville, *J. Organomet. Chem., 66,* 111 (1974).
Reactions of bromo- or iodopentacarbonylmanganese with hexaphenyl-1,4,7,10-tetraphosphadecane.

D. Tertiary Phosphines Containing Donor Groups Other than Phosphorus

200. M. Schäfer and E. Uhlig, *Z. Anorg. Allg. Chem., 407,* 23 (1974).
2-(Diethylphosphinomethyl)pyridine 3d-element complexes and their anomalouw magnetic behavior.

201. E. Uhlig and S. Keiser, *Z. Anorg. Allg. Chem., 406,* 1 (1974).
Palladium(II) complexes of 2-[β-(diphenylphosphino)ethyl] pyridine.

202. T. B. Rauchfuss and D. M. Roundhill, *J. Am. Chem. Soc., 96,* 3098 (1974).
Synthesis and reactions of nucleophilic complexes of rhodium(I) containing *o*-(diphenylphosphino)-*N,N*-dimethylaniline.

203. L. Sacconi, A. Orlandini, and S. Midollini, *Inorg. Chem., 13,* 2850 (1974).
Synthesis, properties, and structural characterization of nonstoichiometric hydrido complexes of nickel with the tetradentate ligand tris(2-diphenylphosphinoethyl)-amine.

204. W. J. Knebel and R. J. Angelici, *Inorg. Chim. Acta, 7,* 713 (1973).
Chromium, molybdenum, and tungsten carbonyl complexes of $Ph_2P(CH_2)_nNR_2$ (R = Me, Et, n = 2, 3) or pyridine donor groups, for example, $Ph_2P(CH_2)_nC_5H_4N$ (n = 1, 2, 3).

205. T. N. Lockyer, *Aust. J. Chem., 27,* 259 (1974).

S-Dealkylation in dimethylformamide was observed for the monochelates of (*o*-methylthiophenyl)diphenylphosphine.

206. S. C. Tripathi, S. C. Srivastava, and A. K. Shrimal, *J. Organomet. Chem., 73,* 343 (1974).
1-(Diphenylphosphino)-2-(diphenylarsino)ethane derivatives of Group VI metal carbonyls.

207. C. A. McAuliffe and R. Pollock, *J. Organomet. Chem., 74,* 463 (1974).
Reactions between *trans*-carbonylchlorobis(triphenylphosphine) iridium and bis-[3-(dimethylarsino)propyl] phenylarsine and -phosphine.

208. K. Suzuki and H. Okuda, *Synth. Inorg. Metal-Org. Chem., 3,* 369 (1973).
Synthesis of fumaronitrile complexes of palladium(O) and platinum(O) containing 1-diphenylphosphino-2-diphenylarsinoethane.

209. G. Kordosky, G. S. Brenner, and D. W. Meek, *Inorg. Chim. Acta, 7,* 605 (1973).
Cobalt(III) complexes of $E(CH_2CH_2CH_2AsMe_2)_3$ (E = P, As, Sb).

210. P. E. Garrou and G. E. Hartwell, *J. Organomet. Chem., 71,* 443 (1974).

211. M. A. Bennett, R. N. Johnson, and I. B. Tomkins, *Inorg. Chem., 13,* 346 (1974).
Pd, Pt, and Ru complexes of olefinic tertiary phosphines.

212. M. A. Bennett, R. N. Johnson, and I. B. Tomkins, *J. Am. Chem. Soc., 96,* 61 (1974).

213. M. A. Bennett, R. N. Johnson, G. B. Robertson, I. B. Tomkins, and P. O. Whimp *J. Organomet. Chem., 77,* C43 (1974).
Coupling and dehydrogenation of the vinyl groups in the coordination sphere of Ru and Rh complexes of olefinic tertiary phosphines.

214. A. J. Carty, G. Ferguson, H. N. Paik, and R. Restivo, *J. Organomet. Chem., 74,* C14 (1974).

215. A. J. Carty, H. N. Paik, and T. W. Ng, *J. Organomet. Chem., 74,* 279 (1974).

216. S. Jacobson, A. J. Carty, M. Mathew, and G. J. Palenik, *J. Am. Chem. Soc., 96,* 4330 (1974).
Fe, Ni, Pd, and Pt complexes of acetylenic phosphines.

217. Y. Nonaka, S. Takahashi, and N. Hagihara, *Mem. Inst. Sci. Ind. Res., Osaka Univ., 31,* 23 (1974); through *Chem. Abstr., 80,* 140678d (1974).
Fe, Co, Rh, Ni, and Pd complexes of diphenylphosphinopolystyrene.

218. W. Beck, R. Höfer, J. Erbe, H. Menzel, U. Nagel, and G. Platzen, *Z. Naturforsch., 29b,* 567 (1974).
Transition-metal carbonyl complexes with Ph_2PCH_2R (R = polystyrene residue).

219. K. R. Mann, W. H. Morrison, Jr., and D. N. Hendrickson, *Inorg. Chem., 13,* 1180 (1974).
Reaction of 1,1′-bis(diphenylphosphino)ferrocene with mercury halides, mercuric cyandie, and other Lewis acids.

220. G. Marr and T. M. White, *J. C. S. Perkin I, 1973,* 1955.
Synthesis of (ferrocenylmethyl)phosphines.

221. C. Elschenbroich and F. Stohler, *J. Organomet. Chem., 67,* C51 (1974).
Bis(diphenylphosphino-h^6-benzene)chromium(O).

222. A. N. Nesmeyanov, K. N. Anisimov, and Z. P. Valueva, *Dokl. Akad. Nauk SSSR, 216,* 106 (1974).
Tris(cyclopentadienyltricarbonylmanganese)phosphine.

223. W. Malisch and M. Kuhn, *J. Organomet. Chem., 73,* C1 (1974).
Phosphido transition-metal complexes with free donor function.

224. H. Nöth and J. Deberitz, *Kem. Kozlem., 40,* 9 (1973); through *Chem. Abstr., 81,* 3205d (1974).
2,4,6-Triphenylphosphorin as *n*- and π-donor ligand in Cr, Mo, and W carbonyl complexes.

IV. Complexes of Phosphites, Phosphonites, and Phosphinites, $R'_{3-n}P(OR)_n$ (n = 1,2,3)

225. Cincinnati Milacron Chemicals, Inc., Ger. Offen. 2,166,425, 11 Apr 1974, U.S. Appl. 84,494, 27 Oct 1970; through *Chem. Abstr., 81,* 79882t (1974).
Synthesis of Ni, Mo, and W, complexes of tricyclohexylphosphite, bis(*o*-phenylene) phenylphosphite, and trineopentylphosphite.

226. M. Y. Darensbourg, D. J. Darensbourg, and D. Drew, *J. Organomet. Chem., 73,* C25 (1974).
Reactions of phenylacetylpentacarbonylmanganese with phosphine and phosphite ligands.

227. F. Le Moigne and R. Dabard, *J. Organomet. Chem., 60,* C14 (1973).
Chiral cyclopentadienylmanganese trimethylphosphite complexes.

228. E. W. Ainscough, T. A. James, S. D. Robinson, and J. N. Wingfield, *J. Organomet. Chem., 60,* C63 (1973).
Ortho-metalation reactions involving some triphenylphosphite complexes of osmium.

229. W. Kläui and H. Werner, *J. Organomet. Chem., 60,* C19 (1973).
Reaction of π-cyclopentadienyl[tris(trimethylstannyl)stannyl] bis(triphenylphosphite)iron with iodine and bromine.

230. L. W. Gosser and G. W. Parshall, *Inorg. Chem., 13,* 1947 (1974).
Co(I) triethylphosphite complexes.

231. M. S. Arabi, A. Maisonnat, S. Attali, and R. Poilblanc, *J. Organomet. Chem., 67,* 109 (1974).
Reaction of $Co_2(CO)_8$ with tertiary phosphites gives several products, which were interconverted by chemical means.

232. H. Neukomm and H. Werner, *Helv. Chim. Acta, 57,* 1067 (1974).
π-Cyclopentadienylrhodium bis(tert. phosphite) complexes.

233. N. I. Yurasova, A. D. Troitskaya, V. V. Sentemov, G. P. Sadakova, Z. L. Shmakova, and T. V. Ryabova, *Zh. Obshch. Khim., 44,* 1650 (1974).

234. A. D. Troitskaya, G. D. Ginzburg, E. A. Zgadzai, V. V. Sentemov, and L. V. Markina, *Zh. Neorg. Khim., 18,* 2682 (1973).

235. L. F. Gogolyukhina, G. A. Levshina, and A. D. Troitskaya, *Tr. Kazan. Khim.-Tekhnol. Inst., 1973,* No. 52, 22; through *Chem. Abstr., 80,* 103320k (1974).

236. L. F. Gogolyukhina, A. D. Troitskaya, and G. A. Levshina, *Zh. Obshch. Khim., 44,* 223 (1974).
Reactions of Rh, Ni, Pd, and Pt thiocyanates with trialkylphosphites.

237. H. E. Shook, Jr., Ger. Offen. 2,353,198, 09 May 1974, U.S. Appl. 300,824, 25 Oct 1972; through *Chem. Abstr., 81,* 51773j (1974).

238. H. E. Shook, Jr., Ger. Offen. 2,353,240, 09 May 1974, U.S. Appl. 300,823, 25 Oct 1972; through *Chem. Abstr., 81,* 51774k(1974).
Nickel(O) tritolylphosphite complexes.

239. C. E. Jones, B. L. Shaw, and B. L. Turtle, *J. C. S. Dalton, 1974,* 992.
O- and C-metallation of 2-alkoxyphenylphosphines by platinum(II).

240. C. A. Tolman and W. C. Seidel, *J. Am. Chem. Soc., 96,* 2774 (1974).
Preparation of olefinbis(tri-*o*-tolylphosphite)nickel complexes.

241. A. D. Troitskaya, G. A. Levshina, I. N. Andreev, K. D. Levshin, and E. V. Chistyakov, *Zh. Obshch. Khim., 44,* 1835 (1974).
Palladium phosphite complexes.

242. G. K. McEwen, C. J. Rix, M. F. Traynor, and J. G. Verkade, *Inorg. Chem., 13,* 2800 (1974).
Preparation of hydridonickel phosphites.

243. W. C. Seidel and L. W. Gosser, *Inorg. Synth., 15,* 9 (1974).
Ethylenebis(tri-*o*-tolylphosphite)nickel(O) and *tris*-(tri-*o*-tolylphosphite)nickel(O).
244. L. F. Gogolyukhina, G. A. Levshina, and A. D. Troitskaya, *Tr. Kazan. Khim.-Tekhnol. Inst., 1973,* No. 52, 28; through *Chem. Abstr., 80,* 103322n (1974).
Reaction of *cis*-dithiocyanatobis(triethoxyphosphine)palladium and bis[thiocyanato(diethylhydrogenphosphite)-(diethylphosphito)palladium] with thiourea.
245. J. J. Mrowca, U.S. 3,776,929, 04 Dec 1973, Appl. 189,834, 18 Oct 1971; through *Chem. Abstr., 80,* 48166n (1974).
Phosphinite, phosphonite, thiophosphinite, dithiophosphonite compounds of palladium and platinum.
246. I. M. Babina, G. D. Ginzburg, and A. D. Troitskaya, *Tr. Kazan. Khim.-Tekhnol. Inst., 1973,* No. 52, 15; through *Chem. Abstr., 80,* 127640r (1974).
Complexes of nickel(II) with triisopropyl- and trifluorobutylphosphite in ethanol and benzene.
247. I. M. Babina, G. D. Ginzburg, and A. D. Troitskaya, *Tr. Kazan. Khim.-Tekhnol. Inst., 1973,* No. 52, 8; through *Chem. Abstr., 80,* 127639x (1974).
Complexes of nickel(II) nitrate, chloride and perchlorate with trialkylphosphites.
248. D. A. Couch and S. D. Robinson, *Inorg. Chim. Acta, 9,* 39 (1974).
Cationic trialkylphosphite derivatives of gold, silver, and the platinum metals.
249. D. A. Couch, and S. D. Robinson, *Inorg. Chem., 13,* 456 (1974).
250. D. A. Couch, S. D. Robinson, and J. N. Wingfield, *J. C. S. Dalton, 1974,* 1309.
Cationic and neutral dialkylphenylphosphonite and alkyldiphenylphosphinite derivatives of copper, silver, gold, and the platinum metals.

V. Complexes of Halophosphines, $R_{3-n}PX_2$ (n = 1, 2, 3).

251. T. Kruck and H. U. Hempel, *Angew. Chem., 86,* 233 (1974).
Hexakis(trifluorophosphine)vanadium hydride.
252. R. A. Head, J. F. Nixon, J. R. Swain, and C. M. Woodard, *J. Organomet. Chem., 76,* 393 (1974).
2,7-Dimethylocta-2,6-diene-1,8-diylruthenium phosphine, fluorophosphine, and carbonyl complexes.
253. M. A. Cairns and J. F. Nixon, *J. Organomet. Chem., 74,* 263 (1974).
Mixed trifluorophosphine(triphenylphosphine)hydrido complexes of cobalt(I), $CoH(PF_3)_{4-n}(Ph_3P)_n$ (n = 1 – 3) and $(\pi-C_4H_7)Co(C_4H_6)PF_3$.
254. M. Kooti and J. F. Nixon, *J. Organomet. Chem., 63,* 415 (1973).
(Phenylazo)phenyl-2C,*N'*-rhodium complexes containing PF_3.
255. O. Stelzer, *Chem. Ber., 107,* 2329 (1974).
Penta- and hexa- coordinated complexes of difluorophenylphosphine with cobalt and nickel halides.
256. G. V. Röschenthaler, R. Schmutzler, and E. Niecke, *Z. Naturforsch., 29b,* 436 (1974).
cis-Tetracarbonylbis[difluoro(trifluorophosphazo)phosphine] molybdenum.
257. A. P. Hagen and E. A. Elphingstone, *J. Inorg. Nucl. Chem., 35,* 3719 (1973).
258. M. Chaigneau and M. Santarromana, *C. R. Acad. Sci., Ser. C, 278,* 1453 (1974).
Reaction of PF_3 with transition-metal oxides.

VI. Complexes of Aminophosphines, $R'_{3-n}P(NR_2)_n$ (n = 1, 2, 3), Diphosphines, R_2P-PR_2, and Related Ligands.

259. D. Sellmann, *Angew. Chem., 85,* 1123 (1973).

Diphosphine transition-metal complexes, $[Cr(CO)_5]_2(P_2H_4)$ and $[(C_5H_5)Mn(CO)_2]_2(P_2H_4)$.
260. J. C. Green, M. L. H. Green, and G. E. Morris, *J. C. S. Chem. Comm., 1974,* 212. Reaction between white phosphorus and bis(π-cyclopentadienyl)molybdenum dihydride forming a MoP_2H_2 system.
261. G. E. Graves and L. W. Houk, *J. Inorg. Nucl. Chem., 36,* 232 (1974). Phosphinohydrazine metal carbonyls.
262. R. B. King and O. Von Stetten, *Inorg. Chem., 13,* 2453 (1974). Metal complexes of some aziridinophosphines related to anticancer drugs.

VII. Homogeneous and Heterogeneous Catalysis by Transition-Metal Phosphine Complexes

263. F. Pruchnik, *Inorg. Nucl. Chem. Lett., 9,* 1229 (1973). 2-Methylallyldichlorohodium(III) catalyses hydrogenation of C=C bonds.
264. B. R. James, L. D. Markham, and D. K. W. Wang, *J. C. S. Chem. Comm., 1974,* 439. Stoichiometric hydrogenation of olefins using $RuHCl(Ph_3P)_3$ and formation of an ortho-metalated ruthenium(II) complex.
265. G. Strathdee and R. Given, *Can. J. Chem., 52,* 2216 (1974).
266. G. Strathdee and R. Given, *Can. J. Chem., 52,* 2226 (1974). Homogeneous catalysis of hydrogen isotope exchange.
267. V. Bazant, M. Capka, H. Jahr, J. Hetflejs, V. Chvalovsky, H. Pracejus, and P. Svoboda, Ger. (East) 100,267, 12 Sep 1973, Appl. 157,601, 08 Sep 1971; through *Chem. Abstr., 80,* 83225c (1974). [Poly(*p*-diphenylphosphinostyrene)[bis(triphenylphosphine)rhodium(I) chloride is an active catalyst for hydrogenation.
268. A. A. Oswald and L. L. Murrell, Ger. Offen. 2,332,167, 10 Jan 1974, U.S. Appl. 265,507, 23 Jun 1972; through *Chem. Abstr., 80,* 83252j (1974). $Ph_2P(CH_2)_8SiCl_3$ was anchored on silica treated with $RhCl_2(CO)_2$ to give $[Ph_2P(CH_2)_8SiCl_3]_2RhCl(CO)$.
269. D. E. Morris and H. B. Tinker, Ger. Offen. 2,359,377, 06 Jun 1974, U.S. Appl. 310,621, 29 Nov 1972; through *Chem. Abstr., 81,* 91053f (1974). Hydroformylation: $[Rh(CO)_3L_2]X$ catalysts ($X = PF_6, BF_4, BPh_4, ClO_4$, $L = Ph_3P$, cyclohexylanisylmethylphosphine).
270. Z. N. Parnes, D. Kh. Shaapuni, M. I. Kalinkin, and D. N. Kursanov, *Izv. Akad. Nauk SSSR, Ser. Khim., 1974,* 1665. Homogeneous catalytic ionic selective reduction of alcohols with H in the presence of $PtCl_2(Ph_3P)_2$ and CF_3CO_2H.
271. N. Kameda and N. Itagaki, *Bull. Chem. Soc. Jap., 46,* 2597 (1973).
272. N. Kameda, Y. Imamura, and M. Takeda, *Nippon Kagaku Kaishi, 1974,* 346; through *Chem. Abstr., 81,* 78296t (1974).
273. E. Uhlig and E. Dinjus, Ger. (East) 101,903, 20 Nov 1973, Appl. 158804, 09 Nov 1971; through *Chem. Abstr., 81,* 154273a (1974).
274. J. S. Yoo and H. Erickson, U.S. 3,755,490, 28 Aug 1973, Appl. 821,134, 01 May 1969; through *Chem. Abstr., 80,* 4087h (1974). Polymerization.
275. J. Beger and H. Reichel, *J. Prakt. Chem., 315,* 1067 (1973). Diene oligomerization.
276. R. De Haan and J. Dekker, *J. Catal., 35,* 202 (1974).
277. N. V. Petrushanskaya, A. I. Kurapova, N. M. Rodionov, and V.Sh. Fel'dblyum, *Zh. Org. Khim., 10,* 1402 (1974). Dimerization of olefins by nickel catalysts.

278. W. E. Tyler and M. B. Dines, U.S. 3,776,972, 04 Dec 1973, Appl. 259,079, 02 Jun 1972; through *Chem. Abstr., 80,* 70972u (1974).
Phosphinecopper complexes capable of recovering olefins from feed streams by complexation.
279. J. Hojo, M. Aramaki, N. Yamazoe, and T. Seiyama, *Kyushu Daigaku Kagaku Shuho, 47,* 332 (1974); through *Chem. Abstr., 81,* 135279 (1974).
Benzaldehyde was oxidized in benzene in the presence of $Ph(Ph_3P)_4$.
280. N. Von Kutepow and F. J. Müller, Ger. Offen., 2,303,271, 08 Aug 1974, Appl. P 2303271.5-42, 24 Jan 1973; through *Chem. Abstr., 81,* 135473z (1974).
Acetic acid and methyl acetate from methanol and CO in presence of $PdCl_2(Ph_3P)_2$.
281. B. L. Haymore and J. A. Ibers, *J. Am. Chem. Soc., 96,* 3325 (1974).
Catalytic reduction of NO by CO to form N_2O and CO_2 in the presence of $[M(NO)_2(Ph_3P)_2]^+$ (M = Ir, Rh) or $IrX_3(NO)(Ph_3P)_2$ (X = Cl, Br).
282. P. Svoboda, T. S. Belopotapova, and J. Hetflejs, *J. Organomet. Chem., 65,* C37 (1974).
Formation of transition metal carbonyl complexes from CO_2.
283. F. Pennella and R. L. Banks, *J. Catal., 35,* 73 (1974).
Inhibiting influence of N_2 to the isomerization of 1-pentene catalyzed by dihydrido(dinitrogen)tris(triphenylphosphine)ruthenium.
284. H. A. Dieck and R. F. Heck, *J. Am. Chem. Soc., 96,* 1133 (1974).
Organophosphinepalladium complexes as catalysts for vinylic hydrogen substitution reactions.
285. Monsanto Co., Brit. 1,349,869, 10 Apr 1974, Appl. 36,234/73 26 May 1971; through *Chem. Abstr., 81,* 25812b (1974).
286. I. Ojima, T. Kogure, and Y. Nagai, *Tetrahedron Lett., 1974,* 1889.
287. W. R. Cullen, A. Fenster, and B. R. James, *Inorg. Nucl. Chem. Lett., 10,* 167 (1974).
288. M. Tanaka, Y. Watanabe, T. Mitsudo, Y. Yasunori, and Y. Takegami, *Chem. Lett., 1974,* 137.
289. Agency of Industrial Sciences and Technology, Japan Kokai 73 99,131, 15 Dec 1973, Appl. 72 33,900, 06 Apr 1972; through *Chem. Abstr., 80,* 95498g (1974).
290. A. J. Solodar, Ger. Offen. 2,312,924, 27 Sep 1793, U.S. Appl. 235,405, 16 Mar 1972; through *Chem. Abstr., 80,* 3672h (1974).
Catalytic reactions with transition-metal complexes of optically active phosphine ligands.
291. S. Ootsuka and K. Tani, Japan. Kokai 73 56,628, 09 Aug 1973, Appl. 71 93,193, 22 Nov 1971; through *Chem. Abstr., 80,* 15070y (1974).
Optical resolution of tertiary phosphine compounds by an asymmetric Pd complex.

VIII. Reactions of Coordinated Ligands

292. T. Kruck, J. Waldmann, M. Höfler, G. Birkenhäger, and C. Odenbrett, *Z. Anorg. Allg. Chem., 402,* 16 (1973).
293. T. Kruck, M. Höfler, and H. Jung, *Chem. Ber., 107,* 2133 (1974).
294. T. Kruck, H. Jung, M. Höfler, and H. Blume, *Chem. Ber., 107,* 2145 (1974).
295. T. Kruck, G. Mäueler, and G. Schmidgen, *Chem. Ber., 107,* 2421 (1974).
Fluorine exchange reactions in metal trifluorophosphine complexes.
296. W. M. Douglas, R. B. Johannesen, and J. K. Ruff, *Inorg. Chem., 13,* 371 (1974).
Reaction of $Fe(CO)_4PF_2Br$ with Ag_2O, Cu_2O gave $Fe(CO)_4PF_2OPF_2Fe(CO)_4$.
297. M. Höfler and M. Schnitzler, *Chem. Ber., 107,* 194 (1974).

The complexes π-$C_5H_5Mn(CO)_2PhPX_2$ (X = CN, NCO, NCS, N_3) were parpared from (π-C_5H_5)$Mn(CO)_2PhPCl_2$ by Cl-pseudohalogen exchange reactions.
298. A. Vizi-Orosz, G. Palyi, and L. Marko, *J. Organomet. Chem., 60,* C25 (1973). Phosphido carbonyl cobalt clusters, $Co(CO)_6P_2$ and $Co_3(CO)_9PS$ were prepared by the reaction of Na[$Co(CO)_4$)] with PX_3 or SPX_3 (X = Cl, Br).
299. W. Ehrl and H. Vahrenkamp, *J. Organomet. Chem., 63,* 389 (1973). Phosphine-bridged binuclear carbonyl complexes of iron and nickel.
300. H. Matsumoto, T. Nakano, and Y. Nagai, *Tetrahedron Lett., 1973,* 5147. Addition of carbon tetrachloride and chloroform to 1-olefins catalyzed by ruthenium(II) complexes.
301. P.-C. Kong and D. M. Roundhill, *J. C. S. Dalton, 1973,* 187. Solvolysis of Pt(O) complexes of Ph_2POR (R = Bu, Pr^i, Me) in ethanol.

IX. Thermodynamic, Kinetic and Mechanistic Studies

302. G. R. Dobson, *Inorg. Chem., 13,* 1790 (1974).
303. W. J. Knebel and R. J. Angelici, *Inorg. Chem., 13,* 627 (1974).
304. B. F. G. Johnson, J. Lewis, and M. V. Twigg, *J. C. S. Dalton, 1974,* 241.
305. M. Basato and A. Poe, *J. C. S. Dalton, 1974,* 456.
306. M. Basato and A. Poe, *J. C. S. Dalton, 1974,* 607.
307. A. Poe and M. V. Twigg, *Inorg. Chem., 13,* 1982 (1974).
308. K. J. Karel and J. R. Norton, *J. Am. Chem. Soc., 96,* 6812 (1974).
309. G. Cardaci, *Inorg. Chem., 13,* 368 (1974).
310. G. Cardaci, *Inorg. Chem., 13,* 2974 (1974).
311. P. C. Ellgen and J. N. Gerlach, *Inorg. Chem., 13,* 1944 (1974). Kinetic studies of substitution reactions on transition metal carbonyls.
312. E. M. Miller and B. L. Shaw, *J. C. S. Dalton, 1974,* 480. Kinetics on oxidative addition reactions of *trans*-$IrCl(CO)(Me_2PR)_2$ (R = Ph, *o*-$MeOC_6H_4$, *p*-$MeOC_6H_4$).
313. D. J. Darensbourg and H. L. Conder, *Inorg. Chem., 13,* 374 (1974). Kinetic studies of phosphine and phosphite exchange reactions of substituted iron tricarbonyl carbene complexes.
314. C. A. McAuliffe and R. Pollock, *J. Organomet. Chem., 77,* 265 (1974). Relative rates of dioxygen uptake by the complexes *trans*-$IrX(CO)(Ph_2PR)_2$ (R = Ph, Me, Et, X = F, Cl, Br, I).
315. R. J. Guschl, R. S. Stewart, and T. L. Brown, *Inorg. Chem., 13,* 417 (1974). Solvent and alkyl substituent effects on the kinetics of base exchange in alkylbis-(dimethylglyoximato)cobalt(III)-trimethylphosphite complexes.
316. D. V. Sokol'skii, Ya. A. Dorfman, and V. S. Emel'yanova, *Kinet. Katal., 14,* 1573 (1973).
317. C. H. Bamford and I. Sakamoto, *J. C. S. Faraday* 1, *70,* 330 (1974).
318. R. G. Pearson and J. Rajaram, *Inorg. Chem., 13,* 246 (1974). Kinetics of redox reactions and oxidative additions.
319. Y. Kubo, A. Yamamoto, and S. Ikeda, *J. Organomet. Chem., 60,* 165 (1973). Kinetic study of the reactions of olefins with $CoH(X_2)(Ph_3P)_3$ (X_2 = N_2, CO).
320. D. A. Sweigart, *Inorg. Chim. Acta, 8,* 317 (1974). Mechanism of ligand substitution on five coordinate iron and cobalt dithiolene complexes.
321. C. G. Grimes and R. G. Pearson, *Inorg. Chem., 13,* 970 (1974). Ligand exchange reactions of the five coordinate complexes $Ni(CN)_2L_3$.

322. C. A. Tolman, W. C. Seidel, and L. W. Gosser, *J. Am. Chem. Soc., 96,* 53 (1974). Thermodynamics of the dissociation of complexes NiL_4.
323. B. R. James and L. D. Markham, *Inorg. Chem., 13,* 97 (1974). Thermodynamics of the reaction $RuCl_2(Ph_3P)_3 \rightleftharpoons RuCl_2(Ph_3P)_2 + Ph_3P$.
324. J. Rimbault and R. Hugel, *Rev. Chim. Miner., 10,* 773 (1973). Distribution equilibria involving Ph_3P and Ph_3PO as ligands in cobalt(II) iodide complexes.
325. I. V. Gavrilova, M. I. Gel'fman, N. A. Kustova, and V. V. Razumovskii, *Zh. Neorg. Khim., 18,* 2856 (1973). Instability constants of platinum(II) complexes in water-acetone solutions.
326. A. D. Troitskaya and V. V. Sentemov, *Zh. Obshch. Khim., 44,* 1625 (1974). Equilibrium constants of nickel(II)thiocyanate trialkylphosphites-ethanol systems.
327. D. A. Redfield, J. H. Nelson, R. A. Henry, D. W. Moore, and H. B. Jonassen, *J. Am. Chem. Soc., 96,* 6298 (1974). Isomerism energetics and mechanisms for palladium(II) phosphine complexes containing 5-methyl- and 5-trifluoromethyltetrazoles.
328. W. J. Louw, *J. C. S. Chem. Commun., 1974,* 353. *Cis-trans* isomerization of dichlorobis(triethylphosphine)platinum.
329. D. J. Thornhill and A. R. Manning, *J. C. S. Dalton, 1974,* 6. Tautomeric equilibria of $[Co(CO)_3L]_2$.
330. P. E. Garrou and G. E. Hartwell, *J. C. S. Chem. Comm., 1974,* 318. Redistribution reactions of organometallic complexes. Carbonyl, halogen, and phosphine ligand exchange between coordinately unsaturated rhodium(I) and iridium(I) complexes.
331. F. Pennella, *J. Organomet. Chem., 65,* C17 (1974). Relative affinities of dinitrogen and of pentene for the dihydridotris(triphenylphosphine)ruthenium moiety.
332. C. A. Tolman, *J. Am. Chem. Soc., 96,* 2780 (1974). Formation constants of (olefin)bis(tri-*o*-tolylphosphite)nickel complexes.
333. A. Musco, W. Kuran, A. Silvani, and M. W. Anker, *J. C. S. Chem. Comm., 1973,* 938. Preference in forming complexes PdL_n with a low-coordination number was determined as $Bu^t{}_2PhP \sim (cyclohexyl)_3P > Pr_3P > (CH_2Ph)_3P > Bu_3P \sim Et_3P \sim Ph_3P > MePh_2P \sim Me_2PhP \sim Me_3P$ by ^{13}C NMR.
334. M. Cusumano, G. Faraone, V. Ricevuto, R. Romeo, and M. Trozzi, *J. C. S. Dalton, 1974,* 490. The kinetics for the reaction $trans\text{-}Pd(NO_2)_2L_2 + Y^- \rightarrow trans\text{-}Pd(NO_2)YL_2 + NO_2^-$ ($L = Pr_3P$, $Y = Cl, N_3, Br, I, SCN, (NH_2)_2CS$) in methanol.
335. G. Cardaci, *J. Organomet. Chem., 76,* 385 (1974). Mechanism of substitution of (π-olefin)tetracarbonyl iron complexes by triphenylphosphine.
336. Yu. N. Kukushkin and L. I. Danilina, *Zh. Neorg. Khim., 19,* 1349 (1974). Reactions of an inner-sphere nitro group in $Rh(CO)NO_2(Ph_3P)_2$.
337. E. Lindner and A. Thasitis, *Z. Anorg. Allg. Chem., 409,* 35 (1974). Comparative investigation on the complex chemical behavior of *N,N*-dimethylacetamide, acetyl-, and (trifluoroacetyl) diphenylphosphine toward metals in low-oxidation states.
338. L. Cassar and M. Foa, *J. Organomet. Chem., 74,* 75 (1974). Mechanism of the nickel-catalyzed synthesis of phosphonium salts.
339. D. V. Sokol'skii, Ya. A. Dorfman, and Z. I. Rogoza, *Zh. Fiz. Khim., 48,* 585 (1974).

Reduction of cerium(IV) complexes by PH_3.
340. I. Bosnyak-Ilcsik, S. Papp, L. Bencze, and G. Palyi, *J. Organomet. Chem., 66,* 149 (1974).
Reaction of low-valent transition-metal complexes and their phosphine derivatives with Karl Fischer reagent.
341. J. Schwartz and J. B. Cannon, *J. Am. Chem. Soc., 96,* 2276 (1974).
Thermal decomposition of bis(triphenylphosphine)(carbonyl)octyliridium.

X. Spectroscopic Investigations

A. NMR and NQR Spectroscopy

342. H. Mahnke, R. J. Clarke, R. Rosanske, and R. K. Sheline, *J. Chem. Phys., 60,* 2997 (1974).
343. R. Mathieu and J. F. Nixon, *J. C. S. Chem. Comm., 1974,* 147.
344. P. R. Hoffman, J. S. Miller, C. B. Ungermann, and K. G. Caulton, *J. Am. Chem. Soc., 95,* 7902 (1973).
345. P. Meakin, R. A. Schunn, and J. P. Jesson, *J. Am. Chem. Soc., 96,* 277 (1974).
346. J. P. Jesson and P. Meakin, *J. Am. Chem. Soc., 96,* 5760 (1974).
347. P. Meakin and J. P. Jesson, *J. Am. Chem. Soc., 95,* 7272 (1973).
348. P. Meakin and J. P. Jesson, *J. Am. Chem. Soc., 96,* 5751 (1974).
349. K. G. Caulton, *J. Am. Chem. Soc., 96,* 3005 (1974).
Inter- and intramolecular ligand exchange in pentacoordinate complexes studied by dynamic NMR spectroscopy.
350. R. J. Cross and N. H. Tennent, *J. C. S. Dalton, 1974,* 1444.
Phosphine exchange in *trans*-chlorobis(methyldiphenylphosphine)[2-(phenazo)-phenyl] platinum(II).
351. P. E. Cattermole, K. G. Orrell, and A. G. Osborne, *J. C. S. Dalton, 1974,* 328.
Preaparation and NMR investigation of stereochemically nonrigid derivatives of dodecacarbonyltetrairidium, $Ir_4(CO)_8L_4$ ($L = MePh_2P$, Me_2PhP, $EtPh_2P$, $P(OPh)_3$, ½ $(Ph_2PCH_2)_2$).
352. K. G. Caulton, *Inorg. Chem., 13,* 1774 (1974).
^{31}P NMR Spectra of $Co(NO)(Ph_3P)_3$, $Rh(NO)(Ph_3P)_3$, and $RhCl_2(NO)(Ph_3P)_2$ indicate the chemical equivalence of the three Ph_3P ligands.
353. D. J. Cole-Hamilton and T. A. Stephenson, *J. C. S. Dalton, 1974,* 754.
Facile optical isomerism reactions in (dimethylphosphino)dithioato complexes of ruthenium(II).
354. E. Lindner and M. Zipper, *Chem. Ber., 107,* 1444 (1974).
355. K. Stanley, R. A. Zelonka, J. Thomson, P. Fiess, and M. C. Baird, *Can. J. Chem., 52,* 1781 (1974).
356. T. Kaneshima, K. Kawakami, and T. Tanaka, *Inorg. Chem., 13,* 2198 (1974).
357. C. H. Bushweller and M. Z. Lourandos, *Inorg. Chem., 13,* 2514 (1974).
Stereodynamics and rotational isomerism.
358. J. P. Fawcett, A. J. Poe, and M. V. Twigg, *J. Organomet. Chem., 61,* 315 (1973).
Positional isomerism of triphenylphosphine nonacarbonylmanganeserhenium.
359. A. G. Ginzburg, L. A. Fedorov, P. V. Petrovskii, E. I. Fedin, V. N. Setkina, and D. N. Kursanov, *J. Organomet. Chem., 73,* 77 (1974).
Protonation of π-cyclopentadienylphosphinemanganese complexes studied by ^{13}C and ^{31}P NMR techniques.
360. D. Dodd, M. D. Johnson, and C. W. Fong, *J. C. S. Dalton, 1974,* 58.
361. S. O. Grim and L. C. Satek, *Z. Naturforsch., 28b,* 683 (1973).

362. S. O. Grim, P. J. Lui, and R. L. Keiter, *Inorg. Chem., 13,* 342 (1974).
363. R. J. Mynott, P. S. Pregosin, and L. M. Venanzi, *J. Coord. Chem., 3,* 145 (1973).
364. D. J. Cardin, B. Cetinkaya, E. Cetinkaya, M. F. Lappert, E. W. Randall, and E. Rosenbert, *J. C. S. Dalton, 1973,* 1982.
365. D. A. Redfield, J. H. Nelson, and L. W. Cary, *Inorg. Nucl. Chem. Lett., 10,* 727 (1974).
366. C. Rüger, A. Mehlhorn, and K. Schwetlick, *Z. Chem., 14,* 196 (1974).
367. C. Blejean and J. L. Chenot, *Bull. Soc. Chim. Fr., 1973,* 2617.
368. D. Rehder and J. Schmidt, *J. Inorg. Nucl. Chem., 36,* 333 (1974).
^{31}P, ^{13}C, ^{19}F, and ^{1}H shifts and discussion of these values with respect to bondings, stereochemistry, and correlation to other data.
369. R. J. Goodfellow and B. F. Taylor, *J. C. S. Dalton, 1974,* 1676.
Sign and magnitude of $^{2}J_{PP}$ in a variety of gold(I), iridium(III), mercury(II), palladium(II), platinum(II) and (IV), and rhodium(III) complexes.
370. K. R. Dixon, K. C. Moss, and M. A. R. Smith, *Inorg. Nucl. Chem. Lett., 10,* 373 (1974).
Inverse, linear correlation of cis- and trans-influence.
371. T. W. Dingle and K. R. Dixon, *Inorg. Chem., 13,* 846 (1974).
NMR spectra of platinum(II) hydrides, AB_2X and AB_2MX spin systems ($A = B = {}^{31}P$, $M = {}^{195}Pt$, $X = {}^{1}H$).
372. T. E. Boyd, T. L. Brown, *Inorg. Chem., 13,* 422 (1974).
^{59}Co NQR spectra of phosphine complexes $CoX(CO)_3L$ and $CoX(CO)_2L_2$ ($L = Bu_3P$, $P(OMe)_3$, $P(OEt)_3$, $P(OPh)_3$, $X = Ph_3Sn$ or other similar group).
373. C. D. Pribula, T. L. Brown, and E. Muenck, *J. Am. Chem. Soc., 96,* 4149 (1974).
Calculated and observed field gradients in $[M(CO)_{5-x}P_x]^n$ complexes.

B. E.S.R. Spectroscopy

374. F. K. Shmidt, V. V. Saraev, G. M. Larin, V. G. Lipovich, and L. V. Mironova, *Izv. Akad. Nauk SSSR, Ser. Khim., 1974,* 2136.
EPR study of the composition and structure of nickel(I) complexes in Ziegler-type catalytic systems.
375. V. V. Saraev, G. M. Larin, F. K. Shmidt, and V. G. Lipovich, *Izv. Akad. Nauk SSSR, Ser. Khim., 1974,* 928.
EPR study of the structure of complexes in the triethylaluminium–dicyclopentadienyltitanium dichloride–phosphine derivative catalytic system.
376. R. Kirmse, W. Dietzsch, and E. Hoyer, *Z. Chem., 14,* 106 (1974).
ESR studies of $Co(S_4C_4H_4)R_3P$.
377. J. A. DeBeer, R. J. Haines, R. Greatrex, and J. A. Van Wyk, *J. C. S. Dalton, 1973,* 2341.
ESR of the cationic derivatives$[(C_5H_5)Fe(CO)Ph_2P]_2^+$.
378. B. B. Wayland and M. E. Abd-Elmageed, *J. Am. Chem. Soc., 96,* 4809 (1974).
379. B. B. Wayland, J. K. Minkiewicz, and M. E. Abd-Elmageed, *J. Am. Chem. Soc., 96,* 2795 (1974).
ESR studies of tetraphenylporphyrinecobalt(II) phosphine complexes.

C. Infrared and Raman Spectroscopy

380. S. J. Cyvin, *Z. Anorg. Allg. Chem., 403,* 193 (1974).
Vibrational frequency shifts of free to complexed trifluorophosphine, tetrakis-(trifluorophosphine)nickel.

381. A. G. Jones and D. B. Powell, *Spectrochim. Acta, 30A,* 1001 (1974).
The vibrational spectra of $(p\text{-FC}_6H_4)_3P$ and its gold(I) halide complexes.

382. H. Minematsu, Y. Nonaka, S. Takahashi, and N. Hagihara, *J. Organomet. Chem., 59,* 395 (1973).
IR and NMR studies of Pd(Maleic anhydride)L_2.

383. D. J. Darensbourg, H. H. Nelson, and C. L. Hyde, *Inorg. Chem., 13,* 2135 (1974).

384. F. T. Delbeke and G. P. Van der Kelen, *J. Organomet. Chem., 64,* 239 (1974).

385. F. T. Delbeke, G. P. Van der Kelen, and Z. Eeckhout, *J. Organomet. Chem., 64,* 265 (1974).
Analysis of the vibrational spectra of substituted transition metal carbonyls.

386. A. D. Troitskaya, V. V. Sentemov, and E. I. Antropova, *Zh. Neorg. Khim., 18,* 3349 (1973).
Infrared absorption spectra of platinum(II) complexes with trialkylphosphites in ethanol.

387. E. A. Allen and W. Wilkinson, *Spectrochim. Acta, 30A,* 1219 (1974).
Vibrational spectra of four coordinate complexes of palladium and nickel with various phosphine ligands.

388. A. D. Troitskaya, G. D. Ginzburg, E. A. Zgadzai, and V. V. Sentemov, *Zh. Neorg. Khim., 19,* 1119 (1974).
Structure of nickel(II) thiocyanate complexes with trialkylphosphites.

389. R. Whyman, *J. Organomet. Chem., 63,* 467 (1973).
Infrared spectroscopic evidence for tricarbonyltriphenylphosphineplatinum and -palladium.

D. ESCA and U.V. Photoelectron Spectroscopy

390. L. J. Matienzo and S. O. Grim, *Anal. Chem., 46,* 2052 (1974).
Interactions of some free phosphorus(III) compounds with gold vapor detected by means of x-ray photoelectron spectroscopy.

E. Mössbauer Spectroscopy

See Ref. 377.

F. Electronic Spectroscopy

391. H. Kato and K. Akimoto, *J. Am. Chem. Soc., 96,* 1351 (1974).
Magnetic circular dichroism spectra of tetrahedral cobalt(II) complexes.

392. A. Merle, M. Dartiguenave, Y. Dartiguenave, J. W. Dawson, and H. B. Gray, *J. Coord. Chem., 3,* 199 (1974).
Electronic structures of dicyanonickel(II) complexes with trimethylphosphine.

393. G. L. Geoffroy, M. S. Wrighton, G. S. Hammond, and H. B. Gray, *J. Am. Chem. Soc., 96,* 3105 (1974).
Electronic absorption and emission spectral studies of square planar rhodium(I) and iridium(I) complexes. Evidence for a charge-transfer emitting state.

G. Various Methods

394. S. Torroni, G. Innorta, A. Foffani, and G. Distefano, *J. Organomet. Chem., 65,* 209 (1974).

Interpretation of the mass spectra of substituted chromium and tungsten carbonyls by means of appearance potential measurements.
395. F. Glocking, T. McBride, and R. J. I. Pollock, *Inorg. Chim. Acta, 8,* 81 (1974). Mass spectroscopic study of platinum alkyl and aryl complexes PtR_2L_2 (L = tertiary phosphine).
396. J. Müller and W. Goll, *J. Organomet. Chem., 69,* C23 (1974). Ion molecule reactions of metall organic complexes. Binuclear ions in the mass spectra of $Ni(PF_3)_4$ and (methoxymethylcarbene)pentacarbonylchromium(O).
397. A. D. Troitskaya, N. P. Burmistrova, G. A. Levshina, and L. F. Gogolyukhina, *Zh. Obshch. Khim., 44,* 1836 (1974).
398. M. P. Brown, R. J. Puddephatt, C. E. E. Upton, and S. W. Lavington, *J. C. S. Dalton, 1974,* 1613. Thermal decomposition studied by differential thermoanalysis, thermogravimetry, or mass spectroscopy.
399. K. Jonas, *J. Organomet. Chem., 78,* 273 (1974). Polarographic studies of diphosphosphine nickel(O) systems with aromatic hydrogen molecules.
400. H. J. Kerrinnes and U. Langbein, *Z. Anorg. Allg. Chem., 406,* 110 (1974). Electrochemical preparation of cobalt complex compounds of acyclic 1,3-diolefins, π-1,3-butadiene[π-butenyl] tributylphosphinecobalt(I).
401. J. G. Norman, Jr., *J. Am. Chem. Soc., 96,* 3327 (1974). SCF-X_α scattered wave calculation of the electronic structure of $Pt(O_2)(PH_3)_2$.
402. B. K. Teo, M. B. Hall, R. F. Fenske, and L. F. Dahl, *J. Organomet. Chem., 70,* 413 (1974). Nonparameterized MO calculations of doubly ligand bridged octacarbonyldimetal complexes with metal-metal interactions.
403. D. M. Allen, A. Cox, T. J. Kemp, L. H. Ali, *J. C. S. Dalton, 1973,* 1899. Photolysis of the complexes $(\pi\text{-}C_5H_5)FeBr(CO)L$ (L = Ph_3P, $P(OPh)_3$).

XI. Crystal and Molecular Structure, Electron Diffraction Studies

404. N. I. Kirillova, A. I. Gusev, A. A. Pasynskii, and Yu. T. Struchkov, *Zh. Strukt. Khim., 15,* 288 (1974). $(\pi\text{-}C_5H_5)NbH_2(CO)(Ph_3P)_2$.
405. P. Meakin, L. J. Guggenberger, F. N. Tebbe, and J. P. Jesson, *Inorg. Chem., 13,* 1025 (1974). $TaH(CO)_2(Me_2PCH_2CH_2PMe_2)_2$.
406. R. M. Kirchner and J. A. Ibers, *Inorg. Chem., 13,* 1667 (1974). $(\pi\text{-}C_5H_5)MoCl[C{=}C(CN)_2][P(OMe)_3]_2$.
407. R. A. Forder and K. Prout, *Acta Crystallogr., 30B,* 2778 (1974). $[(\pi\text{-mesitylene})Mo(Me_2PCH_2CH_2PMe_2)]_2N_2$.
408. M. G. B. Drew and J. D. Wilkins, *J. C. S. Dalton, 1974,* 1654. $[WI(CO)_3(Me_2PhP)_3]^+BPh_4^-$.
409. C. Barbeau and R. J. Dubey, *Can. J. Chem., 52,* 1140 (1974). $(\pi\text{-}C_5H_5)Mn(CO)(Ph_3P)_2 \cdot C_6H_6$.
410. M. Laing, R. Reimann, and E. Singleton, *Inorg. Nucl. Chem. Lett., 20,* 557 (1974). $MnCl(NO)_2(PhP(OMe)_2)_2$.
411. G. Bandoli, D. A. Clemente, U. Mazzi, and E. Tondello, *Cryst. Struct. Commun., 3,* 293 (1974). $TeCl_2[PhP(OEt)_2]_4$.

412. J. A. Jaecker, W. R. Robinson, and R. A. Walton, *J. C. S. Chem. Comm., 1974,* 306. $[ReCl_3(MeCN)(Ph_2PCH_2)]_2$.
413. M. J. Barrow, G. A. Sim, R. C. Dobie, and P. R. Mason, *J. Organomet. Chem., 69,* C4 (1974). π-cyclopentadienyldicarbonyl[bis(trifluoromethyl)phosphinato]iron and π-cyclopentadienyldicarbonyl[bis(trifluoromethyl)phosphine oxidato]iron.
414. H. Vahrenkamp, *J. Organomet. Chem., 63,* 399 (1973). $(\pi\text{-}C_5H_5)Fe_2(CO)_5Me_2P$.
415. F. A. Cotton and J. M. Troup, *J. Am. Chem. Soc., 96,* 4422 (1974). $Fe_2(CO)_7(Ph_2PCH_2PPh_2)$.
416. H. Felkin, P. J. Knowles, B. Meunier, A. Mitschler, L. Ricard, and R. Weiss, *J. C. S. Chem. Comm., 1974,* 44. $(\pi\text{-}C_5H_5)Fe(Ph_2PCH_2CH_2PPh_2)MgBr$.
417. R. Mason and J. A. Zubieta, *J. Organomet. Chem., 66,* 279 (1974). Bis(μ-diphenylphosphido-μ-carbonyl-π-methylcyclopentadienylcarbonyliron)rhodium hexafluorophosphate (2 rhodium-iron).
418. R. Mason and J. A. Zubieta, *J. Organomet. Chem., 66,* 289 (1974). μ[Carbonyl(triphenylphosphine)platino]octacarbonyldiiron (2 platinum-iron, 1 iron-iron)
419. M. Mathew, G. J. Palenik, A. J. Carty, and H. N. Paik, *J. C. S. Chem. Comm., 1974,* 25. $Fe_3(CO)_8[Ph_2PC_4(CF_3)_2]PPh_2$ (trinuclear ferracylbutene complex).
420. V. G. Albano, A. Araneo, P. L. Bellon, G. Ciani, and M. Manassero, *J. Organomet. Chem., 67,* 413 (1974). $Fe(CO)(NO)_2Ph_3P$ and $Fe(NO)_2(Ph_3P)_2$.
421. G. Davey and F. S. Stephens, *J. C. S. Dalton, 1974,* 698. Di-μ-carbonyl-carbonyl[dicarbonyl(methyldiphenylphosphine)-cobaltio-μ-cyclopentadienyliron.
422. A. J. Schultz, R. L. Henry, J. Reed, and R. Eisenberg, *Inorg. Chem., 13,* 732 (1974). $RuCl_3(NO)(MePh_2P)_2$.
423. I. Bernal, A. Clearfield, and J. S. Ricci, Jr., *J. Cryst. Mol. Struct., 4,* 43 (1974). $Ru(CO)[C_2S_2(CF_3)_2](Ph_3P)_2$.
424. D. Hall and R. B. Williamson, *Cryst. Struct. Commun., 3,* 327 (1974). $RuI(CO)(NO)(Ph_3P)_2$.
425. J. C. McConway, A. C. Skapski, L. Phillips, R. J. Young, and G. Wilkinson, *J. C. S. Chem. Comm., 1974,* 327. $[RuH(Ph_3P)_2(\eta\text{-}PhPPh_2)]^+$ (1 Ph ring is bound as an arene to Ru).
426. F. L. Phillips and A. C. Skapski, *J. C. S. Chem. Comm., 1974,* 49. $RuCl_3(Et_2PhP{=}N)(Et_2PhP)_2$ (the $Et_2PhP{=}N^-$ ligand is linearly coordinated to Ru).
427. A. J. F. Fraser and R. O. Gould, *J. C. S. Dalton, 1974,* 1139. $Ru(CS)(Ph_3P)_2$-tri-μ-Cl-$RuCl(Ph_3P)_2$.
428. J. Reed, S. L. Soled, and R. Eisenberg, *Inorg. Chem., 13,* 3001 (1974). $RuCl(SO_4)(NO)(Ph_3P)_2$.
429. A. P. Gaughan, Jr., B. J. Corden, R. Eisenberg, and J. A. Ibers, *Inorg. Chem., 13,* 786 (1974). $Ru(NO)_2(Ph_3P)_2 \cdot \frac{1}{2} C_6H_6$.
430. M. I. Bruce, O. M. Abu Salah, R. E. Davis, and N. V. Raghavan, *J. Organomet. Chem., 64,* C48 (1974).

(π-C_5H_5)Ru($Ph_3P)_2$(C_2Ph)CuCl.
431. A. E. Kalinin, A. I. Gusev, and Yu. T. Struchkov, *Zh. Strukt. Khim., 14,* 859 (1973).
Ru(S_2CH)$_2$(Ph_3P)$_2$.
432. A. C. Skapski and F. A. Stephens, *J. C. S. Dalton, 1974,* 390.
RuH(MeCO_2)(Ph_3P)$_3$.
433. C. A. Ghilardi and L. Sacconi, *Cryst. Struct. Commun., 3,* 415 (1974).
CoH($Ph_2PC_2H_4$)$_3$N.
434. F. S. Stephens, *J. C. S. Dalton, 1974,* 1067.
Di-μ-carbonyldicarbonyl(triethylphosphine)(π-cyclopentadientylnickel)cobalt.
435. W. W. Adams and P. G. Lenhert, *Acta Crystallogr., 29B,* 2412 (1973).
Bis(dimethylglyoximato)(tributylphosphine)(4-pyridyl)cobalt(I).
436. S. Brückner and L. Randaccio, *J. C. S. Dalton, 1974,* 1017.
trans-CoClL$_2$Ph$_3$P (LH = dimethylglyoxime).
437. J. S. Field, P. J. Wheatley, and S. Bhaduri, *J. C. S. Dalton, 1974,* 74.
[(2-(Diphenylphosphino)ethyl)diphenylphosphine oxide] iodonitrosylcobalt(O).
438. A. I. Gusev and Yu. T. Struchkov, *Zh. Strukt. Khim., 15,* 282 (1974).
Rh(O_2CPh)(Ph_3P)$_3$•0.5 C_6H_6.
439. J. A. Muir, M. M. Muir, and A. J. Rivera, *Acta Crystallogr., 30B,* 2062 (1974).
Tetrachloro-μ-dichlorotetrakis(tributylphosphine)dirhodium(III).
440. R. Schlodder and J. A. Ibers, *Inorg. Chem., 13,* 2870 (1974).
[Rh(CO)(Ph_3P_2]$_2$(HCBD) (HCBD = *trans*-1,1,2,3,4,4-hexacyanobutenediide).
441. T. Kashiwagi, N. Yasuoka, N. Kasai, and M. Kakudo, *Technol. Rep. Osaka Univ., 24,* 355 (1974); through *Chem. Abstr., 81,* 160497t (1974).
RhI(C_3H_4)(Ph_3P)$_2$.
442. A. J. Schultz, J. V. McArdle, G. P. Khare, and R. Eisenberg, *J. Organomet. Chem., 72,* 415 (1974).
IrCl_2(CO)(CHF_2)(Ph_3P)$_2$.
443. M. O. Visscher, J. C. Huffmann, and W. E. Streib, *Inorg. Chem., 13,* 792 (1974).
RhCl[P($CH_2CH_2CH{=}CH_2$)$_3$].
444. M. R. Churchill and S. A. Bezman, *Inorg. Chem., 13,* 1418 (1974).
Bis(triphenylphosphine)octakis(phenylethynyl)tetracopperdiiridium (4 Cu-Cu, 8 Cu-Ir).
445. M. J. Bennett, J. L. Pratt, and R. M. Tuggle, *Inorg. Chem., 13,* 2408 (1974).
(η^5-C_5H_5)Ir(CO)Ph_3P.
446. M. R. Churchill and K.-K. G. Lin., *J. Am. Chem. Soc., 96,* 76 (1974).
Ir(C_7H_8)(SnCl_3)(Me_2PhP)$_2$.
447. M. Angoletta, G. Ciani, M. Manassero, and M. Sansoni, *J. C. S. Chem. Comm., 1973,* 789.
[Ir(NO)$_2$Ph$_3$P]$_2$.
448. A. J. Schultz, G. P. Khare, C. D. Meyer, and R. Eisenberg, *Inorg. Chem., 13,* 1019 (1974).
IrCl(CO)(CHF_2)(OCOCF_2Cl)(Ph_3P)$_2$.
449. L. Gastaldi, P. Porta, and A. A. G. Tomlinson, *J. C. S. Dalton, 1974,* 1424.
[1-(Diphenylarsino)-2-(diphenylphosphino)ethane] (O-methylphosphorodithioato)-nickel(II) benzene (2:1).
450. V. Gramlich and C. Salomon, *J. Organomet. Chem., 73,* C61 (1974).
Dichloro[(–)-2,3-O-isopropylidene-2,3-dihydroxy-1,4-bis(diphenylphosphino)butane] nickel(II).
451. C. Krüger and Y. H. Tsay, *Cryst. Struct. Commun., 3,* 455 (1974).
Ni(CO)$_2$(Ph_3P)$_2$.

452. H. Einspahr and J. Donohue, *Inorg. Chem., 13,* 1839 (1974).
$Ni_2(CO)_3][(CF_3)_2PSP(CF_3)_2]_2$.
453. M. Green, J. Howard, J. L. Spencer, and F. G. A. Stone, *J. C. S. Chem. Comm., 1974,* 153.
$Ni(B_7C_2H_9Me_2)(Et_3P)_2$ (a carbadiborallyl nickel complex).
454. L. E. Nikolaeva, A. A. Shevyrev, T. N. Tarkhova, and N. V. Belov, *Kristallografiya, 19,* 516 (1974).
$Ni(SeN_3C_8OH_7)Ph_3P$.
455. H. N. Paik, A. J. Carty, K. Dymock, and G. J. Palenik, *J. Organomet. Chem., 70,* C17 (1974).
sym-$Fe_2(CO)_6(Ph_2PC{\equiv}CPh)_2$ and $Ni_2(CO)_2(Ph_2PC{\equiv}CBu^t)_2$.
456. J. W. Dawson, T. J. McLennan, W. Robinson, A. Merle, M. Dartiguenave, Y. Dartiguenave, and H. B. Gray, *J. Am. Chem. Soc., 96,* 4428 (1974).
$NiBr_2(Me_3P)_3$.
457. F. A. Cotton, B. A. Frenz, and D. L. Hunter, *J. Am. Chem. Soc., 96,* 4820 (1974).
(2,4-Pentanedionato)(triphenylphosphine)ethylnickel(II).
458. M. Zocchi and A. Albinati, *J. Organomet. Chem., 77,* C40 (1974).
[π-Cyclohexylnickelbis(triphenylphosphine)] (+)trichlorozincate(-) (containing a 1,1-trimethylallyl ligand).
459. P. Dapporto, G. Fallani, and L. Sacconi, *Inorg. Chem., 13,* 2847 (1974).
Iodo[1,1,1-tris(diphenylphosphinomethyl)ethane] nickel(I).
460. B. L. Barnett and C. Krüger, *J. Organomet. Chem., 77,* 407 (1974).
(π-Pentenyl)(diisopropylphenylphosphine)methylnickel(II) and (π-peneyl)(dimethylmethylphosphine)methylnickel(II).
461. D. J. Brauer and C. Krüger, *J. Organomet. Chem., 77,* 423 (1974).
$C_2Me_2Ni[(C_6H_{11})_2PCH_2CH_2(C_6H_{11})_2]$.
462. T. N. Tarkhova, L. E. Nikolaeva, M. A. Simonov, A. V. Ablov, N. V. Gerbeleu, and A. M. Romanov, *Dokl. Akad. Nauk SSSR, 214,* 1326 (1974).
$NiLPh_3P$ (L = salicylaldehyde selenosemicarbazone).
463. M. Horike, Y. Kai, N. Yasuoka, and N. Kasai, *J. Organomet. Chem., 72,* 441 (1974).
$Pd(acac)_2Ph_3P{\cdot}0.5\ C_6H_6$.
464. J. A. McGinnety, *J. C. S. Dalton, 1974,* 1038.
$Pd[C_2(CO_2Me)_2](Ph_3P)_2$.
465. S. Jacobson, Y. S. Wong, P. C. Chieh, and A. J. Carty, *J. C. S. Chem. Comm., 1974,* 520.
trans-$Pd(SCN)_2[P(OPh)_3]_2$.
466. A. Immirzi, and A. Musco, *J. C. S. Chem. Comm., 1974,* 400.
$Pd[(C_6H_{11})_3P]_2$ and $Pd(Bu^t{}_2PhP)_2$.
467. K. Okamoto, Y. Kai, N. Yasuoka, and N. Kasai, *J. Organomet. Chem., 65,* 427 (1974).
$Pd(C_3H_4)(Ph_3P)_2$.
468. G. J. Palenik, W. L. Steffen, M. Mathew, M. Li, and D. W. Meek, *Inorg. Nucl. Chem. Lett., 10,* 125 (1974).
$Pd(SCN)_2(Ph_2PCH_2PPh_2)$.
469. M. A. Bennett, P. W. Clark, G. B. Robertson, and P. O. Whimp, *J. Organomet. Chem., 63,* C15 (1973).
Chloro[bis(diphenylphosphino)stilbene] platinum.
470. B. Jovanovic, L. Manojlovic-Muir, and K. W. Muir, *J. C. S. Dalton, 1974,* 195.
(*trans*-$PtCl(SiMePh_2)(Me_2PhP)_2$.
471. J. Große, R. Schmutzler, and W. S. Sheldrick, *Acta Crystallogr., 30B,* 1623 (1974).

$PtCl(PF_2O)(Et_2PhP)_2$.

472. R. F. Stepaniak, and N. C. Payne, *J. Organomet. Chem., 72,* 453 (1974). *trans*-[Methyl(2-oxacyclopentylidene)bis(dimethylphenylphosphine)platinum(II)] hexafluorophosphate.
473. B. W. Davies and N. C. Payne, *Inorg. Chem., 13,* 1848 (1974). Bis(triphenylphosphine)hexafluorobut-2-yneplatinum(O).
474. G. W. Bushnell, K. R. Dixon, and M. A. Khan, *Can. J. Chem., 52,* 1367 (1974). $[PtCl(phen)(Et_3P)_2]BF_4$.
475. N. Brescianti, M. Calligaris, P. Delise, G. Nardin, and L. Randaccio, *J. Am. Chem. Soc., 96,* 5642 (1974). σ-1-(2-phenyl-1,2-dicarbadecahydrododecaboranyl)(tripropylphosphine)(dipropylpropylidenephosphine)platinum(II).
476. S. D. Ittel and J. A. Ibers, *J. Am. Chem. Soc., 96,* 4804 (1974). $[PtCl(HNNC_6H_4F)(Et_3P)_2][ClO_4]$.
477. G. B. Robertson and P. O. Whimp, *Inorg. Chem., 13,* 2082 (1974). Δ-$PtCl[o\text{-}Ph_2PC_6H_4CH{=}CHC_6H_4PPh_2\text{-}o]$.
478. J. M. Baraban and J. A. McGinnety, *Inorg. Chem., 13,* 2864 (1974). $Pt[(C_6H_4NO_2)CHCH(C_6H_4NO_2)](Ph_3P)_2$, 4,4′-dinitro-*trans*-stilbenebis(triphenylphosphine)platinum.
479. P. L. Bellon, M. Manassero, F. Porta, and M. Sansoni, *J. Organomet. Chem., 80,* 139 (1974). *trans*-bis(ethoxycarbonyl)bis(triphenylphosphine)platinum.
480. V. G. Albano and G. Ciani, *J. Organomet. Chem., 66,* 311 (1974). $FePt_2(CO)_5[P(OPh)_3]_3$.
481. B. W. Davis and N. C. Payne, *Can. J. Chem., 51,* 3477 (1973). $[PtMe(MeC{\equiv}CMe)(Me_2PhP)_2]PF_6$.
482. Y. S. Wong, S. Jacobson, P. C. Chieh, and A. J. Carty, *Inorg. Chem., 13,* 284 (1974). *cis*-Thiocyanatoisothiocyanatobis(3,3-dimethylbutynyldiphenylphosphine)platinum-(II).
483. R. F. Stepaniak, and N. C. Payne, *Inorg. Chem., 13,* 797 (1974). *trans*-[(methyl)(methyl-*N,N′*-dimethylaminocarbene)bis-(dimethylphenylphosphine)-platinum(II)] hexafluorophosphate.
484. R. A. Mariezcurrena and S. E. Rasmussen, *Acta. Chem. Scand., 27,* 2678 (1973). Bis(1-ethynylcyclohexanol)bis(triphenylphosphine)platinum.
485. J. S. Field and P. J. Wheatley, *J. C. S. Dalton, 1974,* 702. *trans*-$[Pt(CO)(p\text{-}ClC_6H_4)(Et_3P)_2]PF_6$.
486. D. J. Yarrow, J. A. Ibers, M. Lenarda, and M. Graziani, *J. Organomet. Chem., 70,* 133 (1974). $Pt[C_3H_2(CN)_4](Ph_3P)_2$, a metallocyclobutane complex.
487. L. Manojlovic-Muir, K. W. Muir, and R. Walker, *J. Organomet. Chem., 66,* C21 (1974). *cis*-$PtCl_2(CO)Ph_3P$.
488. M. R. Churchill and K. L. Kalra, *Inorg. Chem., 13,* 1065 (1974). $[CuClPh_3P]_4$.
489. M. R. Churchill and K. L. Kalra, *Inorg. Chem., 13,* 1427 (1974). $[CuBrPh_3P]_4 \cdot 2\ CHCl_3$.
490. A. P. Gaughan, jr., Z. Dori, and J. A. Ibers, *Inorg. Chem., 13,* 1657 (1974). $Cu(BF_4)(Ph_3P)_3$.
491. M. G. Newton, H. D. Caughman, and R. C. Taylor, *J. C. S. Dalton, 1974,* 1031. Trichloro[(2-(diphenylphosphinyl)ethyl)dimethylammonium] copper(II).
492. M. R. Churchill and K. L. Kalra, *Inorg. Chem., 13,* 1899 (1974). $[CuIEt_3P]_4$.

493. W. R. Clayton and S. G. Shore, *Cryst. Struct. Commun., 2,* 605 (1973).
$CuClPh_3P$.

494. D. Coucouvanis, N. C. Baenziger, and S. M. Johnson, *Inorg. Chem., 13,* 1191 (1974).
Bis[bis(triphenylphosphine)silver(I)bis(1,2-dicyano-1,2-ethylenedithiolato)nickelate(II).

495. F. J. Hollander and D. Coucouvanis, *Inorg. Chem., 13,* 2381 (1974).
Tris[bis(triphenylphosphine)silver(I)] tris(dithiooxalato)iron(III).

496. K. Aurivillius, A. Cassel, and L. Falth, *Chem. Scr., 5,* 9 (1974).
Dimeric [bis(2-diphenylphosphino)ethyl)sulfide] chlorosilver

497. N. C. Baenzinger, K. M. Dittemore, and J. R. Doyle, *Inorg. Chem., 13,* 805 (1974).
$AuCl(Ph_3P)_2 \cdot 0.5\ C_6H_6$.

498. S. H. Whitlow, *Can. J. Chem., 52,* 198 (1974).
$Hg(NO_3)_2(Ph_3P)$.

499. K. Aurivillius and L. Falth, *Chem. Scr., 4,* 215 (1973).
$HgL_2P_2Ph_4Et_2S$.

Phosphate Ceramics

A. E. R. Westman

Toronto, Ontario, Canada

This monograph is dedicated to Dr. Horace B. Speakman who, as Director of the Ontario Research Foundation, made it possible for me to spend the school year 1949-1950 at the University of Cambridge, England, where I started my researches on the chromatography of phosphates.

CONTENTS

This contribution was written largely from the viewpoint of the physics and chemistry of ceramic materials. A few methods of preparation of phosphates are included, but methods of manufacture, analytical details, and descriptions of equipment are not. The latter are readily available [855]. Also, the subject matter was restricted to what might be called traditional ceramics. Specialized information on electronic and nuclear materials was omitted because it was not possible to cope with these rapidly expanding fields in this chapter. Hopefully they may be covered in another volume in this series.

The references for this contribution were compiles largely from *Ceramic Abstracts* over the period 1918-1973, inclusive, were submitted by Fellows of the American Ceramic Society in correspondence with me or were culled from my own reference file. They numbered more than 1500, of which about 1000 are quoted herein. For many years ceramists were concerned with the silicates, and in fact the term "silicate chemistry" was commonly used to denote the study of ceramics. Information on phosphate ceramics was often buried in papers devoted primarily to silicates. Consequently there is much information on phosphates in the ceramic literature that is not explicitly included in the titles of the papers, and the usual approach through the decennial index was not rewarding. Therefore a method of scanning the annual subject indices of *Ceramic Abstracts* was adopted. Although laborious, this method yielded a large number of references. With the scanning method used each reference was usually picked up several times, and the method appeared effective insofar as a relation to phosphates could be gleaned from the titles of the papers.

On the other hand, *Ceramic Abstracts* covers such a wide range of journals, that many abstracts dealt with studies in mineralogy, crystallography, and so on, which were beyond the scope of this monograph and are, in fact, covered by other volumes in this series.

In the United States the term "ceramics," at one time restricted to clay products, has been broadened to include the study of glasses and nuclear materials. European practice is to use the term in a more restrictive sense. In particular, the glasses are usually considered by separate societies and in separate journals. The title of this monograph indicates that it is written from a ceramics point of view. The phosporic acids and their polmers are included partly because they show a close relation scientifically to the alkali metal phosphates and polyphosphates and partly because they have achieved some importance in ceramics as bonding and stabilizing materials, particularly in refractories.

The subject of ceramics reaches far back into antiquity, but it might be thought that the use of phosphates would be a comparatively recent development. This to some extent is true. However, small amounts of phosphates have been found in the earliest Egyptian glasses, bone ash from calcined deer bones was available to the early potters, and certain mineral phosphates such as aluminum trimeta-phosphate and various apatites were also used. With the development of the modern phosphate industry a large number of phosphates and phosporic acids have become available and have found their way into ceramic products, notably optical, opal, and semi-conducting glasses, refractory cements, refractories, phosphors, luminors, and nuclear materials. The traditional use of bone ash in English bone china may be contrasted with the modern development of dental and other prosthetic materials based on apatite structures, which is now in progress. The use of phosphates in vitreous enamels has been studied much more extensively in Europe that in the United States, mostly as a replacement for borax.

In spite of these modern developments the tonnage of phosphates used in ceramics is small compared with silicates, which are much cheaper, less volatile, and, in general, more resistant to moisture. As a result, the literature on phosphates in ceramics is less accessible, for the development of knowledge in this field has not been on a planned or logical basis but has tended to expand rapidly in certain directions in which commercial application appeared warranted and to suffer neglect in other directions that might have been of more interest from a scientific viewpoint. However, the contribution of phosphate studies to an overall understanding of ceramic materials, silicones, and borates has been considerable, and in certain cases, for example, glasses, such studies have enabled goals to be reached which could have been achieved only with great difficulty or not at all by other approaches.

A review entitled "Phosphates in Ceramic Ware" was published 30 years ago by Weyl and Kreidl [909] in a series of four articles that dealt with opal glasses, bone china, phosphorus compounds as reducing and refining agents in glasses, and phosphate glasses. Some 66 references were cited. The only review approaching this scope that has been published in the intervening years according to the present search of the literature is one by German [309].

In addition to the Weyl-Kreidl review articles mentioned above, several books or series of articles give a wide coverage of phosphate chemistry and physics and contain information relevant to the topics discussed in this monograph. Many references could have been made to them in the text. Instead they are listed in Table I.

TABLE I. General References

Authors	Titles	Refs.
Corbridge, D. E. C.	*The Structural Chemistry of Phosphorus Compounds*	191
Corbridge, D. E. C.	*The Infra Red Spectra of Phosphorus Compounds*	192
Eitel, Wilhelm	*Silicate Science II*	250
German, W. L.	*The Application of Phosphates in Ceramic Industries*	309
Kalliney, S. Y.	*Cyclophosphates*	442
Liebau, Friedrich	*The Crystal Chemistry of Phosphates*	533
Mackenzie, J. D.	*Modern Aspects of the Vitreous State, Vol. 1*	541
Morey, G. W.	*Properties of Glass*	578
Osterheld, R. K.	*Nonenzymic Hydrolysis at Phosphate Tetrahedra*	647
Thilo, Eric	*Condensed Phosphates and Arsenates*	815
Thilo, Eric	*Chemistry of Condensed Phosphates and Arsenates*	435
Van Wazer, J. R.	*Phosphorus and Its Compounds, Vol. I*	855
Van Wazer, J. R.	*Industrial Chemistry and Technology of Phosphorus and Phosphorus Compounds*	858
Van Wazer, J. R.	*Structure and Properties of Condensed Phosphates*	611
Van Wazer, J. R.	*Principles of Phosphorus Chemistry*	573
Volf, M. B.	*Technical Glasses*	880
Weyl, W. A.	*Phosphates in Ceramic Ware*	904
Weyl, W. A.	*Colored Glasses*	912
Weyl, W. A., and E. C. Marboe	*Constitution of Glasses—A Dynamic Interpretation*	817

In the general plan of this monograph a short review of the phosphate chemistry with a bearing on phosphate ceramics (Section I to III) is followed by sections on phosphate ceramics science, (Sections IV and V); then a review of the application of phosphates in ceramics technology uses a classification based largely on the subject divisions of the American Ceramic Society.

I. Phosphates and Phosphoric Acids—General

A. Compositions and Monenclature

Phosphate nomenclature has only recently reached a state of reasonable standardization. This was partly because of a misunderstanding of the nature of phosphates. Consequently, difficulty is encountered in interpreting some of the earlier work. In this monograph we are concerned largely with inorganic monomers and their condensation polymers for which a considerable degree of standardization has been reached. The latter substances can be divided into linear or chain polymers, cyclic or ring polymers, and branch chain polymers.

Linear or chain polymers can be given the general formula.

$$M_{(n+2)}P_nO_{(3n+1)} \tag{1}$$

where M is a monovalent cation and $n = 0, 1, 2, 3, 4, ...$ The number n denotes the number of phosphorus atoms in the molecule or the "chain length." For $n = 0$ the formula relates to oxides such as H_2O and Na_2O. For $n = 1$ the prefix "ortho" is used. Thus we have orthophosphoric acid, sodium orthophosphate, and so on. For $n = 2$ the prefix "pyro," is used. Thus we have pyrophosphoric acid, sodium pyrophosphate, and so on. For higher values of n the prefixes "tripoly and tetrapoly," are used to give tripolyphosphoric acid and sodium tripolyphosphate.

Cyclic or ring polymers can be given the general formula where M is a monovalent

$$(MPO_3)_n \tag{2}$$

cation and $n = 3, 4, 5, ...$ In this case for $n = 0$ the formula has not meaning and $n =$ 1 or 2 gives formulas for which no compounds have been found. For the other values of n the prefix "meta" is used. Thus we have trimetaphosphoric acid and sodium trimetaphosphate.

Although the above formulas are written in terms of a monovalent cation for convenience, divalent and higher valent cations are encountered with these anions and the formula is adjusted accordingly.

Alternatively, the symbol M may represent one equivalent of cation. It should be noted that although the above system of nomenclature has been widely accepted the system recommended by the International Union of Pure and Applied Chemistry [416] does not use the syllables "poly" and "meta" to distinguish between unbranched chain and ring polymers, respectively. Unbranched chain polymers bear no special designation (although "Catena" may be used) and the syllable "cyclo" is applied to ring polymers. A few examples are given in Table II.

TABLE II. Comparison of Usual and IUPAC Nomenclature

Formula	Usual Designation	IUPAC Designation
Na_3PO_4	Sodium orthophosphate	Sodium monophosphate
$Na_4P_2O_7$	Sodium pyrophosphate	Sodium diphosphate
$Na_5P_3O_{10}$	Sodium tripolyphosphate	Sodium triphosphate
$(NaPO_3)_3$	Sodium trimetaphosphate	Sodium tricyclophosphate
$(NaPO_3)_4$	Sodium tetrametaphosphate	Sodium tetracyclophosphate

In this article the "usual" nomenclature has been adopted to facilitate the use of the information in the references given.

When n is large, formula (1) approaches formula (2) in ultimate or "metaphosphate composition." A sodium glass of this composition, historically known as "Graham's salt." actually has about 7% of its phosphorus in the ring compounds trimeta- and tetrametaphosphate and some larger rings [860a]; the remainder are long chain polyphosphates.

A certain amount of branching is also found as the metaphosphate composition is approached. Van Wazer [861] and Strauss [789], however, have shown that the branching points are readily attacked by water so that in solution branching disappears. This follows from the antibranching rule propounded by Van Wazer (861), which states that when *three* of the oxygens of a PO_4 tetrahedron are shared with other PO_4 tetrahedra essentially all the π character is confined to the nonlinking oxygen. This lack of distribution of π character reduces the resonance energy and lowers the stability.

Because an SiO_4 tetrahedron does not include a nonlinking oxygen, the antibranching rule would not be expected to apply to silicates, and indeed it appears that no such rule has been postulated for silicates. This should be taken into consideration when comparing the constitution of phosphate and silicate simple binary glasses.

When the M/P atomic ratio is less than that corresponding to the metaphosphate composition, the substances are called "ultraphosphates." They have been relatively little explored. Being highly branched, they react vigorously with water so that it is difficult to secure them unchanged in solution and to study them by the methods of filterpaper and column chromatography which have been so useful in investigating the other phosphates.

B. Structure

The structural chemistry of phosphorus compounds has been exhaustively reviewed, to 1964, by Corbridge [191] and their infrared spectra described in detail by the same author [192]. P^{31} NMR studies are reviewed by Van Wazer his associates [862]. The fundamental structural unit of the phosphates is the PO_4 tetrahedron, which consists of a central phosphorus atom surrounded by four oxygen atoms. In writing chemical structural formulas for the phosphates it is customery to indicate five valence bonds with two of them going to one oxygen. It appears likely that the double bond character is smeared over all four bonds to a considerable extent unless one oxygen is terminal; for example there is good X-ray evidence that one of the P—O bonds in orthophosphoric acid is shorter than the other three.

The PO_4 are joined together by sharing corner oxygens to form the various types of polymer described above. The following chemical structural formulas illustrate

Linear or chain

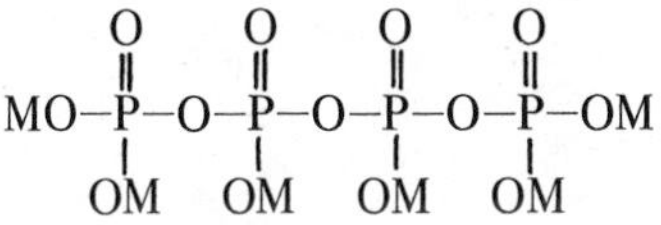

$M_6P_4O_{13}$ Tetrapolyphosphate

1

Cyclic or ring

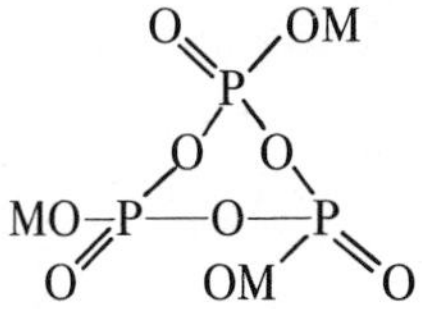

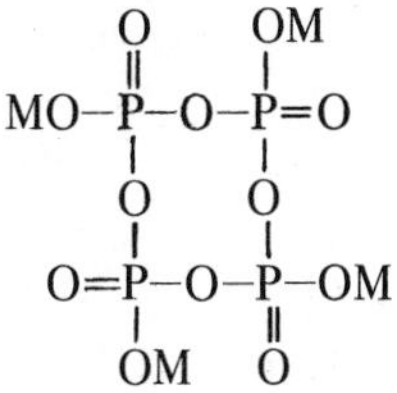

$(MPO_3)_3$ Trimetaphosphate $(MPO_3)_4$ Tetrametaphosphate

2 *3*

Branched chain

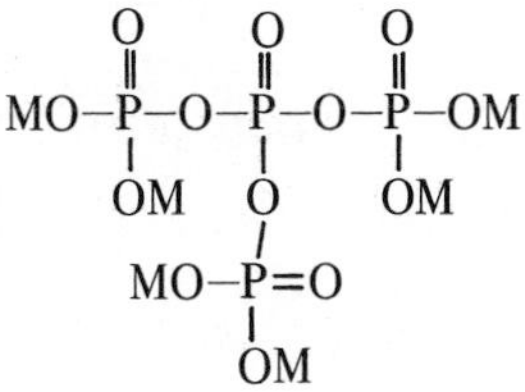

$M_6P_4O_{13}$ Branched chain tetraphosphate

4

the role played by the PO_4 tetrahedron. It will be noticed that the branched-chain tetraphosphate has the same empirical formula as the linear tetrapolyphosphate.

The actual arrangement of the atoms in the polyphosphate chain molecule can be illustrated by Figure 1, which shows the structure of the low-temperature form of

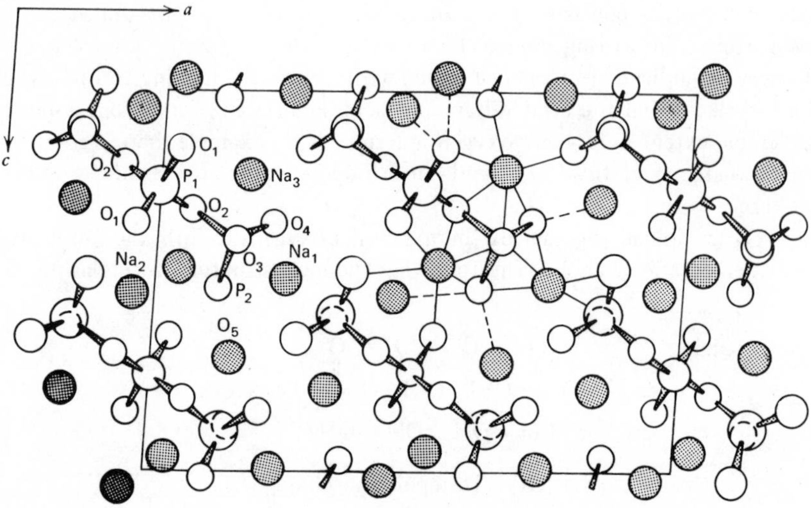

Fig. 1. Structure of sodium tripolyphosphate, phase II, shown in *b* projection. Coordination of sodium ions shown by full lines and coordination of oxygen atoms completed by broken lines. After Davies and Corbridge, *Acta Cryst.* [211].

sodium tripolyphosphate determined by Davies and Corbridge [211]. The tripolyphosphate chain is clearly one of the building units in the crystal. The P—O bond lengths are given as chain 1.61 ± 0.03 Å (inner), 1.68 ± 0.03 (outer), and terminal, 1.50 ± 0.03 Å. When the crystal is dissolved in water, these chains maintain their identity and can be shown to be present by filter-paper chromatography and other chromatographic methods. At room temperature and near neutrality they hydrolyze at a very slow rate to form pyrophosphate chains and orthophosphate molecules. Tripolyphosphate crystals can be dissolved and recrystallized almost indefinitely without affecting the tripolyphosphate chains [689].

C. *PO_4 Tetrahedron*

The PO_4 tetrahedron plays an important role in the structure of the phosphates just as the SiO_4 tetrahedron does in the silicates, and there are many resemblances between silicate and phosphate chemistry. On the other hand, there are striking differences; for example P_2O_5 reacts violently with water but SiO_2 is nearly insoluble. Also, silicates in aqueous solution tend to polymerize and settle out of solution, but polyphosphates tend to hydrolyze and revert to orthophosphate under the same conditions. Thus the P–O–P bridges in the polyphosphates must be considered more reactive under these conditions than the Si–O–Si bridges in the silicates. As a result of filter-paper and column chromatography of the soluble polyphosphates and metaphosphates as well as IR and P^{31}NMR studies, their structures and properties are well understood. We have no comparable wealth of information about the silicates. It appears, however, that the phosphates are more likely to form linear polymers than the silicates. The latter may also be more prone to branching [33].

In the ceramic field, too, this similarity and contrast between phosphates and silicates is also found; for example, P_2O_5 bears little resembalnce to SiO_2 in any of its properties. $AlPO_4$, however, bears a remarkable resemblance to SiO_2, for example, SiO_2 can exist in a number of crystal forms such as alpha quartz, beta quartz, cristobalite, and tridymite. $AlPO_4$ can go through a similar series of transformations so that one has a cristobalite form of $AlPO_4$, and so on.

In general in ceramics the phosphates melt at lower temperatures than the silicates and melt to much less viscous liquids. Thus they may be used as fluxing and bonding materials. They are also more subject to devitrification and volatilization.

The phosphorus-oxygen tetrahedron can be introduced into ceramic products in a number of ways. Phosphorus pentoxide itself is seldom used except in laboratory experiments because in its usual form it is hygroscopic and reacts violently with water or alkalis. Commercial strength (85% H_3PO_4) orthophosporic acid is often used when a liquid is convenient. Bone ash traditionally serves as a source of calcium orthophosphate. Superphosporic acid is used in refractory cements for gunning purposes. A considerable variety of orthophosphates is used to introduce other elements in addition to phosphorus. They include $AlPO_4$, BPO_4, Na_3PO_4, and minerals such as apatite. Unfortunately, references to phosphate material in the literature are often ambiguous; for example, the term "aluminum phosphate" is sometimes used without qualification and probably means aluminum orthosphate. It may also be aluminum trimetaphosphate, a material much favored by ceramists because it introduces alumina, which helps to prevent devitrification and also increases water resistance. A similar result may be obtained by the use of mono-aluminum phosphate, $Al(H_2PO_4)_3$.

Because ceramic materials are usually fired to a relatively high temperature, their properties are determined by their final compositions and ordinarily do not depend on the raw materials used to achieve these compositions.

It has been pointed out that in comparison with the silicates phosphates tend to melt to liquids of lower viscosity and are therefore more prone to devitrification. They are also more prone to attack by water. There appears to have been a considerable amount of development work on the use of phosphates in ceramics which has been abandoned when one or the other of these difficulties has been encountered. Much of this work has not been published. On the other hand, there have been notable cases, for example, bone china, gunning cements, opal glass, and certain

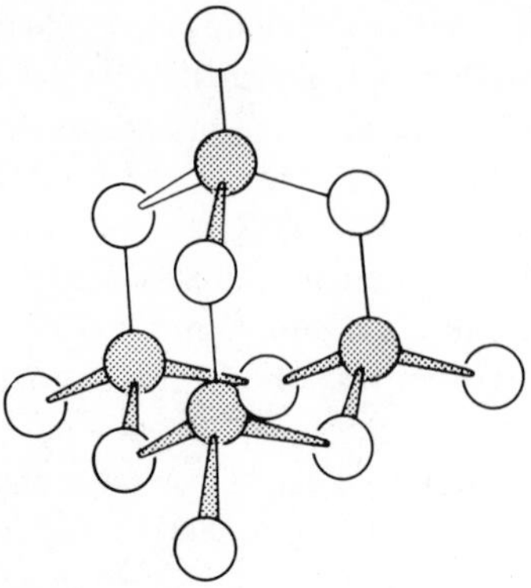

Fig. 2. Molecular structure of P_4O_{10}. From Figure 1.4(*c*), p. 72, Vol. 3, *Topics in Phosphorus Chemistry.*

optical and semiconducting glasses, in which the phosphate compositions have acheived a considerable success. Actually, some of the optical glasses based on phosphates have proved to be superior to the corresponding silicate glasses in resistance to moisture.

In the form of a molecular gas phosphorus pentoxide consists of cage type molecules, each containing four PO_4 tetrahedra, with the phosphorus atoms situated at the corners and having the formula P_4O_{10} (Figure 2). This may be represented by the two-dimensional structural formula (Figure 3). Phosphorus pentoxide in

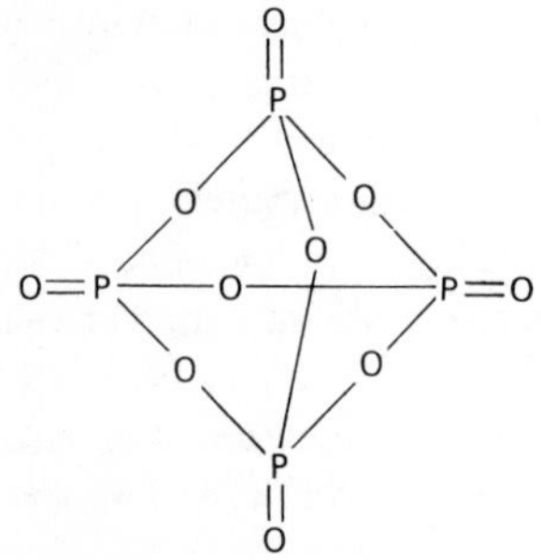

Fig. 3. Structural formula of P_4O_{10}. From Figure 1.5, p. 75, Vol. 3, *Topics in Phosphorus Chemistry.*

the crystalline state can occur in several forms [855]. Hydrolysis of the P_4O_{10} molecule leads in turn to tetrametaphosphate, tetrapoly-phosphate, tripoly-phosphate and orthophosphate, pyrophosphate and orthophosphate, and finally to all orthophosphate. Depending on circumstances, the long-chain phosphates may hydrolyze by splitting off ortho groups as indicated or be converted to trimetaphosphate [338].

D. Structural Theories

In the ceramic literature three general approaches will be found to the structure of phosphate compounds and glasses based on the phosphorus-oxygen tetrahedron: the molecular-orbital, the crystal chemical, and the geophysical.

The molecular-orbital point of view, which is favored by phosphate chemists, regards the PO_4 tetrahedron as a structure held together by covalent bonding. There is a basic bond structure consisting of four sigma bonds due to sp^3 hybridization, going from the phosphorus atom to the four oxygens. In addition, π bonding which utilizes d electrons, is also involved. The structural formulas of compounds containing PO_4 tetrahedra are often written with one double bond per PO_4 tetrahedron, although at least some sharing of this double-bond character with the other three bonds to oxygen is implied. Van Wazer [856] specifies that when three of the oxygens of a given PO_4 tetrahedron are shared with other tetrahedra, as in Figure 2, the π character is confined to the nonlinking oxygen. In all other cases the π character is distributed between all four bonds. The terminal P—O bonds in Figure 2 are shorter than the bonds to bridging oxygens in accordance with the above ideas. This terminal P—O bond has also been considered to have some dative character, written $\overset{\frown}{O2p\pi \text{-} P}3d\pi$, to explain the shortening [161].

Many ceramists have historically adopted a different viewpoint which might be called the crystal chemistry approach and can be described as the application of crystal chemical ideas developed for purely ionic crystals to all crystals and even glasses. Emphasis is placed on atoms, which are considered to be fully ionized. Thus the PO_4 tetrahedron consists of a P^{5+} ion surrounded by four O^{2-} ions in four-fold coordination. A number of properties are assigned to these ions on the basis of crystal chemistry. They include charge, ionic radius, field strength, and polarizability. The field strength is the ratio of the charge to the square of the ionic radius, although Dietzel deviated by using the distance between atom centers [917].

A good historical and descriptive account of this approach is given by Weyl [916]. A more dynamic interpretation is given by Weyl and Marboe in their two-volume text [917]. Although all the modern physical methods such as nuclear magnetic resonance, X-ray diffraction, and infrared spectroscopy agree in establishing that the P—O bonds in the PO_4 tetrahedron are covalent in character and that the P^{5+} cations of crystal chemistry do not exist as such, nevertheless the simpler crystal chemical approach has been useful in a number of circumstances. However, by concentrating on atoms as ions and not as possible parts of molecules, it had the effect, as had the geophysical approach described below, of making it appear that molecules could not be present in a glass. Consequently the paper chromatographic evidence for the presence of polyphosphates in phosphate glasses was received with a great deal of scepticism when it first became available.

One useful attribution of crystal chemistry has been the application of ligand field theory by Douglas and his colleagues at Sheffield University. They have shown that the colors produced by iron and chromium oxides both in glasses and solutions can be understood by the use of this theory [46].

The third approach to ceramics might be called geophysical. It is exemplified by the work of the Geophysical Laboratory of the Carnegie Institute of Washington. It is an approach based on thermodynamics and the phase rule which involves the very accurate measurement of phase equilibria in condensed systems; the data are presented in phase equilibrium diagrams. For the most part the oxides involved are taken as components in the thermodynamic sense, and little attention may be paid to the internal structure of the solid phases and, particularly, the liquids. Chemical compositions are generally given by formulas that merely express the mole ratios of the component oxides. Despite these limitations, there is no doubt that this approach has provided a basis for many advances in the ceramic field, and the production of good phase equilibrium diagrams should be the first step to a study of the ceramic properties of new classes of material.

For many years only a limited number of phase equilibria relating to phosphates were available. In recent years a considerable number has been published. Unfortunately, they deal largely with compounds that are of interest from a luminor or phosphor point of view and gaps still remain in the diagrams that would be of interest in more general ceramic theory.

II. Phosphoric Oxide and Phosphoric Acids

A. *Background*

Before 1950 not much was known about phosphorus inorganic compounds. Orthophosphates, pyrophosphates, tripolyphosphates, and trimeta and tetrametaphosphates were known and the presence of higher polymers was suspected. The latter were usually lumped under the designation "metaphosphate" or "hexametaphosphate." Thus Graham's salt, which is really a glass of the metaphosphate composition $(NaPO_3)_n$, was called "hexametaphosphate," as though it were a single compound. The rapid commercial development of the phosphorus industry, the application of chromatographic methods of analysis, the development of polymer theory, and the introduction of many new physical methods to the study of materials have combined to produce a wealth of information regarding the compounds of phosphorus. This information has been reviewed comprehensively by Corbridge in chapters in Vols. 3 and 6 of *Topics in Phosphorus Chemistry* and by Kalliney & Osterheld in Vol. 7. The present discussion is largely concerned with those phosphorus compounds that are of interest in ceramics. For this reason only inorganic phosphoric compounds that are used as ceramic raw materials, appear in the products, or aid in understanding other compounds which do are included.

B. *Phosphoric Oxide*

Phosphoric oxide, P_2O_5, or phosphorus pentoxide, forms a logical starting point for this discussion. As a gas, it consists of molecules with a formula P_4O_{10} and has the structure shown in Figure 2, a cage type structure formed of four PO_4 tetrahedra, each of which shares three oxygen atoms with the other three. The bond to the fourth or terminal oxygen shows an extreme pi character by employing *d* orbitals.

P_2O_5 is found in at least three crystal forms. The H or rhombohedral form consists of discrete P_4O_{10} molecules as described above and is the usual form. In accordance with Van Wazer's antibranching rule, it is vigorously attacked by water. Its structure has been rather firmly established by a number of physical methods. In particular, the infrared spectra are consistent with the presence of P–O–P and P=O linkages.

There are two orthorhombic forms of P_2O_5. The O form is built of rings of 10 PO_4 tetrahedra, which are linked together to form a continuous three-dimensional structure. The O′ form is a high polymer consisting of infinite sheets composed of rings of six PO_4 tetrahedra, each of which forms linkages to three neighbors. The P/O/P angle is greater than in the H form but the difference between the P–O and P–O–P bond lengths is not so large. The infinite polymer nature of the O and O′ forms results in a much slower reaction with water than is found in the H form. The O form dissolves in water very slowly, even at 90°C. The O′ form also reacts more slowly than the H form, first producing a stiff gel which then slowly dissolves.

The three crystalline forms of P_2O_5 also differ in their melting behavior. The H form melts at 420°C to a metastable liquid of high vapor pressure which presumably consists of P_4O_{10} molecules and rapidly polymerizes to form a glass containing some crystals of the O form. When cooled rapidly to room temperature, this glass has a density that is higher than that of the original crystals, which is unusual because most glasses have a lower density than the corresponding crystals. The O and O′ forms melt at 562 and 580°C, respectively, to produce liquids of considerably lower vapor pressure than the initial liquid produced by melting the H form.

The melting behavior of the different forms of P_2O_5 may be contrasted with that of the different forms of silica. All are molten at 1710°C and all form a viscous liquid that can be easily cooled at room temperature to a glass that has a lower density than the crystal forms.

C. *Phosphoric Acids*

Hydrolysis of P_2O_5 can lead to orthophosphoric acid (H_3PO_4) and some of its polymers, both unbranched chains and cyclic. Higher polymers can be produced by adding P_2O_5 and heating in closed tubes with suitable precautions [404]. A detailed discussion of these points appears in a later section.

The ordinary phosphoric acid of commerce produced by the wet process or the electric furnace process contains about 76% H_3PO_4 by weight or 55% P_2O_5, the remainder being water. This acid is commonly used as a raw material in ceramics as a convenient way of introducing phosphate into enamel or refractory mixes or in making up phosphating solutions for treating steel before vitreous enameling.

"Superphosphoric acid," produced by burning phosphorus at higher rates and temperatures, contains about 76% P_2O_5 by weight, which is equivalent to 105% H_3PO_4. As we show later, it is about 49% ortho-, 42% pyro-, 8% tripoly-, and 1% tetrapoly-phosphoric acid. Although produced primarily for the fertilizer industry, in which it has several advantages, it has been found useful also in the refractories industry in the manufacture of gunning mixes. Polyphosphoric acid, 115% H_3PO_4, is also used in the refractories industry.

It is interesting to note that although we expect all the phosphates that are polymers of orthophosphate to hydrolyze at measurable rates to orthophosphate in the presence of water, yet in these strong acid solutions the reverse is true; for example, a solution with the theoretical H_3PO_4 composition will in fact have 6 mole% each of pyrophosphoric and water as a result of a dynamic equilibrium that occurs. Pure crystalline-H_3PO_4 may melt as high as 42.3°C, but the solution obtained, if given time to equilibrate, will freeze at a much lower temperature; the water and pyroacid, etc., formed depresses the freezing point of the H_3PO_4.

A detailed summary of structural information on orthophosphoric acid is given by Corbridge [191, p. 208]. Structure determinations on the pure crystalline acid of melting point 42.3°C give the structure shown in Figure 4 (Corbridge 111.6).

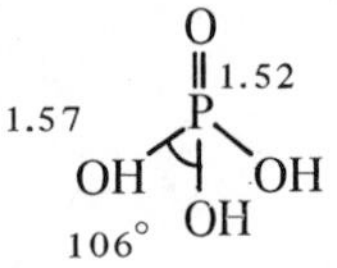

HO/P/O = 105-113°

Fig. 4. Model of H_3PO_4. After Corbridge, *Topics in Phosphorus Chemistry*, [191], p. 208.

The molecule has approximately trigonal symmetry and contains one P—O bond that is shorter than the other three. This short bond corresponds to the phosphoryl oxygen and the remaining oxygens have hydrogens attached. All four oxygens participate in hydrogen bond formation and a continuous sheet structure is built up. In concentrated (86%) solutions radial distribution curves indicate tetrahedral PO_4 groups with mostly one hydrogen bond between any pair of tetrahedra. In more dilute solutions the tetrahedra are bonded to one another indicrectly by water molecules.

III. Phosphates

A. Orthophosphates

Corbridge [191, p. 172] lists some 344 orthophosphates whose structure is known, nearly all of which are inorganic compounds. Not only are many cations and combinations of cations (alkalies, alkaline earths, etc.) included but also many combinations of anions (PO_4,SO_4VO_4, etc). Orthophosphate references in the ceramic literature are concerned largely with calcium orthophosphate [$Ca_3(PO_4)_2$], hydroxy-apatite [$Ca_{10}(OH)_2(PO_4)_6$] and a group of covalent silica-type phosphates of which $AlPO_4$ is of most interest.

Calcium orthophosphate may be used as a replacement of bone ash in the manufacture of chinaware. It may also crystallize from certain glasses or enamels to provide opalescence or opacity. Hydroxyapatite is the main constituent of bone and therefore of interest in the production of bioceramics and bone china. In the presence of traces of fluorine a fluorhydroxy apatite is likely to precipitate from a glass to produce an opalescent effect, according to Weyl [909]. The structure of apatite is described in detail by Corbridge [191, p. 188].

The covalent silica-type phosphates which include $AlPO_4$, BPO_4, $GaPO_4$, $FePO_4$, and $MnPO_4$ constitute a unique class of orthophosphates in which the phosphorus and the other kind of atom are tetrahedrally coordinated by oxygen. Alternate tetrahedra are linked together to form a continuous three-dimensional covalent structure. As an analog, silica could be written $SiSiO_4$. For this reason some authors prefer to call this compound a "mixed oxide" or the "anhydride of a heteropoly acid."

The analogy of $AlPO_4$ to silica is striking. $AlPO_4$ is known to exist in six crystal forms, each of which is paralleled by a corresponding form of silica. These six silica forms are compared with the $AlPO_4$ forms in the following table from Corbridge [191, p. 195]:

$$\text{Quartz} \underset{867^\circ C}{\rightleftharpoons} \text{Tridymite} \underset{1470}{\rightleftharpoons} \text{Cristobalite} \underset{1713}{\rightleftharpoons} \text{Melt}$$

$$SiO_2 \quad \beta \underset{573^\circ C}{\rightleftharpoons} \alpha \quad \beta \underset{117}{\rightleftharpoons} \alpha_1 \underset{163}{\rightleftharpoons} \alpha_2 \quad \beta \underset{220}{\rightleftharpoons} \alpha$$

$$\text{Berlinite } 705 \rightleftharpoons \text{(Tridymite) } 1025 \rightleftharpoons \text{(Cristobalite)} > 1600 \rightleftharpoons \text{Melt}$$

$$AlPO_4 \quad \beta \underset{586^\circ C}{\rightleftharpoons} \alpha \quad \beta \underset{93}{\rightleftharpoons} \alpha_1 \underset{130}{\rightleftharpoons} \alpha_2 \quad \beta \underset{210}{\rightleftharpoons} \alpha$$

In this table the changes from quartz to tridymite to cristobalite to melt involve major changes of structure and are sluggish. The alpha to beta changes in structure are minor and take place more readily. Although silica melts to a viscous liquid, nevertheless it is possible to explore the equilibrium between the liquid and cristobalite and to fix the liquidus at 1713°C. In $AlPO_4$ the exact mechanism of melting comes into question. To parallel silica exactly the pairing of AlO_4 and PO_4 tetrahedra would have to persist through the melting process. On the other hand, it is possible that the pairs break apart or interchange below or at the melting point. Literature references appear to be contradictory. Thus Weyl and Kreidl in 1941 [909] refer to $AlPO_4$ in the glassy state as having the same structure as silica glass, except that the centers of the tetrahedra are not formed by silicon ions but alternately by aluminum and phosphorus ions. Weyl and Marboe, however, in 1961 [916] refer to the experiments of Dietzel and Poegel in 1953 [221] and conclude that at 2000° C Al^{3+} and P^{5+} become free to exchange positions, P_4O_{10} molecules escape, and Al_2O_3 precipitates from the melt. For this reason vitreous $AlPO_4$, which could be called the analog of vitreous silica, does not exist. In a later study (1962) Dietzel and Hinz [220] found that $AlPO_4$ begins to decompose at 1600°C. On rapid heating it melted at 2000°C. A quenched melt showed 90% of the solid in the cristobalite form and flame-sprayed material showed 50 to 58% in the cristobalite form. No glassy $AlPO_4$ was observed, which indicates that the liquid formed by melting $AlPO_4$ must be less viscous than that formed by melting silica.

Beekenkamp and Stevels [65], in studying the structure of some glasses of the composition $M^{111}M^{V}O_4$, tried to include $AlPO_4$, but it decomposed below its melting point and lost P_2O_5. It was possible to melt BPO_4 and some combinations of BPO_4 and $AlPO_4$, but, due to devitrification, no glasses could be obtained on quenching even to 78°K. The composition $75BPO_4$, $25AlPO_4$ (mole%) could be melted but would not vitrify. Addition of 5 mole% SiO_2 enabled glasses to be made. They attributed Kumar's wider range of glass-forming compositions in the B_2O_3-Al_2O_3-P_2O_5 system [510] to silica dissolved from the crucibles he used. It appears that, if $AlPO_4$ could be melted without decomposition, possibly in a P_4O_{10} atmosphere, it would melt at about 2000°C, that it starts to decompose at about 1450°C, and that efforts to produce it in glassy form have so far been unsuccessful.

Among the orthophosphates listed above, which have a $M^{111}M^{V}O_4$ formula, only $AlPO_4$ has been found in an almost complete set of silica structures. YPO_4, $ScPO_4$, and a number of similar compounds have zircon-type structures. The orthophosphates of molybdenum and tungsten have structures based on tetrahedral PO_4 units and octahedral WO_6 and MoO_6 groups.

The orthophosphate $Al(H_2PO_4)_3$, called monoaluminum phosphate, has been recommended as a convenient bond for refractories and has been patented. It disperses in water to give viscous adhesive solutions [333, 334, 335].

B. Condensed Phosphates

The condensed phosphates are chemical compounds which can be considered as arising from the condensation polymerization of orthophosphate, the latter a monomer. Examples of such condensation reactions are shown by the chemical equations (3) and (4). In the first equation the product is a pyrophosphate; in the second, either

$$2Na_2HPO_4 = Na_4P_2O_7 + H_2O \tag{3}$$

$$nNaH_2PO_4 = (NaPO_3)_n + nH_2O \tag{4}$$

a long (theoretically infinite) unbranched chain phosphate or, for n between 3 and about 12, a metaphosphate such as trimetaphosphate.

The condensed phosphates have a long history, starting with the discovery of the glass of the metaphosphate composition by Graham in the 1830s. This was not recognized as a glass and is still referred to as "Graham's salt." Reviews which include the early history of condensed phosphates have been given by Thilo [815], Van Wazer [855, 859], and others. Recent reviews of condensed phosphates include Corbridge [191, 192] and Haiduc [347].

For nearly 120 years after their discovery the chemistry of the condensed phosphates was greatly handicapped by the lack of analytical and separation methods that could deal with mixtures of a number of molecular species of phosphates. Ortho-, pyro-, tripoly-, and possibly trimetaphosphates could be determined. The remainder was usually called "metaphosphate" or even more misleadingly, "hexametaphosphate," with a formula $(M^IPO_3)_6$.

About 1950 the analytical handicap was largely overcome by the introduction of paper chromatography, by means of which linear or unbranched chain phosphates could be determined up to a chain length n (the number of phosphorus atoms per chain) of about 14. In addition, the metaphosphates (cyclophosphates) could be clearly differentiated from the linear phosphates.

The linear condensed phosphates with $n > 4 < 10$, which could be readily separated and determined by paper chromatography, were given the name "oligophosphates" by Thilo, oligo meaning "few." Very long chain phosphates in which the M^1/P ratio is in the neighborhood of unity are sometimes called metaphosphates, although it would be better to save that name for ring phosphates. Alternatively, they may be called HMW, or high molecular weight, phosphates. When the M^1/P ratio is less than unity, the term ultraphosphate is used.

As mentioned earlier, all the condensed phosphate species have skeletons that are formed by linking together PO_4 tetrahedra with shared oxygen atoms so that the skeletons have alternating P and O atoms, whether in chains or rings.

C. *Pyro-, Tripoly-, and Tetrapolyphosphate*

Pyro- and tripolyphosphates are well-known articles of commerce; the latter are used in large quantities as a builder for heavy-duty detergents. Both have complexing and deflocculating properties and may be used in preparing ceramic slips. The structure of the tripolyphosphate is shown in Figure 1. Their phase relations are discussed in Section IV. Some high lead compounds in this system were described later by von Hodenberg, in 1972 [390].

Proof of the existence of tetrapolyphosphate was first shown by paper chromatography [903]. Crystalline sodium tetrapolyphosphate may have been obtained on one occasion by using a method developed by Ostwald to obtain carbonate-free sodium hydroxide, but it appears to be unstable at room temperature and breaks down readily to other phosphates [899]. Crystalline ammonium tetrapolyphosphate can be made readily [899], and quite stable crystals of tetrapolyphosphate can be prepared by using stronger bases such as acridinium and guanidinium [689], as shown by Quimby. The tetrapolyphosphate appears as a stable crystalline phase in the $PbO - P_2O_5$ system, as shown by Langguth, Osterheld, and Karl-Kroupa [522]. A recent discussion of the forces binding the tripoly anion was published in 1971 by Pampuch and Gallus-Olender [651].

D. *Oligophosphates*

The oligophosphates $n > 4$ are principally of interest in connection with the binary phosphate glasses and polyphosphoric acids which are discussed in Section III. As a class they are difficult to obtain in crystalline form and seldom appear in phase equilibrium diagrams. An exception is trömelite, which is a calcium hexapolyphosphate, $Ca_4\,P_6O_{19}$ [925]. Trömelite was reported earlier as a pentapolyphosphate, $Ca_7\,(P_5O_{16})_2$, and is so indicated in Figure 6.

E. *Macromolecular Polyphosphates*

The term "macromolecular poyphosphates" may be used to describe long-chain phosphates, usually in the range of chain lengths of 20 to many thousands. They may result from the thermal condensation of dihydrogen monophosphates according to the equation

$$n(M^1H_2PO_4) \;=\; HO{-}(M^1PO_3)_n{-}OH + (n{-}2)H_2O \qquad (5)$$

the OH groups being at the end of the chains and the number average chain lengths depending on the temperature and the water-vapor pressure according to the equation given by Thilo [p. 44, 815]:

$$\bar{n} \;=\; \exp\left(8.87 - \frac{5100}{RT} - \frac{\log_e p_w}{2}\right) \qquad (6)$$

in which p_w is the water-vapor pressure in torrs.

Most interest attaches to macromolecular polyphosphates in which the M^1/P ratio is in the neighborhood of unity, this having the same overall composition as the metaphosphates or cyclophosphates. In the sodium compounds, in addition to

Graham's glass, which does contain some ring compounds and is soluble in water, four crystalline compounds with the metaphosphate composition have been given the designations $NaPO_3I$, $NaPO_3II$, $NaPO_3III$, and $NaPO_3IV$ in phase equilibrium studies. $NaPO_3I$, which is a trimetaphosphate, exists in two other crystalline forms designated I′ and I″. These six crystal forms, having the composition indicated by $NaPO_3$, can be related as shown in Figure 5, derived from a similar diagram by Van

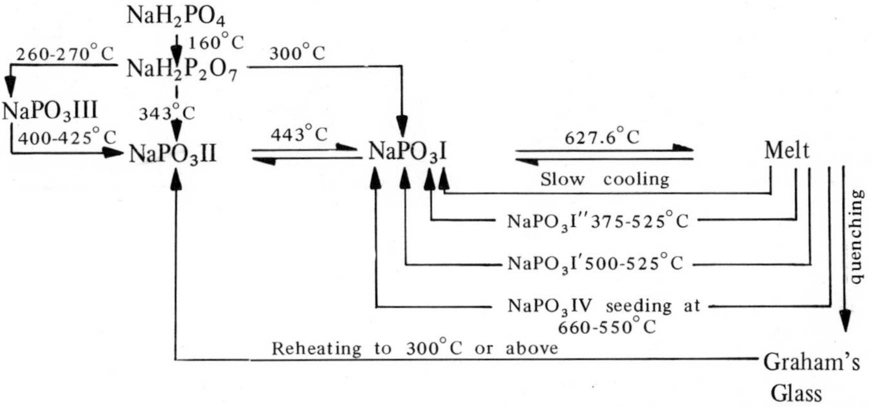

Fig. 5. Condensation Products of NaH_2PO_4. All Forms can be brought to room temperature readily. Forms I, I′, and I″ are trimeta-(tricyclo)phosphates and dissolve readily in water to give supposedly identical solutions. Forms II, III, and IV (sodium Kurrol's salt) are long-chain phosphates which are relatively insoluble in water but dissolve in salt solutions containing cations other than sodium, in the cold, by warming and by boiling respectively, to give highly viscous solutions. After Thio, *Adv. Inorg. Chem. Radiochem.*, [815].

Wazer [859]. They can all be considered to have been formed by the condensation of monosodium dihydrogen orthophosphate.

As indicated, forms II, III, and IV, which are long-chain phosphates, have the unusual property of being relatively insoluble in water, despite the fact that they are sodium salts. This may be because they consist of long chains that can crystallize by a sort of zipper mechanism which releases a considerable lattice energy that precludes solution. The equally unusual property of dissolving in salt solutions containing a cation other than sodium is likely to be due to an ion exchange mechanism that allows the chains to separate individually.

Although all these long-chain crystalline phosphates are formed by PO_4 tetrahedra that share oxygen atoms, a number of arrangements of the tetrahedra are still possible. Some are illustrated in Figure 6, taken from Thilo [815]. The helix arrangement of Kurrol's salt (Figure 6*d*) is of particular interest because it indicates how such a chain could rearrange to form trimetaphosphate, a process discussed later under "hydrolysis."

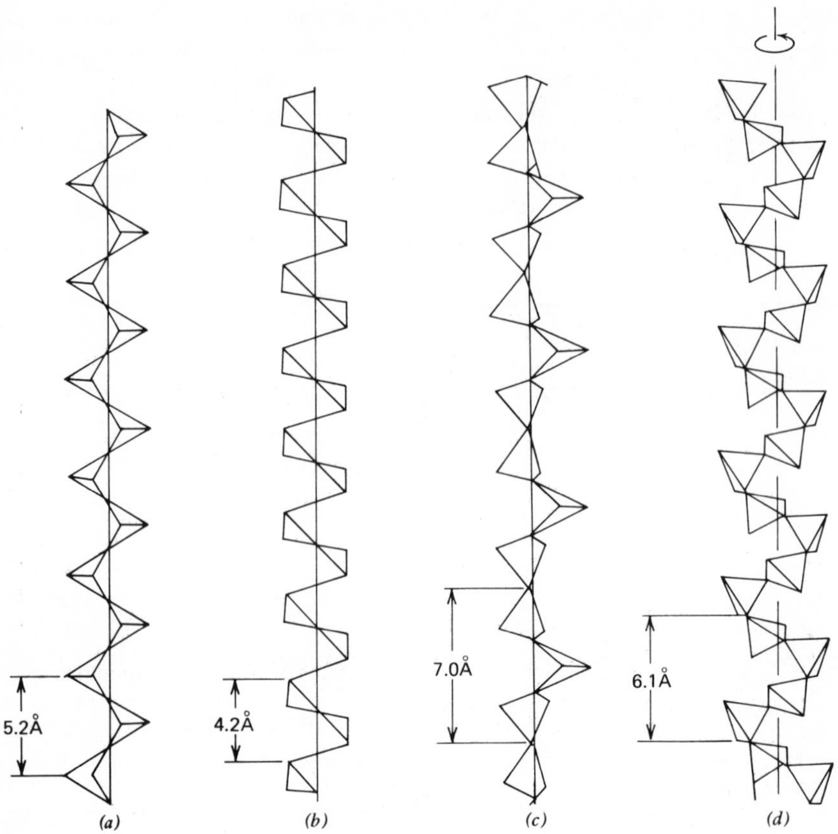

Fig. 6. Various chain types in crystalline macromolecular phosphates: (*a*) $(LiPO_3)_n$, low-temperature form; (*b*) $(RbPO_3)_n$; (*c*) $(NaPO_3)_n$, Maddrell's salt, high temperature form; (*d*) $(NaPO_3)_n$, Kurrol's salt. From Thilo, *Adv. Inorg. Chem. Radiochem.*, [815].

Methods of studying these long-chain phosphates in solution include end-group titration, precipitation by organic solvents, and paper chromatography.

In the titration of any straight-chain phosphate in solution, with a base, it is found that a strongly acid hydrogen is associated with each phosphorus atom along the chain and a weakly acid hydrogen at each end of the chain. By titration between pH 4.5 and 9 it is possible to determine the number of end-group PO_4 tetrahedra and thus arrive at a number average chain length (Van Wazer [858]).

By adding acetone to an aqueous solution of a long-chain phosphate it is possible to produce a fractionation based on chain length, the number average chain length of each fraction being determined by end-group titration (Van Wazer [853]).

Finally, by two-dimensional paper chromatography it is possible to determine the difference ring phosphates which may be present in a solution of Graham's salt; for example, in which the chains are too long to be separated from one another.

F. *Metaphosphates (Cyclophosphates)*

The metaphosphates consist of rings of alternating phosphorus and oxygen atoms. The smallest ring found is the trimetaphosphate ring. Sodium trimetaphosphate may be written $Na_3(P_3O_9)$ to indicate the trimeta ring, but is often simplified to $(NaPO_3)_3$. In phase equilibrium diagrams it is shown as $NaPO_3I$. The most interesting trimetaphosphate in ceramics is the aluminum trimetaphosphate, which is used as a raw material to introduce phosphate and aluminum oxide into the product. The tetrametaphosphate, particularly the aluminum compound $Al_4(P_4O_{12})_3$, has been studied. X-ray studies show that the crystal consists of complexes containing four PO_4 tetrahedra, each of which shares two corners with other tetrahedra and the other two corners with AlO_6 octrahedra. Each AlO_6 octahedron shares its six corners with six P_4O_{12} complexes [659].

True crystalline sodium hexametaphosphate (cyclohexaphosphate) was recovered from Graham's glass by Thilo and Schulke [817], true sodium and lithium hexametaphosphate have been synthesized by Griffith and Buxton [337], and rings as high as the octametaphosphate have been separated and synthesized. None of these higher rings appears to have been of ceramic interest.

A review of the metaphosphates has been published recently by Kalliney under the title "Cyclophosphates" in Vol. 7 of *Topics*.

G. *Hydrolysis*

All the condensed phosphates are subject to hydrolysis in solution, the final product under the right conditions being the orthophosphate. This might appear to have little interest for ceramists. It must be kept in mind, however, when working with phosphate materials. It also has a bearing on the tendency of phosphate glasses to be attacked by moisture, a weakness that has been a handicap for these glasses in the past. It is possible that the same sort of chemical mechanism underlies the attack of moisture on glasses that governs the hydrolysis of condensed phosphate ions in solution. The hydrolysis reaction is the reverse of the condensation reaction given in (3). Essentially it is the disruption of a P—O—P bridge by a water molecule. The same sort of attack of a water molecule on an Si—O—Si bridge is sometimes postulated for silicate glasses in explaining the effect of moisture on their "static fatigue." Static fatigue is the name given to a phenemenon in which a glass, when placed under a constant stress, gradually weakens and breaks. This phenomenon is strongly influenced by moisture and does not occur at liquid nitrogen temperatures, when all moisture is frozen.

The rate of hydrolysis of a condensed phosphate is strongly influenced by the pH.

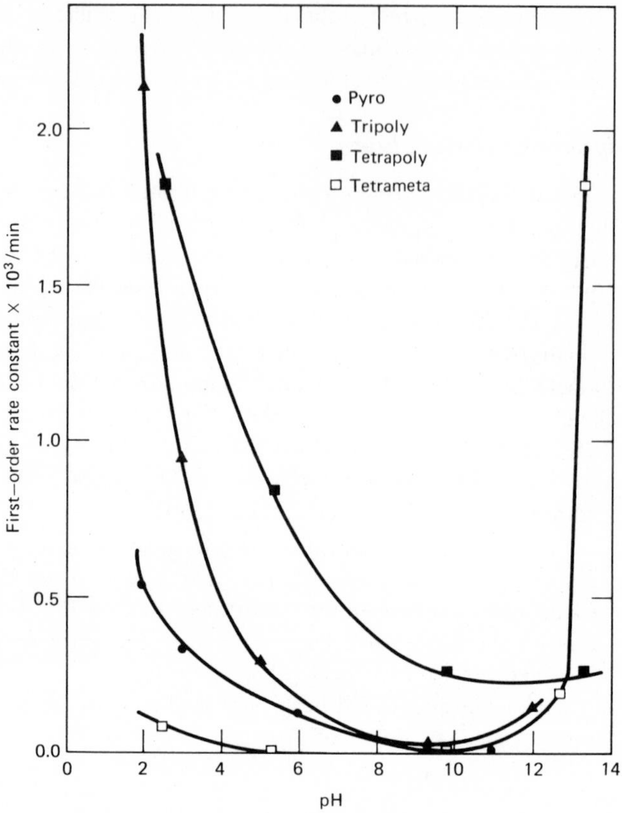

Fig. 7. Effect of pH on the rates of hydrolysis of sodium phosphates. After Crowther and Westman, *Can. J. Chem.* [201].

If the pH is kept constant, first-order kinetics are found to apply. Some rate constants are shown in Figure 7. The linear phosphates are attacked more rapidly under acid conditions and the ring phosphates, under basic conditions. The rapid attack of the tetrametaphosphate under highly basic conditions was shown later by Van Wazer to be due to a sodium ion effect [865].

A complication in studying the hydrolysis of linear phosphates is a conversion to trimetaphosphate which may occur at the same time, particularly in long-chain phosphates near the neutral point. This has been investigated extensively by Thilo [815] and by Griffith and Buxton [338]. This process is not associated with the hydrolytic cleavage of P—O—P bonds, as assumed originally.

H. Ultraphosphates

Ultraphosphates may be defined chemically as species with a lower cation–to–phosphorus ratio than the corresponding metaphosphate composition. In the Na_2O–

P_2O_5 system this means an Na/P ratio of less than one. Ultraphosphates are sometimes defined as having a branched chain structure, but, as shown in Section IB, these are not equivalent definitions. As defined chemically, the ultraphosphates are expected to have branched chain or crosslinked structures. When placed in water, the branching points, which are attacked readily in accordance with Van Wazer's antibranching rule, produce mostly linear phosphates and orthophosphate.

Although ultraphosphate compounds have been reported for the Na_2O–P_2O_5 system, Corbridge [191] concludes that they are not ultraphosphates. The ultraphosphates $CaO.2P_2O_5$ and $2CaO.3P_2O_5$ appear as primary phases in the CaO-P_2O_5 system, shown in Section IV, but they have been questioned by Liebau [533, p. 275]. The ultimate ultraphosphate is P_2O_5 itself, whose various structures have already been discussed in Section II.

In the sodium phosphate glasses, as we approach the metaphosphate composition, there is already some branching taking place. This was studied by Strauss and Treitler [789]. In glasses of higher P_2O_5 content than the metaphosphate composition ultraphosphates undoubtedly are present. A study of these glasses would be of great interest because the silicate glasses of commercial importance have high silica content and might be expected to have much more branching and crosslinking than the sodium phosphate and similar glasses that have been most studied to date.

Unfortunately, the study of alkali metal ultraphosphate glasses presents many difficulties, chief of which is that they react with water to cause local overheating and distortion of the results of chromatographic investigation. A few preliminary experiments in my laboratory in 1958 [756] with glasses made from P_2O_5 and H_3PO_4 showed that the chromatographic analysis of the solution of the glass depended on the solvent used to dissolve it. The widest distribution of polymer chain length was obtained when the glass was dissolved in an ammonium acetate solution. More research is needed in this direction. If we assume that vanadium can substitute for phosphorus in crystal structures, then VPO_5 can be classified as an ultraphosphate. The structure of this compound has been studied recently by Gopal and Calvo [321], Ross and Snyder [722], and Prideaux and Martin [686].

I. Comparison of Phosphates and Silicates

The crystal chemistry of the crystalline phosphates was extensively reviewed by Liebau [533] up to 1965, with 151 references. At the same time, Liebau made an interesting comparison of phosphates and silicates. Although their crystal chemistry was quite similar, they differed in several important respects. In silicates an Si/O ratio of 1:2 was readily reached, as in SiO_2, but in phosphates the maximum P/O ratio was 1:1.25, as in P_2O_5. The P–O bonds in P–O–P bridges in phosphates were calculated from electronegativities to have 30% ionic character. The corresponding Si–O bond was calculated to be 37% ionic.

Liebau ascribed the spreading of the P–O–P bond angle in the condensed phosphates from the theoretical 90° to the observed 122 to 147° to the double-bond character arising from the partial overlap of the orbitals of lone electron pairs on the oxygen with the *d* orbitals of the two phosphorus atoms. This spreading was not found in the silicates, although the Si—O bond was more ionic.

With regard to the types of structure found to date for the crystalline phosphates and silicates, Liebau stressed two differences. No crystalline silicates corresponding to the single-chain oligophosphates had been found. Pyrosilicates were known but not tripolysilicates. This was confirmed in 1969 by Wieker and Hoebbel [924], who separated a number of cyclic silicates by paper chromatography.

On the other hand, the double and triple chains found in the silicates had not been found in the phosphates; for example, the triple-chain silicate $Ba_2Si_3O_8$ had no corresponding phosphate. Liebau was fully aware that more species might be found later in the silicates and phosphates which would require modification of the above conclusions.

IV. Phase Equilibrium Diagrams

A. *General Comments*

Phase equilibrium diagrams are of fundamental importance in ceramic studies. For this reason the American Ceramic Society has issued two volumes entitled "Phase Diagrams for Ceramists" [529], compiled by Levin, Robbins, and McMurdie and very well indexed. One hundred and sixty-four of these diagrams show P_2O_5 as a component, omitting those with H_2O, sulfate, chloride, etc. Many of the phase equilibrium diagrams which list P_2O_5 as a component have been developed because of an interest in slags or in the phosphates that are used in preparing luminescent materials. Table III lists other references to phase equilibrium diagrams found in this survey.

Two factors should be kept in mind in evaluating some of these diagrams. One is that the presence of a small amount of moisture may have a great effect on the stoichiometry in some situations because only a small amount on a wt.% basis may be enough to cause a considerable reduction in, for example, the average chain length of a chain phosphate. The other is the fact that the ultimate analysis of a fairly long-chain oligophosphate may differ very little from the next member in the series because of the high weight of the anion compared with that of the usual cations.

It will be noted, too, that, following tradional practice in this field, compositions of crystalline compounds are usually reported only in terms of the component oxides, and no indication is given of chemical or crystal chemical structure. Also, no indication is given of the constitution of the liquid phases, the designation "liquid" being used. This is discussed further in Section V. Just four diagrams are given in this section:

$Na_2O-P_2O_5$, $CaO-P_2O_5$, $K_2O-Na_2O-P_2O_5$ and $CaO-Al_2O_3-P_2O_5$

B. $Na_2O-P_2O_5$

This system, which is of interest in connection with sodium phosphate glasses, has been the subject of a number of studies, including Morey and Ingerson [579]. Turkdogan and Maddocks [843], and Partridge, Hicks, and Smith [657]. Their conclusions are in good agreement. The diagrams of the first two papers were combined in the American Ceramic Society publication [529, p. 197] to give the diagram reproduced here as Figure 8. This diagram covers the range of compositions from $NaPO_3$(left) to Na_3PO_4 (right), that is, from the metaphosphate composition to the orthophosphate composition.

There are four crystalline phosphates that appear as primary phases in this

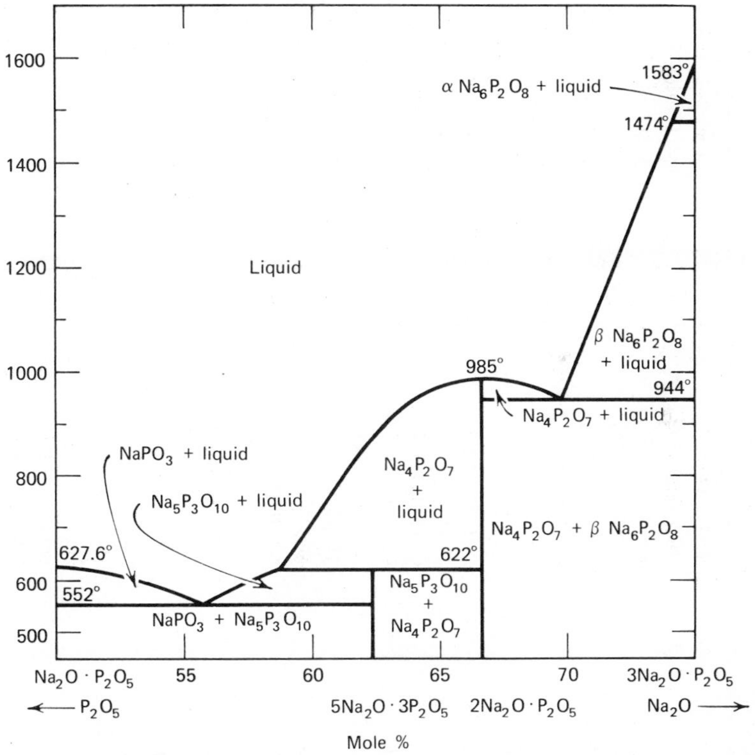

Fig. 8. The system Na PO_3-Na_3PO_4. After Am. Ceram. Soc. *Phase Diagrams for Ceramists,* Levin, Robbins, and McMurdie [529], based on Morey and Ingerson [579] and Turkdogan and Maddocks [843].

TABLE III. Phase Equilibrium Diagram References*

Component Oxides	End Members	Authors	Date	Refs.
$SrO-P_2O_5$	$SrO, SrO.P_2O_5$	Kreidler and Hummel	1967	502
$BaO-P_2O_5$	$BaO, BaO.P_2O_5$	McCauley and Hummel	1968	608
$K_2O-CaO-P_2O_5$	$KPO_3, Ca(PO_3)_2$	Gill and Taylor	1964	311
$Fe_2O_3-Al_2O_3-P_2O_5$	$AlPO_4, FePO_4$	Kobayashi and Kogyo	1966	476
$Cr_2O_3-Al_2O_3-P_2O_5$	$AlPO_4, CrPO_4$	Kobayashi and Kogyo	1966	476
$Na_2O-CaO-P_2O_5$	$Ca_3(PO_4)_2, CaNaPO_4$	Matsuno, Miyhashi, and Ando	1967	554
$CaO-ZnO-P_2O_5$	$Ca_3(PO_4)_2, Zn_3(PO_4)_2$	Kreidler and Hummel	1967	501
$Na_2O-Fe_2O_3-P_2O_5$	$NaPO_3, Fe_2O_3$	Berul and Voskresenskaya	1967	85
$CuO-Tl_2O-P_2O_5$	$Cu(PO_3)_2, TlPO_3$	Lauegt, Scory and Durif	1968	524
$CuO-Rb_2O-P_2O_5$	$Cu(PO_3)_2, RbPO_3$	Lauegt, Scory, and Durif	1968	524
$Na_2O-CaO-P_2O_5$	$Ca_3(PO_4)_2, CaNaPO_4$	Ando and Matsuna	1968	28
$Na_2O-Al_2O_3-P_2O_5$	$Al_2O_3, NaPO_3$	Berul and Voskresenskaya	1968	85
$CaO-SiO_2-P_2O_5$	$3CaO.SiO_2, 3CaO.P_2O_5$	Fix, Hymann and Heinke	1969	276
$Na_2O-WO_3-P_2O_5$	$Na_4P_2O_7, NaPO_3, WO_3$	Bergman and Semenyakova	1970	77
$Na_2O-MoO_3-P_2O_5$	$Na_4P_2O_7, NaPO_3, MoO_3$	Bergman and Semenyakova	1970	77
$ZrO_2-WO_3-P_2O_5$	ZrO_2, WO_3, P_2O_5	Martinek and Hummel	1970	547
$ZrO_2-ThO_2-P_2O_5$	ZrO_2, ThO_2, P_2O_5	Laud and Hummel	1970	523
$CaO-MgO-P_2O_5$	CaO, MgO, P_2O_5	McCauley and Hummel	1971	609
$CaO-BaO-P_2O_5$	$Ca(PO_3)_2-Ba(PO_3)_2$	Bukhalova, Tokman, and Shpakova	1970	147
$CdO-BaO-P_2O_5$	$Cd(PO_3)_2-Ba(PO_3)_2$			
$Na_2O-V_2O_5-P_2O_5$	$NaVO_3-NaPO_3$	Bergman and Sanzharova	1970	76
	$NaVO_3-Na_4P_2O_7$			
	$NaVO_3-Na_3PO_4$			
	KVO_3-KPO_3			
	$KVO_3-K_4P_2O_7$			
	$KVO_3-K_3PO_4$			

Li_2O-RO-P_2O_5	$LiPO_3$-$R(PO_3)_2$	Tokman and Bukhalova	1970	828
K_2O-MoO_3-P_2O_5		Bergman and Semenyakova		78
K_2O-V_2O_5-P_2O_5	KVO_3-KPO_3-$K_4P_2O_7$	Bergman and Sanzharova	1970	75
K_2O-SO_3-B_2O_3-P_2O_5	K_2SO_4-KPO_3-K_3PO_4-$K_2B_4O_7$-$K_4P_2O_7$	Bergman and Matrosova	1970	72
R_2O-B_2O_3-P_2O_5	$R_4P_2O_7$-RBO_2 R_3PO_4-RBO	Bergman and Mikhalkovich	1970	74
Cs_2O-BaO-P_2O_5	$CsPO_3$-$Ba(PO_3)_2$	Tokman and Bukhalova	1970	829

*Additional to "Phase Diagrams for Ceramists" [529].

diagram. From left to right they are sodium trimetaphosphate, called $NaPO_3$ I by Morey, sodium tripolyphosphate, sodium pyrophosphate, and sodium orthophosphate. They exhibit a considerable range of melting points from 622°C., the incongruent melting point for tripoly-, to 1583°C., the melting point of the ortho phosphate. They also differ greatly in the character of the liquidus curves as they approach these melting points. The orthophosphates exhibit the sharp point to be expected from the usual melting point depression when a second species is added. The pyrophosphate shows a more rounded top as though more species were involved. The tripolyphosphate melts incongruently and the pyrophosphate is the primary phase for liquids much higher in sodium content than the tripolyphosphate. This is a reflection of the great stability of the pyrophosphate and its strong crystallizing tendency, which indicates a high lattice energy. The flat curve of the trimetaphosphate shows that a number of species are present in the liquid. It is noteworthy that the one word "liquid" is used for all the liquids in the system, although it appears that they must differ considerably on their constitution. This will be discussed further in Section VI in which the investigations of these liquids will be described.

It should be noted that the tetrapolyphosphate $Na_6P_4O_{13}$, does not appear as a crystalline compound in this diagram nor do the oligophosphates. Lead tetrapolyphosphate appears in the $PbO-P_2O_5$ system [522]. Griffith [336] showed that this could be readily converted to the ammonium salt. Barium tetraphosphate occurs in the $BaO-P_2O_5$ system. The sodium salt in solution can be obtained by the partial hydrolysis of P_2O_5 and can be converted to the ammonium salt on an ion exchange resin. The solution can then be concentrated at 30°C under vacuum. Gartaganis [300] found that a good crop of crystals could be obtained if the solution were not concentrated too much but allowed to evaporate from an open dish. Crystalline sodium tetrapolyphosphate has not been prepared with certainty and appears to be unstable. Gartaganis succeeded on one occasion in preparing some by a method due to Oswald for preparing carbonate-free caustic soda, but the experiment could not be repeated.

C. $CaO-P_2O_5$

A composite diagram from the American Ceramic Society publication [529, Figure 2304] is reproduced here as Figure 9. This system is of interest in connection with metallurgical slags and luminescent phosphates. A number of crystalline phases are indicated, again with a large range of melting points. The same publication shows a number of other diagrams for this system. Unfortunately they do not agree in some respects. The mineral trömelite appears in some [529, Figure 248] as a primary phase. This mineral is also noteworthy because it was described first as a tetrapolyphosphate, then as a pentapolyphosphate [382, 863, 645], C_7P_5 (Fig. 192), and more recently as a hexapolyphosphate [923].

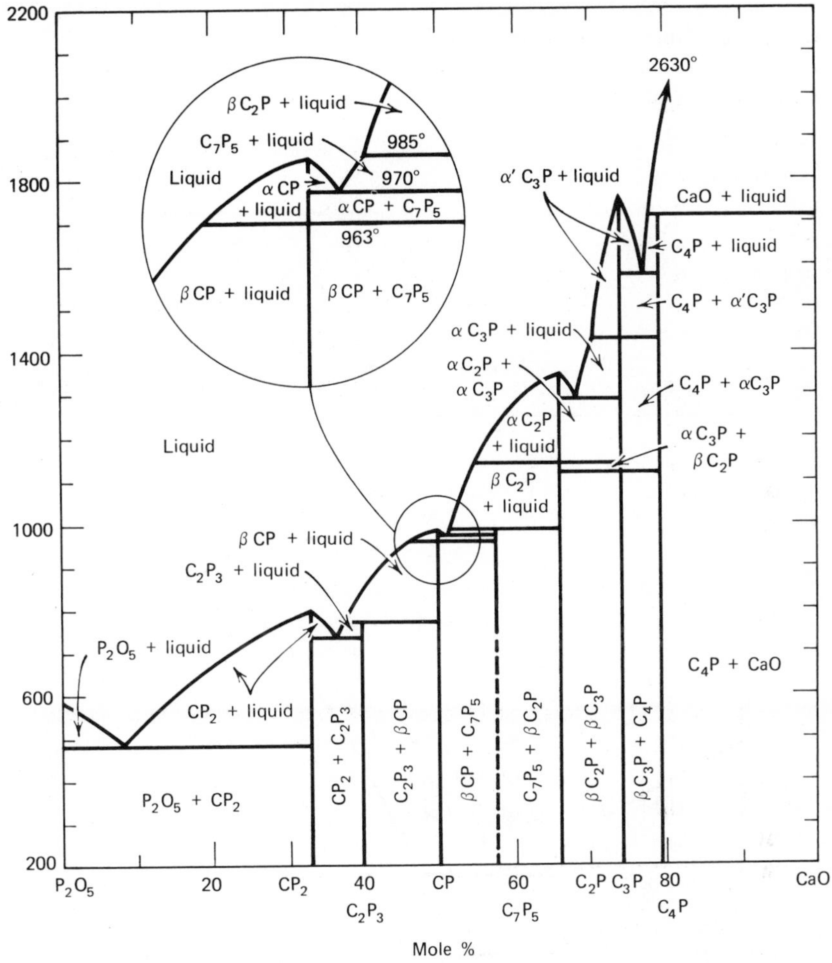

Fig. 9. The system CaO-P_2O_5. C=CaO, P=P_2O_5. After Am. Ceram. Soc. *Phase Diagrams for Ceramists,* Levin, Robbins, and McMurdie [529], based on Kreidler and Hummel, *Inorg. Chem.,* 6(5), 891 (1967); C_7P_5 = pentaphosphate. Note that trömelite is shown as C_7P_5, although it is now considered to be the hexapolyphosphate C_4P_3.

Because no crystalline sodium ultraphosphates have been confirmed, [191, p. 272], some interest attaches to the two compounds $CaO.2P_2O_5$ and $2CaO.3P_2O_5$. From their chemical composition they would be classified as ultraphosphates, but Liebau [533] suggests that they might be acid metaphosphates $CaH_2(PO_3)_4$

and (CaH(PO$_3$)) respectively. The high melting point of calcium orthophosphate, 3CaO.P$_2$O$_5$, is of interest because it has been used for making crucibles.

D. K_2O–Na_2O–P_2O_5

A section of this system is shown in Fig. 10, which is a reproduction of the American Ceramic Society diagram, Fig. 385 [529]. It is of interest in showing that the KPO_3 half of the diagram differs considerably from the $NaPO_3$ side. The occurrence and incongruent melting point of the 3:1 compound $3NaPO_3.KPO_3$ is also of interest.

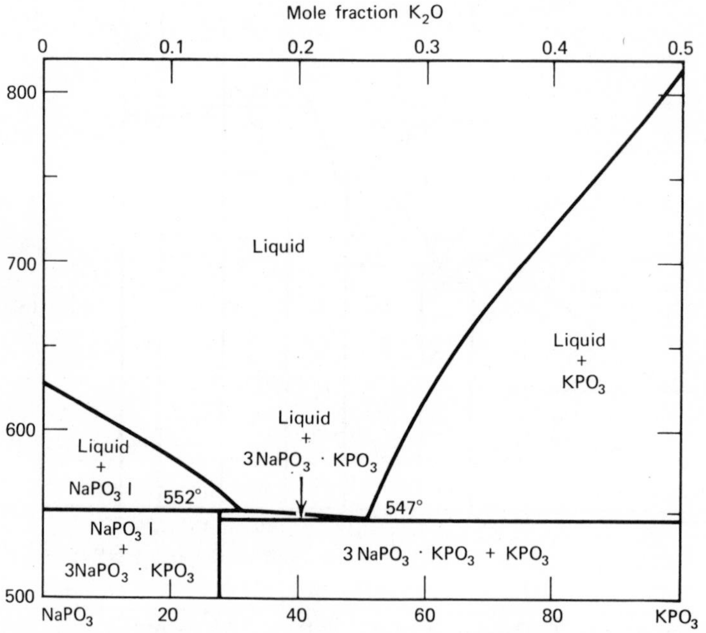

Fig. 10. The system KPO_3-$NaPO_3$. After the Am. Ceram. Soc. Phase Diagrams for Ceramists, Levin, Robbins, and McMurdie [529] based on Morey, *J. Am. Chem. Soc.*, *76*(18), 4725 (1954).

E. CaO–Al_2O_3–P_2O_5

Figure 11 is a reproduction of Figure 640 of the American Ceramic Society [529]. It is an important diagram for the discussion of English bone china and is referred to again under that heading. The letter D has been reinserted before 1450° under $Al_2O_3.P_2O_5$ and before 700° under $Al_2O_3.3P_2O_5$, since these temperatures are those at which noticeable decomposition occurred and are not melting points.

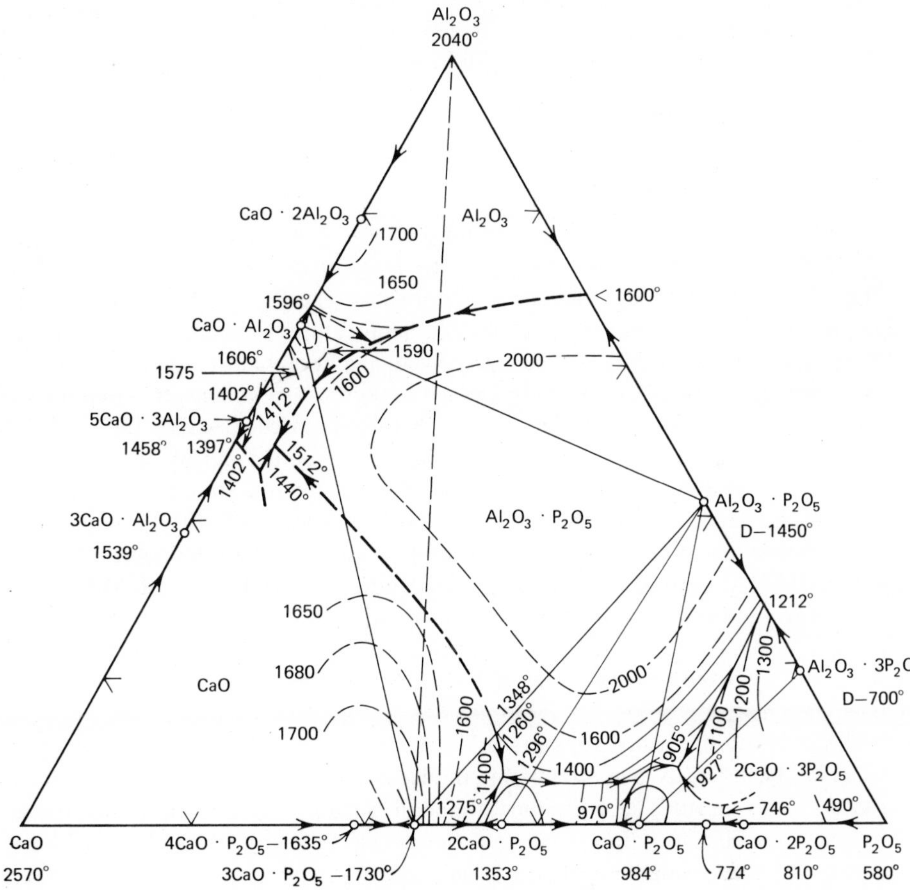

Fig. 11. The system $CaO\text{-}Al_2O_3\text{-}P_2O_5$. After Am. Ceram. Soc. *Phase Diagrams for Ceramists,* Levin, Robbins, and McMurdie, [529], based on St. Pierre [785], Stone, Egan, and Lehr (779a). D indicates onset of decomposition.

These letters were inadvertently omitted when the original was copied by the Society. The Al_2O_3. P_2O_5 is the $AlPO_4$ or aluminum orthophosphate discussed earlier which bears a striking resemblance to SiO_2. As indicated in the caption of Figure 11, the field of $Al_2O_3.P_2O_5$ does not extend so far to the left as the diagram shows. In fact, it does not cross the line joining the Al_2O_3 to $3CaO,P_2O_5$ as shown by St. Pierre (785).

F. $Na_2O–CaO–P_2O_5$

It is noteworthy that only a few binaries in the $Na_2O–CaO–P_2O_5$ system have been published, although from a ceramic point of view this would have been expected to receive early attention. By and large most of the phase diagrams available for phosphates have been developed by slags or luminescent materials in mind.

V. Phosphate Glasses and Polyphosphoric Acids

A. *Nature of the Problem*

In the discussion of phase equilibrium diagrams it was pointed out in Section IV that little attention had been paid to those parts of the diagrams labeled "liquid." In the methods most commonly used in securing data for these diagrams any liquids present in the system at elevated temperatures were quenched to glasses. These liquids were considered to be supercooled, and because little was known about their structure it was believed to be enough to report their composition in terms of oxide components of the system. Although this appeared to satisfy the needs of the thermodynamic approach of these systems, as pointed out by Morey [578], nevertheless glass technologists have studied and debated the structure of glass for at least 150 years. Most of this discussion has been concerned with the soda-lime-silica glass which is produced in the largest volume commercially. According to Morey [577], this composition has been known to mankind for at least 5000 years. The ternary eutectic in this system is such that the glasses can be much above their liquidus temperature when their viscosity is in the working range. This enables them to be worked with little fear of devitrification, that is, separation of a crystalline phase.

The story of the many attempts to establish the "structure of glass" over the years has been largely one of applying the ideas and methods that have proved so successful in the study of crystalline compounds and minerals. By these methods it has been possible to define accurately the detailed dimensions of the crystalline lattice and the positions of all atoms on that lattice. As Liebau has stated [533, p. 267], it is possible in the case of the crystalline phosphates to describe the structure either on the basis of a phosphate anion lattice or a cation lattice. Although modern physical methods have succeeded in accurately defining a short-range order for silicate glasses based on SiO_4 tetrahedra, it is doubtful if they can finally define detailed structure as they have for crystals. It is unlikely that any such structure exists in glass. This does not mean, however, that there may not be a bonding of tetrahedra through shared oxygens to form molecules or polymers which is stronger than the bonds between these molecules or polymers. This kind of information may be said to describe the "constitution" of a glass in contrast to its "structure." It is with the constitution of phosphate glasses and polyphosphoric acids that we are concerned in this section.

Glasses can be formed by heating a crystalline material or a mixture of such materials until a uniform liquid is obtained and then cooling the liquid at such a rate that no evident recrystalization occurs. Thus crystalline sodium tripolyphosphate can be melted by heat and a glass formed by pouring the melt on a cold

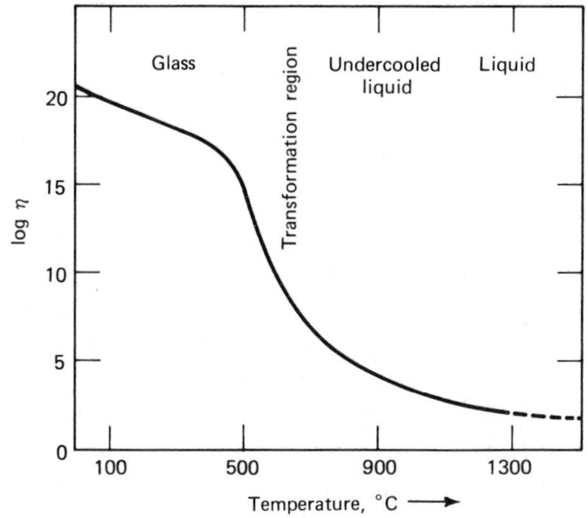

Fig. 12. Viscosity-temperature curve of a plate glass. After Winter, *J. Can. Ceram. Soc.* [928].

copper block. The essential factor is the prevention of recrystallization or devitrification. A glass is more than a supercooled liquid, however, because at some point in the process of cooling it undergoes a second-order transformation in which many of its properties such as thermal expansion undergo a drastic change. Above this glass transition temperature, usually designated T_g, it is supercooled liquid; below T_g it is a glass. T_g may depend on the rate of cooling and its history.

It is more nearly correct to speak of a "transformation" range than a transformation "temperature," as discussed in detail by Winter [928]. Figures 12 to 15, taken from Winter's article, illustrate these ideas. Figure 12 shows how the viscosity of a glass changes as it is heated through the transformation range. Figure 13 shows that the transformation range corresponds to about the same viscosity range for glasses of the different compositions given in Table IV, including a phosphate glass. Figure 14 shows the relation of T_g to the thermal expansion curve of an annealed glass and Figure 15, the relation between T_g and the heating rate.

There has been much speculation about the structural changes occurring in the transformation range. Changes in constitution, for example, polymerization, have sometimes been postulated, but for the sodium phosphate glasses, at least, no change in constitution determined by chromatography was produced by changing the temperature of melting or rate of cooling, and the number average chain length observed agreed with that calculated from the Na/P atomic ratio. Consequently there is no reason to believe that the *constitution* of a sodium phosphate glass at room temperature differs from that of the liquid. For these glasses, at least, then, we must assume that the expansion effects found on heating up to the transformation range involve only atomic vibration effects such as the alternation of a bridging oxygen between two positions, whereas in the transformation range the

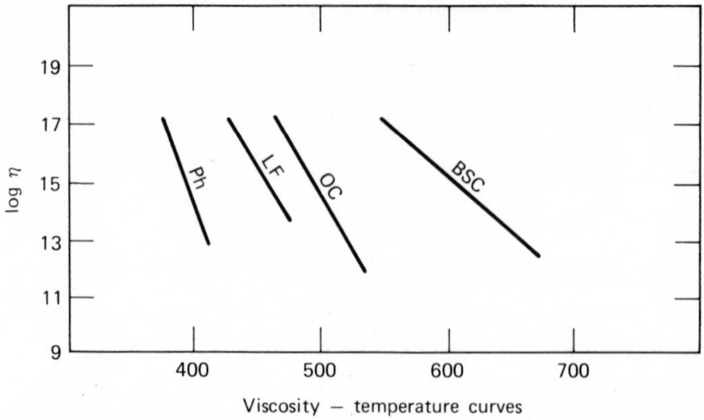

Fig. 13. Viscosity-temperature curves in the transition region. After Winter, *J. Can. Ceram. Soc.* [928]: BSC = borosilicate, OC = ordinary crown, LF = light flint, PH = phosphate.

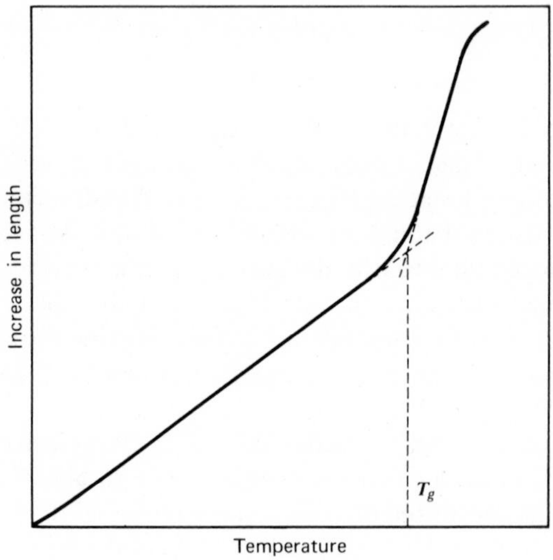

Fig. 14. Typical expansion curve. After Winter, *J. Can. Ceram. Soc.* [928].

TABLE IV. Approximate Composition of Glasses Used for Figure 13

	SiO_2	B_2O_3	Na_2O	K_2O	CaO	PbO	P_2O_5	Al_2O_3
BSC	73	8	10	4	3			
OC	74		10	11	5			
LF	76		12			11		
Ph		14	23	7			34	22

Private communication, Mme Winter.

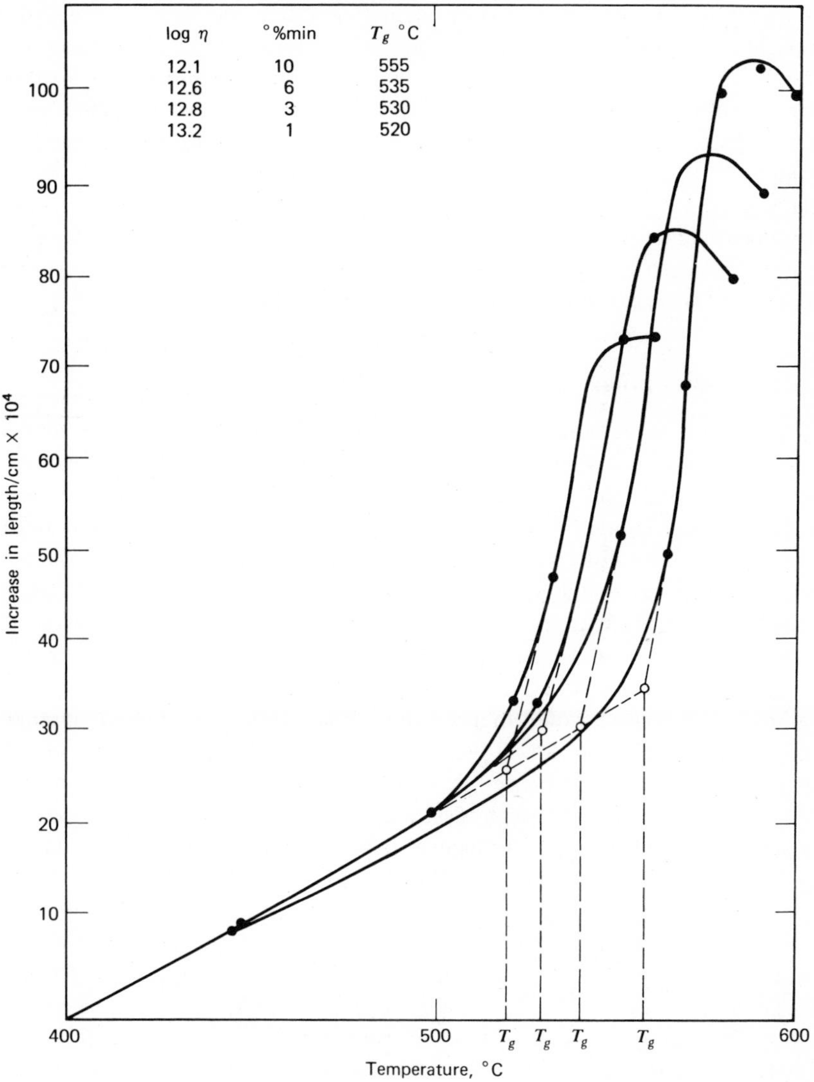

Fig. 15. Effect of the rate of heating on the expansion curves of a crown class. After Winter, *J. Can. Ceram. Soc.* [928].

energy is absorbed by the movement of larger units. Thus a change in structure rather than a change in constitution is involved.

At the same time, a glass differs materially from a crystalline material in a number of ways; for example, it usually has a much lower density than a crystal of the same chemical composition and when subjected to pressure can be inelastically deformed, the deformation being partly removed on reheating.

Because glass has properties intermediate between those of a liquid and a crystal, it is useful to consider it as a state of matter, called the vitreous state [895]. Although for many years it was thought that only a few oxides or mixtures of oxides could be obtained in the vitreous state and the development of glass theory concentrated largely on the commercial soda-lime-silica glass of commerce, in recent years a large number of substances have been put into the vitreous state. The important factor is to avoid devitrification. This avoidance is usually attributed to the development of a high viscosity. It is perhaps more correct, however, to say that the factors that help to avoid crystallization also increase viscosity. An example would be hydrogen bonding which makes it possible to form glasses from a number of substances such as orthophosphoric acid (Dainton) [150, 151, 204, 205].

B. Development of Glass Theory

Early attempts to develop a theory of glass were based on the chemical methods used in the study of crystalline materials. An effort was made to treat glasses as chemical compounds and to assign a chemical formula to a glass. It was soon found that the chemical composition of a glass could be varied in a continuous fashion and that glasses were not compounds of definite proportions.

The next step was to try and demonstrate that compounds of definite proportions existed in a glass. With this in mind, many experiments were done in which the chemical composition of a series of glasses was varied in a systematic manner and some physical property was plotted against that composition in the hopes that some break in the curve would indicate the formation of a chemical compound of the corresponding composition. This method suffered from the difficulty of determining when there was a "break" in a curve and was abandoned for many years after Morey showed that if the physical properties of the glasses in the soda-lime-silica system were plotted on a triangular diagram the lines of equal density, etc., swept across the diagram with no breaks and with no relation to the phase equilibrium diagram for the same system. This was taken to mean that no molecules existed in a glass. Along with the crystal chemistry finding that, for example, a sodium chloride crystal was purely ionic and did not contain any discrete NaCl molecules, this was enough to support the conclusion that molecules could not exist in a glass for a number of years. This conclusion, however, overlooked the fact that the method used would not show the presence of molecules if they were engaged in a dynamic equilibrium so that their relative concentrations changed gradually as the overall composition of the glass was changed. Rather it was a refutation of the crystalite theory of glass structure which postulated the presence of crystals too small to be visible in ordinary light. The physical properties of such a structure might be expected to alter abruptly when a change in the overall composition of a glass changed the primary phase shown on the phase equilibrium diagram.

The crystalite theory was postulated when the first X-ray diffraction data on a glass was obtained by Randall, Rooksby and Cooper [696a]. As X-ray methods became more refined, it was found that the size of the crystals postulated had to be made smaller and smaller until the crystalites needed to be about as small as the unit cell of the crystal considered to be present. At this stage the crystalite theory had been abandoned in Western countries but continued to receive support in Russia. In Russia it has now been modified so that the crystalites are really replaced by small areas showing some long-range order. Meanwhile both groups have paid considerable attention to "microheterogeneities," which can be observed in some specimens of glass under the electron microscope. Important as they may be in commercial glasses, however, it appears that under laboratory conditions these microheterogeneities will disappear if the glass is heated for some time at a high temperature; they therefore appear to be attributable to incomplete melting of the batch materials.

The crystal chemistry theory of glass structure arose largely from the study of crystalline substances in which the bonding between atoms was mostly if not all ionic in character. The atoms in a glass are considered to be completely ionized and conclusions are drawn on the ionic radius of an atom and its charge. Thus the phosphorus atoms in the phosphate glass are given a charge of 5+ and an ionic radius of 0.35 Å, based on measurements of crystals. Ions with a high charge and a small radius have a high field strength. A detailed history of the development of this theory of glass structure has been given by Weyl and Marboe [916]. They conclude their discussion by describing glass structure in terms of three models of liquid structure, namely the Bernal Flawless model, the Frenkel fizzured model, and the Stewart model. The last permits some regularity of structure including some polymers such as chainlike polyphosphate anions. The above account was written in 1961 when the work on phosphate glasses to be described later had been known for some time. The same authors published a two-volume monograph on the subject in 1964 [917]; other publications by Weyl on this subject include [910, 913].

In 1932 Zachariasen published an important paper on crystal chemistry in which he proposed a structure for glass based on a random network. He had noticed the similarity between glasses and crystals in their transparency and strength and proposed that glasses were built up of cation-anion polyhedra, such as SiO_4, in the same way as crystals. However, because there was no periodicity in the glass that could be detected by X-rays, he suggested that the polyhedra were joined together by sharing oxygens, as in crystals, but that the angles at which the polyhedra were joined were not fixed but random in character [952, 953].

The Zachariasen idea of a random network of tetrahedra for silicate glasses was investigated by Warren and his colleagues and the idea of a rigid short-range order within the tetrahedon and a randomly varying angle between tetrahedra was supported by X-ray methods. Thus there arose the Zachariasen-Warren random network theory which dominated discussion of glass structure for many years. [585-586, 883-885]. Unfortunately the supporters of this theory in their enthusiasm took it to mean that there could not be any molecules in the glass; that is, all the tetrahedra were linked into the network and no smaller units containing a limited number of tetrahedra were present. This is in accord with Zachariasen's original proposal, but Warren was always careful to point out that his X-ray results could not be indicative of long-range order; for example, having recently repeated much of his earlier work by using the most modern methods of analysis of his results [585, 586, 885], he then goes on to discuss the possibility of cyclic polymers in a silicate glass. This misinterpretation of Warren's work, however, became so widely accepted that the chromatographic results on sodium phosphate glasses described in the next section were received with considerable scepticism by glass technologists.

As early as 1935 Hägg had taken issue with Zachariasen on the extended network hypothesis, arguing that "large or irregular groups" of tetrahedra could be present in a melt. He suggested that if the number of available oxygen atoms is smaller than necessary for the formation of discrete polyhedral groups with the required coordination, the polyhedra will be joined together, sharing oxygen atoms. Over certain ranges of composition this would allow for the formation of various types of polymer in a melt and presumably in a glass. In the composition range of the usual commercial silicate glasses we would expect a rather extended random network because the silica content is usually in the 70 to 75 wt.% range.

In recent years renewed attempts have been made by Babcock et al. and Robinson [40, 304, 716] to correlate the variation of physical properties with the composition of some silicate glasses with the phase equilibrium diagrams or at least some substructure in the glass. Their findings must be taken into account in evolving a theory of silicate glass structure.

The application of standard analytical procedures by Bell and Jones to sodium phosphate glasses that are soluble in water led to the conclusion that their solutions contained condensed phosphate molecule-ions [67, 68, 432]. Van Wazer in 1948 [861] presented a theory of the molecular structure of sodium phosphate glasses in the range from the sodium tripolyphosphate to the metaphosphate composition. He followed it with experimental data that gave molecular weight distributions that he obtained by using the techniques of polymer or macromolecular chemistry. These data included fractional precipitation with an organic solvent from aqueous solution followed by end-group titration of the fractions. Theoretical chain-length distributions based on the Flory [281] and the Poisson distributions were developed. Van Wazer concluded that the water-soluble sodium phosphate glasses consisted largely of unbranched chain phosphates. As the composition of a metaphosphate was approached, there was evidence that cyclic phosphates were present in appreciable amounts. It was also concluded, however, that the monomer, that is, the orthophosphate would not be found in these glasses both from theory and experiment. As shown later, this conclusion was correct for glasses with a medium or high average chain length but it was possible to produce glasses of short average chain length that contained rather large proportions of orthophosphate.

Our knowledge of phosphate chemistry and ceramics in 1949-1950 could then be summed up by saying that crystalline ortho-, pyro-, and tripolyphosphate were well known, that higher polymers were postulated by Van Wazer but had not been isolated, that Bell, using chemical methods of analysis, had shown that ortho- and tripolyphosphoric acid were present in strong phosphoric acid as well as an unidentified acid, and that Van Wazer had advanced a theory of the constitution of phosphate glasses based on polymer methods of investigation.

C. Chromatography

It was clear that the biggest handicap in the study of phosphates was the lack of methods of analysis that would be applicable to mixtures of a number of species. Consequently investigators turned to chromatography, particularly the filter-paper chromatography developed by Martin and Singe [546a]. In 1949-1950 the prevailing opinion was that paper chromatography was not applicable to phosphates. Preliminary trials had been unproductive. We now know that this was due to two effects: the hydrolysis of the condensed phosphate on the paper, which led to streaking of the spots and complexing of polyphosphates with small amount of impurities in the paper. The latter was particularly important because of the small amounts of phosphates used in paper chromatography. This negative attitude was overcome by Hanes and Isherwood [349a], who developed a paper chromatography method for analyzing for sugar phosphates. They subjected their filter paper to an extended washing procedure, worked with water-miscible solvents, and used apparatus in which the temperature was controlled and the solvent flow was downwards. Both acid and basic solvent mixtures were used.

Using the Hanes and Isherwood apparatus and techniques at Cambridge under the direction of Dr. Hanes, I was able to show that the products of hydrolysis of P_2O_5 could be separated on paper. This project was suggested by Dr. Bernard Raistrick. My colleagues and I continued this investigation of phosphates in the ensuing years [903]. Independently, Ebel and his colleagues studied paper chromatography of condensed phosphates by using the Hanes and Isherwood techniques. He discovered the importance of having a small amount of ammonia present in a trichloracetic acid solvent and introduced the use of two-dimensional paper chromatography for separating unbranched chain phosphates from cyclic phosphates. The effect of the ammonia was attributed to the different phosphate ammonium salts having different solubilities [442, p. 279]. He also used an apparatus in which the solvent mixture ascended the paper by capillarity. These improvements were adopted at Toronto, where operation at 4°C was also found to be advantageous. Ebel also used column chromatography to prepare small amounts of condensed linear phosphates up to the octapolyphosphates [237-243]. Paper chromatography was also studied and used extensively by Thilo and his colleagues [345, 815, 816, 818], by Ando and his group [112], and by Karl Kroupa, Griffith, and others [338, 446, 447]. A review and bibliography are given by Ebel [243].

Column chromatography has undergone a parallel development, as outlined by

Jameson [86, 148, 331, 381, 426, 534, 669], and more recently thin-layer chromatography and electrophoresis have been investigated. Examples of filter paper and column chromatography are given later in this section. Each of these methods has its advocates. The choice depends on the nature of the investigation. A brief review is given by Kalliney [442].

D. Sodium Phosphate Glasses

Early in the development of paper chromatography of condensed phosphates at Toronto [905] it was noticed that chromatograms of the strong phosphoric acids [i.e., those acids containing more P_2O_5 than orthophosphoric acid (72.4 wt.%)] consisted of a series of spots presumably corresponding to higher and higher polymers. A somewhat similar chromatogram was found for a soluble sodium phosphate glass. As the techniques were improved, both classes of material were investigated in detail. It was found that in both cases one was dealing with a dynamic equilibrium

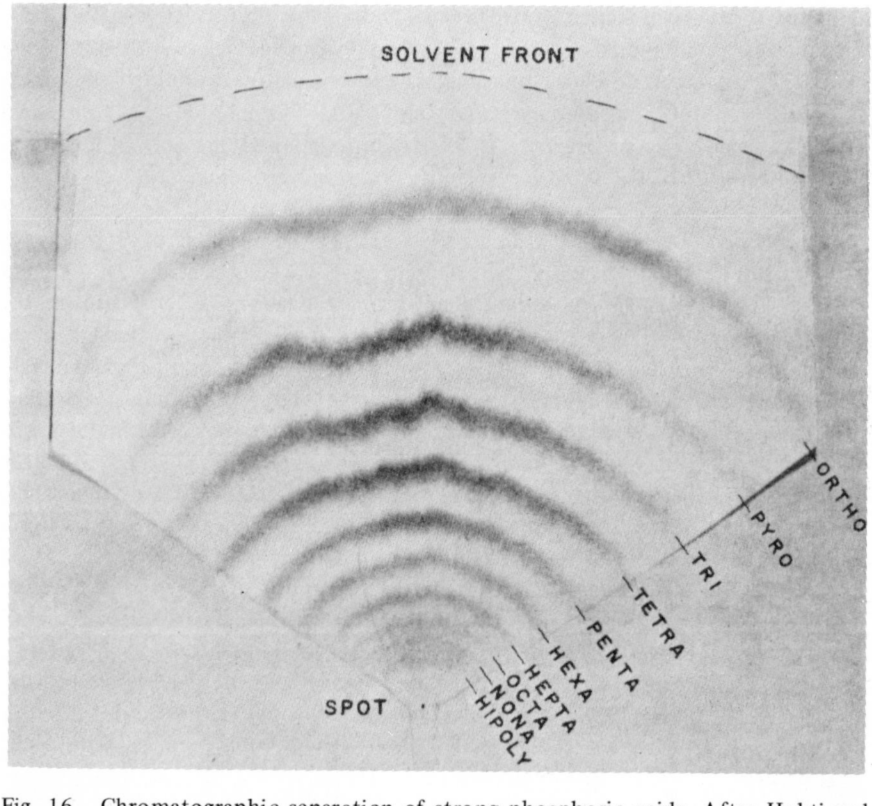

Fig. 16. Chromatographic separation of strong phosphoric acids. After Huhti and Gartaganis, *Can. J. Chem.* [404].

between a number of polymers that were largely unbranched chain polymers. For any given overall composition the equilibrium was quite reproducible and did not depend on the nature of the starting materials, the melting temperature, rate of cooling, or conditions of solution. An illustration of the kind of chromatogram finally achieved is given in Figure 16. In this chromatogram the shape of the paper was such that when the phosphate solution was applied as a thin streak at the bottom the traces of the polymers were stretched out so that more material could be carried without the overlapping of the lines.

The constitutions of a series of soluble sodium phosphate glasses were determined at Toronto by Crowther [897] and redetermined by Gartaganis [898]. It was found that they consisted largely of linear polymers with the general chemical formula

$$M_{(n+2)}P_nO_{(3n+1)} \tag{7}$$

where n is the chain length. In such a system, if M and P are taken to represent the total number of M and P atoms, respectively, the number average chain length will be given by

$$\bar{n} = \frac{2}{[M/P)-1]} \tag{8}$$

If the mole fraction of polymers of chain length n is designated by X_n, Figure 17

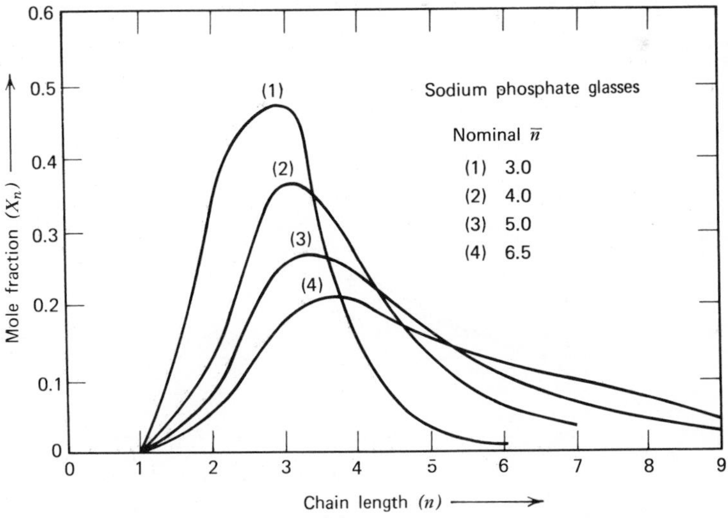

Fig. 17. Constitution diagram for some sodium phosphate glasses. After Westman and Gartaganis, *J. Am. Ceram. Soc.* [898].

shows the way in which the mole fractions of the different polymers varied as the number average chain length was varied, based on the data of Gartaganis [898] for a range of number average chain lengths of 3.0 to 6.5. It will be seen at once that for all four glasses the orthophosphate content was essentially zero. Also, as the number average chain length is changed, the distributions change slowly and a considerable amount of overlapping occurs. One would therefore not expect to find any abrupt changes in the physical properties of these glasses as the *Na/P* ratio was changed.

On the basis of the data shown in Figure 17 it is necessary to conclude that the molten glass consists largely of a number of unbranched polymers that are in dynamic equilibrium with each other or to postulate that they are formed as the glass is dissolved in water. To date none of the results that one would expect from such a postulate have been found experimentally and there appears to be rather general acceptance of the idea that the molten glass does consist largely of unbranched polymers. This is an interesting conclusion in view of the phase equilibrium results discussed in Section IV; for example, one has to picture a tripolyphosphate molecule in a molten glass persisting at a temperature that is much higher than the incongruent melting point of crystalline tripolyphosphate and also picture a number of higher polymers that have not been obtained in crystalline form. There is also the equilibrium between Morey's $NaPO_3I1$, which is sodium trimetaphosphate, and the liquid of the same overall composition at the melting point. The crystalline material is known to be a cyclophosphate, yet it is in equilibrium with a liquid that is largely unbranched chains. This is illustrated in Figure 18, which shows about 3%

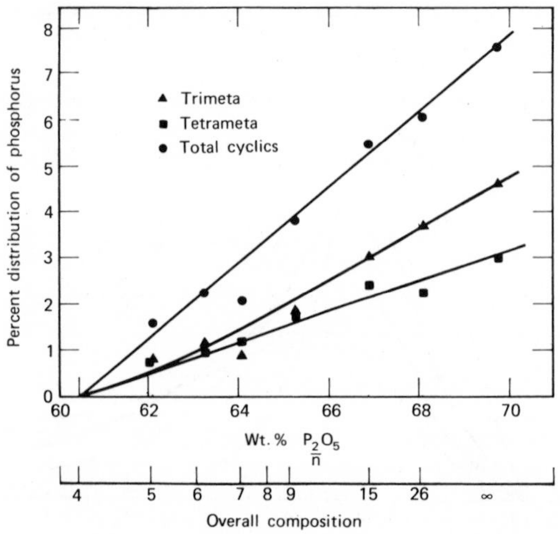

Fig. 18. Cyclic polymers in sodium phosphate glasses. After Westman and Gartaganis, *J. Am. Ceram. Soc.* [898].

trimetaphosphate and 4.6% tetrametaphosphate in a glass of the $NaPO_3$ composition. It appears that at this equilibrium the addition of heat results in trimetaphosphate crystals dissolving in the liquid in which a rearrangement then takes place to restore the normal internal equilibrium between the many polymer species present. The reverse process occurs when heat is subtracted. The system can be treated as a two-component system only because fixing the Na_2O/P_2O_5 ratio of the liquid is sufficient to fix the concentrations of all the polymer species present in the liquid.

If the above picture is true, it might be expected that if trimetaphosphate crystals were both melted and cooled at a very rapid rate it should be possible to produce a glass with a high content of trimetaphosphate polymers. This was tried with the Haller apparatus [519] in which 100-mesh crystals of trimetaphosphate were blown through a gas flame and then cooled rapidly by expanding the flue gases rapidly. In one experiment it appeared that glass of a high trimetaphosphate content was made. The experiment could not be repeated, however, since then experiments on the kinetics of the crystallization of trimetaphosphate from a glass [901] indicate that this is a fast process at the melting point of sodium trimetaphosphate, and it is unlikely that a glass of the trimetaphosphate constitution was achieved. Both production of a glass with the constitution of trimetaphosphate and heating the tripolyphosphate crystal above its melting point remain as challenges to the experimentor. Note that these projects require rapid heating as well as rapid cooling so that the splat technique alone would not suffice. The shock-wave technique might offer possibilities. There is also the possibility of using cations other than sodium or mixtures of cations in an effort to suppress devitrification.

E. Cations Other Than Sodium

A series of studies was instituted in which other alkalis were substituted for sodium. The purpose was to determine whether there was a definite cation effect and also whether glasses of shorter average number length could be made that would contain appreciable amounts of orthophosphate.

A definite alkali effect [898] is illustrated in Figure 19 in which lithium, sodium, and potassium are compared. There is more difference in the constitution of these glasses than could be explained by lack of precision of the chromatographic methods used. As shown in Figure 19, the sodium glasses tend to be intermediate in constitution between the lithium and potassium glasses, but the lithium phosphate glasses tended to have a greater pyrophosphate content than the other two and consequently less of some of the higher polymers.

Calcium and zinc phosphate glasses were investigated by Meadowcroft and Richardson [566, 567]. Although insoluble in water, they could be obtained in solution by the use of acid or calcium complexing agents. Much the same constitutions were found as for the sodium glasses.

The constitutions of mixed alkali phosphate glasses were determined by Murthy et

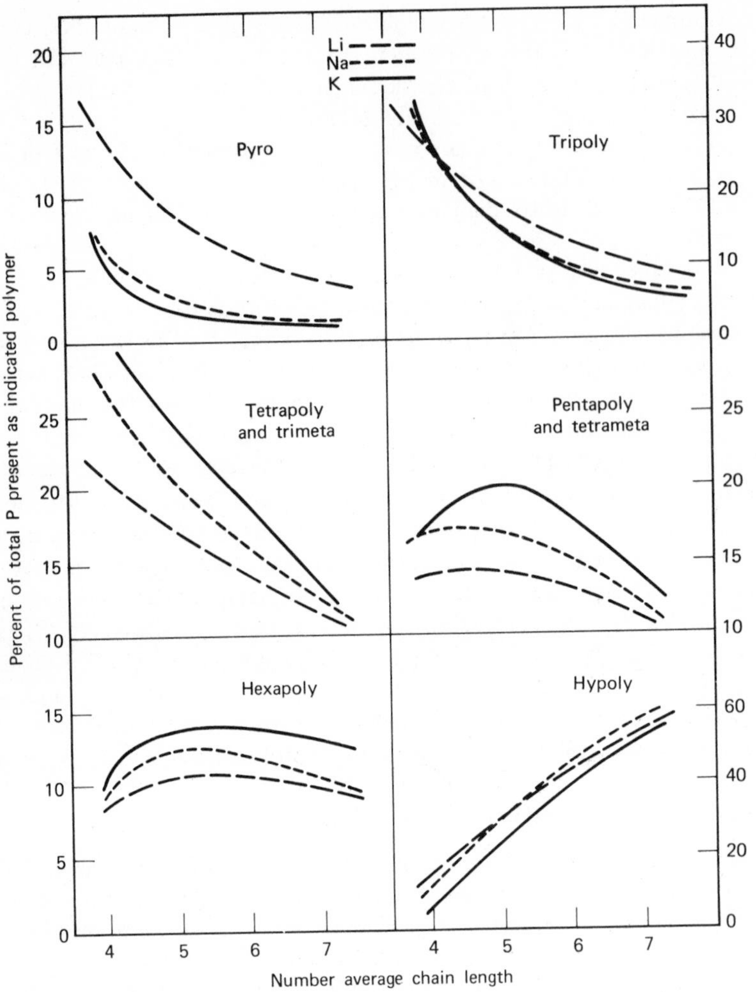

Fig. 19. Illustration of cation effect in phosphate glasses. After Westman and Gartaganis, *J. Am. Ceram. Soc.* [898].

al. and published in a series of papers [603-605, 900]. It was possible in this way to produce glasses with number average chain lengths as low as 1.5 and containing large amounts of orthophosphate.

F. Fluorophosphate Glasses

It was well known that fluorides, when added to molten slags or glasses, greatly reduced their viscosity. It was suspected that this was accomplished by the breaking of an oxygen bridge between two silicon or two phosphorus atoms, the divalent oxygen being replaced by two monovalent fluorines. This subject was investigated by Murthy and associates with filter-paper chromatography [559, 602, 902] and by Williams and associates with other methods [927]. It was found that the addition of fluoride did indeed greatly reduces the chain length of a molten phosphate glass but the heating for even 10 minutes was enought to volatilize all the fluorine either as HF or NaF. The presence of a fluorophosphate ion was also shown.

G. *Polyphosphoric Acids*

In the period 1956 to 1959 both paper and column chromatographic methods were applied to analyses in the P_2O_5-H_2O system. Busch, Ebel, and Blanck in France [148] used column chromatography to prepare specimens of the poly-

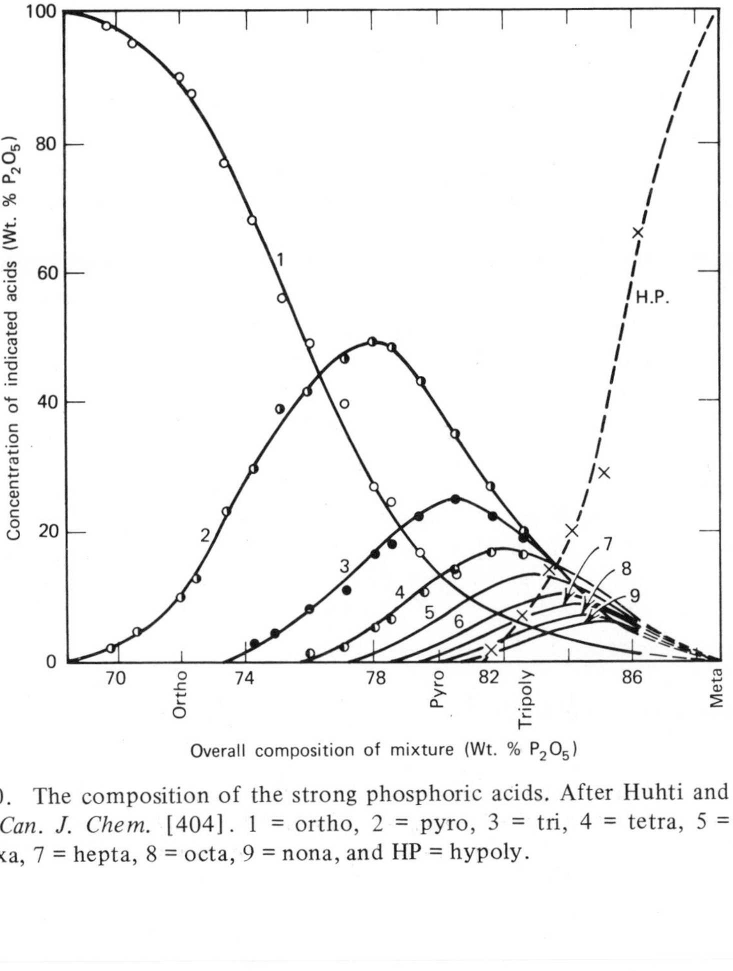

Fig. 20. The composition of the strong phosphoric acids. After Huhti and Gartaganis, *Can. J. Chem.* [404]. 1 = ortho, 2 = pyro, 3 = tri, 4 = tetra, 5 = penta, 6 = hexa, 7 = hepta, 8 = octa, 9 = nona, and HP = hypoly.

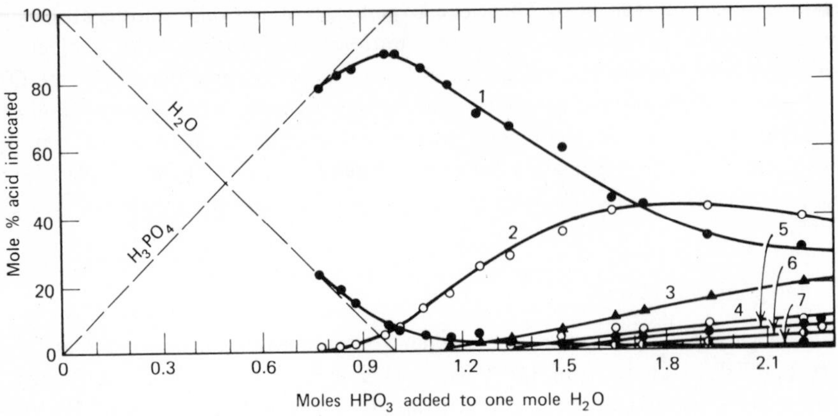

Fig. 21. Computed composition of strong phosphoric acids (mole basis). After Huhti and Gartaganis, *Can. J. Chem.* [404]. 1 = ortho, 2 = pyro, 3 = tri, 4 = tetra, 5 = penta, 6 = hexa, 7 = hepta.

phosphoric acids up to a chain length of eight. Huhti and Gartaganis in Canada [404] and Thilo and Sauer in Germany [816] used paper chromatography to analyze a wide range of acids varying from dilute H_3PO_4 to "strong" acids containing much more P_2O_5 than pure H_3PO_4 does. Column chromatography was used by Ohashi and Sugatami in Japan [644] and Jameson [426] in Scotland for the same purpose. The results of all these investigations were in essential agreement. They may be illustrated by the Huhti and Gartaganis data of Table V and Figures 20 and 21. In Figure 20 it will be seen that the various polyphosphoric acids replace one another as the P_2O_5 content of the solution is increased, thus giving a picture that is similar to that of the sodium acid glasses except that the ortho ion is found in solutions of fairly high average chain length. In Fig. 21 it will be observed that an equilibrium mixture having the overall composition of H_3PO_4 contains about 6 mole % of water and of pyrophosphoric acid. This was confirmed by cryoscopic measurements. Huhti and Gartaganis were able to determine linear polymers up to a chain length of nine. Jameson succeeded in reaching n=14, so that his definition of "highpoly" would be different. His data are discussed more fully in Subsection I-6. No cyclic phosphates were reported in these studies of the strong phosphoric acids.

H. Sodium-Acid Phosphate Glasses

Sodium acid phosphate glasses have compositions intermediate between those of the sodium glasses and the acids. This is illustrated by the composition diagram of the P_2O_5-Na_2O-H_2O system shown in Figure 22 [904]. At the extreme left are the

TABLE V. Composition of Strong Phosphoric Acids.

Composition (wt. % P_2O_5)	Percent of Total Phosphorus as									
	Ortho	Pyro	Tri	Tetra	Penta	Hexa	Hepta	Octa	Nona	"Hy-poly"
68.80	100.00	Trace								
69.81	97.85	2.15								
70.62	95.22	4.78								
72.04	89.91	10.09								
72.44	87.28	12.72								
73.43	76.69	23.31								
74.26	67.78	29.54	2.67							
75.14	55.81	38.88	5.31							
75.97	48.93	41.76	8.23	1.08						
77.12	39.86	46.70	11.16	2.28						
78.02	26.91	49.30	16.85	5.33	1.60					
78.52	24.43	48.29	18.27	6.75	2.26					
79.45	16.73	43.29	22.09	10.69	4.48	1.92	0.08			
80.51	13.46	35.00	24.98	13.99	6.58	3.14	2.84			
81.60	8.06	27.01	22.28	16.99	11.00	5.78	3.72	2.31	1.55	1.28
82.57	5.10	19.91	16.43	16.01	12.64	8.89	6.41	4.11	3.51	6.99
83.48	4.95	16.94	15.82	15.91	12.46	9.71	6.77	5.04	2.99	9.42
84.20	3.63	10.60	11.63	13.05	12.17	9.75	8.19	5.92	4.91	20.16
84.95	2.32	6.97	7.74	11.00	10.45	9.62	8.62	7.85	6.03	29.41
86.26	1.54	2.97	3.31	5.16	5.32	5.54	3.51	3.30	3.30	66.03

Source: After Huhti and Gartaganis, *Can. J. Chem.* [404].
Note: The figures are given to two decimal places for further computation purposes, but the precision may not be better than 1% total phosphorus in some cases.

compositions of the sodium phosphate glasses obtained by ordinary quenching methods, at the extreme right are the compositions of the phosphoric acids which have just been discussed, and in between are the compositions of the sodium acid phosphate glasses, those investigated having compositions along the straight lines shown. The ultraphosphates are above the horizontal line giving the compositions with the metaphosphate ratio.

Filter-paper chromatographic investigation of the sodium acid phosphate glasses showed that they had constitutions that were quite similar to the phosphoric acids. As sodium was substituted for hydrogen, there was no composition at which there was a sudden change in constitution over the range of compositions studied [904].

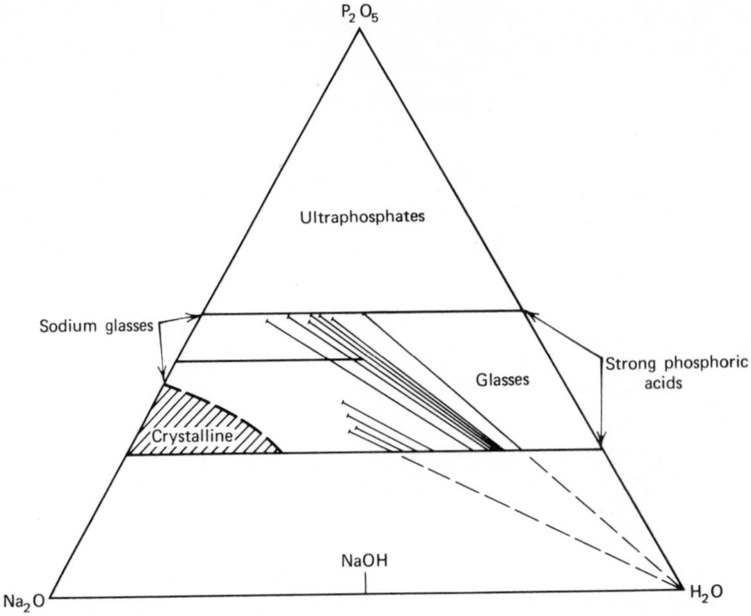

Fig. 22. Glasses in the H_2O-Na_2O-P_2O_5 system. After Westman, Gartaganis, and Smith, *Can. J. Chem.* [904].

I. Distribution Theory

The experimental work that has just been reviewed showed that, in the systems studied and for any particular overall composition, the species present consisted almost wholly of unbranched chain or orthophosphate ions and that a reproducible equilibrium was soon reached between the different species present. It was evident that some sort of dynamic equilibrium was reached by some disproportionation or similar process to which Van Wazer gave the name "random reorganization." It was apparent that a situation was created in which the effect of any reaction occurring at one place in liquid were offset by reverse reactions occurring at some other. Consequently it should be possible to set up a statistical model that would enable the experimental results to be rationalized. However, since a change of cation, for example, from sodium glass to strong acid, obviously changed the chain length distribution, particularly in the shorter chains, it was necessary to have a statistical model that would allow some empirical adjustment of parameters to take care of cation effect.

For the systems with which we are mostly concerned in phosphate ceramics it is sufficient to have a theory based on unbranched chains. Cyclic phosphates are present in only small amounts and can readily be treated separately. In the following paragraphs such a theory, based on a paper by Westman and Beatty in 1966 (896), is presented. This theory was developed over a period of years and owes much to Van Wazer [555, 855], Flory [281], and Jost [433, 434].

1. SYMBOLISM

In the systems under discussion most of the polymers present are unbranched chains that can be represented by the general chemical formula

$$M_{(n+2)}\ P_nO_{(3n+1)} \tag{9}$$

where M represents a monovalent cation and n is the chain length, that is, "the number of phosphorus atoms in the chain." To include M_2O, N can take values 0, 1, 2, 3, and so on.

For the present purpose this can be written in the symmetrical form

$$MO_{1/2}\ \left[\begin{array}{c} O \\ \| \\ O_{1/2}\text{—}P\text{—}O_{1/2} \\ | \\ O \\ | \\ M \end{array}\right]_n\ O_{1/2}M \tag{10}$$

Here the $MO_{1/2}$ groups can be called "chain stoppers" and the bracketed group a "phosphate unit." If the letters M and P are used to represent the number of M and P atoms in a system, then for $M = P$ the systems will consist of one polymer of infinite length or of "wall-to-wall polymers." For $M > P$ the number of molecules will be constant and equal to $(M - P)/2$, since two extra M atoms are needed for chain stoppers for each molecule. Also, the number average chain length $\bar{n}$ will be given by

$$\bar{n} = \frac{2}{[(M/P) - 1]} \tag{11}$$

In a molten glass or in an acid at a suitable temperature consisting of such polymers the molecules are continually reorganizing, keeping the number of atoms, the number of molecules, and therefore the value of $\bar{n}$ constant. This may occur by disproportionation:

$$[\ ]_n + [\ ]_m = [\ ]_{(n+k)} + [\ ]_{(m-k)} \tag{12}$$

where the brackets represent formulas as in (10).

2. FLORY DISTRIBUTION

The Flory distribution, column 2, Table VI, was developed for the case of bifunctional organic monomers. It was applied to sodium phosphate glasses and phosphoric acids by Van Wazer. The fundamental assumption is that in a reorganizing system of linear polymers if a polymer is chosen at random and, starting at a chain stopper, we proceed along the chain, there is a constant probability p that the next unit in the chain is a phosphate unit and not a chain stopper; that is, this probability is independent of the length of the chain. The probability that the next unit is a chain stopper is then $(1 - p)$. The resulting distribution can be written down as shown in column 2, Table VI, where X_n is the mole fraction of molecules of chain length n. The relation between p and $\bar{n}$, as shown at the bottom of column 2, is obtained, in general, from the definitions

$$\sum_{0}^{\infty} X_n = 1 \tag{13}$$

$$\sum_{0}^{\infty} nX_n = \bar{n} \tag{14}$$

For a given M / P ratio $\bar{n}$ can be calculated by using (11), and the ideal Flory distribution the expression for p may also be obtained from (10) with $\bar{n}$ substituted for n by the argument that on the average, starting at one chain stopper, $\bar{n}$ phosphate units (successes) and one chain stopper (failure) have been obtained in $(\bar{n} + 1)$ trials.

The assumptions made in reaching the ideal Flory distribution were reasonable when dealing with polymer systems of very long chains. As shown in column 2 of Table VI, however, it gives a distribution in which the mole fractions constitute a monotonically decreasing geometric series as n increases. Many experimental results, such as those shown in Figure 17, have demonstrated that this is not true for shorter chains, although in the longer chains the mole fractions do approximate to a decreasing geometric series as n increases.

The ideal Flory distribution can be modified to give a good fit to the experimental data by introducing a limited number (usually two) of empirical parameters which are used to modify p for the first few units of a chain. These parameters may be introduced as probability coefficients or as chemical equilibrium constants. In either case it is necessary at present to establish a relation between these parameters and $\bar{n}$ by experiment.

3. PROBABILITY COEFFICIENTS

Jost and Wodtcke [433, 434] introduced the use of multiplyers for p which here are called "probability coefficients" in analogy to activity coefficients, for they modify an ideal situation. Their use is illustrated in column 3, Table VI, in which

TABLE VI. Flory type distributions.

(1) n	(2) Ideal Flory X_n	(3) Flory with Two Probability Coefficients X_n	(4) Flory with Two Equilibrium Constants X_n	(5) Flory with Two Equilibrium Constants and $X_0 = X_1 = 0$ X_n
0	$(1 - p)$	$(1 - a_1 p)$	$X_2 K_1 K_2^2/p^2$	0
1	$p(1 - p)$	$a_1 p(1 - a_2 p)$	$X_2 K_2/p$	0
2	$p^2(1 - p)$	$a_1 a_2 p^2(1 - p)$	X_2	$X_4 K_3 K_4^2/p^2$
3			$X_2 p$	$X_4 K_4/p$
4				
n	$p^n(1 - p)$ $p = \bar{n}/(\bar{n} + 1)$	$a_1 a_2 p^n (1 - p)$* $a_1(a_2 - 1)p^2 + (a_1 + \bar{n})p - \bar{n} = 0$	$X_2 p^{(n - 2)}$* $(K_2 - 1)(1 - \bar{n})p^3 + [\bar{n}(2K_2 - 1 - K_1 K_2^2) + 2(1 - K_2)]p^2 + [K_2(1 - \bar{n}) + 2\bar{n} K_1 K_2^2]p - \bar{n} K_1 K_2^2 = 0$	$X_4 p^{(n - 4)}$† $(K_4 - 1)(3 - \bar{n})p^3 + [\bar{n}(2K_4 - 1 - K_3 K_4^2) + 2(2 - 3K_4 + K_3 K_4^2)]p^2 + [K_4(3 - \bar{n} + 2\bar{n} K_3 K_4 - 4K_3 K_4)]p - \bar{n} K_3 K_4^2 + 2K_3 K_4^2$
			$X_2 = \dfrac{p^2(1 - p)}{(1 - K_2)p^2 + K_2(1 - K_1 K_2)p + K_1 K_2^2}$	$X_4 = \dfrac{p^2(1 - p)}{(1 - K_4)p^2 + K_4(1 - K_3 K_4)p + K_3 K_4^2}$

Source: After Westman and Beatty, *J. Am Ceram Soc.* [896].

*$n \geq 2$.

†$n \geq 4$.

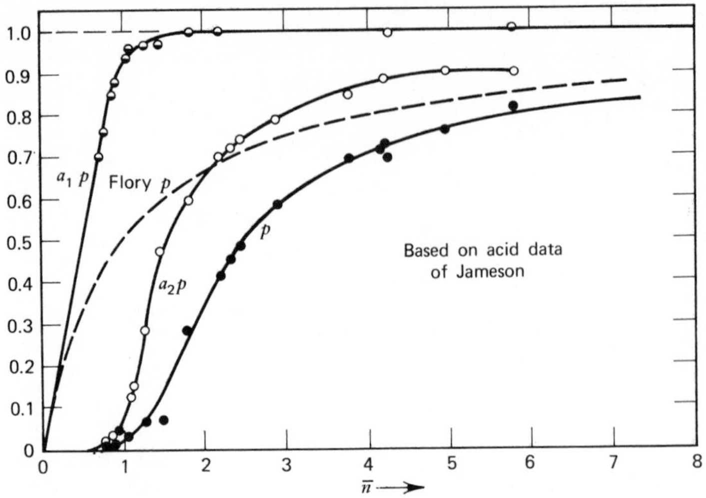

Fig. 23. Observed variations of a_1p, a_2p, and p with $\bar{n}$ compared with that of the Flory p. After Westman, and Beatty, *J. Am. Ceram. Soc.* [896].

two coefficients a_1 and a_2 have been introduced to modify the Flory distribution of column 2. To meet the conditions (13) and (14) the relation between p and $\bar{n}$ becomes the quadratic, shown at the bottom of column 3. Coefficients such as a_1 and a_2 must meet the condition to avoid negative mole fractions.

$$1 \leq a \leq \frac{1}{p} \tag{15}$$

Probably the most precise and extensive data available for testing various theoretical distributions are those obtained by Jameson [426] for the strong phosphoric acids. He used an ion exchange column and determined polymers up to n - 14. His adjusted data are shown in the columns headed a in Table VII. Values of a_1 and a_2 were calculated directly from the first few mole fractions observed and then a_1p, a_2p, p, and the Flory p were plotted against $\bar{n}$ as shown in Figure 23. The relations shown do not lend themselves readily to the formulation of empirical relations nor are the coefficients as suitable for thermodynamic interpretation as the chemical equilibrium constants described next. However, the coefficients have the advantage of requiring an equation for p one lower in degree and of being more directly connected with probability models.

4. EQUILIBRIUM CONSTANTS

It was shown by Huhti and Gartaganis [404] that the dynamic equilibrium between orthophosphoric acid, pyrophosphoric acid, and water for an acid in the

TABLE VII. Jameson acid data*

Acid No. →	1		2		3		4		5	
n	a	c	a	c	a	c	a	c	a	c
0	0.3003		0.2360	0.2358	0.1512	0.1462	0.1154	0.1218	0.0583	0.0609
1	0.6997		0.7565	0.7569	0.8161	0.8273	0.8435	0.8331	0.8212	0.8165
2			0.0076	0.0073	0.0327	0.0258	0.0411	0.0427	0.1177	0.1193
3				0.0001		0.0008		0.0022	0.0028	0.0032
4								0.0001		0.0001
5										
6										
7										
8										
9										
10										
11										
12										
13										
14										
>14										
Adjusted $\bar{n}$	0.6997		0.7716		0.8814		0.9256		1.065	
K_1			0.0030		0.0055		0.0075		0.0109	
K_2									0.185	

TABLE VII. cont.

Acid No. →	6		7		8		9		10	
n	a	c	a	c	a	c	a	c	a	c
0	0.0449	0.0516	0.0328	0.0302	0.0304	0.0316	0.0070			
1	0.8118	0.7986	0.6871	0.6929	0.5119	0.5150	0.4016	0.3988	0.2991	0.2972
2	0.1384	0.1446	0.2594	0.2559	0.3866	0.3773	0.4207	0.4335	0.4139	0.4130
3	0.0049	0.0049	0.0207	0.0194	0.0586	0.0633	0.1230	0.1211	0.1662	0.1705
4		0.0012		0.0015	0.0102	0.0106	0.0355	0.338	0.0704	0.0704
5				0.0001	0.0023	0.0018	0.0082	0.0095	0.0297	0.0291
6						0.0003	0.0031	0.0026	0.0122	0.0120
7							0.0010	0.0007	0.0052	0.0050
8									0.0028	0.0020
9									0.0005	0.0001
10										
11										
12										
13										
14										
>14										
Adjusted $\bar{n}$	1.103		1.268		1.513		1.832		2.192	
K_1	0.0117		0.0161		0.045†					
K_2	0.188		0 205		0.229		0.257		0.297	

Acid No.→	11		12		13		14		15	
n	a	c	a	c	a	c	a	c	a	c
0										
1	0.2812	0.2717	0.2590	0.2554	0.2099	0.2070	0.1540	0.1530	0.1133	0.1313
2	0.3918	0.4005	0.3909	0.3905	0.3298	0.3338	0.2317	0.2385	0.1580	0.1845
3	0.1744	0.1806	0.1766	0.1857	0.1845	0.1933	0.1663	0.1711	0.1212	0.1452
4	0.0841	0.0815	0.0891	0.0883	0.1140	0.1120	0.1178	0.1232	0.0995	0.1145
5	0.0388	0.0368	0.0445	0.0420	0.0705	0.0648	0.0825	0.0885	0.0812	0.0902
6	0.0168	0.0166	0.0214	0.0200	0.0392	0.0375	0.0587	0.0636	0.0635	0.0711
7	0.0075	0.0075	0.0097	0.0095	0.0235	0.0217	0.0448	0.0457	0.0516	0.0560
8	0.0041	0.0034	0.0052	0.0045	0.0139	0.0126	0.0325	0.0329	0.0430	0.0441
9	0.0015	0.0015	0.0026	0.0022	0.0080	0.0073	0.0246	0.0236	0.0354	0.0348
10			0.0009	0.0010	0.0039	0.0042	0.0178	0.0170	0.0293	0.0274
11			0.0001	0.0005	0.0024	0.0024	0.0133	0.0122	0.0238	0.0216
12				0.0002	0.0003	0.0014	0.0099	0.0088	0.0194	0.0170
13				0.0001		0.0008	0.0074	0.0063	0.0153	0.0134
14				0.0001		0.0005	0.0047	0.0045	0.0136	0.0106
>14						0.0006	0.0340	0.0112	0.1320	0.0383
Adjusted $\bar{n}$	2.317		2.420		2.884		4.006		5.067	
K_1										
K_2	0.3060		0.3110		0.3582		0.4602		0.5603	

TABLE VII. cont.

Acid No. →	16		17		18		19		20	
n	a	c	a	c	a	c	a	c	a	c
0										
1	0.1022	0.1252	0.1058	0.1115	0.1000	0.0811	0.1128	0.0431	0.188	
2	0.1293	0.1666	0.1018	0.1232	0.1090	0.0745	0.0924	0.0412	0.076	
3	0.1153	0.1349	0.0903	0.1061	0.0871	0.685	0.1172	0.0394	0.0257	
4	0.0899	0.1092	0.0802	0.0914	0.0762	0.0630	0.0992	0.0377	0.0152	
5	0.0728	0.0885	0.0655	0.0788	0.0744	0.0578	0.0776	0.0361	0.0124	
6	0.0611	0.0716	0.0576	0.0678	0.0681	0.0532	0.0763	0.0346	0.0113	
7	0.0497	0.0580	0.0496	0.0584	0.0595	0.0489	0.0686	0.0331	0.0077	
8	0.0418	0.0470	0.0447	0.0504	0.0501	0.0449	0.0637	0.0316	0.0009	
9	0.0357	0.0380	0.0376	0.0434	0.0475	0.0413	0.0519	0.0303	0.0009	
10	0.0299	0.0308	0.0329	0.0374	0.0418	0.0379	0.0510	0.0290	0.0010	
11	0.0252	0.0249	0.0308	0.0322	0.0378	0.0348	0.0422	0.0277	0.0009	
12	0.0223	0.0202	0.0280	0.0277	0.0353	0.0320	0.0413	0.0266	0.0009	
13	0.0195	0.0164	0.0244	0.0239	0.0302	0.0294	0.0345	0.0254	0.0009	
14	0.0171	0.0132	0.0216	0.0206	0.0279	0.0270	0.0370	0.0243	0.0009	
>14	0.1882	0.0556	0.2294	0.0273	0.1551	0.3056	0.0342	0.5398	0.0010	
Adjusted $\bar{n}$	5.576		7.390		12.286		23.00			
K_1										
K_2	0.6084		0.7796		1.000		1.000			

Source: After Westman and Beatty, *J. Am. Ceram. Soc.* [896], based on Jameson, *J. Chem Soc.* [426].

*a = adjusted observed mole fractions and c = computed mole fractions. For acid 20, $a = 10\ P_n/n$ and is proportional to mole fraction, where P_n is the atomic fraction of the total phosphorus in chains of length n.

†Computed directly from adjusted X_n.

neighborhood of the H_3PO_4 composition could be treated as a mass law equilibrium with an equilibrium constant of approximately 4×10^{-3}. This was confirmed by cryoscopy. It was considered, however, too complicated for acids of higher P_2O_5 content because of the number of equilibrium constants involved and because they could not be related to one another by some common factor such as hydrogen ion concentration.

Later on, however, application of the mass law to multicomponent systems was investigated by Matula, Groenweghe, and Van Wazer [555]. Their studies included branching, rings, and several types of equilibrium constant. Independently, Meadowcroft and Richardson [567] used the "nearest neighbor" equilibrium constants which form part of the present exposition.

In a reorganizing liquid containing k different polymer chain lengths (including zero), all k categories are considered to be in mutual equilibrium. There are $k!/[3!\,(k-3)!]$ possible equilibrium constants, with the categories taken three at a time. There are only k degrees of freedom as far as chain length is concerned, and is reduced to $(k-2)$ if the distribution is expressed in mole fractions and $\bar{n}$ is specified.

There is convenience and logic to some degree in using the "nearest neighbor" equilibrium constants. These are defined by

$$K_n = \frac{X_{(n-1)}X_{(n+1)}}{X_n{}^2} \tag{16}$$

where X is the mole fraction and the subscripts indicate chain length. There are $(k-2)$ such values that can therefore be independent of one another. There is also some reason to believe that equilibrium between the different polymer lengths is maintained by interchange of orthophosphate units, in which case the K_n values correspond to actual reactions in the melt. Although these values are defined in terms of mole fractions rather than activities, this may not be disadvantageous, for if the activity coefficients approximate a geometric series their effect would largely cancel out.

For the ideal Flory distribution shown in column 2, Table VI, the mole fractions form a geometric series with increasing n, so that all the K values are equal to unity. In an actual acid or glass the number of different chain lengths k may be large or infinite in theory. As shown later, however, the observed distributions approach a Flory distribution as n is increased so that the value of K_n can be taken as unity beyond a value of n, which fortunately can be quite low. The way in which the observed K_n values differ from unity for small values of n is shown for a series of sodium glasses and a series of acids by Figure 24 and Figure 25, respectively. For the glasses K_1 and K_2 are zero and K_3 and K_4 are nearly independent of $\bar{n}$, K_3 in particular, which was pointed out to the authors by Van Wazer and his associates. Both are noticeably less than unity, K_3 being about one-

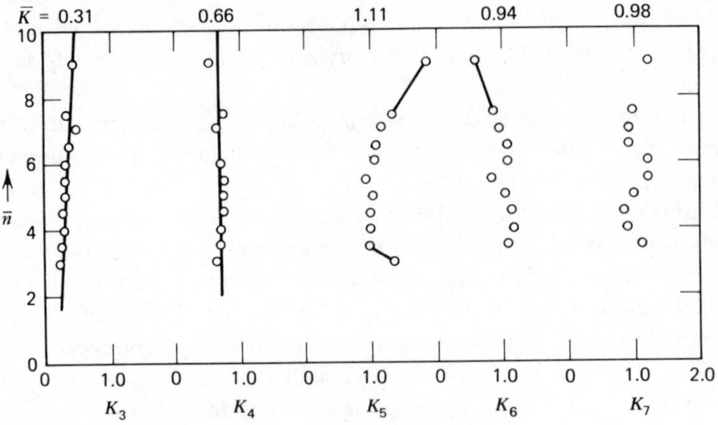

Fig. 24. Observed variations of K_n with $\bar{n}$ for sodium phosphate glasses. After Westman and Beatty, based on Westman and Gartaganis, *J. Am. Ceram. Soc.* [896] and [898], respectively.

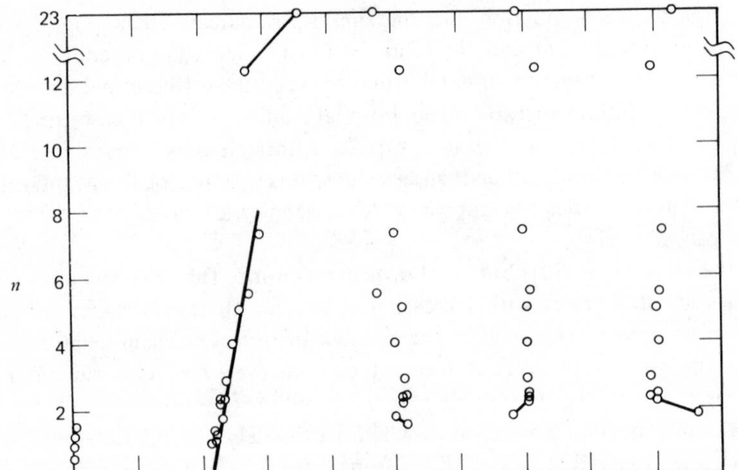

Fig. 25. Observed variations of K_n with n for Jameson Acid Data, K_1 to K_5. After Westman and Beatty, *J. Am. Ceram. Soc.* [896], based on Jameson [426].

third and K_4, about two-thirds. The other K_n values cluster about unity. For the acids K_1 is very low and K_2 is linearly related to $\bar{n}$ for low values of $\bar{n}$ but may become unity for higher values. The other K_n cluster about unity. From this it appears that a reasonable fit to the data could be obtained by deriving two K_n values in each case from the data.

The required modifications of the Flory distribution are shown in Table VI, in which column 4 gives the expressions used for acids and column 5, those for glasses. In each case, given the two K_n values and $\bar{n}$, the cubic equation is solved for p, then X_2 or X_4 is evaluated, as the case may be, and the mole fractions are computed. The use of X_2 or X_4 and p as parameters simplifies the expressions. With this approach iteration methods are avoided and computer time is considerably reduced. Expressions similar to those of columns 4 and 5, Table VI, can be developed for cases in which more than two values of K_n are specified. In fact, they can be written down by symmetry, but for the presently available data they are not necessary.

5. INTERCOMPARISON OF a_n AND K_n

The probability coefficients a_n and the equilibrium constants K_n are shown in Table VI to provide alternative methods for fitting observed molecular weight distributions, given the value of $\bar{n}$. To produce the same distribution the K and a values must be related:

$$K_1 = \frac{[a_2(1 - a_1p)(1 - p)]}{[a_1(1 - a_2p)^2]} \tag{17}$$

$$K_2 = \frac{(1 - a_2p)}{[a_2(1 - p)]} \tag{18}$$

where p is given by the quadratic of column 3 or the cubic of column 4.

For a series of acids or glasses with increasing $\bar{n}$, the K_n values are constant or simply related to $\bar{n}$ and tend to approach unity. They are therefore of more interest, although they require an equation of higher degree. The last is not important if a computer is used. The a values may be of interest in speculating on reasons for departures from the ideal Flory distribution, and they follow the chemical tradition of modifying ideal quantities by coefficients.

6. JAMESON ACID DATA

The Jameson acid data are given in Table VII. Polymers up to $n = 14$ in the strong phosphoric acids were determined by ion exchange column chromatography. The "unreacted" water content in the low n range could be estimated from the chemical analysis and the chromatographic results. Values of K_n could then vary from K_1 to K_{13}. As shown in Figure 25, when these values are plotted against the $\bar{n}$ values, K_1 is small, K_2 is linearly related to $\bar{n}$ to about $\bar{n} = 10$, and the K_n values for subscript n greater than 2 approximate unity. No definite trend with n beyond K_2 is evident. Therefore two empirical K_n values should suffice.

In adjusting the Jameson data, the chromatographic analyses were first adjusted to total 100%. Then, if P_n is defined as the atomic fraction of the total phosphorus occurring in chains of length n, it was found that the sum which should be unity

$$\Sigma_n \; \bar{n} \; \frac{P_n}{n} \tag{19}$$

if the $\bar{n}$ derived from the chemical analysis agreed with the chromatographic results, in fact totaled 1.02 when no hypoly was present. The data were then adjusted so that all acids gave 1.02 for this sum; then division by 1.02 produced values totaling unity. The adjusted data are shown in the columns headed *a* in Table VII.

Use of the adjusted data and the scheme in column 4, Table VI, produces the computed distributions in the column 5 headed *c* in Table VII. The values of k_1 and K_2 were calculated from (20) and (21), based on the adjusted X_n and $\bar{n}$ from the chemical analysis:

$$K_1 = -0.01598 + 0.02470 \; \bar{n} \tag{20}$$

$$K_2 = -0.081941 + 0.094412 \; \bar{n} \tag{21}$$

For most of Table VII the agreement between the observed and computed mole fractions is as close as could be expected from data of this accuracy. For acids 18 and 19 the observed values are somewhat higher than the computed values at low values of $\bar{n}$. For those acids it is difficult to choose a value of K_2 as Fig. 25 shows. The straight line no longer holds and taking $K_2 = 1$ may be overestimating it. Hydrolysis during chromatography can also increase the number of short chains. However, because of the smoothing effect of curve fitting, it is likely that in this table the computed values are more nearly accurate than the observed values.

7. HUHTI AND GARTAGANIS ACID DATA

The Huhti and Gartaganis data (Table V) [404] for the strong acids, obtained by paper chromatography, were analysed in the same way. The data were not so precise or extensive but the result was the same. In fact, the K_n values plotted on the diagram of Fig. 25 showed good agreement with the Jameson data.

The K_1 value for the composition H_3PO_4 corresponds to about 6 mole % unreacted water, which is much higher than that found for the composition H_2SO_4.

On the other hand, the Flory ideal theory would predict an acid nominally H_3PO_4 to be one-half unreacted water, one-quarter orthophosphoric acid, and one-eighth pyrophosphoric acid on a mole fraction basis. In the actual acid it appears that there is more opportunity for water to be eliminated by hydrolysis than to be formed by condensation than the Flory theory anticipates.

8. SODIUM PHOSPHATE GLASS DATA

The most precise and extensive data for soluble sodium phosphate glasses were those of Westman and Gartaganis [898]. No values for X_1 were reported because they were quite small and were neglected at the time this work was done. The K_n versus $\bar{n}$ charts based on these data are shown in Fig. 24. For the range of $\bar{n}$ studied K_3 is linearly related to $\bar{n}$. For higher values of $\bar{n}$ there is no evidence that K_n is different from unity, although the data are evidently less regular than the Jameson data for acids. The K_3 slope approximates the K_1 slope for the Jameson acid data and is much less than the K_2 slope. The K_4 constancy is rather striking when the wide range of compositions covered are considered. A comparison of observed and computed distributions for these data, using K_4 as a constant independent of $\bar{n}$, leads to the same conclusion.

9. OTHER PHOSPHATE GLASS DATA

The data published for mixed alkali phosphate glasses by Murthy and Westman [603] and for sodium phosphate glasses by Murthy [598] and much unpublished data available at Toronto were analyzed by the above method. In all cases the data could be plotted on the diagrams of Figure 24 to give a good fit to the lines.

In the later studies the small amounts of orthophosphate showing in the chromatograms were determined. They usually accounted for less than 1 wt.% of the total phosphorus. Corresponding K_2 values increased rapidly with $\bar{n}$ to as much as 4 or 5, which showed that the small amounts of orthophosphate found in the sodium phosphate glasses over the $\bar{n}$ range from 3 to 9 were probably the result of hydrolysis during the solution and chromatographing of the glasses and were not produced by reorganization in the melt.

For the mixed alkali glasses the data plotted well in the diagram of Figure 24. For the sodium acid glasses [904] the data fitted better in Figure 25 but were somewhat intermediate in character, as might be expected.

When tables like Table VII were prepared, the agreement between observed and calculated values, although as good for some values of $\bar{n}$, was not consistently as good as for the Jameson acid data. This might be expected from the tendency, shown in Figure 24, of the K_n at high subscript n values to deviate from unit.

The conclusion to be drawn from the preceding subsections is that the Flory distribution, column 2, Table VI, can be modified to fit all available data concerned with equilibrium between unbranched linear polymers within the limits of experimental error and that further development of distribution is not necessary for this type of data. When small amounts of cyclophosphates are present, as in the sodium phosphate glasses of larger $\bar{n}$, it is probably good procedure to allow for the effect of the ring formation on the $\bar{n}$ of the unbranched chain polymers and then treat them as though no cyclics were present. In other words, it is likely that a small percentage of cyclics has little effect on the equilibrium between the unbranched chains except by influencing their value of $\bar{n}$. This necessity of shifting $\bar{n}$ for the chain equilibrium to lower values when rings are formed may be the factor that limits ring formation. If so, its effect would be expected to be least or to approach zero as the overall composition approaches the metaphosphate composition because at that composition the formation of rings would have no effect on Na/P ratio of the rest of the melt. As shown in Figure 18, the cyclic content of the sodium phosphate glasses does increase steadily as the metaphosphate composition is approached.

Despite this argument, there is considerable interest in being able to handle chain-ring equilibria and, when data become available for glasses in the ultraphosphate range of compositions, distribution theory that will handle branched chains will be of interest. In the meantime there are many situations in ligand and polymer chemistry, both organic and inorganic, in which a more general distribution theory is required, and great development has occurred in distribution theory since Van Wazer first applied it to the sodium phosphate glasses. This development has taken the form of the generalization of the Flory theory for bifunctional monomers and linear polymers (Table VI, column 2), largely by Flory himself, to include monomers of higher functionality and branched, crosslinked or cyclic polymers. Most of these theories are developed first on a "completely random" basis; that is, the probability that a particular functionality has formed a bond is independent of what has happened to functionalities on the same or other units.

10. GENERALIZATION OF THE FLORY DISTRIBUTION

The ideal Flory distribution in column 2, Table VI, is restricted to bifunctional monomers. This was generalized by Flory [280, 282] to include monomers of functionality higher than 2, which therefore permit branching. (Although the orthophosphate monomer is, on the face of it, trifunctional, yet because of the antibranching rule it acts as a bifunctional monomer over the range of compositions discussed above.) The ideal Flory approach, although it permitted branching, did not permit cyclization and assumed that the probability of a functionality forming a bond was the same at any instant for all unreacted functionalities. This concept was further generalized by Gordon and Scantlebury [323] to cases in which the ease of formation of a bond is influenced by the number of links already formed by the two repeat units to be bonded. This they called the "substitution effect."

In recent years Masson and his associates [919] have reinvestigated the Flory equation for multifunctional monomers under the same condition of no cyclization and have arrived at (4), a somewhat different equation:

$$N_x = \frac{(fx - x)!}{(fx - 2x + 1)!\ x!} (p_{AA})^{x-1}(1 - p_{AA})^{fx-2x+1} \tag{22}$$

where N_x = the mole fraction of x-mer, f = the functionality of the monomer of the form $R - A_f$, $p_{AA} = \alpha f/2\ (f - 1)$, where α = fraction of initially present A groups that have reacted (in consulting the Masson or Flory papers the reader is advised to note carefully the use of two slightly different Ns to denote quite different quantities, in one case a mole fraction and in the other the total number of molecules). The work of the Masson group is discussed further in Subsection J.

For $f = 2$, in which case no branching is possible, (22) reduces to the equivalent of the ideal Flory distribution of Table VI, column 2, if the difference in symbolism is taken into account. In the Westman and Beattie symbolism M_2O is a polymer of zero chain length which is taking part in the dynamic equilibrium. In consequence the term mole fraction is given a different meaning. Like most polymerization equations, (22) was developed for an on going polymerization and is concerned with an instantaneous equilibrium at a time when the fraction α of functional groups has reacted. In a quenched glass or acid the distribution of chain lengths is a stable one that depends on the M/P ratio which determines the distribution in much the same way as α does in a polymerizing system.

There are many ways in which the Flory type of distribution can be generalized. Probably the ultimate has been developed by Van Wazer and his associates in a number of publications which are summed up briefly in a paper by Matula, Groenweghe, and Van Wazer [555]. A short summary is provided by Haiduc [347]. These authors make use of the specialized branch of mathematics known as the theory of graphs and develop an approach applicable to a great variety of what they call "scrambling reactions" in both organic and inorganic reactions. Using the same data for phosphate glasses and phosphoric acids as Westman and Beattie, discussed earlier in this section, they arrive at what are essentially the same conclusions; that is, only a few empirical constants are necessary to get a good fit to the data, the abundance of the longer chains being calculable from a random hypothesis.

In the same article the authors discuss the effect of ring formation to be expected on the molecular distribution. The presence of rings poses the problem of determining the random probability of ring formation. The situation is even more complicated if branching is also present. For phosphate glasses and acids, in which rings are found in only minor amounts, it appears best to treat the rings separately and to consider that they have no effect on the distribution of the unbranched chain molecules except that of changing their M/P ratio when the M/P overall ratio is not unity.

Matula, Groenweghe, and Van Wazer [555] attribute the change from nonrandomness for short chains to randomness for longer chains in the case of phosphate glasses and acids to a change from enthalpy domination of the change in Gibbs' free energy to entropy domination. From this point of view it is interesting to discuss the formation of a glass when a crystal such as sodium tripolyphosphate is melted. In the crystal, shown in Figure 1, the tripoly molecules are packed in a systematic manner and are indistinguishable from one another. When heat equivalent to the lattice energy is supplied, the molecules move in relation to one another. This provides for disproportion and other reactions such as the trading of orthophosphate groups and results in a dynamic equilibrium between a considerable range of chain lengths [curve (1) in Figure 17]. The driving force is the increase in entropy associated with the change from the highly ordered crystal with only one molecular species to a glass with a large number of species that exhibit considerable randomness in their molecular weight distribution. It was postulated that it is only for molecules of very short chain lengths that heat effects are important. In Figure 17 it will also be observed that the systems with the longer number average chain lengths have wider distributions and presumably a higher entropy. The formation of rings in these systems would therefore be expected to decrease the entropy, and this may be the main factor limiting their percentage. In the same way, although a crystal of trimetaphosphate phosphate when melted may for a short time consist wholly of rings, it breaks down quickly to produce a wide range of chain lengths and a limited proportion of rings. If held at about 550°C for some time, however, the whole crystallizes as trimetaphosphate because this releases its lattice energy.

J. Silicate Slags and Glasses

In view of the similarity between PO_4 and SiO_4 tetrahedra, it might be expected that the progress made in understanding phosphate glasses in the last two decades would be parallelled by silicate glasses. This is not the case. Progress in silicates has been held back by the assumption that they tend to polymerize readily in solution and precipitate as gels. Initial attempts by my associates and me to apply paper chromatography to sodium silicates met with some indications that it might be possible but was abandoned. Because a paper chromatographic method for checking various theories regarding silicate glasses was lacking, recourse was had to other methods.

Stevels and his associates [302a, 779, 832], in a study of the electrical porperties of sodium silicate and similar glasses concluded, for example, that sodium silicate glasses with a Na/Si ratio greater than unity differed fundamentally in constitution and gave them the name "invert glasses." These invert glasses were presumed to have the chain constitution found for sodium phosphate glasses.

In studying calcium silicate minerals, it was found that when the mineral was dissolved in an acid solution the calcium was removed from the crystal and the remaining silicate ions quickly polymerized and precipitated. With this in mind, Lentz [528] developed a technique in which trimethylsilyl derivatives of the silicate ions were formed as the calcium was dissolved. This prevented the polymerization of the silicate ions. These derivatives could then be separated by gas-liquid chromatography and identified by high resolution mass spectroscopy. This method was demonstrated for the smaller silicate ions.

Masson and his associates approached the problem of the constitution of calcium silicate and other silicate slags from the thermodynamic viewpoint by using observed changes of activities with composition as an experimental check on speculations about the constitution of the silicate ions. They investigated the Lentz technique and found the conditions for optimum results [937]. They also published a series of papers on the application of polymer theory to silicate melts [919]. The results of their theoretical calculations are compared with observed data on the activities of the silicate ions and also the analyses of the ions that could be determined by the improved Lentz technique. For a number of binery silicate melts the observed results could be represented best in terms of theory in which all chain configurations, including branching, are allowed. However, in the case of $CoO\text{-}SiO_2$ melts at 1450°C the observations were best accounted for by specifying unbranched chains for the silicate ions such as were found for the phosphate ions in phosphate glasses and phosphoric acids.

More recent contributions include [551], [328], [920], and [327].

In view of the contribution that paper chromatography has made to the constitution of phosphate glasses, it is surprising that the paper chromatographic method developed by Wieker and Hoebbel [924] for the condensed silicate ions has not been applied to the constitution of silicate glasses.

K. *Aqueous Glasses*

When Zachariasen [952] laid the basis for the crystal chemistry theory of glass structure, he presented rules for distinguishing glass-forming oxides and listed those he thought were possible glass formers. However, the possibility of cooling a liquid to the glassy state depends to a considerable extent on the rate of cooling, and many materials not on Zachariasen's list of glass formers can be obtained in the vitreous state by adopting extremely fast rates of cooling.

Water is a striking example. Vitreous water was first produced by Washburn not later than 1933 at the U.S. Bureau of Standards [886]. He directed a jet of liquid helium against a thin glass partition and allowed water vapor to enter slowly into a high vaccuum on the other side of the partition. The water vapor condensed on the partition to form an amorphous layer that he called "vitreous water." When this layer was allowed to warm up, it crystallized into ice.

Aqueous solutions of inorganic compounds, including H_3PO_4, can be obtained more readily in the vitreous state, and these "aqueous glasses" have been studied in depth by Dainton and his associates with advanced kinetic and radiation chemistry techniques [150, 151, 204, 205]. Included were concentrated sulfuric acid, hydrofluoric acid, phosphoric acid, and perchloric acid. When these glasses were irradiated, the studies showed that one of the primary radiation chemical species was the electron. A study of rate constants led to the conclusion that the viscosity of glass-forming liquids follows the Tamman-Hesse law, not the Arrhenius law.

A study of diffusion-controlled reactions of the solvated electron with various neutral solutes showed that the rate constants followed the equation $\ln k = A - B/(T - T_0)$, where A, B, and T_0 are characteristic of the medium only. T_0 for aqueous systems is always about 135°K.

A similar study could not be made with acid aqueous glasses because the electrons reacted rapidly with hydrogen ion. However, rate studies of reactions of hydrogen atoms with other solutes, for example, Ag^+, showed a temperature dependence similar to that found for neutral aqueous glasses.

When the viscosity was increased, not by cooling but by the addition of polymeric molecules, it was found necessary to distinguish between bulk viscosity, which is due to structure, and viscosity at the molecular scale, which is characteristic of the small molecules of the medium.

L. *Possible Future Developments*

Because the remaining sections of this contribution are concerned primarily with the technical side of phosphate ceramics, it may be useful to conclude the more scientific sections with a brief discussion of the directions in which the scientific side of phosphate ceramics may be expected to develop.

Much is now known about crystalline and glassy phosphates with, for example, Na/P ratios of more than unity. Little is known, however, about the constitution of the ultraphosphates Na/P<1, except for the limiting case of P_2O_5, and even there much could be learned about glassy P_2O_5. The crystalline ultraphosphates described in the literature have been found on further investigation to be other than ultraphosphates. Ultraphosphate glasses react rather vigorously with water and their investigation by paper chromatography presents problems. Presumably the ultraphosphates are branched and crosslinked and also contain rings. On solution in water it might be expected that the antibranching rule would require that all branching points be hydroyzed and would have to be taken into account in interpreting the data obtained. In 1958 Mrs. Smith [756] investigated glasses made of P_2O_5 and H_3PO_4 in the ultraphosphate range with an H/P ratio of approximately 0.5. She found that the results obtained depended on the solvent in which the glass was dissolved. If water was used, only orthophosphate was found. Evidently the heat generated was enough to ensure almost total reversion to the orthophosphate. When dissolved in an ammonium carbonate solution, orthopyro-, and some tripolyphosphate were found, and in an ammonium acetate solution phosphates as large as nonapolyphosphate were present. There appears to be no reason why the methods developed by Masson for the investigation of slags could not be directed to the composition of glasses of ceramic interest.

The review of phase equilibrium studies in Section IV indicates a need for phase equilibrium studies of systems of more interest in tradional ceramic science; for example, although the Na_2O-CaO-SiO_2 diagram has been investigated a number of times, there appears to be no complete report on the corresponding Na_2O-CaO-P_2O_5 system. Also, little attention has been given to exploring the constitution of the liquid phases, although in phosphates it has been very revealing.

There is the challenge of preparing a sodium phosphate glass whose constitution would consist largely of cyclic phosphates. This would require fast heating and cooling but might easily be within the reach of shockwave techniques.

For the silicate glasses the technique developed by Wieker and Hoebbel [924] should be applied to silicate glasses, starting with the soluble Na_2O-SiO_2 system.

Attempts to apply the Lentz technique to phosphate glasses have not been successful and would bear more investigation. In the meantime a connection between phosphate and silicate glasses might be established by comparing CaO-SiO_2 and CaO-P_2O_5 glasses, using the methods developed by the thermodynamic approach of Masson.

VI. Phosphate Ceramics Technology

A. *Introduction*

The following chapters (VI to XIII) are concerned with the technology of phosphate ceramics. The development of phosphate ceramics science has not always led to the development of a successful technology; in fact, there have been many disappointments and only a few commercial successes. In comparison with silicates, phosphates are expensive, tend to volatilize, are more prone to devitrification, and may have a low resistance to moisture, acids, and other agents. Consequently the development of a new phosphate ceramics technology should be undertaken only with an appreciation of the difficulties likely to be encountered and with plans to circumvent them.

Information regarding phosphate ceramics technology is to a considerable extent buried in the patent literature. Because in recent years patents have been reported in *Ceramic Abstracts* only by title, this constitutes a handicap.

Some evidence of the relative importance attached to the science and technology of phosphate ceramics may be gained from the fact that among the 1319 references bearing directly on these subjects about one-third could be classified as science and two-thirds as technology.

To review the information offered by the 867 references on technology it was necessary to develop a system of classification.

B. *Classification of Technology References*

The classification adopted was based on that used by *Ceramic Abstracts.* Again some estimate of the relative importance attached to these classes in the literature can be gained from the direct references, which were distributed as follows: glasses, 56%; refractories 12%; whitewares, 13%; ceramic metal systems, 6%; cements, limes, and plasters, 7%; abrasives, 1%; structural clay products, one reference only; and other classes, 5%

VII. Glasses, Technical

A. *General*

The phosphate glasses discussed in Chapter V were largely water-soluble or capable of being put into solution by the use of acids, complexing agents, or ion exchange. This was justifiable from a scientific point of view, but to be useful from a ceramic product point of view a considerable degree of chemical durability is necessary. It is true that sodium phosphate glasses of metaphosphate composition are useful as water conditioners and corrosion preventives, but these are chemical rather than ceramic uses. Phosphate glasses in general may have several desirable physical and chemical characteristics, but to make commercial use of these properties it is necessary to develop glass compositions that will make the product durable under the conditions of use. A brief discussion of glass formulation serves as an introduction to this chapter.

1. GLASS FORMULATION

The great bulk of commercial products, such as window and container glasses, are essentially three-component glasses in the Na_2O-CaO-SiO_2 system. The ratios of these three oxides put the composition in the field in which devitrite Na_2O-$3CaO$-$6SiO_2$ is the primary phase. This compound, whose composition lies outside this field, has a very weak crystallizing tendency. Also, the liquidus of these glasses is quite low, but the viscosity is high enough so that they can be worked at temperatures above the liquidus, thus avoiding devitrification. Actually, if all the requirements of fusibility, formability, freedom from devitrification, or liquid-liquid phase separation, and chemical durability are met, the field of possible compositions in this system is rather limited. For this reason there is a close resemblance between the compositions of ancient and modern glasses. If the silica content is raised, the melting and forming are difficult and liquid-liquid phase separation or devitrification may be encountered. If the soda content is raised, chemical duability is decreased and thermal expansion is raised. If the lime content is raised, devitrification is encountered; if it is lowered, solubility is increased. By introducing boric oxide B_2O_3 the alkalies can be greatly reduced or eliminated but fusibility maintained and glasses with high acid resistance and low thermal expansion developed.

Starting with the water-soluble sodium phosphate glasses, it is necessary to increase their chemical resistance by adding alumina Al_2O_3 to decrease the tendency to devitrification and to increase the chemical durability. Also, alkaline earth metal oxides such as calcium oxide and barium oxides may be added to increase further the chemical durability. Other additions for the same purpose may include titanium oxide and zirconium oxide. The formulation of phosphate glasses for commercial purposes is quite an complicated field and has been subject of much patent activity. Usually a particular property of phosphate glass is required, such as UV transmission, and the problem is one of maintaining this property and at the same time formulating a glass that will meet the other requirements of commercial production and use. A detailed review of the formulation of glasses is given by Volf [880].

A few examples from the literature will illustrate our discussion of glass formulation with phosphate glasses. Syritskaya and Yakubik [806] explored the glass region in the system P_2O_5-Al_2O_3-ZnO, thus avoiding any alkalies. They found stable, noncrystalline, and chemically durable glasses in a region of the composition extending from 50 to 73% ZnO on the P_2O_5-ZnO side of the triangle right across the diagram to the range 62 to 73% Al_2O_3 on the P_2O_5-Al_2O_3 side. Stanworth [768] developed a phosphate glass containing 28 to 38% (wt) P_2O_5, 8 to 16 Al_2O_3 13 to 22 B_2O_3, 24 to 34 ZnO + MgO + BaO, and 6 to 10 Na_2O, keeping the mole ratio P_2O_5/Al_2O_3 within the range 1.5 to 3. Weidell [890] developed a "metaphosphate" glass of composition 33 to 41 aluminum metaphosphate, 16 to 21 magnesium metaphosphate, 16 to 21 potassium metaphosphate, 0 to 16.3 B_2O_3, and 2.4 to 20.6 TiO_2, with the requirement that (B_2O_3 + TiO_2) be in the range of 2.4 to 34%. Junge [437] developed glasses based on barium and lithium ultraphosphates. After an alkaline film was removed by washing, these glasses could be formed by the methods used in plastic technology. It was found, however, that the objects formed could not be heat-tempered, which made commercial application impossible.

Volf [880] mentions a number of phosphate glasses classified as fluoride-resistant, high alumina, metal-vapor-resistant, and fiber glass. The fluoride-resistant glasses contained 60 to 80% P_2O_5, the high alumina glasses about 5% P_2O_5, the metal-vapor-resistant glasses 13 to 20% P_2O_5, and the fiberglass 67% P_2O_5. The composition of the last was 2.1 SiO_2, 67.0 P_2O_5, 4.6 B_2O_3, 25.2 CaO and MgO 0.6. Volf remarks that this glass is based on calcium phosphate and is similar to the Corex UV transmitting glass. It will be seen that phosphate glasses used commercially cover a wide range of compositions and in some cases contain rather large amounts of P_2O_5. Some of them are discussed in more detail later. It will be evident that they depart a great deal from the ordinary soda-lime-silica glass used for containers and windows.

2. HISTORY OF GLASS

The origin of glass is discussed by Barag [47] and Gray [329a]. The first glass used by man was "obsidian," a product of volcanic action which occurred naturally. It was chipped and flaked to make arrowheads, knives, and razors, probably about 75,000 B.C.

The Egyptians were the first to make glass vessels, formed around a sand core. This was a difficult process but remained unchanged for about 20 centuries. Roman or Syrian craftsmen invented the blowpipe in the first century B.C.

The development of glassmaking in Europe, in particular the optical glass industry due to the researches of Otto Schott, is outlined by Kiaulehn [457]. Turner [845] wrote six articles reporting on his studies of ancient glasses and glass-making process. Forbes [778] writes on glass throughout the ages. Brill [131] gives the typical compositions of ancient glasses, and Vassus [868] describes about 40 glasses from the Middle Ages. Zschimmer [961] recounts a history of glass research over the 60 years preceding 1927, and Weyl [908] describes the development of optical glasses in a memorial to Otto Schott in 1935. A similar history was provided by French [288].

The earlier glasses were made from sand or quartz; soda from lake deposits or potash from wood ashes was used as a flux. These two constituents were not enough to produce a durable glass, and the success of the glassmaker over many centuries was dependent on the adventitious inclusion of alkaline earths, mainly lime, which occurred naturally in the sand or wood ashes. The ashes of beechwood, particularly beechwood brush, are fairly high in lime. It was only in relatively modern times that lime was added deliberately, for instance, in the development of the Bohemian crystal glass, which was a potash-silica mixture to which limestone was added. Another source of calcium oxide was sea shells.

Brill [131] gives a typical analysis of an ancient East Asiatic glass as high in lead oxide and barium oxide.

Many ancient glasses show a phosphate content as high as 4%. This is particularly true of the glasses of the Middle Ages. This phosphate content was usually the result of the use of wood ashes for potash, beechwood ash (Geilmann, Wilhelm and Jeneman [302]). The percentages were not high enough to have a large effect on the properties of the glass. Wood ash was not the only source of phosphate available to early glassmakers, however. Ganzenmuller [299] considers that phosphate glasses were first made in the fourteenth century and quoted recipes from fourteenth- and fifteenth century treatises which included "burned white bone" and "deer bones" (buffalo or ox). He also refers to an alchemist who hints of "bone glass."

Up to the year 1876 the only glasses available to the manufacturer of optical telescopes and microscopes, apart from the usual commercial glass that was roughly 70% silica, 15% alkali, and 15% limestone, were flint and crown glass. The flint glass was made from silica in the form of flint along with chalk and a considerable addition of lead oxide. The term "flint" is also used in the container-glass industry to indicate a colorless glass in contrast to amber. Crown glass was made from pure quartz by adding chalk and potash. In 1758 [457] Dolland succeeded in making an achromatic telescope by combing lenses of flint and crown glass. However, a colored fringe called the secondary spectrum still remained. Although Harcourt, in the period 1834 to 1870, made a number of new glasses, neither he nor Faraday or Stokes succeeded in solving the problem. In 1876 Abbé, who was a designer of optical instruments, suggested that a new type of glass was needed to solve the problem. In 1879 Otto Schott, who had had some chemical as well as glassmaking experience, started experiments in the cellar of his father's house which were to lead to the tremendous development of optical and special glasses at Jena in Germany. By 1881 he had departed from the usual silica-based glasses and had developed others based on phosphates which successfully got rid of the secondary spectrum. However, these glasses were attacked by atmospheric moisture and could not be used commercially. This disappointment was to be repeated several times in the years to come on other experiments until all research in this direction was discouraged. Schott renewed his efforts with silicate or borate glasses or a combination in the form of a borosilicate glass. By 1883 Jena was producing 44 optical glasses, of which 20 were new. In 1891 sight glasses for boilers, made of two layers of glass of different thermal expansion, were produced; in 1892 and 1893 borosilicate glasses, resistant to temperature change, became available in large quantities for mantle-lamp chimneys all over the world.

The more recent history of glassmaking will become apparent as the references found in this review are discussed. These references were classified under the headings composition, state, structure, manufacture, chemical properties, physical properties, use, and radiation interaction, among which the largest categories were composition, state, structure, and use.

It has been suggested recently that multicomponent glasses might be made without fusion. The glasses would be formed by hydrolysis and polycondensation when heated no higher than the transformation range. This was suggested by Dislich [222].

B. Composition

The following discussion is concerned with references in which the emphasis appears to be mainly on composition in glasses with a P_2O_5 content. In passing, it might be of interest to note that glasses can be made in which elementary phosphorus can be a constituent. Thus Hilton, Jones, and Brau, [384] patented a glass made of germanium tellurium and phosphorus and showing transmission in the 1 to 25 mu range: Hilton and Jones [383] made a similar glass from germanium phosphorus and sulfur.

1. ONE-COMPONENT GLASSES

P_2O_5 by itself can form a glass. In discussing it, Kreidl and Weyl [909] refer to the work of Campbell and Campbell [156] on the allotropy of phosphorus pentoxide. It is concluded that P_2O_5 is unusual because the vitreous form has the lowest solubility in chloroform and the highest density. Its coefficient of expansion and electrical resistance are high and its softening point, low. The optical properties resemble those of silica. The chemical resistance is low, however, and the latter disadvantage can be overcome by adding alumina. Another reference is to Weyl's book on colored glasses [912].

2. TWO-COMPONENT GLASSES

The constitution of the readily soluble binary phosphate glasses has already been discussed in Section V-D. Information regarding the physical properties of sodium phosphate glasses and their dependence on the Na_2O/P_2O_5 ratio was obtained by Fanderlik and Palecek [266] for a number of physical properties. They concluded that the point of inflection occurred at a ratio of 1.1 to 1.2, that is, on the alkaline side of the metaphosphate composition. This does not conflict with the chromatographic results, which indicate the presence of rings and some branching on this side of the metaphosphate composition. Calcium polyphosphate glasses were investigated by Ohashi and Van Wazer [645]. Glasses were formed readily by heating for two hours at 1400°C and quenching. The role of iron in calcium phosphate glasses was investigated by Bishay and Makar [95]. Carpenter and Johnson [185] obtained a patent for a cadmium phosphate glass obtained by heating CdO, $CdCO_3$, or $Cd(NO_3)_2$ with $(NH_4)_3PO_4$. Drake and Scanlon [229] patented a pyrolytic zinc phosphate glass in 1970. Tiwari and Subbarao [825] studied bismuth-phosphate glasses and determined the range of glass formation and physical properties. Homogeneous glasses were formed up to 35 mole% Bi_2O_3. Various physical properties showed maxima, minima, or an abrupt change at about 20 mole% Bi_2O_3. These phenomena were attributed to the change of Bi^{3+} ions from network formers to network modifiers. They were interested in optically transparent shields against gamma rays and neutron flux. Heynes and Rawson [379] report that bismuth oxide behaves similarly to lead oxide. It does not form a glass by itself but will do so in combination with SiO_2, B_2O_3, and P_2O_5. Eipeltauer and Hammer [249] studied the viscous behavior and constitution of binary bismuth-phosphate glasses containing BiO_2 53 to 65%. They concluded that there were at least four but possibly five or more phosphate components, with the chain length increasing with the P_2O_5 content. The existence of the compound $2Bi_2O_3.5P_2O_5$ in the system was indicated by the viscosity isotherms and suggested by the X-ray results. Schulz and Hinz [738] studied the stability of phosphate ions in melts and glasses of a number of binary phosphate systems, including those formed from Pb, Cd, and Zn orthophosphates and those from Na, Li, K, Pb, and Bi pyrophosphates. Molecular weight distributions were determined. They concluded that orthophosphates are stable in all melts except when cations such as Si^{4+} and W^{6+} are present. When cations of such high atomic weight and field strength are present, P_2O_5 is easily split off. High P_2O_5 is an advantage. WO_3 with 12% P_2O_5 forms a deep blue glass. Provance and Wood [687] investigated molybdenum phosphate glasses containing Ag_2O

and K_2O. Kierkegaard, Eistrat, and Rosen-Rosenhall [458] studied both molybdenum and tungsten phosphate glasses.

3. THREE-COMPONENT GLASSES

The following references are concerned with glassmaking characteristics. When some particular use seemed to be the principal consideration, the reference is given later under that use.

Laud and Hummel [523] investigated subsolidus relations in the system ZrO_2-ThO_2-P_2O_5 along the pyrophosphate join. They found that the system contained a number of low-expansion phases "which could be a basis for semirefractory, thermal-shock-resisting ceramic bodies." Matveev, Khodskii, Fisyuk, Bolutenko, and Strugach [557] investigated the systems BaO-TiO_2-P_2O_5, B_2O_3 and SiO_2. The vitrification region at 1400°C and lower was largest for the borate and smallest for the phosphate system. Most glasses had good chemical stability but both microhardness and water resistance decreased with an increase in TiO_2. In all systems density increased with BaO. The region of glass formation in the system Al_2O_3-TiO_2-P_2O_5 was investigated by Syritskaya and Shapovalova [804] who found two glass regions, high phosphate and high titanium. Ohashi and Oshima [643] studied the chemical compositions of crystals and glasses in the systems $NaPO_3$-Na_2SiO_3 and $NaPO_3$-SiO_2. They found some glasses that contained P-O-Si linkages but none in crystalline substances. The silicate ions did not interfere with the filter-paper chromatography of the condensed phosphates. In shifting from P_2O_5 to SiO_2, the structure was transformed successively.

A number of phosphate glasses containing PbO have been described. Silverman, Rothermel, and Sun [750] developed a phosphate glass of the following compositions: one oxide of 6 to 67 MoO_2 or 7 to 84 WO_3, at least one oxide of 7 to 50 Bi_2O_3, 12 to 70 PbO, 13 to 64P_2O_5, and not more than 5% minor oxides. This glass had a lead equivalent greater than 0.33 for 100 kV X-rays. The same authors [723] describe a PbO-WO_3-P_2O_5 glass. Felice de Carli [270] made glasses by mixing sodium metaphosphate, that is, $NaPO_3$, with the oxides of lead, cadmium, and manganese. He concluded that $NaPO_3$ forms orthophosphates and metaphosphates with metallic oxides and that glass beads owe their color to pyrophosphates. This was in 1926, before paper chromatography.

In analogy to borosilicate glasses, there have been a number of studies of borophosphate glasses. Stanworth and Turner [771] studied the possibility of preparing glasses from P_2O_5, Al_2O_3, B_2O_3 and SiO_2 with and without the addition of basic oxides. Glasses containing P_2O_5, B_2O_3, and Na_2O were clear when sufficient Na_2O was present and resistant to atmospheric attack when high percentages of P_2O_5 were present. This is surprising when the attraction of P_2O_5 for water is considered. Other glass makers, however, experimenting with phosphate glasses have said that increasing resistance to moisture could be obtained at times by increasing the P_2O_5 content. The authors also found that a clear glass could not be made from silica and metaphosphoric acid but could be prepared from silicon phosphate. Syritskaya and Kutukova [802] found that the water resistance of a borophosphate glass could be considerably increased by replacing either oxide with La_2O_3. The work of Syritskaya and Yakubik on glasses in the ZnO-Al_2O_3-P_2O_5 system has already been described (VII, A-1). Mazo and Navarro [564] started with a glass containing 65 mole% P_2O_5 and the remainder Li_2O, Na_2O or K_2O. They then replaced the P_2O_5 with B_2O_3 progressively. By paper chromatography they showed a progressive degradation of the structural groups that formed the binary phosphate glasses. Mazo and Fernandex [565] studied the effects of dissolved nitrogen on the properties of alkali borophosphate glasses, with R_2O, 35 mole%, B_2O_3, 17 to 49%, and P_2O_5, 16 to 63%, where R = Li, Na, or k. The effect of bubbling nitrogen through the glass was much smaller for the glasses containing K_2O. The reducing action of NH_3 destroyed the phosphoborate network and produced elemental P and BN. Stanworth [769] introduced RO oxides at the expense of P_2O_5 in Al_2O_3-B_2O_3-P_2O_5 glasses. He was able to make a number of alkali-free glasses with high electrical resistance and high deformation and annealing temperature. Addition of alkali oxides reduced the devitrification tendency but also the electrical resistance. Grossman and Phillips [344] developed a zinc-borophosphate glass.

Beekenkamp [62, 63] and Beekenkamp and Hardeman [64] published three papers on the structure of vitreous borates and borophosphates. They found that the property versus composition curves were different from those of the silicates and suggested improved models based on NMR measurements. They also studied color centers and found that many properties of phosphate glasses show a marked change at the metaphosphate composition, which may be explained by the change from a three-dimensional network to an assembly of chains crosslinked by O-M-O bridges. In general, the properties that showed this effect were related to the bulk rigidity of the glass network. The color centers, however, were not related to the bulk rigidity but to the local rigidity of the structural units. These units alter only gradually as the composition is changed. Consequently there was no sudden change in the color centers at the metaphosphate composition. The analogy in structure between crystalline compounds of the type $AlPO_4$ and silica does not hold in the vitreous state. The resistance to water of some of these glasses could, however, be attributed to a particular borate unit that occurred in the glass. In the Na_2O-B_2O_3-P_2O_5 system NMR studies indicated that glasses with the greatest number of BPO_4 groups show the greatest chemical resistance and glasses with the highest total BO_4 and BPO_4 groups have the highest transformation temperature.

There is much literature on phosphate glasses that also contain V_2O_5. Many of these glasses are of interest from a semiconducting or electrical conducting view point. Kwang-tze, Shu-ming, and You-chi [516] reported on the ESR spectra of the Na_2O-V_2O_5-P_2O_5 system. Grishina and Mel'nik [343] synthesized glasses in the systems V_2O_5-BeO-PbO and Wo_3-MoO_3-P_2O_5 and studied their properties. An ESR study of the structure of vanadiuin V_2O_5-sodium phosphate glasses was made by Nagano, Mochida, Kato, and Seiyama, [619]. Matveev, Khodski, and Fisynk [559] found the glass-forming region in the V_2O_5-BiO_2-P_2O_5 system to be within the boundaries 30 to 85 mole% V_2O_5, 15-50 P_2O_5, and 5-25 Bi_2O_3. The majority of these glasses have low resistivity, coefficients of expansion, softening temperatures, durability to water, and density.

A number of glasses containing uranium oxide have been made, in some cases because of a possible application in reactors. Thus Wirkus and Wilder [930, 931] studied uranium-bearing glasses in both the silicate and phosphate systems. The addition of alumina was found to increase the durability of the phosphate glasses. The best glasses were devitrification-resistant and virtually insoluble in boiling water and most common reagents. HF attacked glasses in both systems and NaOH decomposed the phosphate glasses but not the silicate glasses. A glass containing 30 wt.% UO_2, 62 P_2O_5, and 8 Al_2O_3 had good oxidation and devitrification resistance, a density of 3.30 gm/c, and a linear thermal expansion coefficient of 5.5× 10^{-6}°C. Wilk [926] produced glasses in the system ZnO-UO_2-P_2O_5. In a more general study of the glass-making properties of UO_2 Chakrabarty [164] investigated the systems Na_2O-SiO_2-UO_2, PbO-SiO_2-UO_2, K_2O-B_2O_3-UO_2, K_2O-B_2O_3-UO_2, and Na_2O-P_2O_5-UO_2. Infrared spectra were obtained and the role of uranium was discussed in terms of partial molar refractivity.

4. FOUR OR MORE COMPONENTS

Phosphates are sometimes used in the production of glasses with low softening points. Thus Grimm and Huppert [341] patented a phosphate glass with a low softening point, specifying the composition limits: 34 to 44 P_2O_5, 20 to 80 Al_2O_3, less than 12 B_2O_3, 25 or more alkali, 8 to 21 Na_2O, 5 to 21% K_2O. Kreidl and Weyl [909, 910] produced a phosphate glass with a softening point of 400°C, using aluminum orthophosphate as a vitrifying agent. Water resistance was equal to that of commercial glasses.

Ferrandi, Estrada, and Conde [271] studied the effects of TiO_2 and CaO on the physical properties of a glass with the basic composition Na_2O-25 mole% Al_2O_3-21.75, B_2O_3-35.25, and P_2O_5-18. Dielectric loss was at a minimum with 3% TiO_2 but not altered by substituting CaO for Na_2O.

5. SPECIAL PHOSPHATE GLASSES

A few phosphate glasses with unusual elements present should be mentioned. Golubtsov, Estrop'ev, Kir'yanova, Kondrat'eva, and Khalilev [317] determined the electrical properties of some alkali-free fluorophosphate glasses. The electrical resistance and activation energy for conduction decreased with increasing fluoride content. It was concluded that cationic components were responsible for less than 11% of the transfer of electricity. They suggested that MgF_2 may be incorporated in the glass network to compensate for the decrease in the aluminum metaphosphate content.

Weyl [915, 914] patented a mercury-containing phosphate glass. He also reported on a mercury-lithium phosphate glass compounded in accordance with his ideas on anharmonicity, an Hg-O covalent bond being counteracted by a Li^+ ion.

Biwas and Mukerji [96] prepared a phosphate glass containing ruthenium. They suggested that ruthenium goes into solution in a glass after reacting with a single-bonded oxygen, therefore being more soluble in a phosphate than in a silicate glass.

C. State

1. OPAL AND OPAQUE GLASSES

Most of the references classified under "state" are concerned with opal and opaque glasses, particularly, the former. Ordinarily a glass is transparent because it consists of a single continuous phase containing no internal surfaces at which light can be refracted and so diffused. In this respect it resembles a liquid. Opalescent and opaque glasses differ in being translucent rather than transparent. The translucency may range from a slight milkiness to a white opacity. Such glasses are not transparent because they consist of two or more phases of different refractive index. There is a glassy phase in which finely divided crystals or droplets of a second glassy phase or both are embedded. The presence of crystals may be due to insoluble batch constituents or to devitrification. The droplets are due to immiscibility.

In general it appears that the opal glass due to devitrification is associated with batches containing fluorides and immiscibility opals with batches containing phosphates, although this is not always the case and many opal glasses contain both fluorides and phosphates. With the greatly increased interest in immiscibility there has been more investigation of the phosphate immiscibility type of opal glasses. One authority [99] considers that better opal glasses for certain purposes can be produced by the immiscibility route without fluorine or minor constituents of the apatities. These glasses have higher opacity and do not show roughening of the surfaces: for example, low-expensivity borosilicate opals are now in large-tonnage production for ovenware. The glassy disperse phase can be revealed by the best narrow-angle X-ray diffraction and electron microscope analysis.

In "Phosphate in Ceramic Ware" Weyl [909] devotes the first of the four parts to a discussion of phosphate opal glasses, based mostly on bone ash as a source of phosphate. He states that bone ash opals go far back in the history of glass. Formulas have survived from the fourteenth and fifteenth centuries [299], and bone-ash opals have been manufactured commercially in Bohemia and Silesia since 1670. Fluoride opals based on cryolite were developed when the fertilizing properties of bone ash and guano were recognized, which made them too expensive. Interest in phosphate opals revived when it was found that the borosilicate glasses which Schott developed could not be opacified by cryolite. Weyl emphasised the importance of fluoroapatite formation in the production of opal glasses and therefore the importance of even traces of fluorine. He found that bone ash contained as much as 4% calcium fluoride. The addition of lead oxide was also considered an advantage. The composition of fluoroapatite is $3Ca_3P_2O_8 \cdot CaF_2$ and is thus an orthophosphate combined with a fluoride. It is highly insoluble in water and has a strong crystallizing tendency in glasses.

The literature on phosphate opals and opacity is scattered in time and a lack of agreement between articles is apparent. The following is a brief but necessarily uncritical summary of the references found in this survey:

Fuwa [298] discussed the opacification of glass by means of bone ash, Na_2HPO_4, and $Ca_3(PO_4)_2$, as well as the last with the addition of KNO_3 or As_2O_3. He attributed opalescence to P_2O_3. Zschimmer [962] ascribed the "roughening" of a phosphate opal glass used for lighting purposes to relatively coarse calcium phosphate crystals. Heat treatment rather than composition was considered of prime importance. Körrn [485] gave batch compositions for opal covering glasses which, however, were not included in the abstract. Cauwood [152] reported analyses of opal and alabaster glasses. Phosphate glasses free from fluorine usually contained 2 to 5% phosphoric acid. An anonymous article [13] dicusses the opacification of a silicate and a borosilicate glass. It is reported that the type of alkali had little effect, boric acid had an intensifying action, the RO content had a large effect on the borosilicate glasses, PbO flavored opalescence in the silicate glass, and PbO and BaO had the same effect in the borosilicate. The opalescence is attributed to P_2O_3, probably because of the work of Fuwa. Knapp [473] provided formulas for phosphate opals as well as "devitrification opals." An anonymous article on opal glasses [17] discusses a long list of opacifying agents, including phosphates. A list of all known opacifiers for glass is given [265a] with numerous formulas and a bibliography. Weyl [891] pointed out that for glass used in illumination the degree of transmission could be raised by slowly cooling a phosphate glass, thus increasing the size of the opacifying particles, and obtaining a more uniform distribution. Blau [100] patented a method for controlling the opacity of fluoride opal glasses. Schmidt [735] reviewed the literature on raw materials for opacifying glass. Rooksby [720] used X-rays to study the roles of lead arsenate and lead phosphate in opal glasses. In the phosphate glasses he attributed opacity to the formation of lead orthophosphate, $Pb_3(PO_4)_2$. Blyumen [102] described a milky glass based on the use of apatite. Zschacke [959] ascribed opacity in lead-rich glasses mainly to the presence of minute glass droplets, undissolved constituents and fine gas bubbles, this disagreeing with Rooksby. Zschacke [959] summarized a series of five articles dealing with the action of opacifiers in glasses, glazes, and enamels and recommended phosphates for glasses rich in potassium and lead with the addition of Al_2O_3 and B_2O_3. The best ratio of CaO to P_2O_5 corresponded to the formula $Ca(H_2PO_4)_2$. Charan [166] discussed three methods of achieving opalescence; that is, devitrification, adding insoluble crystals, and immiscibility.

Springer [767] published a bibliography and summary on opaque glasses, and Schweig [740] recommended a furnace temperature of 1150 to 1150°C to avoid roughness of surface in phosphate opal glass. Weyl and Kreidl patented an opalescent glass [918]. The rather wide limits of composition were given as 54 to 66 SiO_2, 0 to 6 Al_2O_3, 12 to 17 Na_2O plus K_2O, 0. to 12 CaO, 0 to 4 BaO, 0 to 5 PbO, 0 to 50 F, 0 to 1 B_2O_3, 2.5 to 5 As_2O_3, and 4 to 9 P_2O_5. They also specified that the total of divalent metals calculated as oxides should not exceed 21%. Bates and Black [53] examined broken surfaces of opal glass at 5000 to 8000 diameters under the electron microscope. They estimated the number of crystals to be 10^7 to $10^8/cm^2$ of surface. The [illegible] o noted bubbles, dendrites, globular structures, and aggregates.

A review of the past and present practice and theory of opaque glass was published by Commons [189]. Hadden [346] discussed the replacement of bone ash by other opacifiers.

Stookey [780, 781] patented an opal glass composition as follows: 50 to 70 SiO_2, 7 to 15 alkali metal oxide, 5 to 25 BaO, 0 to 10 P_2O_5, and at least one of B_2O_3 and Al_2O_3 to a total of not more than 25%. A number of ratios were also specified, thus BaO/P_2O_5 1.5/1 to 4/1 with 9 to 30% BaO + P_2O_5. The light-diffusing particles were stated to be barium phosphate crystals and substantially free from calcium and similar phosphates.

Rooksby [721] used X-ray diffraction to identify the opacifying material in glasses containing lead and arsenic. He reported it was similar to apatite and had a composition of $3Pb_3(AsO_4)_2PbO$. If phosphorus is substituted for arsenic, the isomorphous phosphate is similar to mimetite and pyromorphite.

Kerkhof, Seeliger, and Westphal [453] studied fracture surfaces of opal glass with the electron microscope. They found that surfaces from opposite sides of a crack matched well. Opacity was attributed to tiny embedded crystals, usually in irregular lumps. Merker and Wondratschek [569, 934] studied the crystalline phases in phophate- and arsenate-containing lead glasses and enamels. Two kinds of solid solution were found: a high Na_2BeF_4 type, which was a solid solution of $PbNaPO_4$ and Pb_2SiO_4 or with (Na,K) or K in place of Na or As in place of P, and a low alkali type with solid solutions of $Pb_4A(PO_4)_3$, and $Pb_5(SiO_4)(PO_4)_2$ in which A is alkali and As or V may replace P. Most of the compounds they studied had the apatite structure. They concluded that the explanations of opacification given in the literature were not correct.

Umblia [849] discussed the principles of fluorine and phosphate opal glasses. The main opacifying agents were NaF, CaF_2, and apatite. Rabinovitch and Sil'vestrovich [690] discussed fluoride- and phosphate-containing opal glasses. All opacity was attributed to crystalline substances and a wide variety of fluorides and phosphates were mentioned. Schönborn [736] stated that phosphate opal glasses are typical unmixing dispersoids. With a given concentration of turbidity agent, the total amounts of unmixed material is solely a function of time. If 4.5% bone ash is used in a common borosilicate glass melting above 1400°C the quenched glass will be clear. The thermal treatment on reheating constrols the turbidity.

Coffeen, who has discussed the opacification of enamels, glasses, and glazes [179] reported that the degree of opacification with TiO_2 was a direct function of the volume ratio of the opacifier to the glass phase. As much as 10% of the TiO_2 may dissolve and thus lower the opacity. Glass opacifier systems were tabulated.

In an electron microscope study [851] Vaisfel'd and Rabinovich found approximately one-μ particles like CaF_2 crystals in fluoride opal glasses, which increased in size on heating at 850°C. In phosphate opal glasses droplet-shaped particles of an even smaller size were found. In both, particle size decreased with increasing opacifier content.

Three-, four- and five-component oxide systems containing P_2O_5 were investigated by Das [209] who found that the higher the Na_2O/P_2O_5 ratio, the lower the opacity. Al_2O_3 had a greater effect on the opacity than BaO and CaO. Small amounts of ZnO and F_2 were found to be useful. The furnace atmosphere was unimportant. In a later study of phosphate opacification of glasses Das [210] refers only to crystalline phases determined by X-ray diffraction.

Paoletti and Grammanco [652], reporting on nucleation and crystal growth in opal glasses, used spectrophotometer methods. A recent patent on a phosphate opal glass [454] by Keul gives the composition 66.2 SiO_2, 10.0 B_2O_3, 4.5 Al_2O_3, 5.0 P_2O_5, 1.5 CaO, 0.8 MgO, and 12.0 Na_2O.

Skopina, Paolushkin, and Gurrich [755] reported on the effect of P_2O_5, Cr_2O_3, and Mn_2O_3 on the crystallization of slags. Without these additives the glasses crystallized only on the surface. Harper, James, and Mc Millan [356] described crystal nucleation in lithium silicate glasses. Internal nucleation always occurred in glasses containing P_2O_5 when crystallized at 750°C, but grain size depended on the temperature of the nucleation stage.

Kawazoe, Hasegawa, and Kanazawa [451] reported on critical cooling rates for condensed sodium phosphate melts which decreased with increase in chain length. Occurrence of bond exchange reactions between phosphate anions above 600°C was confirmed by viscosity measurements.

From this review it will be evident that a wide range of compositions can be used for opal glasses and that a number of crystalline materials or immiscibility systems may produce opalescence. Phosphate opals are of recurring interest, with most having been displayed in borosilicate glasses of about 5% P_2O_5 and showing immiscibility. The thermodynamics and phase equilibrium diagrams associated with immiscibility in glasses is exactly parallel, with one exception, to those associated with immiscibility in liquids, for example, phenol-water. With glasses it is possible to cool from the liquid state at such a rate that the ordinary phase equilibrium diagram does not apply, but a second metastable, subliquidus diagram does. This phenomenon results from the high viscosity of the glasses at the liquidus. It has been the subject of a number of papers; Charles [167] is an example.

2. MELTS

A number of references which deal with melts containing phosphates are reviewed briefly.

Löffler [536] reported that "molasses potash" which contained some phosphate made the melting of glass easier than with pure potash. Skeiko, Gitman, and Loichenko [745] reported the linear dependence of the solubility of beryllium oxide in sodium phosphate in the range 700 to 1100°C. For the system Na_2SO_4-$Na_2B_4O_7$-$Na_4P_2O_7$ Bergman, Gasanaliev, and Trunin [80] reported a high melting ternary compound $Na_4P_2O_7.2Na_2SO_4.2Na_2B_4O_7$ with a melting point of 990°C.

In a British patent [764] 6% silica gel was added to $Na_5P_3O_{10}$ fused at 900 to 1000°C to produce pyrophosphates, complex silica-phosphates, and finely divided alpha cristobalite. This shows the limited solubility of silica in phosphate melts.

Müller [591] quenched alkali phosphate glasses from the molten state and examined them by infrared reflection spectra. He confirmed the paper chromatographic finding that glasses with greater than 50% alkali contain no ring phosphates.

Callis, Van Wazer, and Metcalf [155] studied the viscosity of molten sodium phosphates. They found that the flow was Newtonian and similar to that of long-chain hydrocarbons. The flow unit did not exceed about eight P atoms, even though the chain length may be 170 P atoms or greater. Volarovich and Tolstoi [879] determined the viscosity of melts in the metaphosphate-metaborate system $NaPO_3$-$NaBO_2$. The mixture with 60% $NaBO_2$ showed abnormal viscosity. The viscosities of $NaPO_3$ and $NaBO_2$ were very low, just below the melting point. Kingery [463] studied the surface tensions of liquid oxides, including Al_2O_3,B_2O_3, GeO_2, P_2O_5, and SiO_2. Abnormal positive temperature coefficients were found for B_2O_3, GeO_2, and SiO_2 mainly because of changes in the liquid structure with temperature. Presumably no change of this kind occurred with P_2O_5. A similar study was made for the system $Zn(PO_4)_2$-ZnO by Krivovyazov and Voskresenskaya [506]. The gradual shortening of the chains as ZnO was increased led to an increase in surface tension and the appearance of a break in the isotherm.

The relative activity of oxygen ions in molten sodium phosphate was determined from the solubility of sulfur trioxide by Kato, Nishibashi, Nagano, and Mochida [448]. The oxygen ion activity increased exponentially as the Na_2O/P_2O_5 ratio increased. Evaporation of phosphate was appreciable only above 980°C. With sodium silicates Na_2SO_4 separated, but not with sodium phosphate.

Bergman [79] studied the system described in *Ceramic Abstracts* as NaB_2O_3, PO_3, P_2O_7.

Perron and Bell [667] studied the diffusion of phosphorus and sodium in Na_2O-P_2O_5-SiO_2 liquids. The sodium diffusivity was about twice that of the phosphorus. Saito, Goto, and Someno [728] measured the electrical conductivity of a number of melts, including the system PbO-P_2O_5. An increase in the acid oxide produced an increase in the activation energy for conduction and a decrease in the conductivity.

Various systems containing phosphates have been considered as fuels for nuclear reactors. Owens and Mayer [649] measured the molar volumes and surface tensions of melts of mixtures of $NaPO_3$ with UO_2SO_4, Li_2SO_4, Na_2SO_4, or Rb_2SO_4. The properties varied linearly with temperature in all cases, and the volume behavior of all the mixtures was additive. A reduction in the length of the chains of the phosphate anions that constitute molten $NaPO_3$ with increasing sulphate was postulated.

Some studies of liquid silicates have by analogy some interest in regard to phosphate melts. Thus Hess [379] proposed a polymer model for silicate melts with the SiO_4 monomer dominating at low silica compositions and larger branched and ring structures at higher silica compositions. Liquid immiscibility was found at high silica contents. Riebling [708, 709, 711] reported on very accurate measurements of the viscosity and density of a number of silicate melts. His conclusion was that there were structural similarites between an oxide melt and its corresponding glass.

3. DEVITRIFICATION AND IMMISCIBILITY

From the usual glassmaking point of view devitrification is undersirable. Phosphate glasses in general are at a disadvantage because they tend to have a lower viscosity just below their liquidus temperatures than silicate glasses. This may be illustrated by the experiments of Beekenkamp and Stevels [65], who were unsuccessful in cooling melts of the composition 75 mole% BPO_4, 25% $AlPO_4$, but were able to make glass by adding silica to reach the composition BPO_4-71, $AlPO_4$-24, and SiO_2-5 mole%. Veinberg [874] studied the crystallization of K_2O-ZnO-P_2O_5 and Na_2O-ZnO-P_2O_5 glasses. He found that the first gave nondevitrifying and nonhyproscopic glasses in the range K_2O, 15 to 20, ZnO, 40 to 45, and P_2O_5, 40 to 45 mole%. The sodium glasses devitrified if the Na_2O content fell below 30 to 40 mole%, and at that level of alkali they were hygroscopic. Vogel studied the crystallization of phosphate glasses in order to throw light on their structure and was able to grow pure crystals of $Al(PO_3)_3$ from phosphate optical glasses. He concluded that a pure P_2O_5 glass had a reticular structure, whereas a pure HPO_3 glass contained more or less long-chains of PO_4 tetrahedra.

Moriya, Yoji, and Hatano [583] devitrified glasses of the $AlPO_4$ to $Al(PO_3)_3$ composition by reheating them. Nuclei formed only on the surface, and $Al(PO_3)_3$ crystals appeared first. The $AlPO_4$ crystals were in the high temperature cristobalite form.

Although Laird and Bergeron [518] worked with barium borate glasses, the chain folding mechanism which they describe for the growth of $BaO.2B_2O_3$ crystals from a melt suggests that something similar may occur in phosphate melts.

Stranski and Kaischew [788] explained the growth of trimetaphosphate crystals from a melt, solid state, or solution by rearrangement of the other phosphates present.

Shutov and Dertev [748] studied the effect of heat treatment on the crystallization of a glass of the compositoin $Na_2O.P_2O_5.O.1$ ZnO. They used a conductometric method to study the effect of a magnetic field on crystallization.

Sometimes a glass is deliberately devitrified in order to change its electrical conductivity. Kinser [464, 465] made dielectric loss measurements, d-c conductivity measurements and transmission, and electron microscope observations on a glass of composition 55 FeO-45 P_2O_5 (mole%). They were partly devitrified by low-temperature treatments. The dispersed crystalline phase gave rise to Maxwell-Wagner-Sillars heterogeneous dielectric dispersions previously attributed to electron-hopping resonances in these glasses. Moriya and Ueno [582] measured the electrical properties of phosphate glasses containing titanium ions before and after devitrification. Devitrification increased the conductivity, the dielectric constant and tan delta but decreased the activation energy for conduction.

A discussion of metastable liquid immiscibility and subsolidus nucleation was given by Roy [724]. Rindone and Ryder [713] showed that this kind of phase separation could be induced in phosphate glasses by 0.002 to 0.006 wt.% of colloidal platinum. Murthy [598] showed that such separation did not affect the overall constitution of the glass as determined by paper chromatography in sodium phosphate glasses with a range of composition from $\bar{n} = 3$ to 9.

4. MICROHETEROGENEITIES

A number of papers dealing with microheterogeneities in glass have appeared. These have been found in commercial glasses by electron microscope techniques. They appear to be caused by the slow rate at which silicate glasses achieve homogeneity in the melting process due largely to the high viscosity of high silica glasses. Bobkova and Rudakov [140] investigated a range of types of silicate glasses and concluded that the microheterogeneities disappeared on heating at high temperatures and consequently should be considered only as defects in structure but not as characterizing the structure of glass in its equilibrium condition.

5. GLASS CERAMICS

Glass ceramics is a general term applied to purposely and often fully devitrified glasses which have found many uses. The glass may be formed by the usual processes and then devitrified by a special heat treatment. By using a nucleating agent a "shower" of fine crystallites can be produced. The crystallites may be of a compound with a very low thermal expansion coefficient, and their high surface provides strength. The discovery of the first glass ceramic called "pyroceram" has been described by Stookey [782]. Glass ceramics are called "sitalls," particularly in some of the European literature.

Because phosphates are well known agents for producing opalescence, it is not surprising that they have been specified as nucleating agents in the making of glass ceramics. Babcock [39] patented articles bonded by a solder glass in which P_2O_5 could be used as an internal nucleant. Harper and McMillan [355] showed that inclusion of P_2O_5 in a lithium silicate glass resulted in a marked increase in nucleation density. The finest grained glasses also showed liquid-liquid phase separation before crystallization. The P_2O_5 content was usually 1 or 2%, with SiO_2-64-70% and Li_2O-25-36%. The quenched glass was heated rapidly to 500°C, held for one hour for nucleation, and quenched to room temperature. It was then heated rapidly to 750°C, held for one hour, and quenched again. The visible crystals were lithium disilicate.

Nagaoka and Hara [620, 621] published a series of articles on SiO_2-Li_2O-Al_2O_3 glasses nucleated by P_2O_5. The effective range for P_2O_5 was 2 to 4%. A high-volume expansion was attirbuted to the separation of a phosphate glass phase, which could be overcome by the addition of 3% alumina. Crystalline phases found were lithium disilicate, cristobalite, quartz, quartz-spodumene solid solution, and beta-spodumene. Partridge and McMillan [656] also investigated the use of phosphates as nucleants in glass ceramics. A wide range of thermal expansion coefficients could be obtained so that metals and alloys including mild steel and copper, could be matched.

The addition of P_2O_5 to glasses made from tephrite-basalts was investigated by Pavlushkin, Gurevich, and Nikishina [660]. The "Petrositalls" produced had interesting properties.

Phillips and McMillan [673] studied phase separation and crystallization in Li_2O-SiO_2 and Li_2O-SiO_2-P_2O_5 glasses. They produced a fine-grained, high-strength structure which they said was characteristic of phosphate-nucleated glass ceramics.

These investigations were concerned with the use of phosphates as nuceating agents, but because $AlPO_4$ has many resemblances to SiO_2 there was the possibility of using phosphates as a glass-forming constituent as well. This is illustrated by the patent of Petzoldt and Scheidler [669a] of a glass convertible to a glass ceramic. As major constituents this glass contained 20 to 35 SiO_2, 32 to 42 Al_2O_3, and 5 to 17 P_2O_5.

McMillan and Partidge [613] patented a glass ceramic based on a SiO_2-Al_2O_3-Li_2O-MgO glass with 0.5 to 6% P_2O_5 as a nucleating agent. Before this Park [653] had shown that the most distinct phase separation occurred in a glass of composition Na_2O-0.5, B_2O_3-1.5, and SiO_2-6.0, when 0.06 mole of P_2O_5 was added. Mueller [588] determined the partial volumes of SiO_2, $AlPO_4$, $LiAlO_2$, $MgAl_2O_4$, and $ZnAl_2O_4$ for both the glassy and crystalline states of glass ceramics. The differences in partial volumes was attributed to the contracting effect of Li, Mg, and Zn ions and not to the change of Al ions from 6 to 4 coordination.

Carr, El-Bayoumi, and Subramaium [159] and El-Bayoumi and Subramanium [253] studied the nucleation, phase separation, and crystallization of a lead silicate glass containing P_2O_5 and a cerium phosphate glass.

D. *Structure*

1. MACROSTRUCTURE

One reference is the patent of MacAvoy [540], entitled "Method of Making a Cellular Phosphate Glass," which deals with macrostructure of a phosphate glass.

2. ATOMIC ARRANGEMENT

The atomic arrangement in simple binary phosphate glasses from ortho- to metaphosphate composition has already been discussed in detail in the scientific section, in which it was found that for the most part they consisted of long-chain polymers. For glasses that contained more P_2O_5 than required by the metaphosphate composition, that is, glasses of the ultraphosphate composition, little was known. Similarly not much was said about glasses with a number of major components. However, a considerable literature deals with these other types of phosphate glass, and about 40 references were found in which some consideration was given to speculating on their atomic arrangement, although this might not be the main purpose of the investigation described. For the most part these references deal with the variation of physical properties with composition and include nearly all the new methods of physical investigation. Those dealing with the simple binary glasses for the most part confirm the findings of the chromatographic studies described earlier in this monograph. Although the literature on phosphate glasses in general does not lend itself to generalization, a brief review follows:

With regard to structure in general, Gerasimov, Kuznetsov-Felisov, and Kuznetsof [305] discussed what they called the "oriented" structure of phosphate glasses. Bartenev [49], in his book on inorganic glasses, according to a reviewer, makes the statement that some glasses are akin to crosslinked polymers and others to liquid metal alloys, but no comprehensive fundamental theory exists for all inorganic glasses and liquids in general. Wondratschek [933] presents some interesting pictures of cubic, octahedral, tetrahedral, and triangular coordination, with the inside space decreasing in that order. Douglas [225] presents a review of the properties and structures of glasses, referring particularly to silicate but mentioning long-chain phosphate glasses. Eckstein [245] presents a defect arrangement theory of glass structure. As limiting cases he took an ideal crystal and an ideal gas. Intermediate conditions could be approached from either side, that is, by the incipient ordering of a statistical state or the generation of defects in an ordered state.

A number of investigations deal with thermal and mechanical properties. As early as 1930 Tamman and Bandel [810] studied HPO_3 glasses. These had a water content of 7 to 12%. The p,T curves showed a decided increase in expansion at the temperature at which a loss of brittleness occurred. With decreasing water content this temperature increased. Sodium and thallium phosphate glasses were made by Gerasimov, Kuznetsov, Kuznetsov, and Shakhmina [306] in a study of their polymerization and crystallization.

A few papers that discuss the structure of silicate glasses are of interest because reference is made to polymerization or to hypotheses that might be applicable to phosphates. Thus Beyersdorfer [87] discusses silicate glasses in terms of polysilicates held together by van der Waal's forces. He specifies a pentagonal ring formed of SiO_4 tetrahedra as the basic building unit. The theory is supported by changes in density, refractive index, and viscosity with thermal treatment. The Mössbauer effect in certain SnO_2-containing glasses is evidence of the presence of chemical compounds and is inconsistent with a disordered structure. Eisenberg and Takahashi [250] studied the viscoelasticity of "silicate polymers." Demkina [216] who included some lead phosphate glasses in the study of the inner structure of silicate glasses reported that discrete changes in properties were exhibited for compositions lying between definite chemical compounds close to eutectic points. Deeg [213] studied the structure of silicate glasses with mechanic-acoustic techniques. He concluded that the network modifiers were held strongly in the macromolecular network of the glass.

Fitzgerald [277] assembled a vibrational energy absorption spectrum for glass from published anelasticity data. He emphasized the similarity of glass and metals in this aspect and recommended further anelasticity sutdies of glass structure.

Absorption, diffraction, and resonance methods have been applied to the study of a number of phosphate glasses. Danilov [208] studied the spectral absorption of colored phosphate glasses as a function of their color. Co, Ni, Fe, Mn, and U phosphate glasses of the metaphosphate composition were compared with solutions, and spectral absorption varied with fusibility. In a series of papers published from 1939 to 1942 [481-483] Kordes and his colleagues reported on physicochemical studies of the molecular structures of glasses. Some phosphate glasses were included. Regarding binary phosphate glasses, it was concluded that there were two distinct groups: normally built phosphate glasses of the systems P_2O_5 with a CaO, BaO, CdO, and PbO; irregularly built phosphate glasses of the systems P_2O_5 with ZnO, MgO, and BeO. Dispersion measurements provided the most sensitive method of detecting anomalies of structure. In the P_2O_5-ZnO system, by assuming a cristobalite structure, curves of molecular refraction and density against composition could be calculated accurately from ionic radii.

Perry, Wilson, and Kinser [668] determined the oxidation state of iron, copper, vanadium, and titanium in phosphate glasses by heating a specimen in pure oxygen and measuring the oxygen uptake.

Brady [121] used X-ray diffraction methods and radial distribution functions to study the structure of glass of sodium metaphosphate composition. The indicated structure consisted of long chains of PO_4 tetrahedra, crosslinked to other chains by O-Na-O bonds. This agrees with the results of chromatographic and polymer studies. An X-ray study of glasses in the WO_3-P_2O_5 system by Skancke and Kierkegaard [754] indicated the presence in all glasses of aggregates of WO_6 octahedra joined by corners, similar to those existing in $W_2O_3(PO_4)_2$ crystals.

Raman spectra were used by Bobovich [105] to study a large number of binary phosphate glasses and the ternary system Na_2O-TiO_2-P_2O_5. The spectrum obtained for a sodium phosphate glass of metaphosphate composition was the same in general character as that for the same glass in the molten condition and indicated a structure consisting of an almost infinite chain of interlinked PO_4 tetrahedra. When other cations were substituted for sodium, an increase in their coulombic potential strengthens all the bonds and leads to an increase in softening temperature. This rule did not work for Pb^{2+} and Bi^{3+}, which Bobovich attributes to the strongly covalent character of the PbO and BiO bonds. Coincidence of the structure of a crystal and a glass is considered to occur only in cases in which the crystal itself represents a polymer.

Evidence of atomic groups in phosphate glasses was obtained by NMR by Brandenberger [123] and Bray [123a]. Van Wazer and his associates have made extensive contributions to this field [202, 203, 277, 555, 862, 864].

Dielectric and magnetic studies of phosphate glasses have been reported. Hirayama and Rutter [387] made dielectric studies of some borate and phosphate glasses. They found that vitreous P_2O_5 resembled sodium silicate glass in all properties dependent on structure. In his book on colored glasses [912] Weyl states that the coefficients of expansion of this glass is high and the softening temperature, low. The optical properties resemble those of silica glass. Since no sodium ions are present, the electrical resistance is high. However, the chemical resistance is low. The latter disadvantage may be overcome by adding Al_2O_3. Smits and Rutgers [759] studied the behavior of P_2O_5 when heated. At 400°C the vapor pressure of the crystalline material was about 4 atm but dropped to almost zero as the stable glass was formed. This polymerized glass decomposed at higher temperatures to form P_2O_5 vapor that could be condensed into a crystal or a glass according to cooling conditions. The vapor is now known to have the formula P_4O_{10}. The magnetic behavior and microstructure of vanadium phosphate glasses containing 60 to 90 mole% V_2O_5 were studied by Friebele, Wilson, and Kinsar [290]. As-cast glasses with high V_2O_5 concentration separated into two glassy phases, which concentrated the vanadium ions in a vanadium-rich phase. Electron microscope studies showed that the glass must be characterized structurally before other investigations can be attempted.

Williams, Bradley, and Maddocks [927] made a series of studies of binary phosphate and phosphate fluoride glasses. Addition of fluoride caused a decrease in chain length of the glasses when dissolved in water. Surface tension measurements showed linearity when plotted against composition in all cases. Infrared measurements indicated that for glasses of the metaphosphate composition the structure of the potassium glass was quite different from that of the others. Densities and volume of mixing in the binary system $NaPO_3$-KPO_3 showed a slight deviation from ideality. Further studies of fluoride effects in both silicate and phosphate glasses were made by Kumar, Ward, and Williams [511]. In melting under dry conditions fluoride was lost as SiF_4; with moisture present it was lost as HF. When phosphates were present, the loss of fluoride under dry conditions was small. Evstopjev, Petrovsky, and Chalilev [265] studied the structure of glasses based on $BaPO_3F$. Many compositions and properties were studied. They concluded that glass formation was due to the coordination of the polymeric structure due to the donor-acceptor action of the PO_3F group.

When studying the glass-forming properties of alkali molybdates and tungstates, Gossink [326] found that neither MoO_3 or WO_3 would form glasses unless quenched extremely rapidly. When classical glass formers were added, only P_2O_5 gave stable glasses. The crystalline compounds in the glass-forming regions contained infinite chain-type anions. However, the data obtained for the alkali tungstate glasses showed that they contained averagely short chains of WO_4 tetrahedra, which had undergone disproportionation. Alkali molybdate glasses also contained small groups such as MoO_4^{2-}, in addition to which, however, larger groups containing Mo atoms with a coordination number higher than 4 were found. These glasses thus provide an interesting comparison with phosphate glasses, a comparison that Gossink uses to advance his theory that small alkali ions with a higher field strength tend to reduce the size of anionic groups. It is possible that Gossink's conclusions could be checked by chromatographic methods.

A few references on the properites of molten glasses have some bearing on their structure. Callis, Van Wazer, and Metcalf [155a] reported that chains break and reform during measurements of the viscosity of phosphate glasses. Peyches [671] showed by thermal expansion measurements of glass fibers that the structure of such rapidly quenched glass would change at relatively low temperature. Nijjhar and Williams measured the surface tension of molten alkali phosphate mixtures of the metaphosphate composition and made infrared measurements of the quenched glasses [634]. Only slight deviations from ideality were found and no evidence of compound formation. Thakur [814] made an interesting comparison of $NaPO_3$ and KPO_3 melts in studying nucleation and controlled crystallization. The $NaPO_3$ melt, MP 600°C, had a negligible rate of nucleation and readily formed a glass. The KPO_3 melt had a high rate of nucleation and did not form a glass.

Solacolu, Balta, Litianu, and Spuraciu [764] have explained some of the properties of the viteous state such as the structural disorder, phase separation, and increase in viscosity when held at a steady high temperature, by the polymer theory of glass.

Alkali-free glasses with the general formula R_xO_y-68.31, BaO-11.71, La_2O_3-12.83, Tl_2O-1.41, and SrO-5.74 mole% were investigated by Labutina, Pavlushkin, and Artamonova [517]. For R_xO_y they used B_2O_3, GeO_2, P_2O_5, and SiO_2 and found that the cation R had a large effect on structure as revealed by the state of luminescent centers. Johari and Goldstein [428] discussed secondary relaxations of glasses of rigid molecules. Bozomolova, Zhachin, Lazukin and Shapovalova [106] used an EPR study of Cu(II) in glasses of the P_2O_5-Al_2O_3-Cs_2O system. They arrived at a qualitative explanation of certain important relations in the scattering of ions by crystals.

Reisfeld and Boehm [702] discussed the energy transfer between Sm and Eu in phosphate glasses. They suggested that the transfer took place by a dipole-quadrupole mechanism assisted by phonons. A Mossbauer study of ^{57}Fe in an aluminophosphate glass was described by Taragin, Eisenstein, and Haller [811]. Both Fe^{2+} and Fe^{3+} were present and both were octahedrally coordinated. The difference in Debye temperatures was likely due to a difference in sites.

The magnetic properties of manganese phosphate glasses were reported by Friebele, Kinser, and Wilson [289]. Their experiments indicated a small fraction of isolated ions in an ortherwise antiferromagnetically coupled glassy matrix. Dozier and Kinser [227] listed the study of the structure of transition metal phosphate glasses under Current Research.

Nissle and Babcock [636] reported on the relation between stress-optical coefficients and glass composition. They found that such coefficients are linearly related to compositions expressed in mole fractions and are unique for each primary crystallization field. El-Bayoumi and Bishay reported that they had found structural changes associated with certain particular compositoins.

E. *Manufacture*

1. FINING

In the usual process for manufacturing glass the batch is first melted and then held at a high temperature to allow gas bubbles to escape. This is called "fining." Fining agents assist in this process by producing larger bubbles that readily absorb the finer ones as they rise. Weyl and Kreidl [909] found that phosphorus compounds could be used for fining certain glasses, such as selenium ruby, copper ruby,carbon amber, and heat-absorbing, which are melted under reducing conditions.

2. DECOLORIZING

Cousen and Turner [195, 196] showed that the arsenious oxide ordinarily used for decolorizing glass with selenium in a tank furnace could not be replaced with calcium phosphate. Kreidl and Weyl, however [909], refer to Gillinder's work in improving the color of the famous English flint glass by the addition of calcium phosphate. As described later, phosphates have been used as primary batch constituents for a number of commercial glasses required for special purposes, but for ordinary bottle and window glass there does not appear to be a place for them. Rather, their volatility and their tendency to lower chemical resistance mean that they are avoided.

F. Chemical Properties

1. CHEMICAL DURABILITY

The chemical durability of phosphate glasses has always been a matter of considerable importance. As a class, they are not resistant to water and other ordinary reagents, although some resistant phosphate glasses have been developed, and in contrast to silicate glasses they have a inherent resistance to HF. This property is discussed under the next title. The following paragraphs review the references on chemical durability of phosphate glasses in general in chronological order, leaving optical glasses largely to a later section.

Kohlrausch [479] studied the solubility of phosphate glasses, including some Schott phosphate crowns, in cold water. He noted that ordinary glasses gave off alkaline solutions, whereas phosphate glasses produced acid solutions. This was confirmed by Cedivoda [163]. Turner [844], in producing silicate glasses highly resistant to water, investigated the effect of different components and decided that B_2O_3 up to 12% increased the resistance of most glasses. Morey [576] suggested that the attack of water on glass could not be considered in ordinary solubility terms but involved the miscibility of water; the glass was regarded as an undercooled liquid. In silicate glasses he found that small amounts of ZnO or Al_2O_3 were distinctly favorable and that B_2O_3 was a constituent of most of the best glasses. It is now generally considered that the attack of water on glass involves a differential leaching of the alkalies that leaves a layer of siliceous material that governs the rate of further attack. This is diffusion-controlled and is proportional to the square root of the time.

Stanworth and Turner [771] studied the preparation of glasses containing P_2O_5, Al_2O_3, B_2O_3, and SiO_2 and found that those containing Na_2O, B_2O_3, and P_2O_5 were clear when sufficient Na_2O was present and resistant to atmospheric attack even at high percentages of P_2O_5. A clear glass could not be made from silica and metaphosphoric acid alone but could be made by first preparing silicon phosphate.

A series of fluorophosphate glasses was patented by Kodak Ltd. in 1950. Those containing TiO_2 are mentioned as being chemically durable. $Al(PO_3)_3$ was undoubtedly also introduced for the same reason [477].

Dale and Stanworth [206] described a sodium vapor-resistant phosphate glass for use in sodium vapor lights. Mori and Noda [581] studied the effect of adding silica to a phosphate glass. If no alumina were present, silica decreased the durability and the thermal expansion coefficient. Like other investigators, they found that Al_2O_3 increased the stability of phosphate glassware. Takahashi, Iida, Watanabe, and Oguchi [808] reported on glasses of comparable resistance to H_2O and HF to the alumino-silicate glasses. Their glasses were based on the formula xMO (or M_2O), $Al_2O_3.4P_2O_5$, where $x = 1.5$ to 3 and M = Li, Na, K, or Be, Mg, Ca, Sr, Ba, Cd, Zn, or Pb. Recommended were 1.5Be, 3.0Be, 1.5Cd, 2.0Cd or 1.5Zn.

Hurt, Wellington, and Myles (409) discussed the action of alkaline detergents on bottle surfaces. Attack by solutions of these detergents could be largely overcome by adding sodium silicates or phosphates to the solutions.

Shermer, Rynders, Cleek, and Hubbard [746] reported on the use of nonsilicate glasses for pH electrodes. Glasses in the system Na_2O-MgO-P_2O_5 were included. None was satisfactory on the grounds of hygroscopicity and chemical durability, although the two effects were not directly correlated.

Boak, Rapp, Hartley, and Wiends [103] reported on the chemical durability of soda-lime-silica glass in which silica was partly substituted by TiO_2, GeO_2, ZrO_2, and $AlPO_4$. $AlPO_4$ improved the H_2O and alkali resistance but the attack of acid was increased.

Kisilev, Bruevich, Zherdev, Deev, Kudishina, Rozdina, and Korolev [467] described the failure of glass fabrics under the action of heat and phosphoric acid. Clark and Hench [174] discussed the effect of P^{5+}, B^{3+}, and F^- additions on the corrosion of soda-lime-silica glass. F^- blocked the corrosion by stopping the diffusion of Na^+ ions, and P^{5+} influenced the corrosion rate by affecting the conditions for second-phase formation. Morgan [580] patented chemically resistant aluminophosphate glasses.

2. RESISTANCE TO HF

A number of references deal specifically with the resistance of phosphate glasses to HF. A patent by Buck [142] described glasses composed of a fused complex metaphosphate of aluminum and an alkaline earth metal substantially free from silica which was used for bottles, dishes, and stirring rods. Kreidl and Weyl [909] give the composition of two Buck glasses: (1) P_2O_5-77.46, CaO-10.18, Al_2O_3 12.36, and (2) P_2O_5-73.15, CaO-23.77, Al_2O_3-3.08.

Pincus [675, 675a, 676] published papers and patents of HF-resistant glasses based on phosphates. He mentions a glass with P_2O_5 as a major constituent which showed no evidence of attack after being submersed in a bath of HF for 500 hours. This difference between silicate and phosphate glasses is surprising in view of the similarity of the two types. Takahashi, Iida, Wattanabe, and Oguchi [809] also reported on the resistance of phosphate glasses to aqueous HF solutions. Syritskaya and Sergeeva [803] reported on the influence of certain metaphosphates on the properties of glasses in the system La_2O_3-SiO_2-P_2O_5. The composition of a glass with a high chemical resistance to concentrated HF solutions is given as Li_2O-P_2O_5-20, $Al_2O_3.3P_2O_5$-29, $La_2O_3.3P_2O_5$-20, $SiO_2.2P_2O_5$-1, and $PbO.2P_2O_5$-30%. Except for the lead compound, these formulas correspond to the metaphosphate composition. They state that additions of 2.5 to 20% $SiO_2.2P_2O_5$ instead of Al_2O_3-$.3P_2O_5$ have a favorable effect on most properties.

3. SODIUM VAPOR RESISTANCE

Glasses resistant to sodium vapor are needed for sodium vapor lamps. In addition to that of Dale and Stanworth, there is an I.G. Farben A.G. patent on a sodium vapor-resistant phosphate glass [413].

G. Physical Properties

Only a few papers were found that dealt with the physical properties of phosphate glasses in the solid and molten state, particularly with the simple glasses for which the constitution is known from chromatography. The largest group dealt with surface tension.

1. SURFACE TENSION

Callis, Van Wazer and Metcalf [154] published data on the density and surface tension of sodium phosphates in the molten condition. Mole ratios Na_2O/P_2O_5 of 1.0 to 2.15 were covered. The variation of surface tension and density could be given by the McLeod equation with a constant 1.74 ± 0.07. In their studies of phosphate melts and glasses [120, 927] Williams, Bradbury, and Maddocks found that the surface tension of the binary melts Na_2O-P_2O_5, K_2O-P_2O_5, and CaO-P_2O_5 varied linearly with the composition in all cases.

Nijjhar and Williams [634] reported on the surface tensions of a number of molten phosphate mixtures in the system $LiPO_3$-$NaPO_3$, $LiPO_3$-KPO_3, $NaPO_3$-KPO_3, and $NaPO_3$-$Ca(PO_3)_2$. By calculating the surface heats of mixing only slight deviations from thermodynamic ideality were found.

Boyer, Fray, and Meadowcroft [118, 119] reported on the surface tensions and molar volumes of binary phosphates of sodium, lithium, calcium, and zinc, with M_2O, calcium, and zinc, and M_2O/P_2O_5 or MO/P_2O_5 ratios of 1.00 to 1.80. The surface tensions increased monotonically with increasing metal content except for zinc phosphate, in which the surface tension was almost independent of the zinc content. Molar volumes increased linearly with metal content for Na and Li phosphates. For Ca and Zn phosphates the volume remained constant up to ratios of 1.5 or 1.6. In mixtures volumes and surface tensions were ideal in Na-Li, but for mixtures of monovalent and divalent cations large positive excess volumes associated with large negative deviations of the surface tensions were observed. The data were analyzed by using a rigid sphere model. The conclusion was that transport of these phosphates took place by a reorganization process and that long-chain anions found by chromatography may be present on a time average basis.

The surface tension of the melts of the system $Zn(PO_3)_2$-ZnO were published by Krivivazov and Voskresenskaya [506] for compositions between the meta- and orthophosphates of zinc. With increasing ZnO, a gradual shortening of the polyphosphate chains was suggested as the reason for the increase in surface tension.

2. OTHER PHYSICAL PROPERTIES

A phosphate glass with a high coefficient of thermal expansion and good chemical durability is mentioned by Stanworth, [770]: P_2O_5-33.5, Al_2O_3-15.5, PbO-30, B_2O_3-7, Na_2O-14 wt.%. Murthy [600] studied the thermal expansion of sodium phosphate glasses with a view to finding their glass transition points. The data showed that their glass transition temperature was much above the 35°C first estimated by Meadowcroft and Richardson [566] and closer to the Cripp Clark values quoted by the same authors in a later paper [567].

Syritskaya and Shapolova [805] reported the thermal expansion and softening point of glasses in the system Al_2O_3-TiO_2-P_2O_5. Both varied markedly with the composition. The Ti ion strengthened the structure of the high phosphate glasses and weakened that of the high titanium glasses. TiO_2 substituted for P_2O_5 in the high phosphate region decreases thermal expansion but increases it in the high TiO_2 region. These glasses have a low coefficient of expansion because they contain cations with a high charge and relatively small radius. The average softening point of high phosphate glasses exceeds that of high titanium glasses by 80°.

The strength of glass, as of all brittle materials, is a subject of considerable complexity. A summary of the theories has been made by Phillips, [672]. One interesting phenomenon is known as "static fatigue," which is the failure of glass under long-term loading at relatively low stresses. It is known to be due to the attack of moisture and appears likely to be associated with the hydrolysis of the Si-O-Si bonds in ordinary silicate glasses. If this is true, a study of phosphate glasses should be interesting because so much is known regarding the hydrolysis of the P-O-P bond. No references were found to the strength of phosphate glasses as such, but patents by Poole, Snyder, and Boschini [682] and Graham [330] describe the use of potassium phosphates in ion exchange methods for strengthening silicate glasses. By replacing sodium ions with potassium ions a state of compression is produced in the surface of the glass that greatly increases its strength.

In a survey of the thermal conductivities of glasses between −150 and 100°C Ratcliffe [698] included some phosphate glasses. Tsubaki, Hattori, and Tanaka [842] in a study of the thermal conductivity of vitreous sodium phosphates over the range 50 to 160°C, found that the temperature coefficient of thermal conductivity decreased with an increase in the Na_2O/P_2O_5 mole ratio and changed sign from positive to negative at a mole ratio near unity. A similar change in temperature coefficient was found by Trap and Stevels [832] for silicate glasses. They called glasses with a mole ratio Na_2O/SiO_2 greater than unity "invert glasses."

A magnetochemical study of glasses in the system Na_2O-P_2O_5-V_2O_5 was reported by Azarov, Balandina, and Lyutsedarakii [37]. They found that the magnetic moment was due primarily to tri- and tetravalent V ions and that the magnetic properties changed in accordance with changes in phase composition.

Baier, Deeg, and Constabel, [42] published a study of the absorption of orthophosphate on the surface of two ternary borate glasses. Radioactive phosphorus was used with autoradiographs which agreed with cord pattern photographs taken before immersion. The term "cord" is used to designate elongated areas differing in composition from the rest of the glass.

Radioactive fluophosphate glasses containing uranium, plutonium, or thorium were investigated by Sun, Chen-Ysai, and Shan. A stable glass composition was reported [793].

Collins, Rindone, and Mulay [183] searched for superparamagnetism in the iron phosphate system. No precipitation of fine iron phosphate crystals occurred, although such precipitation is found in the silicate system.

Kanazawa, Kawazoe, and Handa [443] reported on the thermal properties and crystallization of a magnesium high-polyphosphate glass. A 10:1 difference in thermal expansion below and above T_g was explained by structural considerations. DTA studies showed that crystallization started at 630°C. Above 660°C it occurred within the glass. The crystals were magnesium cyclophosphate or metaphosphate $Mg(PO_3)_2$. In the same year Kadota, Saito, and Sakaino [439] discussed the stress relaxation mechanism of phosphate glasses. Also, Kadota, Kamiyama, and Sakaino [438] reported the heats of solution of sodium phosphate glasses. A minimum was found near the meta composition.

H. Uses

A considerable number of references were concerned with a particular use of phosphate glasses. Included were optical, 46, semiconduction, 41, UV transmission, 34, color, 28, slags, 14, electric conduction, 8, heat absorption, 7, laser, 7, coating, 5 and a number of minor uses.

1. OPTICAL GLASSES

The term "optical glass" is applied primarily to glasses used for making the optical components of scientific instruments and such widely needed products as spectacles and cameras. Because they must meet stringent requirements with respect to a number of optical properties, many glasses of special composition have been developed by the manufacturers of optical glass. It is natural, therefore, to find them engaged in the manufacture of special glasses for other purposes; for example, laboratory glassware, thermometers, and boiler gauges. Optical glasses are ordinarily exposed to the atmosphere and it is important that they be resistant to attack by water vapor and other gases that would change their surfaces and their transparency. This is one difficulty that must be overcome in the use of phosphate in optical glasses.

History of Optical Glasses. This history has already been outlined in VII-A-2 in the history of glass in general. The history of optical glass in particular has been described by Weyl [908] and the history of optical glass production in the United States, by Glaze [313]. Hovestadt in 1902 [399] published a book entitled

Jene Glass and Its Scientific and Industrial Applications. In referring to this book, Kreidl and Weyl [909] discuss the work of Otto Schott in the use of phosphates to extend the range of optical glasses. By replacing silicates with phosphates Schott produced glasses that had a higher index of refraction than a silicate glass of the same dispersion and so were better for achromizing borate flint glasses. Two typical compositions were given as (1) P_2O_5-70.5, B_2O_3-3.0, K_2O-12.0, MgO-4.0, Al_2O_3-10.0, As-0.51 and (2) P_2O_5-54.0, B_2O_3-3.0, BaO-40.0, Al_2O_3-1.5, As-1.51. Glass (1) was called a light phosphate crown and glass (2) a dense barium phosphate crown. Unfortunately, these glasses lacked weather resistance and were abandoned, despite their valuable properties, and further work on phosphate glasses was discouraged. This discovery of the valuable properties of phosphate glasses and, following that, their poor weather resistance seem to have occurred at intervals of 20 to 30 years in the history of optical glass development.

The many contributions of Otto Schott to the optical glass industry, starting in 1879, are described by Kiaulehn [457] in an interesting account of the relocation of the Jenaer Glaswerk Schott and Genossen at Mainz, West Germany. This book also describes the development of the glass industry in Europe. More recent work on phosphate glasses has permitted them to be used with good results in the design of lenses with flat fields. In addition, in one and possibly more cases phosphate glasses that were able to replace silicate glasses because of their greater moisture resistance have been developed.

Phosphate Glasses and Achromaticity. The development of special glasses for lens systems has been going on for more than 100 years. The primary property of an optical glass for this purpose is its refractivity, usually expressed by n_D, its index of refraction for monochromatic light at the D line, or 589.3 mμ. Unfortunately, the refractive index of a glass does depend on the wavelength of the light, and for ordinary light this variation of index with wavelength, or dispersion, makes it impossible to focus all wavelengths at the same point. This creates a chromatic aberration, which to be corrected requires the use of optical glasses of different dispersions. In this way two wavelengths can be brought to a common focus with conventional achromatization techniques. This gets rid of the primary spectrum but leaves a residual chromatic aberration, known as the secondary spectrum, which is determined by partial dispersion ratios and denoted by P with suitable subscripts. Thus Brewster, Hensler, Rood, and Weidle [128] define P_{FD} as $(n_F - n_D)/(n_F - n_C)$, where C, D, and F indicate the wavelengths 656.3, 589.3, and 486.1 mμ, respectively. A simple achromat with components a and b and FD partial ratios of P_a and P_b has a secondary spectrum measured by $L = [(P_a - P_b)/(\nu_a - \nu_b)]f$, where f is the focal length and ν is the reciprocal relative dispersion or Abbé number and equals $(n_D - 1)/(n_F - n_C)$. These authors found that for silicate glasses a plot of the partial dispersion ratios as a function of the Abbé number gave a smooth curve that approached a straight line. For glasses of the same Abbé number the borates have partial dispersion ratios lower than the silicates and the phosphates have higher ratios. A judicious choice of a borate and a phosphate glass minimized L and superior optical instruments such as microscopes were developed on this basis.

The glass systems from which suitable glasses were developed were PbO-La_2O_3-Al_2O_3-B_2O_3 for low, PbO-Al_2O_3-B_2O_3 and CaO-Al_2O_3-B_2O_3 for imtermediate, and alkali Al_2O_3-B_2O_3 for high Abbé number borates. Phosphate glasses of varying Abbé number were made by adding PbO to the system Al_2O_3-MgO-K_2O-P_2O_5. Stability and durability had to be kept in mind.

Wright [936] discussed the relation between composition and optical properties of a large number of optical glasses which included some phosphate glasses. He remarked that the phosphate glasses differed only slightly from the borosilicates and barium crowns and that this did not outweigh the practical disadvantage of weathering instability. Nicolardot [632] proposed a nomenclature and classification of optical glasses. Sun and Huggins [794] patented a beryllium borophosphate glass with an n_D between 1.505 and 1.53 and a ν between 67 and 72.

Brewster, Kreidl, and Pett [125] published information on the use of lanthanum and barium in glass-forming systems. For an equal increase in refractive index of high-index, low-dispersion crown glasses BaO and LaO decrease the ν value less than other constituents. Both impair chemical resistance. Small additions of Al_2O_3, ZrO_2, ThO_2 and particularly TiO_2, Cb_2O_5, and Ta_2O_5 increase the chemical resistance of Ba and La glasses more than does silica. La_2O_3 forms glasses with phosphoric acid and BaO. A pure lanthanum metaphosphate glass exists, which has a refractive index of 1.60 and a ν value of 60, and high-index, low-dispersion crown glasses were produced. Phosphate glasses require Al_2O_3 for stability, which modifies their optical properties. Kreidl [489] patented a high-index phosphate glass which had the composition barium metaphosphate -42-75% and lanthanum metaphosphate-25-58%. Weissenberg and Meinert [892] applied for a patent on an optical crown glass that contained 20 to 55% Mg, Ca, Sr, or Ba metaphosphate and 45 to 80% Mg, Ca, Sr, or Ba orthophosphate. Broemer and Meinert [133] claimed an optical glass of high refraction, medium dispersion, abnormal partial dispersion, and excellent chemical stability. It contained 45 to 70 wt.% of lead borate or phosphate combined with boric acid, aluminum oxide, and tantalum, zinc, cadmium, zirconium, and alkaline earth, or rare earth oxides. Hensler [370a] showed that the ultraviolet absorption of some phosphate glasses was influenced by the oxidizing or reducing character of the melting conditions. Expressed in moles, the glasses contained $Al(PO_3)_3$-12, MgO-2 to 5, Li_2O-2 to 5. Hensler [371] discussed the transitions determining the optical dispersion in glass.

Walter [424] patented an optical crown glass based on phosphates. Two compositions were given: (1) boron orthophosphate, 10 to 90, lanthanum metaphosphate, 0 to 80, barium and/or strontium pyrophosphate, 0 to 78 mole%; (2) boron orthophosphate, 12 to 38%, aluminum metaphosphate, 2 to 25%, and/or lanthanum metaphosphate, 5 to 65%, and barium pyrophosphate, stontium pyrosphate, barium orthophorphate, and/or strontium orthophosphate, 20 to 80%. A Baush and Lomb patent [127a] claims a titanium phosphate superflint glasses with unusually low Abbé values of about 40 and refractive indices (n_D) of 1.6 to 1.75. They are composed of TiO_2-7 to 35, Al_2O_3-1 to 10, alkali-metal oxide-1 to 20, alkali-earth oxide-1 to 10, P_2O_5- 30 to 60, B_2O_3-3 to 18, and As_2O_3-3 to 30.

Fluorophosphate Optical Glasses. An interest in fluorophosphate optical glasses has developed, particularly in recent years. Thus Kodak Ltd., [477] obtained patents on these glasses. Chemical stability was claimed for one containing TiO_2. Another contained $Al(PO_3)_3$ in amounts of 15 to 40%. Another patent mentions a composition of 32 wt.% LiF and 68 wt.% $Al(PO_3)_3$ [236]. Broemer and Meinert [134] patented a fluorophosphate optical glass, and Toda [827] took out a German patent on optical fluorophosphate crowns. Broemer, Meinert, and Spincie [137] patented an optical phosphate glass with anomalous partial dispersion. A later patent by Broemer and Meinert [135] refers to a fluorophosphate glass with anomalous partial dispersion. A later patent by Broemer and Meinert [135] refers to a fluorophosphate glass with this property. Tzumitani and Toda [418] patented a stable fluorophosphate optical glass, and Levina and Shershev [530] published an infrared study of glasses in the system $KBeF_3$-KPO_3.

Golubtsov, Khalilev, Evstrop'v, and Ivanova [318] reported on the regions of glass formation and certain properties of glasses in the systems $Al(PO_3)_3$-BaF_2-RF_2, where R = Mg, Ca, or Sr. They found that the width of the glass-forming region decreased with the decrease in the field force of R. BaF_2 is a glass former that promotes stabilization. Ogita [640] patented an amorphous fluorophosphate optical glass free from yellow coloring.

From the brief abstracts available it is not clear whether the interest in fluorophosphate glasses derives from their apparent moisture resistance or their special optical properties.

Other Phosphate Optical Glasses. A few other references to phosphate optical glasses illustrate the wide range of composition studied.

Tillyer [823] in discussing ophthalmological glasses, remarks that phosphate and borate glasses have been stabilized and some remarkable new properties, discovered. Softening temperatures as low as 160°C have been reached by using beryllium fluoride. Johnson [430] patented a phosphate glass, and Weyl and Kreidle [500] discussed the production of phosphate glasses, starting with "tetraphosphoric acid," a commercial mixture of various phosphoric acids, and aluminum metaphosphate. A table of compositions and properties for 16 phosphate glasses is given.

A phosphate-glass batch patented by Gelstharp [303] shows the following composition: $Al(PO_3)_3$-50 to 100 wt.%, NaH_2PO_4-0 to 26%, and $MgHPO_4$-O to 27%. Weissenberg, Bredow, and Meinert [893] patented an optical flint glass based on lead orthophosphate. Composition ranges were $Pb_3(PO_4)_2$-47.6 to 94.3, $Ca(PO_3)_2$-

4.8 to 9.0, In PO_4-0 to 47.6, and Na_3AlF_6-0 to 1. These glasses had a high n_D and a low ν value. It is claimed that aluminum phosphate or indium phosphate increases the chemical stability of the glass. Jahn [425] patented a lead phosphate optical glass.

Robinson and Fournier reported on the coordination of Yb^{3+} in phosphate, silicate, and germanate glasses [715], and Deeg and Young [214] patented a method for producing phosphate optical glasses. A colorless P_2O_5 glass with an anomalous dispersion in the short wavelength region was patented by Izumitani and Masuda [417].

Optical Absorption. Phosphate optical glasses have found some use as color filters. Hoxie and Werner [400] patented a color filter which in a 6-mm thickness simulated a color temperature of about 3500K. It was made from a phosphate glass with of composition P_2O_5-45 to 80, divalent metal oxide-0 to 20, Al_2O_3-5 to 25, SiO_2-0 to 30, and B_2O_3-0 to 20. It contained 1 to 5% ferrous oxide and 0.005 to 0.015 cobalt oxide.

Landry, Fournier, and Young [521] reported electron spin resonance and optical absorption studies of Cr^{3+} in a phosphate glass, and Jahn [423] published an investigation of the violet coloration of phosphate glasses containing titania. The cations K^+, Mg^{2+}, Na^+, and Ba^{2+} were compared. The relative bond strengths between the different cations and oxygen ions were used to predict the degree of color expected. Landry [520] published a similar study of Mo^{3+} in a phosphate glass.

Fournier and Bartram [285] described the inhomogeneous broadening of the optical spectra of Yb^{3+} in phosphate glass. Weber, Sharp, and Miller [888] discussed the optical spectra, relaxation, and energy transfer of Eu^{3+} and Cr^{3+} in an europium phosphate glass. They decided that nonradiative transfer involved a combination of energy migration through the Eu^{3+} system to fast relaxing Cr^{3+} ions which act as quenching centers. Mukerji [589] reported on the absorption spectra of ruthenium in borosilicate, phosphate, and aluminophosphate glasses. Ru^{6+} and Ru^{4+} or mixtures of the two were found.

2. SEMICONDUCTING GLASSES

Two modes of conduction of electricity are found in glasses: ionic and electronic or semiconducting. In ionic conduction the current is due to the movement of ions such as Na^+ and the electrical characteristics are similar to those of electrolytic conduction. In electronic conduction the current is carried by electrons. A general account of electronic conduction is given by Trapp and Stevels [834] who cite an example of a 6-mole% Fe_3O_4 glass that conducts by ions over the whole temperature range and a 15-mole% Fe_3O_4 glass that is an ionic conductor at high temperatures and an electronic conductor at low.

Some glasses are natural semiconductors; for example, the glasses based on the chalcogenides S, Se, and Te. Others which are made to be semiconducting by artificial means include those that have transition metal oxides in glass produced by directed crystallization, hot pressing of powders, or injection of electrons. The base glasses may be silicates, phosphates, or borates. The vanadium glasses were usually combined with phosphates to extend their zones of vitrification. Trapp and Stevels conclude with a discussion of the use of V_2O_5-P_2O_5 glasses in microchannel disk amplifiers. In an earlier article [833] they described a range of semiconducting glasses based on iron. An iron boroaluminate glass was found to have advantages.

In a discussion of the progress made in glass science Kreidl [494] comments that more than 200 papers were published in one year on amorphous semiconductors. Featured was an effort to adapt a band structure model to them.

Vanadium Phosphate Glasses. Weyl and Marboe [916], Part III, refer to work reported by Rawson [699] which showed that small additions of P_2O_5 to the semiconducting V_2O_5 form a glass on cooling. Baynton, Rawson, and Stanworth [711] reported on the semiconducting properties of glasses in the systems BaO-V_2O_5-P_2O_5 and Na_2O-BaO-V_2O_5-P_2O_5, containing 50 to 87 mole% V_2O_5. Manakata, Kawamura, Asahara, and Iwamoto [594] published a study of high vanadium phosphate glasses. The conductivity was presumed because of the valence exchange between V^{4+} and V^{5+}. A large amount of V^{4+} was found. When specific resistance was plotted against the BaO-P_2O_5 ratio, a minimum was found which shifted toward higher values of the ratio as the temperature rose.

Nador [616] published the properties of glasses in the system V_2O_5-P_2O_5. Munakata [593] reported on the electrical conductivity of high vanadium phosphate glass.

Balta, and Valea [44] described the semiconducting properties of some glasses from the system V_2O_5-P_2O_5-As_2O_3 with more than 60% V_2O_5. Hamblen, Weidel, and Blari [349] described the preparation of ceramic semiconductors from high vanadium glasses. The glasses contained 75 to 85 wt.% of V_2O_5 and metaphosphates of Ba, Pb, Li, Na, Cd, V, and K. Good quality reproducible glasses were made in 500 g melts. A great change in specific electrical resistance occurred when the glasses were heat-treated to produce varying amounts of devitrification. It was found that the thermal history below the liquidus but above the transformation point greatly effected electrical conductivity.

Nador [617] reported on electron spin resonance in the system V_2O_5-P_2O_5. Matveev, Mel'nik, and Glasova [560] studied the effect of adding CdO to a vanadium phosphate glass and found that CdO does not play an independent role in increasing the conductivity of this system because its effect is related to the V_2O_5/P_2O_5 ratio in the glass. This finding led to a further theory of glass formation and electrical conduction. A Blair, Hamblen, and Weidel patent [98] gives V_2O_5-metal metaphosphate glass compositions. Trap described phosphovanadate glasses and their use in display panels, some of which contained additions of PbO, MoO_3, or WO_3 [831].

Matveev, Khodskii, and Fisynk [558] published a detailed description of semiconducting glasses in the system V_2O_5-Bi_2O_3-P_2O_5. The electrical conductivity varied from 10^{-4} to 10^{-8}/ohm (cm), and the temperature coefficient of resistance varied from 2.57 to 5.46%. The energy of activation was 2.57 to 5.46 eV. The glasses started to soften at 280 to 470°C, and densities ranged from 3.2 to 4.2 g/cm^3. The coefficient of thermal expansion values were in the range 67.3 to 102.8 $\times$ 10^{-7}/deg.

Gamma ray or neutron irradiation had no effect on the electrical properties of *n*-type V_2O_5-P_4O_{10} semiconducting glasses. The internal friction of V_2O_5-P_2O_5 glasses in the kilohertz region and 100 to 400°K. was studied by Field [272]. He reported that he had found a relaxation peak below room temperature that was strongly dependent on composition and showed a relation between internal friction and electrical conductivity.

Caley and Murthy [153] investigated the electrical conductivity of glasses in the system P_4O_{10}-V_2O_5 and P_4O_{10}-WO_3 and found that tungsten phosphate glass was sensitive optically and electrically to small substitutions of V_2O_5. Caley published a comparison of the electrical conductivities and optical spectra of tungsten-vanadium-phosphate glasses. The conductivities were stongly dependent on the WO_3 and V_2O_5 contents and could be accounted for by the optical absorption spectra. The phenomena were attributed to the Redox effects of the V and W ions.

Electron and nuclear magnetic resonance studies of semiconducting phosphate glasses were made by Lynch, Sayer, Siegel, and Rosenblatt with parmagnetic V-phosphate and Mo-phosphate glasses. They reported that the mobilty was temperature-dependent and that the line shape of an average paramagnetic site was unchanged over a temperature range of 77 to 300°K. The ESR studies showed strongly exchanged, narrowed interaction with V^{4+}/V^{5+} and Mo^{5+}/Mo^{6+} ratios and line shapes. NMR studies of ^{51}V and ^{31}P glasses verified these conclusions [539].

Friebele, Wilson, and Kinser reported on the magnetic behavior and microstructure of vanadium phosphate glasses [290]. A transition temperature near −75°C was observed. As cast, the glasses with high V_2O_5 concentrations separated into two glassy phases; the vanadium ions were concentrated in the vanadium-rich phase. Attempts to reduce V^{5+} to lower valance states produced phase separation in the glass and weak magnetic behavior. Explanations for these phenomena were advanced. Electron microscope results indicated that a glass must be characterized structurally before other investigations can be attempted.

"Polaronic hopping" was advanced as a means of explaining electric conduction in glasses prepared from V_2O_5-P_2O_5 melts containing 20% P_2O_5 by Sayer and Lynch; Mansingh, Reyes, and Rosenblatt [732, 733]. Conductivity effects are attributed to electron hopping in a disordered lattice, and conventional semiconductor theory is used to interpret the frequency-dependent properties. Possible application of the glasses involves barrier layer effects, catalytic fuel-cell reactions, and thermoelectric properties. Sayer and Lynch [732] reported on impurity effects in semiconducting glasses. They added V_2O_5, CuO, and MoO_3 as impurities to a tungsten phosphate semiconducting glass from which they derived dramatic effects and found evidence for direct optical effects of polarons in phosphate glasses.

France and Hooper [286] described a nuclear magnetic resonance study of a semiconducting vanadium phosphate glass, and Schaake and Hench [734] reported on polaronic effects in potassium phosphate glasses. Regan and Drake [700] discussed vanadium phosphate glass threshold switching devices. Under Current Ceramic Research Winters and LaCourse [929] listed "Carrier Mobilities in V_2O_5-P_2O_5 Glasses" and Tatom and Kinser listed "Vanadium Phosphate Glasses" [812].

Adler's book entitled, *Amorphous Semiconductors,* devotes several chapters to vanadium phosphate glasses [2].

Iron Phosphate Glasses. Hansen [350] discussed the semiconduction of a phosphate glasses of composition FeO-55, P_2O_5-45 mole%, based on measurements made at 25 and 400°C. The resistance increase varied with the Fe^{3+}/Fe total ratio with a minimum at 0.5. Conduction was *p*-type when this ratio was lower than 0.38 and *n*-type at higher ratios. Hansen suggested that carrier mobility was a thermally activated diffusion process such as that proposed for crystalline metal oxides and that the activation energy for conduction was related to the mobility rather than the formation of carriers.

Hansen and Splann [351] discussed the dielectric constant and loss factor for the same glass. They found an ac dispersion with temperature dependence similar to that of dc conductivity. The dielectric constant was independent of frequency and temperature below the dispersion region. The same glass was studied by Kinser [465], who reported dielectric loss measurements, dc conductivities, and transmission electron microscopic observations. It was found that the glass partially devitrified after low-temperature treatments. Further studies on this glass were published [227] by Dozier, Wilson, Friebele, and Kinser. Changes in electrical properties were related to the Fe^{3+}/Fe_{total} ratio and to devitrification induced by heat treatment. The electrical relations were typical of a amorphous semiconductor, and $FePO_4$ was the primary crystalline phase for all compositions studied. The d-c resistivity changes were due primarily to devitrification.

Dozier, Wilson, Friebele and Kisner [228], in their discussion of the electrical properties of a glass with composition FeO-55, P_2O_5-45, related changes to the ratio Fe^{3+}/Fe_{total} and to devitrification induced by heat treatment. Log resistivity versus $10^3/T$ gave a straight-line plot with two sections of different slope. This typified an amorphous semiconductor. Vaughan, Wilson, and Kinser [870] reported a study of the Mössbauer effect in iron phosphate glasses. Octahedral coordination and short-range magnetic ordering was indicated at 77°K. Precipitation of at least two crystalline phases was shown on heat treatment. Vaughan and Wilson [869] listed other Mössbauer studies.

3. ULTRAVIOLET TRANSMITTING GLASSES

A considerable literature deals with the use of phosphate glasses for their ultraviolet transmitting properties. In this they are superior in general to the silicate glasses and, provided that they are weather resistant and can meet cost criteria, they can be quite useful. The high point of interest appears to have developed in the late twenties and early thirties following the discovery about 1925 of the physiological effects of ultraviolet light.

Other Phosphate Glasses. Boynton, Rawson, and Stanworth [56] reported on a wide range of oxide systems. Glass formation was found with Li_2O-MoO_3, K_2O-MoO_3, and BaO-MoO_3. The MoO_3-P_2O_5 system also produced glasses, provided that the P_2O_5 content was higher than 11%. The same was true for the WO_3-P_2O_5 system. In stabilizing experiments P_2O_5 was found to be a most promising additive.

Kurkjian [513] described the use of P_2O_5-SiO_2 films on semiconductor devices.

Sayer and Austin [731] submitted a paper that treated the high field conduction of glasses of the transition metal phosphates in general. They postulated a field-dependent hopping process between sites separated by a distance four to six times the average ion spacing. This was attributed to enhancement of the field by the microscopic structure of the glass.

Kreidl and Weyl [909] refer to the work of Eder and Valenta [246], who in 1894 found that the phosphate crowns of Otto Schott were superior to other optical glasses in ultraviolet transmission and were less expensive than pure silica glass for this purpose. Zschimmer reporting on experiments that related the ultraviolet transmission of glasses to their chemical composition, found that commercial glasses based on the absence of ferric oxide lost their ultraviolet transmission after prolonged exposure to sunlight [909, 960]. Hood [403] described a new ultraviolet transmitting glass and obtained a patent on a phosphate glass with a small ferric oxide content. This was probably the "Corex" glass referred to by Kreidl and Weyl [909] as being an almost pure calcium glass with minor additions of boric oxide, alumina, and silica. The chemical resistance of this glass they described as imperfect but sufficient for many purposes.

I.G. Farbenind, A.G. [411] patented a glass that was described as highly resistant to water and permeable to ultraviolet light. It was composed largely of barium metaphosphate but calcium and magnesium metaphosphates might also have been present. Grimm and Huppert [340] patented a glass in which silica was replaced wholly or in part by Al_2O_3 and P_2O_5, in the proportion of more than 1 mole of Al_2O_3 to 1 mole of P_2O_5. High UV transparency, high softening temperature, and low coefficient of expansion were claimed. Reference was also made [19] to glasses manufactured by the I.G. method in which silica was replaced entirely by aluminum orthophosphate. A similar reference [18] claims a glass of high resistance to water with the composition: K_2O-10, CaO-12, Al_2O_3- 32.5, and P_2O_5-45.5%. I.G. Farben patents [408a, 412] covered ratios of P_2O_5/Al_2O_3 between 1.1 and 3.2 and the addition of ammonium phosphate to the usual batches of transparent phosphate glass.

Coblentz and Stair [175] studied the changes in the transmission of UV light when a phosphate-lime glass (Corex A) was exposed to light. Only light of wavelengths shorter than 290 mμ caused a depreciation in transmission.

Huppert and Wolff [408b] give the composition of a UV transparent glass as K_2O-12.5, CaO-5, BaO-2.5, MgO-7.5, B_2O_3-17.5, Al_2O_3-28.8, and P_2O_5-26.2%. Gentil [303a] reported that the UV transmission of the I.G. phosphate glasses was nearly equal to quartz glass and superior to the silicate glasses. Hoffman [392] reported on the coloring effects of radiation on phosphate glass, and Kaufmann and Bungartz [449] reported that barium was better than calcium in a phosphate glass. Previous I.G. Farben patents were widened to include aluminum, calcium, magnesium, and aluminum metaphosphate [413a].

Hoffman [391] reported that the red coloring of phosphate glasses by ultraviolet rays was due to phosphorus atoms. Kreidl and Weyl [500] report that P_2O_5, which was first used to improve the color of heavy flint glasses, was later recognized as a valuable constituent of optical glasses that permitted excellent UV transmission. Poor chemical resistivity could be largely overcome by the addition of Al_2O_3. Pincus [674] patented an optical glass with aluminum metaphosphate as its main constituent. The composition range was P_2O_5-45 to 80, Al_2O_3-8 to 25, SiO_2 1 to 30 wt.%, with a specific composition of 68, 16, and 16 respectively. It could be worked and annealed by conventional equipment and was resistant to temperature changes. As little as 0.1% alkali improved its working properties. The introduction of FeO resulted in a heat-absorbing glass with a low coefficient of expansion. If both ferric and ferrous iron were present, the glass absorbed both IR and UV rays. A Science New Letter [21] refers to the use of phosphate UV-transmitting glasses in windowpanes of hospitals, in optical glass, and in electric insulators. Holly [393] describes a black UV-transmitting glass that is almost opaque to visible light. A fluorescent phosphate glass was patented in 1946 [800].

Stevels [775, 776] said that as the network forming ions were varied in the direction B^{3+}-Si^{4+}-P^{5+} it was found that the phosphate glasses have the greatest UV transmission because absorption is displaced toward shorter wavelengths as the potential of liberation of an electron (Madelung potential) increases.

Stein [773] discussed the Kordes theory of UV transmission of binary phosphate glasses. The difference between phosphate and silicate glasses was attributed to a phosphate glass with free corner oxygen atoms as well as oxygen bridges [484]. "Uviol" phosphate glasses were the subject of a paper by Manasevich [545].

Kordes and Nieder [484] published data on the transmission limit of about 15 binary phosphate glasses. Some showed extension of this limit to shorter wavelengths. In each of the systems P_2O_5-ZnO and P_2O_5-CdO two composition ranges of high UV transparency were attributed to the polarization properties of the nonnoble, gaslike zinc and cadmium ions. In comparing borate, silicate, and phosphate glasses the curves of physical properties versus composition were explained on the basis of the ratio of "corner oxygens" to "bridging oxygens."

Duffy [230] discussed the UV transparency of sodium phosphate glasses that contained metal impurities. He concluded that metal ions dissolved in phosphoric acid or in sodium phosphate glass exist as phosphate complexes, although the nature of the phosphorus (V) oxyanion ligands was probably different in each case. The phosphate ligands attached to the metal ion in the glass may be linked to the phosphorus-oxygen network of the glass.

4. COLORED GLASSES

The principal reference on the subject of colored glasses in general is Weyl's *Coloured Glasses* [912], which reviews the history of the subject up to mid-1949 and provides a great deal of information on borate and phosphate glasses. Much may be learned by comparing the colors of aqueous solutions, crystalline compounds, and glasses of approximately the same composition, and Weyl makes full use of this approach. In what is possibly the simplest case, the color of a material by transmitted light may be due to colloidal particles that are essentially opaque. The color then depends only on the particle size distribution, the longer wavelengths being transmitted in preference to the shorter wavelengths. All other colors are due to absorption of energy by electrons moving to different energy levels. Most of these colors may be due to ions of the transition metals in which valence electrons can be shifted into higher energy orbits. Ions of the rare earths allow electron transitions in their inner protected orbits. Many metallic ions allow electron transitions between two ions of the same element; for example, iron and copper glasses. When actual metallic particles are precipitated in a glass, the color may be due to the light absorption properties of the metal in question.

All colors dependent on electron transitions are influenced by the environment of the group responsible for them. Thus the color of the same ion may be different in silicate, borate, and phosphate glasses and may also change noticeably with variations in the amount and nature of the other constituents of the glass; for example, phosphate glasses containing ferrous ions show less color than the corresponding silicate glasses but retain their heat-absorbing properties and are therefore sometimes preferred for this purpose.

Phosphate crystals, aqueous solutions, and glasses are convenient hosts for many ions, particularly transition metal ions. Consequently a considerable literature deals with colors of phosphate glasses. In the time since the publication of Weyl's book particular attention has been paid to ligand field theory. A transition metal ion in a glass or crystal is surrounded by anions arranged in octahedral or tetrahedral symmetry. Ligand field theory is concerned with calculating the new energy levels of the transition metal ion from the interaction of the electrostatic field from the nearest neighbor anions, called "ligands." Douglas and his associates at Sheffield University have made good use of ligand field theory in explaining colors in glasses, crystals, and solutions [54]. Finally mention should be made of a series of glasses developed by Knudsen [474], which owed their color not to selective absorption but to selective dispersion of light. These glasses were two-phased. The refractive index versus wavelength curves for the two phases separately intersected within the visible region. At this wavelength there was no scattering of light and consequently the wavelength was transmitted at little loss.

Colored Phosphate Glasses. References of a general character found in this survey are reviewed in the following paragraphs. More specific colors are then discussed.

Cohn [181] studied the cause of color produced by the fusion of borax or sodium metaphosphate with metallic oxides. He concluded that the colors were due to the formation of pyrophosphates in the metaphosphate glasses. No orthophosphates were formed as long as the beads were transparent. Long patented a violet-colored phosphate glass that contained a relatively small percentage of TiO_2 and was substantially free from silica [537a].

Weyl [907] discussed the theoretical basis of glass coloring and concluded that the color of cobalt glass depended more on the number of "ligands" with which the Co atom was combined and less on its chemical nature. This explained the formation of blue and red cobalt glasses.

Bhatnager, Khosla, and Chand [89] reported their studies of the nature of the multivalent ions of Mn, Co, and Ni in phosphate glasses using magnetic susceptibilities. They found Mn in reduced colorless glasses to be divalent but trivalent in pink-violet colored glasses. Co blue and Ni yellow were bivalent. On reduction, Co blue was unaffected, but Ni yellow evolved opaque masses that were ferromagnetic in nature and therefore consisted of metallic nickel. Armistead [31a] patented a glass colored by Cu and Ce of the following composition: P_2O_5- 45 to 80%, Al_2O_3-8 to 25, SiO_2-1 to 30, CuO-0.1 to 5, and CeO_2-0.1 to 3. The weight percent ratio of P_2O_5/Al_2O_3 was between 3/1 and 6/1.

Lindroth [535] reported that the color of glass depended on (1) ionic equilibria, (2) complex coordination, and (3) dispersed colloidal elements in a glassy matrix. In (3) the color was a function of particle size and absorption spectrum. Mukerji and Biswas [588] have made an interesting comparison of phosphate and silicate glasses by studying the oxidation states of ruthenium, which are contrasted in the two types of glass. Edwards, Paul, and Douglas [248] made a similar study of iron in RO-P_2O_5 glasses, in which R = Mg, Ca, Sr, or Ba. The magnesium phosphate glasses differed from the others. Reisfeld, Hormodaly, and Barent [704] used Ce^{3+} as a probe of the crystal field and the nature of the impurity-ligand bond in borate and phosphate glasses.

Amber Glasses. Amber-colored glasses are of considerable importance commercially for beverage bottles, medical prescription bottles, and signal lights. The amber color of beverage bottles has traditionally been called a "carbon-sulfur amber" because both sulfur and carbon are usually added to the glass to produce the color. For many years there was disagreement about the source of the color, but more recent work shows that ferric ions play a large role.

Moore and Prasad [575] reported on a spectrophotometric study of the state of iron in silicate and borosilicate glasses. They concluded that ferrous iron produced a rich blue, ferric ion, a rich amber, and ferroferric ion, a perfect neutral gray. A colorless glass was found to have ferric ions that were probably associated with three adjacent SiO_4 tetrahedra. Brewster and Kreidl [126] discussed the change in color of iron-containing glasses as the composition of the glass was changed. Weyl [911] reported that color could be the result of the presence of two states of valency of the same element. Ferrous and ferric ions together produce no color in sulfate, slight color in phosphate, blue in silicate, and gray in borate glasses. In order to throw light on the color produced by iron compounds in phosphate glasses, Ram, Bose, and Kumar [694, 695] studied ferriphosphate complexes in solution by polarographic methods. They advanced an explanation of the suppression of iron colors in phosphate glasses. Kumar [509] investigated a wide range of phosphate glasses, including some containing lead oxide. P_2O_5 was found to shift the spectrophotometric absorption edge due to amber toward shorter wavelengths. Some showed excellent chemical durability, low deformation temperature, and little visible absorption due to iron. Bamford [45] published the results of a study of the relation of the magnetic properties of iron to its coloring effect in sodium phosphate glasses. He found that ferrous and ferric ions were subject to octahedral ligand fields. The color was dependent on the valency condition of the Fe ions determined by the temperature variation of their magnetic susceptibilities. Mössbauer, optical, and electron paramagnetic resonance spectra were used by Kurkjian and Sigety [514] to study the coordination of Fe^{3+}. They found it to be tetrahedral in silicate glasses but octahedral in phosphate glasses. In studies of the role of Fe^{3+} in amber glass Zaman and Douglas [955] concluded that the chromophore was an Fe^{3+} ion surrounded tetrahedrally by three oxygens and one sulfur. In discussing the structure and technology of glasses, Douglas [226] reported that absorption bands showed the different oxidation states and coordination of an ion surrounded by oxygens. With melts in equilibrium and a given partial pressure of oxygen the state of oxidation increases in the order Li, Na, and K and with the total amount of alkali present. Poole [681] accepted Douglas' conclusion that in a carbon-sulfur amber glass the color center is a ferric ion coordinated by both oxygen and sulfur ions. Harding and Ryder [354] concluded that the amber color of silicate glasses had an intensity proportional to the product $(Fe^{3+}) \times (S^{2-})$. The color center contained one sulfide ion for each ferric ion. Extreme oxidation or reduction removed one or another of the essential components. Harding [353] described the effect of the base glass composition on the color of amber glass. He found that the S^{2-}/SO_4^{2-} balance was not wholly a function of melt oxygen availability. Different alkali and alkaline earth cations had a different effect. Amber color centers contained Fe^{3+} and S^{2-} and the transmitted color correlated with the levels of these species. The color centers involved only 13% of the Fe^{3+} and 5 to 6% of the S^{2-} in the glass.

Uranium Glasses. Kreidl and Colbert [496] remarked on the unusual colors produced by uranium in glass. Murthy and Ciric [601], using UO_2, made a number of glasses, including many in the system Na_2O-B_2O_3-UO_2. Numerous shades of red were encountered.

5. SLAGS

Slags play an important role in metallurgical processes and consequently a number of papers in the metallurgical literature deal with them. Because slags are essentially glasses and often contain considerable amounts of phosphates, a few references may be of interest. Slags in general are discussed in a publication of the Institute of Mining and Metallurgy [415] entitled "Physical Chemistry of Melts." Toop and Samis [838] proposed some new concepts regarding the ionic constituents of silicate slags which included mixed heteropoly acid formation. Phase equilibria relating to slags were described by Muan and Osborn [587]. Activities in liquid slags have received attention. Thus Carter, [169] studied four ternary systems, including FeO-CaO-P_2O_5, Toop and Samis [830] measured the activities of ions in silicate melts, and Holmquist [394] calculated the activity of sodium oxide in sodium silicate melts in equilibrium with sodium sulfate and a controlled atmosphere at 1200°C.

Several references are concerned with the effect of adding fluorides to melts. Thus Kumar, Ward, and Williams [511] reported that fluorides in acid slags broke oxygen bridges but in alkaline slags had little effect. Both silicates and phosphates were used. Behrendt and Wentrup [66] reported that a ternary eutectic existed between lime, silico-carnotite, and "tetraphosphate."

Trömel [836] investigated the CaO-P_2O_5-FeO system near the CaO corner in an effort to overcome difficulties encountered in producing good soluble slags in the basic Bessemer process.

Kristofferson [505] published a paper on phosphates and silicophosphates in slags, and Olsen and Maetz [639] studied the FeO-CaO-P_2O_5 system. Over part of the diagram the melt separated into two layers.

Of particular interest to ceramists is the series of articles, published by Masson and his associates, dealing with slags from the viewpoint of thermodynamics, activities and ionic distributions, vapor-phase chromatography and using the Lentz technique and molecular size distributions [328, 550-553, 758, 919, 937].

6. OTHER ELECTRICAL PROPERTIES

Semiconducting glasses were discussed under Subsection H-2. References to other electrical properties of phosphate glasses follow.

Kumar [510] discussed the electrical properties of glasses in the system Al_2O_3-B_2O_3-P_2O_5. Included were the dc conductivity, power factor, and dielectric loss. Beekenkamp and Stevels [65], referring to a 1956 paper by Kumar, suggest that Kumar's glasses dissolved silica from the crucibles which broadened the glass-forming region. More general approaches to the question of the conductivity of glasses and ceramics were made by Myuller [606] and Adler [3]. The dielectric strength and dielectric constant of phosphate glasses were discussed by Patashinsky [658] and Kobayashi and Yokota [475], respectively. The latter investigated the system TeO_2-$Ba(PO_3)_2$-RPO_3, in which R represented Li, Na, or K. The softening temperatures were 380 to 430°C, the dielectric constant, 15 to 26 at 1 Mc/sec, and the dielectric loss, 0.002 to 0.008.

Surface conducting glasses were studied by Green and Blodgett [332]. Glasses containing Pb, Bi, or Sb oxides became electrically conducting after several hours of reduction in hydrogen. The surface conductivity was both stable and reproducible. In a U. S. patent abstracted in 1962 Hensler [370] gave the composition of a glass that could be reduced to produce a conducting surface as silver phosphate, -8 to 25, aluminum phosphate, -30 to 75, alkali-metal phosphate, -10 to 30, and barium phosphate up to 25 wt. %. At 300 to 400°C a reducing atmosphere for two to three hours resulted in a resistance of less than 10 $ohms/cm^2$. A similar patent was issued in 1964 [369]. Golubtsov, Evstrop'ev, Kir'yanova, Kondrat'eva, and Khalolev [317] reported on the electrical properties of alkali-free fluorophosphate glasses. Anionic conductivity was indicated. Cationic components did not exceed 11% of the electric transfer. MgF_2 entered the glass network as if to compensate for the decrease in aluminum metaphosphate content.

7. HEAT-ABSORBING GLASSES

Heat-absorbing glasses have a number of uses. One of technical importance is in projection lanterns for colored slides in which freedom from color is a requirement. Most of the heat-absorbing glasses depend on ferrous iron to absorb the heat rays and here phosphates have some advantage because the absorption is just outside the visible range. Berger [70] patented a colorless, heat-absorbing glass which contained at least 25% (P_2O_5 + B_2O_3) and 0.2% FeO. He pointed out the advantage of ferrophosphorus and other phosphorus compounds as reducing agents. Conti [190] described the results obtained by the Corning, Jena, and Boileau-Mercier glass works in producing heat-absorbing glasses. "Phosphoric" glasses were included. Tillyer, Moulton, and Gunn [824] patented a heat-absorbing glass of a barium aluminum phosphate type containing 1 to 10% $Fe_3(PO_4)_2.8H_2O$.

Weyl and Kreidl [909] in their review of phosphates in ceramic ware, described the use of ferrophosphorus and other phosphorus compounds in the production of heat-absorbing and copper ruby glasses. Pincus [674] patented a UV-transmitting glass that was also heat-absorbing. The main component was aluminum metaphosphate. The composition range specified was P_2O_5-45 to 80, Al_2O_3-8 to 25, and SiO_2-1 to 30wt.%, with a specific composition of 68, 16, and 16, respectively. Golubeva and Prok [316] reported on a heat-insulating glass containing P_2O_5 and Al_2O_3. A high stability toward water and acids was claimed. An iron-bearing phosphate glass [22] which absorbed 90% IR and transmitted 85% light was reported in 1946. It resisted weathering without surface treatment. Mook and Ricker [637] also patented a heat-absorbing glass, which, however, was based on silicates rather than phosphates. The IR spectra and absorption characteristics of divalent metal phosphate glasses containing residual water were reported by Naruse [624] and Naruse, Abe, and Inoue [625]. The glass systems P_2O_5-Al_2O_3-ZnO-R_2O and P_2O_5-Al_2O_3-B_2O_3-R_2O were studied. Satisfactory IR shielding power was obtained with FeO greater than 3.2%, R_2O less than 5%, and metallic silicon as a reducing agent.

8. LASER GLASSES

A general discussion of laser glasses had been published by Snitzer [761]. Laser glasses depend on the presence of ions of the rare earth elements neodymium, ytterbium, and erbium, with neodymium getting the most mention. Snitzer says that neodymium has been made to lase in a variety of glasses but the requirements of high efficiency, durability, and ease of manufacture have been met best by alkali-alkaline earth-silicate glasses. In an earlier paper, however, Snitzer, Woodcock, and Segre [762] give an example in which an erbium-silicate glass laser was improved by shifting to a phosphate glass with the composition P_2O_5-61.3, ZnO-14.7, Al_2O_3-8.4, Nd_2O_3-0.2, Yb_2O-14.9, Er_2O_3-0.5 wt. %.

Depaolis and Mauer [217] patented a phosphate glass for laser use in which the composition was La_2O_3-14, BaO-23, P_2O_5-60, and Nd_2O_3-3%. Kreidl speculates that the high oxygen content of phosphate glasses due to the high valency of phosphorus allows a more symmetrical accomodation of the Nd^{3+} ion. Young and Snitzer [951] published a discussion of the laser ions of neodymium, ytterbium and erbium, and Mazelsky and Ohlmann [561] described as a laser a crystalline fluorapatite material doped with neodymium and manganese.

Hirayama, and Melamed [386] have patented some laser phosphate glass compositions, and Mazelsky and Ohlmann [562] have patented a laser containing neodymium-doped calcium fluoroapatite. Reisfeld and Eckstein [703] reported on the absorption and emission spectra of Tm^{3+} and Er^{3+} in borate and phosphate glasses. They considered that the high luminescence efficiency of Tm^{3+}indicated laser possibilities. Graf [329] patented a laser of a phosphate base laser glass. Weber, Damen, Danielmeyer, and Tofield [889] reported that laser action had been observed in both crystal and glass of the unusual compound NdP_5O_{14}, which they called an ultraphosphate.

9. COATINGS

A limited use has been found for phosphate coatings on technical glassware. Navias [627] patented an electrical discharge device with a silica glass envelope coated on the inside surface by heating such oxides as P_2O_5, B_2O_3 or PbO, thus lessening a wall-charging tendency. Van Bakel and Gast [852] patented a phosphate coating for incandescent lamps which transmitted 95% of the light but prevented the filament from shining through. Repsher [705] patented a discharge device which had a thin exterior coating of a transparent phosphate. Snow and Deal [763] studied the polarization phenomena and other properties of phosphosilicate glass films on silicon. Such films were compared with thermally produced silica layers. A nondestructive method for the analysis of glassy phosphosilica films by secondary K alpha emission was described by Pink and Lyn [677]. Bonniaud and Sombret [111] described the use of glass as a coating for radioactive waste. Phosphate and silicophosphate glasses were tried. A means of stabilizing the volatile fluorides and both semicontinuous and continuous processes were evolved. Kennedy, [452] patented a method for reactive sputter deposition of a phosphosilicate glass.

10. OTHER USES

Phosphate glasses are mentioned in the literature surveyed in connection with a number of other minor uses; for example, solder or low-melting glasses are important in vacuum and electronic technology and as components of low-melting enamels. Commercial glasses in this category, described by Umblia [850], included some P_2O_5 glasses. The process of bonding articles together by an internally nucleated devitrified solder glass was described by Babcock [38]. The composition given was PbO-70 to 73, B_2O_3-15 to 29, and 0.1 to 10% of at least one internal nucleating agent (Cr_2O_3, P_2O_5 or V_2O_5), with Cr_2O_3 not exceeding 1%. Grimm and Huppert patented a low-melting phosphate glass [341] whose composition was P_2O_5-34 to 44, Al_2O_3-20 to 80, less then 12% B_2O_3, at least 25% alkali containing 8-21% Na_2O, 5-21% K_2O. Kreidl and Weyl [909] with regard to a similar glass, suggest that if part of the B_2O_3 were replaced by ZnO a resistant phosphate glass with a softening point of 400°C could be obtained. Kreidl and Weyl [500a] gave a comprehensive account of the development of low-melting glasses, including some low-melting phosphate glasses. A phosphate glass electrode with a good selectivity for alkaline earth ions which could be used for measuring the activity of Ca^{2+} in natural systems containing Mg^{2+} was described by Truesdell and Pommer [839]. A nontoxic phosphate glass patented by Gordon [324] had a composition of alkali-metal oxide 25 to 58, P_2O_5-29 to 68, Al_2O_3-0 to 20, with total impurities less than 4%.

Phosphate glasses have been mentioned in connection with nuclear reactors both as fuel and a method of waste disposal. With fuel use in mind, Owens and Meyer [649] reported on the molar volumes and surface tensions of fused metaphosphate-sulfate systems. Ideal behavior on mixing was observed. A phosphate glass process for the disposal of high-level radioactive wastes was described by Tuthill, Weth, Emma, Strickland, and Hatch [847]. It was demonstrated on a pilot-plant scale with platinum crucibles. Tuthill, Strickland, and Weth [846] investigated a 99% aluminum crucible as a substitute for platinum crucibles. Corrision resistance was satisfactory but thermal shock resistance was too low. Bonniaud and Sombret [111] described the use of glass as a coating for radioactive wastes. Phosphates and silicophosphate glasses were tested. Infrared-transmitting glasses were discussed by Foulstich [268]. A fluophosphate glass was found to be almost free from absorption over the whole region from 0.32 to 4.1 mu, with strong steep absorption in the infrared at 4.1. Kanazawa, Kawazoe, and Ikeda [444] reported on the properties of some calcium silicophosphate and some magnesium silicophosphate glasses. Poole, Snyder, and Boschini [682] patented a tripotassium phosphate treatment for strengthening glass. Dumesnil, Hewitt, and Bozarth [233] patented low-expansion, low-melting zinc phosphovanadate glass composition suitable for seals to alumina over the range 380 to 480°C. Roberts [714] reported on the preparation of glasses in the system FeO-K_2O-P_2O_5. Glasses with up to 30% iron and suitable for alkaline farm soils were evolved.

I. Interaction with Radiation

A large literature has been built up, particularly in recent years, on the interaction of radiation with glass. Historically, it was noted many years ago that window glasses which had been decolorized with vanadium, over a period of years acquired a slight purplish color. This was called solarization. Radiation can have many other effects on glass, depending on the type of radiation and the composition of the glass. These effects can often be studied with a view to arriving at the nature of the structure of the glass. In addition to the study of these effects, much attention has been paid to the use of glasses, particularly silver phosphate glasses as a means for determining X-ray dosimetry. There has also been much interest in the use of special glasses as shielding from particular types of radiation.

1. EFFECTS

References dealing with the effects of radiation on glasses, particularly phosphate glasses, are reviewed first.

Sun and Kreidl [795] published a comprehensive article on the coloration of glass by radiation which included its use on dosimeters. They also reported that the K-Ba-Al phosphate glass developed by Weyl and others has been an outstanding contribution. Progress has been made in the understanding of color center formation in simple crystals, but little knowledge of this kind has been available for glasses. Coloration in glasses containing neither polivalent ions nor easily reducible metallic ions could logically be explained by color center formation. Direct proof of the formation of color centers in glasses was lacking, but the parallelism between crystals and glasses with respect to color formation was quite general. It was suggested that the structure of a glass was not so simple as Warren's picture of perfectly random orientation.

Kreidl [490] discussed the interaction of glass with high energy radiation. The wide use of atomic energy required optical glasses that could be used under high-energy radiation. Of special interest were dosimetry, shielding, and the causes and prevention of color changes. For high energy radiation the effect depended on the density and the component nuclei of the glass. For most practical sources a maximum lead content afforded the best shielding power against high-energy radiation. For silicate glasses a dark brown color was produced; with phosphate glasses, a striking red. Ordinary absorption starts at much shorter wavelengths than for silica glasses. Color centers consisted of an electron trapped in a local deficiency in negative charge with respect to the next positive neighbors. In phosphate glasses coloration could be prevented by the addition of about 1% cerium oxide. Phosphate glasses with a few percent of silver would detect gamma radiation over wide energy and intensity ranges. The formation of color centers in glass exposed to gamma radiation has been discussed by Kreidl and Hensler [497], and Kreidl, Hensler, and Blair have discussed radiation damage to glasses [499]. Stevels [777] presented a number of interesting structure diagrams and pointed out the remarkable differences between quartz crystal and quartz glass after irradiation. By studying the network defects present in these materials before and after irradiation by X-rays and fast neutrons it was possible to draw up "operational equations" that described the processes induced by these radiations. Kreidl [493] contributed a section dealing with the effect of high-energy radiation to a book on glass manufacture. Alers [8], reporting on the effects of gamma radiation on cerium-bearing phosphate glass, concluded that the cerium remained in the cerous condition and that the height of the radiation-induced absorption peak was inversely proportional to the amount of cerium present. Bishay [94] discussed the gamma-ray-induced coloring of some phosphate glasses in considerable detail. The effect of composition, melting conditions, and small additions was investigated. Under reducing conditions the UV-induced absorption increased and the visible-range-induced absorption decreased. The reverse was found if K^+ was replaced by Na^+ or Li^+. Berger, Robinstein, Kurkjian, and Treptow [71] reported on the Faraday rotation of rare-earth (III) phosphate glasses. The rotation was ascribed primarily to strong electric dipole transitions that involved the rare-earth 4f electrons. A simplified equation for an effective transition wavelength was given. Prediction of rotations of ions, molecules, etc., on the basis of relative magnitudes of magnetic susceptibility and concentration was not meaningful.

Nakai [622] published an electron paramagnetic study of gamma-ray-irradiated phosphate glasses. He suggested that the paramagnetic centers are connected with the peculiar structure of phosphate glasses and considered trapped holes in the glass network as well as trapped electrons in the alkaline or alkaline-earth ions. In phosphate glasses modified with such ions as Ca_2^+, Ba_2^+, or Li^+, which can coordinate with the tetrahedral atom in "end groups," the weak resonance is easily detected. A comprehensive review on radiation effects in quartz, silica, and glasses was published by Lell, Kreidl, and Hensler [527]. Among 290 references, nine were concerned with phosphate glasses. The general conclusion was that the study of radiation effects on solid material was an important and almost indispensable tool for exploring the defect structure in solids, which in turn helped in the interpretation of radiation phenomena. A similar study of complex glasses was published by Kreidl and Lell [498]. Feldmann and Treinin [269] reported on I^-photosensitized reactions in metaphosphate glass. In their study of the photo-oxidation of Cl^- and Br^- optical and electron spin resonance methods were used. They postulated an energy transfer from an excited I^- to a phosphate polymer to a metaphosphate color center. The primary processes were discussed and an exciton-type mechanism was proposed for the energy transfer.

Becherescu, Menessy, and Cristea [58] discussed the relation between the structure of phosphate glasses and their behavior under irradiation. Their glasses contain 40 to 60 mole % P_2O_5 and the remainder, Na_2O. All became a rose red when irradiated with beta or gamma rays. Glassy $(NaPO_3)_n$ was colored by radiation but the crystalline material was not. Bershov, Martirosyan, Zavadovskaya, and Starodubtsev [84] reported on the relation between point defects and the radiation breakdown of aluminophosphate glasses. The PO_3^{2-} and PO_4^{2-} ions are interpreted within the framework of the theory of molecular orbitals. Sandoe, Sarkies, and Parke [730] published the absorption and fluorescence spectra due to Er^{3+} in a wide range of silicate, phosphate, and borate glasses.

2. DOSIMETRY

Kreidl [491, 492] issued preliminary and final reports on phosphate glass dosimeters. Kreidl and Blair [495] discussed recent developments in glass dosimetry, and Blair [97] patented a phosphate glass composition for a dosimeter: aluminum metaphosphate-50, magnesium metaphosphate-25, lithium metaphosphate-25, and 8 parts by weight of silver metaphosphate. Cropper [199] discussed gamma-ray induced coloration in glass systems. He postulated an initial effect that produced free electrons and trapped holes and quoted the data of Schulman et al. on silver phosphate glass. A phosphate glass for dosimetry of X and gamma rays was patented by Bedier, Carpentier, Francois, Brand-Clement, and Meneret [61]. Its composition was given as Li_2O-1.5 to 7.7, BeO-1.3 to 6.6, Na_2O-0 to 7.6, Al_2O_3-2.9 to 8.5, Ag_2O-1.5 to 9.3, and P_2O_5-71.8 to 80.5%.

Ernst Leitz, G.m.b.H. (260) patented a silver activated phosphate glass. Both borate and phosphate glasses were investigated by Becherescu, Menessy, and Christea, who measured the influence of X-radiation on the optical absorption. They reported [57] on three glasses with the following compositions: (1) P_2O_5-50 to 66, Na_2O-33 to 50 (one 45 K_2O); (2) B_2O_3-67 to 95, Na_2O-5 to 33; and (3) B_2O_3-67 to 95, BaO-5 to 33 (5BaO + Na_2O) to (30 BaO + 3 mole % Na_2O). Both gamma and X-radiation was used. The P_2O_5 glasses, which were 10 to 20 times more sensitive and reddish, showed high sensitivity and stability at 40°C that would permit their use as dosimeters. Boulos and Kreidl [117] reported on the structure and properties of silver borate glasses. They made borate glasses from 0 to 35 mole % Ag_2O but found it difficult to melt silicate glasses containing appreciable amounts of silver because of their high melting temperatures. A relatively large amount of silver, that is, an Ag/P ratio of unity, could be obtained in phosphate glasses. There was some atomic silver in the high silver glasses. Yokota and Nakajima [950] patented glass materials for a silver-activated phosphate glass dosimeter. Kaes [440] discussed the influence of the base glass on the radiophotoluminescence of silver-activated dosimeter glasses. A cation gave higher sensitivity and more rapid decay when added as metaphosphate than as oxide. Oxide additions reduced the half-width of the luminescent band with decreasing cation valence. Kaes and Scharmann [441] reported on the melting and heat-treatment effect on the dosimeter properties of silver-activated phosphate glasses. Broemer, Meinert, and Preuss [136] patented a silver-activated phosphate glass batched with nitrate.

3. SHIELDING

Rothermel, Sun, and Silverman [723] published an article that dealt with a radiation-absorbing glass based on the PbO-WO_3-P_2O_5 system and patented a phosphate glass with a lead equivalent for 100 kv X-rays greater than 0.33. The composition covered contained one oxide of 6 to 67 MoO_3 or 7 to 84 WO_3 and at least one oxide of 5-50 Bi_2O_3, 12 to 70 PbO, or 13-64 P_2O_5, with not more than 5% minor oxides. Brewster and Kreidl [127] published a comprehensive article on radiation-absorbing glasses which included a large number of elements and their oxides as Pb, Ta (Nb, Th), B_2O_3, P_2O_5, and GeO_2. Two other series included Cd, Ti, Zr, B_2O_3, and Pb, Cd, B_2O_3. Glasses of high X-ray thermal neutron and combined X-ray and thermal neutron absorptivity with satisfactory technological properties were developed. The absorption of high-energy electromagnetic radiation and particles depended on the density and absorptivities of the elements in the glass. Over a wide range of compositions the density is predictable from the composition. Silicate glasses of high lead content are used but chemical durability could be improved. Reference is made to the Rothermal, Sun, and Silverman glasses.

VIII. Refractories

A. General

The refractories industry is important not only to ceramics, but its products are essential to the production of metals, Portland and other cements, and a number of inorganic chemicals. For nearly all of these industries better refractories make higher temperatures, quicker reactions, and longer runs possible. Thus there is constant striving to improve present refractories and to develop new and better ones. Kraner [487] described in detail the development of better refractories for the basic oxygen steel process and emphasized the important part played by minor impurities; for example, the deleterious effect of small amounts of alumina on the properties of lime-bonded silica bricks. He did not mention the use of phosphates but in a private communication to me he stated that phosphates and phosphoric acid were being used in a wide variety of refractories, particularly those containing substantial amounts of alumina. Although it is difficult to arrive at any estimate of the importance of phosphates in refractories, a considerable literature deals with the subject, particularly in patents. Some authors praise, others condemn the introduction into refractory products. They have, however, certainly been found useful in certain special cases and indications are that they might find more uses if, as suggested by Kraner, more attention were paid to the effect of minor impurities. One refractories manufacturer in 1971 obligingly scanned his patent file on phosphates and phosphoric acid and sent me a list of 155 patents covering a wide variety of materials and products. Most frequent were patents for bonding alumina, magnesia, and silica and for the manufacture of a wide variety of plastic refractories, cements, and gunning mixes. As shown later, phosphates have sometimes been good stabilizing agents for certain refractories.

B. Refractory Phosphates

A number of references are concerned with the possible use of phosphates as refractories in their own right; for example, calcium orthophosphate has been used for making crucibles for special purposes. Most of the references are concerned with aluminum orthophosphate, $AlPO_4$, which is isostructural with silica, $SiSiO_4$, and might be expected to be highly refractory. As shown in Sections III and IV, however, this compound starts to decompose at temperatures well below its possible melting point, which greatly limits its usefulness.

Huettenlocher [401] reported that $AlPO_4$ and SiO_2, which could be written $SiSiO_4$, were isostructural. Hummel [406] made a direct comparison of various forms of silica with the corresponding forms of $AlPO_4$. He concluded that $AlPO_4$ more closely resembles silica structurally than any other natural or artificial mineral. In the quartz types of $AlPO_4$ he found that aluminum and phosphorus layers alternate in planes parallel to the basal plane. A similar report on BPO_4

by Hummel and Kapinski [407] appeared. $AlPO_4$ vaporized without decomposition near 1450°C. Tromel and Olsen [837] reported that crucibles of pure $3CaO.P_2O_5$, that is, $Ca_3(PO_4)_2$, melted at 1820°C and would withstand ferrous oxide melts at 1600°C, iron low-phosphate slags, and liquid iron.

Dietzel and Pögel [221] reported that the melting point of $AlPO_4$ was about 2000°C or about 300° higher than silica. At 2000°C they considered that Al^{3+} and P^{5+} ions were able to exchange positions so that P_4O_{10} escaped and Al_2O_3 precipitated. Weyl [916] concluded that an $AlPO_4$ glass could not be made. Possibly it could be done under high pressure of P_4O_{10}.

Harrison, McKinstry, and Hummel [358] published a paper on high-temperature zirconium phosphates. They established a "normal" zirconium pyrophosphate ZrP_2O_7 which had a reversible inversion at 300°C and dissociated to a zirconyl compound, tentatively $(ZrO)_2P_2O_7$, at 1550°C with the loss of P_2O_5 as vapor. The zirconyl compound was stable to about 1600°C and had a low thermal expansion coefficient.

Perloff [666] described the temperature inversions of anhydrous gallium orthophosphate. Berlinite, high-cristobalite, and low cristobalite forms were found but no tridymite forms.

Pepperhoff [664] published infrared spectra that could be used for the identification of a number of phosphate compounds in slags and refactory systems. Floerke and Lachenmayr used differential thermal analysis and X-ray methods to investigate in detail the various forms of $AlPO_4$ [279].

Allen [9] described how an aluminum phosphate foam was used to solve a problem for the National Aeronautic and Space Administration. Christyakova, Sivkina, Sadkov, Khashkovskaya, and Povsheva [172] described several aluminum orthophosphates. Laud and Hummel [523] reported on subsolidus relations in the system ZrO_2-ThO_2-P_2O_5. They found a number of low expansion phases which they suggested could be used as a basis for semirefractory, thermal-shock-resisting ceramic bodies. Alper and McNally [11] patented a fused cast refractory from the MgO-P_2O_5 system. Peri [665] described $AlPO_4$ as a mixed oxide of Al and P, and the Al and P atoms alternating regularly in linked oxide tetrahedra. He also showed that $AlPO_4$ could be prepared readily as a stable gel with a high surface area. Chiola, Smith, and Vanderpol [170] patented a refractory metal phosphate and phosphide coatings for refractory metal leads, and Hloch, Medic, and Kohlhaas [388] patented a process for the manufacture of condensed aluminum phosphates.

Although the phosphates have not found many uses as refractories in their own right, they have been studied and used as bonding materials for refractories and other ceramic products. Many of the references do not refer to a specific refractory.

C. Phosphate Bonding–General

The first reference is a patent by Lefranc [526] which describes a ceramic binder made by treating "calcic phosphate" with enough phosphoric acid to make an acid phosphate but not enough to permit free acid. A fundamental study of phosphate bonding in refractories was reported in a series of articles by Kingery [462]. He reviewed 33 references that dealt with siliceous material and phosphoric acid, oxides and phosphoric acid, and direct addition or formation of acid phosphates. Except for dental cements, no bond mechanisms had been established and no systematic studies of bonding properties had been made except for metaphosphates. For cold-setting properties he found that a weakly basic or amphoteric cation of moderately small ionic radius is required. For a refractory mortar he found that bonding with monoaluminum and monomagnesium phosphate produced mortars that were equivalent or superior to commercial mortars. In contrast to many commercial mortars no intermediate weak temperature zone was found with the phosphate bonds. Data on the reactions between some 40 oxides and phosphate bonds were presented in a table in Kingery's second article. His third article discussed phosphate absorption by clay as well as bond migration; his fourth dealt with monoaluminum phosphate and monomagnesium phosphate. Gregor [334] reported on a new bond for refractories. This was formed from aluminum hydrogen phosphates, the most useful containing 1.5 to 1.85 moles of Al_2O_3 to 3.0 moles of P_2O_5.

Veale [871] patented a method of impregnating clay firebrick of 5 to 30% porosity with hot fluid phosphoric acid and heating to 500°F. Sheets, Bulloff, and Duckworth [744] described a process for bonding Al_2O_3, ZrO_2, mullite, beryllia, or SiC with phosphoric acid or aluminum phosphate. Eubanks and Moore [264] described an investigation of aluminum phosphate coatings for the thermal insulation of air frames.

Bechtel and Ploss [59, 60] reported on the bonding of ceramics with monoaluminum phosphate $Al(H_2PO_4)_3$, a material that seems to have been preferred by a number of investigators. They bonded fireclay, bauxite, chrome, and silica and obtained excellent strengths in the intermediate temperature range of 600 to 1000°C without bad effects on the high-temperature properties. A book on refractories by Harders and Kienow [352] contains many references to phosphates and phosphoric acid. For certain products approaching the mullite composition sintering could be promoted at low temperatures by phosphates, phosphoric acid, or mineralizers. In a discussion of calcium silicates reference is made to the use of stabilizers such as phosphate, borate, chromium oxide compounds, iron oxide, and V_2O_5 to protect against "dusting" produced by the shift to the gamma modification.

Preuser [684] discussed the chemical setting of refractories by dilute phosphoric acid. Considerable strength was produced at 200 to 300°C. Reinhart [701] published a comprehensive review of binders for ceramic bodies.

Gilham-Dayton [310] reported on the phosphate bonding of refractory materials. He found that unfired silica brick bonded with phosphates were unsatisfactory because of the quartz inversion but that similar mullite bricks were useful. Cold-setting, phosphate-bonded ramming mixes of chrome-magnesite and magnesite were described. Bechtel and Ploss [60] reported on the use of monoaluminum phosphate solution for bonding ceramic materials and described the phase changes and softening that attended the heating of a number of phosphates mixed with clay. The highest refractoriness under load was shown by the mixture corundum -77 to 80, clay -15, $Al(OH)_3$-2 to 5, and $Al_2O_3.3P_2O_5$, that is $Al(PO_3)_3$, -2.5 wt. %.

Long [538] patented a ceramic material consisting of a finely divided refractory and a binder of an eutectic mixture of a refractory oxide and a metal pyrophosphate. Rashkovan, Kuzminskaya, and Kopeikin, [697] described the thermal changes occurring in an aluminophosphate binder. They listed the conversions for a binder with P_2O_5/Al_2O_3 = 2.3 when heated to 1500°C. Taniguchi described refractory specialties bonded by aluminum phosphate. A similar paper was published by Chakravorthy [165]. In his dictionary of ceramics Dodd [223] states that the term "aluminum phosphate" usually refers to the orthophosphate $AlPO_4$. Its bonding powers are attributed to reaction with basis oxides or alumina in the refactory. It preserves its strength in the intermediate temperature range 500 to 1000°C, in which organic bonds are destroyed and ceramic bonding has not yet developed.

Herbst and Lyon [372] patented a refractory composition consisting of H_2O-1 to 10, H_3PO_4-1 to 8, polycarboxylic acid -1 to 8, and alumina or alumina mixed with 3 to 15% refractory clay. Chvatal [173] patented a binding agent for refractories which was stable and highly viscous, based on monophosphates of a number of metals and additional materials. Also Kupzog, Koltermann, and Bartha [512] reported on the binding of corundum and mullite refractories with monoaluminum phosphate. With corundum cold strength increased remarkably after heating to 300°C but fluctuated considerably up to 1300°C with mullite; the high-temperature strength was maximum at 500°C, fluctuated between 500 and 1000°C, and then decreased rapidly. The latter deterioration could be reduced by the addition of 10% good bonding clay. Khoroshavin and D'yachknov [456] reported that increasing concentrations of phosphoric acid increases the setting of $MgO.Al_2O_3$, chromite, fused alumina, dunite, and dinas but hardly affects the setting of MgO, Cr_2O_3, magnesite, fused alumina, dunite, and quartzite. Miller [572] patented some heat-initiated, phosphate-bonded compositions.

Palfreyman [650] published data on the hot strength of high alumina refactories. Values of more than 2000 psi were found for fused mullite specimens bonded with ground calcined alumina and 5 of 85% H_3PO_4. At all temperatures above 1000 to 1400°C the fused mullite specimens were considerably stronger than those containing Al_2O_3 grogs. Some large variations in strength with changes in temperatures were attributed to polymorphic changes in the $AlPO_4$ formed. Fernandez [258] discussed the use of phosphate bonding agents in the manufacture of shaped and unshaped refactories. Magnesite, chromite, corundum, quartzite, and SiO_2 bricks were tested. The temperature of deformation decreased and the porosity decreased with increases in $(H_2PO_4)_2Mg$, $PO_4H_2NH_4$, and $PO_4H(NH_4)_2$.

Bartha, Lehmann, and Koltermann [50] discussed the bonding of a number of refactory materials with H_3PO_4. With corundum cold strength increased with treatment temperature to approximately 400 Kg/cm_2, but above 250°C the hot strength was reduced by the formation of $AlPO_4$. With mullite and $Al(H_2PO_4)_3$ hot strength increased with temperature as $Al(PO_3)_3$ was formed but dropped when $AlPO_4$ was formed above 700°C. With alpha tridymite at 200 to 300°C pyrophosphoric acid and a SiO_2-P_2O_5 compound formed, but decomposition at 1000°C gave little cold or hot strength. With MgO decomposition of the $Mg_3(PO_4)_2$ at 700°C ruined the hot strength. With spinel formation of $Mg_2P_2O_7$ above 400°C gave an increase in cold strength and a good hot strength to 1000°C but formation of $Mg_3(PO_4)_2$ above 1000°C lowered the strength. Ohba, Hiragushi, and Kitamara [646] showed that brickwork in a wall did not expand so much on heating as might have been expected. This was due to the restraining action of frictional forces. Expansion values were 90% of theoretical with phosphate bonding and 60% of theoretical with air-setting mortar. Mackintosh, Ford, and White [542] discussed the influence of phosphates on the bonding of oxide bodies at high temperatures.

D. Phosphate Bonding-Specialized

1. ALUMINA-(CORUNDUM)

The bonding of alumina by phosphates has been a favorite subject for investigation. Matveev and Rabukhin [556] reported on the bonding of corundum with a mixture of H_3PO_4 and $Al(OH)_3$, when heated to 1650°C. The highest cold and fired strength were obtained using 20% of the mixture, regardless of its composition ratio. Klyucharov and Skoblo [471] dealt with the composition of the hardening products when corundum was bonded with an aluminophosphate binder. At normal temperatures the hardening was due to the formation of variscite. The acid/alumina ratio that produced the most variscite gave the highest strength. Yutina, Zhukova, and Lysak [951a], reported on the reaction of phosphoric acid with some forms of alumina. Even at 120°C finely dispersed corundum reacted with H_3PO_4 to produce aluminophosphates. The high strength of compressed corundum-H_3PO_4 concretes was apparently due to the acid phosphates that form at 100 to 120°C.

King [460] patented a high-alumina refractory. This was made by binding a size-graded, high-alumina material containing 1 to 10% $Al(OH)_3$ with a phosphate-bonding ingredient and oxalic acid. The last increased storage life by retarding the breakdown of the phosphate bond. Christyakova, Sivkina, Sadkov, and Khashkoskaya [171] described the effect of corundum on the phases formed by heating in the Al_2O_3-P_2O_5-H_2O system. Stable amorphous products formed in this system at 20 to 80°C. With increasing Al_2O_3 both the amorphous material and its stability increased. On cooling this phase crystallized to $AlPO_4$ (Berlinite). An increase in Al_2O_3 also increases the thermal stability because of the formation of various modifications of $AlPO_4$.

Bidard [90] patented the use of aluminum monoorthophosphate as a fugitive binder for porous alumina articles. Eti and Hall [263] discussed control of the premature hardening in phosphate-bonded, high-alumina articles. The mixture investigated was tabular alumina, various meshes -14 to -325, -82%: raw kyanite, 35 mesh, -9%; 85% phosphoric acid -4%; 10% solution polyvinal alcohol -0.5%, and water -4.5 wt. %. Premature hardening could be controlled by inhibitors such as acetyl acetone, 5-sulfosalicylic acid, or dextrin, used at about half the level of the H_3PO_4.

After a detailed study of the volume stability of phosphate-bonded, high-alumina brick, Baab and Blackwood [38] concluded that they were inferior to conventionally bonded, high-alumina brick in which high-temperature volume stability is a requirement. They investigated 85% Al_2O_3 brick with P_2O_5 contents of 3.10 and 3.26% and 90% Al_2O_3 brick with 5.20% P_2O_5. Volume instability on initial heating was attributed to liquid formation at relatively low temperatures. The phosphate-bonded brick had some advantages such as cold strength.

O'Hara, Duga, and Sheets [642] described the stabilization of phosphate-bonded high-alumina brick by the addition of Cr_2O_3. Up to 2500°F they found that phase inversions in the bond were significantly reduced by Cr_2O_3 and concluded that the fine particulate material of the Al_2O_3-bond interface was most likely the cristobalite form of $AlPO_4$. The Cr_2O_3 retained amorphous phases over a broader temperature range. A detailed account is given of the phases formed on firing. It is noteworthy that they concluded that $AlPO_4$ is stable to at least 3200°F. Fisher [274] patented a refractory mortar consisting of high-alumina materials, graphite, phosphate binder, methyl cellulose, and water. Halpin [348] reported on tests of phosphate-bonded, 85% alumina brick intended for blast furnace linings. Tests with K_2CO_3 at 954°C showed outstanding resistance to cracking.

2. SILICA

As early as 1923 Hugill and Rees [403] reported on the influence of H_3PO_4 and phosphates on the rate of inversion of quartz in silica brick manufacture. They concluded that no advantage was derived by adding either pure P_2O_5 or basic slag. In contrast [143] Budnikoff concluded that a phosphate bond greatly accelerated the formation of tridymite and deserved plant scale trials. Deadmore and Machin [212] stated that fine-grained, high purity St. Peter silica sand bonded with phosphates of calcium and aluminum made a satisfactory silica refractory suitable for some furnace applications.

Veale and West [872] patented a bond for silica refractories consisting of H_3PO_4 and calcium phosphate, with P_2O_5-4 to 15% and H_2O/H_3PO_4 equal or greater than 1.3. By firing to at least 680°F a silicon pyrophosphate bond was formed.

Borodai and Edvokimova [113] described a refractory composition that used an alumino-phosphate bond. Quartz bonded in this way with less than 20% bond endured prolonged heating at 700°C without change. When fired at 150 to 900°C, the specimens had insignificant phosphocristobalite. Of interest in connection with silica refractories was Heilmann's investigation of the polymorphous silica phases in the system $(NaPO_3)_x$-Na_2SiO_3 [365]. He found tridymite and cristobalite. A maximum of 0.17% P_2O_5 was incorporated in the cristobalite. The nucleating and stabilizing abilities of silicon diphosphate found in this system were investigated. Yates [948] patented molding powders consisting of colloidal silica coated with a trivalent metal phosphate.

3. MAGNESITE (CHROMITE)

Harris and Kelly [357] described a basic insulating refractory containing 40 and 60% chromite. High porosity was obtained by the action of H_3PO_4 and ammonium lignin sulfonate on MgO. With 70% porosity the refractory weighed 70 lb/ft^3. Dess [218] patented a phosphate-bonded basic refractory. The following mixture was compressed: chrome ore + 50 to 80% MgO + 1 to 6% concentrated (70 to 98%) H_2SO_4 + 1 to 8% concentrated (50 to 90%) phosphoric acid. DiBello and Pradel [219] described the performance of phosphate-bonded refractory magnesias. They decided that amorphous sodium phosphate glasses are best for binding a basic refractory aggregate.

Staron [772], discussing the bonding of basic refractory brick, states that polymerized, long-chain alkali phosphates gave much higher strength than oxysulfates in the range 700 to 1200°C. The volume stability of magnesite and magnesite-chromite materials depended on the type of silicate present at 1650 to 1750°C. The CaO/SiO_2 mole ratio was important, the phosphate increasing or decreasing shrinkage, depending on whether this ratio was less than or greater than unity, respectively. The phases present after firing were determined.

Ved, Zharov, Bocharov, and Overko [873] described structure formation during the hardening of the magnesia binder in the presence of phosphates. They concluded that phosphate additives greatly improve the mechanical properties of the MgO binder. New compounds were formed and optimum structures obtained with 3 to 5% iron phosphate and kaolin-phosphate additives. Treffner and Foessel [835] patented a phosphate-bonded basic refractory composition consisting of dead-burned, low-silica magnesite combined with a sodium polyphosphate and a calcium compound. High strength was obtained at 2300 to 2700°F. In another paper [283] they state that the optimum binder addition corresponded to a calculated $CaO/(P_2O_5+SiO_2)$ ratio in the total mixture of 0.90 to 1.05. Under these conditions the strengths at 2700°F were expected to exceed 1500 psi.

4. OTHER MATERIALS

Hummel and Henry [406a] reported on the physical properties of some refractory phosphates of beryllium, cerium, and zirconium. Nickerson [631] patented an oxidation-resistant article consisting of a carbonaceous matrix of carbon or graphite with 4 to 20 wt. % of oxyacids or ammonium salts of phosphorus, plus a zinc halide and boric acid.

Bonding of dolomite was discussed by Shishkina, Babin, and Koka [747] in connection with the service of water-resistant dolomite brick in steel making. With 0.7 to 1.0% P_2O_5 in the dolomite charges, they found no increase in the phosphorus content of the steel during direct contact.

Bremser and Nelson [124] discussed the phosphate bonding of zirconia, and Duma [231] described similar bonding of fired clay.

The bonding of mullite by aluminum and chromium-phosphate was described by deAza and Pecci [35]. The viscosity, bond stability, and nature of thermal conversions of the solid residues of aluminum-chromium phosphate binders were reported by Bromberg, Kasatkina, Kopeikin, Kuz'minskaya, Rashkovan, and Tananaev [138].

Sudakas and Sinichkina [792] reported that $PbO\text{-}B_2O_3$ mixtures with orthophosphoric acid set instantly but that control of setting times could be obtained by using $ZnO\text{-}B_2O_3$ and $CdO\text{-}B_2O_3$ mixes. Aleksandrova and Maslennikova [7] described heatproof concrete based on phosphate bonds. Tsubaki and Tanaka [841] reported on the effect of phosphoric and phosphates on the strength and other properties of thermal insulators in the vermiculite-bond clay system. The marked strength increase at 800 and 1000°C obtained by adding $Mg(H_2PO_4)_2$ was attributed to the formation of bridging bonds with the MgO and Al_2O_3 components in the bond clays.

Visser [877] reported that the alkali penetration of fireclay bricks in blast furnaces could be reduced materially by impregnating the bricks with phosphate. Bricks containing 39 to 91% Al_2O_3 were treated with a 50% solution of $AlPO_4$. Nekrasov and Alexandrova [629] described refractory alumina silicate concretes with phosphate binders. Addition of alumina or alumina hydrate lowers the hardening temperature from 200 to 100°C.

Owen, Visser, and von Laar [648] patented the impregnation of firebricks with phosphates as a means of obtaining alkali resistance, described by Visser, above.

Watanabe and Watanabe [887] patented an inorganic liquid refractory composition consisting of a metal nitride and H_3PO_4 or ammonium salts. Yavorsky [949] patented an aluminum phosphate-bonded refractory, and Niesz [633] listed a project concerned with the phosphate bonding of lightweight inorganic thermal insulation.

E. Stabilization

Difficulties are encountered occasionally in refractories and cements because of the inversion of one crystalline form to another, often accompanied by a considerable change in volume. A classic case is the "dusting" of dicalcium silicate in Portland cement. Such inversions can often be prevented or minimized by the addition of other materials which act as stabilizers. There are several references to the use of phosphates for this purpose.

A Belgian patent, P612,195, to Basic Inc. [52] describes an agglomerated dolomite refractory consisting of dead-burned dolomite and containing coarse material passing 6.73 mm combined with fine material passing 0.207 mm, the latter being stabilized by 0.5 to 2.0% basic calcium triphosphate to prevent the inversion of beta to alpha calcium disilicate.

Svikis and Phillips [797] described the processing of North American kyanite concentrates into volume-stable, dense, and highly refractory aggregates. They had found that phosphoric acid and aluminum phosphates were highly effective as additives that caused no reduction in refractoriness.

Nadachowski [615], reporting on the stabilization of dicalcium silicate in magnesite-dolomite products, found that a phosphate stabilizer of about 1.5% P_2O_5 prevented the disintegration of $3CaO.SiO_2$ and $2CaO.SiO_2$ and decomposed only on contact with solid carbon.

Svikis [796] gave a further report on the properties of phosphate-stabilized refractory materials made from Canadian kyanite concentrate. Addition of 20 to 30% calcined alumina and 2 to 4% H_3PO_4 substantially improved the refractory properties. For a maximum bond at high temperatures the H_3PO_4 should be kept to a minimum.

Miki, Osaka, and Shikenjo [570] reported on nonslaking dolomite refractories. With a cone 27 mixture of dolomite, silica, and dunite dusting due to $2CaO.SiO_2$ was completely prevented by the addition of 1.0 to 2.0% apatite. The apatite addition was not so effective in the dolomite-alumina system.

Alvarez-Estrada, Demetrio, and Boia discussed the effect of P_2O_5 on the sintering of magnesia in a Bolivian publication [27].

F. Cements, Mortars, and Ramming and Gunning Mixes

A number of cements, mortars, and ramming and gunning mixes are used in the steel and other high-temperature industries. Of particular importance are gunning mixes which can be used to effect hot repairs to a furnace and thus avoid shutdowns. Such mixes must have the property of sticking to a hot refractory and forming a stable bond. For this purpose mixes containing polyphosphates are particularly useful.

Kuzell [515] published an article that described the Clarkdale method of hot patching operating furnaces. It involved the pneumatic spraying of an aqueous suspension of finely divided refractory particles.

Herold and Burst [375] discussed the use of metaphosphates in refractory mortars. They added 1 to 5% metaphosphates to a standard fireclay mortar and reported on the strength, fusion point, volume, and porosity.

Phosphate-bonded alumina, castable refractories were described by Gitzen, Hart, and MacZura [312] who used sintered Al_2O_3 grog bonded with phosphoric acid. They found that a high bond strength developed at 650°F and that the product had excellent serviceability in the range of 3400°F.

Precopio and Bateman [683] mention ceramic cements useful at Mach 3. MgO and fiberglass mica-base phosphate cements were used to make electrical insulating systems at 350 to 500°C in high-speed aircraft.

High-temperature refractory concretes with a metaphosphoric bonding agent were patented by Barta and Prochazka [48]. No firing was required. High refractoriness, thermal shock resistance, and mechanical strength at high temperatures were claimed.

Gunning mixes with phosphate bonds were described by Tseitlin and Tarasova [840]. Phosphate bonds were used with siliceous, semiacid fireclay and kaolin-fireclay bodies. The materials gave seven to nine times the life of the sodium silicate-bonded materials previously used.

Rickles [707] studied the effect of heat on a number of inorganic cements to 1650°C. X-ray, DTA, and vapor diffusion techniques were used on aluminum, boron, and zirconium phosphates. Monoaluminum phosphate gave DTA peaks at 200, 500, and 1500°C, corresponding to the formation of aluminum metaphosphate, aluminum orthophosphate, and $Al_2O_3 + P_2O_5$, respectively.

Heilich, Rohr, and Hart [364] published a paper on the pneumatic gunning of refractory concrete. Bash, Tuminov, and Rakhimbaev [51] discussed the making of plugging cement from phosphorus slag. Granular phosphorus slag in 2 to 10% amount was used as an activator in plugging bore holes at 100 to 150°C.

Peek and Koehme published papers that described ramming mixes of acid and basic refractories. Monoaluminum phosphate was used as a binder [663].

A mold-dressing containing alkaline earth metal hydroxyl apatite was patented by Tuvell [848]. The ignition loss at 800°C for one hour was below 1.1%. The mean particle diameter was 1 to 10μ.

In an article that discussed steel plant refractories Kappmeyer and Hubble [445] give the composition of a phosphate-bonded gunning mix: SiO_2-46, Al_2O_3-40, Fe_2O_3-2, TiO_2-2, MgO-5, P_2O_5-2 and alkali-2. After heating cycles of 500 to 1750°F they make the following comparison of the phosphate-bonded and clay-bonded mixes: 1 cycle, no change versus heavy cracks; 150 cycles, small cracks versus heavy cracks; 250 cycles, increased cracking versus spalled off; 500 cycles, badly cracked. In a comparison of five types of blast-furnace stockline refractories the phosphate-bonded, 85% alumina refractory was superior to high superduty and dense mullite bricks and only slightly inferior to 90% dense alumina and SiC brick in a cyclic spalling test. The phosphate-bonded brick had about twice, the strength at 2000°F as the other refractories.

G. Filters

A ceramic filter that could be classed as a refractory because it was fired to 1500°C was described by Rodina [717]. It consisted of 90% Zr concentrate with 10% of an aluminophosphate binder with an Al_2O_3/P_2O_5 ratio of 1/3. It was stated as having high spalling resistance, a rupture strength of 60 kg/cm^2, and a pore diameter of 35μ.

H. Refractories for Phosphates

Two references were found to refractories that were resistant to phosphate melts. Fisk [275] described a high lime crucible formed by bonding CaO with TiO_2. The fired bond consisted of $3CaO.2TiO_2$. Magai and Ando [618] reported that forsterite refractories made from peridotite and seawater magnesia resisted the corrosion of fused phosphates. The latter tended to diffuse through a forsterite crucible without the bond.

IX. Whitewares

In ceramic terminology the term "whitewares" includes such products as tableware (china, porcelain, stoneware, semivitreous ware, and earthenware), sanitary ware, electrical porcelain, and glazes and enamels for ceramic ware. A typical unit consists of an inner part or "body" and an outer coating or "glaze," which is essentially a glass. The term "enamel" is not strictly defined but is usually applied to a glaze that is opaque and usually in a decorative design.

At present the principal use of phosphates in whitewares takes the form of "bone ash", which is essentially calcium orthophosphates and which is required in high percentages as a raw material in the manufacture of English bone china. Some apatite is employed in a similar way in Japan. Minor uses of phosphates include dispersants in the preparation of "slips" or dispersions in water of body materials for casting ware in plaster molds. Another potentially important use is in the manufacture of "bioceramics" for bone replacement in the practice of surgery.

A general reference on whitewares is the book by Jackson [420].

A. Bone China

1. HISTORY

Bone china has an interesting history for which many references can be given.

A review of individual papers in this field would be beyond the scope of this chapter. Instead, author, date, and reference number are listed and followed by a short review:

Wood, 1901 [935], Brown, 1926 [139], Honey, 1928 [395], Wenham, 1933 [894], Vernay, 1934 [875], Grimson, 1937 [342], Eardley, 1937 [234], Weyl, 1941 [909], Anon, 1945 [20], Svec, 1949 [798], Sprackling, 1952 [778], Anon, 1956 [25], Jackson, 1969 [420].

English bone china, which may contain as much as 50% bone ash, was an outgrowth of the earnest endeavors of European potters in the seventeenth and eighteenth centuries to find the "secret" of the Chinese porcelain that was being imported into Europe in large quantities. Ultimately, examination of an unfired porcelain body from China revealed the fact that the raw materials were a white clay or kaolin and chinastone, the latter a granite consisting largely of orthoclase. An intensive search of Europe for "china clay" revealed deposits in Saxony, near Limoges in France, and notably in Cornwall, England, where "Cornish stone," a partly decomposed feldspar similar to chinastone in its ceramic properties, was also found. In 1768 William Cooksworthy took out a patent for a porcelain made from china clay and chinastone, and similar "hard-paste" porcelains were developed on the Continent.

Soft-paste porcelains were developed in England by adding fluxing materials to the usual porcelain bodies——first glassy frits, then soapstone, and finally bone ash. The last resulted in the English bone china developed in 1800 by Josiah Spode, Jr.

2. BONE ASH

Bone ash was originally made by the calcination of deer bones, but presently the bone comes from the cattle-producing countries. It consists of about 90% calcium orthophosphate and 10% calcium carbonate. Its calcination and grinding must be carefully controlled to produce the best results. In addition to being an important ingredient in a bone china body, it may also be used in the glaze and as a replacement for flint in bedding potteryware in saggers; the latter are refractory vessels that support the ware when it is stacked in the kilns [16].

German [308] published a detailed account of the preparation and use of bone ash. Jackson and Holdcroft [422] discussed the fluxing effect of bone ash in English china. Jobling [427] noted the effect of calcining conditions on the physical properties of bone ash, and Majunder [543] recommended grinding it until 80% passed through 0.01 mm. Jackson [420] gave a detailed account of the preparation and use of bone ash.

3. BODIES

A typical bone china body consists of china clay -25, Cornish stone -25, and bone ash 50%. It is not very plastic and requires great skill in forming and handling. The formed piece is fired to a high temperature of 1200 to 1240°C and has a short firing range. The "bisque" or "biscuit" is then dipped in a suspension of the glaze materials and fired to 1070 to 1100°C. The decoration is applied overglaze and may require several firings to different temperatures to produce the desired result.

In contrast to bone china, porcelain is usually made from china clay 60, ballclay -10, quartz 15, and feldspar 15%. The bisque is fired to the low temperature of 900°C, dipped in glaze suspension, dried, and fired to 1400 to 1500°C under reducing conditions. The soft bisque body absorbs the glaze readily and this and the high firing temperature produce a translucent product. The decoration is applied to the bisque as an underglaze which limits the colors that can be used.

In contrast to the rather cold appearance of porcelain, bone china has a very white, soft, warm tone, and because the decoration is overglaze a wide range of colors can be used.

Several advantages may be attributed to the use of bone ash in a bone china body. Like other phosphates it acts as a flux to produce translucency and strength at a lower temperature. In addition, as suggested by Weyl, it may result in the precipitation of apatite in finely divided crystals. Finally, in the discussion of glass colors it was pointed out that phosphates had the property of reducing the color due to traces of iron and this may help in producing a white body, an effect investigated by Parmelee [654].

Bone china bodies have been discussed by Edwards [247], Jackson [419], Krause and Schlegeelmilch [488], and Elliott [255].

4. MANUFACTURING METHODS

Detailed descriptions of manufacturing methods for bone china have been provided by Moore [574], German [307], Anon, [23], Brain [122], Bruce and Wilkinson [141], Mulroy [592], and Jackson [420].

5. CONSTITUTION

The constitution of bone china has been the subject of a number of investigations. Klein [470] studied the microstructure of bone china in thin sections. As quoted by Weyl [909], he found that the china consisted almost entirely of isotropic inhomogeneous material in two substances, both extremely fine grained and intimately interwoven. The refractive index of one substance was approximately 1.53 and that of the other, approximately 1.61. The material with the lower refractive index resembled the glassy phase found in ordinary vitrified feldspar-clay-quartz bodies.

Cronshaw [198] published a microscopic study of a bone china body. Features were the large amount of isotropic material, the small size of the crystals, and a matrix crowded with ill-defined microlites. Krause and Schlegelmilsch [488] reported on the structure and properties of bone china, but more definite information was provided by Massazza [549] who used chemical, metallurgical, and X-ray techniques. The chemical analysis showed SiO_2-34.65, Al_2O_3-16.78, P_2O_5-17.11, CaO-28.71, MgO-0.41, and alkalies -2.34. Massazza concluded that the china was chiefly tricalcium phosphate, anorthite ($CaO.Al_2O_3.$-$2SiO_2$), and glass, with some dicalcium phosphate and quartz. Klause [469] also reviewed the subject.

St. Pierre published a series of papers on the constitution of bone china based on an extensive study of phase equilibria, reactions occurring on firing, and microscopic and X-ray studies of fired bodies. He concluded that the finished commercial product would consist of the crystalline phases beta tricalcium phosphate and anorthite surrounded by a siliceous glass [783, 784, 786, 787]. This finding was in agreement with the preliminary findings of earlier workers but did not support Weyl's hypothesis that some forms of apatite would be crystalline phases. Thus it appears that most of the phosphate ends up in the crystalline phase beta tricalcium phosphate and not in the glassy bond. St. Pierre's statement that the crystalline phases are surrounded by a siliceous glass has sometimes been misinterpreted to mean a reference to the glaze. It was the unglazed bodies that were investigated.

B. Other Bodies

Apart from bone china and the bioceramics discussed later, there appears to have been little use of phosphates in other ceramic bodies. A few references were found.

Thompson and Parmelee [819] investigated the effect of B_2O_3, P_2O_5, and Fe_3O_4 on the thermal expansion of Florida kaolin-quartz mixtures. Jackson [421] patented a ceramic composition consisting of a base material containing kaolin, ball clay, and china clay -4 to 28, wollastonite -70 to 90, and boron phosphate -2 to 10%. Improved dielectric properties were claimed.

Kirchner [466] patented a low-expansion ceramic body consisting of a sintered mixture of uranium dioxide and uranium phosphate. Whitaker patented the use of phosphatic waste materials for the production of ceramic ware. The analysis of the waste was phosphate -26, Al_2O_3-24, Fe_2O_3-4, CaO-8, SiO_2-15.

Honma and Honma [396] described what were apparently sodium borphosphate glasses made by heating mixtures of sodium dihydrogen phosphate and boric acid.

C. Glazes, Enamels, and Pigments

General references to ceramic glazes are Parmelee's book [655], and two later editions, and the abstracts by Koenig and Earhart [478]. Parmelee gives no references to phosphates in glazes but remarks that they are used to only a limited extent and tend to cause opacity and raise the index of refraction. Koenig and Earhart give several references which are incorporated in the following brief review.

Stull [791] reported on a cheap enamel for stoneware. Expressed in the conventional glaze formula, this enamel has the composition

0.2 ZnO	0.40 to 0.55 Al_2O_3	2.5 SiO_2
0.6 CaO		0.2 P_2O_5
0.2 K_2O		0.8 B_2O_3

Such formulas express the composition of the fired glaze in moles: the first, or RO, column adds up to unity; the second, or R_2O_3, column with some authors would include the B_2O_3 and the P_2O_5; the last is then the RO_2 column. Stull puts the P_2O_5 and B_2O_3 in the last column as a way of indicating the acid oxides.

Ramsden [696] described a study of chromium red glazes. The effect of P_2O_5 was to destroy the crimson red color. This is similar to the effect of phosphates on the color of glasses containing iron oxide. Fritz, in a discussion of phosphate glasses and glazes [292], reported that glazes rich in PbO did not develop the typical yellow shade in the presence of phosphates. Some water-soluble glazes became chemically resistant when fired to the body because of a combination of phosphate glass with alumina in the body.

Pearson [661] described the quality and formulas of a number of glazes, including one described as phosphoric acid, opaque. In discussing cobalt and nickel colors [558], Mellor mentions "phosphate blue." Presumably this was made by calcining cobalt oxide with P_2O_5 and possibly some alumina.

Earhart [235] described the use of phosphate opacifying agents in sanitary ware glazes. Pearson [662] discussed phosphoric acid opaque glazes. He considered them better for translucent glazes than for opaque. A nonweathering, fine matte glaze could be obtained with potassium metaphosphate-100, lead phosphate-95, calcium phosphate-10, and alumina-15. Another recipe was feldspar-43, boric acid-37, bone ash-17, and ZnO-3. The boric acid improved the weathering resistance of the phosphates.

Hughan [402] described early Chinese ceramic glazes, which included the "sang-de-boeuf" and "flambé" reds. The latter differed from the former in having its purple effects, which Hughan attributed to the presence of iron and phosphate.

National Lead Co. [626] patented a glass-decorating enamel in which discoloration due to the PbO-TiO_2 reaction was prevented by adding 0.01 to 0.10 parts of P_2O_5 for each part of PbO. An enamel with a bright blue-white color after firing on glass was obtained from a glaze with the following parts by weight: Pb_3O_4-160.6, SiO_2-54.0, Na_2CO_3-3.6, H_3BO_3-9.4, TiO_2-8.0, Al_2O_3-8.2, and NaH_2-$PO_4.H_2O$-30.4.

Bierbrauer [92, 93], reporting on stoneware glaze experiments with bone ash, gave 19 formulas for glazes in which calcium phosphate was a constituent. Bopp [112] informed me that he has used as much as 2% bone ash in a copper turquoise matte glaze at cone 2.

A pink pigment for highly fired porcelain was described by Hirano, Iida, and Terasaki [385]. Two formulas were given: $MnHPO_4$-10 to 40, Al_2O_3-60 to 90, and $Mn(NO_3)_2$-33, $AlPO_4$-9, and $Al(OH)_3$-58%. The pigment was fired to cones 3 to 6 and then pulverized.

D. Dispersion and Adsorption

In many ceramic processes clays, often admixed with nonclay raw materials, are handled in the form of aqueous suspension of "slip" for casting ware in plaster molds, applying glaze materials to whitewares, or applying enamels to metals. When deflocculation of the clays and other materials is required, phosphates have proved to be useful. A number of references have been made to the use of phosphates for this purpose. Most of them deal with the effect of phosphates on clays from a dispersion or adsorption point of view.

Plotnikov and Natansson [679] published a paper on the effect of soluble phosphates on the viscosity of clay suspensions. Among the phosphates tested, $(NH_4)_3PO_4$ and Na_2HPO_4 had the most effect. They were adsorbed on the kaolin and the maximum adsorption was reached at low concentrations. The mechanism of phosphates fixation by monmorillonitic and kaolinitic clays was reported by Coleman [82] to be a replacement of OH by H_3PO_4 below pH 5.

Van Wazer and Besmurtnuk [860] described the action of phosphates on kaolin suspensions. A number of phosphates were used with sodium and calcium kaolin. They concluded that it was unlikely that deflocculants greatly superior to the polyphosphates would be found. Zapp and Goller [956] discussed the effects of sodium metaphosphates on the particle-size distribution of elutriated kaolins. They found that 0.1% of sodium metaphosphate sharply reduced the viscosity of the slip so that coarse particles settled quickly and quartz, feldspar and mica could be removed.

Goldsztaub, Henin, and Wey [315] reported on the adsorption of sodium and calcium phosphates in montmorillonite and kaolinite. The maximum was at pH 4 to 5. They decided that the adsorption was due to aluminum ions on the exterior surface of the microcrystals. Engelhardt and Smolinski [257] discussed the reaction of polyphosphates with clay minerals. They concluded that kaolinite and montmorillonite adsorb sodium tripolyphosphate from aqueous solution in accordance with the Freundlich isotherm. After adsorption a slow decomposition of clay minerals takes place to give SiO_2 and aluminum polyphosphates in solution.

Rolfe, Miller and McQueen [718] published a paper that described the dispersion characteristics of montmorillonite, kaolinite, and illite clays and their control with phosphate dispersants. The adsorption of phosphate ions in Egyptian clays was described by Tobia and Milad [826]. Removal of very fine particles from the clays increased the phosphate adsorption and altered the character of the adsorption curves. They also concluded that colloidal particles had an inhibiting effect on adsorption by noncolloidal particles.

Bergseth [81] reported on the combination with and release from clay minerals of phosphates through anion exchange treatment with ^{32}P. He concluded that the removal of adsorptively bound phosphate ions was the greater, the greater the charge and polarizability of the exchangeable anions.

Bidwell, Jepson, and Toms [91] compared the adsorption of polyacrylic and polyphosphate on kaolinite in water. With sodium pyrophosphate, aluminum phosphate complexes were formed; with sodium polyacrylic, adsorption occurred at the positively charged edges of the kaolinite. Joyce and Worrall [436] reported on the adsorption of polyanions by two kaolinite clays. They suggested that the mechanism of deflocculation involves more than the supposed masking of positive edge charges. With monoionic clays, the H form adsorbed more readily than the Na form, which accords with the view that the Na form is more highly ionized.

Schwiete [741] studied the action of polyanionic deflocculation of the components of fine ceramic bodies. He concluded that polyphosphates liquify better than the usual electrolytes and also do not attack gypsum molds appreciably. Colwell [184] described phosphate sorption by iron and aluminum oxides. He found that sorption varied greatly with pH. Price [685] reported that crystallite growth of beryllium oxide powders was inhibited by adsorbed phosphate.

Hoover and Sinkovitz [398] patented the use of aminodiphosphonates for dispersing clays, pigments, and the like.

E. Bioceramics

Biological hydroxyapatites and closely related compounds form the mineral and inorganic phase of tooth enamel, dentine, and bone. Considerable interest has been shown in the possibility of forming hydroxy- and similar apatite structures by ceramic techniques. There is the hope that not only can suitable implants take the place of lost bone in the body but also that porous apatite structures will be invaded by the body fluids and gradully changed to bone. There are the problems of preparing the desired apatites, of molding them into the desired shapes with a specified porosity, and the long time tests to determine their compatibility with body tissue. Metal and plastic implants are at best inert, but the bioceramics hold out the hope that they will provide close adhesion of body tissue.

Most of the references deal with the study of apatites, their analysis and structure and their preparation and forming.

Perloff and Posner [664] described the preparation of pure hydroxyapatite crystals. Fucher and Ring [273], used infrared methods for determining fluorapatite in hydroxy apatite. Negas and Roth [628] in their study of the dehydoxylation of strontium and barium hydroxyapatite found that the crystals underwent cell parameter contractions at high temperature but neither type could be completely dehydrated. Levitt, Crayton, Monroe, and Condrate [531] reported that the chemical composition of hydroxyapatite was $Ca_{12}(PO_4)_4(OH)_2$ and that forming by hot pressing showed promise. (Corbridge, p. 189 [191] gives $Ca_{10}(OH)_2(PO_4)_6$). They concluded that whether apatite shapes could be used better as permanent orthopedic prostheses or as templates to be determined by long-term tests. McClellan and Lehr [610] published a crystal chemical investigation of natural apatites, and Elliot [254] reviewed progress in their chemistry, crystal chemistry, and structure. Kreidler and Hummel [503] published the lattice parameters of about 35 apatites.

Botz, Rechter, Reynolds, and Bazell [116] patented a reinforced porous ceramic bone prosthesis. In their 1971 book *Ceramics in Severe Environments* Kriegel and Palmer [504] include the human body in their list of environments. Hulbert, Klawitter, and Bowman [405] published a history of ceramic orthopedic implants, and Heide and Hoffmann [363] published a review of the problems encountered in finding ceramics compatible with living tissue. Campbell, Burroughs, and Cochran, [157] patented a surgically implantable prosthetic joint.

Brigham and Vickery [129] patented a dental prosthesis model base composition containing calcium fluoride. The patent described a dry blend of refractory oxide -40 to 50, alkali phosphate -6 to 8, and akaline earth fluoride -42 to 47%. This blend was mixed with a 40% SiO_2 sol in the proportion 4 to 5 parts of blend to one part sol. These authors also patented a castable-refractory die composition suitable for use with porcelains to make dental prostheses [130]. Sipe and Whittmore [752] listed a current ceramic research item as "Ceramics as Dental Bone Implants" and Rasmussen [680] listed "Materials for Dental and Medical Applications–Tissue Ingrowth and Osteosis."

Bokros and Ellis [108] patented a prosthetic, blood-circulating device that had a pyrolytic carbon-coated blood-contacting surface. Mountvala [584] discussed the surface properties of bioceramic vascular prostheses.

Avnimelech, Moreno, and Brown [34] reported the solubility product of finely divided hydroxy apatite $(Ca)^5(PO_4)^3(OH)$ to be $6.3 \pm 2.1 \times 10^{-59}$ at 25°C. Bortz and Onesto [114] reported on plasma and flame-sprayed bioceramics for use in hard or soft tissue organs; for example, plasma sprayed ZrO_2 tubular elements. Brown, Kenner, Schnittgrund, and Frakes [140] listed a current research project as "Delayed Fracture and Aging of Orthopedic Ceramics in Vitro and in Vivo." Ten more references were obtained from the 1973 Ceramic Abstracts, which indicates recent interest in the subject of bioceramics. They are listed in the next paragraph.

Franklin [287] listed "Properties of Calcium Apatites" under current ceramic research. Griffiths [339] listed "Ceramics as Prosthetic Materials," Hauth and Crayton [360], "Fabrication of Calcium Fluoroapatite by Hot Pressing," Hench, Paschall, Paschall, and McVay submitted a paper at the American Ceramic Society's annual meeting entitled, "Histological Responses at Bioglass and Bioglass-

Ceramic Interfaces." Fibroblast compatability was improved by altering surface morphology, by precorrosion, or by forming a composite of bioglass with Al_2O_3. However, stimulation of bone growth required a higher surface activity of Na^+, Ca^{2+}, and P^{5+} ions. Marcus, Lukas, Pollack, and Hench [546] listed a project concerned with making artificial corneas from biocompatible glasses. Other current research projects included one by McVay, Rodebush, Paschall, Paschall, Allen, Piotrowski, and Hench [614] entitled, "Medical, Histological, and Tissue Culture Evaluation of Bioglass-Ceramics." Miller, Piotrówski, and Allen [571] listed studies of the biomechanical behavior of bioglass-ceramic prosthesis. Sipe and Whittemore [753] reported that they were working on ceramics as dental bone implants. Tennery and Driskell [813] submitted a paper on SEM studies of bone-ceramic interfaces in calcium phosphate resorbable ceramics in dogs. A clearly defined interface could not be readily identified, but the ceramic was resorbed as bone growth proceeded. The SEM results were corroborated by ground section histology techniques.

F. Molds and Raw Materials for Whitewares

The Japanese have shown a great interest in the use of by-product gypsum from phosphoric acid manufacture for making molds for whiteware. References include papers by Murakami and his associates [596, 597] and Yamada and his associates [938-945]. The general conclusion is that by-product gypsum can be used for molds for slip casting.

Vetter [876] discussed the use of lithium minerals as raw materials for the ceramic industry. Included were the phosphate minerals amblygonite, triphylite, and dilithium sodium phosphate.

X. Ceramic Metal Systems

This heading from Ceramic Abstracts covers enamels on metals usually referred to as "vitreous" or "porcelain", and also such products as "cermets." However, only references to enamels were found in this survey.

A. General

The standard work in this field is Andrews *Porcelain Enamels* [29]. There is also a book by Vargin [866], a review by Danielson, [207] on opacifying agents, and bibliographies published by the American Ceramic Society [178, 429]. Both Andrews and Vargin state that phosphates have found little use as major constituents of enamels but point out that minor additions of phosphates enhance the opacity of titania-opacified enamels. It is noteworthy that papers that suggest the substitution of phosphates for borax in enameling were published in Europe and at times when there was difficulty in securing borax. A review article on phosphate enamels was published in 1954 [24]. Phosphoric acid has been used for the pickling and pretreatment of iron and steel for enameling purposes.

B. Iron and Steel

1. PICKLING AND PRETREATMENT

Phosphoric acid and phosphates may be used merely to clean the steel for enameling or to build up a phosphate coating on the steel which is advantageous in the subsequent enameling.

An anonymous article [26] described a new method of spray phosphoric acid pickling. Satisfactory adherence was obtained when solution containing 20 to 30% H_3PO_4 was sprayed on sheets of iron used for single-coat enameling.

Knanishu [472] recommended grit blasting for cold phosphating and vapor degreasing for hot phosphating processes. Fussel and Hadley [297] reported that phosphate coating produced a significantly rougher metal-oxide interface under the conditions used. Farbwerke Hoechst [267] patented a process for producing finely crystalline phosphate layers on metal surfaces. The Pyrene Co. [688] also patented a phosphate coating for steel.

2. COMPOSITIONS

As early as 1904 Hermsdorf and Wagner [373, 374] obtained German patents on the use of phosphates, particularly calcium phosphate, to increase refractoriness and give opacity in enameling iron. Two frits were used, one containing fluorides. According to Weyl, these processes could result in the formation of apatite. Blumenberg [101] patented a flux for enamel compositions that contained alkalimetal boron phosphate. Zschacke [958] found that it was not possible to replace borax completely with phosphates in enamels and glazes. Löffler [537], discussing phosphate enamels in a German patent, states that the composition must be regulated so that it will not contain appreciable amounts of alkaline earth and fluorine at the same time. Weyl would expect fluoroapatite under these conditions.

Frost and Commons [293] patented enamel frits that contained 7 to 15% zirconium silicate and 0.5 to 4% aluminum phosphate. Kautz [450] described a silica-free, acid-resisting, cover-coat enamel of the aluminophosphate type. The batch composition was KNO_3-4.6, K_2CO_3-3.2, borax (cryst)-13.2, Na_2CO_3-7.3, cryolite-4.8, ZnO-3.7, $Al(OH)_3$-17.9, monoammonium orthophosphate-39.6 and ZnO-5.7%. The mill batch was frit-100, clay-4, sodium nitrite-0.5, zirconium opacifier-4.0, barium molydbate-1.5, and antimony oxide-1.5%.

Huppert [408] reviewed the applications of phosphate-bearing raw materials and discussed the future of phosphate enamels based on experiences with phosphate glasses. Agarov and Chistova [36], in their discussion of phosphate enamels, reported on the properties of a number of glasses in the Na_2O-SiO_2-P_2O_5-Al_2O_3-B_2O_3 system. Keeping the Na_2O and B_2O_3 at 20 and 10 mole %, respectively, they varied the other three constituents. Glasses containing 40 to 50 mole % P_2O_5 and 10 to 29 mole % of silica and alumina had a low melting point. More alumina prevented glass formation and more silica greatly increased the viscosity. A number of additions enabled the thermal expansion to be varied from 70×10^{-7} to 140×10^{-7}.

Iler [414] patented a vitreous coating for metal that contained colloidal silica and a water-soluble phosphorus compound. The SiO_2:P_2O_5 ratio was 41:10.

Foster and Seidel patented a phosphate glass coating for electrical steel sheets [284]. The coating consisted of P_2O_5-60 to 70%, MnO-10 to 14, Al_2O_3-9 to 12, and not more than 5% of other compatible bodies. Pevzner, Dzhavukteyan, and Mishel [670] described acid-resistant glass-crystalline coatings for metals with high thermal expansion factors (copper and austenitic steels). The P_2O_5 content should be no greater than 4 to 5%, the firing temperature no greater than 840 to 880°. Wieczerek and Cook [922] discussed fluoride-free, direct-on, titania cover enamels that show reflection values greater than 85% and other favorable factors. The reflectance values were markedly influenced by the proportion of P_2O_5 and Al_2O_3 as minor additions.

3. OPACITY

Phosphates are related to opacity in two ways. They may produce opacity by crystallizing in the enamel or may assist other opacifiers, particularly TiO_2. High opacity is important in enamels because it permits thinner coats which have greater resistance to chipping.

In a series of five articles [959] Zschacke and his associates discussed the action of opacifiers in glazes and enamels for glasses. With phosphates, opacification and reflectance were best when the mole ratio CaO/H_3PO_4 corresponded to $Ca(H_2PO_4)_2$, with reheating to 852°. The best melting and best white were obtained with Na_2HPO_4 and $CaCO_3$ as raw materials. For low-melting enamels barium phosphates had a good opacifying action. Leadless enamels containing 10 to 20% B_2O_3 were made opaque only with phosphates or with Na_2SiF_6. Similar general articles were published by Coffeen and associates [179-180]. Bahnsen and Bryant [41] patented frits containing 10 to 20% ZrO_2 and 1 to 7.5% P_2O_5.

When TiO_2 is used as an opacifier, it is more effective if the crystals of TiO_2 in the fired enamel are in the anastase rather than the rutile form. This was shown by Friedberg and Peterson [291] and by Shannon and Friedberg [742]. Baldwin [43] stated that ZrO_2-opacified enamels were also improved by mill additions of materials containing P_2O_5.

Other evidence for the use of phosphates in TiO_2 opacified enamels is furnished by Kopelman and Wainer in whose patent a superopaque enamel frit of the following composition was claimed: Na_2O-7 to 11%, B_2O_3-5 to 11, SiO_2-35 to 55, Al_2O_3-3 to 9, ZnO-2 to 6, sodium aluminum fluoride-7 to 11, TiO_2-8 to 14, and P_2O_5-2 to 5, with Na_2O, B_2O_3 and TiO_2 less than 30%. Niklewski and Ashby [635] reported that the stability of the color of TiO_2-opacified enamels was increased by the addition of P_2O_5.

Cowan [197], who made a quantitative study with an X-ray spectrometer of the opacifying crystals in titania enamels, concluded that the reflectance stability of P_2O_5 containing enamels could not be explained by the quantities of the opacifying crystals present. This subject was also investigated by Eppler and Spencer-Strong [259].

C. *Aluminum*

Phosphates have been of some interest in formulating enamels for aluminum because low-melting enamels are required. Donahey, Morris, and Sweo [224] discussed phosphate base glasses as enamels for aluminum and its alloys. Leadless enamels of good adherence and excellent surface were developed in a wide range of colors. Good hardness and class H (Porcelain Enamel Institute) acid resistance were achieved. However, improved weatherability would be required for continued outdoor exposure. Ground and cover coats were fired at or below 1000°F in 3 to 5 minutes.

Allen [10] patented a vitreous enamel, based on phosphates, for aluminum and aluminum alloys. The following composition range was covered: P_2O_5-39.1 to 48.2 wt.%, Al_2O_3-20.8 to 28.1, Na_2O-18.25 to 22.8, TiO_2-2.4 to 2.7, B_2O_3-0 to 8, K_2O-0 to 15, Li_2O-0 to 3.5, and ZrO_2-0 to 4.0. It was said that Li_2O would provide resistance to soap and detergents. Bezborodov, Mazo, and Kaminskaya [88] described enamels for aluminum based on the K_2O-PbO-B_2O_3-SiO_2 system, with P_2O_5 partly replacing the PbO and SiO_2. A selected composition was K_2O-24.02, PbO-19.02, Al_2O_3-13.03, P_2O_5-36.27, and SiO_2-7.66, which at 540° gave a lustrous white but blistered surface. Chemical resistance could be improved by adding Al_2O_3. A white enamel had the following composition: K_2O-8, Na_2O-6, Li_2O-3, PbO-25, Al_2O_3-13.03, P_2O_5-29.31, SiO_2-7.66, B_2O_3-8, and MoO_3-2% as a mill addition. Barium nitrate-0.25 to 0.5% was added to prevent the reduction of PbO in colored enamels. The enamel was fired to 540 to 580°. The rather high percentages of phosphates in these formulas are noteworthy. Budnikov, Azarov, et al. [144] reported on the evolution of gases during the reaction of phosphate enamels with aluminum.

Pretreatment of aluminum before enameling with frits, including phosphate frits, was discussed by Yamada and Tomino [945] and by Yamada [944]. Anodic oxidation was included.

XI. Abrasives and Abrasive Wheels

References to the use of phosphates in grinding wheels were difficult to secure. There is little in the literature that deals with grinding wheels and what there is does not mention phosphates. Undoubtedly some phosphates could be introduced into the bond in vitrified grinding wheels, but with the cost of phosphates and their tendency to be attacked by water, which most grinding wheels must withstand, it appears unlikely that much use has been made of them. There is an extensive patent literature, but patents are abstracted by title only in *Ceramic Abstracts*, and in view of the considerations just stated a search of the patents did not appear justified. The following is, in the main, a brief review of the general literature on bonded abrasives.

Gormley published a series of five articles [325] that dealt with technical aspects of vitrified grinding wheel manufacture. Phosphates are not mentioned in the abstracts of these articles. Kingery, Sidheva, and Waugh reported on the structure and properties of vitrified bonded abrasives. They did not mention phosphates. Coes article [177] and book [176] on abrasives do not mention phosphates.

The only reference to phosphates in connection with abrasives found in this survey was in Mazilianskas' article [563] in which he discussed frits for ceramic bonds. A typical frit composition for a variety of abrasives was the following: flint -30%, potash spar -25, boric acid -25, petalite -10, zircon -5, and calcium phosphate -5. In addition, a patent [743] was issued to Sheets and O'Hara for a phosphate-bonded grinding wheel.

XII. Cements, Limes, and Plasters

A. *General*

Some cements depend on the hydraulic properties of phosphates or the combing properties of phosphoric acid for their setting properties. In addition, a number of references have been made to the relation of phosphates to the manufacture and properties of cements such as Portland, which depend on the setting properties of other compounds. A large literature, mostly Japanese, deals with the properties and uses of by-product gypsum which occurs in the acid process for phosphate manufacture.

B. *Hydraulic Properties of Phosphates*

An early reference is that of Schröder [737] who patented a process for producing a phosphoric acid and an aluminum phosphate cement. Ershov and Basman [262] reported on the bonding characteristics of calcium phosphates in the system $CaO-P_2O_5$. Monocalcium phosphate, formed at 600°C., had no bonding properties. Di-, tri-, and tetracalcium phosphates, formed at 800, 1000, and 1200°C, respectively, had bonding properties, the tri-calcium phosphate being the most active. These phosphates hardened rapidly when mixed with water and had optimum strength after three days. Rickles [707] studied the changes that occurred when the cementitious materials, aluminum, boron, and zirconium phosphate, were heated. Monoaluminum phosphate changed to aluminum metaphosphate, then to aluminum orthophosphate, and finally to Al_2O_3 and P_2O_5. Boron orthophosphate sublimed at 1235°C, and zirconium orthophosphate changed to zirconyl pyrophosphate at 1520°C. Funtikov, Kolbasov, and Kaznetsova [296] reported that a mixture of phosphorus slags, lime, and Portland cement could produce a strong rocklike material. Hloch, Kohlhaas, Medic, and Neises [389] patented a hardener for water-glass cements which consisted of condensed $AlPO_4$ prepared by subjecting acid $AlPO_4$ to a two-stage heat treatment. Hloch, Medic, and Kohlaas [388] patented a process for the manufacture of condensed aluminum phosphates.

C. Phosphate Cements

Ceramic Abstracts provides some references to industrial and dental phosphate cements. The latter could be studied more effectively by searching the literature on dentistry.

1. INDUSTRIAL CEMENTS

Pavlish [659a] patented cementitious compositions consisting of magnesium oxychloride and magnesium oxysulfate cements with a water soluble alkali-metal phosphate as an additive claimed to control, retard and resist after-expansion. About 0.5% was most effective. Bondley and Knoll [109] patented conductive cements consisting of powdered aluminum silicate and aluminum phosphate and a conductive filler. These were mixed with dilute H_3PO_4. When fired to 600°C, they were good to 700°C for glass and metal parts assembly.

Golynko-Vol'fson and Sudakas [319], commenting on the binding properties of phosphates systems which consisted of 75% H_3PO_4 and 25% oxide powders, concluded that with decreasing ionic potential the reaction speeded up and vice versa. Golynko Vol'fson and Sudakas [320] discussed the binding properties of a mixture of TiO_2 and H_3PO_4 when hardened at 260 to 300°C. High strength, good stability in neutral and acid media, heat resistance to 1050 to 1100°C, and high volume electric resistance were obtained. It was a colloidally dispersed product. Heating to 770 to 880°C resulted in crystalline $5TiO_2.2P_2O_5$.

2. DENTAL CEMENTS

Crowell [200], in discussing dental cements based on zinc oxide, stated that oxychloride cements were superseded by phosphate cements which, in turn, were superseded by silicate cements. Heynemann [378] patented dental cements based on H_3PO_4 with oxides, phosphates, or silicates and a compound of lithium. Eberley [244] patented dental cement powders consisting of SiO_2 and Al_2O_3 fused together. These powders, when mixed with finely ground calcined basic oxides and H_3PO_4, set at room temperature. Ruff, Friedrich, and Ascher [725] stated that both "silicate" and "phosphates" cements harden by the formation of phosphate. Thomsen [820] patented a dental cement consisting of 145 parts of weight of silicate cement and 40 to 60 parts of pure H_3PO_4. Volland, Paffenbarger, and Sweeney [881] described a dental cement consisting of a powder of calcined ZnO, modified by MgO and other oxides, and a liquid of aqueous H_3PO_4 buffered by ZnO, $Al(OH)_3$, or both. The formula for a high-strength dental silicate cement was given by Budnikov and Gol'denberg [145] as calcined Al_2O_3-23.2, quartz sand -33.3, synthetic cryolite-36.0, and $CaHPO_4$-7.5. This was fired to 1350°C, ground, and mixed with a liquid containing 8 parts Al_2O_3, 9 parts ZnO, and 83 parts boiling H_3PO_4.

DeMent [215] patented a cold-setting dental cement that contained ZnO and 20 to 80% natural tooth enamel powder mixed with a phosphoric acid of specific gravity 1.5 to 1.85. About 1969 a group of Japanese investigators published a series of at least nine papers, of which two were abstracted. The latter dealt with the relevant properties of zinc orthophosphate [380].

D. Portland Cement

1. MANUFACTURE

The raw material for Portland cement manufacture are commonly fed to the rotary cement kiln in the form of an aqueous slurry. Phosphates may be used as a dispersant for this slurry and may also be added to the mix to affect the properties of the clinker.

Romig and Kester [719] reported on the use of sodium tripolyphosphate as a deflocculating agent in a wet-process cement slurry. Van Wazer [854] used Stormer viscosity data to show the effect of phosphate on limestone-shale slurries of different consistancies. Schumacher [739] discussed the same subject and stressed the fuel savings and increased production resulting from the reduction of the water content of cement slurries.

Although American specifications for Portland cement set a low limit on phosphates, several references advocate the addition of phosphates to the raw mix. Hartmann and Hagermann [359] discussed the importance of phosphates in Portland cement. They followed the changes in calcium phosphates on heating. Zimanovskaya and Vodzinskaya [957] discussed the influence of fluorine in the presence of phosphates on the formation and crystallization reactions of clinker minerals. The burned product contained a solid solution of calcium orthophosphate in calcium orthosilicate. In the range of 3 to 5% fluorine stabilized the tricalcium silicate phase. Kukolev and Mel'nik [508] reported on the microstructure of Portland cement klinker with additions of Cr_2O_3, P_2O_5, V_2O_5 and BaO. The additions had little effect on the mineralogical composition and P_2O_5, V_2O_5 and BaO had little effect on the grain size.

Sanada and Miyazawa [729] reported that additions of P_2O_5 and Cr_2O_3 increased the burnability of the charge. The soundness of the cement was improved by adding 2 to 3% wt. of $CaSO_4.2H_2O$ or $MgSO_4$. Arapova, Pakhomova, and Demkina [31] reported on the use of phosphorites to increase the strength of Portland cement. Maximum strength was achieved with 0.2% P_2O_5 in the clinker.

Tien, Butt, and Vorob'eva [822] reported on apatite as a mineralizer during the firing of Portland cement clinker. An optimum amount (1 to 3%) in the raw mix decreased the temperature of clinker formation by 50 to 100° and increased the crystallization of minerals in the clinker. As much as 5% inhibits the binding action of the CaO and decreases hydraulic activity. Akatsu, Maeda, and Ikeda [4] described the effect of Cr_2O_3 on the strength and color of Portland cement clinker. Budnikov, Entin, and Babin [146] reported on the effect of sodium tripolyphosphate on Portland cement slurries. The liquifying effect was associated with exchange adsorption in colloid disperse systems and does not depend on the mineralogical composition of the slurry unless it contains calcium or sulfate ions. Ryba, Korab, and Jirku [726] described the influence of P_2O_5 on Portland cement clinker. The optimum was 0.5% P_2O_5, the allowable, not more than 2% P_2O_5. Above 2% there was rapid deterioration but only a slight influence on setting time.

Butt, Vorokeva, and Van Tien [149] investigated the effect of Vietnamese apatite on cement manufacture. The apatite was better than calcium orthophosphate and, along with some CaF_2, accelerated the formation of tricalcium silicate and increased its hydraulic activity. Kryzhanovskaya, Shchetkina, and Svirskaya [507] investigated the effect of phosphorus slags and found that the P_2O_5 content should be no greater than 2%. After 90 to 180 days of curing the strength decreased. Khnykin, Timashev, Malinin, and Ryazin, [455] also investigated the effect of P_2O_5 on the phase composition of Portland cement clinker. No solid solution was observed between dicalcium silicate and P_2O_5. As P_2O_5 increased, the tricalcium silicate, dicalcium silicate, and tricalcium aluminate decreased.

2. SETTING PROPERTIES

A number of references that dealt with the effect of phosphates in a cement clinker on its setting properties were found. Koyanagi [486] reported that small quantities of monocalcium phosphate retarded the setting of cement by preventing the hydration of calcium aluminate. Elsner von Gronow [256], in discussing small admixtures in cement, said that phosphates occurred to about 0.3%. Ershov [261] stated that small amounts of P_2O_5 in cement had an intense effect on speed of hardening and increase in strength; P_2O_5 in 0.2 to 0.3% amounts tripled the 24-hr strength; $3CaO.P_2O_5$ and $4CaO.P_2O_5$, when fired to 1100-1200°C, hydrated intensively. In opposition to some of the above statements Nurse [638] reported that P_2O_5 in raw materials may cause difficulties in firing, erratic setting, and slow strength development. As the percentage of P_2O_5 rises to 2.25%, the early strength decreases to less than the British Standard requirements. The cement is also affected more by the water-cement ratio.

Simonovskaya and Shpunt [751] reported that calcium phosphates in cement do not reduce quality. The strength loss and destruction of cement shapes containing 1 to 2% P_2O_5 were caused by large amounts of free CaO in the clinker because the experiments were made with charges with a high saturation coefficient. Steinour [774] stated that the P_2O_5 in the rock was usually less than 0.37% and insignificant; 2 to 2.5% P_2O_5 can be tolerated in the clinker with a proper mix design and still produce a low free-lime product with a good percentage of tricalcium silicate. The phosphate, however, must not be left in a soluble form. The product may have a low one-day strength and greater sensitivity to increased water. Cherkinskii and Koroleva [169] also discussed the influence of phosphates on the hardening of cement suspension. Sychev and his associates [799], discussed the use of Kara-Tau phosphorites in cement manufacture. They claimed that the cement industry should be prepared to use the waste as well as nonstandard and raw phosphoric ore with 6 to 15% P_2O_5.

Ryba, Korab, and Jirku [726] reported the optimum P_2O_5 as 0.5%, the allowable as not larger than 2%. Above 2% rapid deterioration was encountered, but only a slight influence was exerted on the setting time.

In a somewhat related article Thomson [821] described the effect of a sodium "hexametaphosphate" aqueous solution on set cement at room temperature. After 30 to 40 minutes contact the cement suddenly disintegrated and silica was thrown out of solution.

E. *Aluminous Cement*

The following references deal mainly with the manufacture of aluminous cement as a by-product of phosphorus manufacture. Akiyama [5, 6] discussed the preparation of aluminous cement from aluminum phosphate or limestone and calcium phosphate ore. Patents in this field were awarded [12, 410, 14, 15, 461, 630].

F. *By-Product Gypsum*

The usual acid process for the manufacture of phosphoric acid has gypsum as a by-product. A large literature, mostly Japanese, deals with the properties of this gypsum and its possible uses. One use discussed is for molds for ceramic slip casting [596]. A partial list of references follows: Netuka [630], Heinerth [366], Yamada et al. [939, 941], Murakami, Tanaka, and Bukan [597], Murakami and Tanaka [596], Yamada [943], Maki [544], Ognyanova and Kramer [641], Martusevicius and Martynaitis [548], Berezovskii [69], Yamada, Mizutani, and

Nagai [942], Hegner [362], Yamada, Ho, and Tamai [938], and Kitchen and Skinner [468]. Tabikh and Miller [807], and Berry [82, 83].

XIII. Ceramic-Related Products

A number of uses for phosphates, although not strictly ceramic, have some ceramic interest. This section deals with them briefly.

A. Phosphate Binders

Phosphates have been used as binders for a wide variety of materials. Heating is usually involved in establishing the bond. Some examples follow.

Rhodes and Gainsbury [706] patented a jewelers investment composition consisting of 100 3.5μ silica and 34 ml of 0.25 wt. % of sodium phosphate glass. Nakamura and Kawai [623] published articles on the use of phosphates in the manufacture of Co-Fe ferrite magnets. These phosphates not only promoted sintering but also increased the coercive force of the magnet by precipitating in the crystal grain or at the grain boundary.

McDaniel [612] patented a reconstituted mica sheet made of ground mica and 5 to 50% aluminum phosphate. The sheet was heated to higher than 230°C. Heyman [377] patented a phosphate-impregnated mica sheet, and Comeforo and various associates [185-188] discussed phosphate-bonded synthetic mica and talc.

Audrieth [32] patented a method for improving or making clay products by adding 0.5 to 5 wt. % sodium metaphosphate, sodium pyrophosphate, or sodium polyphosphate and firing to 1800 to 1900°F. The product had a uniform red color, no white bloom and increased compressive strength.

Vondracek [882] discussed phosphate-bonded asbestos composites for structural and electrical applications, and Chase and Copeland [168] described fibrous-reinforced aluminum phosphate laminates. Gebbett, Hourd, and Bloomfield [301] patented molded polymeric products made by mixing alkali-metal cyclophosphates with a variety of oxides and silicates. Abolins and Lukes [1] patented inorganic molding compositions consisting of asbestos and orthophosphoric acid.

Rodina [717] described ceramic filter materials with an aluminophosphate binder. Ten percent of binder ($Al_2O_3/P_2O_5=1/3$) was used with zirconia concentrate and fired to 1550°C. Sicka [749] patented a foamed ceramic made from fly ash and phosphoric acid, and McCarthy and Lovette [607] described a hot pressing method for fixing solid radioactive wastes in a phosphate glass.

B. Corrosion Inhibition

Long-chain polyphosphates in aqueous solution have the property of complexing with multivalent cations, for example, Ca^{2+}, and so can be used in water softening.

They also have the property of stabilizing a thin layer of calcium carbonate on a metal surface when present in small concentrations, a phenemenon known as a "threshold" treatment. This has been explained by Raistrick [691, 692, 693] on the basis that the molecular structure of the polyphosphates is such that it fits into the carbonate lattice and keeps carbonate nuclei from growing. This mechanism is accepted by Corsaro, Ritter, Hrubik, and Stephens [194].

Boies, Bregman, and Newman [107] patented a corrosion inhibitor that consisted of a molecularly dehydrated phosphate glass of the empirical formula $Na_{12}P_{10}O_{31}$. This was added to nitrate fertilizers in amounts of 100 to 1000 pm.

King [459] patented a water-soluble glass that consisted of P_2O_5-33.3, MO-2.5 to 17.5, and A_2O-32.5 to 100 mole %, where M is a divalent metal and A is a monovalent metal. Munter [595] patented a method of stabilizing an alkali-metal phosphate glass in water, and Fuchs [294], patented glassy sodium polyphosphates with a $(Na_2O+H_2O)/P_2O_5$ ratio of 1.10 to 1.065 with terminal groups on the polyphosphates chain of not less than 75 mole % OH; the remainder were ONa groups. Fuchs also patented slowly soluble phosphate glasses [295]. The composition range was P_2O_5-46.5-48.5, Na_2O-35.5 to 44.5, Al_2O_3-1 to 4.25, and CaO-4 to 15 mole %. The fusion temperature was less than 1380°F and the dissolving rate 25 to 125 mg/liter. Wirth and Robertson [932] patented polyphosphate glasses for water treatment to prevent corrosion. Yasutake and Fujita [946] patented a process for preparing sodium metaphosphate. Vogel [878] patented a method of forming a phosphate glass to use in the threshold treatment of corrosion.

C. *Bearings and Lubricants*

In 1962 Strub [790] patented a bearing consisting of polytetrafluoroethylene in which granular particles of a phosphate glass were uniformly dispersed.

Syritskaya [801] described aluminum phosphate glasses that were used as lubricants for metal extrusion. Godron [314] patented the use of phosphate glasses for extruding metals, for fertilizers, for low sintering glasses, and for aluminum enamels. Zager and Schneider [954] described the use of glass as a lubricant during the extrusion of steel.

D. *Other Uses*

A few other uses may be of interest. Wexler, Garfinkel, Jones, Hasegawa, and Krinsky [906] evaporated under vacuum a thin film of potassium metaphosphate on a glass blank and used it as the sensor of a hygrometer for radiosonde work. Pitrot [678] patented a composite lead phosphate-lead silicate pigment. The composition in moles was PbO-2 to 8, P_2O_5-1, SiO_2-0.75 to 7.5 moles/mole PbO. Phosphates are common additives to the caustic soda used in return-bottle soaker-washers. Fletcher, Kier, Johnson, and Slingsby [278] reported on the resistance of glasses to some industrial detergents and detergent additives. Wickham [921] suggested the use of lead pyrophosphate instead of PbO as a flux for crystal growth which caused less attack on refractories and heating elements. Hanes and Shaner [361] patented some rare earth oxide compositions coated with phosphates. Some interest has also been shown in molten phosphate reactor fuels and in the use of phosphoric acid in the manufacture of colored roofing granules. Yates [947] patented a positively charged fibrous cerium phosphate. Lee and Moos [525] described the magnetic properties of $TbPO_4$, a canted antiferromagnet. Brixner, Bierstedt, Jaep, and Barkley [132] discussed pure single crystals of alpha-$Pb_3(PO_4)_2$. The beta-to-alpha phase transition occurred at 180°C.

Bortz, Nakamura, and Aleshin [115] submitted a paper that discussed the manufacture of lightweight aggregate, brick, block, tile, pipe, and insulating concrete from Florida phosphate slime.

References

1. Visvaldis Abolins and R. M. Lukes, U. S. 3,322,549; *Ceram. Abstr.* **46P**(9), 225j (1967).
2. David Alder, *Amorphous Semiconductors,* C. R. C. Press, Cleveland, Ohio, 1971; *Ceram. Abstr.,* **51** (10), 262c (1972).
3. David Alder, *Am. Ceram. Soc. Bull.,* **52** (2), 154 (1973).
4. K. Akatsu, Katsusuke Maeda, and Isoroku Ikeda, *Semento Gijutsu Nempo,* **24**, 39 (1970), Cem. Assoc. Jap (English trans.), 20-23; *Ceram. Abstr.,* **51** (9), 214f (1972).
5. Kei-ichi Akiyama, *Waseda Appl. Chem. Soc. Bull.,* **15** (1), 1 (1938); *Cer. Abstr.,* 18, A (1), 8 (1939).
6. Kei-ichi Akiyama, *Waseda Appl. Chem. Soc. Bull.,* **15** (4), 15 (1938); *Cer. Am. Abstr.,* **19A** (4), 86 (1940).
7. G. N. Aleksandrova and M. G. Maslennikova, *Beton Zhelezobeton* (10), 24 (1972); *Ceram. Abstr.,* **52** (9), 190d (1973).
8. P. B. Aleirs, *J. Opt. Soc. Am.,* **51**(11), 1251 (1961); *Ceram. Abstr.,* **41**(3), 57i (1962).
9. A. C. Allen, *Ceram. Ind.,* **83**(2), 52, 56 (1964); *Ceram. Abstr.,* **43**(11), 315f (1964).
10. R. P. Allen, U.S. 2,866,713 (1958); *Ceram. Abstr.,* **38P**(5), 124f (1959).
11. A. M. Alper and R. N. McNalley, U.S. 3,600,206 (1971); *Ceram. Abstr.,* **50P**(11), 272g (1971).
11a. T. Ando, J. Ito, S. H. Ishii, and T. Soda, *Bull. Chem. Soc., Japan,* **25**, 78 (1952).
12. Anon., Brit. 263,124, *Rock Prod.,* **30**(17), 82 (1927); *Ceram. Abstr.,* **6P**(11), 500 (1927) (1927).
13. Anon., *Ceram. Ind.,* **10**(3), 314 (1928); *Ceram. Abstr.,* **7A**(5), 273 (1928).
14. Anon., Brit., 267,518, *Rock Prod.,* **31**(2), (1928); *Ceram. Abstr.,* **7P**(4), 216 (1928).
15. Anon., Ger. 483,399 (1927), *Tonind. Ztg.,* **53**(98), 1724 (1929); *Rock Prod.,* **33**(19) 86 (1930); *Ceram. Abstr.,* **10P**(2), 101 (1931).
16. Anon., *Times Eng. Supp.,* **30**(40), 1932; *Ceram. Abstr.,* **11A**(9), 496 (1932).
17. Anon., *Glass,* **9**(3), 96 (1932); *Ceram. Abstr.,* **11**(11), 561 (1932).
18. Anon, *Ind. Chim.,* **20**(235), 626 (1933); *Ceram. Abstr.,* **13A**(2), 84 (1934).
19. Anon., *Ind. Chim.,* **20**(239), 949 (1933); *Ceram. Abstr.,* **13A**(10), 255 (1934).
20. Anon., *Pottery Gaz.,* **70**(819), 511 (1945); *Ceram. Abstr.,* **25A**(2), 37 (1946).
21. Anon., *Sci. News.,* **47**(6), 84 (1945); *Ceram. Abstr.,* **24A**(6), 106 (1945).

22. Anon., *New Equip. Dig.*, **11**(1), 64 (1946); *Am. Glass Rev.*, **65**(13), 18 (1945); *Ceram. Abstr.*, **25A**(4), 64 (1946).
23. Anon., *Pottery & Glass*, **32**(7), 195 (1954); *Pottery Gaz.*, **79**(926), 1165 (1954); *Ceram. Abstr.*, **33**(11), 194a (1954).
24. Anon., *Ceramics*, **6**, 12 (1954).
25. Anon., *Roy. Doulton Mag.*, (9), 10 (1956); *Ceram. Abstr.*, **36**(3), 64e (1957).
26. Anon., *Met. Prod. Mfg.*, **15**(12), 36 (1958); *Ceram. Abstr.*, **38**(4), 96b (1959).
27. Demetrio Alvarez-Estrada and P. D. Botia, *Bol. Soc. Esp. Ceram.*, 8(4), 423 (1969); *Ceram. Abstr.*, **50**(8), 197a (1971).
28. Jumpei Ando and Seiichi Matsuno, *Bull. Chem. Soc. Japan*, **41**(2), 342 (1968); *Ceram. Abstr.*, **48**(4), 127d (1969).
29. A. I. Andrews, *Porcelain Enamels*, 2nd ed. (1961), Garrard, Champaign, Ill., 1961.
30. L. M. Angus-Butterworth, *Endeavour*, **VI**(23), 112-118 (1947).
31. A. S. Arapova, V. I. Pakhomova, and L. M. Demkina, *Tsement*, (1), 10 (1969); *Ceram. Abstr.*, **50**(6), 144b (1971).
31a. W. H. Armistead, U.S. 2,532,386 (1950); *Ceram. Abstr.*, **30P**(8), 136e (1951).
32. L. F. Audrieth, U.S. 2,880,099 (1959); *Ceram. Abstr.*, **38P**(8), 207d (1959).
33. A. Audsley and J. Aveston, *J. Chem. Soc.*, **1962**, 2320; *Ceram. Abstr.*, **43**(3), 74d (1964).
34. Y. Avnimelech, E. C. Moreno, and W. E. Brown, *J. Res. Natl. Bur. Stand.*, **77A**(1), 149 (1973); *Ceram. Abstr.*, **52**(9), E. 214, (1973).
35. S. deAza and C. R. Pecci, *Bol. Soc. Esp. Ceram.*, 7(3) 321 (1968); *Ceram. Abstr.*, **50**(6), 148d (1971).
36. K. P. Azarov and E. M. Chistova, *Zh. Prikl. Khim.*, **31**(10), 1602 (1958); *Ceram. Abstr.*, **41**(3), 566 (1962).
37. K. P. Azarov, V. V. Balandina, and V. A. Lyutsedarskii, *Zh. Priklad. Khim.*, **34**(11), 2560 (1961); *Ceram. Abstr.*, **41**(10), 236c (1962).
38. K. A. Baab and J. M. Blackwood, *Bull. Am. Ceram. Soc.*, **50**(7), 607 (1971); *Ceram. Abstr.*, **50**(8), 199g (1971).
39. C. L. Babcock, U.S. 3,063,198 (1962); *Ceram. Abstr.*, **42**(3), 69b (1963).
40. C. L. Babcock, *J. Am. Ceram. Soc.*, **51**(3), 163 (1968); *ibid.*, **56**(2), 55 (1973).
41. W. J. Bahnsen and E. E. Bryant, U.S. 2,324,812 (1943).
42. Ernst Baier, Emil Deeg, and H. G. Constabel, *Glastech., Ber.*, **39**(3), 136 (1966); *Ceram. Abstr.*, **46**(10), 253e (1967).
43. W. J. Baldwin, U.S. 2,483,393 (1949), U.S. 2,500,231 (1950); *Ceram. Abstr.*, **29P**(9), 177i (1950).
44. Petru Balta and Valriu Velea, *Bull. Inst. Politeh, Bucur.*, **25**(4), 55 (1963); *Ceram. Abstr.*, **43**(8), 213d (1964).
45. C. R. Bamford, *Phys. Chem. Glasses*, **3**(2), 54 (1962); *Ceram. Abstr.*, **42**(2), 41d (1963).
46. C. R. Bamford, *Phys. Chem. Glasses.*, **3**(6), 189 (1962).
47. D. P. Barag, *IX Intern. Congr. Glass, Artistic Historical Commun.*, **B11**, 183 (1971).
48. R. Barta and S. Prochazke, *Stavivo*, **39**(8), 282 (1961); *Ceram. Abstr.*, **43**(9), 249f (1964).
49. G. M. Bartenev, *The Structure and Properties of Inorganic Glasses*, translated from the Russian by F. F. Jaray, Wolters-Noordhoff, Groningen, The Netherlands, 1970.
50. P. Bartha, H. Lehmann, and M. Koltermann, *Ber. Dtsch. Keram, Ges.*, **48**(3), 111 (1971); *Ceram. Abstr.*, **P50**(10), 249f (1971).
51. S. M. Bash, A. D. Tumanov, and Sh. M. Rakhimbaev, *Tsement*, **33**(6), 15 (1967); *Ceram. Abstr.*, **48**(1), 3h (1969).
52. Basic Inc., Belgian P612, 195.

53. T. F. Bates and M. V. Black, *Glass Ind.*, **29**(9), 487-492, 516-518 (1948); *Ceram. Abstr.*, **29**(3), 42h (1950).
54. T. Bates and R. W. Douglas, *Trans. Soc. Glass Technol.*, **XLIII**, 289 (1959).
56. P. L. Baynton, H. Rawson, and J. E. Stanworth, *Nature*, **178**(4539), 910 (1956); *Ceram. Abstr.*, **36**(3), 57e (1957).
57. D. Becherescu, J. Menessy, and V. Christea, *Glas. Emaill. Keram. Tech.*, **20**(10), 341 (1969); *Ceram. Abstr.*, **49**(9), 214g (1970).
58. D. Becherescu, I. Menessy, and V. Gristea, *Glas. Emaill. Keram.-Tech.*, **22**(8), 281 (1971).
59. H. Bechtel, and G. Ploss, *Ber. Dtsch. Keram. Ges.*, **37**(8), 362 (1960); *Ceram. Abstr.*, **40**(6), 141h (1961).
60. H. Bechtel, and G. Ploss, *Ber. Dtsch. Keram, Ges.*, **40**(7), 399 (1963); *Ceram. Abstr.*, **43**(6), 159i (1964).
61. Renée Bedier, Serge Carpentier, Henri Francois, A. M. Brand-Clement, and Jean Meneret, U.S. 3,294,700 (1966); *Ceram. Abstr.*, **46P**(3), 69i (1967).
62. P. Beekenkamp, Doctorate Thesis, Technical High school, Eindhoven, Holland (1965), Colour Centres, Borate, Phosphate and Borophosphate Glasses.
63. P. Beekenkamp, *Klei en Keramiek*, **17**(8), 230 (1967); *Ceram. Abstr.*, **47**(5), 128i (1968).
64. P. Beekenkamp, and G. E. G. Hardeman, *Verres Réfract.*, **20**(6), 419 (1966); *Ceram. Abstr.*, **47**(7), 196h (1968).
65. P. Beekenkamp and J. M. Stevels, *Phys. Chem. Glasses*, **4**(6), 229 (1963); *Ceram. Abstr.*, **43**(6), 156e (1964).
66. G. Behrendt and Wentzuph, *Arch. Eisenhuttenwes.*, **7**, 95 (1933); *Ceram. Abstr.*, **34A**(6), 163 (1934).
67. R. N. Bell, *Ind. Eng. Chem.*, **39**(2), 136 (1947); *Ceram. Abstr.*, **26A**(9), 188d (1947).
68. R. N. Bell, *Ind. Eng. Chem. Anal. Ed.*, **19**, 97 (1947).
69. V. I. Berezovskii, *Zh. Prikl. Khim.*, **38**(8), 1687 (1965); *Ceram. Abstr.*, **48**(8), 274e (1969).
70. Edwin Berger, U.S. 1,961,603 (1934); *Ceram. Abstr.*, **13p**(8), 207 (1934).
71. S. B. Berger, C. B. Rubinstein, C. R. Kurkjian, and A. W. Treptow, *Phys. Rev.*, **133**, 3A, A723 (1964).
72. A. G. Bergman and V. A. Matrosova, *Zh. Neorg. Khim.*, **15**(6), 1703 (1970); *Ceram. Abstr.*, **52**(1), 29a (1973).
73. A. G. Bergman and L. N. Mikhalkovich, *Zh. Neorg. Khim.*, **15**(6), 1677 (1970); *Ceram. Abstr.*, **52**(1), 28g (1973).
74. A. G. Bergman and L. N. Mikhalkovich, *Zh. Neorg. Khim.*, **15**(8), 2270 (1970); *Ceram. Abstr.*, **52**(1), 28g (1973).
75. A. G. Bergman and Z. I. Sanzharova, *Zh. Neorg. Khim.*, **15**(4), 1139 (1970); *Ceram. Abstr.*, **52**(1), 28j (1973).
76. A. G. Bergman and Z. I. Sanzharova, *Zh. Neorg. Khim.*, **15**(6), 1708 (1970); *Ceram. Abstr.*, **52**(1), 25d (1973).
77. A. G. Bergman and L. V. Semenyakova, *Russ. Inorg. Chem.*, May (1970), 711.
78. A. G. Bergman and L. V. Semenyakova, *Zh. Neorg. Khim.*, **15**(8), 2287 (1970); *Ceram. Abstr.*, **52**(1), 28f (1973).
79. A. G. Bergmann, A. M. Gasanaliev, and A. S. Trunin, *Zh. Neorg. Khim.*, **14**(6), 1681 (1969); *Ceram. Abstr.*, (5), 124f (1972).
80. A. G. Bergman, A. M. Gasanaliev, and A. S. Trunin, *Zh. Neorg. Khim.*, **14**(5), 1422 (1969); *Ceram. Abstr.*, **51**(1), 25j (1972).
81. H. Bergseth, *Kolloidn Z.Z. Polym.*, **215**(1), 52 (1967); *Ceram. Abstr.*, **48**(4), 149a (1969).
82. E. E. Berry, Jr., *Appl. Chem. Biotechnol.*, **22**(6), 667 (1972); *Ceram. Abstr.*, **52**(8),

166g (1973).
83. E. E. Berry, Jr., *Appl. Chem. Biotechnol.*, **22**(6), 673 (1972); *Ceram. Abstr.*, **52**(8), 187a (1973).
84. L. V. Bershov, V. O. Martirosyan, E. K. Zavadovskaya, and V. A. Starodubtsev, *Izv. Akad. Nauk SSSR, Neorg. Mater.*, 7(3), 476 (1971); *Ceram. Abstr.*, **52**(6), 124j (1973).
85. S. I. Berul and N. K. Voskresenskaya, *Izv. Akad. Nauk SSSR, Neorgan. Mater.*, 3(3), 534 (1967); *Ceram. Abstr.*, 48(8), 297e (1969).
86. John Beukenkamp, William Rieman, III and Siegfied Lindenbaum, *Anal. Chem.*, **26**, 505 (1954).
87. Paul Beyersdorfer, *Glas.Emaill. Keram. Tech.*, **20**(12), 417 (1969); *Ceram. Abstr.*, **49**(9), 214j (1970).
88. M. A. Bezborodov, E. E. Mazo, and V. S. Kaminskaya, *Steklo Mat. Keram.*, **17**(1), 35 (1960); *Ceram Abstr.*, 39(9), 205f (1960).
89. S. S. Bhatnager, B. D. Khosla, and Ram. Chand, *J. Indian Chem. Soc.*, **17**, 515 (1940); *Chem. Abstr.*, **35**, 1197 (1941); *Ceram. Abstr.*, **23A**(5), 87 (1944).
90. Jean-Claude Bidard, U.S. 3,538,202 (1970); *Ceram. Abstr.*, **P50** (3), 79F (1971).
91. J. I. Bidewll, W. B. Jepson, and G. L. Toms, *Clay Miner*, 8(4), 445 (1970); *Ceram. Abstr.*, **50**(10), 264d (1971).
92. Gebhard Bierbrauer, *Keram. Z.*, **10**(1), 14 (1958); *Ceram. Abstr.*, **37**(7), 177f (1958).
93. Gebhard Bierbrauer, *Keram. Z.*, **10**(6), 26 (1958); *Ceram. Abstr.*, **38**(1), 16d (1959).
94. Adli M. Bishay, *J. Am. Cer. Soc.*, **44**(11), 545 (1961); *Ceram. Abstr.*, **41**(1), 4h (1962).
95. A. M. Bishay, and L. Maker, *J. Am. Ceram. Soc.*, **52**(11), 605 (1969); *Ceram. Abstr.*, **49**(1), 5d (1970).
96. S. R. Biswas, and J. Mukerji, *Cent. Glass Ceram. Res. Inst. Bull. (India)*, **15**(4), 99 (1968); *Ceram. Abstr.*, (10), 338i (1969).
97. G. E. Blair, U.S. 2,999,819 (1961); *Ceram. Abstr.*, **41p**(3), 59a (1962).
98. G. E. Blair, D. P. Hamblen, and R. A. Weidel, U.S. 3,278,317 (1966); *Ceram. Abstr.*, **46P**(1), 6f (1967).
99. H. H. Blau, private communication.
100. H. H. Blau, U.S. 2,132,399 (1938).
101. Henry Blumenberg, Jr., U.S. 1,601,231 & 2 (1926); *Ceram. Abstr.*, **5P**(12), 424 (1926).
102. L. M. Blyumen, *Stekolnaya Prom.*, **15**(6), 21 (1939); *Ceram. Abstr.*, **19A**(3), 64 (1940).
103. T. Boak, G. F. Rapp, H. T. Hartley, and B. E. Wiens, *Bull. Am. Ceram. Soc.*, **47**(8), 727 (1968).
104. M. M. Bobkova and V. V. Rudakov, Translated from *Steklomat. Keram.*, **6**, 11 (1967).
105. Ya. S. Bobovich, *Optics Spectrosc., USSR*, English translation **13**(4), 274 (1962).
106. L. D. Bogomolova, V. A. Zhachin, V. N. Lazukin, and N. F. Shopovalova, *Dokl. Akad. Nauk. SSSR*, **198**(4), 805 (1971); *Ceram. Abstr.*, **52**(11), 245i (1973).
107. D. B. Boies, J. I. Bregman, and T. R. Newman, U.S. 3,067,024 (1962); *Ceram. Abstr.*, **42P**(3), 69e (1963).
108. J. C. Bokros and W. H. Ellis, U.S. 3,685,059 (1972); *Ceram. Abstr.*, **52**(1), 15e (1973).
109. R. J. Bondley and M. E. Knoll, U.S. 2,888,406 (1959); *Ceram. Abstr.*, **38P**(9), 242j (1959).
110. Pierre Bonneman-Bémia, *Ann. Chim., 11e Ser.*, 15-16 (1941).
111. R. Bonniaud and C. Combret, *Verres Refract.*, **25**(2), (1971); *Ceram Abstr.*, **51**(1), 268f (1972).
112. Harold Bopp, private communication.
113. F. Ya Borodai and T. M. Evdovimova, *Izv. Akad. Nauk SSSR Neorg. Mater.* **5**(8), 1406 (1969); *Ceram. Abstr.*, **51**(8), 195e (1972).
114. S. A. Bortz and E. J. Onesto, *Bull. Am. Ceram. Soc.*, **52**(12), 898 (1973).

115. S. A. Bortz, H. H. Nakamura, and Eugene Aleshin, *Bull. Am. Ceram. Soc.*, **52**(4) 411 (1973); *Am. Ceram. Soc. Prog.*, 1-S-73.
116. S. A. Bortz, H. L. Rechter, W. E. Reynolds, and Seymour Bazell, U.S. 3,662,405 (1972); *Ceram. Abstr.*, **51P**(8), 198i (1972).
117. E. N. Boulos and N. J. Kreidl, *J. Am. Ceram. Soc.*, **54**(8), 368 (1971).
118. A. J. G. Boyer, D. J. Fray, and T. R. Meadowcroft, *Phys. Chem. Glasses* 8(3), 96 (1967); *Ceram Abstr.*, **47**(2), 57d (9168).
119. A. J. G. Boyer, D. J. Fray, and T. R. Meadowcroft, *J. Phys. Chem.*, **71**(5), 1442 (1967); *Ceram. Abstr.*, 48(4), 158g (1969).
120. B. T. Bradbury, and W. R. Maddocks, *J. Soc. Glass Tech.*, **42**, 325t-336t (1959).
121. W. George Brady, *J. Chem. Phys.*, **28**(1), 48-50 (1958); *Ceram. Abstr.*, **37**(7), 170h (1958).
122. E. W. Brain, *Pottery and Glass*, **37**(6), 457 (1959); *Ceram. Abstr.*, **38**(10), 264h (1959).
123. J. R. Brandenberger, Master's thesis under P. J. Bray, at Brown University (1964).
123a. P. J. Bray, *The Structure of Glass*, 7, 52 (1966), Consultants Bureau, New York.
124. A. H. Bremser and J. A. Nelson, *Bull. Am. Ceram. Soc.*, **46**(3), 280 (1967); *Ceram. Abstr.*, **46**(4), 97a (1967).
125. G. F. Brewster, N. J. Kreidl, and T. G. Pett, *Trans. Soc. Glass Tech.*, **31**, 153 (1947).
126. G. F. Brewster and N. J. Kreidl, *J. Soc. Glass Tech.*, **35**(167), 332 (1951).
127. G. F. Brewster and N. J. Kreidl, *J. Am. Ceram. Soc.*, **35**(10), 259 (1952).
127a. G. F. Brewster and R. A. Weidel, U.S. 3,490,928 (1970); *Ceram. Abstr.*, **49**(5), 115b (1970).
128. G. F. Brewster, J. R. Henser, J. L. Rood, and R. A. Weidel, *Appl. Opt.*, **5**(12), 1891 (1966); *Ceram. Abstr.*, **47**(9), 264b (1968).
129. Kristin Brigham and R. C. Vickery U.S. 3,647,488 (1972); *Ceram. Abstr.*, **51**(6), 136a (1972).
130. Kristin Brigham and R. C. Vickery, U.S. 3,649,732 (1972); *Ceram. Abstr.*, **51**(6), 136d (1972).
131. R. H. Brill, *IX Int. Cong. Glass*, Versailles, Section BI (1971).
132. L. H. Brixner, P. E. Bierstedt, W. F. Jalp, and J. R. Barkley, *Mater. Res. Bull.*, 8(5), 497 (1973); *Ceram. Abstr.*, **52**(9), 207h (1973).
133. Heinz Broemer and Norbert Meinert, U.S. 3,043,702, Ger. App. 17.9.59.
134. Heinz Broemer and Norbert Meinert, U.S. 3,492,136 (1970); *Ceram. Abstr.*, **49P**(5), 114a (1970).
135. Heinz Broemer and Norbert Meinert, U.S. 3,671, 276 (1972); *Ceram. Abstr.*, **51P**(9), 219f (1972).
136. Heinz Broemer, Norbert Meinert, and H. J. Preuss, U.S. 3,740,241 (1973); *Ceram. Abstr.*, **52**(10), 223a (1973).
137. Heinz Broemer, Norbert Meinert, and Johann Spincie, U.S. 3,597,245 (1971); *Ceram. Abstr.*, **50P**(11), 271d (1971).
138. A. V. Bromberg, A. G. Kasatkina, V. A. Kopeikin, A. I. Kuz'minskaya, I. L. Rashkovan, and I. V. Tananaev, *Izv. Akad. Nauk SSSR Neorg. Mater.*, 5(4), 805 (1969); *Ceram Abstr.*, **50**(9), 234g (1971).
139. Leslie Brown, *Ceramist*, 7(5), 297(1926); *Ceram. Abstr.*, **5A**(5), 160 (1926).
140. S. D. Brown G. H. Kenner, G. D. Schnittgrund, and J. T. Frakes, *Current Ceram. Res. Bull. Am. Ceram. Soc.*, **52**(2), 233 (1973).
141. R. H. Bruce and W. T. Wilkinson, *Trans. Brit. Ceram. Soc.*, **65**(5), 233 (1966); *Ceram. Abstr.*, **45**(10), 275i (1966).
142. E. C. Buck, U.S. 1,570,202 (1926); *Ceram. Abstr.*, **5P**(3), 81 (1926).
143. Peter Budnikoff, *Feuerfest*, **4**(9), 136 (1928); *Ceram. Abstr.*, **8A**(3) 22 (1929).

144. P. P. Budnikov, K. P. Azarov et al., *Steklomat. Keram.*, **18**(12), 23 (1961); *Ceram. Abstr.*, **45**(2), 30e (1966).
145. P. P. Budnikov and I. G. Gol'denberg, *Zh. Prikl. Khim.* **17**(7-8), 417 (1944); *Ceram. Abstr.*, **27**(2), 28g (1948).
146. P. P. Budnikov, Z. B. Entin, and G. A. Babin, *Kolloidn. Zh.*, **32**(3), 333 (1970); *Ceram. Abstr.*, **52**(11), 241j (1973).
147. G. A. Bukhalova, I. A. Tokman, and V. M. Shpakova, *Zh. Neorg. Khim.*, **15**(6), 1691 (1970); *Ceram. Abstr.*, **52**(1), 24i (1973).
148. Norbert Busch, J. P. Ebel, and Monique Blanck, *Bull. Soc. Chim., (Fr.)*, 486 (1957).
149. Yu. M. Butt, M. A. Vorobeva, and Bui Van Tien, *Trans. Mosk. Khim., Technol. Inst.*, (68), 169 (1971); *Ceram. Abstr.*, **52**(8), 165j (1973).
150. G. V. Buxton, F. C. R. Cattel, and F. S. Dainton, *Trans. Faraday Soc.*, No. 579, **67**(3), 687 (1971).
151. G. V. Buxton, F. S. Dainton, T. E. Lantz, and F. P. Sargent, *Trans Faraday Soc.*, **66**(12), 2962 (1970); *Ceram. Abstr.*, **50**(9), 219j (1971).
152. R. H. Caley, *J. Can. Ceram. Soc.*, **39**, 7 (1970); *Ceram Abstr.*, **51**(8), 190d (1972).
153. R. H. Caley, and M. K. Murthy, *J. Am. Ceram. Soc.*, **53**(5), 254 (1970); *Cer. Abstr.*, **49**(6), 137j (1970).
154. C. F. Callis, J. R. Van Wazer, and J. S. Metcalf, *J. Am. Chem. Soc.*, 77(6), 1468 (1955); *Ceram. Abstr.*, **36**(2), 49a (1957).
155. C. F. Callis, J. R. Van Wazer, and J. S. Metcalf, *J. Am. Chem. Soc.*, 77(6), 1468 (1955).
155a. C. F. Callis, J. R. Van Wazer, and J. S. Metcalf, *J. Am. Chem. Soc.*, 77(6), 1471 (1955).
156. A. N. Campbell and A. J. R. Campbell, *Trans. Faraday. Soc.*, **31**, 1567 (1935).
157. W. B. Campbell, J. E. Burroughs, and J. K. Cochran, U.S. 3,638,243 (1972); *Ceram. Abstr.*, **51**(5), 111f (1972).
158. H. W. Carpenter and P. D. Johnson, U.S. 3,084,055, Appl., 15.11.60.
159. Sandra Carr, Osama El-Bayoumi, and K. N. Subramanian, *Bull. Am. Ceram. Soc.*, **52**(4), 386 (1973); *Am. Ceram. Soc. Prog.*, 48-G-73.
160. P. T. Carter, *Roy. Tech. Coll. Met. Club J.*, **1955-1956**(8) 43; *Abstr. J. Iron Steel Inst. (London)*, **186**(3), 362 (1957); *Ceram. Abstr.*, **36**(10), 236b (1957).
161. A. J. Carty, *Chem.* 13/12 *News*, **41**, 6 (1972), University of Waterloos, Waterloo, Canada.
162. J. D. Callwood, J. H. Davidson, and V. Dimbleby, *J. Soc. Glass Tech.*, **12**(45), 7 (1928); *Ceram. Abstr.*, **8A** (1), 22 (1929).
163. Franz Cedivoda, *Chem. Ltg.*, **5**, 347 (1901).
164. M. R. Chakrabarty, *Can. Uranium Res. Found. Rept.* 1962) ORF62-2; *Ceram. Abstr.*, **42**(9), 239i (1963).
165. S. K. Chakravorthy, *Ceramics (Coll. Ceram. Tech.) (Calcutta)*, 8, 30 (1966); *Ceram. Abstr.*, **47**(5), 133a (1968).
166. R. Charan, *Trans. Indian Ceram. Soc.*, **2**(2), 85 (1943); *Ceram. Abstr.*, **23A**(9), 150 (1944).
167. R. J. Charles, *VIII Int. Glass Congr. London*, **34**, 29 (July 1968).
168. V. A. Chase and R. L. Copeland, *Am. Ceram. Soc.*, 1965 Mtg., Symposium #2.
169. Yu. S. Cherkinski and A. T. Koroleva, *Izv. Akad. Nauk SSSR Neorg. Mater.* **4**(10), 1825 (1968); *Ceram. Abstr.*, **50**(4), 98f (1971).
170. Vincent Chiola, J. S. Smith, and C. D. Vanderpol, U.S. 3,721,852 (1973); *Ceram. Abstr.*, **52**(7), 147d (1973).
171. A. A. Christyakova, V. A. Sivkina, V. I. Sadkov, and A. P. Khashkovskaya, *Izv. Akad. Nauk SSSR Neorg. Mater.*, **5**(10), 1738 (1969); *Ceram. Abstr.*, **51**(9), 234g (1972).
172. A. A. Christyakova, V. A. Sivkina, V. I. Sadkov, A. P. Khashkovskaya, and L. G. Polvysheva, *Izv. Akad. Nauk. SSSR Neorg. Mater.*, **5**(3) 536 (1969); *Ceram. Abstr.*, **50**(9), 239e (1971).

173. Theodor Chvatal, U.S. 3,329,516 (1967); *Ceram. Abstr.*, **46**(10), 259d (1967).
174. A. E. Clark, Jr., and L. L. Hench, *Bull. Am. Ceram. Soc.*, **52**(4), 379 (1973); *Am. Ceram. Soc. Prog.*, 1-G-73.
175. W. W. Coblentz and R. Stair, *Proc. Natl. Acad. Sci.*, **20**(12), 630 (1934); *Ceram Abstr.*, **15A**(4) 119 (1936).
176. Loring Coes, *Abrasives*, Springerverlag, New York, 1971.
177. J. Coes, Jr., *Abrasives Applied Mineralogy*, **I**, Springer Verlag, New York, 1971; *Ceram. Abstr.*, **51**(6), 153c (1972).
178. W. W. Coffeen, *Bibliog. Enamel Division. Am. Ceram. Soc.*, **1959**; *Ceram. Abstr.*, **40B**(4), 105d (1961).
179. W. W. Coffeen, *Ceram. Ind.*, **I,70**(4), 120, (1958); II *ibid.*, (5), 76, 81; *Ceram. Abstr.*, **38**(6), 153f (1959).
180. W. W. Coffeen and D. V. Van Gordon, *Ceramics (Sao Paulo)*, **4**(13), 16 (1958); *Ceram. Abstr.*, **38**(2), 51i (1959).
181. Theodore Cohn, *Chem. News.*, **129**, 32-35 (1924); *Ceram. Abstr.*, **A**(1) 7, 1925.
182. R. Coleman, *Soil Sci. Soc. Am. Proc.*, **9**, 72-78 (1944); *Ceram. Abstr.*, **26A**(1), 20 (1947).
183. D. W. Collins, G. E. Rindone, and L. N. Mulay, *J. Am. Ceram. Soc.*, **54**(1), 52 (1971).
184. J. D. Colwell, *Aust. J. Appl. Sci.*, **10**(1), 95-103 (1959); *Ceram. Abstr.*, **39**(7), 176i (1960).
185. J. E. Comeforo, R. A. Hatch, R. A. Humphry, and Wilhelm Eitel, *J. Am. Ceram. Soc.*, **36**, 286 (1953).
186. J. E. Comeforo, J. G. Breedlove, and Hans Thurnauer, *J. Am. Ceram. Soc.*, **37**(4) 191 (1954).
187. J. E. Comeforo, *J. Am. Ceram. Soc.*, **37**, 427 (1954).
188. J. E. Comeforo, U.S. 2,704,261 (1955).
189. C. H. Commons, Jr., *Bull. Am. Ceram. Soc.*, **27**(9), 337 (1948); *Ceram. Abstr.*, **27**(10), 221b (1948).
190. M. Conti, *Ind. Vetro Ceram.*, **8**(9), 355 (1935); *Ceram. Abstr.*, **15A**(3), 88 (1936).
191. D. E. C. Corbridge, *Topics on Phosphorus Chemistry*, Vol. 3., Interscience, New York, 1966, pp. 57-435.
192. D. E. C. Corbridge, *Topics in Phosphorus Chemistry*, Vol. 6, Interscience, New York, p. 235, 1969.
193. See 329a.
194. G. Corsaro, H. S. Ritter, W. Hrubik, and H. L. Stephens, *J. Am. Water Works Assoc.*, **48**, 683 (1956); *Abstr. J. Appl. Chem. (London)*, **7**(1), 1 (1957); *Ceram. Abstr.*, **36**(9), 217f (1957).
195. Arnold Cousen and W. E. S. Turner, *Meet. Soc. Glass Technol.*, University of Leeds, November 19, 1924; *Ceram. Abstr.*, **4**(2), 34 (1925).
196. Arnold Cousen and W. E. S. Turner, *Glass Ind.*, **7**(9), 213 (1926); *Ceram. Abstr.*, **5**(11), 348 (1926).
197. R. E. Cowan, *Bull. Am. Ceram. Soc.*, **35**(2), 53-56 (1956); *Ceram Abstr.*, **35**(4), 71j (1956).
198. H. B. Cronshaw, *Trans Ceram. Soc. (Engl)*, **17**, 153 (1917-1918).
199. W. H. Cropper, *J. Am. Ceram. Soc.*, **45**(6), 293 (1962); *Ceram. Abstr.*, **41**(7), 164c (1962).
200. W. S. Crowell, *Trans. Am. Inst. Chem. Eng.*, **19**, 19 (1927); *Ceram. Abstr.*, **7A**(9), 596 (1928).
201. Joan Crowther and A. E. R. Westman, *Can. J. Chem.*, **34**, 969 (1956).
202. M. M. Crutchfield, G. H. Dungan, and J. R. Van Wazer, *Topics in Phosphorus Chemistry*, Vol. 5, Interscience, New York, 1967, p. 1.

203. M. M. Crutchfield, C. H. Dungan, J. H. Letcher, V. Mark, and J. R. Van Wazer, *Topics in Phosphorus Chemistry,* Vol. 5, Interscience, New York, 1967, p. 237.
204. F. S. Dainton, *Berichte der Bunsen-Ges.,* **75**(7), 608 (1971).
205. F. S. Dainton and F. T. Jones, *Radiation Res.,* **17**(3), 388 (1962).
206. A. E. Dale and J. Stanworth, *J. Soc. Glass Technol.,* **35**, 185 (1951).
207. R. R. Danielson, *J. Can. Ceram. Soc.,* **23**, 40-42 (1954); *Ceram. Abstr.,* **37**(6), 142h (1958).
208. V. P. Danilov, *Dokl. Akad. Nauk. SSSR,* **48**, 109 (1945).
209. C. R. Das, *Cent. Glass Ceram. Res. Inst. Bull (India),* **12**(2), 58 (1965); *Ceram. Abstr.,* **45**(3), 61a (1966).
210. C. R. Das, *Cent. Glass Ceram. Res. Inst. Bull. (India),* **14**(3), 85 (1967); *Ceram. Abstr.,* **47**(6), 169c (1968).
211. D. R. Davies and D. E. C. Corbridge, *Acta. Cryst.,* (11), 315 (1958).
212. D. L. Deadmore and J. S. Machin, *Illinois State Geol. Surv. Circ.,* No. 335 (1962); *Ceram. Abstr.,* **42**(8), 214i (1963).
213. Emil Deeg, *Glastech. Ber.,* **31**, 1, 85, 124, 229 (1958).
214. E. W. Deeg and R. W. Young, U.S. 3,964,179 (1972); *Ceram. Abstr.,* **52**(3), 64d (1973).
215. Jack DeMent, U.S. 2,549, 180 (1951); *Ceram. Abstr.,* **31P**(1), 8f (1952).
216. L. I. Demkina, *Steklomat. Keram,* **11**(2), 10 (1954); *Ceram. Abstr.,* **34**(10), 177g (1955).
217. P. F. De Paolis and P. B. Mauer, U.S. 3,250,721, (1966); *Ceram. Abstr.,* **45P**(7), 184c (1966).
218. H. M. Dess, U.S. 3,227,567 (1966); *Ceram. Abstr.,* **45**(4), 101c (1966).
219. P. M. DiBello, and A. M. Pradel, *J. Can. Ceram. Soc.,* **37**, 13-14 (1968); *Ceram. Abstr.,* **50**(1), 13C (1971).
220. A. Dietzel and Ingeborg Hinz, *Ber. Dtsch. Keram. Ges.,* **39** (12), 569 (1962); *Ceram. Abstr.,* **42**(7), 183h (1963).
221. A. Dietzel and H. J. Poegel, *Naturwissenschaften,* **40**, 604 (1953).
222. Helmut Dislich, *Glastech. Ber.,* **44**(1), (1971); *Ceram. Abstr.,* **51**(7), 160a (1972).
223. A. E. Dodd, *Dictionary of Ceramics,* George Newnes, London, 1967, p. 10.
224. J. W. Donahey, G. J. Morris, and B. J. Sweo, *The International Enamelist,* **1**(4), 6 (1951).
225. R. W. Douglas, *Progr. Ceram. Sci.,* **I**(5), 200 (1961).
226. R. W. Douglas, *VIII Int. Congr. Glass Soc. Glass Technol.,* Sheffield, England, 23 (1969).
227. A. Dozier and D. L. Kinser, *Current Ceram. Res. Bull. Am. Cer. Soc.,* **52**(2), 238 (1973).
228. A. W. Dozier, L. K. Wilson, E. J. Friebele, and D. L. Kisner, *J. Am. Ceram. Soc.,* **55**(7), 373 (1972); *Ceram. Abstr.,* **5**(8), 189d (1972).
229. C. Drake and I. Scanlan, *Ceramics,* 2024613 (1970).
230. J. A. Duffy, *Phys. Chem. Glasses,* **13**(3), 65 (1972).
231. Gyorgy Duma, *Epitoanyag,* **20**(12), 450 (1968); *Ceram. Abstr.,* **50**(4), 113i (1971).
232. W. H. Dumbaugh, Jr., G. B. Carrier, and J. E. Flannery, *Bull. Am. Ceram. Soc.,* **52**(4), 383 (1973); *Am. Ceram. Soc. Prog.,* 24-C-73.
233. M. E. Dumesnil, R. R. Hewitt, and J. L. Bozarth, U.S. 3,650,778 (1972); *Ceram. Abstr.,* **51**(7), 162g (1972).
234. D. Eardley, *Pottery Glass Rec.,* **19**(10), 261 (1937); *Ceram. Abstr.,* **17A**(1), 26 (1938).
235. W. H. Earhart, *Bull. Am. Ceram. Soc.,* **20**(9), 312 (1941); *Ceram. Abstr.,* **20**(11), 267 (1941).
236. Eastman Kodak Co., Swiss 270, 499 (1951); *Abstr., Chem. Zent.,* **122II**(11), 1649 (1951); *Ceram. Abstr.,* **32P**(4), 60h (1953).
237. J. P. Ebel and Y. Volmer, *C. R. Acad. Sci., Paris,* **233**, 415 (1951).
238. J. R. Ebel, *C. R. Acad. Sci., Paris,* **234**, 621, 732 (1952).
239. J. R. Ehel, *Bull. Soc. Chem. (Fr.),* 991, 1085, 1089, 1096 (1953).
240. J. P. Ebel and J. Colas, *C. R. Acad. Sci., Paris,* **239**, 173 (1954).

241. J. P. Ebel, and J. Colas, *Bull. Soc. Chim. (Fr.)*, 1087 (1955).
242. J. P. Ebel, and N. Busch, *C. R. Acad. Sci., Paris*, **242**, 647 (1956).
243. J. P. Ebel, *Bull. Soc., Chim. (Fr.)*, 1663 (1968).
244. N. E. Eberly, U.S. 1,671,104 (1928); *Ceram. Abstr.*, **7P**(8), 516 (1928).
245. B. Eckstein, *Vortr. Fachausschussber I D.G.G.*, Würzburg, Germany (April 27, 28, 1961).
246. J. M. Eder and E. Valenta, Denkschrift der Mathem, *Naturwiss, Kl., Akad. Wiss., Wien*, **61**, 285 (1894).
247. H. W. Edwards, *Trans. Ceram. Soc. (Engl.)*, **3**, 32 (1903-1904).
248. R. J. Edwards, A. Paul, and R. W. Douglas, *Phys. Chem. Glasses*, **13**(5) 137 (1972); *Ceram. Abs.*, 52(6), 124h (1973).
249. Eduard Eipeltauer and Elizabeth Hammer, *Glastech. Ber.*, **39**(6), 294 (1966); *Ceram. Abstr.*, 47(6), 169g (1968).
250. A. Eisenberg and K. Takahashi, *J. Non-Cryst. Solids*, 3(3), 279 (1970).
251. Wilhelm Eitel, *Silicate Science II: Glasses, Enamels, Slags*, Academic, New York, 1964, p. 379, *Non-Silicate Glasses.*
252. Osama El-Bayoumi and A. M. Bishay, *Bull. Am. Ceram. Soc.*, **52**(4), 390 (1973); *Am. Ceram. Soc. Prog.* 72-G-73.
253. Osama El-Bayoumi, and K. N. Subramanian, *Bull. Am. Ceram. Soc.*, **52**(4), 386 (1973); *Am. Ceram. Soc. Prog.*, 50-G-73.
254. J. C. Elliot. *Calcif. Tissue Res.*, 3(4), 293 (1969); *Ceram. Abstr.*, **49**(4), 106b (1970).
255. N. Elliott, *Pottery Glass Rec.*, **19**(11), 289 (1937); *Ceram. Abstr.*, **17A**(2), 77 (1938).
256. H. Elsner von Gronow, *Zement-Kalk-Gips*, **1**(5), 85 (1948); *Ceram. Abstr.*, **29**(4), 65f (1950).
257. W. von Engelhardt and A. von Smolinski, *kolloid. Z.*, **151**(1), 47 (1957).
258. Fernandez Enrique de Mariguel, *Bol. Soc. Esp. Ceram.*, 9(3), 311 (1970); *Ceram. Abstr.*, **50**(6), 149g (1971).
259. R. A. Eppler and G. H. Spencer-Strong, *J. Am. Ceram. Soc.*, **52**(5), 263 (1969); *Ceram. Abstr.*, **48**(7), 246h (1969).
260. Ernst Leitz, G. H., Brit. 1,169,312 (1969); *Ceram. Abstr.*, **49**(10), 246d (1970).
261. L. D. Ershov, *Tsement*, **21**(4), 19 (1955); *Ceram. Abstr.*, **35P**(5), 90f (1956).
262. L. D. Ershov and R. M. Basman, *Ukr. Khim. Zh.*, **21**(6), 783 (1955); *Ceram. Abstr.*, **35**(10), 205g (1956).
263. Ersin Eti, and W. B. Hall, *Bull. Am. Ceram. Soc.*, **50**(7), 604 (1971); *Ceram. Abstr.*, **50**(8), 197b (1971).
264. A. G. Eubanks, and D. G. Moore, *NASA Tech Note D-106* (1959); *Ceram. Abstr.*, **40**(5), 111i (1961).
265. K. S. Evstopjev, G. T. Petrovsky, and W. D. Chalilev, *Verres Réfract.*, **25**(4), 161 (1971).
265a. L. F. Rev. Belge *ind. verrières, céram., email*, 4(2)26-29, (3)57-58, (4)77-78 (1933).
266. M. Fanderlik and M. Palecek, *Silik. J.*, **5**(7), 127 (1966); *Ceram. Abstr.*, **47**(6), 167f (1968).
267. Farbwerke Hoechst Akt. Ges., Brit. 1,005,964 (1965); *Ceram. Abstr.*, **46P**(10), 250e (1967).
268. M. Faulstich, Eric Schott (1959), *Beitrage zur angewandten Glasforschung*, Wissenschaftliche Verlags-gesellschaft M. B. H. Stuttgart.
269. T. Feldmann and A. Treinin, *J. Phys. Chem.*, **72**(11), 3768 (1968); *Ceram. Abstr.*, **48**(5), 173c (1969).
270. Felica de Carli, *Atti V. II Congr. Chim. pura appl.*, 1146-1150 (1926); *Ceram. Abstr.*, 7,A(9), 642 (1928).
271. V. A. Ferrandis, D. A. Estrada, and C. S. Conde, *Rev. Sci. Appl. (Madrid)*, **13**, 97 (1959); *Abstr. J. Appl. Chem. (London)*, **9**(10) 11-354 (1959); *Ceram. Abstr.*, **39**(2), 31b (1960).
272. M. B. Field, *J. Appl. Phys.*. **40**(6), 2628 (1969); *Ceram. Abstr.*, **48**(11), 369f (1969).

273. R. B. Fischer and C. E. Ring, *Ann. Chem.*, (3), 431 (1957); *Ceram. Abstr.*, **36**(11), 279i (1957).
274. R. E. Fisher, U.S. 3,649,313 (1972); *Ceram. Abstr.*, **51**(6), 135g (1972).
275. H. G. Fisk, *J. Am. Ceram. Soc.*, **34**(1), 9 (1951); *Ceram. Abstr.*, **30**(2), 29i (1951).
276. W. Fix, H. Hymann, and R. Heinke, *J. Am. Ceram. Soc.*, **52**(6), 346 (1969); *Ceram. Abstr.*, **48**(8), 199a (1969).
277. J. V. Fitzgerald, *J. Am. Ceram. Soc.*, **34**(10) 314 (1951).
278. W. W. Fletcher, E. S. Kier, P. G. Johnson, and B. Slingsby, *Glass Technol.*, **3**, 195 (1962).
279. O. W. Floerke and H. Lachenmayr, *Ber. Dtsch. Keram. Ges.*, **39**(1), 55 (1962); *Ceram. Abstr.*, **46**(5), 135e (1967).
280. P. J. Flory, *J. Am. Chem. Soc.*, **63**, 3083 (1941).
281. P. J. Flory, *J. Am. Chem. Soc.*, **64**, 2205 (1942).
282. P. J. Flory, *Principles of Polymer Chemistry,* Cornell University Press, Ithaca, New York, (1953).
283. A. H. Foessel and W. S. Treffner, *Bull. Am. Ceram. Soc.*, **49**(7), 652 (1970).
284. Karl Foster and Joseph Seidel, U.S. 3,620,779 (1971); *Ceram. Abstr.*, **15**(3), 54f (1972).
285. J. T. Fournier and R. H. Bartram, *J. Phys. Chem. Solids,* **31**12), 2615 (1970); *Ceram. Abstr.*, **52**(5), 98c (1973).
286. P. W. France and H. O. Hooper, *J. Phys. Chem. Solids,* **31**(6), 1307 (1970); *Ceram. Abstr.*, **52**(4), 80i (1973).
287. A. D. Franklin, *Curr. Ceram. Res. Bull., Am. Ceram. Soc.*, **52**(2), 229 (1973).
288. J. W. French, *Proc. Roy Phil. Soc. Glasg.*, **52**, 113-37 (1924); *J. Soc. Chem. Ind.*, **44**, B170 (1925); *Ceram. Abstr.*, **4**(6), 157 (1925).
289. E. J. Friebele, D. L. Kinser, and L. K. Wilson, *Bull. Am. Ceram. Soc.*, **52**(4), 384 (1973); *Am. Ceram. Soc. Prog.*, 37-G-73.
290. E. J. Friebele, L. K. Wilson, and D. L. Kinser, *J. Am. Ceram. Soc.*, **55**(3), 164 (1972).
291. A. L. Friedberg and E. A. Peterson, *J. Am. Ceram. Soc.*, **33**, 17 (1950).
292. Helene Fritz, *Keram. Rundsch.*, **26**, 195, 201, 205 (1918).
293. L. J. Frost and C. H. Commons, U.S. 2,339,260 (1944).
294. R. J. Fuchs, U.S. 3,127,238 (1964); U.S. 3,130,002 (1964); *Ceram. Abstr.*, **43P**(7), 190b (1964).
295. R. J. Fuchs, U.S. 3,338,670 (1967); *Ceram. Abstr.*, **46P**(11), 279f (1967).
296. V. I. Funtikov, V. M. Kolbasov, and M. N. Kuznetsova, *Trans. Mosk. Khim. Tekhnol. Inst.*, **1966**, No. 50, 140; *Ceram. Abstr.*, **47**(6), 165h (1968).
297. L. E. Fussell and R. H. Hadley, *J. Am. Ceram. Soc.*, **41**(3), 81 (1958); *Ceram. Abstr.*, **37**(4), 89j (1958).
298. Kitzuso Fuwa, *J. Jap. Ceram. Assoc.*, **32**(382), 431 (1924); *Ceram. Abstr.*, **5**(1), 18 (1926).
299. Ganzenmüller, *Glashütte,* **68**(51), 873 (1938); *Ceram. Abstr.*, **18A**(7), 178 (1939).
300. P. A. Gartaganis and A. E. R. Westman; see E. J. Griffith, *J. Inorg. Nucl. Chem.*, **26**, 1381 (1964), for a discussion of a private communication on crystalline ammonium tetrapolyphosphate. See also Joan Crowther and A. E. R. Westman, *Can. J. Chem.*, **34**, 977 (1956).
301. D. Gebbett, A. Hourd, and P. R. Bloomfield, Brit. Pat 1,018,401; *Chem. Abstr.*, **64**, 9407b (1966).
302. Wilhelm Geilmann and Hans Jenemann, *Glastech, Ber,* **26**(9), 259 (1953); *Ceram. Abstr.*, **34**(4), 67j (1955).
302a. R. J. H. Gelsing, H. N. Stein, and J. M. Stevels, *Phys. Chem. Glasses,* **7**(6), 185 (1966).
303. Frederick Gelstharp, U.S. 2,294,844 (1942); *Ceram. Abstr.*, **21**(10), 212 (1942).
303a. Karl Gentil, *Umschau,* (39), 341 (1935); *Abstr. Chem. Zent,* II, 3821 (1935); *Ceram. Abstr.*, **15A**(5), 147 (1936).
304. A. N. Georoff and C. H. Babcock, *J. Am. Ceram. Soc.*, **56**(2), 97 (1973).

305. V. V. Gerasimov, L. I. Kuznetsov-Felisov, and E. V. Kuznetsov, *Izv. Akad. Nauk SSSR Neorg. Mater.*, **4**(10), 1819 (1968); *Ceram. Abstr.*, **50**(4), 99j (1971).
306. V. V. Gerasimov, L. I. Kuznetsov, E. V. Kuznetsov, and T. B. Shakhmina, *Izv. Akad. Nauk SSSR Neorg. Mater.*, **5**(6), 1062 (1969); *Ceram. Abstr.*, **50**(9), 219e (1971).
307. W. L. German, *Trans. Brit. Ceram. Soc.*, **51**(3), 198 (1952); *Ceram. Abstr.*, **32**(3), 45b (1953).
308. W. L. German, *Ceramics,* **4**(43), 56 (1952); *Ceram. Abstr.*, **32**(7), 123g (1953).
309. W. L. German, *Ceramics,* **6**(66), 252 (1954); *Ceram. Abstr.*, **37**(4), 100h (1958).
310. P. A. Gilham-Dayton, *Trans. Brit. Ceram. Soc.*, **62**(11), 895 (1963); *Ceram. Abstr.*, **43**(6), 162a (1964).
311. J. B. Gill and R. M. Taylor, *J. Chem. Soc. Suppl.* **1964**, No. 2, 5905; *Ceram. Abstr.*, **46**(8), 209d (1967).
312. W. H. Gitzen, L. D. Hart, and G. Maczura, *Bull. Amer. Ceram. Soc.*, **35**(6), 217 (1956); *Ceram. Abstr.*, **35**(8), 167h (1956).
313. Francis W. Glaze, *Bull. Amer. Ceram. Soc.*, **32**(7), 242 (1953); *Ceram. Abstr.*, **32**(9), 170h (1953).
314. Y. Godron, *Can.*, 779, 708, 3-1-1966.
315. S. Goldsztaub, S. Heńin, and R. Wey, *Clay Min. Bull.*, **2**(12), 162 (1954) (Fran); *Ceram. Abstr.*, **35**(2), 43h (1956).
316. N. V. Golubeva and I. M. Prok, *Zh. Prikl. Khim. (Leningrad)*, **17**(7-8), 422 (1944); *Ceram. Abstr.*, **24A**(8), 141 (1945).
317. L. A. Golubtsov, K. K. Evstrop'ev, T. N. Kiryanova, V. S. Kondrat'eva, and V. D. Khalilev, *Izv. Akad. Nauk SSSR Neorg. Mater.*, 7(2), 248 (1971); *Ceram. Abstr.*, **52**(6), 123c (1973).
318. V. D. Golubtsov, K. S. Khalilev, K. S. Evstrop'ev, and E. N. Ivanova, *Zh. Prikl. Khim. (Leningrad)*, **44**(1), 180 (1971); *Ceram. Abstr.*, **52**(1), 8f (1973).
319. S. L. Golynko-Vol'fson and L. G. Sudakas, *Zh. Prikl. Khim. (Leningrad)*, **38**(7), 1466 (1965); *Ceram. Abstr.*, **48**(7), 245f (1969).
320. S. L. Golynko-Vol'fson and L. G. Sudakas, *Izv. Akad. Nauk SSSR Neorg. Mater.*, **2**(2), 343 (1966); *Ceram. Abstr.*, **47**(1), 11a (1958).
321. R. Gopal and C. Calvo, *J. Solid State Chem.*, **5**(3), 432 (1972); *Ceram. Abstr.*, **52**(4), 91d (1973).
322. M. Gordon, *Recent Adv. Theory Thermal Degradation* S.C.I.; see also *J. Phys. Chem.*, **64**, 1929; *Proc. Roy Soc. (London)*, **A258**, 215-236 (1960).
323. M. Gordon and G. R. Scantlebury, *Trans. Faraday, Soc.*, **60**, 604 (1964).
324. Y. G. Gordon, U.S. 3,248,234 (1966); *Ceram. Abstr.*, **45P**(7), 183b (1966).
325. M. W. Gormley, *Bull. Am. Ceram. Soc.*, **37**(2), 77;(3), 144(4), 189;(5), 210;(6), 283 (1968).
326. R. G. Gossink, Doctorate thesis, Technische Hogeschool, Eindhoven, The Netherlands (1971), Properties of Vitreous and Molten Alkali Molybdates and Tungstates.
327. J. Gotz and C. R. Masson, *Int. Congr. Glass Sci. Tech. Commun.*, 9(1), 261 (1971).
328. J. Gotz, W. D. Jamieson, and C. R. Masson, *Proc. Int. Comm. Glass, Toronto,* **69** (1969).
329. R. E. Graf, U.S. 3,731,226 (1973); *Ceram. Abstr.*, **52**(8), 170e (1973).
329a. Beverly Gray, reference lost.
330. P. W. L. Graham, U.S. 3,473,906 (1969); *Ceram. Abstr.*, **49P**(2), 38g (1970).
331. J. A. Grande and John Keukenkamp, *Anal. Chem.*, **28**(9), 1496 (1956); *Ceram. Abstr.*, **36**(3), 72e (1957).
332. R. L. Green and K. B. Blodgett, *J. Am. Ceram. Soc.*, **31**(4), 89 (1948).
332a. S. Greenfield and M. Clift, Pergamon Press, Oxford, to be published, June 1975.
333. H. H. Gregor, U.S. 2,460,344 (1949); *Ceram. Abstr.*, **28**(8), 197c (1949).
334. H. H. Gregor, *Brick Clay Rec.*, **117**(2), 63, 68 (1950); *Ceram. Abstr.*, **31**(10), 183f (1952).
335. H. H. Gregor, U.S. 2,538,867 (1951); *Ceram. Abstr.*, **30P**(9), 169b (1951).

336. E. J. Griffith, *J. Inorg. Nucl. Chem.*, **26**, 1381 (1964).
337. E. J. Griffith and R. L. Buxton, *Inorg. Chem.*, **4**, 549 (1965).
338. E. J. Griffith and R. L. Buxton, *J. Am. Chem. Soc.*, **89**(12), 2884 (1967).
339. R. Griffiths, *Curr. Ceram. Res. Bull. Am. Ceram. Soc.*, **52**(2), 224 (1973).
340. H. G. Grimm and Paul Huppert, *Ceramics,* **589**, 035 (1933); *Ceram. Abstr.*, **13P**(8), 207 (1934).
341. Hans Grimm and Paul Huppert, U.S. 2,227,082 (1940); *Ceram. Abstr.*, **20P**(3), 69 (1941).
342. J. F. Grimson, *Pottery Glass Rec.*, **19**(11), 291 (1937); *Ceram. Abstr.*, **17A**(2), 76 (1938).
343. N. P. Grishina and M. T. Mel'nik, *Stekloobraznoe Sostoyanie, Minsk,* **S63**(4), 74 (1964).
344. D. G. Grossman and C. J. Phillips, *J. Am. Ceram. Soc.*, **47**(9), 471 (1964); *Ceram. Abstr.*, **43**(10), 275g (1964).
345. H. Grunze and Eric Thilo, *Die Paperchromatographie der kondensierten Phosphate,* Akademie Berlin, 1955.
346. R. Haddan, *J. Soc. Glass Technol.*, **33**, 175 (1964).
347. Ionel Haiduc, *The Chemistry of Inorganic Ring Systems,* Vols. 1 and 2, Wiley, New York, 1970.
348. J. A. Halpin, *Ber. Dtsch., Keram. Ges,* **49**(7), 219 (1972); *Ceram. Abstr.*, **52**(4), 82i (1973).
349. D. P. Hamblen, R. A. Weidel, and G. E. Blair, *J. Am. Ceram. Soc.*, **46**(10), 499 (1963); *Ceram. Abstr.*, **42**(11), 310d (1963).
349a. C. S. Hanes and F. A. Isherwood, *Nature,* **164**, 1107 (1949).
350. K. W. Hansen, *J. Electrochem. Soc.*, **112**(10), 994 (1965); *Ceram. Abstr.*, **46**(11), 277j (1967).
351. K. W. Hansen and M. T. Splann, *J. Electrochem. Soc.*, **113**(9), 895 (1966); *Ceram. Abstr.*, **47**(8), 231g (1968).
352. Friederich Harders and Sigismund Kienow, *Feuerfestkunde: Herstellung, Eigenschaften and VerwendungFeuerfester Baustoffe,* Springer-Verlag, Berlin, 1960.
353. F. L. Harding, *Glass Technol.*, **13**(2), 43 (1972); *Ceram. Abstr.*, **51**(11), 268i (1972).
354. F. L. Harding and R. J. Ryder, *J. Can. Ceram. Soc.*, (39), 59 (1970); *Ceram. Abstr.*, **51**(8), 189d (1972).
355. H. Harper and P. W. McMillan, *Phys. Chem. Glasses,* **13**(4), 97 (1972).
356. H. Harper, P. F. James, and P. W. McMillan, *Discuss. Faraday Soc.*, (56), 206 (1970); *Ceram. Abstr.*, **52**(2), 35b (1973).
357. H. M. Harris and H. J. Kelly, *Bull. Am. Ceram. Soc.*, **37**(7), 307 (1958); *Ceram. Abstr.*, **37**(9), 237e (1958).
358. D. E. Harrison, H. A. McKinstry, and F. A. Hammel, *J. Am. Ceram. Soc.*, **37**(6), 277 (1954); *Ceram. Abstr.*, **33**(7), 134j (1954).
359. H. Hartmann and H. Hagermann, *Zement-Kalk-Grips,* **6**(3), 81 (1953); *Ceram. Abstr.*, **33**(1), 20b (1954).
360. W. Hauth and P. H. Crayton, *Curr. Ceram. Res. Bull. Am. Ceram. Soc.*, **52**(2), 223 (1973).
361. J. W. Haynes and K. H. Shaner, U.S. 3,607,371 (1971); *Ceram. Abstr.*, **51P**(1), 14c.
362. Pavel Hegner, *Silikaty,* **11**(1), 17 (1967); *Ceram. Abstr.*, **47**(10), 296b (1968).
363. H. Heide and U. Hoffmann, *Ber. Dtsch. Keram. Ges.*, **49**(6), 185 (1972); *Ceram. Abstr.*, **52**(2), 41b (1973).
364. R. P. Heilich, F. J. Rohr, and L. D. Hart, *Bull. Am. Ceram. Soc.*, **46**(7), 674 (1967); *Ceram. Abstr.*, **46**(8), 180g (1967).
365. Robert Heilmann, *Glastech. Ber.*, **43**(5), 183 (1970).
366. Erich Heinerth, U.S. 2,851,335 (1958); *Ceram. Abstr.*, **38P**(1), 26b (1959).
367. L. L. Hench, H. F. Paschall, M. Paschall, and J. McVay, *Bull. Am. Ceram. Soc.*, **52**(4), 432 (1973); *Am. Ceram. Soc. Prog.*, 7-S3-73.
368. J. R. Hensler, *J. Am. Ceram. Soc.*, **38**, 423 (1955).

369. J. R. Hensler, U.S. 3,118,788 (1964); *Ceram. Abstr.*, **43P**(5), 119d (1964).
370. J. R. Hensler, U.S. 2,999,339 (1956); *Ceram. Abstr.*, **41P**(3), 59g (1962).
370a. J. R. Hensler, *Advances in Glass Technology*, Part 2, Plenum, New York, 1963, p. 25.
371. J. R. Hensler, *Proc. VII Int. Congr. Glass, Brussels*, (July 1965).
372. H. J. Herbst and J. E. Lyon, U.S. 3,316,110 (1967); *Ceram. Abstr.*, **46**(8), 182C (1967).
373. L. Hermsdorf and R. Wagner, Ger. 166,672 (1904).
374. H. Hermsdorf and R. Wagner, Ger. 179,997 (1904).
375. P. G. Herold and J. F. Burst, *J. Missouri University School Mines Met., Bull. Tech*, Series 18(2), 34pp (1964); *Ceram. Abstr.*, **27B**(11), 246h (1948).
376. P. C. Hess, *Geochim. Acta 1971*, **35**(3), 289; *Phys. Chem. Glasses*, **12**(5), 1971, Abstr. 652.
377. M. D. Heyman, U.S. 2,865,426 (1958); *Ceram. Abstr.*, **38P**(3), 83c (1959).
378. Hans Heynemann, Ger. 520,138 (1927); *Ceram. Abstr.*, **10P**(8), 545 (1931).
379. M. S. R. Heynes and H. Rawson, *J. Soc. Glass Tech.*, **41**(203), 347 (1957); *Ceram. Abstr.*, **38**(6), 151j (1959).
380. Setsuo Higashi, Koji Morimoto, Akiya Satomura, Kozo Horie, Misaki Anzai, Masao Toriumi, Rokuro Yamamoto, and Taro Ozaki, *J. Nippon University School Dent.*, **11**(2), 55 (1969); *Ceram. Abstr.*, **49**(4), 88f (1970).
381. C. E. Higgins and W. H. Baldwin, *Anal. Chem.*, **27**, 1780 (1955).
382. W. L. Hill, G. T. Faust, and D. S. Reynolds, *Am. J. Sci.*, 242, 457, 542 (1944).
383. A. R. Hilton, Jr. and C. E. Jones, Jr., U.S. 3,370,964 (1968); U.S. 3,370,965 (1968); *Ceram. Abstr.*, **47P**(6), 170e (1968).
384. A. R. Hilton, Jr., C. E. Jones, Jr., and M. J. Brau, U.S. 3,338,728 (1967); *Ceram. Abstr.*, **46P**(11), 278f (1967).
385. Kosuke Hirano, Rihei Iida, and Koji Terasaki, *Rep. Imp. Fuel Res. Inst., Japan*, **17**, 56 (1936); *Abstr. Chem. Zent.*, **11**, 2423 (1936); *Ceram. Abstr.*, **16A**(2), 67 (1937).
386. Chikara Hirayama and N. T. Melamed, U.S. 3,549,554 (1970); *Ceram. Abstr.*, **50P**(4), 102c (1971).
387. Chikara Hirayama and M. M. Rutter, *J. Am. Ceram. Soc.*, **42**(8), 367 (1959); *Ceram. Abstr.*, **38**(9), 230e (1959).
388. Albert Hloch, Nikolay Medic, and Rudolf Kohlhaas, U.S. 3,650,683 (1972); *Ceram. Abstr.*, **51**(7), 175h (1972).
389. Albert Hloch, Rudolf Kohlhaas, Nikolay Medic, and Helmut Neises, U.S. 3,445,257 (1969); *Ceram. Abstr.*, **48P**(8), 274b (1969).
390. R. von Hodenberg, *Ber. Dtsch. Keram, Ges*, **49**(8), 243 (1972); *Ceram. Abstr.*, **52**(4), 91c (1973).
391. Josef Hoffmann, *Sprechsaal*, **70**(41), 517-519(42), 529-531 (1937); *Ceram. Abstr.*, **17A**(5), 180 (1938).
392. J. Hoffmann, *Glass*, **1A**(12), 519, 522 (1937); *Ceram. Abstr.*, **17**(5), 180 (1938), **18A**(1), 15 (1939).
393. J. G. Holley, U.S. 2,398,530 (1946); *Ceram. Abstr.*, **25P**(6), 104 (1946).
394. Stig Holmguist, *Amer. Ceram. Soc., Glass Div., Fall Meet.* (October 1966).
395. W. B. Honey, *Pottery Gaz.*, **53**(610), 644ii (1928); *Ceram. Abstr.*, **7A**(6), 381 (1928).
396. Kou Honma and Kouji Honma, *Nippon Kagaku Kaishi*, (5), 856 (1972); *Ceram. Abstr.*, **52**(6), 138b (1973).
397. H. P. Hood, *Glass Ind.*, **7**(12), 287 (1926); *Ceram. Abstr.*, **6**(2), 55 (1927).
398. M. F. Hoover and G. D. Sinkovitz, U.S. 3,713,859 (1973); *Ceram. Abstr.*, **52**(5), 105j (1973).
399. H. Hovestadt, *Jena Glass and Its Scientific and Industrial Applications*, MacMillan, London, 1902.
400. J. P. Hoxie and A. J. Werner, U.S. 3,113,033 (1963); *Ceram. Abstr.*, **43P**(4), 86i (1964).
401. H. F. Huettenlocher, *Z. Krist*, **90**, 508 (1935).

402. H. R. Hughan, *Ceram. Age,* **56**(2), 40 (1950).
403. W. Hugill and W. J. Rees, *Trans. Ceram. Soc. (Engl.),* **23**, 304-361 (1923); *Ceram. Abstr.,* **4**(10), 280 (1925).
404. Anna-Lisa Huhti and P. A. Gartaganis, *Can. J. Chem.,* **34**, 785 (1956).
405. S. F. Hulbert, J. J. Klawitter, and J. L. Bowman, *Mater. Res. Bull.,* 7(11), 1239 (1972); *Ceram. Abstr.,* **52**(3), 66a (1973).
406. F. A. Hummell, *J. Am. Ceram. Soc.,* **32**(10), 320 (1949); *Ceram. Abstr.,* **28**(11), 268f (1949).
406a. F. A. Hummel and E. C. Henry, *Penn State College, School Min. Ind., Memo Rept. 7* (August 1946), PB 60,660; *Ceram. Abstr.,* **28B**(2), 45a (1949).
407. F. A. Hummel and T. A. Kupinski, *J. Am. Chem. Soc.,* **72**, 5318 (1950).
408. P. A. Huppert, *Cer. Age,* **61**(2), 32 (1953); *Ceram. Abstr.,* **32**(9), 157e (1953).
408a. Paul Huppert and Hans Wolff, U.S. 2,077,481 (1934); *Ceram. Abstr.,* **16P**(7), 205 (1937).
408b. Paul Huppert and Hans Wolff, Ger. 620,347 (1935); *Chem. Abstr.,* **30**, 830 (1936); *Ceram. Abstr.,* **15P**(8), 237.
409. N. A. Hurt, R. Wellington, and L. H. Myles, *Ind. Chemist* (December 581, 1953).
410. I. G. Farbenind, A. G., Brit. 287,036 (1928); *Ceram. Abstr.,* **7P**(12), 815 (1928).
411. I. G. Farbenind, A. G., Fr. 751,524 (1933); *Ceram. Abstr.,* **13P**(10), 259 (1934).
412. I. G. Farbenind, A. G., Ger. 596,471 (1934), addition to 580,295; *Ceram. Abstr.,* **14P**(5), 113 (1935).
413. I. G. Farbenind, A. G., Ger. 654,925 (1935); *Chem. Ztg.,* **61**(103-104), 1018 (1937); *Ceram. Abstr.,* **17P**(9), 306 (1938).
413a. I. G. Farbenind, A. G., Ger. 634,698; 634,699; 636,035 (1936); *Ceram. Abstr.,* **16P**(8), 245 (1937).
414. R. K. Iler, U.S. 3,041,205 (1962); *Ceram. Abstr.,* **41P**(10), 234f (1962).
415. Institute of Mining and Metallurgy, London, *Chem. Eng. News,* **31**(15), 1575 (1953); *Ceram. Abstr.,* **32B**(9), 170C (1953).
416. I.U.P.A.C. Commission on the Nomenclature of Inorganic Chemistry, *Pure and Applied Chem.,* **28**, 27 (1971).
417. Tetsuro Izumitani and Isao Masudo, U.S. 3,751,272 (1973); *Ceram. Abstr.,* **52**(11), 245a (1973).
418. Tetsuro Izumitani and Siichi Toda, U.S. 3,656,976 (1972); *Ceram. Abstr.,* **51P**(8), 194b (1972).
419. C. Jackson, *Trans. Ceram. Soc. (Engl.),* **27**(3), 151 (1928); *Ceram. Abstr.,* **8A**(7), 481 (1929).
420. George Jackson, *Introduction to Whitewares,* Maclaren, London, 1969.
421. W. M. Jackson, II, U.S. 2,726,963 (1955); *Ceram. Abstr.,* **35P**(4), 79b (1956).
422. W. Jackson and A. D. Holdcroft, *Trans. Ceram. Soc. (Engl.),* **4**, 6 (1904-1905).
423. Walter Jahn, *Glastech. Ber.,* **39**(3), 118 (1966); *Ceram. Abstr.,* **46**(10), 252j (1967).
424. Walter Jahn, U.S. 3,455,707 (1969); *Ceram. Abstr.,* **48P**(10), 339f (1969).
425. Walter Jahn, *Ceram. Abstr.,* **50P**(9), 221j (1971).
426. R. F. Jameson, *J. Chem. Soc.,* 752 (February 1959).
427. A. Jobling, *Trans. Brit. Ceram. Soc.,* **67**(11), 511 (1968); *Ceram. Abstr.,* **48**(6), 228j (1969).
428. G. P. Johari and Martin Goldstein, *J. Chem. Phys.,* **53**(6), 2372 (1970); *Ceram. Abstr.,* **51**(7), 161d (1972).
429. G. H. Johnson, Ceramic-Metal Systems and Enamel Bibliography and Abstracts 1963-1964, *Am. Ceram. Soc.,* Columbus, Ohio, *Ceram. Abstr.,* **44B**(8), 246d (1965).
430. G. W. Johnson, Brit. 498,049 (1939); *Ceram. Abstr.,* **18P**(4), 100 (1939).
431. J. T. Jones and M. F. Berard, Ceramics-Indistrial Processing and Testing, Iowa State University Press, Ames, Iowa, 1972.

432. L. T. Jones, *Ind. Eng. Chem. Anal.*, **14**, 536 (1942).
433. Karl-Heinz Jost, *Makromol. Chem.*, **55**, 203 (1962).
434. Karl-Heinz Jost and Friederick Wodtke, *Makromol. Chem.*, **53**, 1 (1962).
435. Karl-Heinz Jost, Horst Worzala, and Eric Thilo, *Z. anorg. allgem. Chem.*, **325**(1-2), 98 (1963).
436. I. H. Joyce and W. E. Worrall, *Trans. Brit. Ceram. Soc.*, **69**(5), 211 (1970); *Ceram. Abstr.*, **51**(1), 9h (1972).
437. A. E. Junge, U.S. Patent 3,485,646 (1969).
438. Kazuya Kadota, Koji Kamiyama, and Teruo Sakaino, *Yokyo Kyokai Shi,* **80**(1), 1 (1972); *Ceram. Abstr.*, **52**(6), 123c (1973).
439. Kazuya Kadota, Shogo Saito, and Teruo Sakaino, *Yogyo Kyokai Shi,* **80**(5), 179 (1972); *Ceram. Abstr.*, **52**(8), 169b (1973).
440. H. H. Kaes, *Glastech. Ber.*, **45**(6), 234 (1972); *Ceram. Abstr.*, **52**(3), 61e (1973).
441. H. H. Kaes and Arthur Scharmann, *Glastech. Ber.*, **45**(5), 182 (1972); *Ceram. Abstr.*, **52**(3), 62a (1973).
442. S. Y. Kalliney, *Topics in Phosphorus Chemistry,* Vol. 7, Interscience, New York, 1972, p. 255.
443. Takafumi Kanazawa, Hiroshi Kawazoe, and Minoru Handa, *Kogyo Kyokai Shi,* **80**(7), 263 (1972); *Ceram. Abstr.*, **52**(10), 221h (1973).
444. Takafumi Kanazawa, Hiroshi Kawazoe, and Masayoshi Ikeda, *Yogyo Kyokai Shi,* 78(4), 121 (1970); *Ceram. Abstr.*, **51**(3), 57h (1972).
445. K. K. Kappmeyer and D. H. Hubble, *Bull. Am. Ceram. Soc.*, **51**(7), 568 (1972).
446. Editha Karl-Kroupa, *Anal. Chem.*, **28**(7), 1091 (1956); *Ceram. Abstr.*, **36**(3), 75i (1957).
447. Editha Karl-Kroupa, J. R. Van-Wazer, and C. H. Russell, *Scott's Standard Methods of Chemical Analysis,* Vol. 1, 6th ed., Chapter 35, Van Nostrand, New York, 1962.
448. Akio Kato, Reiko Nishibashi, Masamitsu Nagano, and Isao Mochida, *Ceram. Abstr.*, **51**(5), 123c (1972).
449. Waldemar Kaufmann and Everhard Bungartz, U.S. 2,031,958 (1936); *Ceram. Abstr.*, **15**(5), 149 (1936).
450. K. Kautz, *J. Am. Ceram. Soc.*, **28**, 88 (1945).
451. Hiroshi Kawazoe, Nasashi Hasegawa, and Takafumi Kanazawa, *Yogyo Kyokai Shi,* **80**(6), 251 (1972); *Ceram. Abstr.*, **52**(9), 209h (1973).
452. T. N. Kennedy, U.S. 3,743,587 (1973); *Ceram. Abstr.*, **52**(10), 222g (1973).
453. Frank Kerkhof, Robert Seeliger, and Walter Westphal, *Glastech. Ber.*, **28**(7), 262 (1955); *Ceram. Abstr.*, **36**(10), 231c (1957).
454. Robert Keul, U.S. 3,498,801 (1970); *Ceram. Abstr.*, **49P**(7), 168d (1970).
455. Yu. F. Kynykin, V. V. Timashev, Yu. S. Malinin, and V. P. Ryazin, *Trans. Mosk. Khim-Tekhnol. Inst.*, (68), 164 (1971); *Ceram. Abstr.*, **52**(9), 190h (1973).
456. L. B. Khoroshavin, P. N. Dyachkov et al., *Ogneupory,* **33**(3), 40 (1968); *Ceram. Abstr.*, **47**(9), 267g (1968).
457. Walther Kiaulehn, *Der Zug der 41 Glasmacher,* Jenaer Glasswerk Schott and Genossen, Mainz, 1959.
458. Peder Kierkegaard, Keji Eistrat, and Astrid Rosen-Rosenhall, *Acta Chem. Scand.*, **18**(10), 2237 (1964).
459. C. S. King, U.S. 2,370,472 (1945); *Ceram. Abstr.*, **24P**(4), 82 (1945).
460. D. F. King, U.S. 3,284,218 (1966); *Ceram. Abstr.*, **46P**(2), 44e (1967).
461. George King, U.S. 2,859,124 (1958); *Ceram. Abstr.*, **38**(2), 36a (1959).
462. W. D. Kingery, *J. Am. Ceram. Soc.*, **73**(8), 239 (1950); *ibid.*, 242,247; *Ceram. Abstr.*, **29**(9), 182i (1950).
463. W. D. Kingery, *J. Am. Ceram. Soc.*, **42**(1), 6 (1959); *Ceram. Abstr.*, **38**(2), 67g (1959).
464. D. L. Kinser, *Int. Comm. Glass Proc.*, **59** (1969).

465. D. L. Kinser, *J. Electrochem. Soc.*, **117**(4), 546 (1970); *Ceram. Abstr.*, **51**, 218i (1972).
466. H. P. Kirchner, U.S. 3,126,349 (1964); *Ceram. Abstr.*, **43P**(7), 192b (1964).
467. B. A. Kiselev, V. N. Bruevich, Yu. V. Zherdev, I. S. Deev, V. A. Kudishina, N. A. Rozdina, and A. Ya. Korolev, *Izo. Akad. Nauk SSSR Neorg. Mater.*, 7(3), 472 (1971); *Ceram. Abstr.*, **52**(6), 123e (1973).
468. D. Kitchen and W. J. Skinner, *J. Appl. Chem. Biotechnol.*, **21**(2), 53 (1971); *Ceram. Abstr.*, (11), 268f (1971).
469. H. Klause, *Keram. Z.*, 8(7), 331 (1956); *Ceram. Abstr.*, **36**(1), 136 (1957).
470. A. A. Klein, *Bur. Stand. Tech. Paper No. 80* (December 8, 1916).
471. Ya. V. Klyucharov and L. I. Skoblo, *Zh. Prikl. Khim.*, **38**(3), 520 (1965); *Ceram. Abstr.*, **47**(7), 200i (1968).
472. J. B. P. Knanishu, Rep. 133953, U.S. Gov't. Dept. *Res. Rep.*, **30**(5), 340 (1958); *Ceram. Abstr.*, **40**(9), 205i (1961).
473. O. Knapp, *Glashütte*, **60**, 755 (1930); *Ceram. Abstr.*, **A**(10), 760 (1931).
474. Erik Knudsen, *Kolloid-Z.*, **69**(1), 35 (1934); *Ceram. Abstr.*, **15**(7), 202 (1936).
475. Keyi Kobayashi and Ryosuke Yokota, *Yogyo Kyokai Shi*, **72**(7), 116 (1964); *Ceram. Abstr.*, **46**(5), 118e (1967).
476. Taneo Kobayashi, *Kogyo Kyokai Shi*, **74**(6), 190 (1966); *Ceram. Abstr.*, **48**(1), 37e (1969).
477. Kodak Ltd., Brit. 649,512; 649,513; 649,733; 649,734 (1950); *Ceram. Abstr.*, **31P**(2), 23b (1952).
478. J. H. Koenig and W. H. Earhart, *Literature Abstracts of Ceramic Glazes* (1951).
479. F. Kohlrausch, *An. Phys.*, **44**, 577 (1891).
480. E. Kordes, *Glastech. Ber.*, **17**(3), 65-76 (1939).
481. E. Kordes, *Abstr. Angew. Chem.*, **53**(41-42), 480 (1940); *Ceram. Abstr.*, **21A**(1), 10 (1942).
482. E. Kordes, *Z. Phys. Chem.*, **B50**, 194 (1941); *Ceram. Abstr.*, **22A**(8), 135 (1943).
483. E. Kordes, W. Vogel, and Ruth Felerowsky, *Z. Electrochem.*, **57**(4), 282 (1953); *Ceram. Abstr.*, **32**(11), 189g (1953).
484. Ernst Kordes and Rudolf Nieder, *Glastech. Ber.*, **41**(2), 41 (1968); *Ceram. Abstr.*, **49**(8), 190e (1970).
485. G. Körrn, *Keram Rundsch.*, **35**, 463,581,671 (1927); *Ceram. Abstr.*, **8A**(11), 789 (1929).
486. K. Koyanagi, *J. Soc. Chem. Ind. (Japan)*, **33**(7), 276 (1930); *Tonind. Ztg.*, **54**(98), 1534 (1930); *Ceram. Abstr.*, **10A**(3), 171 (1931).
487. H. M. Kraner, *Bull. Am. Ceram. Soc.*, **50**(7), 598 (1971).
488. Otto Krause and Schlegelmilsch, *Ber. Dtsch. Keram. Ges*, **13**(10), 437 (1932); *Ceram. Abstr.*, **12A**(2), 68 (1933).
489. N. J. Kreidl, U.S. 2,544,460 (1951); *Ceram. Abstr.*, **31P**(2), 22h (1952).
490. N. J. Kreidl, *III, Int. Congr. Glass, Venice*, **1**(7), 263 (1953).
491. N. J. Kreidl, No bsr 63242 NE01551 Task 3, 2, 1953.
492. N. J. Kreidl, Prelim. Rep. High level Dosimeter G1 (1954). Final Rep. No bsr 57016; NE051551, 2, 3 57010; NE051551, 3.7.
493. N. J. Kreidl, *Tooley's Handbook of Glass Manufacture*, Ogden, 1960.
494. N. J. Kreidl, *Verres Réfract.*, **25**(4/5), 155 (1971).
495. N. J. Kreidl and G. E. Blair, *Nucleonics*, **14**(3), 82 (1956).
496. N. J. Kreidl and W. Colbert, *J. Opt. Soc. Am.*, **35**(1), 731 (1945).
497. N. J. Kreidl and J. R. Hensler, *J. Am. Ceram. Soc.*, **38**(12), 423 (1955).
498. N. J. Kreidl and E. Lell, *Interaction of Radiation with Solids*, Plenum, New York, 1967.
499. N. J. Kreidl, J. R. Hensler, and G. E. Blair, *Irradiat. Damage Glass Confr.*, AT930-1-1321 (1954-57).
500. N. J. Kreidl and W. A. Weyl, *J. Am. Ceram. Soc.*, **24**(11), 372 (1941).

500a. N. J. Kreidl and W. A. Weyl, *Glass Ind.*, **23**(9), 335(10), 384(11), 426, 441(12), 465, 480 (1942).
501. E. R. Kreidler and F. A. Hummel, *Inorg. Chem.*, **6**, 524 (1967).
502. E. R. Kreidler and F. A. Hummel, *Inorg. Chem.*, **6**(5), 884 (1967); *Ceram. Abstr.*, **48**(11), 386e (1969).
503. E. R. Kreidler and F. A. Hummel, *Am. Min.*, **55**(1-2), 170 (1970); *Ceram. Abstr.*, **49**(7), 181f (1970).
504. W. W. Kriegel and Hayne Palmour, Eds., *Materials Science Research,* Vol 5, Plenum New York, 100 1971; Ceramics in Severe Environments, *Ceram. Abstr.*, **51**(8), 212 (1972).
505. Kristoffersen, *Tid. Kjemi Bergwes. Met.*, **11**(4), 46 (1951); *Ceram. Abstr.*, **30**(9), 167g (1951).
506. E. L. Krivovazov and N. K. Voskresenskaya, *Izv. Akad. Nauk SSSR Neorg. Mater.*, **5**(10), 1734 (1969); *Ceram. Abstr.*, (51), 239a (1972).
507. I. A. Kryzhanovskaya, T. Yu. Shchetkina, and Yu. L. Svirskaya, *Tsem.*, (6), 10 (1971); *Ceram. Abstr.*, **52**(2), 32d (1973).
508. G. V. Kukolev and M. Y. Mel'nik, *Dokl. Akad. Nauk SSSR,* **132**(1), 168 (1960); *Ceram. Abstr.*, **43**(7), 186b (1964).
509. S. Kumar, *Cent. Glass Ceram. Res. Inst. Bull. (India),* **6**(1), 13 (1959); *Ceram. Abstr.*, **39**(9), 207e (1960).
510. S. Kumar, *Cent. Glass Ceram. Res. Inst. Bull. (India),* 7(3), 117 (1960); *Ceram. Abstr.*, **40**(7), 161a (1961).
511. D. Kumar, R. G. Ward, and D. J. Williams, Discuss. *Faraday. Soc.*, **32**, 147 (1961).
512. E. Kupzog, M. Koltermann, and P. Bartha, *Ber. Dtsch. Keram. Ges.*, **44**(9), 445 (1967); *Ceram. Abstr.*, **47**(7), 200g (1968).
513. C. R. Kurkjian, *Phys. Rev.*, **133**, 723 (1964).
514. C. R. Kurkjian and E. A. Sigety, *Phys. Chem. Glasses,* **9**(3), 73 (1968); *Ceram. Abstr.*, **48**(1), 6d (1969).
515. C. R. Kuzell, *Amer. Inst. Min. Met. Eng. Tech. Pub. 995 Met. Tech.*, **6**(2), (1939); *Ceram. Abstr.*, **18A**(6), 155 (1939).
516. Hsu Kwang-tze, Chen Shu-ming, and Tang K'o You-chi, *Hsuch Tung Pao,* 1963, No. 252; *Sci. Abstr. China Chem. Technol.*, **2**(1), 4 (1964); *Ceram. Abstr.*, **44**(4), 129a (1965).
517. L. V. Labutina, N. M. Pavlushkin, and M. V. Artamonova, *Trans. Mosk. Khim. Tekhnol. Inst.*, (63), 3 (1969); *Ceram. Abstr.*, **52**(1), 5d (1973).
518. J. A. Laird and C. G. Bergerson, *J. Am. Ceram. Soc.*, **53**(9), 482 (1970).
519. B. M. Landis, Thesis, University of Toledo (1956); supervision, W. Haller.
520. R. J. Landry, *J. Chem. Phys.*, **48**(3), 1422 (1968); *Ceram. Abstr.*, **48**(1), 7b (1969).
521. R. J. Landry, J. T. Fournier, and C. G. Young, *J. Chem. Phys.*, **46**(43), 1285 (1967); *Ceram. Abstr.*, **47**(5), 127i (1968).
522. R. P. Langguth, R. K. Osterheld, and E. Karl-Kroupa, *J. Phys. Chem.*, **60**, 1335-1336 (1956).
523. K. R. Laud and F. A. Hummel, *J. Amer. Ceram. Soc.*, **54**(8), 407 (1971).
524. M. Lauegt, M. Scory, and A. Durif, *Mat. Res. Bull.*, **3**(12), 963 (1968); *Ceram. Abstr.*, **48**(9), 331f (1969).
525. J. N. Lee and H. W. Moos, *Solid State Commun.*, **9**(13), 1139 (1971); *Ceram. Abstr.*, **52**(9), 212d (1973).
526. J. G. A. Lefranc, U.S. 2,099,367 (1937); *Ceram. Abstr.*, **17**(2), 77 (1938).
527. Eberhard Lell, N. J. Kreidl, and J. R. Hensler, *Progress Ceramic Science,* Vol. 4, Pergamon, Oxford, 1966.
528. C. W. Lentz, M.S.S., January 28, 1964, Dow Corning Corp., *Inorg. Chem.*, **3**, 574 (1964).
528a. J. H. Letcher and J. R. Van Wazer, *Topics in Phosphorus Chemistry,* Vol. 5, Interscience, New York, 1967, p. 75.

529. E. M. Levin, C. R. Robbins, and H. F. McMurdie, *Phase Diagrams for Ceramists,* Am. Ceram. Soc. (Diagrams 1-2066) (1964), Supplement (Diagrams 2067-4149) (1969).
530. M. E. Levina and B. S. Shershev, *Zh. Prikl. Khim. (Leningrad),* **41**(9), 2068 (1968).
531. S. R. Levitt, P. H. Crayton, E. A. Monroe, and R. A. Condrate, *J. Biomed. Mater. Res.,* **3**, 683 (1969).
532. R. M. Levy and J. R. Van Wazer, *J. Chem. Phys.,* **45**, 1824 (1966).
533. Friedrich Liebau, *Fortshr. Miner., Stuttgart,* **42**(2), 266 (1966).
534. Siegfried Lindenbaum, T. V. Peters, and William Riemann III, *Anal. Chim. Acta,* **11**, 530 (1954).
535. Stig Lindroth, *Glas. Emaill. Keram. Tech.,* **4**(1), 1 (1953); English translation *ibid.*(4), 117; *Ceram. Abstr.,* **32**(9), 158e (1953).
536. Johannes Löffler, *Glastech. Ber.,* **12**(10), 332 (1934); *Ceram. Abstr.,* **16A**(5), 146 (1937).
537. J. Löffler, Ger. 693,743 (1939).
537a. Bernard Long, U.S. 1,749,823 (1930); *Ceram. Abstr.,* **9P**(7), 522 (1930).
538. R. A. Long, U.S. 3,131,073 (1964); *Ceram. Abstr.,* **43P**(8), 230i (1964).
539. G. F. Lynch, M. Sayer, S. L. Segel, and G. Rosenblatt, *J. Appl. Phys.,* **42**(7), 2587 (1971); *Ceram. Abstr.,* **51**(5), 105h (1972).
540. T. C. MacAvoy, U.S. 3,342,572 (1967); *Ceram. Abstr.,* **47**(1), 9d (1968).
541. J. D. MacKenzie, *Modern Aspects of the Vitreous State,* Vol. 1, Butterworth, Washington, D. C., 1960; *Ceram. Abstr.,* **40B**(8), 201f (1961).
542. G. H. Mackintosh, W. F. Ford, and J. White, *Curr. Ceram. Res. Bull., Am. Ceram. Soc.,* **52**(2), 237 (1973).
543. B. L. Majunder, *Chem. Age. (India),* Ser. 7, 105 (April 1953); *Ceram. Abstr.,* **32**(11), 194a (1953).
544. Iwao Maki and Yuichi Suzukawa, *Yogyo Kyoshi Shi.,* **71**(3), 54 (1963); *Ceram. Abstr.,* **43**(4), 103a (1964).
545. T. G. Manasevich, *Steklo Mat. Trudy Inst. Stekla,* (1), 79 (1967); *Ceram. Abstr.,* **49**(7), 167c (1970).
546. R. Marcus, G. Lukas, F. Pollack, and L. L. Hench, *Curr. Cer. Res. Bull., Am. Ceramic Soc.,* **52**(2), 233 (1973).
546a. A. J. P. Martin and R. L. M. Synge, *Biochem.,* **35**, 91 (1941).
547. A. Martinek and F. A. Hummel, *J. Am. Ceram. Soc.,* **53**(3), 159 (1970); *Ceram. Abstr.,* **49**(4), 106e (1970).
548. M. Martusevicius and M. Martynaitis, *Liet., T. S. R. Aukst. Mokyklu Mosklo Darb., Chem. Chem. Technol.,* **6**, 195 (1965); *Ceram. Abstr.,* **45**(1), 2b (1966).
549. F. Massazza, *Ceramica,* (1), 43 (1954); *Ceram. Abstr.,* **33**(5), 93f (1954).
550. C. R. Masson, *J. Am. Ceram. Soc.,* **51**, 134 (1968).
551. C. R. Masson, *J. Iron Steel Inst. (London),* **210**, 89-96 (1971); *Chem. Met. Iron Steel, The Iron and Steel Institute, (London),* **3**, 11 (1973).
552. C. R. Masson, I. B. Smith, and S. G. Whiteway, *Can. J. Chem.,* **48**(1), 201 (1970).
553. C. R. Masson, I. B. Smith, and S. G. Whiteway, *Can. J. Chem.,* **48**(9), 1456 (1970).
554. Seiichi Matsuno, Toshiki Miyahashi, and Jumpei Ando, *Kogyo Kogoku Zasshi,* **70**(10), 1638 (1967); *Ceram. Abstr.,* **49**(3), 82i (1970).
555. D. W. Matula, L. C. D. Groenweghe, and J. R. Van Wazer, private communications; also published in part in *J. Chem. Phys.,* **41**(10), 3105 (1964).
556. M. A. Matveev and A. I. Rabukhim, *Keram. Z.,* **16**(12), 770 (1964); *Ceram. Abstr.,* **44**(6), 172d (1965).
557. M. A. Matveev, L. G. Khodskii, G. K. Fizyuk, A. I. Bolutenko, and L. S. Strugach, *Izv. Akad. Nauk SSSR Neorgan, Mater.,* **2**(6), (1966); *Ceram. Abstr.,* **47**(2), 38h (1968).
558. M. A. Matveev, L. G. Khodskii, and G. K. Fisyuk, *Izv. Akad. Nauk SSSR Neorg. Mater.,* **4**(1), 163 (1968); *Ceram. Abstr.,* **49**(10), 244e (1970).

559. M. A. Matveev, L. G. Khodskii, and G. K. Fisyuk, *Inorg. Mater. Consultants Bur. Transl.*, **4**(1), 135 (1968); *Phys. Chem. Glass, Abstr.*, **790**, 12(5) (1971).
560. M. A. Matveev, M. T. Mel'nik, and M. P. Glasova, *Steklo Mat. Keram.*, **22**(12), 12 (1965); *Ceram. Abstr.*, **46**(8), 176i (1967).
561. Robert Mazelsky and R. C. Ohlmann, U.S. 3,504,300 (1970); *Ceram. Abstr.*, **49P**(7), 173c (1970).
562. Robert Mazelsky and R. C. Ohlmann, U.S. 3,617,937 (1971); *Ceram. Abstr.*, **51P**(3), 65h (1972).
563. Stasys Mazilianskas, *Ceram. Age*, **40** (April 1958).
564. J. L. O. Mazo and J. M. F. Navarro, *Bol. Soc. Esp. Ceram.*, **9**(6), 721 (1970); *ibid.*, **10**(1), 37; *Phys. Chem. Glasses, Abstr.*, **12**(5), 655 (1971); 789, *ibid.*
565. J. L. O. Mazo and J. M. F. Navarro, *Bol. Soc. Esp. Ceram.*, **10**(3), 325 (1971); *Ceram. Abstr.*, **51**(4), 80g (1972).
566. T. R. Meadowcroft and F. D. Richardson, *Trans. Faraday Soc.*, **59**(487), 1564-1571 (1963); *Ceram. Abstr.*, **46**(5), 136i (1967).
567. T. R. Meadowcroft and F. D. Richardson, *Trans. Faraday Soc.*, **61**(505), 54 (1965); *Ceram. Abstr.*, **45**(1), 5d (1966).
568. J. W. Mellor, *Trans. Eng. Ceram. Soc.*, **36**, 1 (1936-1937).
569. L. Merker and H. Wondratschek, *Silikat.*, 8(9), 373 (1957); *Ceram. Abstr.*, **37**(10), 293b (1958).
570. Kazuyuki Miki, *Osaka Kogyo Gijustu Skikenjo Kiho*, **5**, 50 (1954); *ibid.*, 166; *Ceram. Abstr.*, **38**(1), 13c (1959).
571. G. Miller, G. Piotrowski, and W. C. Allen, *Curr. Ceram. Res. Bull., Am. Ceram. Soc.*, **52**(2), 233 (1973).
572. R. E. Miller, U.S. 3,427,174 (1969); *Ceram. Abstr.*, **48**(5), 178h (1969).
573. K. Moedritzer, G. M. Burch, J. R. Van Wazer, and H. K. Hofmeister, *Inorg. Chem.*, **2**, 1152 (1963).
574. Bernard Moore, *Trans. Ceram. Soc. (Engl)*, **5**, 37 (1905-1906).
575. S. Moore and S. N. Prasad, *J. Soc. Glass Tech.*, **33**(155), 336T (1949); *Ceram. Abstr.*, **29**(10), 199i (1950).
576. G. W. Morey, *Ind. Eng. Chem.*, **17**, 389-392 (1925).
577. G. W. Morey, *Sci. Mon.*, *XLII*, 541 (1936).
578. G. W. Morey, *Properties of Glass*, Reinhold, New York, 1954.
579. G. W. Morey and Earl Ingerson, *Am. J. Sci.*, **242**(1), 1 (1944); *Ceram. Abstr.*, **23A**(7), 129 (1944).
580. D. W. Morgan, U.S. 3,746,556; *Ceram. Abstr.*, **52**(10), 222f (1973).
581. Yoshi Mori and Akira Noda, *Bull. Osaka Ind. Res. Inst.*, **4**(4), 181 (1953); *Ceram. Abstr.*, **34**(1), 5h (1955).
582. Yoshiro Moriya and Tsutomu Ueno, *Osaka Kogyo Gijutsu Shikensho Kiho.*, **15**(1), 27 (1964); *Ceram. Abstr.*, **43**(9), 247e (1964).
583. Taro Moriya, Yoji Akao, and Naobumi Hatano, *Yogyo Kyokai Shi*, **68**(774), 145 (1960); *Ceram. Abstr.*, **40**(9), 208b (1961).
584. A. J. Mountvala, *Bull. Am. Ceram. Soc.*, **52**(5), 479 (1973).
585. R. L. Mozzi and B. E. Warren, *J. Appl. Cryst.*, **2**(4), 164 (1969); *Ceram. Abstr.*, **49**(4), 106d (1970).
586. R. L. Mozzi and B. E. Warren, *J. Appl. Cryst.*, **3**(3), 251 (1970).
587. Arnulf Maun and E. F. Osborn, *Phase Equilibria Among Oxides in Steel-Making*, Addison Wesley, Reading, Mass., 1965; *Ceram. Abstr.*, **44B**(7), 221f (1965).
588. Gerd Mueller, *Glastech. Ber.*, **45**(6), 221 (1972); *Ceram. Abstr.*, **52**(3), 61g (1973).
589. J. Mukerji, *Glass Technol.*, **13**(5), 135 (1972); *Ceram. Abstr.*, **52**(6), 122f (1973).
590. J. Mukerji and S. R. Biswas, *Glass Techn.*, **12**(4), 107 (1971); *Ceram. Abstr.*, **51**(2), 32e

(1972).

591. Klaus-Peter Müller, *Glastech. Ber.*, **42**(3), 83-89 (1969).

592. B. Mulroy, *Ceramics,* **17**, 208 (1966).

593. Motosuke Munakata, *Solid-State Electron,* **1**(3), 159 (1960).

594. Motsuke Munakata, Susumu Kawamura, Jumpei Asahara, and Masajiro Iwamoto, *Kogyo Kyokai Shi,* **67**(766), 344 (1959); *Ceram. Abstr.*, **39**(11), 253h (1960).

595. C. J. Munter, Can. 445,880 (1947); *Ceram. Abstr.*, **27P**(9), 204a (1948).

596. Keiichi Murakami and Hirobumi Tanaka, *Sekko to Sekkai,* **61**, 273 (1962); *Ceram. Abstr.*, **42**(2), 35d (1963).

597. Keiichi Murakami, Hirobumi Tanaka, and Bukan Sudo, *Sekko to Sekkai,* **54**, 207 (1961); *Ceram. Abstr. II,* **41**(3), 54a (1962); *VI,* **41**(3), 54b (1962).

598. M. K. Murthy, *J. Am. Ceram. Soc.*, **44**(8), 412 (1961); *Ceram. Abstr.*, **40**(10), 237d (1961).

599. M. K. Murthy, *J. Am. Ceram. Soc.*, **46**(11), 558 (1963).

600. M. K. Murthy, *O.R.F. Phosphate Res. Rep. 63-1,* internal report, Thermal expansion properties of sodium phosphate glasses.

601. M. K. Murthy and J. Ciric, *Can. Uranium Res. Found. Rep. No. R-8* (1961); *Ceram. Abstr.*, **41**(10), 237c (1962).

602. M. K. Murthy and A. Mueller, *J. Am. Ceram. Soc.*, **46**(11), 530 (1963); M. K. Murthy, *ibid.*, 558; *Ceram. Abstr.*, **43**(1), 3j (1964); I, II, **43**(9), 248a (1964).

603. M. K. Murthy and A. E. R. Westman, *J. Am. Ceram. Soc.*, **45**(9), 401 (1962); *Ceram. Abstr.*, **41**(10), 235b (1962).

604. M. K. Murthy and A. E. R. Westman, *J. Am. Ceram. Soc.*, **49**(6), 310-311 (1966).

605. M. K. Murthy, M. L. Smith, and A. E. R. Westman, *J. Am. Ceram. Soc.*, **44**(3), 97 (1961); *Ceram. Abstr.*, **40**(4), 84i (1961).

606. R. L. Myuller, *Electrical Conductivity of Vitreous Substances,* Plenum, New York, p. 10011.

607. G. J. McCarthy and Maribeth Lovette, *Bull. Am. Ceram. Soc.*, **51**(8), 655 (1972).

608. R. A. McCauley and F. A. Hummel, *Trans. Brit. Ceram. Soc.*, **67**(12), 619 (1968); *Ceram. Abstr.*, **48**(6), 236b (1969).

609. R. A. McCauley and F. A. Hummel, *J. Electrochem. Soc. Solid State Sci.*, **118**(5), 755 (1971).

610. G. H. McClellan and J. R. Lehr, *Am. Min.*, **54**(9-10), 1374 (1969); *Ceram. Abstr.*, **49**(3), 79d (1970).

611. J. F. McCullough, J. R. Van Wazer, and E. J. Griffith, Jr., *Am. Chem. Soc.*, 78, 4528 (1956).

612. W. T. McDaniel, Jr., and P. N. Sales, U.S. 2,760,879 (1956); *Ceram. Abstr.*, **36P**(1), 15e (1957).

613. P. W. McMillan and Graham Partridge, U.S. 3,647,489 (1972); *Ceram. Abstr.*, **51**(6), 132g (1972).

614. J. McVay, M. Rodebush, M. Paschall, H. A. Paschall, W. C. Allen, G. Piotrowski, and L. L. Hench, *Curr. Ceram. Res. Bull. Am. Ceram. Soc.*, **52**(2), 233 (1973).

615. F. Nadachowski, *Pr. Inst. Hutn.*, **10**(5), 237 (1958); *Abstr. J. Iron Steel Inst. (London),* **192**(2), 176; *Ceram. Abstr.*, **38**(9), 238c (1959).

616. B. Nador, *Steklo Mat. Keram.*, **17**(10), 18 (1960).

617. B. Nador, *Acta Chim. Acad. Sci. Hung.*, **40**(1), 1 (1964) (in English); *Ceram. Abstr.*, **44**(1), 24i (1965).

618. Shoichiro Nagai and Jumpei Audo, *J. Ceram. Assoc. Japan.*, **61**(680), 64 (1953); *Ceram. Abstr.*, **32**(6), 104j (1953).

619. Masamitsu Nagano, Isao Mochida, Akio Kato, and Tetsuro Seiyama, *Yogyo Kyokai Shi,* **79**(8), 270 (1971); *Ceram. Abstr.*, **51**(6), 131j (1972).

620. Kinnosuke Nagaoka and Mikio Hara, *Osaka Kogyo Gijutsu Shikensho Kiho,* **11**(2), 115 (1960); *Ceram. Abstr.,* **43**(5), 118c, d (1964).
621. Kinnosuke Nagaoka, Mikio Hara, and Hirokichi Tanaka, *Osaka Kogyo Gijutsu Shikensho Kiho,* **12**(2), 48, 150 (1961); **12**(3), 292 (1961), and **13**(2), 105 (1962).
622. Yasuo Nakai, *Bull. Chem. Soc. Japan.,* **38**(8), 1308 (1965); *Ceram. Abstr.,* **46**(5), 118i (1967).
623. Hiroshi Nakamura and Noboru Kawai, *Funtaioyobi Funma Tsuyakin,* II, **6**(1), 25 (1959); *Ceram. Abstr.,* **40**(1), 15e (1961). For Part I., *Ceram. Abstr.,* **39**(7), 163e (1960).
624. Akira Naruse, *Yogyo Kyokai Shi,* **69**(790), 351 (1961); *Ceram. Abstr.,* **46**(4), 92h (1967).
625. Akira Naruse, Yoshihiro Abe, and Hiroyoshi Inoue, *Yokyo Kyokai Shi,* **76**(2), 36 (1968); *Ceram. Abstr.,* **49**(10), 243b (1970).
626. National Lead Co., Brit. 724,374 (26.8.52 U.S.); Appl. 24.8.53; Pub. 16.2.55.
627. Louis Navias, U.S. 2,292,151 (1942).
628. T. Negas and R. S. Roth, *J. Res. Natl. Bur. Stand.,* **72A**(6), 783 (1968); *Ceram. Abstr.,* (6), 233j (1969).
629. K. D. Nekrassov and G. N. Alexandrova, *Bull. Soc. Fr. Ceram.,* (98), 21 (1973); *Ceram. Abstr.,* **52**(11), 246j (1973).
630. V. Netuka, *Chem. pruomysl,* **7**, 181 (1957); *Abstr. J. App. Chem. (London),* 8(8), ii-157 (1958); *Ceram. Abstr.,* **38P**(2), 36j (1959).
631. J. D. Nickerson, U.S. 2,906,632 (1959); *Ceram. Abstr.,* **39P**(3), 62f (1960).
632. P. Nicolardot, *Rev. Gen. Colloid,* **4**, 9, 39 (1926); *Chem. Ind.,* **45B**, 407 (1926); *Ceram. Abstr.,* **5A**(8), 246 (1926).
633. D. E. Niesz, *Bull. Am. Ceram. Soc.,* **52**(2), 211 (1973).
634. R. S. Mijjhar and D. J. Williams, *Trans. Faraday Soc.,* **64**(7), 1784 (1968); *Ceram. Abstr.,* **48**(9), 330j (1969).
635. B. K. Niklewski and R. H. Ashby, *Inst. Vitreous Enamellers Bull.,* **4**(5), 7 (1954); *Ceram. Abstr.,* **37**(9), 228j (1958).
636. T. R. Nissle and C. L. Babcock, *J. Am. Ceram. Soc.,* **56**(11), 596 (1973).
637. G. C. Nook and R. W. Ricker, U.S. 2,397,195 (1946); *Ceram. Abstr.,* **25P**(5), 83 (1946).
638. R. W. Nurse, *J. Appl. Chem. (London),* **2**(12), 708 (1952); *Ceram. Abstr.,* **34**(9), 156e (1955).
639. W. Oelsen and H. Maetz, *Mitt. Kaiser Wilhelm Inst. Eisenforsch, Düsseldorf,* **23**(12), 195 (1941); *Abstr. Iron Steel Inst. (London),* **145**(1), 179A (1942); *Ceram. Abstr.,* **23A**(11), 200 (1944).
640. Nobruyo Ogita, U.S. 3,743,492 (1973); *Ceram. Abstr.,* **52**(10), 222a (1973).
641. E. Z. Ognyanova and G. L. Kramer, *Tsement,* **30**(3), 4 (1964); *Ceram. Abstr.,* **46**(9), 215d (1967).
642. M. J. O'Hara, J. J. Duga, and H. D. Sheets, *J. Bull. Amer. Ceram. Soc.,* **51**(7), 590 (1972); *Ceram. Abstr.,* **51**(8), 196C (1972).
643. Shigeru Ohashi and Fumio Oshima, *Bull. Chem. Soc. Japan,* **36**(11), 1489 (1963).
644. Shigeru Ohashi and H. Sugatami, *Bull. Chem. Soc. Japan,* **30**, 864 (1957).
645. Shigeru Ohashi and J. R. Van Wazer, *J. Am. Chem. Soc.,* **81**, 830 (1959).
646. Hirosha Ohba, Keisuke Hiragushi, and Tomoyasu Kitamara, *Taikabutsu,* **23**(167), 554 (1971); *Ceram. Abstr.,* **52**(1), 13h (1973).
647. R. K. Osterheld, *Topics in Phosphorus Chemistry,* Vol. 7, Interscience, New York, 1972, pp. 103-255.
648. A. J. Owen, Reier Visser, and Jacobus van Laar, U.S. 3,708,317 (1973); *Ceram. Abstr.,* **52**(4), 84a (1973).
649. B. B. Owens and S. W. Mayer, *J. Am. Ceram. Soc.,* **47**(7), 347 (1964); *Ceram. Abstr.,* **43**(8), 236a (1964).
650. M. Palfreyman, *Bull. Am. Ceram. Soc.,* **49**(7), 638 (1970).

651. Roman Pampuch and Joanna Gallus-Olender, *Pol. Akad. Nauk Oddzial Krakowie Pr. Kom. Ceram. Ceram.*, (17), 77 (1971); *Ceram. Abstr.*, **52**(4), 916 (1973).
652. G. Paoletti and F. Giammanco, *Verres Réfract.*, **23**(6), 679 (1969); *Ceram. Abstr.*, **51**(7), 160e (1972).
653. Y. W. Park, *J. Korean Ceram. Soc.*, **7**(3), 83 (1970); *Ceram. Abstr.*, **52**(9), 192e (1973).
654. C. W. Parmelee, *Trans. Am. Ceram. Soc.*, **8**, 236 (1906).
655. C. W. Parmelee, *Ceram. Glazes, Indus. Pub.*, 1st ed. (1948); C. W. Parmelee 3rd ed. Completely revised and enlarged by C. G. Harman, Cahners Publishing Co. Boston, copyright 1973.
656. G. Partridge and P. W. McMillan, *Glass Tech.*, **4**(6), 173 (1963); *Ceram. Abstr.*, **43**(6), 155d (1964).
657. E. P. Partridge, Victor Hicks, and G. W. Smith, *J. Am. Chem. Soc.*, **63**(2), 454 (1941); *Ceram. Abstr.*, **21A**(2), 51 (1942).
658. A. M. Patashinsky, Bachelor's Thesis, Massachusetts Institute of Technology, 1939, Harvard College Library, *Breakdown Voltage of Phosphate Glass.*
659. Linus Pauling and J. Sherman, *Miner. Petrog. Abt. A.Z. Krist.*, **96**, 481 (1937); *Chem. Abstr.*, **31**, 829 (1937); *Ceram. Abstr.*, **19A**(5), 126 (1940).
659a. A. E. Pavlish, U.S. 2,745,759 (1956); *Ceram. Abstr.*, **35P**(10), 207d (1956).
660. N. M. Pavlushkin, Ts. N. Gurevich, and L. I. Nikishina, *Trans. Mosk. Khim. Tekhnol. Inst.*, (59), 36 (1969); *Ceram. Abstr.*, **50**(6), 145h (1971).
661. B. M. Pearson, *Ceram. Age.*, **10**(3), 92 (1927).
662. B. M. Pearson, *Ceramic Age,* Part II, **54**(5), 325 (1949).
663. Walter Peek and Volker Koehne, *Sprechsaal Keram. Glas. Emaill. Silik.*, **100**(21), 830 (1967), **101**(12), 507 (1968); *Ceram. Abstr.*, **49**(1), 9e (1970).
664. Werner Pepperhoff, *Arch. Eissenhuttenwes.*, **29**(3), 153 (1958); *Ceram. Abstr.*, **37**(9), 253h (1958).
665. J. B. Peri, *Faraday Soc. Discuss.*, No. 52, 55 (September 1971).
666. Alvin Perloff, *J. Am. Ceram. Soc.*, **39**(3), 83-88 (1956); *Ceram. Abstr.*, **35**(4), 851 (1956).
667. P. O. Perron and H. B. Bell, *Trans. Brit. Ceram. Soc.*, **66**(8), 347 (1967); *Ceram. Abstr.*, **47**(5), 151a (1968).
668. C. H. Perry, L. K. Wilson, and D. L. Kinser, *Bull. Am. Ceram. Soc.*, **52**(4), 380 (1973); *Am. Ceram. Soc. Prog.*, 15-G-73.
669. T. V. Peters and William Rieman III, *Anal. Chim. Acta,* **14**, 131 (1956).
669a. Juergen Petzolt and Herwig Scheidler, U.S. 3,642,504 (1972); *Ceram. Abstr.*, (5), 107g (1972).
670. B. Z. Pevzner, S. G. Dzhavuktsyan, and V. E. Mishel, *Steklo Mat. Keram.*, (4), 17 (1972); *Ceram. Abstr.*, **52**(2), 35c (1973).
671. I. Peychès, *Trans. Soc. Glass Technol.*, **36**, 164 (1952).
672. C. J. Phillips, *Am. Sci.*, **53**(1), 20 (1965).
673. S. V. Phillips and P. W. McMillan, *Glass Technol.*, **6**(2), 46 (1965); *Ceram. Abstr.*, **44**(8), 228g (1965).
674. A. G. Pincus, U.S. 2,359,789 (1944).
675. A. G. Pincus, *Bull. Am. Ceram. Soc.*, **24**(1), 40 (1945); *Sci. Illust.*, **3**(1), 30, 61 (1948); *Ceram. Abstr.*, **28**(5), 129e (1949).
675a. A. G. Pincus, reference by Bacon. See 674.
676. A. G. Pincus, *Ceram. Age.*, **53**(5), 260-264 (1949).
677. F. X. Pink and V. Lyn, *Electrochem. Technol.*, **6**(7-8), 258 (1968); *Ceram. Abstr.*, **48**(4), 120j (1969).
678. A. R. Pitrot, U.S. 2,822,285 (1958); *Ceram. Abstr.*, **37P**(5), 129d (1958).
679. W. A. Plotnikov and E. M. Notansson, *Zh. Prikl. Khim.*, **6**, 839 (1933); *Ceram. Abstr.*, **14A**(2), 53 (1935).

680. Gary Pollard, Norman Smyrl, and J. P. Devlin, *J. Phys. Chem.*, **76**(13), 1826 (1972).
681. J. P. Poole, *Int. Comm. Glass, Proc.*, 169 (1969).
682. J. P. Poole, H. C. Snyder, and M. A. Boschini, U.S. 3,607,172 (1971); *Ceram. Abstr.*, **51P**(1), 5f (1972).
683. F. M. Precopio and J. R. Bateman, *Ceram. Age,* **77**(2), 38 (1961); *Ceram. Abstr.*, **41**(3), 77h (1962).
684. Erhard Preuser, *Silikattech,* **12**(2), 81 (1961); *Ceram. Abstr.*, **41**(10), 239 (1962).
685. G. H. Price, W. I. Stuart, and D. G. Walker, *J. Am. Ceram. Soc.*, **51**(9), 481 (1968); *Ceram. Abstr.*, **47**(10), 319c (1968).
686. P. Prideau and R. Martin, *Curr. Cer. Res. Bull. Am. Ceram. Soc.*, **52**(2), 223 (1973).
687. J. D. Provance and D. C. Wood, *J. Am. Ceram. Soc.*, **50**(10), 516 (1967).
688. Pyrene Co. Ltd., Brit. 1,176,066 (1970); *Ceram. Abstr.*, **P50**(1), 5d (1971).
689. O. T. Quimby, *J. Phys. Chem.*, **58**, 603 (1954).
690. E. M. Rabinovitch and S. I. Sil'vestrovich, *Zh. Prikl. Khim.*, **32**(8), 1690 (1959); *Ceram. Abstr.*, **41**(8), 187i (1962).
691. Bernard Raistrick, *J. Roy Coll. Sci.*, **XIX**, 9-27 (1949).
692. Bernard Raistrick, *Faraday Soc. Discuss.*, **5**, 234 (1949).
693. Bernard Raistrick, *Chem. Ind.*, 408-4-14, 10 May (1952).
694. Atma Ram, A. K. Bose, and S. Kumar, *J. Sci. Ind. Res. (India),* **13B**(3), 217 (1954); *Ceram. Abstr.*, **36**(1), 7a (1957).
695. Atma Ram, A. K. Bose, and S. Kumar, *J. Sci. Ind. Res. (India),* **15B**(2), 78 (1956).
696. C. E. Ramsden, *Trans. Eng. Ceram. Soc.*, **12**, 239 (1912-1913).
696a. J. T. Randall, H. P. Rooksby, and B. S. Cooper, *J. Soc. Glass Technol.*, **14**(54), 219 (1930).
697. I. L. Rashkovan, L. N. Kuziminskaya, and V. A. Kopeikin, *Izv. Akad. Nauk SSSR Neorg. Mater.*, **2**(3), 541 (1966); *Ceram. Abstr.*, **47**(2), 46j (1968).
698. E. H. Ratcliffe, *Glass Technol.*, **4**(4), 113 (1963); *Ceram. Abstr.*, **43**(1), 4f (1964).
699. H. Rawson, *Trans. IV Congrès. Int. Verre,* 62 (1956).
700. M. Regan and C. F. Drake, *Mater. Res. Bull.*, **7**(12), 1559 (1972); *Ceram. Abstr.*, **52**(4), 81d (1973).
701. Friedrich Reinhart, *Glas. Emaill. Keram. Tech.*, **13**(1), 11 (1962); *Ceram. Abstr.*, **41**(8), 197g (1962).
702. Renata Reisfeld and L. Boehm, *J. Solid State Chem.*, **4**(3), 417 (1971); *Ceram. Abstr.*, **52**(1), 7c (1973).
703. R. Reisfeld and Y. Eckstein, *J. Solid State Chem.*, **5**(2), 174 (1972); *Ceram. Abstr.*, **52**(2), 35A (1973).
704. Renata Reisfeld, J. Hormodaly, and Baruch Barnett, *Chem. Phys. Lett.*, **17**(2), 248 (1972); *Ceram. Abstr.*, **52**(10), 219g (1973).
705. R. W. Repsher, U.S. 3,023,337 (1962); *Ceram. Abstr.*, **41P**(7), 165e (1962).
706. E. C. Rhodes and P. E. Gainsbury, U.S. 2,681,860 (1954); *Chem. Abstr.*, **49**(7), 4962f (1955); *Ceram. Abstr.*, **37P**(9), 240f (1958).
707. R. N. Rickles, *J. Appl. Chem.*, **15**(2), 74 (1965); *Ceram.Abstr.*, **46**(10), 248d (1967).
708. E. F. Riebling, *Can. J. Chem.*, **42**, 2811 (1964).
709. E. F. Riebling, *Rev. Int. Hautes Temp. Réfract.*, **4**(1), 65-76 (1967); *Ceram. Abstr.*, **47**(1), 7d (1968).
710. E. F. Riebling, P. E. Blaszyk, and D. W. Smith, *J. Am. Ceram. Soc.*, **50**(12), 641 (1967).
711. E. F. Riebling, *J. Am. Ceram. Soc.*, **51**(3), 143 (1968); *Ceram. Abstr.*, **47**(4), 93d (1968).
712. J. G. Riess and J. R. Van Wazer, *Inorg. Chem.*, **5**(2), 178 (1966); *Ceram. Abstr.*, **47**(5), 150h (1968).
713. G. E. Rindone and R. J. Ryder, *Glass Ind.*, **38**(1), 29, 51 (1957); *Ceram. Abstr.*, **40**(2), 32C (1961).
714. G. J. Roberts, *Bull. Am. Ceram. Soc.*, **52**(4), 383 (1973); *Amer. Ceram. Soc. Progr.*,

26-G-73.

715. C. C. Robinson and J. T. Fournier, *J. Phys. Chem. Solids,* **31**(5), 895 (1970); *Ceram. Abstr.,* **52**(4), 80h (1973).

716. H. A. Robinson, *J. Am. Ceram. Soc.,* **52**(7), 392 (1969).

717. T. I. Rodina, *Steklo Mat. Keram.,* **26**(1), 24 (1969); *Ceram. Abstr.,* **50**(1), 12i (1971).

718. B. N. Rolfe, R. F. Miller, and I. S. McQueen, *U.S. Geol. Survey Profess. Paper,* 1960, No. 334-G, U.S. Gov't Printing Office, Washington, D. C.; *Ceram. Abstr.,* **39**(11), 268d (1960).

719. John Romig and Bruce Kester, *Rock Prod.,* **56**(5), 64 (1953); *Ceram. Abstr.,* **32**(8), 135C (1953).

720. H. P. Rooksby, *J. Soc. Glass Technol.,* **23**, 76 (1939); *Ceram. Abstr.,* **21A**(8), 167 (1942).

721. H. P. Rooksby, *Analyst,* 77(920), 759 (1952); *Ceram. Abstr.,* **32**(4), 58f (1953).

722. E. Ross and R. Snyder, *Curr. Ceram. Res. Bull. Am. Ceram. Soc.,* **52**(2), 223 (1973).

723. J. J. Rothermel, K-H Sun, and Alexander Silverman, *J. Am. Ceram. Soc.,* **32**(5), 153 (1949).

724. R. Roy, *J. Am. Ceram. Soc.,* **43**(12), 670 (1960).

725. O. Ruff, C. Friedrich, and E. Ascher, *Z. Angew. Chem.,* **43**(42), 1081-1087 (1930); *Ceram. Abstr.,* **10A**(6), 409 (1931).

726. Juraj Ryba, Otokor Korab, and Juraj Jirku, *Stavivo,* **48**(6), 167 (1970); *Ceram. Abstr.,* **51**(3), 53j (192), **51**(9), 214f (1972).

727. Ya. I. Ryskin and G. P. Stabitskaya, *Opt. Spectrosc.,* 8, 320-324 (1960).

728. Hiroshi Saito, Kazuhiro Goto, and Mayumi Someno, *Tetsu to Hagane,* **55**(7), 539 (1969); *Ceram. Abstr.,* **49**(1), 23i (1970).

729. Yoshiaki Sanada and Kiyoshi Miyazawa, *Sekko to Sekkai,* **45**, 56-59 (1960); *Ceram. Abstr.,* **39**(7), 154i (1960).

730. J. N. Sandoe, P. H. Sarkies, and S. Parke, *J. Phys.,* **D5**(10), 1788 (1972); *Ceram. Abstr.,* **52**(5), 98e (1973).

731. M. Sayer and I. G. Austin, *Can. Ceram. Soc. Progr., 72nd. Ann. Mtg.* (February 1974).

732. M. Sayer and G. F. Lynch, *Can. Ceram. Soc. 1973 Ann. Meet., Toronto.*

733. M. Sayer, A. Mansingh, J. M. Reyes, and G. Rosenblatt, *J. Appl. Phys.,* **42**(7), 2857 (1971).

734. H. F. Schaake and L. L. Hench, *J. Non-cryst. Solids.,* **2**, 292 (1970); *Ceram. Abstr.,* **51**(6), 132d (1972).

735. Rudolf Schmidt, *Sprechsaal,* **71**(48), 585 (1938); *Ceram. Abstr.,* **18A**(7), 178 (1939).

736. H. Schönborn, *Silik. Tech.,* **10**, 390-400 (1959).

737. Willi Schröder, Ger. 490,803, *Tonind. Ztg.,* **54**(34), 579 (1930); *Ceram. Abstr.,* **9**(12), 1024 (1930).

738. Ingeborg Schulz and Wilhelm Hinz, *Glastech. Ceram.,* **29**(8), 319 (1956); *Ceram. Abstr.,* **37**(1), 12a (1958).

739. C. P. Schumacher, *Nonmet. Min. Process.,* 3(12), 22 (1962); *Ceram. Abstr.,* **42**(11), 302e (1963).

740. Bruno Schweig, *Glass,* **19**(12), 315 (1942); **20**(1), 5 (1943); *Ceram. Abstr.,* **22A**(5), 70 (1943).

741. Schwiete, *Euro-Ceram.,* 9(6), 151 (1959); *Ceram. Abstr.,* **40**(1), 23a (1961).

742. R. D. Shannon and A. L. Friedberg, *University of Illinois, Eng. Exp. Sta. Bull. # 456* (1960).

743. H. D. Sheets, Jr., and M. J. O'Hara, U. S. 3,619,151 (1971).

744. H. D. Sheets, Jr., J. J. Bulloff, and W. H. Duckworth, *Brick Clay Rec.,* **133**(1), 55-57 (1958).

745. I. N. Sheiko, E. B. Gitman, and V. Ya. Lichenko, *Ukr. Khim. Zh.,* **38**(4) 305, (1972), *Ceram. Abstr.,* **51**(8), 210e (1972).

746. H. F. Shermer, G. F. Rynders, G. W. Cleek, and Ronald Hubbard, *J. Res. Natl. Bur. Stand.*, **52**(5), 251-258 (1954); *Ceram. Abstr.*, **38**(1), 9a (1959).
747. V. I. Shishkina, P. N. Babin, and P. A. Koka, *Izv. Akad. Nauk. Kazakh,* SSSR *Ser. Gorn. Dela, Met. Stroimat.* (5), 169 (1955); *Ceram. Abstr.*, **35**(8), 168f (1956).
748. N. I. Shutov and N. K. Dertev, *Steklomat.* (1) 77 (1969); *Ceram. Abstr.*, **50**(9), 220f (1971).
749. R. W. Sicka, U. S. 3,625,723 (1971); *Ceram. Abstr.*, **51P**(3), 60b (1972).
750. Alexander Silverman, J. J. Rothermel, and K.-H. Sun, U. S. 2,518,194. (1950); *Ceram. Abstr.*, **30P**(4), 64e (1951).
751. R. E. Simanovskoya and S. Ya. Shpunt, *Dokl. Akad. Nauk. SSSR,* 101(5), 917 (1955); *Ceram. Abstr.*, **35**(4), 70g (1956).
752. J. J. Sipe and O. J. Whittemore, Jr., *Bull. Am. Ceram. Soc.*, **51**(2), 214 (1972). Current ceramic research item.
753. J. J. Sipe and O. J. Whittemore, Jr., *Curr. Ceram. Res. Bull. Am. Ceram. Soc.*, **52**(2), 238 (1973).
754. Anne Skancke and Peder Kierkegaard, *Ark. Kem.*, **27**(3), 197 (1967); *Ceram. Abstr.*, **47**(6), 169i (1968).
755. L. V. Skopina, N. M. Pavlushkin, and Ts. N. Gurevich, *Trans. Mosk. Khim. Tekhnol. Inst.*, (63), 67 (1969); *Ceram. Abstr.*, **52**, 6a (1973).
756. M. J. Smith, *Chem. Rep. No. 5805,* Ontario Research Foundation, Sheridan Park, Ontario (1958).
757. M. J. Smith, *Anal. Chem.*, **31**, 1023 (1959).
758. I. B. Smith and C. R. Masson, *Can. J. Chem.*, **49**(5), 683 (1971).
759. A. Smits and A. J. Rutgers, *Chem. Soc. (London),* **75**, 2573 (1924).
760. D. M. Smyth, T. B. Tripp, and G. A. Shirn, *J. Electrochem. Soc.*, **113**(2), 100 (1966); *Ceram. Abstr.*, **47**(7), 207b (1968).
761. Elias Snitzer, *Bull. Am. Ceram. Soc.*, **52**(6), 516 (1973).
762. Elias Snitzer, R. F. Woodcock, and J. Segne, *IEEE J. Quant. Electron.*, QE, **4**(5), 360 (1968).
763. E. H. Snow and B. E. Deal, *J. Electrochem. Soc.*, **113**(3), 263 (1966); *Ceram. Abstr.*, **47**(7), 207f (1968).
764. S. Solacolie, P. Balta, G. Jitianu, and C. Spuraciu, *Epitoayag,* **25**(3), 94 (1973); *Ceram. Abstr.*, **52**(10), 221C (1973).
765. South African Iron & Steel Ind. Corp. Ltd., B. P. 898,303, App. 14.4.59, Pub. 6.6.62.
766. Helen Sprackling, *Craft Horizons,* **12**(5), 20 (1952); *Ceram. Abstr.*, **32**(1), 1 (1953).
767. L. Springer, *Sprechsaal,* **76**, 182 (1943); *Ceram. Abstr.*, **25A**(9), 157 (1946).
768. J. E. Stanworth, U. S. 2,390,191 (1945); *Ceram. Abstr.*, **25P**(2), 34 (1946).
769. J. E. Stanworth, *J. Soc. Glass. Technol.*, **30**, 381 (1946).
770. J. E. Stanworth, U. S. 2,441,853 (1948); *Chem. Abstr.*, **42**, 5192c.
771. J. E. Stanworth and W. E. S. Turner, *J. Soc. Glass Technol.*, **21**(87), 368 (1937); *Ceram. Abstr.*, **17A**(9), 304 (1938).
772. J. Staron, *Ber. Dtsch. Keram. Ges.*, **46**(7), 369 (1969); *Ceram. Abstr.*, **49**(3), 68g (1970).
773. H. Stein, *Vortr. Fachausschussber. I D.G.G.* Würzburg, Germany (April 27, 28, 1961).
774. H. H. Steinour, *Pit and Quarry,* **50**(3), 93-101 (1957); *Ceram. Abstr.*, **37**(4), 86j (1958).
775. J. M. Stevels, *Verres Réfract.*, (4-14 February 1948); *Ceram. Abstr.*, **27**(?), 222 (1948).
776. J. M. Stevels, *Proc. XI Congr. Pure Appl. Chem.*, **5**, 519 (1947).
777. J. M. Stevels, *NRC Conf. Non-Crystalline Solids,* Alfred, New York (September 3-5, 1958).
778. J. M. Stevels, *Philips Techn. Rev.*, **22**,(9/10), 281 (1960-1961).
779. J. M. Stevels and H. J. L. Trap, *Int. Glass Congr.*, Munich (1959).

779a. P. E. Stone, E. P. Egan, and J. R. Lehr, *J. Am. Ceram. Soc.*, **39**(3), 89 (1956).
780. S. D. Stookey, U. S. 2,559,805 (1951); *Ceram. Abstr.*, **31P**(4), 61d (1952).
781. S. D. Stookey, Can. 480,668 (1952); *Ceram. Abstr.*, **32P**(2), 23g (1953).
782. S. D. Stookey, *Int. Sci. Technol.*, **40** (July 1962).
783. P. D. S. St. Pierre, *J. Am. Ceram. Soc.*, **37**(6), 243 (1954); *Ceram. Abstr.*, **33**(7), 127E (1954).
784. P. D. S. St. Pierre, *J. Am. Ceram. Soc.*, **38**(6), 217 (1955); *Ceram. Abstr.*, **34**(7), 126f (1955).
785. P. D. S. St. Pierre, *J. Am. Ceram. Soc.*, **39**(4), 147 (1956); *Ceram. Abstr.*, **35**(5), 105f (1956).
786. P. D. S. St. Pierre, *J. Am. Ceram. Soc.*, **39**(10), 361 (1956); *Ceram. Abstr.*, **35**(1), 250i (1956).
787. P. E. Stone, E. P. Egan, Jr., and J. R. Lehr, *J. Am. Ceram. Soc.*, **39**(13), 389 (1956); *Ceram. Abstr.*, **35**(5) 105f (1956).
788. I. N. Stranski, and R. Kaischew, *Phys. Z.*, **36**., 393 (1935).
789. U. P. Strauss and T. L. Treitler, *J. Am. Chem. Soc.*, 77, 1473 (1955); ibid, 78, 3553 (1956).
790. Rene' Strub, U. S. 3,067,135 (1962); *Ceram. Abstr.*, **42P**(3), 69c (1963).
791. P. E. Sturrock, *Diss. Abstr.*, **21**, 2090-2091 (1961).
792. L. G. Sudakas and L. M. Sinickina, *Izv. Akad. Nauk SSSR Neorg. Mater.*, 7(3), 537 (1971); *Ceram. Abstr.*, **52**(6), 121a (1973).
793. Kuan-Han Sun, Ching-Tan Chen-Tsai, and Yueh Shan, U. S. 3,373,116 (1968); *Ceram. Abstr.*, **47P**(6), 171e (1968).
794. K-H Sun and M. L. Huggins, U. S. 2,415,661 (1947); *Ceram. Abstr.*, **26P**(4), 84b (1947).
795. K-H Sun and N. J. Kreidl, *Glass Ind.*, **33**(10), 511 (11), 589 (12), 651 (1952)., *Ceram. Abstr.*, **32**(11), 187i (1953).
796. V. D. Svikis, *Bull. Am. Ceram. Soc.*, **38**(5), 264 (1959); *Ceram. Abstr.*, **38**(7); 183h (1959).
797. V. D. Svikis and J. G. Phillips, *Bull. Am. Ceram. Soc.*, **35**(8), 305 (1956); *Ceram. Abs.*, **35**(10), 213g (1956).
798. Svec, J. J., *Crockery Glass J.*, **145**(6), 114 (1949); *Ceram. Abstr.*, **31**(5), 90j (1952).
799. M. Sychev, V. I. Korneev, N. S. Sirazitdinov, A. F. Yugai, M. S. Yanovskaya, G. N. Kas'yanova, and V. K. Balitskii, *Tsem.* (8), 7 (1969); *Ceram. Abstr.*, **51**(8), 188a (1972).
800. Sylvania Electric Products, Inc., Australia, 140,322 (1946); *Ceram. Abstr.*, (2), 23e (1952).
801. Z. M. Syritskaya, *Proc. All. Union Conf. Glassy State, 3rd, Leningrad* (English translation), 1956/1960, 295-298.
802. Z. M. Syritskaya and E. S. Kutukova, *Steklomat.* (3), 34 (1968); *Ceram. Abstr.*, **50**(1), 8h (1971).
803. Z. M. Syritskaya and G. N. Sergeeva, *Steklomat. Trudy Inst. Stekla,* No. 2 30 (1965); *Ceram. Abstr.*, **45**(8), 211f (1966).
804. Z. M. Syritskaya and N. F. Shapovalova, *Steklomat. Inform. Mat. Inst. Stekla,* No. 1, 1-6 (1964); *Ceram. Abstr.*, **44**(10), 277a (1965).
805. Z. M. Syritskaya and N. F. Shapolova, *Steklomat. Trudy Inst. Stekla,* (2), 111 (1966); *Ceram. Abstr.*, **47**(10), 299f (1968).
806. Z. M. Syritskaya and V. V. Yakubik, *Steklomat. Keram.*, **17**(2), 18 (1960); *Ceram. Abstr.*, **41**(8), 186f (1962).
807. A. A. Tabikh and F. M. Miller, *Cem. Concr. Res.*, **1**(6), 663 (1971); *Ceram. Abstr.*, **51**(5), 104a (1972).
808. Kentaro Takahashi, Kenji Iida, Yuichi Watanabe, and Maki Oguchi, *J. Ceram. Assoc. Japan*, **61**(690), 621 (1953); *Ceram. Abstr.*, **33**(5), 87c (1954).

809. K. Takahashi, K. Iida, Y. Wattanabe, and M. Oguchi, *Bull. Fac. Engg.*, Yokohama National University, **4**, 69-76 (1955).
810. G. Tamman and G. Z. Bandel, *Z. anorg. allgem. Chem.*, **192**(2), 139 (1930); *Ceram. Abstr.*, **10A**(3), 179 (1931).
811. M. F. Taragin, J. C. Eisenstein, and W. Haller, *Phys. Chem. Glasses*, **13**(5), 149 (1972); *Ceram. Abstr.*, **52**(6), 124f (1973).
812. J. Tatom and D. L. Kinser, *Curr. Ceram. Res., Bull. Amer. Ceram. Soc.*, **52**(2), 238 (1973).
813. V. J. Tennery and T. D. Driskell, *Bull. Am. Ceram. Soc.*, **52**(4), 430 (1973); *Am. Ceram. Soc. Prog.* 6-S3-73.
814. R. L. Thakur, *Cent. Glass Ceram. Res. Inst. Bull.*, **10**(2), 51 (1963).
815. Eric Thilo, *Adv. Inorg. Chem. Radiochem.*, **4**, 1 (1962).
816. Eric Thilo and R. Sauer, *J. Prakt. Chem.*, **4**, 324-348 (1957).
817. Eric Thilo and U. Schülke, *Angew. Chem., Int. Ed.*, **2**(12), 742 (1963).
818. Eric Thilo and W. Wieker, *Z. anorg. allgem. Chem.*, **277**, 27 (1954).
819. C. L. Thompson and C. W. Parmelee, *J. Am. Ceram. Soc.*, **22**(5), 170 (1939); *Ceram. Abstr.*, **18A**(6), 155 (1939).
820. J. E. Thomsen, U. S. 1,792,200 (1931); *Ceram. Abstr.*, **10P**(5), 328.
821. R. T. Thomson, *Analyst*, **61**, 320 (1936); *Ceram. Abstr.*, **15A**(11), 323 (1936).
822. B. V. Tien, Y. M. Butt, and M. A. Vorob'eva, *Trans. Mosk-Teckhnol. Inst.* (63), 165 (1969); *Ceram. Abstr.*, **52**(1), 1j (1973).
823. E. D. Tillyer, *J. Opt. Soc. Am.*, **28**(1-4) (1938); *Ceram. Abstr.*, **17A**(11), 351 (1938).
824. E. D. Tillyer, H. R. Moulton, and T. A. Gunn, U. S. 2,226,418 (1940).
825. A. N. Tiwari and E. C. Subbarao, *J. Am. Ceram. Soc.*, **53**(5) 258 (1970); *Ceram. Abstr.*, **49**(6), 137g (1970).
826. S. K. Tobia and N. E. Milad, *J. Chem. U.A.R.*, **6**(1), 11 (1963); *Ceram. Abstr.*, **45**(6), 165a (1966).
827. S. Toda, *Ger.* 2,024,613 (1970).
828. I. A. Tokman and G. A. Bukhalova, *Zh. Neorg. Khim.*, **15**(3), 881 (1970); *Ceram. Abstr.*, **52**(1), 26a (1973).
829. I. A. Tokman and G. A. Bukhalova, *Zh. Neorg. Khim.*, **17**(2), 544 (1972); *Ceram. Abstr.*, **52**(9), 213b (1973).
830. G. W. Toop and C. S. Samis, *Trans. Met. Soc. AIME*, **224**,878 (1962).
831. H. J. L. Trap, Netherlands 6,601,167 (1966), 6,612,854 (1966), 6,807,093 (1968), and 7,012,522 (1970).
832. H. J. L. Trap and J. M. Stevels, *Glastech. Ber.*, **32K**, V1, 31-52 (1959).
833. H. J. L. Trap and J. M. Stevels, Verres *Réfract.*, **16**, 337 (1962).
834. H. J. L. Trap and J. M. Stevels, *Verres Réfract.*, **25**, 176 (1971).
835. W. S. Treffner and A. H. Foessel, U. S. 3,522,063 (1970); *Ceram. Abstr.*, **49**(11), 279g (1970).
836. G. Trömel, *Mitt. Kaiser Wilhelm Inst. Eisenforsch., Düsseldorf*, **63**, 21 (1943); *Abstr., J. Iron Steel Inst. (London)*, 147(1), 158A (1943); *Ceram. Abstr.*, **23A**(11), 202 (1944).
837. G. Trömel and W. Olsen, *Arch. Eisenhüttenwes.*, **23**, 17-20 (1952); *Chem. Abstr.*, 25.6.52/5496.
838. G. W. Troop and C. S. Samis, *Can. Met. Quart.*, **1**(2), 129 (1962).
839. A. H. Truesdell and A. M. Pommer, *Science*, **142** (3597), 1292 (1963); *Ceram. Abstr.*, **43**(3), 55b (1964).
840. L. A. Tseitlin and T. E. Tarasova, *Ogneupory*, **29**(4), 177 (1964); translated in *Refract. J.*, **40**(9), 361 (1964); *Ceram. Abstr.*, **46**(4), 107i (1967).
841. Takayuki Tsubaki and Masami Tanaka, *Yogyo Kyokai Shi*, **80**(3), 105 (1972); *Ceram. Abstr.*, **52**(7), 148e (1973).

842. Takayuki Tsubaki, Makoto Hattori, and Masami Tanaka, *Kolloid N.Z.Z. Polym.*, **225** (2), 112 (1968); *Ceram. Abstr.*, **48**(4), 97f (1969).
843. E. T. Turkdogan and W. R. Maddocks, *J. Iron Steel Inst.*, **172**, 6 (1952).
844. W. E. S. Turner, *J. Soc. Chem. Ind.*, **43**, 1262-1263 (1924); *Ceram. Abstr.*, **4**(3), 64 (1925).
845. W. E. S. Turner, *J. Soc. Glass Technol.* 38, 436 (1954): **38**, 445; **40**, 39; **40**, 162; **40**, 277; **43**, 262 (1959).
846. E. J. Tuthill, Gerald Strickland, and G. G. Weth, *Ind. Eng. Chem. Process Des. Dev.*, **8**(1), 36 (1969; *Ceram. Abstr.*, **48**(5), 178a (1969).
847. E. J. Tuthill, G. G. Weth, L. C. Emma, Gerald Strickland, and L. P. Hatch, *Ind. Eng. Chem. Process Des. Dev.*, **6**(3), 314 (1967); *Ceram. Abstr.*, **46**(10), 253a (1967).
848. M. E. Tuvell, U. S. 3,379,541 (1968); *Ceram. Abstr.*, **47P**(7), 204d (1968).
849. E. Umblia, Glastek. Tidskr., **12**(1), 5 (1957); *Ceram. Abstr.*, **36**(6), 129j (1957).
850. E. Umblia, Tidskr. Glastek., **18**(5), 122 (1963); *Ceram. Abstr.*, **43**(2), 29d (1964).
851. N. M. Vaisfel'd and E. M. Rabinovich, *Zh. Prikl. Khim.*, **35**,(11), 2393 (1962); *Ceram. Abstr.*, **43**(3), 54d (1964).
852. H. A. Van Bakel and T. A. M. Gast, U. S. 2,843,504 (1958); *Ceram. Abstr.*, **38P**(1), 10g (1959).
853. J. R. Van Wazer, *J. Am. Chem. Soc.*, 72, 644 (1950).
854. J. R. Van Wazer. *Rock Prod.*, **56**(12), 104 (1953); *Ceram. Abstr.*, **33**(4), 63b (1954).
855. J. R. Van Wazer, *Phosphorus and Its Compounds*, Vol. I, Interscience, New York, 1958.
856. J. R. Van Wazer, *Monsanto Technol. Rev.*, **4**(1), 11 (1959).
857. J. R. Van Wazer, *Bull. Soc. Chim Fr. 5e Sér.* (4-6), 1732 (1968).
858. J. R. Van Wazer, *Industrial Chemistry and Technology of Phosphorus and Phosphorus Compounds–A Survey*, Wiley, New York, 1969.
859. J. R. Van Wazer, *Kirk-Othmer Encyc. Chem. Techn.*, Vol. 15, pp. 232-276: *Phosphoric Acids and Phosphates*, Interscience, New York.
860. J. R. Van Wazer and E. Besmurtnuk, *J. Phys. Colloid Chem.*, **54**, 89 (1950); *Ceram. Abstr.*, **32**(2), 33i (1953).
860a. J. R. Van Wazer and E. J. Karl-Kroupa, *J. Am. Chem. Soc.*, **78**, 1772 (1956).
861. J. R. Van Wazer and K. A. Holst, *J. Am. Chem. Soc.*, **72**(2), 639 (1950).
862. J. R. Van Wazer and J. H. Letcher, in *Topics in Phosphorus Chemistry*, Vol. 5, Interscience, New York, 1967, p. 169.
863. J. R. Van Wazer and S. Ohashi, *J. Am. Chem. Soc.*, **80**, 1010 (1958).
864. J. R. Van Wazer, C. F. Callis, J. N. Shoolery, and R. C. Jones, *J. Am. Chem. Soc.*, **58**, 5715 (1956).
865. J. R. Van Wazer, E. J. Griffith, and J. F. McCullough, VII, *J. Am. Chem. Soc.*, 77(2), 287 (1955); *Ceram. Abstr.*, **36**(2), 48j (1957).
866. V. V. Vargin, *Technology of Enamels*, Maclaren, London, 1967.
867. V. V. Vargin and T. S. Tsekhomskoya, *Zh. Prikl. Khim.*, **33**(12), 2633 (1960); *Ceram. Abstr.*, **41**(9) 208g (1962).
868. C. D. Vassas, *IX Int. Congr. Glass*, Versailles, Section BIII (1971).
869. J. G. Vaughan and L. K. Wilson, *Curr. Ceram. Res., Bull. Am. Ceram. Soc.*, **52**(2), 238 (1973).
870. J. G. Vaughan, L. K. Wilson, and D. L. Kinser, *Bull. Am. Ceram. Soc.*, **52**(4), 384 (1973); *Am. Ceram. Soc. Prog.*, 35-G-73.
871. J. H. Veale, U. S. 2,805,174 (1957); *Ceram. Abstr.*, **37P**(1), 18c (1958).
872. J. H. Veale and H. F. West, U. S. 2,802,750 (1967); *Ceram. Abstr.*, **36P**(11), 271c (1957).
873. E. I. Ved, E. F. Zharov, V. K. Bocharov, and A. P. Overko, *Izv. Vyssh. Ucheb. Zaved. Khim. Khim. Tekknol.*, **13**(8), 1193 (1970); *Ceram. Abstr.*, **50**(4), 103j (1971).

874. T. I. Veinberg, *Zh. Prikl., Khim.*, **32**(8), 1685 (1959); *Ceram. Abstr.*, **41**(8), 186b (1962).
875. A. S. Vernay, *Arts. Decor.*, **41**(2), 10 (1934); *Ceram. Abstr.*, **13A**(8), 215 (1934).
876. Hans Vetter, *Euro-Ceram.* 8(5), 119 (1958); *Ceram. Abstr.*, **37**(9), 248b (1958).
877. R. Visser, *Tonind. Ztg. Keram. Rundsch.*, **96**(7), 182 (1972); *Ceram. Abstr.*, **52**(4), 82h (1973).
878. Hans Vogel, U. S. 3,720,505 (1973); *Ceram. Abstr.*, **52**(7), 147h (1973).
879. M. Volarovich and D. Tolstoi, *C. R. Acad. Sci.* (USSR), **269** (1932); *Ceram. Abstr.*, **13A** (4), 102 (1934); *Sprechsaal*, **66**,(48), 823 (1933).
880. M. B. Volf, *Technical Glasses*, Pitman, London, 1961.
882. C. H. Vondracek, *Am. Ceram. Soc. Meet.*, Philadelphia (May 1965).
883. B. E. Warren, *Chem. Revs.*, **26**(2), 237 (1940); *Ceram. Abstr.*, **19**(8), 187 (1940).
884. B. E. Warren and J. Biscoe, *J. Am. Ceram. Soc.*, **21**, 259 (1938).
885. B. E. Warren and R. L. Mozzi, *J. Appl. Cryst.* **3**, 259 (March 1970); *J. Appl. Cryst.* g., **2**(4), 164 (1969); *Ceram. Abstr.* (4), 106d (1970).
886. E. W. Washburn, *Can. Chem. Met.*, **17**, 209, 226, 227 (1933). An account of Washburn's visit to Toronto during which he told me and others about his work on vitreous water at the Bureau of Standards in Washington. Washburn died shortly after this visit.
887. Ryuji Watanabe and Massayuki Watanabe, U. S. 3,709,723 (1973); *Ceram. Abstr.*, **52**(4), 84a (1973).
888. M. J. Weber, E. J. Sharp, and J. E. Miller, *J. Phys. Chem. Solids*, **32**(10), 2275 (1971); *Ceram. Abstr.*, **52**(5), 98h (1973).
889. H. O. Weber, T. C. Damen, H. G. Danielmeyer, and B. C. Tofield, *Appl. Phys. Lett.*, **22**(10), 534 (1973); *Ceram. Abstr.*, **52**(10), 237e (1973).
890. R. A. Weidell, U. S. 3,328,181 (1967); *Ceram. Abstr.*, **46P**(9), 220f (1967).
891. Weigl, *Glashutte*, **64**(1), 9-10 (1934); *Ceram. Abstr.*, **13A**(8), 204 (1934).
892. Gustav Weissenberg and Norbert Meinert, U. S. 2,996,391, App. 30.6.54.
893. Gustav Weissenberg, Heinz Bredow, and Norbert Meinert, U. S. 2,723,203 (1955); *Ceram. Abstr.*, **35P**(4), 75e (1956).
894. Edward Wenbam, *Connoisseur*, **92**(387), 310-317 (1933); *Ceram. Abstr.*, **13A**(2), 29 (1934).
895. A. E. R. Westman, *Modern Aspects of the Vitreous State*, Vol. I., J. D. MacKenzie, Ed., Butterworth, Washing D. C., 1960, p. 63.
896. A. E. R. Westman and R. Beatty, *J. Am. Ceram. Soc.*, **49**(2), 63 (1966); *Ceram. Abstr.*, **45**(3), 61d (1965).
897. A. E. R. Westman and Joan Crowther, *J. Am. Ceram. Soc.*, **37**(9), 420 (1954); *Ceram. Abstr.*, **33**(10), 176i (1954).
898. A. E. R. Westman and P. A. Gartoganis, *J. Am. Ceram. Soc.*, **40**(9), 293 (1957); *Ceram. Abstr.*, **36**(10), 230j (1957).
899. A. E. R. Westman and P. Gartaganis, unpublished work at Ontario Research Foundation: 1. Preparation of $Na_6P_4O_{13}$ by Ostwald's method for pure NaOH. 2. Crystallization of ammonium tetrapoly.
900. A. E. R. Westman and M. K. Murthy, *J. Am. Ceram. Soc.*, **44**(3), 97(10), 475 (1961); *Ceram. Abstr.*, **40**(11), 258i (1961).
901. A. E. R. Westman and M. K. Murthy, Symposium on Nucleation and Crystallization in Glasses and Melts, *Am. Cer. Soc.*, Columbus, Ohio (1962).
902. A. E. R. Westman and M. K. Murthy, *J. Am. Ceram. Soc.*, **47**(8), 375 (1964).
903. A. E. R. Westman and A. E. Scott, *Nature*, **168**, 740 (1951).
904. A. E. R. Westman, P. A. Gartaganis, and M. J. Smith, *Can. J. Chem.*, **37**, 1764 (1959).
905. A. E. R. Westman, A. E. Scott, and J. T. Pedley, *Chem. in Can.*, **4**, 189 (1952).
906. Arnold Wexler, S. B. Garfinkel, F. E. Jones, Saburo Hasegawa, and Albert Krinsky, *J. Res. Natl. Bur. Stand.*, **55**(2), 71 (1955); *Ceram. Abstr.*, **35**(1), 18h (1956).

907. W. Weyl, *Keram. Rundsch.*, **41**(15), 188 (1933); *Ceram. Abstr.*, **13A**(5), 113 (1934).
908. W. Weyl, *Angew. Chem.* **48**, 677-678 (1935); *Ceram. Abstr.*, **17**(5), 177 (1938).
909. W. A. Weyl, *J. Am. Ceram. Soc.*, **24**,(7) 221, (8) 245 (1941); W. A. Weyl, and N. J. Kreidl, *J. Am. Ceram. Soc.*, **24**,(11), 337 (1941). See also 500.
910. W. A. Weyl, *J. Chem. Educ.*, **27**(9), 520 (1950); *Ceram. Abstr.*, **30**(4), 73j (1951).
911. W. A. Weyl, *J. Phys. Colloid. Chem.*, **55**, 507 (1951); *Ceram. Abstr.*, **31**(8), 152b (1952).
912. W. A. Weyl, Coloured Glasses, *Pub. Soc. Glass Technol.*, Sheffield (1951).
913. W. A. Weyl, *Round Table Discussion, Dtsch. Glastech. Ges.*, Tübingen, Germany, May (1956).
914. W. A. Weyl, *VIII Int. Congr. Glass*, Brussels, 1965, Paper M-I-1.
915. W. A. Weyl, U. S. 3,499,774 (1970); *Ceram. Abstr.*, **49P**(7), 168e (1970).
916. W. A. Weyl and E. C. Marboe, *The Glass Industry*, August-December (1960), January-April (1961).
917. W. A. Weyl and E. C. Marboe, Interscience, New York, 1964. *Ceram. Abstr.*, **44**(3), 107b (1965).
918. W. A. Weyl and N. J. Kreidl, U. S. 2,394,502 (1946); *Ceram. Abstr.*, **25P**(4), 65 (1946).
919. S. G. Whiteway, I. B. Smith, and C. R. Masson, *Can. J. Chem.*, **48**(1), 33, (1970).
920. S. G. Whiteway, I. B. Smith, and C. R. Masson, *Can. J. Chem.*, **51**, 1422 (1973).
921. D. G. Wickham, *J. Appl. Phys.*, **33**(12), 3597 (1962); *Ceram. Abstr.*, **42**(5), 147h (1963).
922. I. W. Wieczorek and R. L. Cook, *Bull. Am. Ceram. Soc.*, **52**(4), 364 (1973), *Am. Ceram. Soc. Prog.* 5-C-73.
923. W. Z. Wieker, *Electrochem.*, **64**, 1047 (1960).
924. W. Wieker and D. Hoebbel, *Z. anorg. allgem. Chem.*, **366**, 139 (1969).
925. W. Wieker, A. R. Grimme, and E. Thilo, *Z. anorg. allgem. Chem.*, **330**, 78-90 (1964).
926. H. P. Wilk, *Szklo Ceram.*, **19**(3), 76 (1968); *Ceram. Abstr.*, **48**(4), 97e (1969).
927. D. J. Williams, B. T. Bradbury, and W. R. Maddocks, *J. Soc. Glass Technol.*, **43**(213), 308-24T (1959); *Ceram. Abstr.*, **39**(6), 135a (1960).
928. Winter, A., *J. Can. Ceram. Soc.*, **41**, 27 (1972).
929. Winters, R., and W. La Course, *Curr. Ceram. Res. Bull. Am. Ceram. Soc.*, **52**(2), 223 (1973).
930. C. D. Wirkus and D. R. Wilker, IS-267, *Ames Laboratory R & D Report*, U. S. Atomic Energy Commission (February 1961).
931. C. D. Wirkus and D. R. Wilder, *J. Nuclear Met.*, **5**(1), 140 (1962).
932. L. R. Wirth, Jr., and R. S. Robertson, U. S. 3,432,428 (1969); *Ceram. Abstr.* **48P**(6), 213a (1969).
933. H. Wondratschek, *Glashütten-Hanbuch Hüttentechn*, Vereinigung der deutschen Glasindustrie (HVG), 1965.
934. H. Wondratschek and L. Merker, *Naturwissenschaften*, **43**, 494 (1956); *Ceram. Abstr.*, **40**(5), 129e (1961).
935. H. E. Wood, *Trans. Ceram. Soc. (Engl.)*, 1,21 (1901).
936. F. E. Wright, *J. Am. Ceram. Soc.*, **3**(10), 783 (1920).
937. F. F. H. Wu, J. Gotz, W. D. Jamieson, and C. R. Masson, *J. Chromatogr.* (Chrom. 4632), **48**, 515 (1970).
938. Tamotsu Yamada, Kamame Ito, and Tatsuyasu Tamai, *Sekko to Sekkai* **102**, 251 (1969); *Ceram. Abstr.*, **49**(7), 162g (1970).
939. Tamotsu Yamada, Tetsuya Iwato, Shiyiro Abo, and Shoichero Nagai, *Sekko to Sekkai*, **40**, 17 (1959); *Ceram. Abstr.*, **38**(8) 201b (1959).
940. Tamotsu Yamada, Teruhisa Kawose, and Shoichiro Nagai, *Sekko to Sekkai*, **53**, 164 (1961); *Ceram. Abstr.*, **41**(3), 54b (1962).
941. Tamotsu Yamada, Naohisa Koike, Ito Masatako, and Shoiciro Nagai, *Sekko to Sekkai*, **42**, 19 (1959); *Ceram. Abstr.*, **39**(3), 52h (1960).

942. Tamotsu Yamada, Chihiro Mizutani, and Shoichior Nagai, *Sekko to Sekkai*, **85**, 198 (1966); *Ceram. Abstr.*, **48**(4), 123e (1969).
943. Tamotsu Yamada, Masatoshi Miyama, and Shoichiro Nagai, *Sekko to Sekkai*, **67**, 223 (1963); *Ceram. Abstr.*, **44**(4), 109 (1965).
944. Toshio Yamada, *Yogyo Kyokai Shi,* **67**(768), 399 (1959); *Ceram. Abstr.*, **39**(11), 252g (1960).
945. Toshio Yamada and Shinichiro Tomino, *Yogyo Kyokai Shi,* **67**(757), 16 (1959); *Ceram. Abstr.*, **38**(6), 150j (1959).
946. Yoshito Yasutake and Yosisige Fujita, U. S. 3,458,279 (1969); *Ceram. Abstr.*, **48P**(11), 382h (1969).
947. P. C. Yates, U. S. 3,615,807 (1971); *Ceram. Abstr.*, **51P**(2), 46b (1972).
948. P. C. Yates, U. S. 3,650,783 (1972).
949. P. J. Yavorsky, U. S. 3,730,744 (1973), *Ceram. Abstr.*, **52**(8), 171j (1973).
950. Ryosuke Yokota and Saburo Nakajima, U. S. 3,607,321 (1971); *Ceram. Abstr.*, **51**(1) 4h (1972).
951. C. G. Young and E. Snitzer, *IEEE. J. Quant. Mech.*, **5**(360), QE4 (1968).
951a. A. S. Yutina, ZD Zhukova, and S. V. Lysak, *Izv. Akad. Nauk SSSR Neorg. Mater.*, 2 (11), 2020 (1966); *Ceram. Abstr.*, **47**(9), 268e (1968).
952. W. H. Zachariasen, *J. Am. Chem. Soc.*, **54**, 3841 (1932).
953. W. H. Zachariasen, *J. Chem. Phys.*, **3**(3), 162 (1935).
954. Ludvik Zagar and Gerd Schneider, *Glastech. Ber.*, **41**(11), 446 (1968); *Ceram. Abstr.*, **49**(9), 214h (1970).
955. M. Zaman and R. W. Douglas, *VIII Int., Congr. Glass, Soc. Glass Technol.* 257 (1969).
956. Friederich Zapp and S. Goller, *Tonind. Ztg, Keram. Rundsch.*, **76**(17/18), 264 (1952); *Ceram. Abstr.*, **32**(9), 162g (1953).
957. R. E. Zimanovskaya and Z. V. Vodzinskaya, *Tsement*, **21**(5), 12 (1955); *Ceram. Abstr.*, **35**(5), 89a (1956).
958. F. H. Zschacke, *Keram. Rundsch.*, **42**(52), 644 (1934); *Ceram. Abstr.*, **15A**(11), 324-325 (1936).
959. F. M. Zschacke et al., *Keram. Rundsch.*, **49**, 197-201 (1941); **49**, 265 (1941). **49**, 331-332, 351-353 (1941); **49**, 443-445 (1941); **50**, 21-23, 31-34 (1942).
960. E. Zschimmer, *Phys. Z.* **8**, 611 (1907).
961. E. Zschimmer, *Sprechsaal*, **60**(52), 1016-1022 (1927); *Ceram. Abstr.*, **7A**(4), 220 (1928).
962. E. Zschimmer, K. Hesse, and L. Stoess, *Sprechsaal* (58), 513, 529 (1925); *Chem. Zent. I*, 207 (1926); *Ceram. Abstr., Glass* 5, *A* (12), 392 (1926).
963. R. Zsigmondy, *Dinglers Polytech. J.*, **273**, 29 (1889).

ESR of Phosphorus Compounds

P. Schipper, E. H. J. M. Jansen, and H. M. Buck

Eindhoven University of Technology
Department of Organic Chemistry
Eindhoven, The Netherlands

CONTENTS

Phosphorus compounds are involved in a variety of free radical reactions. A survey of the literature in this field through 1964 has been reported by Walling and Pearson [143]. A division was made in phosphorus-centered radicals and radicals of various structural types. The phosphorus-centered radicals were classified according to phosphinyl radicals of type $R_2\dot{P}$ and phosphoranyl radicals of type $R_4\dot{P}$. The same classification was also used by Bentrude in a review up to 1966 which is comprised of the ESR and chemistry of phosphorus free radicals [15]. Over the last 10 years an increasing number of ESR studies has been performed on phosphorus compounds. Techniques of generation have been improved to the extent that a variety of radicals, whose existence earlier was only presumed as reaction intermediates, could be observed. The ESR of the phosphorus compounds has its own intrinsic interest because it provides more insight into the structure and nature of bonding of phosphorus. This review is concerned with the ESR of phosphorus compounds which have the unpaired electron on phosphorus as well as on neighbor atoms. Our primary emphasis is to correlate the phosphorus splittings with the spin distribution in the phosphorus compounds. Therefore a classification has been made according to the nature of the orbital to which the unpaired electron is formally confined.

In phosphorus π radicals the unpaired electron may be located in a phosphorus p or d orbital, in a p orbital on the ligands, or in both. In this class of radicals the isotropic phosphorus splitting is a consequence of spin polarization. Phosphorus splittings smaller than 80 gauss were observed. When phosphorus is substituted in a β position to a π radical center, a hyperconjugative mechanism appears to be important. In this class of compounds phosphorus splittings are observed 250 gauss. In phosphorus-centered σ radicals the unpaired electron resides in an orbital on phosphorus with a substantial $3s$ contribution. Much larger phosphorus splittings are observed here: 300 to 1300 gauss. This review is based on literature appearing up to mid-1974.

I. Phosphorus π Radicals

A. *Interpretation of ^{31}P Hyperfine Splittings*

The isotropic hyperfine splitting for an unpaired electron in the $3s$ orbital of phosphorus has been calculated to be 3640 gauss [145, 146]. In phosphorus σ radicals, in which the unpaired electron resides in an s- and p-hybridized orbital, the amount of s character can be derived directly from the observed phosphorus coupling constant. The radicals in this section, however, have a structure in which the unpaired electron is formally confined to π orbitals on phosphorus and its ligands. The observed phosphorus splitting for radicals of this structure results from a spin polarization mechanism. In π radicals it is customary to link the observed hyperfine splitting not to the s character of the unpaired electron but to the σ-π polarization parameters for the system concerned; for instance, in proton hyperfine splittings in π-carbon systems the relation between the proton hyperfine splitting and the π spin density on the adjacent carbon atom is given by the McConnell relation

$$a_i = Q\rho_i \tag{1}$$

where a_i is the observed hyperfine splitting of the proton attached to the ith carbon atom and ρ_i is the unpaired electron denisty at the ith carbon. Q is usually about 25 gauss, but depends on the particular system. In nuclei that can accommodate π-electron spin density directly on the atom concerned a two-term equation has to be considered. As a general model for such systems, the theoretical formulation for the relation between ^{13}C splittings and spin densities in carbon π radicals, given by Karplus and Fraenkel [83], can be used. This expression is

$$a_{^{13}C} = (S_C + \sum_X Q^C_{CX})\rho_C + \sum_X Q^C_{XC}\rho_X \tag{2}$$

where S_C covers the polarization of 1*s* electrons and the Q^C_{CX} terms cover spin polarization in the C–X sigma bonds for the spin density on the carbon atom concerned. The Q^C_{XC} terms represent the polarization in the X–C sigma bonds from spin density on neighboring atoms X. Similar formulas are known to be applicable to other nuclei such as ^{14}N, ^{17}O, and ^{19}F. For ^{31}P (2) takes the form

$$a_{^{31}P} = Q^P_{PP}\rho^\pi_P + \sum_X Q^P_{XP}\rho^\pi_X \tag{3}$$

in which

$$Q^P_{PP} = S^P + Q^P_{PX}$$

Hunter and Symons [73] showed the validity of a similar equation for a wide range of radicals with various nuclei, including phosphorus.

Although many Q parameters have been evaluated for nuclei such as ^{13}C and ^{14}N, reports in the literature for ^{31}P are rather scarce. In some cases it was possible to estimate Q values from experimental phosphorus splittings. A value of 80 gauss has been derived for Q^P_{PP} from the observed phosphorus splitting of $\dot{P}H_2$ [111]. In this radical the unpaired electron resides in a p_z orbital on phosphorus. A positive sign has been indicated [73]. The adjacent atom term Q^P_{CP} has been determinated to be 40.68 gauss, which is the a_P of $H_2\dot{C}-P^+Ph_3$. In this radical the unpaired electron is located at the methylene carbon atom [99, 24]. Similarly, a value of 33 gauss was derived for Q^P_{OP} from the a_p of PO_4^{2-} [73, 134]. From the point of view that spin polarization induces a negative spin density on the atoms linked to the radical center, a negative sign has been indicated for the adjacent term Q^{P}_{XP} [9]. Insertion of the values, for Q^{P}_{PP} and Q^{P}_{CF} in (3) leads to a quantitative solution for the relation between phosphorus hyperfine splittings and spin densities for a

$$a_{^{31}P} = 80\rho^{\pi}_{P} - 40\sum_{C}\rho^{\pi}_{C} \tag{4}$$

for a phosphorus π radical in which phosphorus is linked to carbon atoms. Although (4) may be valid for a wide range of phosphorus π radicals, possibly with accommodated Q values for particular systems, its application has been reported in only a few cases [24, 73, 142].

B. Compounds Containing Disubstituted Phosphorus

1. PHOSPHINYL RADICALS, $R_2P^{\bullet}$

A number of phosphinyl radicals such as $\dot{P}H_2$, $\dot{P}F_2$, and $\dot{P}Cl_2$ have been trapped in solid matrices. The structure of these radicals could be derived from the observed hyperfine splittings.

The $\dot{P}H_2$ radical was generated by γ irradiation of PH_3 in a krypton matrix at 4.2°K [111]. The ESR spectrum showed isotropic splittings of 80 gauss for phosphorus and 18 gauss for hydrogen. Because an electron in a 3*s* orbital of phosphorus has a theoretical coupling of 3640 gauss, the orbital of the unpaired electron has a negligible amount of 3*s* character. Therefore, it was concluded that the odd electron occupies a phosphorus 3*p* orbital perpendicular to the molecular plane.

Kokoszka and Brinckman [89, 90] reported the ESR of $\dot{P}Cl_2$, generated by photolysis of PCl_3 at 77°K. An anisotropic phosphorus splitting, which varied from 28 to 269 gauss, was observed. The anisotropic *g* value changes from 2.001 to 2.021. Since the *x* and *y* components of the *g* tensor are identical a Cl—P—Cl bond angle of 90° was suggested. The observed phosphorus coupling constants $A_{\parallel}(P) = 269$ gauss and $A_{\perp}(P) = 28$ gauss can be written in terms of an isotropic and an anisotropic contribution according to

$$\begin{aligned} A_{\parallel} &= A + 2B \\ A_{\perp} &= A - B \end{aligned} \tag{5}$$

The isotropic component *A* of the coupling arises from *s*-orbital spin density of the valence shell and the anisotropic component *B* from spin density in the *p* orbital. Thus the phosphorus 3*p*-orbital spin density can be estimated from

$$\rho^{\pi}_{P} = B(P)_{exp}/B(P)_{ref} \tag{6}$$

Reference values have been calculated by Atkins and Symons [6] and Hurd and

Coodin [74]. Because the sign of the couplings given by Kokoszka and Brinckman were uncertain, definite values for A(P) and B(P) could not be given. However, assuming various signs, they derived a spin density of 0.61 to 0.78 in the phosphorus 3p orbital. From these data a molecular geometry was depicted similar to that of the $\dot{P}H_2$ radical. The ESR spectrum of $\dot{P}Cl_2$ was also reported by Wei and Current

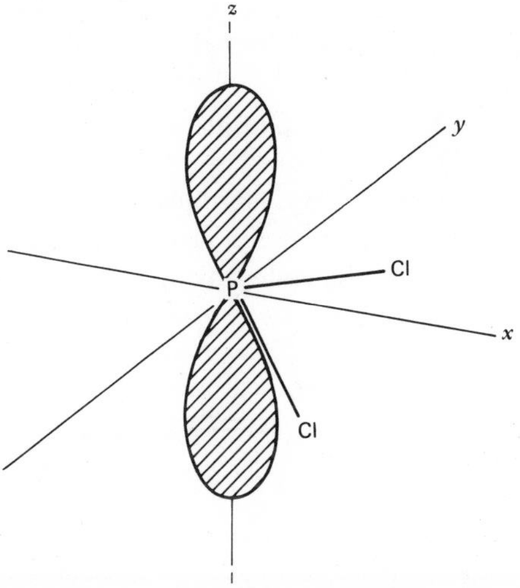

[150], who generated the PCl_2 radical in noble gas and nitrogen matrices by thermolysis of PCl_3 at 1000°C. Definite values could be determined for the phosphorus splitting: $A_{\parallel}$(P) = 293 gauss and $A_{\perp}$(P) = −30.5 gauss. According to (5), isotropic and anisotropic contributions were calculated: A(P) = 77 gauss and B(P) = 107.5 gauss. The reference value of 125 gauss for B(P) [74] leads to a spin density of 0.82 in the phosphorus 3p orbital.

Similar calculations were performed for the $\dot{P}F_2$ radical by Nelson et al. [119], who trapped the radical in a xenon matrix and observed an isotropic ESR spectrum at 77°K and an anisotropic spectrum at 4.2°K. The isotropic phosphorus splitting a_P = 84.6 gauss was obtained directly from the spectrum. From the observed anisotropic splittings $A_{\parallel}$(P) = 308.1 gauss and $A_{\perp}$(P) = −27.2 gauss the anisotropic contribution B(P) has been calculated to be 111.7 gauss. According to the procedure used for $\dot{P}Cl_2$, a phosphorus 3p spin density of 0.83 was calculated. The $\dot{P}F_2$ radical was observed earlier by Morton et al. [144] on irradiation of ND_4PF_6 with isotropic splittings for A(P) = 36.0 gauss and A(F) = 60.5 gauss. Similar values were found by Wei et al. [151], who generated the radical both by photolysis of PF_2H and by passing P_2F_2 over a heated wire. These values, however, were probably in error because of matrix effects [119].

Kilcast and Thomson (87, 138) calculated the polarisation parameter Q_{PP}^P by comparing the preceding isotropic phosphorus coupling constants with spin densities, obtained by CNDO calculations. Q_{PP}^P values were estimated for $\dot{P}H_2$, 80 gauss and $\dot{P}Cl_2$, 97.5 gauss. The isotropic phosphorus coupling for $\dot{P}F_2$, found by Nelson et al. (119), gave a Q_{PP}^P of 90.3 gauss. The similarity of this value to the previous ones was concluded (138) to be in support of the work of Nelson.

Ingold et al. [63] generated the $(EtO)_2P$ and $(Me_2N)_2P$ radicals in solution and obtained similar values for the isotropic phosphorus hyperfine splittings as found for the radicals mentioned above: 78.5 and 70.2 gauss, respectively. Therefore the same structure was proposed.

The generation of divalent organo phosphinyl radicals has also been reported. Schmidt et al. [131, 132] obtained diphenylphosphinyl radicals by photolysis of tetraphenyl diphosphines and diphenylphosphines and trapped the products in a coldfinger at 77°K. A broad signal was observed with a weakly anisotropic g factor of 2.009. Wong and Wan [152] generated the diphenyl phosphinyl radical (1) by UV irradiation of tetraphenyldiphosphine in benzene at 77°K.

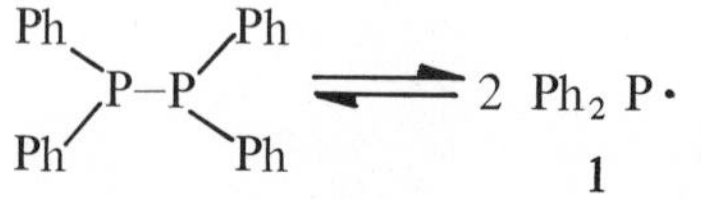

$$\underset{Ph}{\overset{Ph}{>}}P{-}P\underset{Ph}{\overset{Ph}{<}} \;\rightleftharpoons\; 2\;Ph_2\,P\cdot$$

$$\mathbf{1}$$

The spectrum could be explained by assuming coupling of the odd electron with phosphorus (a_P = 12 gauss) and three equivalent protons, a_H = 5.8 gauss. A radical structure was suggested in which the C–P–C angle is close to 90°, similar to the inorganic phosphinyl radicals, and one of the rings is twisted out of the plane of the phosphorus $3p$ orbital. By using the McConnell relation for the observed proton splittings a spin density of 70% was estimated in one of the phenyl rings. From this it has been concluded that 30% of the spin resides on phosphorus, but because the phosphorus splitting may also be due to spin polarization the observed a_P value is *not* necessarily connected with a spin density on phosphorus (4).

2. PHOSPHORIN RADICAL IONS

The phosphorin radical ions are another class of disubstituted phosphorus radicals. These radicals have been studied by Dimroth over a wide range of substituted compounds of type 2 [41, 42, 43]. Radical cations, anions, and even in one

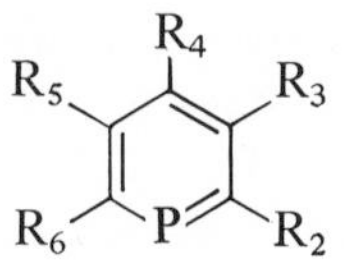

case a radical trianion could be generated. The radical cations were prepared chemically by oxidizing agents such as $AuCl_3$ or $Hg(OAc)_2$ in DMF and lead tetrabenzoate or -acetate in benzene and also by a suitable oxidant such as 2,4,6-triphenoxyl (**3**) in benzene [42].

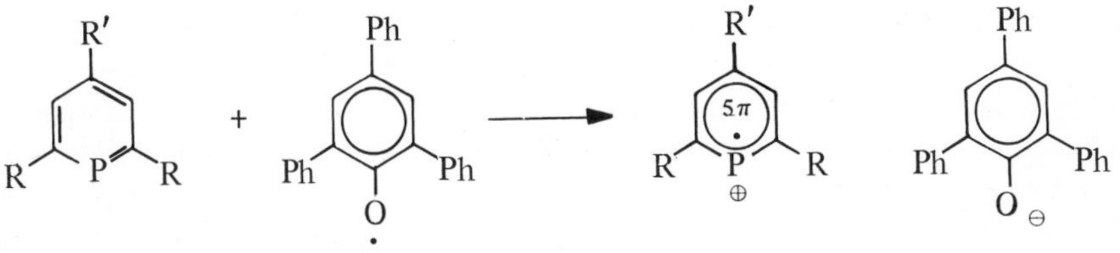

3

Well-resolved ESR spectra were obtained of the phosphorin cation radicals (**4**) and (**5**) [41]. The ESR spectrum of (**4**) could be explained, assuming a_p = 23.3, $a_H^{3,5}$ = 2.3, $a_H^p = a_H^o$ = 2.3, a_H^m = 0.7 gauss (Ph at the 4 position), and $a_H^p = a_H^o$ = 0.46, a_H^m = 0.23 (Ph at the 2 and 6 positions). These assignments could be made by comparison with the pentadeuterophenyl substituted analog (**6**), which has a similar a_p of 23.3 and shows a twofold triplet splitting of 2.4 gauss caused by the two-ring protons (at the 3 and 5 positions).

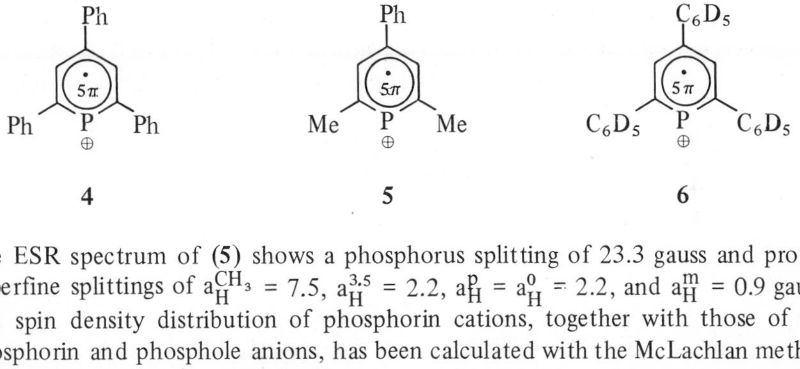

4 **5** **6**

The ESR spectrum of (**5**) shows a phosphorus splitting of 23.3 gauss and proton hyperfine splittings of $a_H^{CH_3}$ = 7.5, $a_H^{3,5}$ = 2.2, $a_H^p = a_H^o$ = 2.2, and a_H^m = 0.9 gauss. The spin density distribution of phosphorin cations, together with those of the phosphorin and phosphole anions, has been calculated with the McLachlan method by Thomson and Kilcast [138]. Phosphorus spin densities of 0.2 to 0.3 were estimated for phosphorin cation radicals with various substituents. By comparing these values with the experimental phosphorus splitting constants (22 to 26 gauss) and using the one-term equation

$$a_P^{exp} = Q_P^{eff} \cdot \rho_P^{calc} \tag{7}$$

a value of 98 gauss was obtained for Q_P^{eff}, which seems to be in good accord with the Q value of $\cdot PH_2$ being 80 gauss. As outlined in Section A, however, a two-term equation has to be applied when phosphorus takes part in the π system. Therefore the observed phosphorus splitting constant of radical (**5**) (23.2 gauss) can be better understood by using (4). The reported methyl splitting (7.5 gauss) indicates a spin density of 0.3 ($Q_H^{C-CH_3}$ = 25 gauss) on each of the carbons adjacent to phosphorus. By substituting this value in (4) and assuming a near zero spin density on phosphorus, we can calculate an a_P of about 24 gauss, which is in good agreement with the reported value of 23.2 gauss. The almost unit spin density on those positions, which can be calculated from proton splittings by using the McConnell relation (1), also indicates little spin density on phosphorus. In contrast with the phosphorin cations, the ESR spectra of the radical anions are much less resolved and show only a phosphorus doublet splitting [45]. The reduction with alkali metals of (**7**) gives, in stages, the monoanion radical 32.4 gauss, the diamagnetic dianion and the paramagnetic trianion, 4.6 gauss. The phosphorus doublet splitting constant of the trianion could be assigned by deuteration of the phenyl substituents [40, 41].

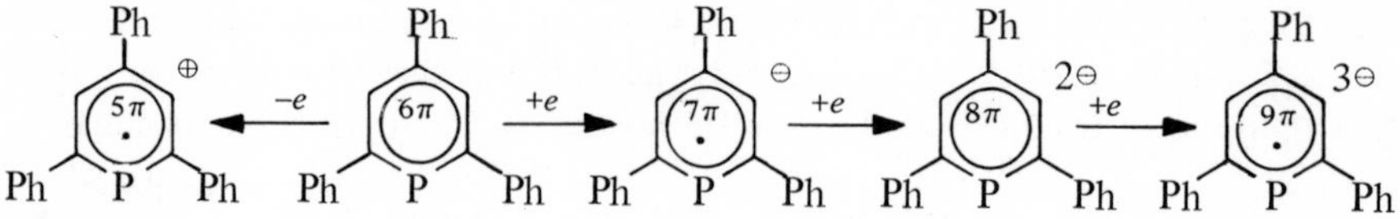

7

Phosphorus σ-π polarization constants were also calculated by Jongsma et al. [79]. They generated the radical ion of (**8**) with potassium in THF. A well-resolved ESR spectrum was obtained from which all hyperfine splittings could be derived. The

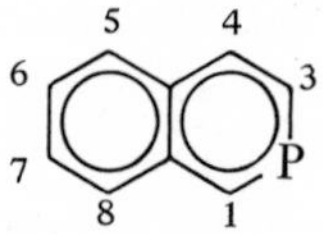

8

hydrogen-splitting constants were computed by using the HMO method in combination with the McConnell relation:

Position	1	P	3	4	5	6	7	8
a^{exp}	7.23	23.6	0.50	3.62	3.52	1.00	0.50	3.62
a^{calc}	5.65	–	0.37	3.51	3.24	2.18	0.92	4.09

With a similar one-term equation (7), used by Thomson and Kilcast [138], a Q_P^{eff} of 118 gauss has been determined. These calculations were also performed for the radical anion of (**9**). With a calculated ρ_P of 0.29 and an observed a_P of 33.5 gauss a 116-gauss Q_P^{eff} value was inferred.

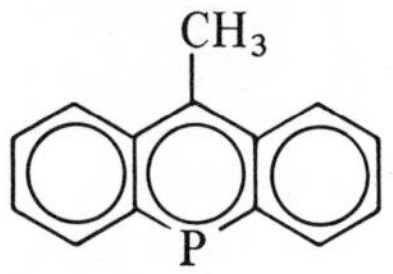

9

Though these Q_P^{eff} parameters seem to be adequate in estimating the π-spin density on phosphorus, the σ-π polarization of the C–P σ bonds should also be taken into account and is of real importance when there is appreciable spin density on the carbon atoms adjacent to phosphorus.

C. Compounds Containing Trisubstituted Phosphorus

1. PHOSPHINIUM CATION RADICALS, $R_3\dot{P}^{+}$

There are few reports regarding the ESR of phosphinium π radicals. Lucken et al. [100] considered the occurrence of $Ph_3P^{\dot{+}}$ as an intermediate species in the reaction of triphenylphosphine with chloranil. Its existence could not be established, however. Boekestein et al. [21] proved by means of spin-trapping techniques the existence of the tris(dimethylamino)phosphine radical cation as an intermediate species in the reaction of tris(dimethylamino)phosphine with 9,10-phenanthrenequinone. ESR of phosphinium radicals has been reported by Bersohn [18]. A phosphorus splitting of 2.32 gauss was observed, but the method of generation was not mentioned. Substituted triphenylphosphinium radicals were generated by Tomaschewski [140] by treatment of the corresponding phosphine with silver perchlorate. The ESR of the tris(*p*-dimethylaminophenyl)phosphinium radical showed interaction of the unpaired electron with phosphorus (a_p = 12 gauss) and three equivalent nitrogens (a_N = 6.2 gauss), which indicates delocalization of the odd electron over phosphorus and the three phenyl rings.

2. PHOSPHINE FREE RADICALS, $R_2P{-}\dot{C}R'_2$

Lucken [97] reported the generation of the radical $(HOCH_2)_2P{-}\dot{C}HOH$ by hydrogen abstraction from tris(hydroxymethyl)phosphine by hydroxyl radicals in aqueous solution. The hyperfine splittings (a_P = 29.1 gauss, a_H = 14.8 gauss) are somewhat smaller than those of the phosphobetaine radical $(HOCH_2)_3P^+ - \dot{C}HOH$ (a_P = 33 gauss, a_H = 17 gauss), but, from the similarity, it was concluded that no appreciable delocalization of the unpaired electron onto the phosphorus nucleus takes place. In both cases solely a spin polarization mechanism has been proposed to account for the observed phosphorus splitting. A similar observation was made by Begum et al. (9), who generated $Et_2P - \dot{C}HCH_3$ (a_P = 32 gauss, a_H = 22 gauss) and $Et_3P^+{-}\dot{C}HCH_3$ (a_P = 40 gauss, a_H = 25 gauss) by γ irradiation of the corresponding phosphines and phosphonium salts at 77°K. The authors considered the smaller hyperfine splittings of the phosphine free radical to be within experimental error. A similar sequence in hyperfine splittings was observed earlier however, by Muller et al. [118] for radicals **10** (a_P = 6.8 gauss) and **11** (a_P = 16.8 gauss).

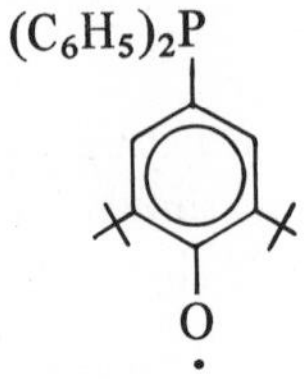

10

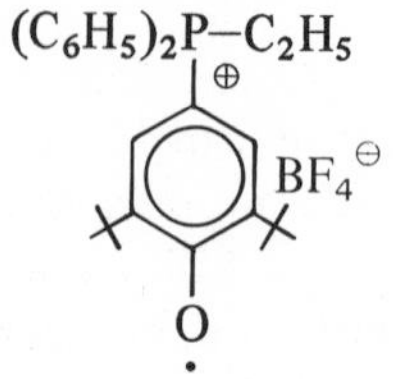

11

The smaller hyperfine splittings in phosphine free radicals may be explained by delocalization of the unpaired electron onto the phosphorus nucleus. Such delocalization gives a positive contribution to the isotropic negative phosphorus splitting (4); hence the magnitude should be less than for the phosphobetaine radical cations.

3. PHOSPHINE ANION RADICALS, $R_3P^{\overline{\cdot}}$

A number of alkali metal reductions of aryl phosphines have been reported. Usually aryl-cleaved products were observed. Britt and Kaiser [22] showed that reaction of triphenyl phosphine with alkali metals leads not to the triphenylphosphine radical anion but to the radical (**12**) formed by phenyl cleavage according to the scheme

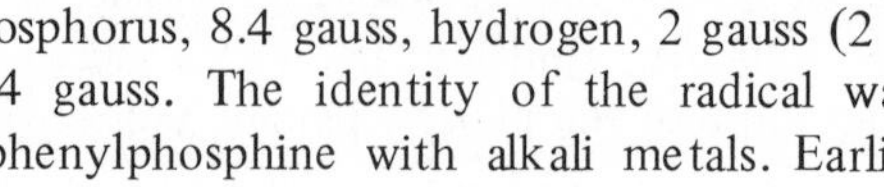

$$Ph_3P + 2M \longrightarrow Ph_2P{-}M + PhM$$

$$Ph_2PM + M \longrightarrow \underset{\mathbf{12}}{Ph_2PM^{\overline{\cdot}}} + M^+$$

The ESR spectrum of (12), generated by reduction with potassium, shows hyperfine splittings of phosphorus, 8.4 gauss, hydrogen, 2 gauss (2 H), 0.8 gauss (4 H), and potassium, 0.24 gauss. The identity of the radical was confirmed by its generation from diphenylphosphine with alkali metals. Earlier, radical (**12**) was erroneously assigned to the triphenylphosphine anion radical (Hanna [65], Kabachnik [81]). Similar phenyl cleavage has been observed in the reaction of phenylbiphenylene phosphine with alkali metals [23]. The ESR of the resulting metal biphenylene phosphine anion radical (**13**) showed a phosphorus splitting of 8.5 gauss. A second free radical, which shows a triplet (8.8 gauss), probably arising from two phosphorus atoms, was formed in this reaction. The structure of this radical was assigned to radical anion (**14**).

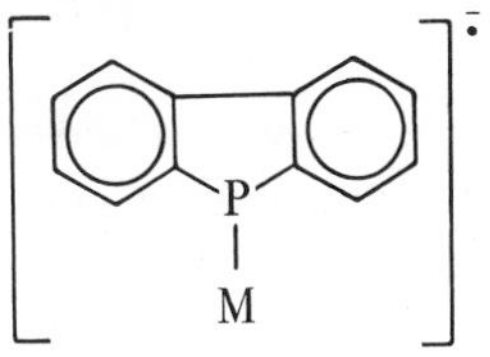

13

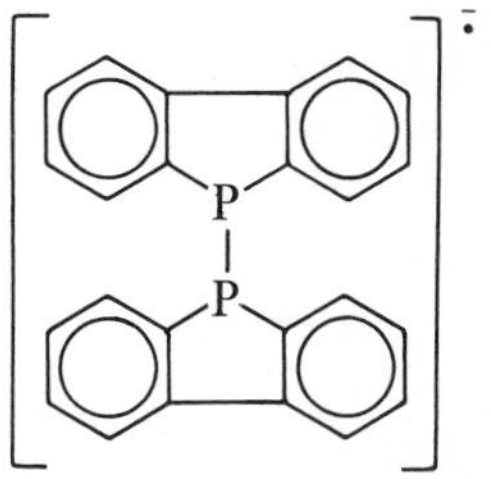

14

Aryl cleavage has been further reported in the alkali metal reduction of 4,4′-bis-(diphenylphosphine)biphenyl [70] and tris(1-naphtyl)phosphine [71]. Uncleaved products were claimed in the reaction of alkali metals with phospholes by Kilcast and Thomson [86], who studied the ESR of the radical ions of phospholes (**15**), (**16**) and (**17**).

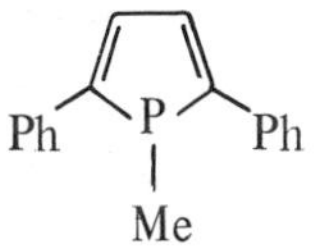

15

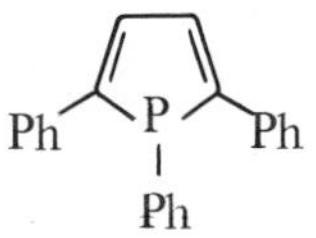

16

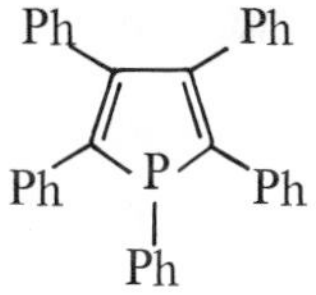

17

The increase in the a_P going from (**15**) (23.5 gauss), to (**16**) (26.5 gauss), to (**17**) (31.3 gauss), despite the more delocalized system, has been explained by assuming that the phenyl rings were twisted out of plane with consequent increasing localization of the unpaired electron in the phosphole ring. However, a much smaller a_P (15 gauss) was observed by Dessy and Pohl [39], who generated the radical anion of (**17**) by electrolytic reduction. Gerson et al. [55] generated the radical anion of dimethylphenyl phosphine (**18**) both by reaction with alkali metals and electrolysis of the parent compound. On electrolysis, the ESR spectrum showed a temperature-dependent phosphorus splitting of 8.28 to 8.67 gauss in the range of −40 to −60°C, whereas the proton splittings $a_H^{CH_3} = 0.78$, $a_H^P = 9.06$ (1 H), $a_H^o = 3.31$ (2 H), $a_H^m = 0.39$ (2 H) were not affected by change in temperature. From the spin population in the benzene ring it was concluded that the dimethylphosphine substituent is

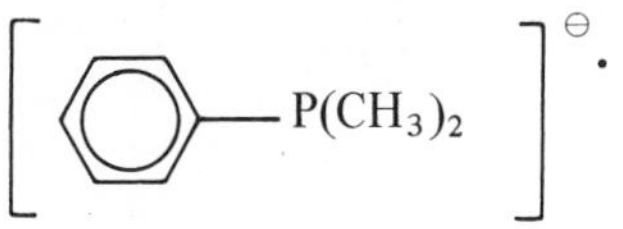

18

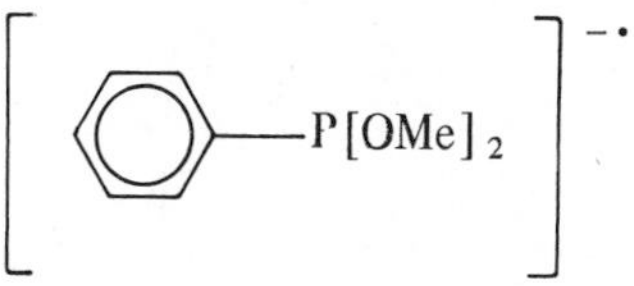

19

electron attracting. This effect has been attributed to electron delocalization of the vacant $3d$ levels of phosphorus.

Similar hyperfine splittings (a_P = 9.0, a_H^p = 5.9, a_H^o = 3.9 gauss) were also observed by Boekestein et al. [20] in the electrochemical reduction of dimethyl phenylphosphonite (**19**). The electrochemical reduction appeared to be a more fruitful method in the generation of anion radicals of parent phosphines. Ilyasov [76] reported the ESR of the triphenylphosphine radical anion in which a phosphorus splitting of 3.3 gauss was observed. The delocalization of the unpaired electron over phosphorus and the three phenyl rings has been established by subsequent introduction of pentadeuterophenyl groups [77]. Prokovjev [122] reported the ESR of radical anions of substituted ethylbenzoates (**20**) and (**21**). Phosphorus splittings of 3.75 to 5.20 gauss were observed for (**20**), whereas much lower splittings (1.1 gauss), which indicate lower spin densities on the adjacent carbon atom, were observed for (**21**). Wallace

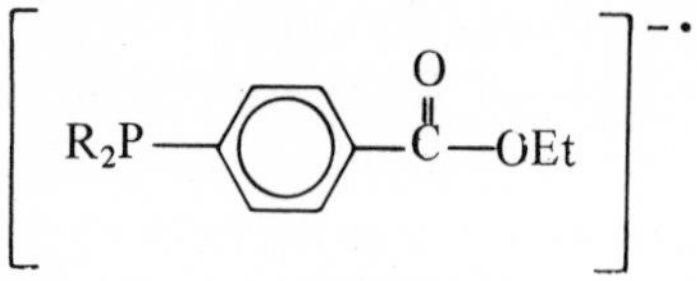

20

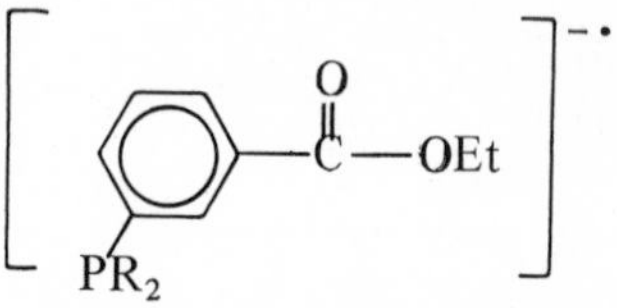

21

et al. [142] reported an ESR study of the anion radicals of cyclopolyphosphines (**22**) and (**23**). Well-resolved ESR spectra were obtained from which the following hyperfine splittings could be derived: a_P = 18.0, $a_F^{C-CF_3}$ = 18.6, $a_F^{P-CF_3}$ = 3.93 gauss for (**22**) and a_P = 10.9 (2 P), 0.5, $a_F^{C-CF_3}$ = 18.05, $a_F^{P-CF_3}$ = 8.85 (2 CF_3), 0.15 gauss for (**23**). From these experimental results carbon spin densities of 0.39 were calculated for both compounds by using the McConnell relation with $Q_F^{CF_3}$ = 47 gauss. The spin densities on phosphorus were estimated by using the two-term equation of Hunter and Symons [73]. Assuming a negative sign for the observed phosphorus splitting, the phosphorus spin density turned out to be −0.07 and 0.03 for (**22**) and (**23**), respectively. Thus the unpaired electron is localized mainly on the ethylenic carbons of both radicals.

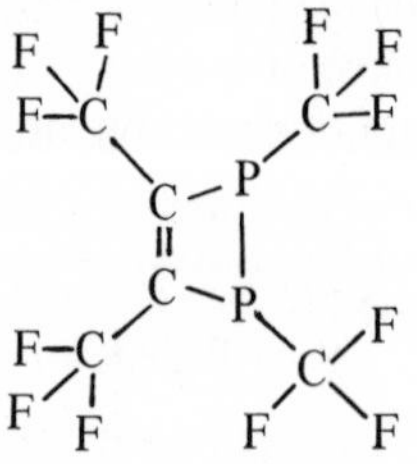

22

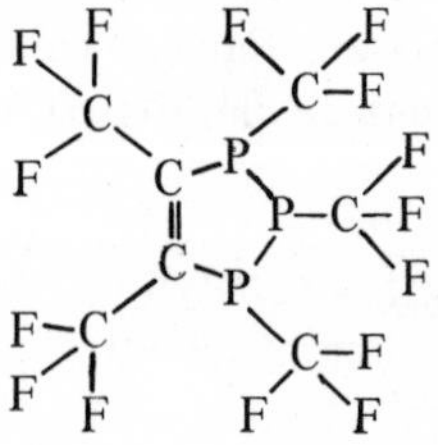

23

D. Compounds Containing Tetrasubstituted Phosphorus

1. PHOSPHOBETAINE CATION RADICALS, $R_3\overset{+}{P}{-}\dot{C}R'_2$

The ESR study of phosphobetaine radical cations is particularly relevant to the problem of p_π-d_π bonding to phosphorus. A number of these radicals were trapped in solid matrices. The X-irradiation of a single crystal of carboxymethyltriphenylphosphonium chloride has been shown to yield the corresponding oriented radical cation (**24**) (Lucken and Mazeline [98]).

$$Ph_3\overset{+}{P}{-}CH_2CO_2H\,Cl^- \xrightarrow{\text{X-ray}} Ph_3\overset{+}{P}{-}\dot{C}HCO_2H\,Cl^-$$

24

A slightly anisotropic phosphorus splitting that varied from 32 to 36 gauss was observed, from which an isotopic component of 34.6 gauss was obtained. From the similarity of the proton splittings in **24** to those of other radicals of type $R{\cdot}\dot{C}H{\cdot}CO_2H$, in which R=H, OH, and CO_2H, it was concluded that delocalization of the unpaired electron to phosphorus by π bonding is negligible. A similar result [99] was reported for the triphenylphosphoniummethylide radical cation $Ph_3\overset{+}{P}{-}\dot{C}H_2$ (**25**) (a_P^{iso} = 40.6 gauss).

A positive sign was indicated for the isotropic phosphorus splittings of **24** and **25**. Because spin polarization of the σ electrons in the C–P bond induces a negative spin density on phosphorus, a negative sign has to be expected in the absence of π bonding. Therefore the mechanism of the interaction of the phosphorus nucleus with the electron spin remained unclear.

Symons et al. [9, 101] generated phosphobetaine radical cations of type $R_2\dot{C}{-}\overset{+}{P}R_3$, and related structures $R_2\dot{C}{-}\overset{+}{P}HR_2$ and $R_2\dot{C}{-}P(O)R_2$ in which R = alkyl, by γ irradiation of the corresponding phosphonium salts, phosphines, and phosphine oxides in sulfuric acid at 77°K. As in the observations of Lucken and Mazeline [98, 99], no significant p_π-d_π bonding to phosphorus could be detected.

It has been pointed out [9] that the results obtained by Lucken and Mazeline [98, 99] are more in agreement with a negative sign for the isotropic phosphorus-splitting constant, which supports a spin polarization mechanism.

The ESR of phosphobetaine radical cations generated in liquid phase has also been reported. Radicals of the type $R_3\overset{+}{P}{-}\dot{C}HOH$ and related structures, $R_2P(O){-}\dot{C}HOH$ and $R_2\overset{+}{P}H{-}\dot{C}HOH$ in which R = CH_2OH, were formed on hydrogen abstraction of the parent compounds by hydroxyl radicals in a rapid flow system (Lucken [97]). As with the radical cations trapped in solid matrices, the proton-coupling constants do not indicate any appreciable delocalization of the unpaired electron onto the phosphorus atom.

Earlier, an ESR study had been reported (Müller [118]) of 2,6-di-*t*-butylphenoxy radicals substituted at the 4 position with tri- and tetravalent phosphorus. The spectra show a parallel increase in a_H^m and a_P, which was interpreted as being

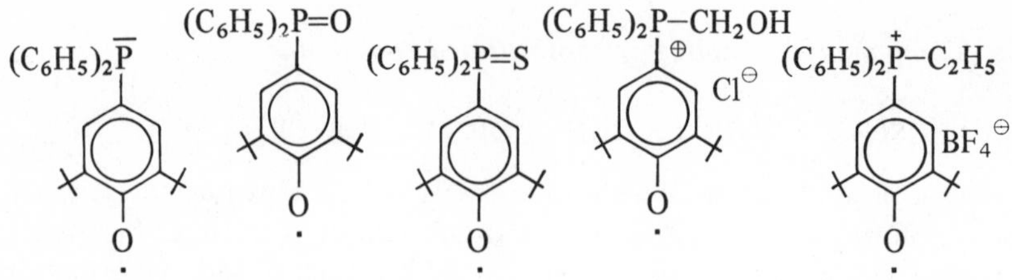

a_P	6.8	14.7	15.4	15.1	16.8
a_H^m	1.6	2.0	2.0	2.15	2.35

due to the increasing electrophilicity of the substituted phosphorus.

The mechanism of the interaction of the phosphorus nucleus and the electron spin in type $R_2\dot{C}{-}\overset{+}{P}R_3$ radical cations has been illustrated by the work of Buck et al. [24]. The radical cation (**26**) of triphenylphosphonium diphenylmethylide was generated by electrolytic oxidation of the methylide or by reduction of the corresponding dipositive ion.

$$Ph_2C = PPh_3 \rightleftharpoons Ph_2\dot{C}{-}\overset{+}{P}Ph_3 \rightleftharpoons Ph_2\overset{+}{C}{-}\overset{+}{P}Ph_3$$

26

The ESR spectrum of (**26**) is given in Figure 1. A phosphorus doublet splitting of

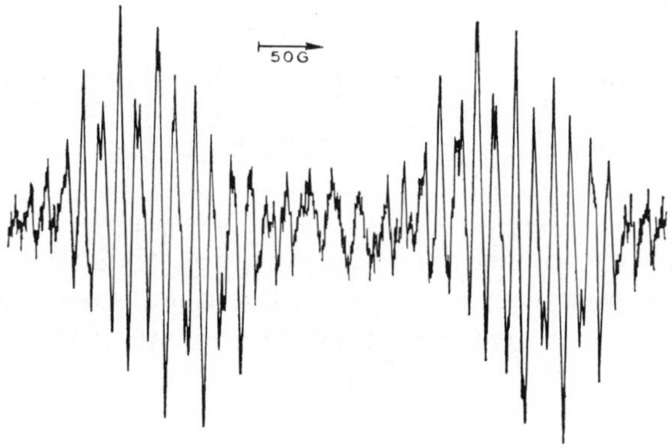

Fig. 1. The ESP spectrum of $(C_6H_5)_2\dot{C}{-}P^+(C_6H_5)_3$ in benzonitrile.

26.6 gauss was observed. The similarity between the proton coupling constants of (**26**) and the parent neutral radical $(C_6H_5)_2\dot{C}H$ also indicated no delocalization of the unpaired electron onto the triphenylphosphonium part of the molecule. From the point of view that phosphorus hyperfine splitting arises from spin polarization in the C–P σ bond, the spin density at the central carbon atom has been calculated; an equation similar to McConnell's relationship was used:

$$a_P = |Q^P_{CP}| \rho^\pi_C \tag{8}$$

The $|Q|$ value has been taken to be 40.68 gauss, which is the phosphorus coupling constant of $H_2\dot{C}-\overset{+}{P}Ph_3$ [99]. From the splitting constants and spin densities, which are presented in tabular form, it appears that the triphenylphosphonium group has

The splitting constants and spin densities of $(C_6H_5)_2\dot{C}-P^+(C_6H_5)_3$

	ortho		meta		para		Central carbon	
a_H	2.71	(3.05)	1.17	(1.22)	2.97	(3.05)		(8.36)
a_P							26.60	
ρ_c	0.100	(0.113)	-0.043	(-0.045)	0.110	(0.113)	0.654	(0.602)

*The values in parentheses refer to $(C_6H_5)_2\dot{C}H$.

little influence on the spin distribution of the parent neutral radical. The same observations were made for radicals **(27)** and **(28)** [24].

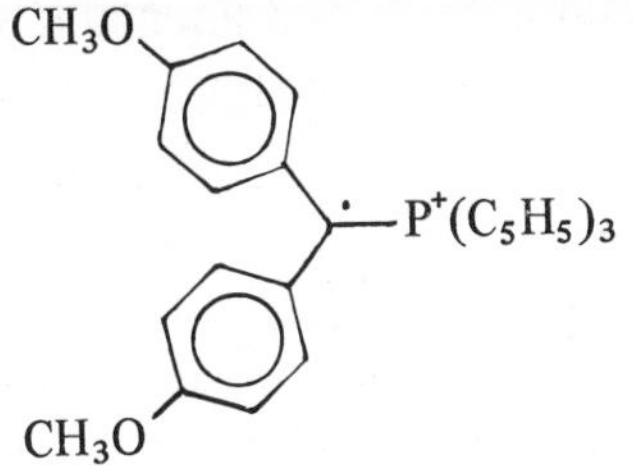

27

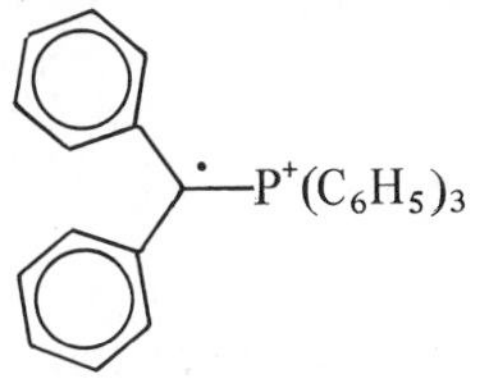

28

The method of stabilizing radicals by introduction of the triphenylphosphonium group has been applied by Kooistra et al. [91] in an ESR study of radical analogs of Chichibabin's and Schlenk's hydrocarbons.

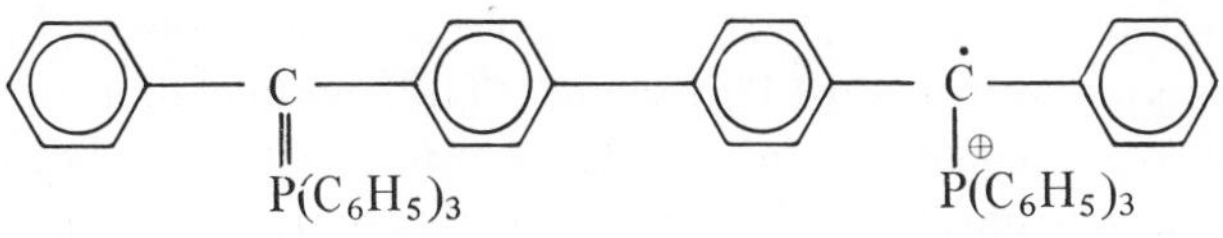

29

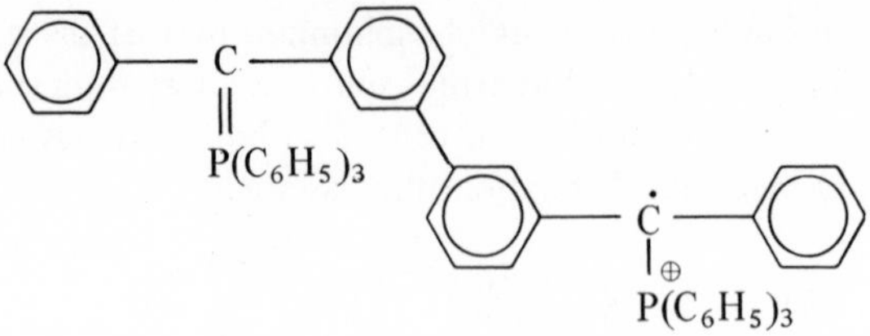

30

The ESR spectrum of (**29**), which consists of a triplet of quintets, could be explained by assuming coupling of the odd electron with the two phosphorus nuclei (a_p = 11.71 gauss). The quintet has been explained by coupling of the four biphenyl protons. Zero spin density was indicated for the outer phenyl rings. In radical (**30**) the spin density is delocalized over one-half the molecule and interaction is found for only one phosphorus atom. An a_P of 26.62 gauss was observed close to the value of the phosphorus splitting in (**26**).

Phosphobetaine radical cations were also obtained by oxidation of phosphorus hetero aromatics [41, 44]. 1,1-Dimethoxy-2,4,6-triphenyl-λ-5-phosphorin (**31**) was oxidized to its radical cation either electrolytically or by leadtetraacetate.

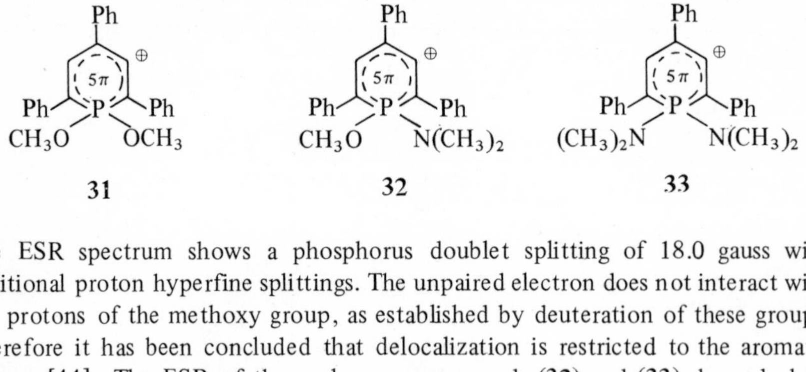

31 **32** **33**

The ESR spectrum shows a phosphorus doublet splitting of 18.0 gauss with additional proton hyperfine splittings. The unpaired electron does not interact with the protons of the methoxy group, as established by deuteration of these groups. Therefore it has been concluded that delocalization is restricted to the aromatic system [44]. The ESR of the analogous compounds (**32**) and (**33**) showed phosphorus splittings of 20.9 and 22.3 gauss, respectively.

2. PHOSPHOBETAINE FREE RADICALS, $R_3\overset{+}{P}-\overset{\dot{-}}{C}R'_2$

In phosphobetaine free radicals a tetrahedrally coordinated phosphorus atom is linked to a radical anion. These radicals are stereoisomers of phosphoranyl radicals, in which phosphorus has a trigonal bipyramidal (TBP) configuration by use of its 3*d* orbitals in the σ-bond base structure. In this conformation the unpaired electron occupies an equatorial position. The appreciable amount of *s* character of the orbital of the unpaired electron results in phosphorus splittings of 600 to 1300 gauss (IIIB).

Phosphoryanyl radicals in liquid phase with one or more aryl ligands appear to be unstable and to isomerize to a phosphobetaine structure [20]. When dimethylphenyl phosphonite was photolyzed in the presence of di-*t*-butyl-peroxide, a radical was observed (**34**) with a low phosphorus doublet splitting (a_P = 9.7 gauss). From the proton hyperfine splittings (a_H^P = 9.7 gauss, a_H^m = 0.9 gauss, a_H^0 = 5.5 gauss) it was concluded that the odd electron is located in the phenyl ring. Thus the phosphorus splitting arises from spin polarization in the P—C phenyl σ bond. When the phenyl ring is replaced by a methoxy group the phosphorus compound maintains a TBP configuration.*

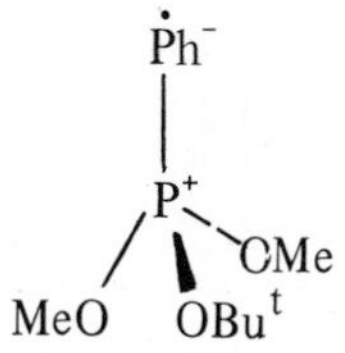

34

Earlier, Rothuis et al. [125, 127] reported a similar stereoisomerization of radicals generated by photolysis of bis(2,2′-biphenylene)hydrogenphosphorane [67].

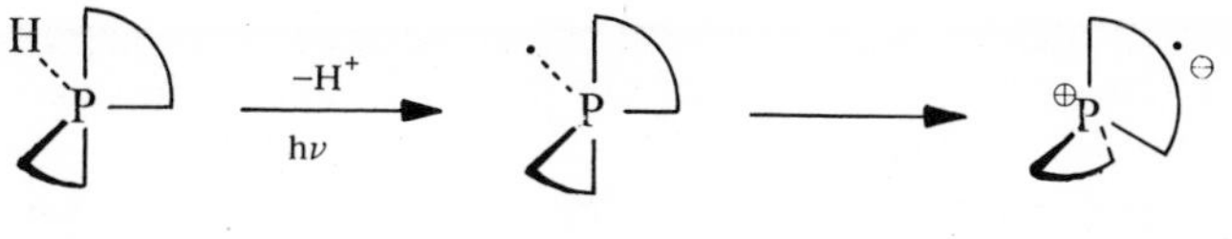

35 **36** **37**

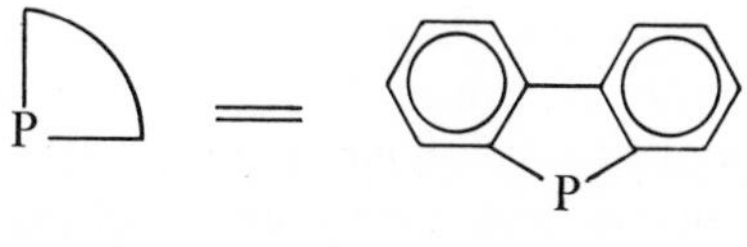

Because phosphorane (**35**) has a TBP configuration, the initial species formed (**36**) will have a similar geometry. However, a large a_P (600 to 1300 gauss), representative of a TBP configuration, was not observed. Therefore a stereoisomerization to structure **37** was indicated. Radical **37** could also be generated by electrolytic reduction of the corresponding phosphonium salt. From the observed proton hyperfine splittings a_H = 0.85 (4H), 1.70 gauss (2H), 2.55 gauss (2H) and a phosphorus splitting of 18 gauss it was concluded that the spin is delocalized in one biphenyl system. A spin density of 0.364 on the two carbons linked to phosphorus has been calculated by using relation (8) with Q_{CP}^P = 40.68 gauss [128]. Similar results were obtained for related radicals substituted with electron donating groups such as OCH_3 (**38**) and $N(CH_3)_2$ (**39**) groups. The phosphorus splittings of compounds **37**, **38** and **39** appeared to be temperature dependent as shown in the accompanying table.

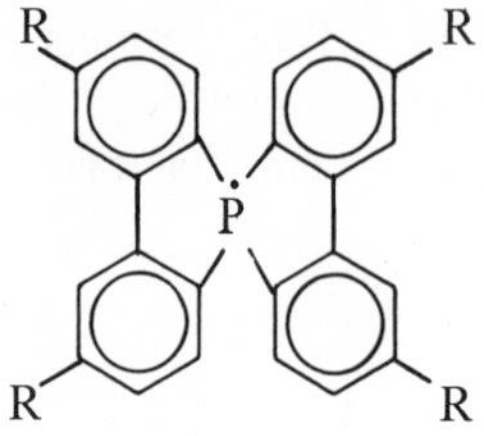

Temp.	$a_P(R{=}H)$	$a_P(R{=}OCH_3)$	$a_P(R{=}N(CH_3)_2)$
20°	18.6 G	21.6 G	22.9[a] G
0°	17.4	20.3	21.3[a]
−20°	16.7	19.3	19.6[a]
−40°	15.6	18.3	17.8[a]
−60°	14.7	17.1[a]	16.8[a]
−80°	14.2	16.3[a]	15.5[a]
−100°	13.2[a]	15.7[a]	
−120°	12.9[a]		

[a] After irradiation of the diamagnetic solution

At lower temperatures a decrease in radical concentration was observed. This process appeared to be reversible. Dimerization was postulated according to

$$2(\mathrm{bi}\phi)_2\mathrm{P}\cdot \underset{\Delta \text{ or } h\nu}{\overset{-\Delta}{\rightleftharpoons}} [(\mathrm{bi}\phi)_2\mathrm{P}]_2$$

Electron-donating substituents, as in **38** and **39**, appear to destabilize the monomeric form [125, 127].

Preferential localization of the unpaired electron in the aromatic ligands has been observed in the unsymmetrically substituted radicals **40** to **44**, which are formed during electrolytic reduction of the corresponding phosphonium salts. As inferred from proton splittings, the odd electron in radicals **40** to **43** is delocalized in the biphenyl system. In radical **44**, however, the unpaired electron appeared to be completely localized in the *p*-nitrophenyl group. In this ligand a unit spin density has been calculated from the hyperfine splittings: a_P = 6.5 gauss, a_N = 13.0 gauss, and a_H = 3.3 gauss, with Q_P = 40.68 gauss and Q_N and Q_H both 27 gauss. The localization of the unpaired electron in the *p*-nitrophenyl group has been explained

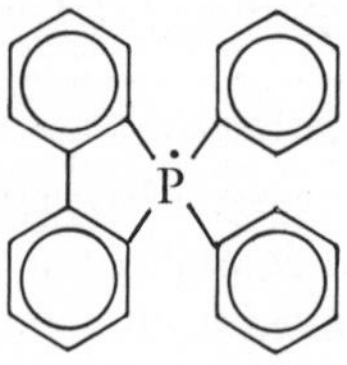

40

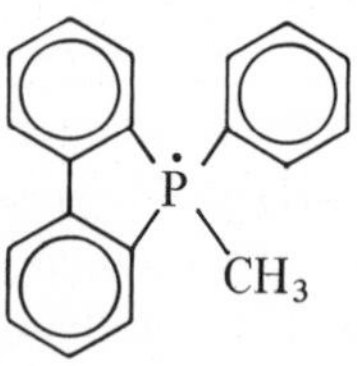

41

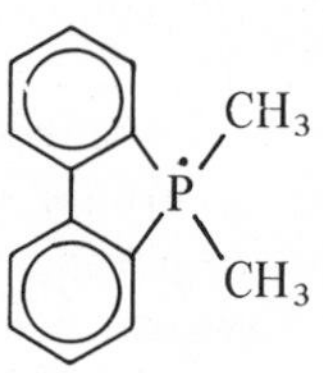

42

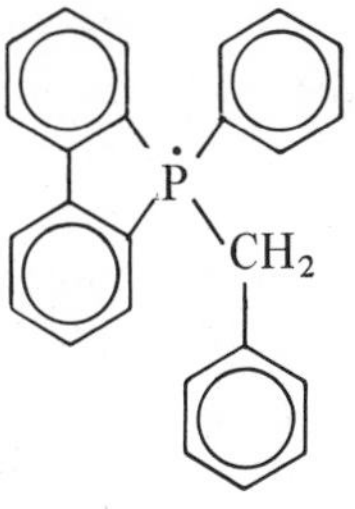

43

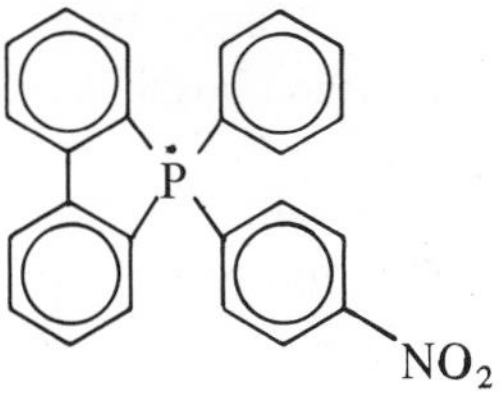

44

by a higher electron affinity of this ligand relative to the biphenylene system and the phenyl group. The spin distributions in compounds **40** to **44** were supported by PPP-UHF calculations, in which all five 3*d*-orbitals of phosphorus (tetrahedral configuration) participate in the π electron system. These calculations indicate a spin density on phosphorus as given in tabular form for compound **40** [126]. Delocalization of the odd electron to the phosphorus by p_π-d_π bonding was also observed in

Calculated odd electron density

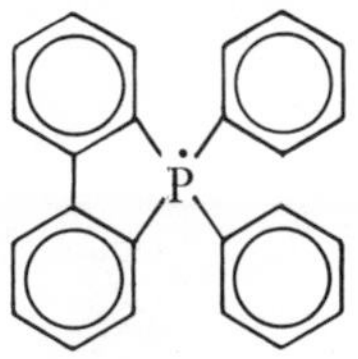

Phosphorus	Biphenylene	Phenyl
+0.20	+0.85	-0.025

compound **45**, generated by electrolytic reduction of the corresponding phosphonium salt [76, 125]. The hyperfine splittings $a_P = 4.05$ gauss, $a_H = 2.70$ gauss (4H), and $a_H = 1.35$ gauss (8H) indicate delocalization of the unpaired electron over the phosphorus atom and the four phenyl rings. An odd electron density of 0.4 has been calculated on the phosphorus atom and of 0.15 in each of the phenyl groups [125].

The observed small a_P is in accord with the π spin density on phosphorus, which gives a positive contribution to the negative isotropic coupling due to spin polarization in the four C–P σ bonds (4). Delocalization of spin to the phosphorus nucleus by π bonding appears to be more important than in the phosphobetaine

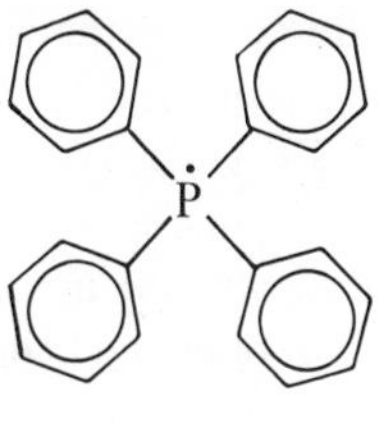

45

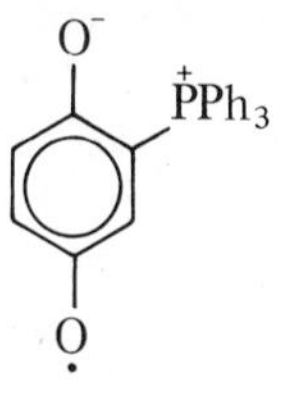

46

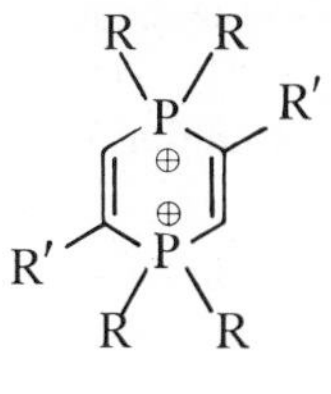

47

radical cations. In the latter species the unpaired electron occupies a nonbonding orbital, whereas in the phosphobetaine free radicals the odd electron resides in the lowest antibonding level of the aryl ligand which is energetically more proximate to the phosphorus 3*d* levels.

Preferential localization of the odd electron had been reported earlier for the semiquinone phosphobetaine radical **46** [95], formed during the reaction of triphenylphosphine with *p*-benzoquinone. It was concluded that the unpaired electron is largely confined to the semiquinone part of the molecule and that it is best described as a substituted semiquinone. In contrast to this observation, the unpaired electron in radical cations of (**47**) appears to occupy a molecular orbital that includes the two phosphorus atoms and the four ring carbons [124]. From the reported proton splittings (a_H = 3 gauss) a spin density of about 0.1 on the ring carbons can be deduced, leaving 0.3, in a *d* orbital on each of the phosphorus nuclei. Using relation (4), we can calculate a phosphorus splitting according to: $-40 \times 0.2 + 80 \times 0.3 = 16$ gauss, which is in excellent agreement with the reported phosphorus splittings of 13.2 to 16 gauss for various substituted compounds. The value of 80 gauss in (4) is derived from $\dot{P}H_2$, in which the unpaired electron resides in a *p* orbital on phosphorus. Thus the result of the preceding calculation indicates that the value for the polarization parameter Q^P_{PP} does not vary significantly for an electron either in a *p* orbital or a *d* orbital on phosphorus.

3. PHOSPHINE OXIDE ANION RADICALS, $R_3PO^{\overline{\cdot}}$

Alkyl- and arylphosphine oxides react with alkali metals to give paramagnetic solutions of "phosphyls":

$$R_3P{=}O + M \longrightarrow R_3\dot{P}{-}OM$$

Kabachnik [80] and Cowley and Hnoosh [26] have generated triphenyl phosphyl by treatment of triphenylphosphine oxide with potassium in tetrahydrofuran. Both report a phosphorus doublet splitting of 5.25 gauss. Different assignments were made for the aromatic proton splittings: Kabachnik assigned values of 3.4 gauss for three para protons and 1.75 gauss for six ortho protons, whereas Cowley and Hnoosh indicate nine equivalent proton splittings of 1.75 gauss. The latter authors observed an additional potassium splitting of 0.875 gauss. These assignments were supported by a HMO spin density calculation [26]. Ilyasov et al. [75] generated the triphenylphosphine oxide free radical by electrolytic reduction of the parent compound. The ESR spectrum could be explained by hyperfine splittings of phosphorus and the aromatic protons at a ratio of constants $a_P : a^p_H : a^o_H = 3{:}2{:}1$; which is in agreement with the observations of Kabachnik. A phosphorus doublet splitting of 6.4 gauss was observed. The higher value for the phosphorus doublet splitting compared with those of the phosphyls has been ascribed to the free anion character of this radical [75]. Conjugation of the unpaired electron over phosphorus and the three phenyl rings has been demonstrated by subsequent introduction of pentadeuterophenyl groups. On each introduction of a C_6D_5 group, the number of proton splittings was reduced [77].

Alkyl phenyl-substituted phosphine oxide anion radicals have also been reported. Potassium reduction of phenyl dimethyl phosphine oxide resulted in the formation of $(C_6H_5)(CH_3)_2\dot{P}(O)^-$. The 29-line spectrum has been explained by a phosphorus splitting of 8.75 gauss, three aromatic proton splittings of 3.5 gauss, and methyl splittings of 0.875 gauss [26]. Kabachnik [80, 81] published the spectrum of phenyl dimethyl phosphyl together with those of diphenyl phosphyl and tri-*n*-butyl phosphyl. The ESR spectrum of tri-*n*-butyl phosphyl reveals a single line with a poorly defined hyperfine structure consisting of 12 components that could be explained, assuming coupling of the odd electron with the phosphorus and six α protons. The small phosphorus splitting constant has been interpreted by localization of the odd electron in a *d* orbital on phosphorus which should result in a weak interaction. Such localization, however, would result in a splitting of the same order of magnitude as for a *p* orbital on phosphorus, which is 80 gauss (D2). Therefore delocalization over phosphorus and oxygen seems to be more reasonable. In this case the opposite signs of the polarization constants Q^P_{PP} and Q^P_{PO} are responsible for the observed small a_P (4). Delocalization of the odd electron over phosphorus and nitrogen by p_π-d_π bonding has been observed in $[(CH_3)_2N]Ph_2\dot{P}(O)^-$ [26]. The ESR spectrum showed an a_P of 7.2 gauss and a nitrogen splitting of 4.9 gauss. In contrast to this observation $[(CH_3)_2N]_3P(O)$ gives an ESR singulet when treated with alkali metals [50, 120]. More dilute solutions exhibit a 35-line ESR spectrum [25] which has been ascribed to a solvated electron with four equivalent, nearest neighbor solvent molecules, tetrahedrally arranged, with an a_P of 4.91 gauss, and six next-nearest neighbor solvent molecules, octahedrally arranged with an a_P of 1.83 gauss.

Phosphole oxide anions (**48**) and (**49**) where generated by Thomson and Kilcast [139] by reduction of the parent phosphine oxides with lithium, potassium, or sodium in both 1,2-dimethoxy ethane and tetrahydrofuran.

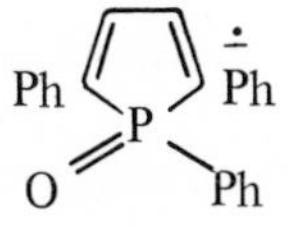

48

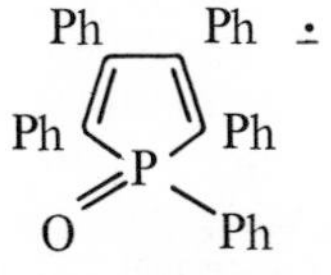

49

The observed phosphorus splitting depends on the alkali metal used. For compound **48** the a_P varied from 14.95 to 16.22 gauss, for **49** from 15.25 to 16.40 gauss. An additional hyperfine structure was observed, characteristic of the metal used. The dependence of the phosphorus splitting and the additional fine structure on the alkali metal involved have been ascribed to association of the gegenion with the phosphorus nucleus via the phosphorus d orbitals [139].

E. Phosphorus-Substituted Nitroxide Free Radicals

To identify unstable, short-lived free radicals that occur as intermediates in reactions of organo phosphorus compounds the technique of spin trapping has been used. Nitroso compounds are often employed as scavengers.

Adevik and Lagercrantz [2] reported the spin trapping by 2-methyl-2-nitrosopropane of the radical ion PO_3^{2-}, formed by γ irradiation of sodium phosphite. The resulting nitroxide radical (50) is stable enough to be detected by ESR spectroscopy.

$$\left[\begin{array}{c} O_3P-\underset{\underset{O\cdot}{|}}{N}-C(CH_3)_3 \end{array} \right]^{2-}$$

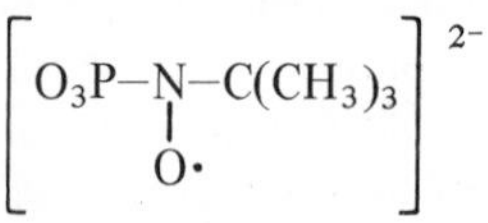

50

The six-line ESR spectrum could be interpreted by interaction of the unpaired electron with phosphorus, a_P = 12.0 gauss, and nitrogen, a_N = 13.4 gauss. By using the same scavenger a number of phosphonyl and phosphinyl radicals have been trapped in aqueous solution [82]. The ESR spectra of the resulting radicals showed phosphorus splittings of 6 to 14 gauss.

Gagnaire et al. [51] reported the ESR spectrum of radical 52, generated by oxidation of 51 with nitrobenzoic acid. The ESR spectrum shows a phosphorus

51   **52**

splitting of 12.8 gauss and a nitrogen splitting of 10.5 gauss, which were noted to be different from those of related structure 50 [51].

Boekestein et al. [19] reported the occurrence of the tris(dimethylamino)-phosphinium radical as an intermediate species in the reaction of tris(dimethylamino)phosphine with bis(p-nitrophenyl)methyl bromide in tetrahydrofuran or dimethyl formamide. The intermediate phosphinium radical, trapped with 2-methyl-2-nitrosopropane, yielded spin adducts 53 and 54 as a result of

$$t\text{-Bu-N-}\overset{+}{\underset{\cdot O}{\text{P}}}[N(CH_3)_2]_3$$

$$\overset{\ominus Br}{t\text{-Bu-N-}\underset{\cdot O(CH_3)_2}{\overset{\oplus}{N}}\text{-P}[N(CH_3)_2]_2}$$

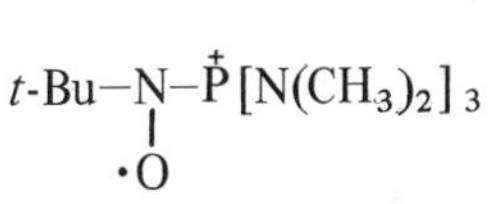

53   **54**

trapping at phosphorus and nitrogen, respectively. The structures of **53** ($a_N = 11.9$ gauss, $a_P = 14.8$ gauss) and **54** ($a_N = 18.4$ gauss and 0.85 gauss) were derived from the hyperfine splittings. Structure **54** was also observed in the reaction of tris-dimethylamino)phosphine with 9,10-phenanthrenequinones in the presence of 2-methyl-2-nitrosopropane [21]. A second species was detected, which has been ascribed to the trapped dimer phosphorus radical **55**, as inferred from the hyperfine splittings $a_P = 11.5$ and 3.1 gauss and $a_N = 12.2$ gauss.

$$[(CH_3)_2N]_3P + \cdot\overset{\oplus}{P}[N(CH_3)_2]_3 \longrightarrow [(CH_3)_2N]_3\overset{\oplus}{P}\text{-}\overset{\cdot}{P}[N(CH_3)_2]_3 \longrightarrow$$
$$+N=O$$
$$\longrightarrow [(CH_3)_2N]_3\overset{\oplus}{P}\text{-}\underset{\cdot O\text{-}N}{P}[N(CH_3)_2]_3$$
$$\underset{+}{\cdot O\text{-}N}$$

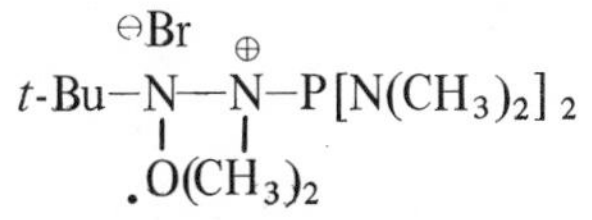

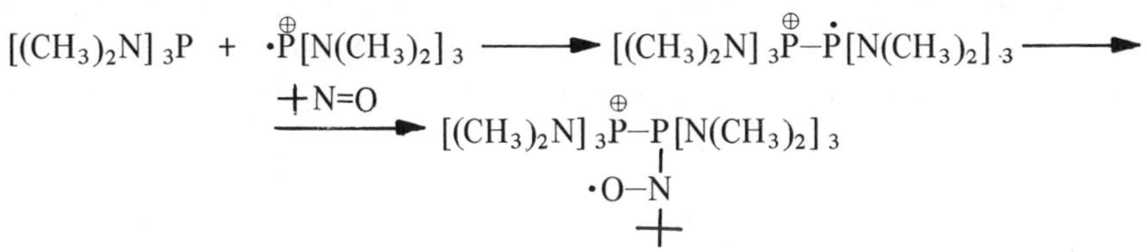

55

II. β-Phosphorus Substituted π Radicals

A. Radical Type $\overset{\cdot}{R}\text{-}O\text{-}P$

The ESR spectra of a number of phosphorus-substituted aromatic radicals, in which phosphorus is separated from the aromatic moiety by oxygen, have been reported. Gulick and Geske [64] generated a series of nitro aromatic anion radicals, para substituted with various phosphate groups by electrolytic reduction. From the similarity of the nitrogen and aromatic proton splittings to those of the unsubstituted nitrobenzene anion radical it was concluded that the phosphorus substituent has not significantly perturbed the spin distribution in the nitrophenyl ring.

The observed phosphorus splitting appeared to increase on introduction of steric restrictions, which is obvious for **57** (10.5 gauss) compared with **56** (7.92 gauss). This indicates [64] that the coupling is favored by conformations in which phosphorus is out of the plane of the aromatic ring. The apparent angular dependence of the coupling to phosphorus has been proposed to be of the form

$$a_P = C \langle\cos^2\theta\rangle$$

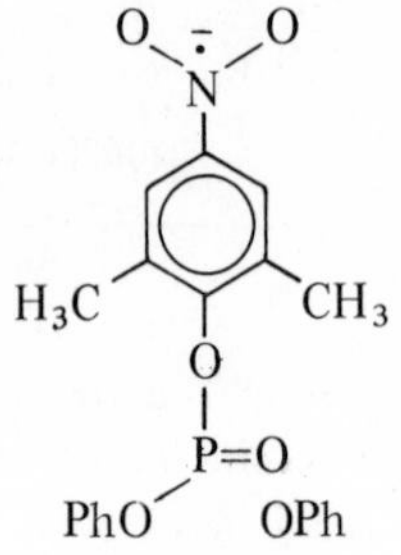

57

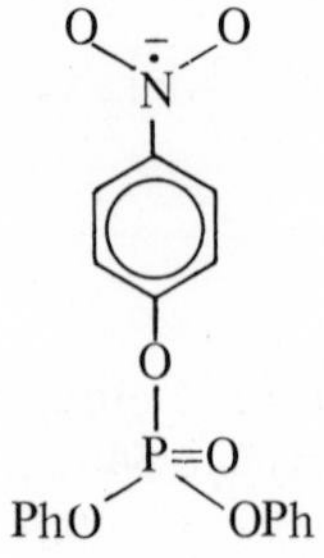

56

in which θ is the dihedral angle between to O—P bond and the axis of the p_z orbital associated with the aromatic system. This angular dependence has been interpreted as possible evidence of a hyperconjugative coupling mechanism. A similar phenomenon was noted previously by Allen and Bond [3] who studied the ESR of semiquinone phosphate free radicals **58** and **59**, which were generated by oxidizing quinol phosphates with $KMnO_4$.

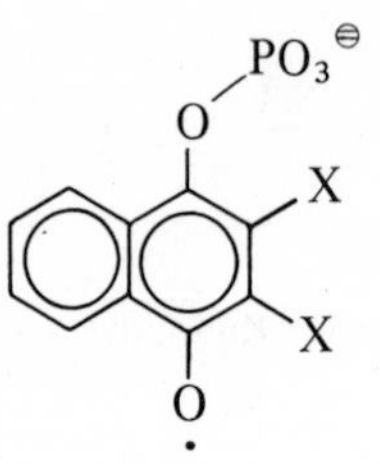

58

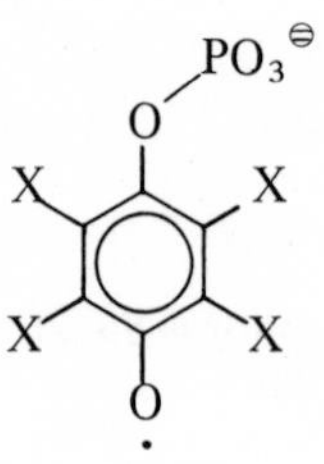

59

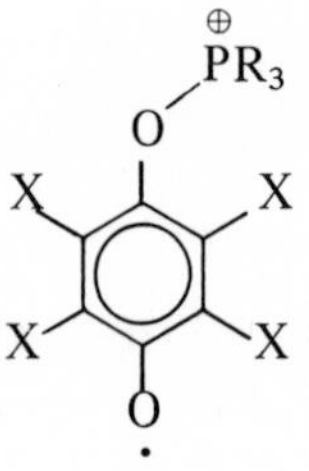

60

Large phosphorus splittings (17 to 20 gauss) were observed only when X was a bulky group like methyl or chloro, whereas a much smaller splitting (2 gauss) was noted when X was hydrogen. The relatively large phosphorus splitting has been explained by assuming a hyperconjugative mechanism that involves overlap of the 3*s* orbital of phosphorus with the carbon π system. Similar radicals were

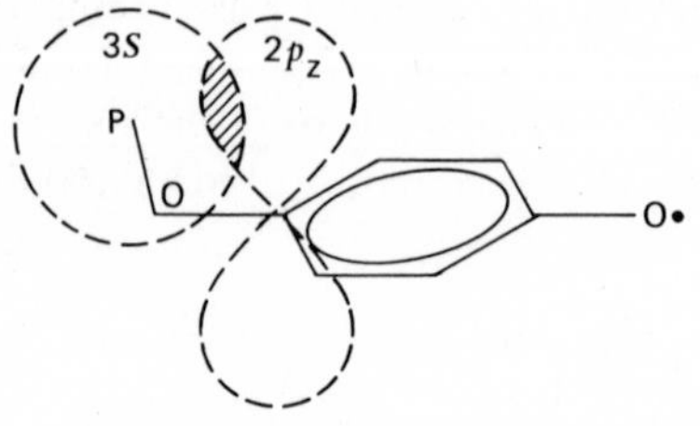

proposed as intermediates in the reaction of phosphines with tetrahalosemiquinones (Cl, Br, I) [100]. Phosphorus hyperfine splittings were observed in the range of 1.80 to 2.42 gauss. Because the a_P depends on the nature of the quinone used, it was concluded that the observed species were not identical to the triphenyl phosphonium radical. Therefore the structure of the paramagnetic species was assigned to **60**. The observed a_P's, however, are very small in comparison with the phosphorus splitting of the closely related structure **59**, which shows an a_P of 20 gauss for X = Cl. This was further confirmed by the phosphorus splitting of the paramagnetic species **61** (19.1 gauss), formed during the reaction of diethyl acetyl phosphoramidite with chloranil at −70°C [123]. A similar adduct radical (**62**) (a_P = 11.2 gauss) was assigned as an intermediate in the reaction of tris(dimethylamino)-phosphine with 9,10-phenanthrenequinones [21].

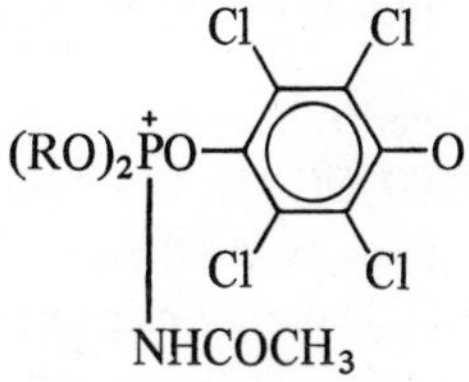

61

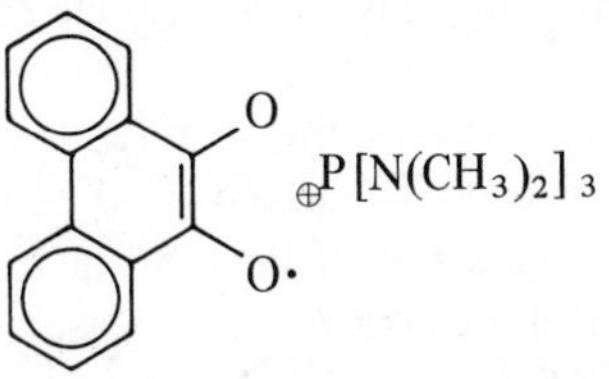

62

Angular dependence of coupling to phosphorus has also been established for alkyl π radicals separated from the phosphorus by oxygen. A number of these radical types have been generated in aqueous solution by the action of hydroxyl radicals on alkyl phosphates. Lucken [96] reported the ESR of radicals of types $R_1R_2P(O)O\dot{C}H_2$ and $R_1R_2P(O)O\dot{C}HCH_3$, where R_1 is alkoxy and R_2 is H, OH, alkyl and alkoxy. In spite of the fact that the carbon 2 p_z spin population is equal in both radical types, the ^{31}P coupling of the latter (11.2 to 14.6 gauss) appeared to be 1.5 times larger than that of the former (7.8 to 10.5 gauss). This has been explained by steric hindrance of the methyl substituents that forces the P–O bond out of the nodal plane of the carbon $2p_z$ orbital, thus increasing the extent of hyperconjugation of the phosphate group. An increase in a_P on the imposition of more steric restrictions was also indicated by the results of Metcalf and Waters [106] for the radicals $Me\dot{C}HOPO(OEt)_2$ and $Me_2\dot{C}OPO(H)OCHMe_2$. The phosphorus splitting of the latter (31.9 gauss) is much larger than that for the former, less substituted radical (14.1 gauss). Consequently, still smaller phosphorus splittings (4.5 to 5 gauss) were observed for radicals of type $R\dot{C}HOPO_3$ derived from monoalkyl phosphates [130].

Symons et al. [11] reported an ESR study of the radical $H_2\dot{C}OPO(OMe)_2$, generated in a solid matrix by γ irradiation of frozen trimethyl phosphate at 77°K. The value of the proton hyperfine splitting (a_H^{iso} = 19.5 gauss) is close to previously reported values for this radical in liquid phase (20.4 gauss [96] and 20.1 gauss [106]), whereas the value for the phosphorus splitting (a_P^{iso} = 4.0 gauss) differs markedly; 10.5 [96] and 10.3 gauss [106]. The differences have been explained by assuming various geometries. The radical in solid matrix is supposed to be planar. In this configuration the observed a_P rises from spin polarization in the P—O σ bond because of the positive spin density on oxygen by direct π overlap, which induces a negative spin density on phosphorus. Assuming a value of 30 gauss for the polarization constant Q_{PO}, a spin density of 0.14 has been calculated on oxygen, which leaves 0.86 on carbon. The larger phosphorus splitting in solution has been explained by assuming a hyperconjugative mechanism caused by an out-of-plane twisting of the phosphate group with respect to the nodal plane of the carbon 2 p_z orbital. Hyperconjugation should give a positive contribution to the phosphorus splitting [11]. Because the sign of the a_P is unknown, the mechanism of the phosphorus coupling, either spin polarization or hyperconjugation or both, remained uncertain.

The mechanism of the interaction of the unpaired electron with the phosphorus nucleus has been elucidated for the radical $CH_3\dot{C}HOP(O)_2OC_2H_5$. Ezra and Bernard [47] obtained this radical by γ irradiation of magnesium diethylphosphate at 77°K after annealing to 160°K. Isotropic values of +4.5 and +6.7 gauss were determined at 210 and 270°K, respectively. The radical structure was established by deuterium and ^{13}C substitution. A spin density of 0.88 could be derived for the α carbon atom. From the remaining spin density of 0.12 on oxygen a phosphorus splitting of −3.6 gauss, due to spin polarization, was established by comparison with the PO_4^{2-} radical [73, 135]. Therefore the hyperconjugation contributes +8.1 gauss to the isotropic phosphorus coupling of +4.5 gauss [47].

B. Radical Type $\dot{R}-C-P$

Conformational preferences were found in β-phosphorus substituted alkyl π radicals in which phosphorus is separated from the radical center by a tetrahedral-carbon. Symons et al. [137, 102, 103] reported the ESR spectra of **63** and **64**,

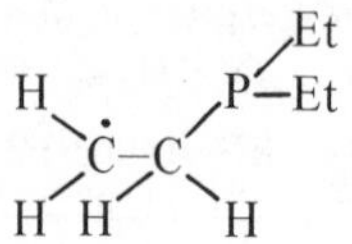

63

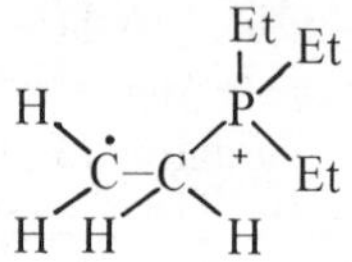

64

generated in solid matrix by γ irradiation of powdered samples of the corresponding phosphines and phosphonium salts at 77°K. The ESR spectrum of **63** showed a large, slightly anisotropic phosphorus splitting from 230 to 295 gauss, from which an isotropic coupling of 251 gauss has been derived. This large coupling was ascribed to delocalization of the unpaired electron to phosphorus, resulting from a preferred conformation in which the phosphorus nucleus eclipses the carbon $2p_z$ orbital occupied by the unpaired electron [102, 103]. A similar phenomenon was found in radical **64**, with a tetravalent phosphorus. A smaller phosphorus coupling constant ($a_P^{iso} = 120$ gauss) has been observed, however. The greatly enhanced coupling in radical **63** with respect to **64** was ascribed to a homoconjugation effect involving the phosphorus lone-pair electrons. Because the lone-pair orbital is strongly hybridized, it results in an increase in the isotropic coupling [137]. The mechanism of spin delocalization

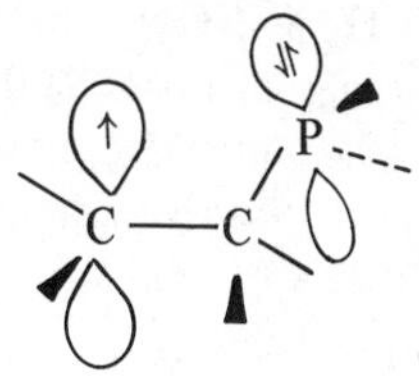

onto the tetravalent phosphorus in radical **64** is unclear. Symons [103] pointed out that an interaction involving p_π-d_π overlap is unlikely because of the large isotropic character of the coupling. Alternatively, a hyperconjugative interaction involving the C—P sigma bond has been proposed. This mechanism was also suggested to account for the large phosphorus splitting in the phosphorus-substituted cyclohexadienyl radical **65** [109]. Radical **65** was observed along with the σ radical PhṖO(OH) on γ

H

P—

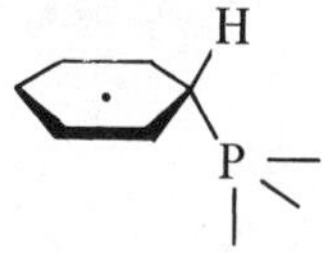

65

irradiation of phenylphosphonic acid PhPO(OH)H at 77°K. The nearly isotropic ESR spectrum of **65** showed proton couplings of 40 gauss (1H) and 10 gauss (3H) and an a_P of 200 gauss. Damerau and Lassman [27, 28, 29, 93] generated radicals of the type **66** from the parent trichloro compounds by reaction with $\cdot CO_2^-$ radicals

$$R_1O \underset{R_1O}{\overset{}{\diagdown}} \overset{\overset{O}{\parallel}}{\underset{}{P}} - \overset{\overset{OR_2}{\mid}}{\underset{}{CH}} - \dot{C}Cl_2$$

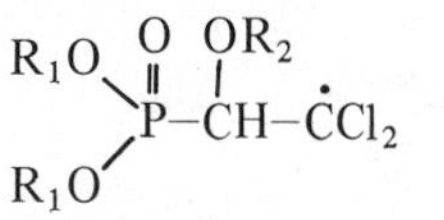

66

in aqueous solution. The phosphorus coupling constants (60 to 69 gauss) appeared to depend on the size of R_1 and R_2 and, moreover, varied with temperature. Therefore it was concluded that the coupling to phosphorus is angular-dependent and is probably hyperconjugative [27].

A number of β-phosphorus-substituted alkyl radicals have been generated in the liquid phase by addition of phosphorus centered radicals to unsaturated compounds. The radicals are characterized by large phosphorus splittings in the range of 30 to 180 gauss, which were commonly attributed to conformational preferences.

Beckwith [7] reported the reaction of the radicals $H\dot{P}O_2$, $\dot{H}PO_3$, and $\dot{P}O_3^{2-}$ with unsaturated compounds of the type $R_2C = CR'_2$, where R,R′ is H, CH_3, and CO_2H and nitroalkane acianions ($CH_2 = NO_2$). The ESR spectra of the adducts show phosphorus splittings in the range of 60 to 100 gauss. The coupling to phosphorus was discussed in terms of a conjugative mechanism involving phosphorus *3d* orbitals. Gilbert et al. [56] performed similar reactions. The phosphorus splittings of the adducts (25 to 75 gauss) appeared to depend on conformation, as inferred from their variation with temperature.

Davies et al. [32] reported the addition of dialkoxyphosphonyl radical $(RO)_2\dot{P}O$ (R = alkyl) to terminal olefins. The ESR spectra show phosphorus splittings of 94 to 112 gauss, consistent with β-phosphorus substituted alkyl radicals. Similar radicals (**67**) (a_P = 77 – 122 gauss) were obtained on addition of spirophosphoranyl radicals to olefins [62].

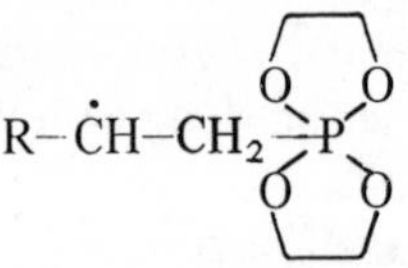

67

Intramolecular addition reactions were observed by Davies et al. [36]. Phosphoranyl radicals containing 2-alkenoxy substituents cyclize to give oxaphosphetans (**68**), with the unpaired electron centered on an exocyclic carbon atom. Phosphorus splittings of 176 to 199 gauss were observed for various substituents. When 3-alkenoxy substituents were present, oxaphospholan radicals (**69**) were obtained (a_P^{iso} = 127 – 136 gauss).

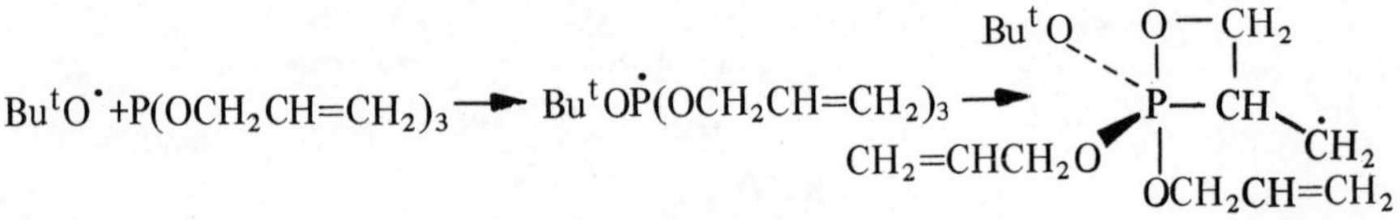

68

$$R\dot{P}(OCH_2CH_2CH{=}CH_2)_3 \longrightarrow$$

(structure 69)

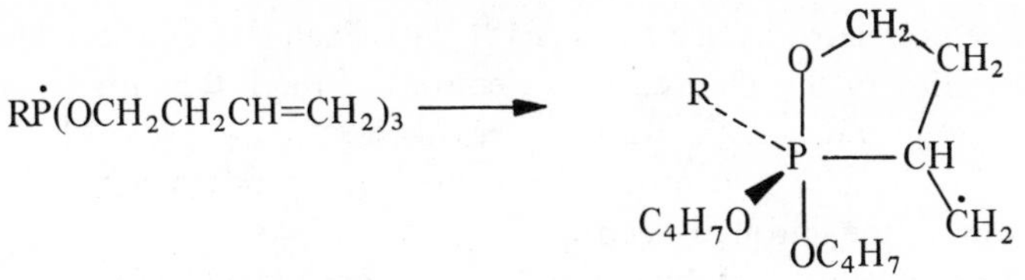

69

III. Phosphorus-Centered σ-Radicals

A. Tetrahedral Phosphorus Radicals

Particularly in the last few years, many ESR studies concerning tetrahedral phosphorus radicals have been reported. These radicals can be classified into four types: $\dot{P}R_3^+$, $R_2\dot{P}O$, $R\dot{P}O_2^-$, and $\dot{P}O_3^{2-}$. All have structures in which the unpaired electron is located in a hybrid orbital on phosphorus with appreciable s character, resulting in phosphorus splittings in the range of 400 to 800 gauss. The major factor for the differences in the a_p's is the relative electronegativity of the ligands, as pointed out by Symons et al. [8, 12].

The 3s character of the orbital containing the unpaired electron can be estimated by comparison of the observed isotropic phosphorus splitting constant with the Hartree-Fock value of 3640 gauss for an electron in the valence 3s orbital of phosphorus [6, 145, 146]. The 3p character can be inferred from the anisotropic component $B(P)$ of the phosphorus splitting by comparison with the corresponding value $B(P)^{ref}$, calculated for an isolated atom [6]. A reference value of 103 gauss [145, 146] is commonly used.

According to this method, p/s ratios have been calculated for many compounds. From the results it appeared that the radicals have a pyramidal structure in which the odd electron occupies an approximate sp^3 orbital on phosphorus. In a few cases hyperfine splittings were supported by CNDO and UHF/RHF calculations [1, 59, 68, 87].

1. PHOSPHINIUM RADICALS, $R_3\dot{P}^+$

Compounds identified as phosphinium radicals by their ESR spectra have been prepared by exposure of the corresponding trialkyl phosphines to γ-rays after solution in sulfuric acid. Using this method, Symons et al. generated the radicals $\dot{P}H_3^+$, $\dot{P}Me_3^+$, $\dot{P}Et_3^+$, $\dot{P}Bu_3^+$, and $\dot{P}(OH)_3^+$ [8, 9, 14, 58], although the assignment of the $\dot{P}(OH)_3^+$ radical was uncertain [58]. In addition, dimeric radicals of type $[R_3P{-}PR_3]^{+\bullet}$ were observed [104]. For the $\dot{P}R_3^+$ radicals isotropic phosphorus splittings were reported of 795 (R = OH), 517 (R = H), 393 (R = Me), 384 (R = Et), and 360 gauss (R = Bu). This decrease in a_p^{iso} and consequent increase in the p/s ratio (from 3.2 to 9.6) of the phosphorus hybrid orbital containing the unpaired electron were correlated by the authors [8] with Paulings electronegativity theory [121]. According to this theory, the σ electrons of the P-R bonds are increasingly withdrawn from phosphorus with increasing electronegativity of the ligands. This results in a smaller 3*s* contribution to the P-R σ bonds, hence in an increasing 3*s* population for the unpaired electron [8].

Formation of dimeric phosphinium radicals was postulated by Lyons and Symons [104] according to (**70**).

$$\dot{P}R_3^+ + PR_3 \longrightarrow [R_3P\text{—}PR_3]^{+\cdot}$$

70

From the ESR spectra, however, the existence of **70** could not be confirmed because the central line of the expected triplet (due to two equivalent phosphorus atoms) was obscured by unidentified species. Two identical phosphorus splittings of 454.7 gauss were assigned to compound **70** with R = Et and of 462 gauss with R = Bu. From anisotropic components *p*/*s* ratios of 3.3 and 3.4, respectively, were estimated. Calculations [104] indicate that PR_3 groups are more pyramidal than the monomeric cation radicals $\dot{P}R_3^+$. As expected, the dimer species $[R_3P\text{—}PR_3]^{+\cdot}$ have an intermediate structure, which was concluded form the 3*s* and 3*p* character.

By means of UHF and RHF calculations Aarons et al. [1] predicted the geometries and the hyperfine coupling constants of $\dot{P}H_3^+$, $\dot{P}F_3^+$, and $\dot{P}Cl_3^+$, as shown in tabular form. Only those of $\dot{P}H_3^+$ could be compared with experimental results.

Radical	a_P^{iso}	a_X	Angle out of plane	Experimental Results
$\dot{P}H_3^+$	309.9	7.0	15°30′	a_P^{iso}=517; $a_H \leqslant 10$; θ=14°30′
$\dot{P}F_3^+$	958.2	41.4	19°48′	
$\dot{P}Cl_3^+$	555.6	-1.4	18°	

Another recent calculation on $\dot{P}H_3^+$ was reported by Gorlov et al. [59], using the UHF-INDO method. s-Orbital spin densities were computed of 0.12 for phosphorus and of 0.096 for hydrogen, from which an a_p of 442 gauss and an a_H of 48 gauss can be derived which are in good agreement with the experimental results.

2. PHOSPHONYL RADICALS, $R_2\dot{P}O$

ESR studies concerning phosphonyl radicals, measured both in solid state and solution, revealed isotropic phosphorus splittings of 350 to 750 gauss. The lowest values for these phosphorus splittings (355 to 375 gauss) in these series were determined for phosphonyl radicals with alkyl substituents. The $Me_2\dot{P}O$ radical

(a_P^{iso} = 375 gauss) has been generated by Begum and Symons [14] on γ irradiation of Me_3PO at 77°K. At 140°K the ESR spectrum showed well-resolved proton splittings of 5.6 gauss. A p/s ratio of 7.6 was calculated for the orbital with the unpaired electron.

Alkyl- and alkoxy-substituted phosphonyl radicals were measured in solution by Davies et al. [30]. In this study a relation was shown between the isotropic phosphorus splitting constants of phosphonyl radicals of type $R_2\dot{P}O$ and phosphoranyl radicals of type $\dot{P}R_4$ (Section B) and the $J_{P\text{-}H}$ coupling constants derived from the NMR spectra of the protic parents, $R_2P(O)H$ and HPR_4.

$$a_P/\text{gaus} = 1.52 \cdot J_{P\text{-}H}/\text{Hz} - 350 \tag{9}$$

The values of the phosphorus splittings were not reported, but from the plot of (9) istropic phosphorus splittings of about 355 gauss can be estimated for $Et_2\dot{P}O$ and $(n\text{-}C_6H_{13})_2\dot{P}O$ and a value of about 450 gauss for $Et(EtO)\dot{P}O$.

When alkyl was replaced by alkoxy groups, an increase in a_p was observed. Dialkoxyphosphonyl radicals appeared to have isotropic phosphorus splittings of 671 to 700 gauss. These compounds were generated in solutions by UV irradiation of the parent tetraalkyl pyrophosphites (**71**) or dialkyl phosphonates (**72**) in the presence of di-*t*-butyl peroxide [32]:

$$Bu^tO{-}OBu^t \xrightarrow{h\nu} 2\ Bu^tO\cdot$$

$$Bu^tO^\cdot + \underset{\textbf{71}}{(RO)_2POP(OR)_2} \longrightarrow Bu^tP(OR)_2 + O\dot{P}(OR)_2$$

$$Bu^tO^\cdot + \underset{\textbf{72}}{(RO)_2P(O)H} \longrightarrow Bu^tOH + (RO)_2\dot{P}O$$

Isotropic phosphorus splittings were reported in the range of 685 to 700 gauss. Similar splittings were determined in the solid state by Kerr et al. [84] for phosphonyl radicals with MeO- (684 gauss) and EtO-substituents (671 gauss), which were obtained in a pure state on γ irradiation of phosphorus esters in glassy solutions of methyl bromide. From the anisotropic splittings the p/s ratio was calculated to be about 3.0, consistent with a sp^3 hybridization.

Various ESR studies have been reported on phenyl-substituted phosphonyl radicals in solid matrices. On X-irradiation of single crystals of diphenylphosphine oxide the diphenylphosphonyl radical $Ph_2\dot{P}O$ could be identified by means of its ESR spectrum which shows an isotropic phosphorus splitting of 390 gauss at room temperature [53]. From a comparison with the π radical $Ph_2NO^\cdot$ it was concluded that in $Ph_2\dot{P}O$ no delocalization of the unpaired electron onto the phenyl rings takes place by p_π-d_π

bonding. Similar results were reported [52] for the sulfur analog $Ph_2\dot{P}S$ (a_P^{iso} = 351 gauss). Irradiation of a single crystal of phenyl phosphonic acid at 77°K gave rise to the radical $Ph\dot{P}(O)OH$ (a_P^{iso} = 540 gauss) [54]. The anisotropic proton splittings, 15 to 17.4 gauss, could be assigned by comparing with the dueterated compound $Ph\dot{P}(O)OD$.

Mishra and Symons [108] observed the radicals $Ph\dot{P}(O)Cl$ and $Ph\dot{P}(S)Cl$ formed on chloride ion loss by the parent anion radicals $Ph\dot{P}OCl_2^-$ and $Ph\dot{P}SCl_2^-$ (Section B). These compounds were identified by comparing with similar radicals like $H\dot{P}O_2^-$ [112] and $Ph\dot{P}O_2^-$ [54].

The highest value for the isotropic phosphorus splittings in this series of phosphonyl radicals was observed in the dihalogen substituted radical $\dot{P}OCl_2$. This radical has been generated by Begum and Symons [13] on γ irradiation of $POCl_3$ at 77°K. Anisotropic splittings of 692 to 893 gauss were observed for phosphorus and of 15 to 24 gauss for the two equivalent chlorine nuclei. At 100°K all hyperfine splittings became isotropic (a_P^{iso} = 759 and a_{Cl} = 18 gauss).

3. ANION RADICALS, TYPE $R\dot{P}O_2^-$

A number of ESR studies has been reported of this type of radicals with various substituents. Isotropic phosphorus splittings lie in the range of 430 to 700 gauss.

The radical $H\dot{P}O_2^-$ (**73**) has been identified in γ-irradiated hypophosphites in several ESR studies. Morton [112] obtained this radical on γ irradiation of ammonium hypophosphite. The ESR spectrum revealed an anisotropic phosphorus splitting from 439 to 606 gauss. A remarkably large isotropic proton splitting of 82 gauss, corresponding to 16% spin in the hydrogen 1*s* orbital, was observed. An irregular tetrahedral structure was inferred from the 3*p*:3*s* ratio of 3.93 for the orbital containing the unpaired electron. Similar results were reported by Atkins et al. [4] for (**73**), generated by γ irradiation of several hypophosphites at various temperatures. $H\dot{P}O_2^-$ was generated in aqueous solution on hydrogen abstraction from hypophosphorus acid by hydroxyl radicals in a rapid flow system [7]. The isotropic hyperfine splittings of 484 and 90 gauss for phosphorus and hydrogen respectively, are consistent with the spectral parameters of this radical in solid matrix [4]. In a similar way the radical $HO\dot{P}O_2^-$ (**74**) (a_P^{iso} = 658 gauss) has been generated from H_3PO_3 as precursor [7].

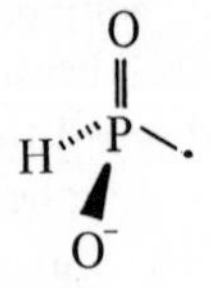

73

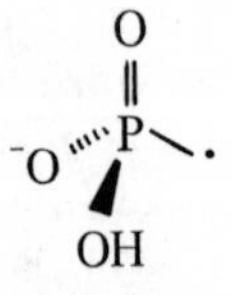

74

A CNDO calculation for $H\dot{P}O_2^-$ by Kilcast and Thomson [87], predicted isotropic hyperfine splittings of 530 and 125 gauss for, phosphorus and hydrogen respectively, which are in good agreement with the experimental results. Morton [112] reported an ESR study of the dimeric phosphorus compound $O_2\dot{P}-PHO_2^{2-}$, which was formed by the reaction of an HPO_2^- radical with a $H_2PO_2^-$ anion. A radical with a similar structure, $O_2\dot{P}-PO_3^{3-}$, was formed on γ irradiation of $P_2O_4^{6-}$ (114).

Geoffrey and Lucken [54] obtained $Ph\dot{P}O_2^-$ (a_P^{iso} = 490 gauss) on X irradiation of various polycrystalline salts of phenylphosphonic acid. The anisotropic parameters, which depend on the kind of gegenion, indicate a nonplanar structure in which the unpaired electron is located on the phosphorus atom (p/s ratio = 4.3). No delocalization onto the phenyl ring was observed.

Kerr et al. [84] reported an ESR study of radicals in γ-irradiated diethylphosphite. Not only the bulk esters but also the glassy solutions of the esters in methylbromide and in 2-methyl-tetrahydrofuran (MTHF) were studied. The radical types formed depend on the nature of the matrix used. $EtO\dot{P}O_2^-$ was exclusively observed in the MTHF matrix. From the hyperfine principle values a $3p/3s$ orbital spin density ratio of 2.98 was derived for the unpaired electron on phosphorus in accord with a tetrahedral configuration. A similar result was reported for the $EtO\dot{P}O_2^-$ radical (a_P^{iso} = 673 gauss) in γ-irradiated magnesium diethyl phosphate [46].

A relatively large value for the phosphorus splitting (a_P^{iso} = 690 gauss), consistent with the high electronegativity of the substituent, was found for $F\dot{P}O_2^-$ [12], which earlier was erroneously assigned to the radical $\dot{P}F^{\pm}$ [115]. The pyramidal $F\dot{P}O_2^-$ has proved to be structurally similar to the isoelectronic $\dot{P}O_3^{2-}$ (Section 4). The isotropic fluorine coupling (165 gauss) was found to be close to the a_F of the pyramidal $\dot{C}F_3$ radical (144 gauss). Similar results were reported by Fessenden [48]. CNDO calculations [87] performed for this radical predicted hyperfine splittings (a_P^{iso} = 552 and a_F = 138 gauss) which are in reasonable agreement with the experimental values.

4. THE HYPOPHOSPHATE RADICAL, $\dot{P}O_3^{2-}$

In a number of studies the radical $\dot{P}O_3^{2-}$ has been identified in irradiated solids by its ESR spectrum. Isotropic phosphorus splittings were found in the range of 550 to 800 gauss. The pyramidal structure has been inferred from the p/s ratios, which varied from 2.7 to 3.3, depending on the nature of the host crystal. Published work includes γ irradiation of disodium orthophosphite ($Na_2HPO_3 \cdot 5H_2O$) (72), calcite ($CaCO_3$) with phosphorus impurities [133], alkali phosphates (P_2O_5/Mg_2O) [149], and hypophosphates ($(NH_4)_2H_2P_2O_6$) [114], (Na_2HPO_3) and ($MHPO_3$), in which M = Mg, Ca, Sr, Ba, and Cd [136]. In addition, X irradiation studies of magnesium phosphite ($MgHPO_3 \cdot 6H_2O$) [66] and ammonium fluorophosphate ($(NH_4)_2PO_3F \cdot H_2O$) [68] were reported, and an ESR study of the $\dot{P}O_3^{2-}$ radical in the liquid phase [7] has been published.

Calculations concerning the $\dot{P}O_3^{2-}$ radical have been performed by Herring et al. [68], who discussed the electronic structure in terms of the semiempirical LCAO extended Hückel MO method. Principal values of 745.7, 597.9, and 597.9 gauss for the ^{31}P hyperfine interaction tensor were calculated, which are in good accord with the experimental results. A CNDO calculation performed by Kilcast and Thomson [87] predicted an isotropic phosphorus splitting of 744 gauss.

B. Phosphoranyl Radicals, $R_4\dot{P}$

1. PHOSPHORANYL RADICALS IN IRRADIATIED SOLIDS

The most extensively investigated phosphoranyl radical in solid matrix is the $\dot{P}F_4$ radical. The first study was reported by Morton [115], who identified the radical by ESR in γ-irradiated NH_4PF_6 at 295°K. The isotropic ESR spectrum revealed a phosphorus splitting of 1346 gauss and four equivalent fluorine splittings of 196 gauss. It was concluded that $\dot{P}F_4$ has a tetrahedral or square-pyramidal configuration in which the unpaired electron is largely confined to the 4*s* atomic orbital of phosphorus. Subsequently, Atkins and Symons [5] obtained an ESR spectrum with similar features on γ irradiation of ammonium and potassium hexafluorophosphate at 300°K, whereas at 210°K only one broad line was observed. The radical species involved was similarly assigned to $\dot{P}F_4$. The experimental results were explained in terms of a trigonal bypiramidal (TBP) structure, similar to PF_5, with the unpaired electron accommodated in an *s*- and *p*-hydridized orbital on phosphorus. The equivalence of the four fluorine atoms was ascribed to a rapid pairwise interconversion of two sets of nonequivalent fluorine atoms. The broad line at 210°K seemed to support this interpretation, because at this temperature the inversion rate is apparently equivalent to the hyperfine splitting frequency. Two distinct sets of nonequivalent fluorine atoms were observed by Fessenden and Schuler [49] for a radical formed on γ irradiation of PF_3 in SF_6 matrix at −175°C. The spectral parameters a_P^{31} = 1330 gauss, $a_{F_1}^{iso}$ = 282 gauss (2F), and $a_{F_2}^{iso}$ = 59 gauss (2F) were assigned to $\dot{P}F_4$ in a TBP configuration undergoing slow inversion. In a recent communication Mishra and Symons [110] concluded that the species with four equivalent fluorine atoms, reported in earlier studies, is not the $\dot{P}F_4$ radical but almost certainly $\dot{P}F_5^-$. In this species, as with $\dot{S}F_5$, four fluorine atoms are equivalent whereas one fluorine atom has an almost zero hyperfine coupling constant. The latter could be observed at room temperature under high resolution. The previously reported [5] temperature-dependent spectrum appeared to rise from a restriction of the overall rotation of the radical. On reinvestigation [110], the four strongly coupled fluorine atoms remained equivalent over a temperature range of 77°K to room temperature.

A theoretical study of the structure of $\dot{P}F_4$ has been reported by Higuchi [69]. By assuming C_{2v} symmetry and the participation of phosphorus 3*d* orbitals he calculated the isotropic hyperfine splittings of phosphorus and the two kinds of

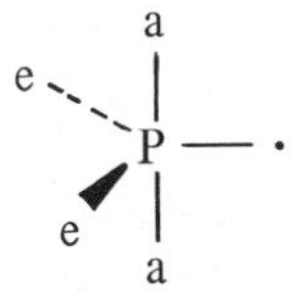

fluorine nuclei by valence bond or molecular orbital approaches as a function of $\angle F_{ap}PF_{ap}$ and $\angle F_{eq}PF_{eq}$. On comparing the calculated hyperfine splittings with the values observed by Fessenden and Schuler [49], he estimated angles of 174 and 109° between the apical and equatorial PF bonds, respectively. The calculations indicate a slightly distorted TBP structure in which the fluorines with the larger hyperfine splitting constants occupy the apical positions.

Nelson et al. [119] reported angles of 180 and 102°, which were obtained by comparing Higuchi's calculated data with the isotropic components of the observed spectral data of $\dot{P}F_4$ in γ-irradiated PF_3 at 4.2°K ($a_{\dot{P}}^{iso}$ = 1391 gauss, $a_{F_{ap}}^{iso}$ = 380 gauss, and $a_{F_{eq}}^{iso}$ = 45 gauss). The structurally similar $\dot{P}Cl_4$ radical, generated by UV irradiation of neat PCl_3 at −196°C, has been reported by Kokoszka and Brinckman [89, 90]. The isotropic ESR spectrum exhibited a phosphorus splitting of 1214 gauss and two sets of nonequivalent chlorine splittings of 62.5 and 7.5 gauss. Several organo halophosphines were similar irradiated, but only the ESR spectrum of irradiated CH_3PCl_2 showed a large doublet splitting, which was assigned to $(CH_3)\dot{P}Cl_3$. The large phosphorus splitting of 1077 gauss indicated a structure analogous to $\dot{P}Cl_4$. The decrease in phosphorus splitting with respect to $\dot{P}Cl_4$ has been ascribed to the decrease in electronegativity of the CH_3 substituent in comparison with chlorine. CNDO calculations, performed by Kilcast and Thomson [87], for $\dot{P}Cl_4$ and $\dot{P}F_4$, predicted phosphorus splittings of 968 and 1330 gauss, respectively, which are in fair agreement with the experimental values. The correlation for the chlorine and fluorine splittings, as inferred from the *s*-orbital spin densities, was concluded to be poor [87]. In a recent study of Gillbro and Williams [57] however, it is pointed out that the spin densities in the ligand *p* orbitals have to be considered. Indeed, the ratio of the calculated *p* orbital spin densities of the apical and equatorial ligands [87] are more in agreement with the experimental results.

Begum and Symons [13] observed the phosphorus oxychloride radical anion ($POCl_3^-$) in γ-irradiated $POCl_3$ at 77°K. The isotropic ESR spectrum revealed a phosphorus splitting of 1359 gauss and chlorine splittings of 67 (2 Cl) and 15 gauss (1 Cl). From the similarity of the hyperfine splittings to those of $\dot{P}Cl_4$ it was concluded that $\dot{P}OCl_3^-$ has a TBP structure in which the two strongly coupled chlorines occupy apical positions. Kerr and Williams [85] obtained similar results for $\dot{P}OCl_3^-$.

The anisotropic ESR spectrum of $\dot{P}OCl_3^-$ was observed by Gillbro and Williams [57] on γ irradiation of single crystals of phosphorus oxychloride. From the anisotropic hyperfine splittings of the two, magnetically equivalent, apical chlorines (25 to 70 gauss) spin densities of 0.29 in each of the ligand $3p_\sigma$ orbitals were estimated. The nearly isotropic phosphorus splitting of 1371 gauss corresponds to a spin density of 0.38 in the 3*s* orbital on phosphorus. Thus the unpaired electron is largely confined to the phosphorus 3*s* orbital and the *p* orbitals of the apical ligands, as indicated by the sum of the spin densities on these positions (0.96).

The phenyl substituted phosphoranyl radicals $Ph\dot{P}(O)Cl_2^-$ and $Ph\dot{P}(S)Cl_2^-$ were produced on γ irradiation of phenyl phosphonic dichloride and phenylphosphonothionic dichloride [108]. The phosphorus splittings of 1000 and 973 gauss, respectively, are consistent with a TBP configuration, in which the two chlorines occupy apical positions. The similarity of the chlorine splittings, 51 and 43 gauss, respectively, to those of the apical chlorines in $\dot{P}Cl_4$ (62 gauss), supported this conclusion. No delocalization of the unpaired electron onto the phenyl ring has been observed.

A related structure was depicted [107] for radical **75** observed on γ irradiation of hexachlorocyclotriphosphazene as inferred from the isotropic hyperfine splittings [a_P = 1290 and a_{Cl} = 52 gauss (2 Cl)]. The unpaired electron is entirely confined to one phosphorus atom.

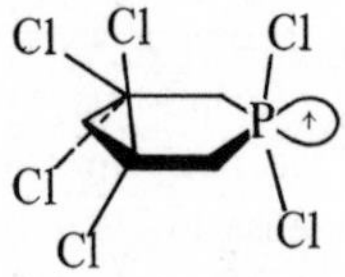

75

The influence of ligand electronegativity on a_P^{iso} of phosphoranyl radicals has been discussed by Symons et al. [58]. It was concluded that replacement of H or Me by –OR groups leads to an increase in the isotropic phosphorus coupling for both apical and equatorial sites. In agreement with this observation, a value of 610 gauss was found for the isotropic component of the phosphorus splitting of $Me_3\dot{P}O^-$ in γ-irradiated Me_3PO [14], whereas a value of 860 gauss was observed for $(MeO)_3\dot{P}O^-$ [58]. However, the radicals $\dot{P}H_4$ and $\dot{P}O_4^{4-}$ show an anomalous behavior with respect to this trend.

The $\dot{P}H_4$ radical, produced on photolysis of PH_3 in krypton at 10°K, has been identified by McDowell et al. [105]. The isotropic ESR spectrum revealed a phosphorus splitting of 973 gauss and four equivalent hydrogen splittings of 21 gauss. The equivalence of the hydrogen atoms was ascribed to tunneling motions. Because of the high value of the a_P and the small hydrogen splittings, Symons suggested the structure $P_2H_6^+$ as an alternative for the observed species [58].

The radical species in γ-irradiated phenacite (Be_2SiO_4) was assigned to $\dot{P}O_3^{2-}$ by Lozylowsky et al. [94]. However, the phosphorus splitting constant (a_P^{iso} = 1156 gauss) is exceptionally large with respect to $\overset{\cdot}{P}O_3^{2-}$ in other host crystals (550 to 800 gauss). Symons [135] reconsidered the reported ESR parameters of this radical and assigned the species to the phosphoranyl radical $\overset{\cdot}{P}O_4^{4-}$. The relatively high value of the phosphorus splitting with respect to the phosphoranyl radical with more electronegative substituents [i.e., $(MeO_3)\overset{\cdot}{P}O^-$] has been discussed in terms of Be^{2+} coordination to oxygen in the host phenacite [58].

In contrast to $\overset{\cdot}{P}O_4^{4-}$, the $\overset{\cdot}{P}O_4^{2-}$ radical has a tetrahedral structure with the spin density located on oxygen, which results in a small phosphorus splitting (30 gauss) [73] (Chapter 1). Isostructural with $\overset{\cdot}{P}O_4^{2-}$ are $\overset{\cdot}{P}O_2Cl_2$ (a_P^{iso} = 44 gauss [13]), $\overset{\cdot}{P}O_2F_2$ (a_P^{iso} = 43.8 gauss [10]), $\overset{\cdot}{P}O_3F^-$ (a_P^{iso} = 39.1 gauss [10]), and $(OH)_2\overset{\cdot}{P}O_2$ (a_P^{iso} = 39.0 gauss [134]).

2. PHOSPHORANYL RADICALS IN SOLUTION

Phosphoranyl radicals in solution have been the subject of intensive ESR studies during the last few years. The first study was reported by Kochi and Krusic [88], who detected a phosphorus centered radical (a_P = 618 gauss) in the reaction of photolytically generated t-butoxy radicals with trimethylphosphine. The ESR spectrum was assigned to the phosphoranyl adduct $(CH_3)_3\overset{\cdot}{P}OBu^t$.

In a similar way a variety of phosphoranyl radicals has been generated by reaction of alkyl, thiyl, and alkoxy radicals with trivalent phosphorus compounds. Published work includes alkyl alkoxyphosphoranyl radicals $R_n\overset{\cdot}{P}(OR')_{4-n}$, where R=H, alkyl, and R'=alkyl [31, 34ab, 35, 63, 88, 92, 147]; alkenoxyphosphoranyl radicals $R_n\overset{\cdot}{P}(OR')_{4-n}$, where R=alkyl, alkoxy, and R'=alkenyl [36]; chlorophosphoranyl radicals $Bu^tO(R)_n\overset{\cdot}{P}Cl_{3-n}$ and $Bu^tO(RO)_n\overset{\cdot}{P}Cl_{3-n}$, where R=alkyl [35, 61]; amino-substituted phosphoranyl radicals $Bu^tO(R)_n\overset{\cdot}{P}(NR_2')_{3-n}$ and $Bu^tO(RO)_n\overset{\cdot}{P}(NR_2')_{3-n}$, where R and R'=alkyl [37, 38]; and the thiyl-substituted compounds $MeS\overset{\cdot}{P}Bu_3^n$ and $MeS\overset{\cdot}{P}(OEt)_3$ [33]. Phosphoranyl radicals may undergo α- or β-scission reactions, according to the scheme

$$RO^\bullet + PXYZ \longrightarrow RO\overset{\cdot}{P}XYZ \overset{\alpha}{\underset{\beta}{<}} \begin{matrix} ROPYZ + X^\bullet \\ \\ O{=}PXYZ + R^\bullet \end{matrix}$$

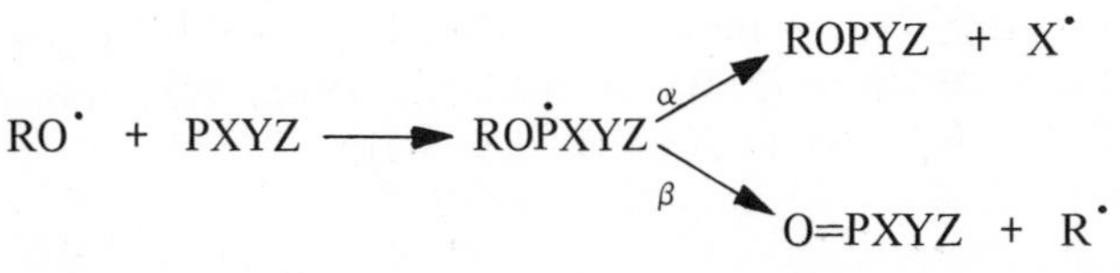

The kinetics of these scission processes have been examined by ESR [34b, 35, 60, 63, 147] and in product studies [16, 17]. As a consequence of α scission, not only mono-t-butoxy substituted radicals but also the disubstituted compounds were observed [37, 38, 61, 92]:

$$ROPYZ + RO^\bullet \longrightarrow (RO)_2\overset{\cdot}{P}YZ$$

In some cases cleaved alkyl radicals (resulting from β scission) were observed, which can react with the parent phosphine or phosphite PXYZ, according to [31, 35]

$$^{\bullet}R' + PXYZ \longrightarrow R'\dot{P}XYZ$$

Tetraalkoxyphosphoranyl radicals that appeared to react with oxygen produced radical species with small phosphorus splittings (~9 gauss) [34b, 148]. This doublet has been assigned to phosphoranylperoxy radicals (**76**):

$$(RO)_4P^{\bullet} + O_2 \longrightarrow (RO)_4POO^{\bullet}$$

76

The structure of phosphoranyl radicals in solution, as inferred from the large phosphorus splittings in the range of 600 to 1200 gauss, has been discussed in terms of a distorted TBP configuration in which the unpaired electron occupies an equatorial position, analogous to Higuchi's calculated structure for $\dot{P}F_4$ [69]. A marked preference of the ligands for one of the stereochemically different sites in the TBP configuration has been observed in various ESR studies. Rules concerning their preference were evaluated for the isostructural pentacoordinated phosphoranes by Muetteries et al. [116, 117] and Trippett [141]. The major factor for the difference in apicophilicity appears to be the electronegativity of the ligands. The most electronegative groups preferentially occupy apical sites. Another factor is the ability of the ligand to enter into p_π-d_π bonding with phosphorus, which will be most efficient when the donor atom is in an equatorial position. When the phosphorus atom in a phosphorane is incorporated into a four- or five-membered ring, the ring prefers to span apical and equatorial positions.

The results obtained for phosphoranyl radicals suggest that their structures are governed by the same preference rules.

In phosphoranyl radicals of type $R_nP(OR')_{n-4}$, where R=H or alkyl, the alkoxy group preferentially occupies an apical position. for instance $H_3\dot{P}OBu^t$ was assigned to structure 77, as inferred from the two distinct hydrogen splittings of 139 (1 H) and 10 gauss (2 H). A similar structure (**78**) was depicted for $(CH_3)_3\dot{P}OBu^t$ [92] [a_H = 4.6 (1 Me) and 2.8 (2 Me)].

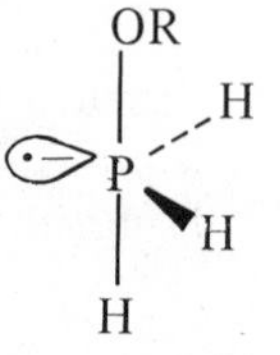

77

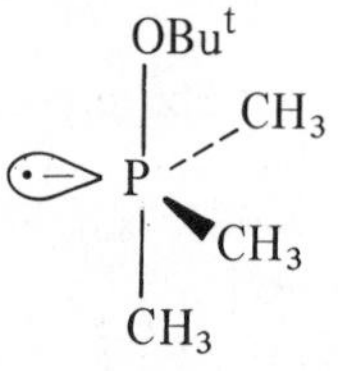

78

Griller and Roberts [61] concluded from an ESR study of chlorophosphoranyl radicals that the apicophilicity of chlorine is greater than that of an alkoxy group, as indicated by structures **79**, **80**, and cyclic structure **81**.

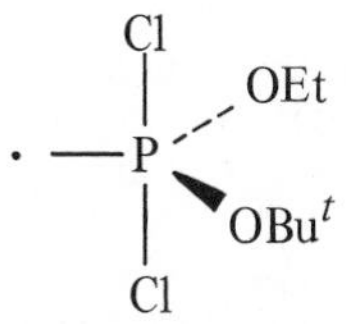

79

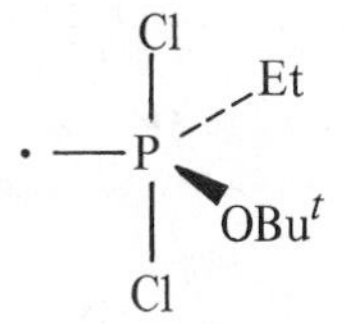

80

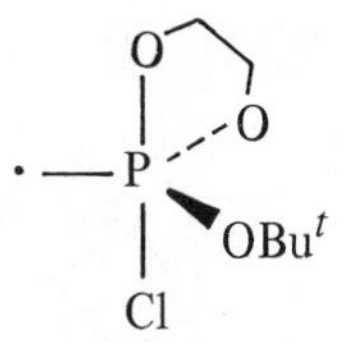

81

In amino alkoxy-substituted phosphoranyl radicals [37] the amino groups generally occupy the equatorial positions, although the reversed order was observed for **82** at $-150°C$. When the temperature was raised to $-120°C$, an isomerization to **83** could be detected.

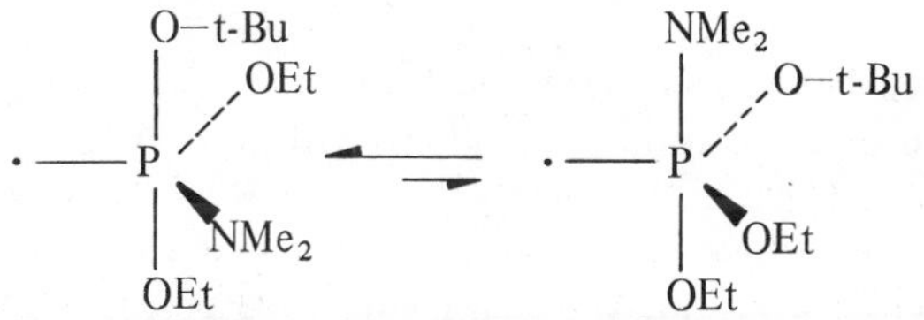

83 **82**

A rapid exchange of apical and equatorial ligands, probably by pseudorotation, has been observed in the amino phosphoranyl radicals **84** and **85**, in which phosphorus is incorporated in a five-membered ring [38].

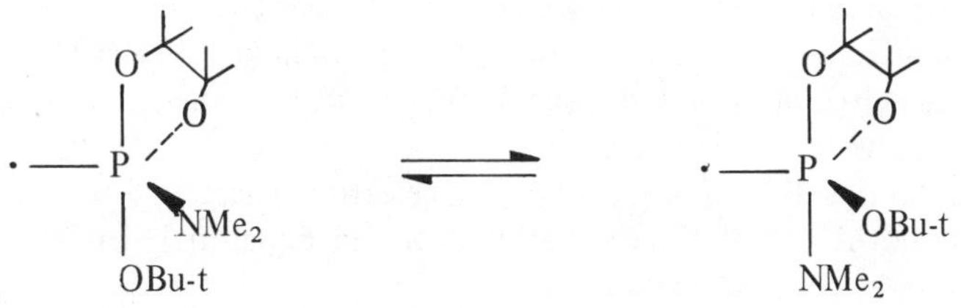

84 **85**

At $-120°C$ the ESR spectrum showed the features of radical **84** (a_P = 849 gauss, a_N = 9.5 gauss), whereas at $-110°C$ radical **85** (a_P = 784 gauss, a_N = 24 gauss) predominated. On increasing the temperature, line-broadening effects indicated an increasingly rapid isomerization.

A similar isomerization was not observed in spirophosphoranyl radicals **86** where (R and R′=H, CH_3), produced by Griller and Roberts [62] on photolysis of the protic parents in the presence of di-*t*-butyl peroxide.

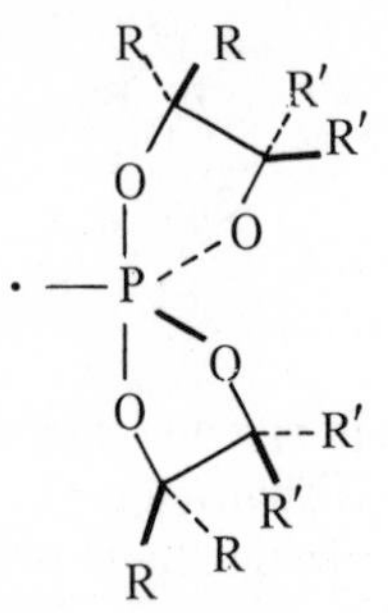

86

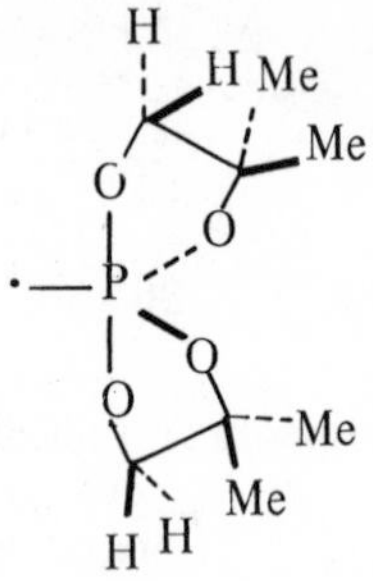

87

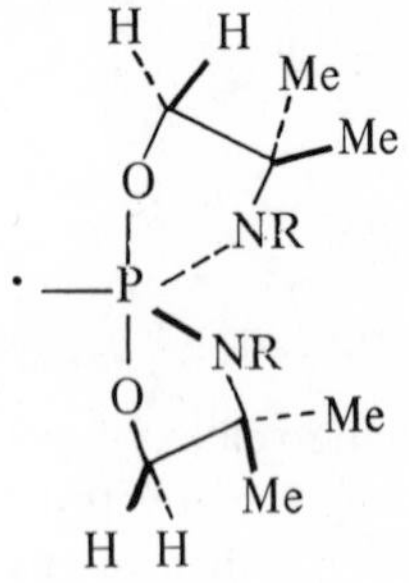

88

The spectra remained unchanged even up to +120°C, which indicated that compounds of **86** type are strongly stabilized with respect to their acyclic analog. Phosphoranyl radical **86** in which R = H and R′ = CH_3, was assigned (87) based on a comparison with the nitrogen containing radical **88**. In the latter species the nitrogen atoms are known to occupy equatorial positions. As a consequence, the primary alkoxy group resides in apical positions. The similar magnitude of the proton splittings of **87** and **88** (3.5 gauss) indicated an analogous configuration. This conformational preference was ascribed to the smaller electronegativity and the greater p_π-d_π bonding ability of the tertiary relative to the primary alkoxy group. Consequently, the latter group will have the greatest apicophilicity.

The effect of electronegativity of ligands both in axial and equatorial positions on the a_P^{iso} has been discussed recently by Gillbro and Williams [57]. A theoretical model was proposed in which the unpaired electron resides in a molecular orbital, which was represented as a Rundle three-center nonbonding orbital [129] involving the apical ligand orbitals and the symmetry allowed 3*s* and 3*p* orbitals of phosphorus in an antibonding combination.

According to this description, the unpaired electron is largely distributed over the p_σ orbitals of the apical ligands but not of the equatorial, irrespective of the electronegativities of these ligands.

This was supported by the large spin densities on the apical chlorine atoms in $\dot{P}OCl_3^-$ (III B2) and by the large splittings of the apical hydrogens, in comparison with the equatorial, as in **77** [92]. Consequently, the phosphorus splitting will be more affected by changes in electronegativity of the apical ligands. This conclusion was supported with other experimental data [57]. Replacement of the two equatorial chlorines in $\dot{P}Cl_4$ [90] by alkoxy groups as in $Bu^tO(EtO)\dot{P}Cl_2$ [61] results in a relatively small decrease of a_P^{iso} from 1217 to 1145 gauss ($POCl_3$, a_P^{iso} = 1371 gauss, exhibited an anomalous trend [57]), However, further replacement of the two apical chlorine atoms by ethoxy groups as in $Bu^tO\dot{P}(OEt)_3$ [34b, 35, 61] shows a significant decrease in the a_P^{iso} from 1145 to 890 gauss.

Acknowledgement

We express our sincere thanks to Mrs. H. A. M. Rothuis for helpful assistnace in searching the literature. We also thank Mrs. P. Meyer for typing the manuscript.

TABLE 1. Disubstituted-Phosphorus π Radicals

Compound	a_P^{iso},G	$a_P(\parallel)$,G	$a_P(\perp)$,G	Other Splittings,G	3s Character	3p Character	Refs.
1. Phosphinyl radicals							
$\dot{P}H_2$	80	–	–	$a_H = 18$	0.022	–	111
	82.3	–	–	$a_H = 18.0$	–	–	105
$\dot{P}Cl_2$	33	–	–	–	–	–	89
	–	269	28	$a_{Cl}(\parallel) = 15$ $a_{Cl}(\perp) = 0$	small	0.61-0.78	89,90
	77	293	−30.5	$a_{Cl}(\parallel) = 17.5$ $a_{Cl}(\perp) < 0.3$	–	0.83	150
	–	272.5	22.5	$a_{Cl}(\parallel) = 16$ $a_{Cl}(\perp) = 0$	–	–	13
$\dot{P}F_2$	–	308.1	−27.2	$a_F(\parallel) = 125.5$ $a_F(\perp) = -13.5$	0.024	0.83	119
	84.6	–	–	$a_F^{iso} = 32.5$			119
	36.0	–	–	$a_F^{iso} = 60.5$	–	–	144
	–	307	−83.0	$a_F(\parallel) = 127$ $a_F(\perp) = 33.5$	0.013	0.91	151
$(EtO)_2\dot{P}$	78.5	–	–	–	–	–	63
$(Me_2N)_2\dot{P}$	70.2	–	–	$a_N = 4.2$ $a_H = 2.1$	–	–	63
$\dot{P}Ph_2$	12	–	–	$a_H = 5.8(3H)$	–	–	152

Compound	a_P,G	Other Splittings,G	Refs.
2. Phosphorin radical ions			
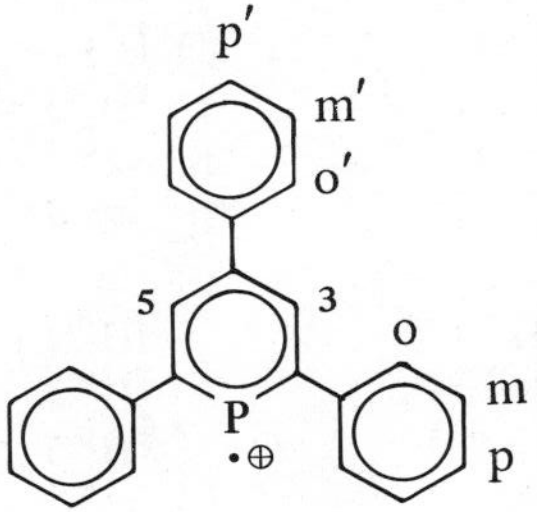	23.2 24.2	a_H = 2.3(p',o',3,5) a_H = 0.7 (m') a_H = 0.46 (p,o) a_H = 0.23 (m) —	41,42 138
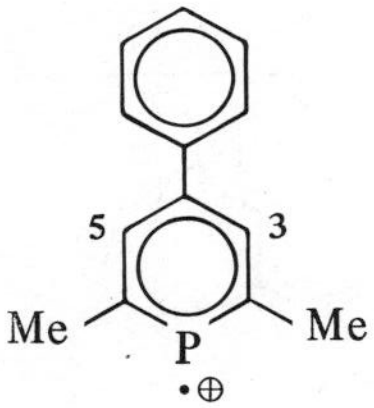	23.2	a_H = 7.5 (methyl) a_H = 2.2 (p,o,3,5) a_H = 0.9 (m)	41
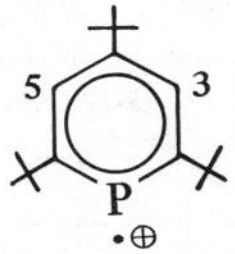	26.9 26.7	a_H = 2.9 (3,5) a_H = 0.8 (t-Bu)	41,43 138

TABLE I. cont.

Compound					a_P,G	Other Splittings,G	Refs.
2. Phosphorin radical ions (*continued*)							
R_2	R_3	R_4	R_5	R_6			
C_6D_5	H	C_6D_5	H	C_6D_5	23.2		41,43
C_6D_5	H	C_6H_5	H	C_6D_5	23.2		41,43
C_6H_5	C_6H_5	C_6H_5	H	C_6H_5	24.8		43
C_6H_5	C_6H_5	C_6H_5	C_6H_5	C_6H_5	26.0		43
p-ClC_6H_4	H	C_6H_5	H	C_6H_5	22.6		43
p-ClC_6H_4	H	*p*-ClC_6H_4	H	*p*-ClC_6H_4	22.4		43
C_6H_5	H	*p*-$CH_3OC_6H_4$	H	C_6H_5	21.6		43
p-$CH_3OC_6H_4$	H	C_6H_5	H	C_6H_5	21.7		43
p-$CH_3OC_6H_4$	H	*p*-$CH_3OC_6H_4$	H	C_6H_5	21.6		43
p-$CH_3OC_6H_4$	H	C_6H_5	H	*p*-$CH_3CO_6H_4$	22.3		42,43
C_6H_5	H	*t*-Bu	H	C_6H_5	23.7		41,43
t-Bu	H	C_6H_5	H	*t*-Bu	23.7		41,43
					24.1		138
t-Bu	H	*p*-$CH_3OC_6H_4$	H	*t*-Bu	23.3		43

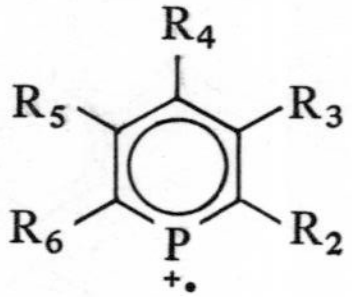

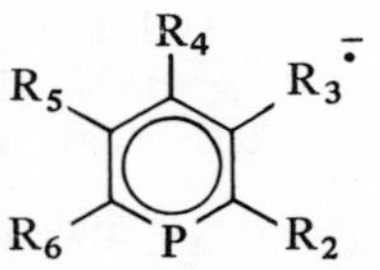

	R_2	R_3	R_4	R_5	R_6			
	C_6H_5	H	C_6H_5	H	C_6H_5	32.4		41,45
						32.9		138
						4.6 (3-)		41,45
	C_6D_5	H	C_6D_5	H	C_6D_5	32.4		41
	C_6D_5	H	C_6H_5	H	C_6D_5	32.4		41
	C_6H_5	H	*t*-Bu	H	C_6H_5	31.8		41
	t-Bu	H	C_6H_5	H	*t*-Bu	30.5		41
						30.4		138
	t-Bu	H	*p*-$CH_3OC_6H_5$	H	*t*-Bu	29.1		41
	t-Bu	H	*t*-Bu	H	*t*-Bu	27.8		41
						26.9		138
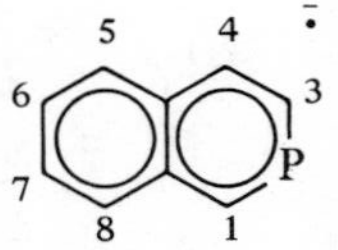						23.6	a_H = 7.23 (1) a_H = 3.62 (4,8) a_H = 3.52 (5) a_H = 1.00 (6) a_H = 0.50 (3,7)	79
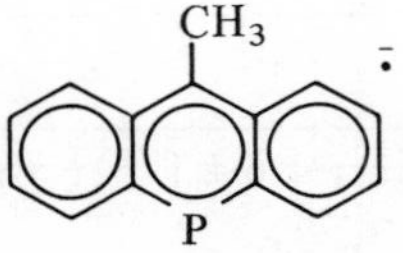						33.5	a_H = 6.5 (methyl)	79

TABLE II. Trisubstituted-Phosphorus π Radicals

Compound	a_P ,G	Other Splittings ,G	Refs.
1. Phosphinium radicals, $R_3\dot{P}^{+}\cdot$			
$\dot{P}Ph_3^+$	2.32	—	18
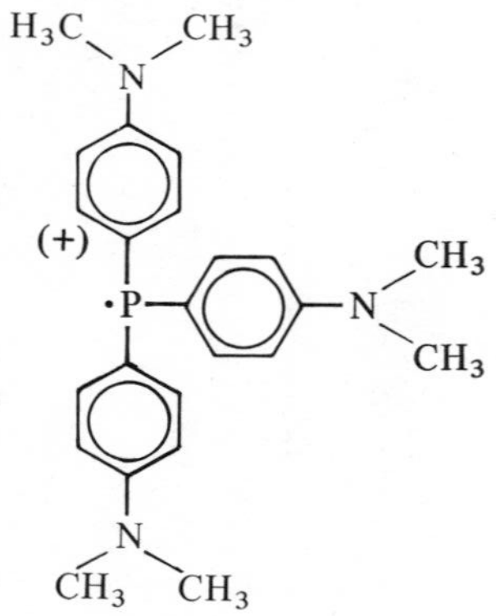	12	$a_N = 6.2$	140
2. Phosphine Free radicals, $R_2P-\dot{C}R'_2$			
$(HOCH_2)_2P-\dot{C}HOH$	29.1	$a_H = 14.8$	97
$CH_3\dot{C}H-P(Et)_2$	32	$a_H = 22$ (4 H)	9
$C_3H_7\dot{C}H-P(Bu)_2$	32	$a_H = 20$ (3 H)	9

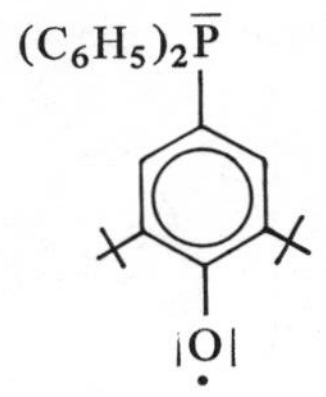	6.8	a_H = 1.6 (*m*)	118
3. Phosphine anion radicals, $R_3P^{\overline{\cdot}}$			
$Ph_2PK^{\overline{\cdot}}$	8.4	a_H = 2.0 (*p,o*) a_H = 0.8 (*m*) a_K = 0.24	22
$PPh_3^{\overline{\cdot}}$	8.36	a_H = 2.3 a_H = 0.23 (12 components)	65
$Ph_3PK^{\overline{\cdot}}$	8.5	a_H = 2.3 (4H)	81
$(\rho\text{-}DC_6H_4)_3PK^{\overline{\cdot}}$	8.5	a_H = 2.1 (2H)	81
$Ph_2(CH_3)PK^{\overline{\cdot}}$	—	a_H = 5.65 (4H)	81
		a_H = 2.1 (4H)	
$Ph(CH_3)_2PK^{\overline{\cdot}}$	8.5	a_H = 2.9 (*p,o*) a_H = 1.2 (methyl)	81
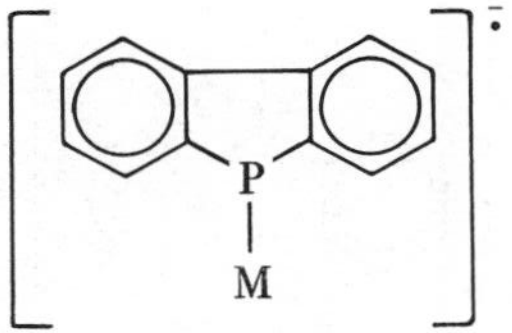 M = Li, Na, K, Cs	8.5	a_H = 2.4 (2 H)	23

TABLE II. cont.

Compound	a_P, G	Other Splittings, G	Refs
3. Phosphine anion radicals, $R_3P^{\bar{\cdot}}$ (*continued*)			
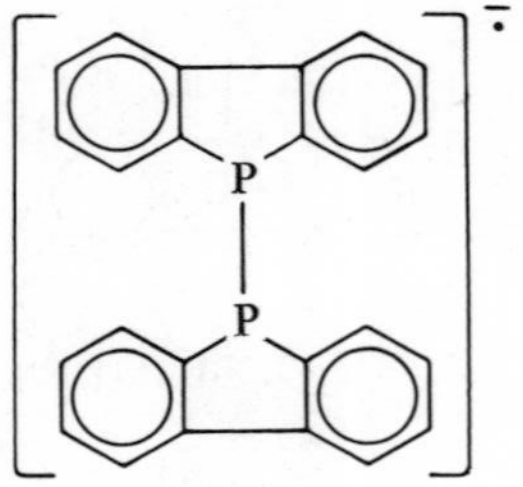	8.8 (2 P)	–	23
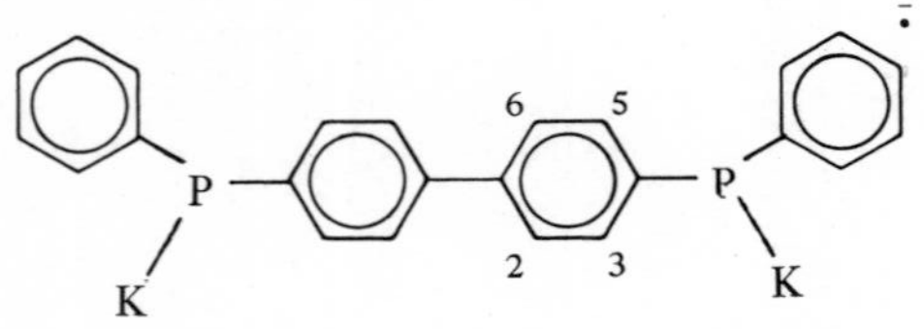	3.05 (2 P)	a_H = 2.05 (*p,o*) a_H = 1.02 (3,3′,5,5′) a_H = 0.4 (2,2′,6,6′)	70
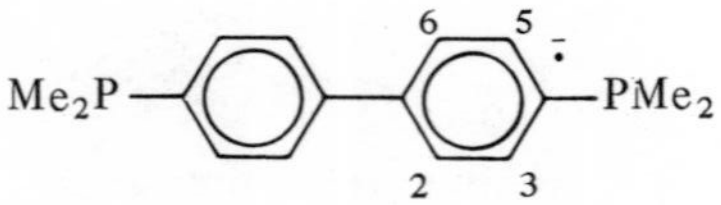	5.7 (2 P)	a_H = 2.2 (3,3′,5,5′) a_H = 0.44 (2,2′,6,6′) a_K = 0.88	26
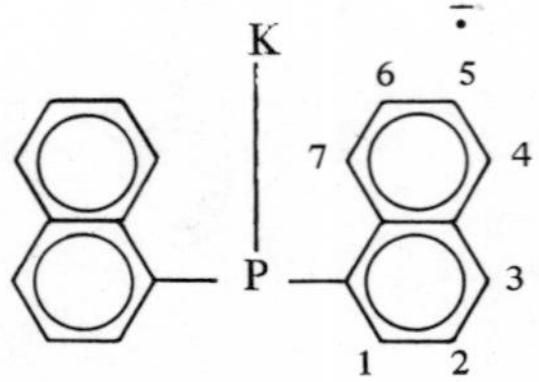	7.2	a_H = 3.60 (1,1′,3,3′) a_H = 2.40 (5,5′,7,7′) a_H = 1.20 (2,2′) a_H = 0.40 (6,6′)	71

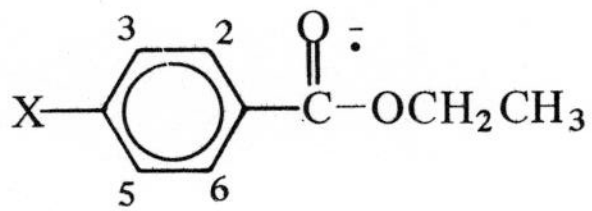

$X = PH_2$	3.75	$a_H = 14.0\ (PH_2)$ $a_H = 3.75\ (2,6)$ $a_H = 1.00\ (CH_2)$	122
$X = PMe_2$	4.60	$a_H = 3.80\ (2,6)$ $a_H = 1.00\ (CH_2)$	122
$X = PEt_2$	5.20	$a_H = 3.80\ (2,6)$ $a_H = 1.00\ (CH_2)$ $a_H = 0.35\ (PEt_2)$	122
$X = P(i\text{-}Pr)_2$	3.80	$a_H = 3.90\ (2,6)$ $a_H = 1.00\ (CH_2)$ $a_H = 0.30\ (P(i\text{-}Pr)_2)$	122

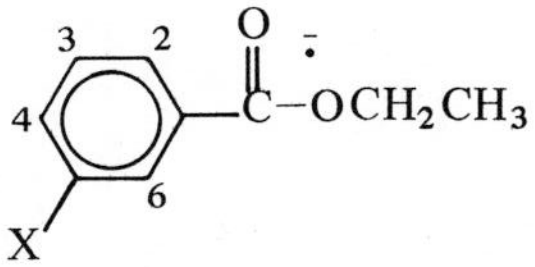

$X = PH_2$	1.1	$a_H = 8.5\ (4)$ $a_H = 4.8\ (2)$ $a_H = 3.6\ (6)$ $a_H = 1.1\ (3, CH_2)$	122
$X = PMe_2$	1.3	$a_H = 8.2\ (4)$ $a_H = 4.8\ (2)$ $a_H = 3.8\ (6)$ $a_H = 1.1\ (3,CH_2)$	122

TABLE II. cont.

Compound	a_P ,G	Other Splittings ,G	Refs.
3. Phosphine anion radicals, $R_3P^{\bar{\cdot}}$ (*continued*)			
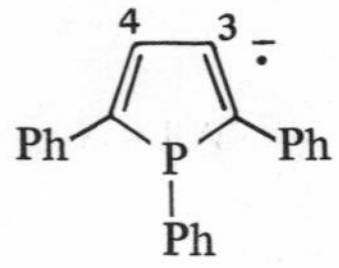	26.5-28.1	a_H = 2.26 (3,4,*p*,*o*) of the 2 and 5 Ph rings a_H = 0.43 (*m*)	86,138
	28.7	—	78
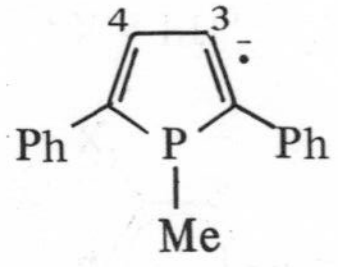	23.5-28.0	a_H = 2.48 (3,4,methyl) a_H = 1.31 (*p*,*o*) a_H = 0.62 (*m*)	86,138
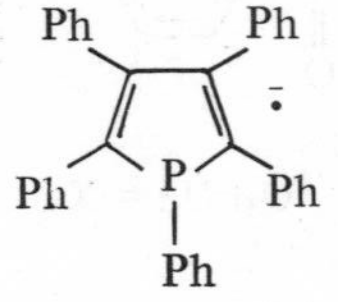	31.3	—	86
	15	—	39
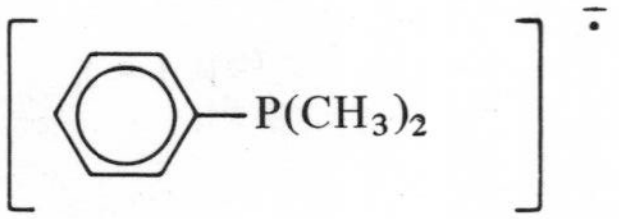	8.28-8.67	a_H = 9.06 (*p*) a_H = 3.31 (*o*) a_H = 0.39 (*m*) a_H = 0.78 (methyl)	55
	7.2-7.9	a_H = 8.9 (*p*) a_H = 3.5 (*o*) a_H < 0.4 (*m*) a_H = 0.9 (methyl)	55

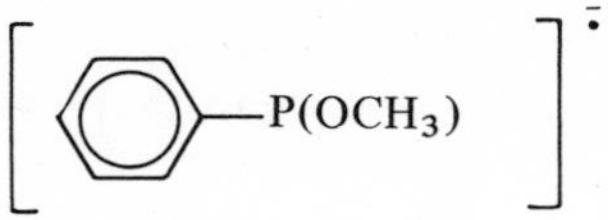	9.0	$a_H = 5.9$ (*p*) $a_H = 3.9$ (*o*)	20
$PPh_3^{\dot{-}}$	3.3	$a_H = 2.2$ (*p*) $a_H = 1.1$ (*o*)	76
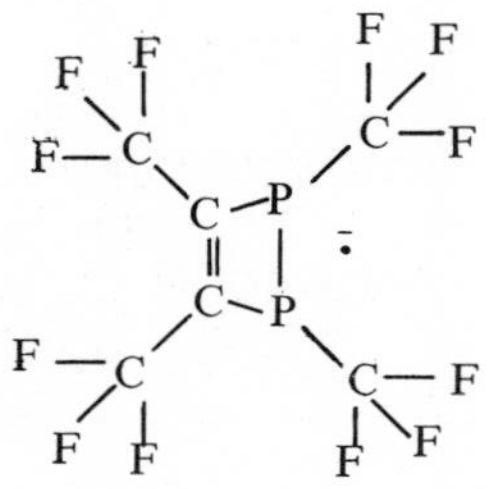	18.0	$a_F = 18.60$ ($C-CF_3$) $a_F = 3.93$ ($P-CF_3$)	142
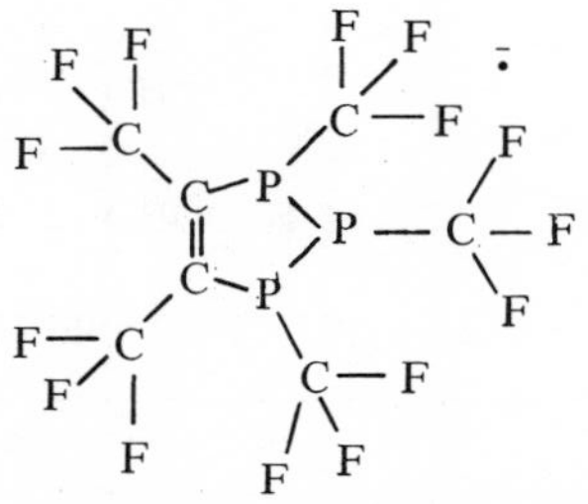	10.90 (2 P) 0.5 (1 P)	$a_F = 18.05$ ($C-CF_3$) $a_F = 8.85$ ($C-P-CF_3$) $a_F = 0.15$ ($P-P-CF_3$)	142

TABLE III. Tetrasubstituted-Phosphorus π Radicals

Compound	a_P, G	Other Splittings, G	Refs.
1. Phosphobetaine cation radicals, $R_3\overset{+}{P}-\dot{C}R'_2$			
$Ph_3\overset{+}{P}-\dot{C}HCO_2H$	37.2, 34.3, 32.5 (iso) 34.7	a_H = 30.4, 19.0, 9.0 (iso) 19.4	98
$Ph_3\overset{+}{P}-\dot{C}H_2$	41.8, 38.2, 36.8 (iso) 40.4	a_H = 27.5, 18.9, 16.8 (iso) 21.2	99
$Me(H)\dot{C}-\overset{+}{P}(H)(Et)_2$	36	a_H = 24 (4 H) a_H = 30 (1 H)	9
$CH_3\dot{C}H-\overset{+}{P}Et_3$	41	a_H = 25	9
$CH_3(CH_2)_2\dot{C}H-\overset{+}{P}Bu_3$	40	a_H = 25	9
$\dot{C}H_2-\overset{+}{P}Bu_3$	41	a_H = 25	9
$CH_3(CH_2)_2\dot{C}HP(O)Bu_2$	36	a_H = 23 (3 H)	9
$Me(PhCO)\dot{C}-\overset{+}{P}Ph_3$	31	a_H = 26 (methyl)	101
$(HOCH_2)_2(H)\overset{+}{P}-\dot{C}HOH$	25.7	a_H = 17.5 (C–H) a_H = 8.4 (P–H)	97
$(HOCH_2)_3\overset{+}{P}-\dot{C}HOH$	33.8	a_H = 17.6	97
$(HOCH_2)_2(O)P-CHOH$	27.2	a_H = 180 (C–H) a_H = 2.1 (OH)	97
$H_2\dot{C}-P(O)Me_2$	21	a_H = 38	14
$H_2\dot{C}-\overset{+}{P}(OH)Me_2$	20	a_H = 40	14

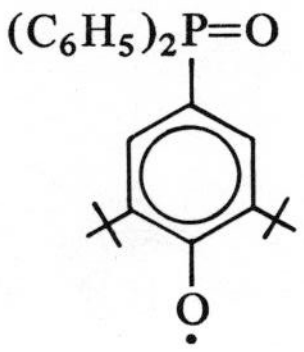	14.7	$a_H = 2.0$ (*m*)	118
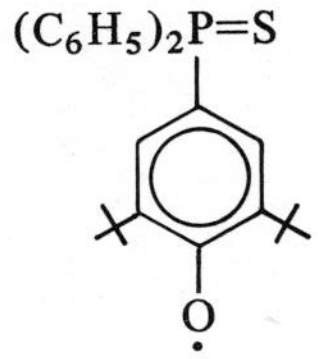	15.4	$a_H = 2.0$ (*m*)	118
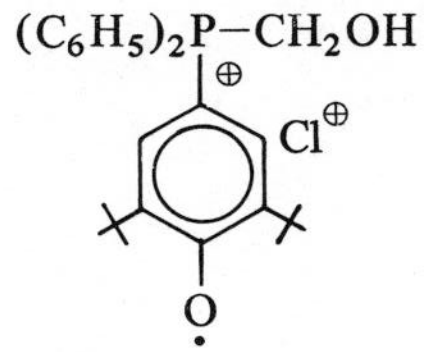	15.1	$a_H = 2.15$	118
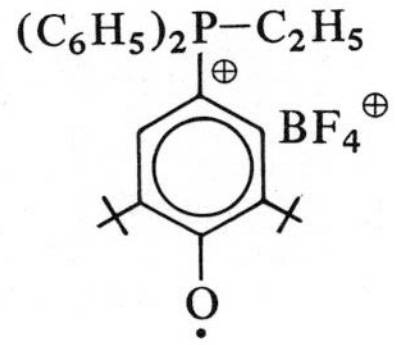	16.8	$a_H = 2.35$	118

TABLE III. cont.

Compound	a_P, G	Other Splittings, G	Refs.
1. Phosphobetaine cation radicals, $R_3\overset{+}{P}-\dot{C}R'_2$ (*continued*)			
$Ph_2\dot{C}-\overset{+}{P}Ph_3$	26.60	a_H = 2.97 (*p*,2 H) a_H = 2.71 (*o*,4 H) a_H = 1.17 (*m*,4 H)	24
$(p\text{-}CH_3OC_6H_4)_2\dot{C}-\overset{+}{P}Ph_3$	24.75	a_H = 2.5 (*o*,4 H) a_H = 0.72 (*m*,4 H) a_H = 0.46 (methoxy)	24
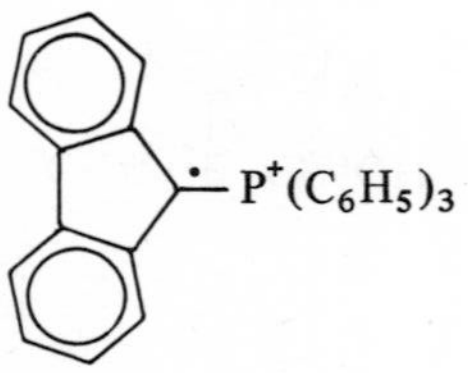	22.25	a_H = 3.75 (3) a_H = 1.78 (5) a_H = 0.50 (4,6)	24
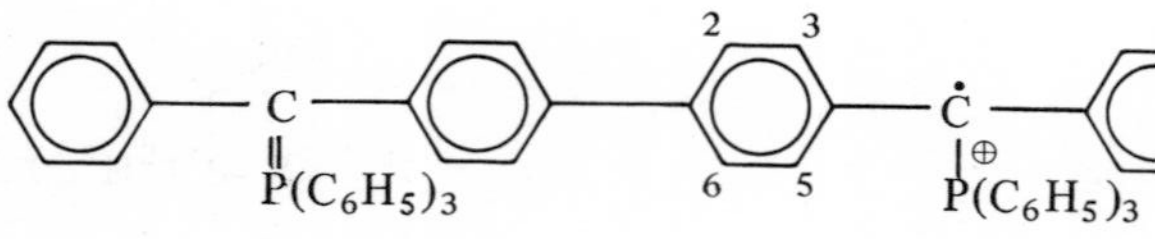	11.71 (2 P)	a_H = 1.77 (3,3′,5,5′)	91
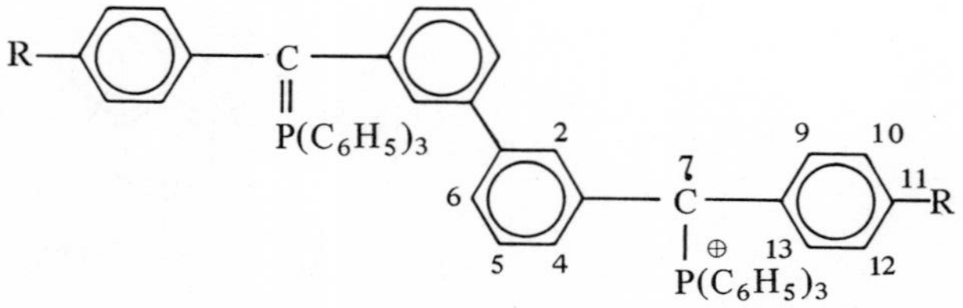			

R = H	26.62	a_H = 5.82 (11) a_H = 4.41 (9,13) a_H = 2.33 (2,4) a_H = 1.08 (6)	91
R = t-Bu	26.05	a_H = 3.92 (9, 13) a_H = 1.43 (2,4) a_H = 1.12 (6,10,12) a_H = 0.92 (5)	91
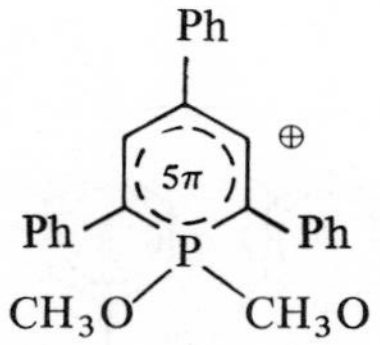	18.0	Total expansion 35.9	41,44
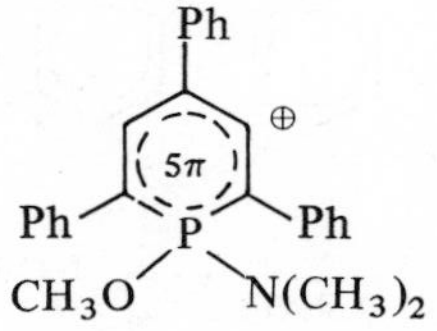	20.9	Total expansion 45.5	41
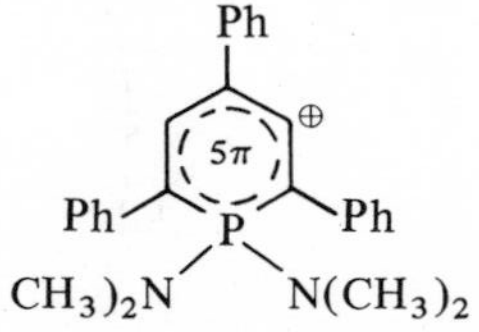	22.3	Total expansion 50	41

TABLE III. cont.

Compound	a_P, G	Other Splittings, G	Refs.
2. Phosphobetaine free radicals, $R_3\overset{+}{P}-\overset{\bullet}{\bar{C}}R'_2$			
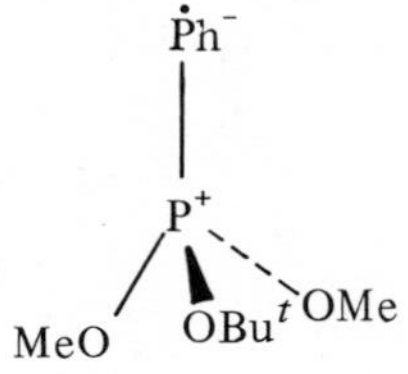	9.7	$a_H = 9.7$ (*p*) $a_H = 5.5$ (*o*) $a_H = 0.9$ (*m*)	20
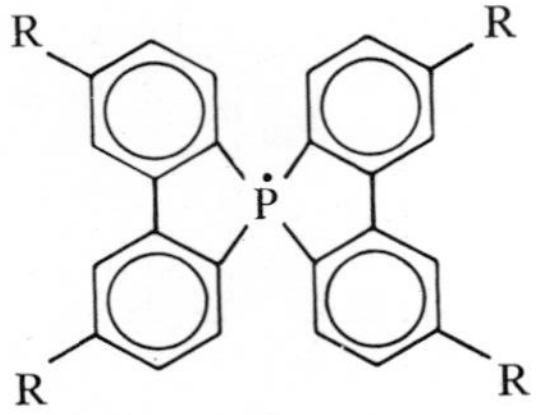			
R = H	12.9-18.6	–	127
	14.8-19.4	$a_H = 2.55$ (2 H) $a_H = 1.70$ (2 H) $a_H = 0.85$ (4 H)	125,128
	17.9	–	67
R = OCH_3	15.7-21.6	–	125,127
R = $N(Me)_2$	15.5-22.9	–	125,127
R = F	10.3-15.5	–	125

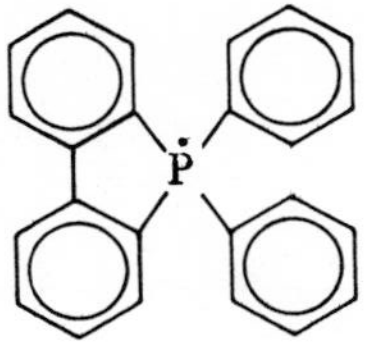	9.5	$a_H = 2.4$	125,126
	8.5	—	125,127
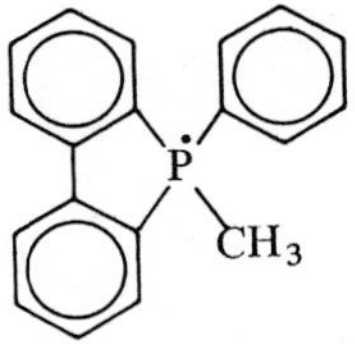	14.6	—	126
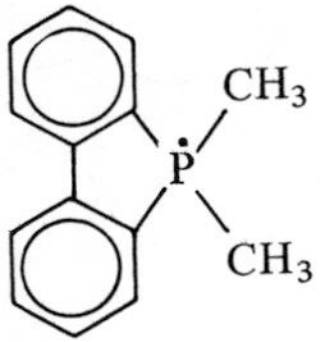	17.4	$a_H = 2.5$	125,126
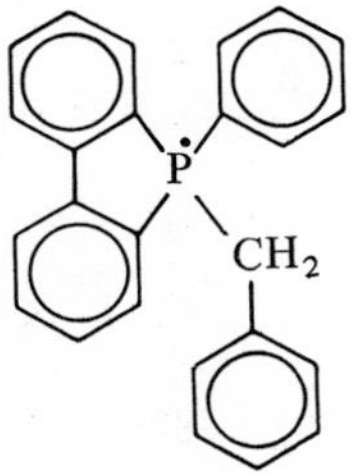	9.5	$a_H = 2.5$	125,126

TABLE III. cont.

Compound	a_P,G	Other Spittings,G	Refs.
2. Phosphobetaine free radicals, $R_3\overset{+}{P}-\overset{\cdot -}{C}R'_2$ *(continued)*			
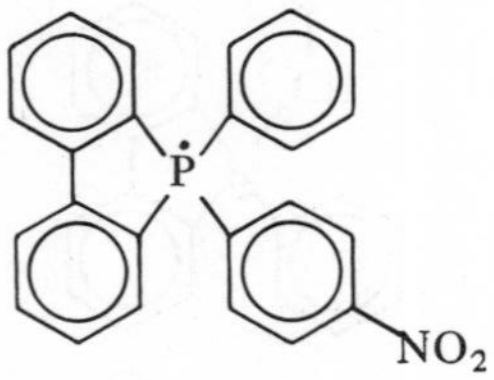	6.5	a_N = 13.0 a_H = 3.3	125,126
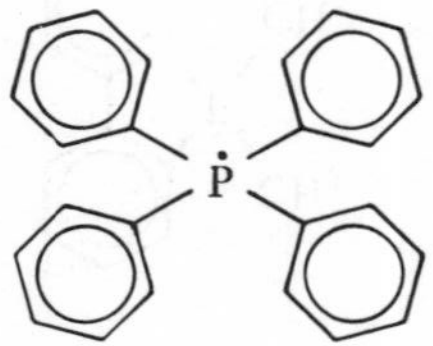	4.0	a_H = 2.7 (*p*) a_H = 1.35 (*o*)	125,127
	3.9	—	76
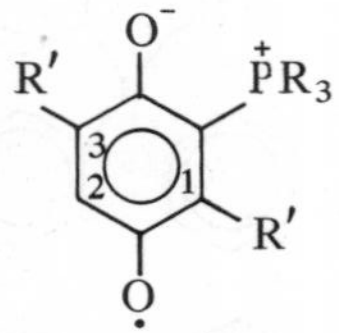			
R R′			
Ph H	4.70	a_H = 5.80 (1) a_H = 1.70 (3) a_H = 1.55 (2)	95

Bu	H	5.00	a_H = 5.15 (1) a_H = 1.70 (2,3)	95
CH_2CH_2CN	H	4.50	a_H = 6.00 (1) a_H = 1.55 (3) a_H = 1.25 (2)	95
Ph, $(NMe_2)_2$	H	4.00	a_H = 5.55 (1) a_H = 1.55 (2,3) a_N = 0.2-0.3	95
Ph, $(NEt_2)_2$	H	3.85	a_H = 5.60 (1) a_H = 1.50 (2,3) a_N = 0.2-0.3	95
Bu	Bu	4.40	a_H = 1.60 (2)	95
Bu	Ph	4.30	a_H = 1.60 (2)	95

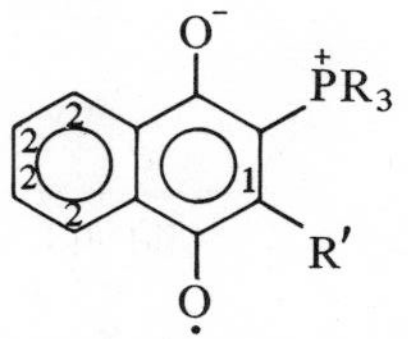

R	R′			
Ph	H	5.70	a_H = 6.70 (1) a_H = 0.2-0.4 (2)	95
Bu	H	6.25	a_H = 6.25 (1) a_H = 0.1-0.3 (2)	95
Ph, $(NMe_2)_2$	H	6.05	a_H = 6.05 (1)	95
Bu	CH_3	5.45	a_H = 6.4 (methyl) a_H = 0.1-0.2 (2)	95

TABLE II. cont.

Compound	a_P,G	Other Splittings,G	Refs.
2. Phosphobetaine free radicals, $R_3\overset{+}{P}-\dot{\bar{C}}R'_2$ (*continued*)			
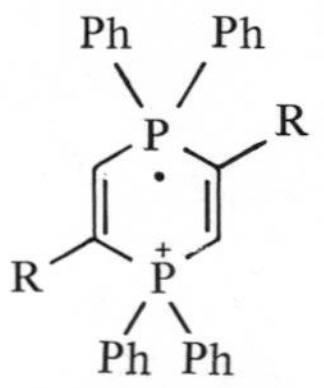			
R = Me	15.8	a_H = 4.8 (methyl)	124
		a_H = 2.5	
R = Bu	16.0	a_H = 2.8	124
R = Ph	16.0	a_H = 3.8	124
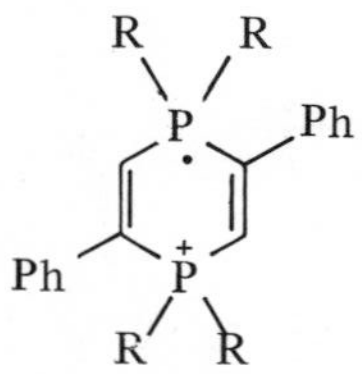			
R = Et	14.2	a_H = 4.8	124
R = p-$CH_3OC_6H_4$	16.4	a_H = 3.8	124
R = p-FC_6H_4	16.3	a_H = 3.7	124

3. Phosphine oxide anion radicals, $R_3PO^{\overline{\cdot}}$

$Ph_3POK^{\overline{\cdot}}$	5.25	$a_H = 3.5$ (*p*) $a_H = 1.75$ (*o*)	80,81
	5.25	$a_H = 1.75$ (*p,o*) $a_K = 0.875$	26
$Ph_2P(O)K^{\overline{\cdot}}$	7.9	$a_H = 2.6$ (*o,p*) $a_K = 0.4$	26
$(Bu)_3POK^{\overline{\cdot}}$	—	a = 0.37 (12 components)	80,81
$Ph(Me)_2POK^{\overline{\cdot}}$	3.36	$a_H = 4.6$(*p,o*) $a_H = 1.6$ (methyl)	80,81
$Ph_2MePOK^{\overline{\cdot}}$	10		80,81
$Ph_3PO^{\overline{\cdot}}$	6.4	$a_H = 4.2$(*p*) $a_H = 2.1$(*o*) $a_H = 0.15$(*m*)	75
	4.4	$a_H = 3.0$(*p*) $a_H = 1.5$(*o*)	76
β–Ph=CH(OEt)$_2$PO$^{\overline{\cdot}}$	7.8	$a_H = 4.8$(1H) $a_H = 1.9$(2H)	75
$Ph(Me)_2PO^{\overline{\cdot}}$	8.75	$a_H = 3.5$(*p,o*) $a_H = 0.875$(methyl)	26
$(CH_3)_2NPh_2PO^{\overline{\cdot}}$	7.2	$a_N = 4.9$ $a_H = 2.44$(*p,o*) $a_H = 0.2$(methyl)	26
$[(CH_3)_2N]_3PO + e^-$	4.91 (4 P) 1.83 (6 P)		25
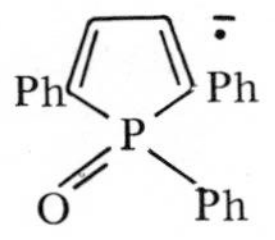	14.95-16.22		139

TABLE III. cont.

Compound	a_P,G	Other Splittings,G	Refs.
3. Phosphine oxide anion radicals, $R_3PO^{\overline{\cdot}}$ (*continued*)			
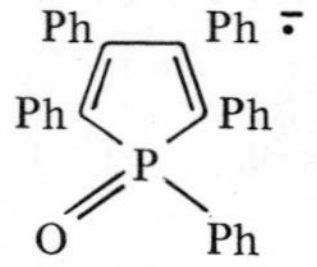	15.25-16.40	–	139
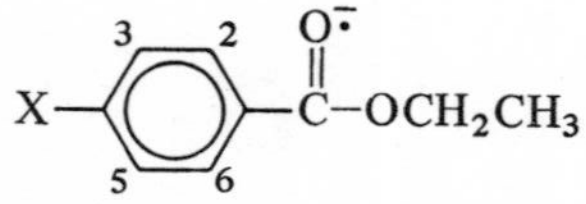			
X			
$(CH_3)_2PO$	12.80	a_H = 3.20 (2) a_H = 1.00 (3,CH_2) a_H = 0.60 [$P(O)(CH_3)_2$]	122
$(C_2H_5)_2PO$	13.10	a_H = 3.20 (2) a_H = 0.95 (3,CH_2) a_H = 0.30 [$P(O)(C_2H_5)_2$]	122
$(C_2H_5)_2PS$	13.50	a_H = 3.00 (2) a_H = 1.00 (3,CH_2) a_H = 0.60 [$P(S)(C_2H_5)_2$]	122
$(i\text{-}C_3H_7)_2PO$	13.40	a_H = 3.20 (2) a_H = 1.10 (3,CH_2)	122

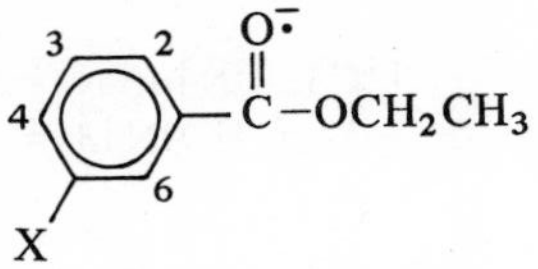

X		$a_H(2)$	$a_H(3)$	$a_H(4)$	$a_H(6)$	$a_H(CH_2)$	
$(CH_3)_2PO$	–	5.4	1.0	8.8	2.8	1.0	122
$(C_2H_5)_2PO$	–	5.5	0.95	8.8	2.7	0.95	122
$(i\text{-}C_3H_7)_2PO$	–	5.6	1.0	8.8	2.65	1.0	122
$[(CH_3)_2N]_2PO$	–	5.8	1.0	8.9	2.5	1.0	122

4. Inorganic phosphorus π radicals

	a_P^{iso}	$a_P(\parallel)$	$a_P(\perp)$		
$\dot{P}O_4^{2-}$	30	–	–	–	73,134
	19.4	20.1	18.8,18.6	–	133
$Cl_2\dot{P}O_2$	44	43	44	$a_{Cl}(\parallel) = 6.5$	13
				$a_{Cl}(\perp) = 3.5$	
$F_2\dot{P}O_2$	43.8	–	–	$a_F = 52.5$	10
$\dot{P}O_3F^-$	39.1	–	–	$a_F = 8.0$	10
$(OH)_2\dot{P}O_2$	39.0	41.0	38.0	–	134

TABLE IV. Phosphorus-Substituted Nitroxide Free Radicals

Compound	a_P,G	a_N,G	Refs.
$R{-}N(O\cdot){-}C(CH_3)_3$			
R = PO_3^{2-}	12.0	13.4	2,82
R = $POH(O)^-$	10.6	12.3	82
R = $P(OCH_3)_2(O)$	13.8	10.3	82
R = $P(OEt)_2(O)$	13.1	9.9	82
R = $P(OBu)_2(O)$	13.1	9.9	82
R = P(Ph)(OH)(O)	10.4	12.0	82
R = PPh_2	11.4	13.2	82
R = P(H)Ph	6.5	12.0	82
R = $P(S)(CH_3)_2$	10.5	11.5	82
R = 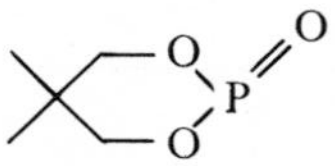	12.8	10.5	51
R = $\overset{+}{P}[N(CH_3)_2]_3$	12.0,14.8	10.3,11.9	19
R = $\overset{+}{N}(CH_3)_2{-}P[N(CH_3)_2]_2$		18.4,0.85	19,21

R = $[(CH_3)_2N]_3\overset{+}{P}$-$P[N(CH_3)_2]_3$	11.5,3.1	12.2	21
R = $\overset{+}{P}(OEt)_3$	11.2	13.3	21
R = 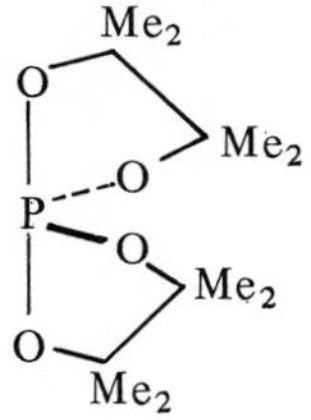	1.73	14.7	62

TABLE V. β-Phosphorus-Substituted π Radicals

1. Radical type $\dot{R}$—O—P

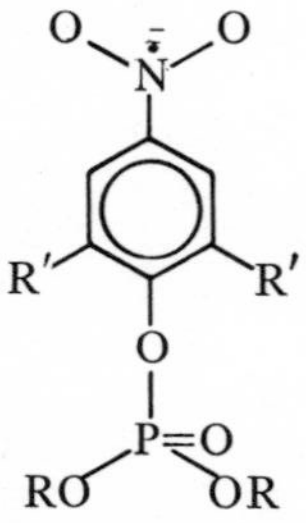

Compound: R	R′	a_P, G	Other Splittings, G: $a_H(o)$	$a_H(m)$	a_N	Refs.
Et	H	7.05	3.64	1.13	10.43	64
		7.28	3.44	1.13	9.93	64
Ph	H	7.92	3.42	1.12	9.96	64
		8.35	3.47	1.12	9.50	64
p-tolyl	H	7.90	3.46	1.12	10.14	64
		8.00	3.45	1.12	9.89	64
o-tolyl	H	8.27	3.46	1.12	10.15	64
		8.39	3.46	1.12	9.87	64
Ph	Me	10.50	3.39	1.03	10.50	64
		11.06	3.40	1.02	9.75	64

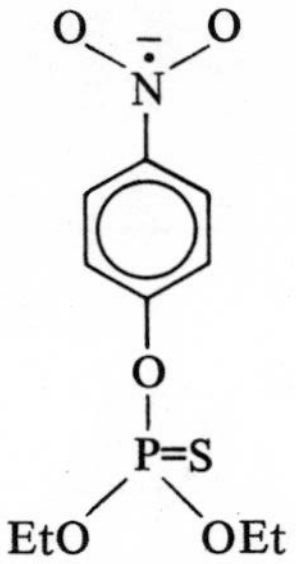	10.29	3.47	1.13	10.27	64
	10.84	3.46	1.12	9.71	64
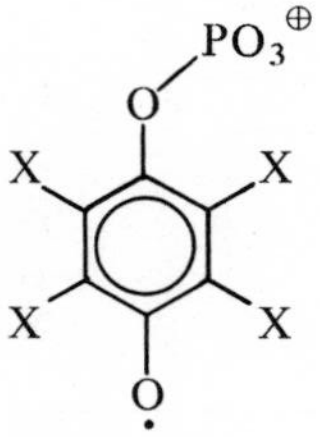					
X = H	1.3		a_H = 4.1 (triplet)		3
X = Me	17.2		a_H = 5.0 (septet)		3
			a_H = 0,95 (quintet/septet)		
X = Cl	20		—		3

TABLE V. cont.

Compound		a_P,G	Other Splittings,G	Refs.
1. Radical type $\dot{\mathbf{R}}$**—O—P** (*continued*)				
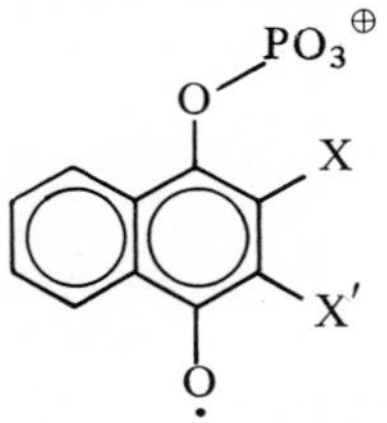				
X	X′			
H	H	2.0	a_H = 5.2 (triplet)	3
Me	H	19.1	a_H = 2.4 (triplet)	3
Me	Me	18.9	a_H = 7.8 (quartet)	3
Me	Cl	18.5	a_H = 2.4 (triplet)	3
			a_H = 8.2 (septet)	
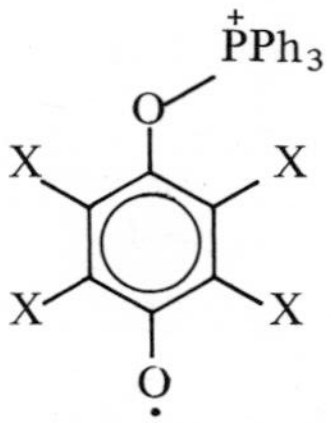				
X				
Cl		2.42	—	100

Br	2.04		—	100
I	1.80		—	100
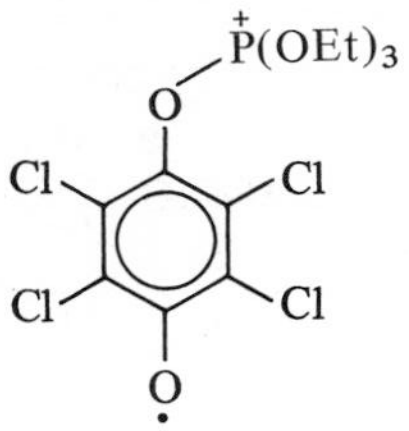	1.83			100
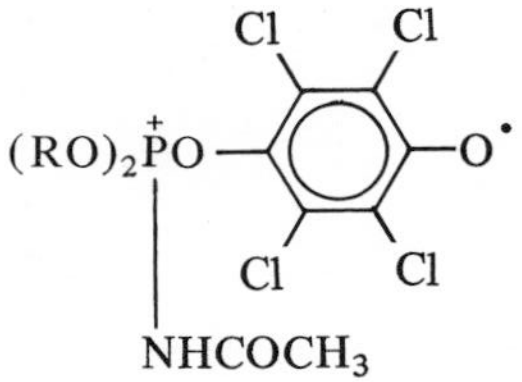	19.1		—	123
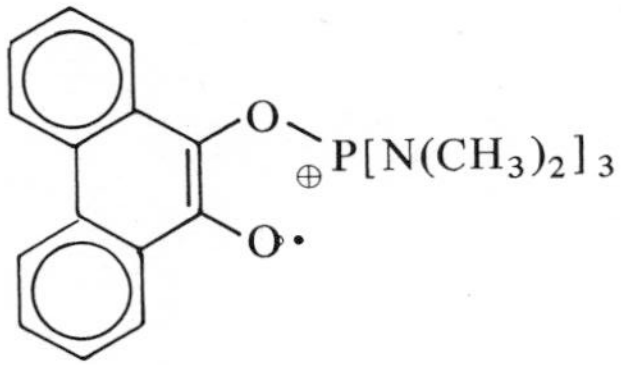	11.2		—	21
$R_1R_2P(O)O\dot{C}H_2$				
R_1 R_2		$a_H(\alpha H)$	$a_H(\beta H)$	
H OMe	12.4	20.3	—	106
	12.4	20.4	—	96

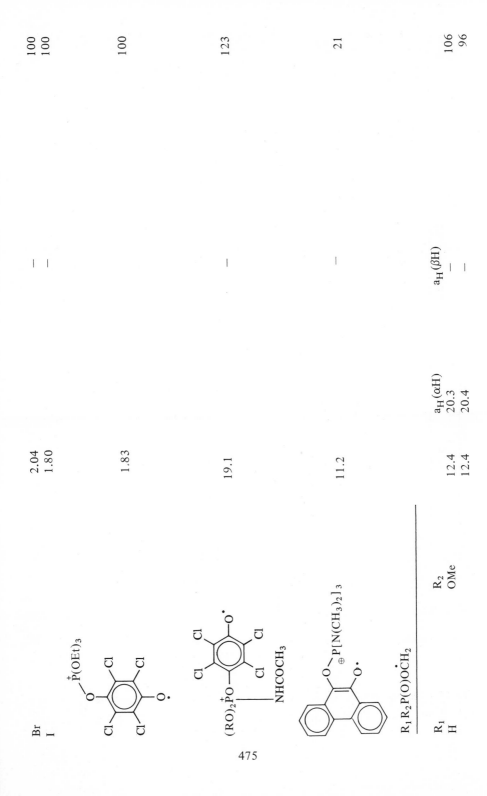

TABLE V. cont.

Compound		a_P,G	Other Splittings,G			Refs.
1. Radical type $\dot{R}$—O—P (*continued*)						
			$a_H(\alpha H)$	$a_H(\beta H)$		
OMe	OMe	4.0	19.5	—		11
		10.5	20.4	—		96
		10.3	20.1	—		106
Me	OMe	10.4	20.3	—		106
		10.5	20.1	—		96
OH	OMe	7.8	19.2	—		96
$R_1R_2P(O)O\dot{C}HCH_3$						
R_1	R_2					
H	OEt	17.7	17.7	24.6		106
		17.4	17.4	24.0		96
OH	OEt	11.2	17.4	23.8		96
H	$OCHMe_2$	31.9	—	22.0		106
Me	OEt	14.5	18.0	24.5		106
Et	OEt	13.8	18.0	24.2		96
OMe	OEt	14.1	18.0	24.5		96
OEt	OEt	14.1	18.1	24.6		106
		14.2	17.9	24.4		96
$Et\dot{C}HOP(O)(H)OPr$		17.1	17.1	23.1		106
$\dot{C}H_2OPO_3^-$		5.70	18.62			130
$\dot{C}HOPO_3^{2-}$ | CHOH | CH_2OH		5.55	18.25	11.33	a_H = 0.30 (γH) a_H = 0.60 (OH)	130

$\dot{C}HOPO_3^{2-}$–CHOH–CO_2^-	4.37	18.20	11.32	a_H = 0.32 (OH)	130
$\dot{C}HOPO_3^{2-}$–$CH_2NH_3^+$	4.90	19.00	13.11	a_N = 9.51 a_H = 0.30 (NH_3)	130
$\dot{C}HOPO_3^{2-}$–$CHNH_3^+$–CO_2^-	4.69	18.29	13.51	a_N = 7.15 a_H = 0.14 (NH_3)	130
$CH_2OPO_3^{2-}$–•CH–CHO	0.52	18.34	27.17(CH_2) 1.58(CHO)		130
$CH_2OPO_3^{2-}$–•CNH_2–CO_2^-	0.97		10.85	a_N = 5.69 a_H = 2.85, 1.87 (NH_2)	130
$CH_3\dot{C}HOP(O)_2OEt$	4.5-6.7	16.9	23.4		47
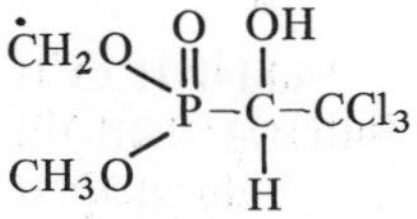	10.6	20.4			29
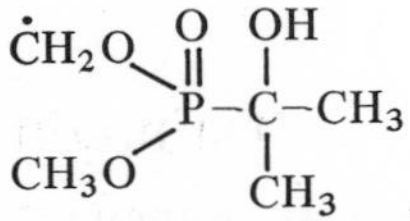	8.5	20.2			29
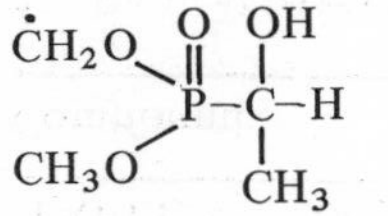	9.0	20.2			28

TABLE V. cont.

Compound		a_P,G	Other Splittings,G			Refs.
2. Radical type $R_2\dot{C}-C-P$			$a_H(\alpha H)$	$a_H(\beta H)$		
$H_2\dot{C}CH_2PEt_2$		251	15			103
		251	20	14		102
$H_2\dot{C}CH_2\overset{+}{P}Et_3$		120	14			103
$Et\dot{C}HCH_2\overset{+}{P}(H)Bu_2$		150				103
$H_2\dot{C}CHMe\overset{+}{P}Ph_3$		110	20	13		103
		200			a_H = 40 a_H = 10 (3H)	109
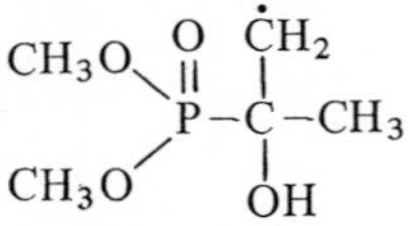		78.8		22.2		
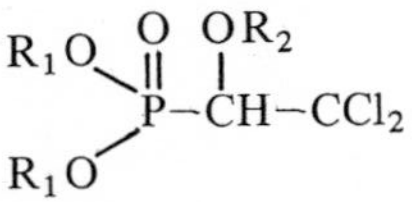						
R_1	R_2				a_{Cl}	
Me	H	66.0		9.0	3.6	27,28
Et	H	65.2		9.1	3.6	27,28
n-Pr	H	65.0		9.0	3.6	27,28
i-Pr	H	64.5		9.1	3.6	27,28
Me	OCMe	61.5		7.2	3.6	27,28
Me	OCEt	61.5		7.2	3.6	27,28
Me	OCPr	60.9		7.2	3.6	27,28

Me	Me	68.9		7.2	3.6	27,28
Me	Et	65.7		9.1	3.6	27,28
$^{-}O_2P(H)CH(CO_2H)\dot{C}HCO_2H$		64.5	19.9	10.4	a_H = 3.5 (PH)	7
$^{-}O_2P(OH)CH(CO_2H)\dot{C}HCO_2H$		65.3	20.2	8.5	—	7
$^{-}O_2P(H)CH(CO_2^-)\dot{C}HCO_2^-$		64.1	19.8	6.05	a_H = 3.0 (PH)	7
		64.0	19.8	6.0	a_H = 3.0 (PH)	56
$^{2-}O_3PCH(CO_2^-)\dot{C}HCO_2^-$		55.5	19.5	4.5	—	7
		54.7	19.4	4.3	—	56
$^{-}O_2P(H)CH_2\dot{C}HCO_2H$		74.5	19.8	17.5	a_H = 2.8 (PH)	7
$^{-}O_2P(OH)CH_2\dot{C}HCO_2H$		75.6	20.1	17.8	—	7
$^{-}O_2P(H)CH_2\dot{C}HCO_2^-$		75.5	20.0	17.6	a_H = 1.6 (PH)	7
		75.4	20.0	17.6	a_H = 1.7 (PH)	56
$^{2-}O_3PCH_2\dot{C}HCO_2^-$		56.4	19.7	21.0	—	7
		55.9	19.6	21.1	—	56
$^{-}O_2P(H)CH_2\dot{C}MeCO_2H$		73.5	—	22.0 (3 H)	a_H = 2.5 (PH)	7
				13.2 (2 H)		
$^{-}O_2P(OH)CH_2\dot{C}MeCO_2H$		81.5	—	22.0 (3 H)	—	7
				13.3 (2 H)		
$^{-}O_2P(H)CH_2\dot{C}MeCO_2^-$		78.5	—	22.0 (3 H)	a_H = 1.5 (PH)	7
				13.8 (2 H)		
$^{-}O_2P(H)CHMe\dot{C}HCO_2^-$		75.6	19.8	10.1	a_H = 1.6 (PH)	56
$^{2-}O_3PCH_2\dot{C}MeCO_2^-$		67.1	—	21.2 (3 H)	—	7
				15.5 (2 H)		
$^{2-}O_3PCHMe\dot{C}HCO_2^-$		49.8	19.7	5.4 (1 H)	—	56
				0.8 (3 H)		
$^{-}O_2P(H)CMe_2\dot{C}HCO_2H$		64.7	19.8	—	a_H = 2.8	7
$^{-}O_2P(H)CH(CO_2H)\dot{C}Me_2$		64.0	—	21.0 (6 H)	a_H = 2.7	7
				6.0 (1 H)		
$^{-}O_2P(OH)CMe_2\dot{C}HCO_2H$		70.0	19.6	—	—	7

TABLE V. cont.

Compound	a_P,G	Other Splittings,G				Refs.
2. Radical type $\dot{R}-C-P$ (*continued*)						
		$a_H(\alpha H)$		$a_H(\beta H)$		
$^{-}O_2P(OH)CH(CO_2H)\dot{C}Me_2$	81.0	—		23.5 (6 H)	—	7
				6.7 (1 H)		
$^{-}O_2P(H)CMe_2\dot{C}HCO_2^-$	66.0	19.7		—	a_H = 1.6 (PH)	7
$^{-}O_2P(H)CH(CO_2^-)\dot{C}Me_2$	63.0	—		23.1 (6 H)	a_H = 2.9 (PH)	7
				7.0 (1 H)		
$^{2-}O_3PCMe_2\dot{C}HCO_2^-$	65.0	19.5		—	—	7
$^{-}O_2P(H)CH_2\dot{C}Me_2$	94.0	—		23.3 (6 H)	a_H = 2.2 (PH)	7
				15.0 (2 H)		
$^{-}O_2P(OH)CH_2\dot{C}Me_2$	105	—		23.3 (6 H)	—	7
				14.7 (2 H)		
$^{2-}O_3PCH_2\dot{C}Me_2$	89.0	—		23.0 (6 H)	—	7
				15.0 (2 H)		
$^{-}O_2P(H)CH_2\dot{C}HOEt$	75.7	15.0		16.5	a_H = 1.5 (γH)	7
$^{-}O_2P(OH)CH_2\dot{C}HOEt$	80.7	15.3		16.4	a_H = 1.5 (γH)	7
$^{2-}O_3PCH_2\dot{C}HOEt$	66.7	15.0		17.7	a_H = 1.5 (γH)	7
$^{2-}O_3PCH(OEt)\dot{C}H_2$	46	22		18	—	7
$^{2-}O_3PCH_2\dot{C}HCOMe$	61	18		17	a_H = 1.6 (γH)	7
$^{2-}O_3PCH_2\dot{C}HOAc$	65.9	19.0		20.5	a_H = 1.6 (γH)	56
$^{-}O_2P(H)CH_2\dot{C}HCN$	65.6	20.0		16.0	a_H = 2.5 (PH)	56
					a_N = 3.4	
$^{2-}O_3PCH_2\dot{C}HCN$	62.0	19.5		16.1	a_N = 3.4	56
		$a_H(\beta H)$	$a_H(\gamma H)$	a_N		
$^{2-}O_3PCH_2NO_2^{\dot{-}}$	31.4	10.7	—	24.9	—	56
	31.6	10.6	—	25.2	—	7

$^{2-}O_3PCHEtNO_2^{\bullet -}$	36.8	2.2	0.85 0.25	23.8	—	56
$^{2-}O_3PCHMeNO_2^{\bullet -}$	27.0	3.2	0.3	25.6	—	56
$^{2-}O_3PCHMe_2NO_2H^{\bullet}$	10.9	—	—	27.6	a_H = 3.1 (OH)	56
$^{-}O_2P(H)CH_2NO_2^{\bullet -}$	29.2	10.1	—	25.1	a_H = 1.5 (PH)	56
	29.2	10.17	—	25.2	a_H = 1.48 (PH)	7
$^{-}O_2P(H)CHMeNO_2^{\bullet -}$	26.8	6.1	0.4	25.5	a_H = 1.4 (PH)	56
$^{-}O_2P(H)CHEtNO_2^{\bullet -}$	34.7	2.7	0.5	24.3	a_H = 2.0 (PH)	56
$^{-}O_2P(H)CMe_2NO_2^{\bullet -}$	16.9	—	0.3	26.3	a_H = 1.0 (PH)	56
$^{-}O_2P(Ph)CH_2NO_2^{\bullet -}$	33.4	10.1	—	24.5	—	56
$^{-}O_2P(Ph)CHMeNO_2^{\bullet -}$	30.4	3.7	0.3	24.8	—	56
$^{-}O_2P(Ph)CHEtNO_2^{\bullet -}$	38.7	2.2	0.9	23.3	—	56
		a_H(NH)	a_H(βH)	a_N		
$^{2-}O_3PCHMeN(H)\dot{O}$	28.5	14.4	12.1	14.4	—	56
$^{2-}O_3PCHEtN(H)\dot{O}$	35.3	14.4	6.0	14.4	—	56
$^{-}O_2P(H)CH_2N(H)\dot{O}$	38.6	13.9	10.7	13.9	a_H = 2.3 (PH)	56
$^{-}O_2P(H)CHMeN(H)\dot{O}$	37.1	13.9	10.2	13.9	a_H = 2.4 (PH)	56
$^{-}O_2P(H)CHEtN(H)\dot{O}$	38.7	13.9	7.2	13.9	a_H = 2.4 (PH)	56
$^{-}O_2P(Ph)CHMeN(H)\dot{O}$	40.6	13.9	9.1	13.9	—	56
$^{-}O_2P(Ph)CHEtN(H)\dot{O}$	41.8	13.9	5.9	13.9	—	56
		a_H(αH)		a_H(βH)		
$(EtO)_2P(O)CH_2\dot{C}Me_2$	112.5			13.5(P-CH_2) 22.9(Me)		32
$(EtO)_2P(O)CH_2\dot{C}HCH_2Pr$	93.0	21.5		16.8(P-CH_2) 27.4(CH_2Pr)		32
$(EtO)_2P(O)CH_2\dot{C}HCH_2Bu$	94.1	21.5		16.5(P-CH_2) 27.4(CH_2Bu)		32

TABLE V. cont.

Compound		a_P,G	Other Splittings,G		Refs.
2. Radical type $\dot{R}-C-P$ (*continued*)					
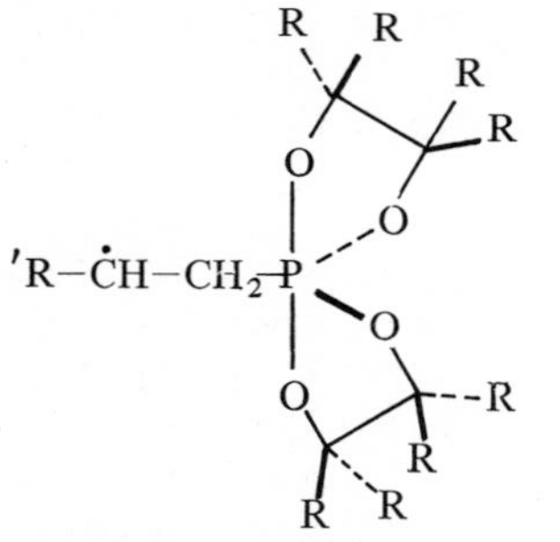			$a_H(\alpha H)$	$a_H(\beta H)$	
R	R′				
H	Bu^n	122.5	20.8	16.0(P-CH_2) 26.8(n-Bu)	62
H	Bu^t	104.5	20.1	20.1	62
Me	Bu^n	112.4			62
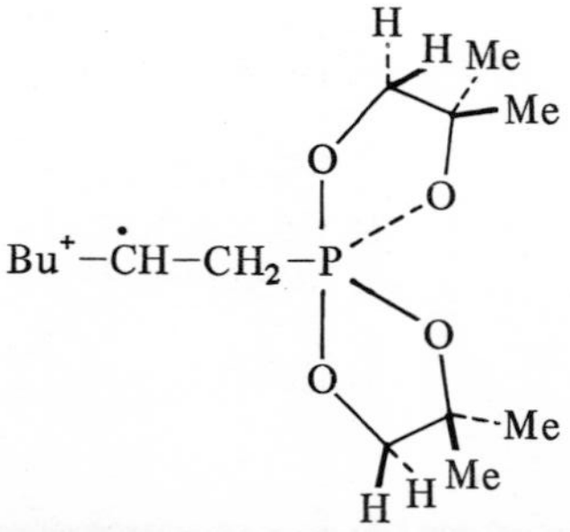		98.0	20.6	18.3	62

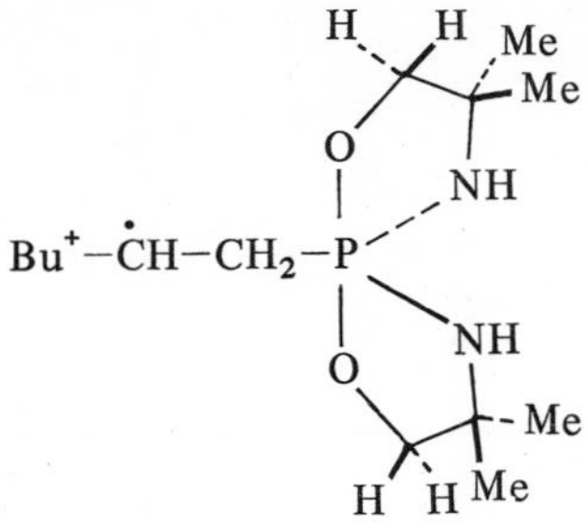

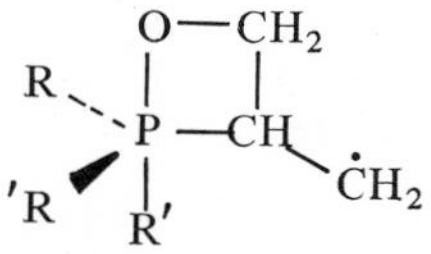

R	R′				
OBu^t	$CH_2=CHCH_2O$	179	20.5	20.5	36
OEt	$CH_2=CHCH_2O$	182	20.0	20.0	36
Me	$CH_2=CHCH_2O$	180	20.0	20.0	36
OBu^t	OEt	176	20.0	20.0	36
OEt	OEt	181	20.0	20.0	36
Me	OEt	180	20.0	20.0	36

TABLE V. cont.

Compound		a_P,G	Other Splittings, G		Refs.
2. Radical type $\dot{R}-C-P$ (*continued*)					
			$a_H(\alpha H)$	$a_H(\beta H)$	
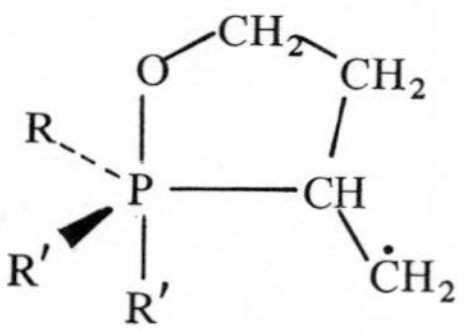					
R	R′				
OEt	$CH_2=CHCH_2CH_2O$	127	21.0	25.0	36
Me	$CH_2=CHCH_2CH_2O$	136	21.0	25.0	36
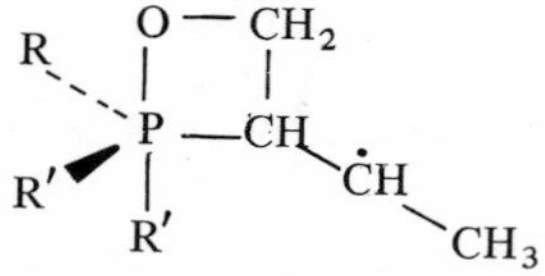					
R	R′				
OBu[t]	$CH_3CH=CHCH_2O$	190	19.9	14.3(1H) 23.0(3H)	36
OEt	$CH_3CH=CHCH_2O$	199	19.9	14.3(1H) 23.0(3H)	
Me	$CH_3CH=CHCH_2O$	195	20.0	14.3(1H) 23.0(3H)	36

TABLE VI. Tetrahedral Phosphorus Radicals

Compound	a_P^{iso},G	$a_P(\parallel)$,G	$a_P(\perp)$,G	Other Splittings,G	3s Character	3p Character	Refs.
1. Phosphinium radicals, $R_3\dot{P}^{+}\cdot$							
$\dot{P}H_3^+$	517	706	423		0.142	0.917	8
$\dot{P}Me_3^+$	—	600	290		0.108	1.00	14
$\dot{P}Et_3^+$	384	588	283		0.106	0.985	8
$\dot{P}Bu_3^+$	360	556	262		0.099	0.951	8
$\dot{P}(OH)_3^+$	795	930	728		—	—	58
$Et_3\dot{P}{-}PEt_3^+$	454.7 (2 P)	540	412		0.125	0.414	104
$Bu_3\dot{P}{-}PBu_3^+$	462 (2 P)	550	418		0.127	0.427	104
$(Me_2PS)_2^{\dot{-}}$	490 (2 P)	553	459		0.135	0.304	104
2. Phosphonyl radicals, $R_2\dot{P}O$							
$Me_2\dot{P}O$	375	535	295	$a_H(\parallel) = 20$ $a_H(\perp) = 17.5$	0.103	0.78	14
	—	433	375,310	$a_H^{iso} = 5.6$	—	—	14
$(MeO)_2\dot{P}O$	700	—	—	—	—	—	32
	690.7	802	635	—	—	—	58
	684	795	629	—	$p/s = 2.86$		84
$(EtO)_2\dot{P}O$	687	—	—	—	—	—	32
	671	784	614	—	$p/s = 2.98$		84
$(n\text{-}BuO)_2\dot{P}O$	688	—	—	—	—	—	32
$(i\text{-}PrO)_2\dot{P}O$	685	—	—	—	—	—	32
$Ph_2\dot{P}O$	390	514	332,324	—	0.11	0.60	53
$Ph_2\dot{P}S$	351	472	290,285	—	0.09	0.58	52
$Ph\dot{P}(O)OH$	540	667	476	$a_H(\parallel) = 17.5$	0.15	0.61	54

TABLE VI. cont.

Compound	a_P^{iso},G	$a_P(\parallel)$,G	$a_P(\perp)$,G	Other Splittings,G	3s Character	3p Character	Refs.
2. Phosphonyl radicals, $R_2\dot{P}O$							
				$a_H(\perp) = 15$			
$Ph\dot{P}(O)OD$	—	666	476	—	—	—	54
$Ph\dot{P}(O)Cl$	—	≥ 550	ca. 450	—	—	—	108
$\dot{P}hP(S)Cl$	430	550	370	—	—	—	108
$\dot{P}OCl_2$	—	893	692	$a_{Cl}(\parallel) = 24$	—	—	13
	—			$a_{Cl}(\perp) = 15$			
	759			$a_{Cl}^{iso} = 18$	—	—	13
3. Anion radicals, type $R\dot{P}O_2^-$							
$H\dot{P}O_2^-$	495	606	439	$a_H^{iso} = 82$	0.14	0.55	112
$H\dot{P}O_2^- \cdot X^+$							
X = Mg	520.7	642.2	459.9	$a_H^{iso} = 92.6$	0.15	0.61	4
X = Ca	511.8	632.9	451.3	$a_H^{iso} = 88.0$	—	—	4
X = Na	472	586	415	$a_H^{iso} = 88.6$	—	—	4
$H\dot{P}O_2^-$	484	—	—	$a_H^{iso} = 90.0$	—	—	7
$HO\dot{P}O_2^-$	658	—	—	—	—	—	7
$O_2P{-}\dot{P}HO_2^{2-}$	—	487	337,329	$a_P(\parallel) = 188$	—	—	112
				$a_P(\perp) = 129,125$			
				$a_H^{iso} = 32$			
$O_2P{-}\dot{P}O_3^{3-}$	414	523	350	$a_P(\parallel) = 157$	—	—	114
				$a_P(\perp) = 124$			
				$a_P^{iso} = 188$			

$Ph\dot{P}O_2^- \cdot X^+$							
X = Na	433	545	377	—	0.12	0.54	54
	505	628	444	—	0.14	0.59	54
X = K	499	616	440	—	0.13	0.56	54
X = Mg	514	640	451	—	0.14	0.61	54
X = Ca	510	630	450	—	0.14	0.58	54
X = Sr	490	608	432	—	0.13	0.57	54
X = Ba	471	588	413	—	0.13	0.56	54
$MeO\dot{P}O_2^-$	556	649	510	—	p/s = 2.94		84
$EtO\dot{P}O_2^-$	557	651	510	—	p/s = 2.99		84
$EtO\dot{P}O_2^-$	—	780	624,615	—	0.18	0.52	46
$F\dot{P}O_2^-$	690	807	630,620	a_F = 177,207,121	0.19	0.57	12
				a_F^{iso} = 165			
$F\dot{P}O_2^-$	698.5			a_F^{iso} = 168.6	—	—	48
4. The hypophosphate radical, $\dot{P}O_3^{2-}$							
	—	699	540	—	—	—	12
	571	—	—	—	—	—	7
	595	702.5	540.6		0.165	0.527	72
	699	819.1	639.1		0.194	0.584	133
	671.4-784.3	—	—		0.2	0.6	149
	751	870	702,696		0.208	0.55	114
	675	789	618		—	—	66
	—	650.8	507,503		0.178	0.484	68
	627.4	739.3	571.4		p/s = 3.1		113

TABLE VI. cont.

Compound	a_P^{iso},G	$a_P(\parallel)$,G	$a_P(\perp)$,G	Other Splittings,G	3s 3p Character	Refs.
4. The hypophosphate radical, $\dot{P}O_3^{2-}$						
$\dot{P}O_3^{2-}.X^+$						
X = Na	—	706.5	540.2		p/s = 3.2	136
X = Mg	—	786	615		3.2	136
X = Ca	—	814.9	640		3.1	136
X = Sr	—	752	582		3.2	136
X = Ba	—	682.6	523		3.3	136
X = Cd	—	778	611		3.2	136

TABLE VII. Phosphoranyl Radicals in Irradiated Solids

Compound	a_P^{iso},G	$a_P(\parallel)$,G	$a_P(\perp)$,G	Other Splittings,G	Refs.
$\dot{P}F_4$	1346	—	—	a_F = 196	115
	1330	—	—	—	5
	1354.4			a_F = 197.6	48
	1321.5			a_F = 293.0 (ap)	
				a_F = 59.8 (eq)	
	1330	—	—	a_F = 282 (ap)	49
				a_F = 59 (eq)	
	1391	1508	1328	$a_F(\parallel)$ = 408 (ap)	119
				$a_F(\perp)$ = 366 (ap)	
				a_F^{iso} = 380 (ap)	
				a_F^{iso} < 45 (eq)	
$\dot{P}Cl_4$	1206	—	—	a_{Cl} = 62 (ap)	89
				a_{Cl} = 7 (eq)	
	1214	—	—	a_{Cl} = 62,5 (ap)	90
				a_{Cl} = 7,5 (eq)	
	1217	—	—	—	85
	1248	—	—	a_{Cl} = 52	13
$CH_3\dot{P}Cl_3$	1077	—	—	—	89,90
$\dot{P}OCl_3^-$	1371	—	—	a_{Cl} = 40 (ap)	57
				a_{Cl} = 18 (eq)	
	1367	—	—	a_{Cl} = 67 (ap)	85
				a_{Cl} = 14 (eq)	
	1359			a_{Cl} = 67 (ap)	13
				a_{Cl} = 15 (eq)	
$Ph\dot{P}(O)Cl_2^-$	1000	1150	925	a_{Cl}^{iso} = 51	108
$Ph\dot{P}(S)Cl_2^-$	973	1120	900	a_{Cl}^{iso} = 43	108

TABLE VII. cont.

Compound	a_P^{iso},G	$a_P(\parallel)$,G	$a_P(\perp)$,G	Other Splittings,G	Refs.
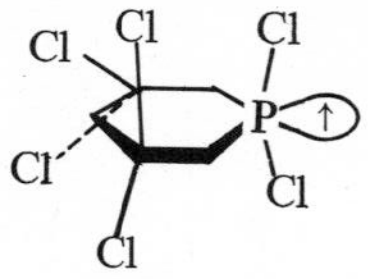	1290	—	—	$a_{Cl} = 52$	107
$Me_3\dot{P}O^-(H^+)$	610	738	548	—	14
$(MeO)_3\dot{P}O^-$	860	970	805	—	58
$\dot{P}H_4$	973.8	—	—	$a_H = 21.0$	58,105
$\dot{P}O_4^{4-}$	1156	1270	1099	—	58,94,135

TABLE VIII. Phosphoranyl Radicals in Solution

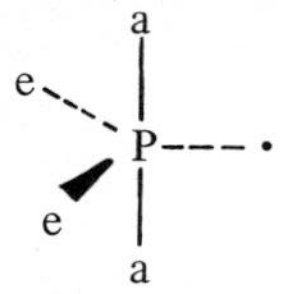

Compound				a_P,G	Other Splittings,G	Refs.
ap	ap	eq	eq			
OBu^t	H	H	H	626.7	a_H = 139.6 (ap)	92
					a_H = 10.8 (eq)	
OBu^t	H	H	Me	631.5	a_H = 139.6 (ap)	92
					a_H = 9.76 (eq)	
					a_H = 4.05 (methyl)	
OBu^t	OBu^t	H	H	672.5	a_H = 0.6	92
OBu^t	H	Me	Me	631.4	a_H = 142.9	92
					a_H = 3.45 (methyl)	
OBu^t	D	Me	Me	630.3	a_D = 21.3	92
					a_H = 3.40 (methyl)	
OBu^t	OBu^t	H	Me	703.3	a_H = 6.84	92
					a_H = 3.15 (methyl)	
OBu^t	Me	Me	Me	618.7	a_H = 4.6 (ap)	92
					a_H = 2.8 (eq)	
				618	—	88
OBu^t	OBu^t	Me	Me	713.7	a_H = 2.5	92
				712.4	a_H = 2.5	92
				701	—	63

TABLE VIII. cont.

Compound				a_P,G	Other Splittings,G	Refs.
OBu^t	OBu^t	Et	Et	713.9	a_H = 2.3	92
				705	a_H = 2.3	31
				704	—	37
OBu^t	OBu^t	*i*-Pr	*i*-Pr	714.6	a_H = 1.7	92
OBu^t	OBu^t	*t*-Bu	*t*-Bu	705		63
$R\dot{P}(OR')_3$						
R	R′					
OBu^t	Me			887		147
				891.5		34b
OBu^t	Et			889		147
				887.1-893.1		34b
				890		61
				891		34a
OBu^t	Et			889		35
OBu^t	*i*-Pr			900		147
				902.3		34b
OBu^t	Bu^t			918		147
				920.0		34b
OBu^t	Bu^n			886.7-891.9		34b
OBu^t	c-C_5H_9			902.3		34b
OBu^t	n-C_5H_{11}			904.2		34b
OBu^t	$ClCH_2CH_2$			908.5		34b
OBu^t	CF_3CH_2			940.2-943.1		35
OEt	Me			884.0		35
OEt	Et			884.5-886.2		35

OEt	*i*-Pr	889.5	35
OEt	*n*-Bu	887.9	35
OEt	*t*-Bu	907.4-908.6	35
OEt	neo-C_5H_{11}	899.2	35
OEt	cyclo-C_5H_9	892.5-893.5	35
OEt	*n*-C_{12}-H_{25}	889.3	35
OEt	*c*-C_5H_9, Me,Me	885.7-885.8	35
OEt	*c*-C_5H_9, Et,Et	886.8-888.0	35
OEt	$ClCH_2CH_2$	912.1	35
OEt	CF_3CH_2	953.8-954.4	35
Me	Me	784.3-779.6	35
Me	Et	782.5-786.7	35
Me	*i*-Pr	780.6-787.4	35
Me	*n*-Bu	784.2-789.2	35
Me	*t*-Bu	884.0	35
Me	neo-C_5H_{11}	795.9-803.1	35
Me	cyclo-C_5H_9	788.6-795.8	35
Me	*c*-C_5H_9, Me,Me	781.8-785.3	35
Me	*c*-C_5H_9, Et,Et	783.8-789.9	35
Me	$PhCH_2$, Et,Et	791.2	35
OBu^t	CH_2=$CHCH_2$	897	36
OBu^t	CH_2=$CHCH_2$ Et,Et	884	36
OEt	CH_2=$CHCH_2$,Et,Et	865	36
OBu^t	CH_2=$CHCH_2CH_2$	894	36
OEt	CH_2=$CHCH_2CH_2$	893	36
OBu^t	CH_2=$CHCH_2CH_2CH_2$	893	36
OEt	CH_2=$CHCH_2CH_2CH_2$	890	36
Me	CH_2=$CHCH_2CH_2CH_2$	794	36

TABLE VIII. cont.

Compound				a_P,G	Other Splittings,G	Refs.
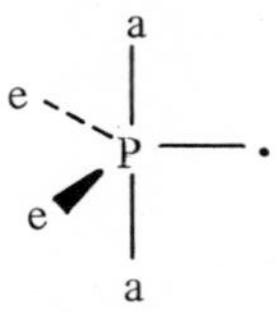						
ap	ap	eq	eq			
Cl	OBu^t	OEt	OEt	1034-1037	$a_{^{35}Cl}$ = 47.2 $a_{^{37}Cl}$ = 39.3	61
Cl	OBu^t	OBu^t	Et	943	$a_{^{35}Cl}$ = 30 $a_{^{37}Cl}$ = 25	61
Cl	Cl	OBu^t	OEt	1145	$a_{^{35}Cl}$ = 34.1 $a_{^{37}Cl}$ = 28.1	61
Cl	Cl	OBu^t	Me	1023	$a_{^{35}Cl}$ = 30.9 $a_{^{37}Cl}$ = 25.8	61
Cl	Cl	OBu^t	Et	1003	$a_{^{35}Cl}$ = 31.3 $a_{^{37}Cl}$ = 26.0	61
Cl	OBu^t	Et	Et	794	$a_{^{35}Cl}$ = 38.9 $a_{^{37}Cl}$ = 32.2	61
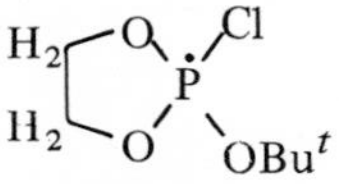				1033	$a_{^{35}Cl}$ = 42.8 $a_{^{37}Cl}$ = 35.9 a_H = 1.3 (2 H)	61
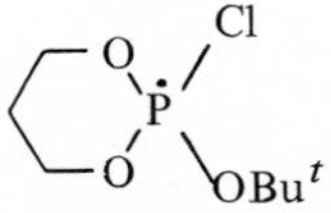				1070	$a_{^{35}Cl}$ = 44.8 $a_{^{37}Cl}$ = 37.1	61

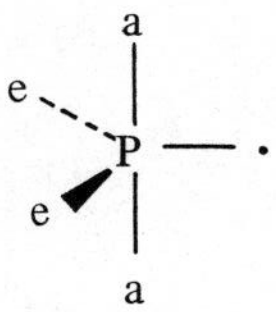

ap	ap	eq	eq			
OBu^t	OEt	NMe_2	NMe_2	785	—	37
OBu^t	OBu^t	NMe_2	NMe_2	771	—	37
OBu^t	OEt	OEt	NMe_2	841	$a_N = 2.8$	37
OEt	NMe_2	OBu^t	OEt	697	$a_N = 12.7$	37
OBu^t	OBu^t	Et	NMe_2	740	—	37
OBu^t	NMe_2	Et	Et	581	$a_N = 12.6$	37
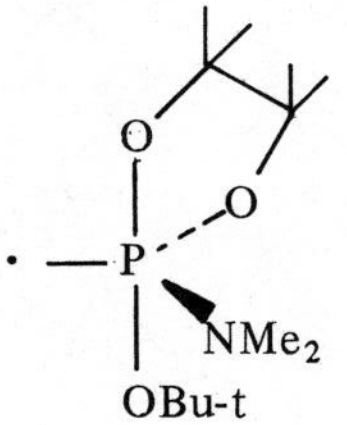				849	$a_N = 9.5$	38
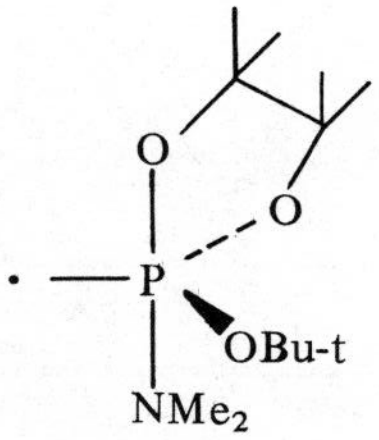				784	$a_N = 24$	38

TABLE VIII. cont.

Compound	a_P,G	Other Splittings,G	Refs.
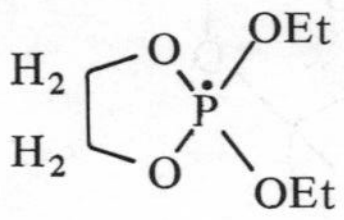	906.3-908.4	a_H = 1.8 (3 H)	35
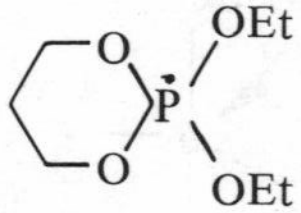	921.0-930.2	–	35
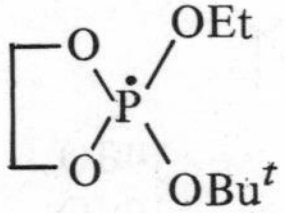	906.5-909.5	–	35
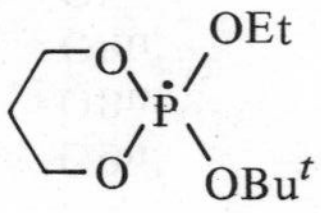	947.0		35
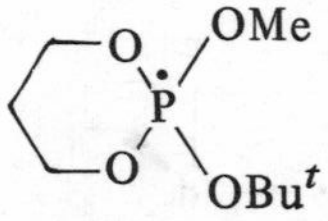	945		147

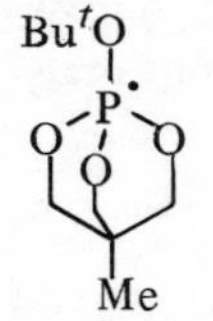	973.2	—	34b
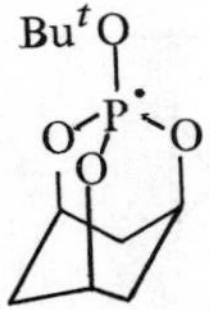	955.7	—	34b
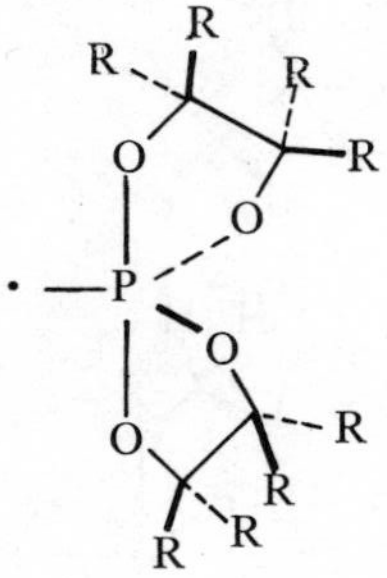			
R = H	910-920	a_H = 3.4 (3 H) a_H = 1.0 (3 H)	60,62
R = Me	911-916	—	60,62

TABLE VIII. cont.

Compound	a_P,G	Other Splittings,G	Refs.
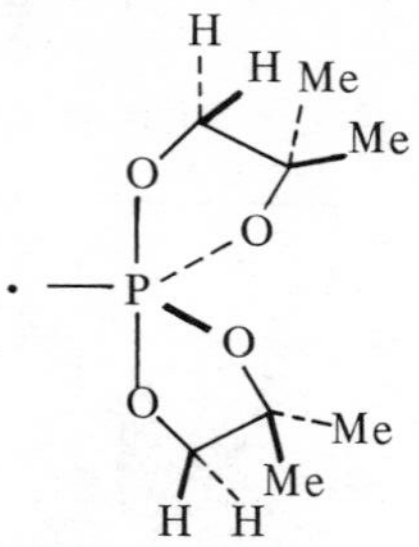	902-910	a_H = 4.1-3.5 (2H)	62
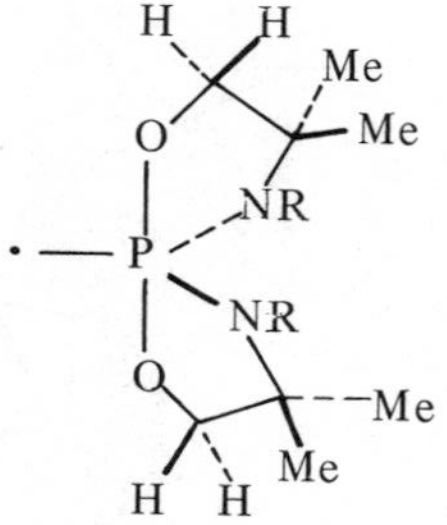			
R = H	807	a_H = 3.6 (3 H) a_N = 6.1 (2 N)	62
R = D	805-809	a_H = 3.5 (3 H) a_N = 6.1 (2 N)	62
$MeS\dot{P}(Bu^n)_3$	635.5	—	33
$MeS\dot{P}(OEt)_3$	748.1		33

Phosphoranylperoxy radicals		
$Bu^tO(MeO)_3POO\cdot$	9.2	148
$Bu^tO(EtO)_3POO\cdot$	9.5	148
	9.45	34b
$Bu^tO(i\text{-}PrO)_3POO\cdot$	8.9	148
$(Bu^tO)_4POO\cdot$	9.0	148
$(EtO)_4POO\cdot$	9.25	34b

References

1. L. J. Aarons, I. H. Hillier, and M. F. Guest, *J. Chem. Soc., Faraday II,* **1974**, 167.
2. G. Adevik, and C. Lagercrantz, *Acta Chem. Scand.,* **24**, 2253 (1970).
3. B. T. Allen, and A. Bond, *J. Phys. Chem.,* **68**, 2439 (1964).
4. P. W. Atkins, N. Keen, and M. C. R. Symons, *J. Chem. Soc.,* **1963**, 250.
5. P. W. Atkins, and M. C. R. Symons, *J. Chem. Soc.,* **1964**, 4363.
6. P. W. Atkins, and M. C. R. Symons, The Structure of Inorganic Radicals, Elsevier, Amsterdam, 1967.
7. A. L. J. Beckwith, *Aust. J. Chem.,* **25**, 1887 (1972).
8. A. Begum, A. R. Lyons, and M. C. R. Symons, *J. Chem. Soc., A,* **1971**, 2290.
9. A. Begum, A. R. Lyons, and M. C. R. Symons, *J. Chem. Soc., A,* **1971**, 2388.
10. A. Begum, S. Subramanian, and M. C. R. Symons, *J. Chem. Soc., A,* **1970**, 1323.
11. A. Begum, S. Subramanian, and M. C. R. Symons, *J. Chem. Soc., A,* **1970**, 1334.
12. A. Begum, S. Subramanian, and M. C. R. Symons, *J. Chem. Soc., A,* **1971**, 700.
13. A. Begum, and M. C. R. Symons, *J. Chem. Soc., A,* **1971**, 2065.
14. A. Begum, and M. C. R. Symons, *J. Chem. Soc., Faraday II,* **69**, 43 (1973).
15. W. G. Bentrude, *Ann. Rev. Phys. Chem.,* **18**, 283 (1967).
16. W. G. Bentrude, J. J. L. Fu, and P. E. Rogers, *J. Am. Chem. Soc.,* **95**, 3625 (1973).
17. W. G. Bentrude, E. R. Hansen, W. A. Khan, T. B. Min, and P. E. Rogers, *J. Am. Chem. Soc.,* **95**, 2286 (1973).
18. M. Bersohn, quoted in W. G. Bentrude, *Ann. Rev. Phys. Chem.,* **18**, 315 (1967).
19. G. Boekestein, and H. M. Buck, *Rec. Trav. Chim.,* **92**, 1095 (1973).
20. G. Boekestein, E. H. J. M. Jansen, and H. M. Buck, *J. Chem. Soc., Chem. Comm.,* **1974**, 118.
21. G. Boekestein, W. G. Voncken, E. H. J. M. Jansen, and H. M. Buck, *Rec. Trav. Chim.,* **93**, 69 (1974).
22. A. D. Britt, and E. T. Kaiser, *J. Phys. Chem.,* **69**, 2775 (1965).
23. A. D. Britt, and E. T. Kaiser, *J. Org. Chem.,* **31**, 112 (1966).
24. H. M. Buck, A. H. Huizer, S. J. Oldenburg, and P. Schipper, *Phosphorus,* **1**, 97 (1971).
25. H. L. J. Chen, and M. Bersohn, *J. Am. Chem. Soc.,* **88**, 2663 (1966).
26. A. H. Cowley, and M. H. Hnoosh, *J. Am. Chem. Soc.,* **88**, 2595 (1966).
27. W. Damerau, and G. Lassmann, and Kh. Lohs, *J. Magn. Res.,* **5**, 408 (1971).
28. W. Damerau, and G. Lassmann, and Kh. Lohs, *Z. Chem.,* **11**, 182 (1971).
29. W. Damerau, G. Lassmann, and Kh. Lohs, *Z. Naturforsch.,* B**25**, 152 (1970).
30. A. G. Davies, R. W. Dennis, D. Griller, K. U. Ingold, and B. P. Roberts, *Mol. Phys.,* **25**, 989 (1973).
31. A. G. Davies, R. W. Dennis, D. Griller, and B. P. Roberts, *J. Organomet. Chem.,* **40**, C33 (1972).
32. A. G. Davies, D. Griller, and B. P. Roberts, *J. Am. Chem. Soc.,* **94**, 1782 (1972).
33. A. G. Davies, D. Griller, and B. P. Roberts, *J. Organomet. Chem.,* **38**, C8 (1972).
34a. A. G. Davies, D. Griller, and B. P. Roberts, *Angew. Chem.,* **83**, 800 (1971).
34b. A. G. Davies, D. Griller, and B. P. Roberts, *J. Chem. Soc., Perkin II,* **1972**, 993.
35. A. G. Davies, D. Griller, and B. P. Roberts, *J. Chem. Soc., Perkin II,* **1972**, 2224.
36. A. G. Davies, M. J. Parrott, and B. P. Roberts, *J. Chem. Soc., Chem. Comm.,* **1974**, 27.
37. R. W. Dennis, and B. P. Roberts, *J. Organomet. Chem.,* **43**, C2 (1972).
38. R. W. Dennis, and B. P. Roberts, *J. Organomet. Chem.,* **47**, C8 (1973).
39. R. E. Dessy, and R. L. Pohl, *J. Am. Chem. Soc.,* **90**, 1995, (1968).
40. K. Dimroth, *Chimie Organique du Phosphore,* Paris 1969, Edition du Centre National de la Recherche Scientifique, Paris **182**, 139 (1970).

41. K. Dimroth, *Fortschr. Chem. Forsch.*, **38**, 1, (1973), Springer Verlag.
42. K. Dimroth, N. Greif, H. Perst, F. W. Steuber, W. Sauer, and L. Duttka, *Angew. Chem.*, **79**, 58 (1967).
43. K. Dimroth, N. Greif, W. Städe, and F. W. Steuber, *Angew. Chem.*, **79**, 725 (1967).
44. K. Dimroth, A. Hettche, W. Städe, and F. W. Steuber, *Angew. Chem.*, **81**, 784 (1969).
45. K. Dimroth, and F. W. Steuber, *Angew. Chem.*, **79**, 410 (1967).
46. F. S. Ezra, and W. A. Bernhard, *J. Chem. Phys.*, **59**, 3543 (1973).
47. F. S. Ezra, and W. A. Bernhard, *J. Chem. Phys.*, **60**, 1711 (1974).
48. R. W. Fessenden, *J. Magn. Res.*, **1**, 277 (1968).
49. R. W. Fessenden, and R. H. Schuler, *J. Chem. Phys.*, **45**, 1845 (1966).
50. G. Fraenkel, S. H. Ellis, and D. T. Dix, *J. Am. Chem. Soc.*, **87**, 1406 (1965).
51. D. Gagnaire, A. Rassat, J. B. Robert, and P. Ruelle, *Tetrahedron Lett.*, **43**, 4449 (1972).
52. M. Geoffroy, *Helv. Chim. Act.*, **56**, 1552 (1973).
53. M. Geoffroy, and E. A. C. Lucken, *Mol. Phys.*, **22**, 257 (1971).
54. M. Geoffroy, and E. A. C. Lucken, *Mol. Phys.*, **24**, 335 (1972).
55. F. Gerson, G. Plattner, and H. Bock, *Helv. Chim. Act.*, **53**, 1629 (1970).
56. B. C. Gilbert, J. P. Larkin, R. O. C. Norman, and P. M. Storey, *J. Chem. Soc., Perkin II*, **1972**, 1508.
57. T. Gillbro, and F. Williams, *J. Am. Chem. Soc.*, **96**, 5032 (1974).
58. I. S. Ginns, S. P. Mishra, and M. C. R. Symons, *J. Chem. Soc. Dalton*, **1973**, 2509.
59. Y. I. Gorlov, I. I. Ukrainsky, and V. V. Penkovsky, *Theor. Chim. Act.*, **34**, 31 (1974).
60. D. Griller, and B. P. Roberts, *J. Organomet. Chem.*, **42**, C 47 (1972).
61. D. Griller, and B. P. Roberts, *J. Chem. Soc., Perkin II*, **1973**, 1339.
62. D. Griller, and B. P. Roberts, *J. Chem. Soc., Perkin II*, **1973**, 1416.
63. D. Griller, B. P. Roberts, A. G. Davies, and K. U. Ingold, *J. Am. Chem. Soc.*, **96**, 554 (1974).
64. W. M. Gulick, Jr., and D. H. Geske, *J. Am. Chem. Soc.*, **88**, 2928 (1966).
65. M. W. Hanna, *J. Chem. Phys.*, **37**, 685 (1962).
66. M. W. Hanna, and L. J. Altman, *J. Chem. Phys.*, **36**, 1788 (1962).
67. D. Hellwinkel, *Chem. Ber.*, **102**, 528 (1969).
68. F. G. Herring, J. H. Hwang, W. C. Lin, and C. A. McDowell, *J. Phys. Chem.*, **70**, 2487 (1966).
69. J. Higuchi, *J. Chem. Phys.*, **50**, 1001 (1969).
70. M. H. Hnoosh, and R. A. Zingaro, *Can. J. Chem.*, **47**, 4679 (1969).
71. M. H. Hnoosh, and R. A. Zingaro, *J. Am. Chem. Soc.*, **92**, 4388 (1970).
72. A. Horsfield, J. R. Morton, and D. H. Whiffen. *Mol. Phys.*. **4**, 475 (1961).
73. T. F. Hunter, and M. C. R. Symons, *J. Chem. Soc., A*, **1967**, 1770.
74. C. M. Hurd, and P. Coodin, *J. Phys. Chem. Solids*, **28**, 523 (1967).
75. A. V. Il'Yasov, Yu. M. Kargin, Ya. A. Levin, B. V. Mel'nikov, and V. S. Galeev, *Izv. Akad. Nauk SSSR, Ser. Khim.*, **12**, 2841 (1968).
76. A. V. Il'Yasov, Yu. M. Kargin, Ya. A. Levin, I. D. Morozova, B. V. Mel'inkov, A. A. Vafina, N. N. Sotinkova, and V. S. Galeev, *Izv. Akad. Nauk SSSR, Ser. Khim.*, **4**, 770 (1971).
77. A V. Ilyasov, Ya. A. Levin, and I. D. Morosova, *J. Mol. Struct.*, **19**, 671 (1973).
78. E. G. Janzen, W. B. Harrison, and C. M. Dubose, Jr., *J. Organomet. Chem.*, **40**, 281 (1972).
79. C. Jongsma, H. G. de Graaf, and F. Bickelhaupt, *Tetrahedron Lett.*, **14**, 1267 (1974).
80. M. I. Kabachnik, *Tetrahedron*, **20**, 655 (1964).
81. M. I. Kabachnik, V. V. Voevodsskii, T. A. Mastryukova, S. P. Solodovnikov, and T. A. Melenteva, *Zh. Obshch. Khim.*, **34**, 3234 (1964).
82. H. Karlsson, and C. Lagercrantz, *Act. Chem. Scand.*, **24**, 3411 (1970).

83. M. Karplus, and G. K. Fraenkel, *J. Chem. Phys.*, **35**, 1312 (1961).
84. C. M. L. Kerr, K. Webster, and F. Williams, *Mol. Phys.*, **25**, 1461 (1973).
85. C. M. L. Kerr, and F. Williams, *J. Phys. Chem.*, **75**, 3023 (1971).
86. D. Kilcast, and C. Thomson, *Tetrahedron*, **27**, 5705 (1971).
87. D. Kilcast, and C. Thomson, *J. Chem. Soc., Faraday II*, **68**, 435 (1972).
88. J. K. Kochi, and P. J. Krusic, *J. Am. Chem. Soc.*, **91**, 3944 (1969).
89. G. F. Kokoszka, and F. E. Brinckman, *Chem. Comm.*, **1968**, 349.
90. G. F. Kokoszka, and F. E. Brinckman, *J. Am. Chem. Soc.*, **92**, 1199 (1970).
91. C. Kooistra, J. M. F. van Dijk, P. M. van Lier, and H. M. Buck, *Rec. Trav. Chim.*, **92**, 961 (1973).
92. P. J. Krusic, W. Mahler, and J. K. Kocki, *J. Am. Chem. Soc.*, **94**, 6033 (1972).
93. G. Lassmann, W. Damerau, and Kh. Lohs, *Z. Naturforsch.*, B **24**, 1375 (1969).
94. H. Lozykowski, R. G. Wilson, and F. Holuj, *J. Chem. Phys.*, **51**, 2309 (1969).
95. E. A. C. Lucken, *J. Chem. Soc.*, **1963**, 5123.
96. E. A. C. Lucken, *J. Chem. Soc., A*, **1966**, 1354.
97. E. A. C. Lucken, *J. Chem. Soc., A*, **1966**, 1357.
98. E. A. C. Lucken, and C. Mazeline, *J. Chem. Soc., A*, **1966**, 1074.
99. E. A. C. Lucken, and C. Mazeline, *J. Chem. Soc., A*, **1967**, 439.
100. E. A. C. Lucken, F. Ramirez, V. P. Catto, D. Rhum, and S. Dershowitz, *Tetrahedron*, **22**, 637 (1966).
101. A. R. Lyons, G. W. Neison, and M. C. R. Symons, *J. Chem. Soc., Faraday II*, **68**, 807 (1972).
102. A. R. Lyons, and M. C. R. Symons, *Chem. Comm.*, **1971**, 1068.
103. A. R. Lyons, and M. C. R. Symons, *J. Chem. Soc., Faraday II*, **68**, 622 (1972).
104. A. R. Lyons, and M. C. R. Symons, *J. Chem. Soc., Faraday II*, **68**, 1589 (1972).
105. C. A. McDowell, K. A. R. Mitchell, and P. Raghunathan, *J. Chem. Phys.*, **57**, 1699 (1972).
106. A. R. Metcalfe, and W. A. Waters, *J. Chem. Soc., B*, **1967**, 340.
107. S. P. Mishra, and M. C. R. symons, *J. Chem. Soc., Chem. Comm.*, **1973**, 313.
108. S. P. Mishra, and M. C. R. Symons, *J. Chem. Soc., Dalton*, **1973**, 1494.
109. S. P. Mishra, and M. C. R. Symons, *Tetrahedron Lett.*, **41**, 4061 (1973).
110. S. P. Mishra, and M. C. R. Symons, *J. Chem. Soc., Chem. Comm.*, **1974**, 279.
111. R. L. Morehouse, J. J. Christiansen, and W. Gordy, *J. Chem. Phys.*, **45**, 1747 (1966).
112. J. R. Morton, *Mol. Phys.*, **5**, 217 (1962).
113. J. R. Morton, *J. Phys. Chem. Solids*, **24**, 209 (1963).
114. J. R. Morton, *Mol. Phys.*, **6**, 193 (1963).
115. J. R. Morton, *Can. J. Phys.*, **41**, 706 (1963).
116. E. L. Muetterties, W. Mahler, and R. Schmutzler, *Inorg. Chem.*, **2**, 613 (1963).
117. E. L. Muetterties, W. Mahler, K. J. Packer, and R. Schmutzler, *Inorg. Chem.*, **3**, 1298 (1964).
118. E. Müller, H. Eggensperger, and K. Scheffler, *Ann*,. **658**, 103 (1962).
119. W. Nelson, G. Jackel, and W. Gordy, *J..Chem. Phys*,. **52**, 4572 (1970).
120. H. Normant, T. Cuvigny, J. Normant, and B. Angelo, *Bull. Soc Chim..Fr.*, 3441 (1965).
121. L. Pauling, *J. Chem. Phys.*, **51**, 2767 (1969).
122. A. I. Prokof'ev, S. P. Solodovnikov, I. G. Malakhova, E. N. Tsvetkov, and M. I. Kabachnik, *J. Gen. Chem. USSR*, **43**, 2601 (1973), *Zh. Obshch. Khim.*, **43**, 2621 (1973).
123. A. N. Pudovik, E. S. Batyeva, A. V. Il'yasov, V. D. Nesterenko, A. Sh. Mukhtarov, and N. P. Anoshina, *Zh. Obshch. Khim.*, **43**, 1451 (1973).
124. R. D. Rieke, R. A. Copenhafer, A. M. Aguiar, M. S. Chattha, and J. C. Williams, jun., *J. Chem. Soc., Chem. Comm.*, **1972**, 1130.

125. R. Rothuis, Preparation and ESR of Phosphorus Spiro Compounds, Thesis, Eindhoven, The Netherlands, 1974.
126. R. Rothuis, J. J. H. M. Font Freide, J. M. F. van Dijk, and H. M. Buck, *Rec. Trav. Chim.*, 93, 128 (1974).
127. R. Rothuis, J. J. H. M. Font Freide, and H. M. Buck, *Rec. Trav. Chim.*, **92**, 1308 (1973).
128. R. Rothuis, T. K. J. Luderer, and H. M. Buck, *Rec. Trav. Chim.*, **91**, 836 (1972).
129. R. E. Rundle, *Surv. Progr. Chem.*, **1**, 81 (1963).
130. A. Samuni, and P. Neta, *J. Phys. Chem.*, 77, 2425 (1973).
131. U. Schmidt, F. Geigen, A. Müller, and K. Markau, *Angew. Chem.*, **75**, 640 (1963).
132. U. Schmidt, K. Kabitzke, K. Markau, and A. Müller, *Chem. Ber.*, **99**, 1497 (1966).
133. R. A. Serway, and S. A. Marshall, *J. Chem. Phys.*, **45**, 4098 (1966).
134. S. Subramanian, M. C. R. Symons, and H. W. Wardale, *J. Chem. Soc., A*, **1970**, 1239.
135. M. C. R. Symons, *J. Chem..Phys.*, **53**, 857 (1970).
136. M. C. R. Symons, *J. Chem. Soc., A*, **1970**, 1998.
137. M. C. R. Symons, *Chem. Phys. Lett.*, **19**, 61 (1973).
138. C. Thomson, and D. Kilcast, *Chem. Comm.*, **1971**, 214.
139. C. Thomson, and D. Kilcast, *Chem. Comm.*, **1971**, 782.
140. G. Tomaschewski, *J. Prakt. Chem.*, **33**, 168 (1966).
141. S. Trippett, 2nd. *IUPAC Conf. Phys. Org. Chem.*, Noordwijkerhout, The Netherlands, 1974.
142. T. C. Wallace, R. West, and A. H. Cowley, *Inorg. Chem.*, **13**, 182 (1974).
143. C. Walling, and M. S. Pearson, in *Topics in Phosphorus Chemistry*, Vol. 3, M. Grayson and E. J. Griffith, Eds., Interscience, New York, 1966, pp. 1-56.
144. J. K. S. Wan, J. R. Morton, and H. J. Bernstein, *Can. J. Chem.*, **44**, 1957 (1966).
145. R. E. Watson, and A. J. Freeman, *Phys. Rev.*, **123**, 521 (1961).
146. R. E. Watson, and A. J. Freeman, *Phys. Rev.*, **124**, 1117 (1961).
147. G. B. Watts, D. Griller, and K. U. Ingold, *J. Am. Chem. Soc.*, **94**, 8784 (1972).
148. G. B. Watts, and K. U. Ingold, *J. Am. Chem. Soc.*, **94**, 2528 (1972).
149. R. A. Weeks, and P. J. Bray, *J. Chem. Phys.*, **48**, 5 (1968).
150. M. S. Wei, J. H. Current, and J. Gendell, *J. Chem. Phys.*, **57**, 2431 (1972).
151. M. S. Wei, J. H. Current, and J. Gendell, *J. Chem. Phys.*, **52**, 1592 (1970).
152. S. K. Wong, and J. K. S. Wan, *Spectrosc. Lett.*, **3**, 135-138 (1970).

Subject Index

Topics in Phosphorus Chemistry

Cumulative Index, Volumes 1–9